VOICE COMPRESSION AND COMMUNICATIONS

VOICE COMPRESSION AND COMMUNICATIONS

Principles and Applications for Fixed and Wireless Channels

Lajos Hanzo
Department of Electronics and Computer Science
University of Southampton, U.K.

F. Clare A. Somerville
Department of Electronics and Computer Science
University of Southampton, U.K.
currently with Bell Labs U.K.

Jason P. Woodard
Department of Electronics and Computer Science
University of Southampton, U.K.
currently with Ubinetics, U.K.

IEEE SERIES ON
**DIGITAL
& MOBILE
COMMUNICATION**

John B. Anderson, *Series Editor*

IEEE PRESS

WILEY-
INTERSCIENCE

A JOHN WILEY & SONS, INC., PUBLICATION
New York • Chichester • Weinheim • Brisbane • Singapore • Toronto

For ordering and customer service, call 1-800-CALL-WILEY.

Library of Congress Cataloging-in-Publication Data is available.

ISBN 0-471-15039-8

Printed in the United States of America.

10 9 8 7 6 5 4 3 2 1

Dedicated, with sincere thanks, to our supportive spouses and to the contributors to the field listed in the author index.

Books of Related Interest from the IEEE Press

WIRELESS VIDEO COMMUNICATIONS: Second to Third Generation Systems and Beyond
Lajos Hanzo, P. J. Cherriman, Jurgen Streit
2001   Hardcover   1,056 pp   IEEE Order No. PC5880   ISBN 0-7803-6032-X

SINGLE- AND MULTI-CARRIER QUADRATURE AMPLITUDE MODULATION: Principles and Applications
A John Wiley & Sons, Ltd. book published in cooperation with IEEE Press.
Lajos Hanzo, William Webb, and Thomas Keller
2000   Hardcover   712 pp   IEEE Order No. PC5864   ISBN 0-7803-6015-X

INTELLIGENT SIGNAL PROCESSING
Simon Haykin and Bart Kosko
2001   Hardcover   573 pp   IEEE Order No. PC5860   ISBN 0-7803-6010-9

SOFTWARE RADIO TECHNOLOGIES: Selected Readings
Joseph Mitola III and Zoran Zvonar
2001   Hardcover   483 pp   IEEE Order No. PC5871   ISBN 0-7803-6022-2

Contents

Chapter 14 Mixed-Multiband Excitation   501

Preface, Motivation and The Speech Coding Scene

In the era of third-generation (3G) wireless personal communications standards, despite the emergence of broad-band access network standard proposals, the most important mobile radio services are still based on voice communications. Even when the predicted surge of wireless data and Internet services becomes a reality, voice will remain the most natural means of human communication, although it may be delivered via the Internet, predominantly after compression.

This book is dedicated mainly to voice compression issues. Error resilience, coding delay, implementational complexity, and bitrate are also at the center of our discussions, characterizing many different speech codecs incorporated in source-sensitivity matched wireless transceivers. Here we attempt a rudimentary comparison of some of the codec schemes treated in the book in terms of their speech quality and bitrate in order to provide a roadmap for the reader with reference to Cox's work [1, 2]. The formally evaluated mean opinion score (MOS) values of the various codecs described in this book are shown in Figure 1.

Observe in the figure that a range of speech codecs have emerged over the years. These codecs attained the quality of the 64 kbps G.711 PCM speech codec, though at the cost of significantly increased coding delay and implementational complexity. The 8 kbps G.729 codec is the most recent addition to the International Telecommunications Union's (ITU) standard schemes; it significantly outperforms all previous standard ITU codecs in terms of robustness. The performance target of the 4 kbps ITU codec (ITU4) is also to maintain this impressive set of specifications. The family of codecs designed for various mobile radio systems include the 13 kbps Regular Pulse Excited (RPE) scheme of the Global System of Mobile communications known as GSM, the 7.95 kbps IS-54, and the IS-95 Pan-American schemes, the 6.7 kbps Japanese Digital Cellular (JDC) and 3.45 kbps half-rate

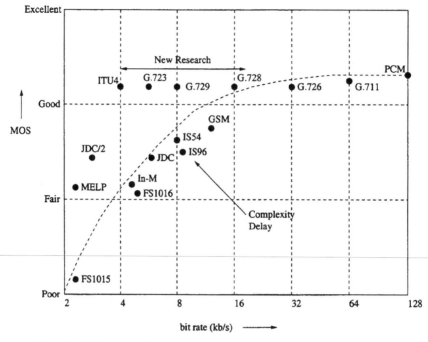

Figure 1   Subjective speech quality of various codecs. Cox *et al.* [1] © IEEE, 1996.

JDC arrangement (JDC/2). These exhibit slightly lower MOS values than the ITU codecs. Let us now consider the subjective quality of these schemes in a little more depth.

The 2.4 kbps U.S. Department of Defense Federal Standard codec known as FS-1015 is the only vocoder in this group, and it has a rather synthetic speech quality, associated with the lowest subjective assessment in the figure. The 64 kbps G.711 PCM codec and the G.726/G.727 Adaptive Differential PCM (ADPCM) schemes are waveform codecs. They exhibit a low implementational complexity associated with a modest bitrate economy. The remaining codecs belong to the hybrid coding family and achieve significant bitrate economies at the cost of increased complexity and delay.

Specifically, the 16 kbps G.728 backward-adaptive scheme maintains a similar speech quality to the 32 and 64 kbps waveform codecs, while also featuring an impressively low, 2 ms delay. This scheme was standardized during the early 1990s. The similar, but significantly more robust, 8 kbps G.729 codec was approved in March 1996 by the ITU. Its standardization overlapped with G.723.1 codec developments. The G.723.1 codec's 6.4 kbps mode maintains a speech quality similar to the G.711, G.726, G.727, G.728, and G.728 codecs, while its 5.3 kbps mode exhibits a speech quality similar to the cellular speech codecs of the late 1980s. Work is currently under way to standardize a 4 kbps ITU scheme, which we refer to here as ITU4.

In parallel to the ITU's standardization activities, a range of speech coding standards have been proposed for regional cellular mobile systems. The standardization of the 13 kbps RPE-LTP full-rate GSM (GSM-FR) codec dates back to the second half of the 1980s, representing the first standard hybrid codec. Its complexity is significantly lower than that of the more recent Code Excited Linear Predictive (CELP)-based codecs. As Figure 0.1 shows, there is also a similar-rate Enhanced Full-Rate GSM codec (GSM-EFR), which matches the speech quality of the G.729 and G.728 schemes. The original GSM-FR codec's development was followed a little later by the release of the 7.95 kbps Vector Sum Excited Linear Predictive (VSELP) IS-54 American cellular standard. Due to advances in the field, the

7.95 kbps IS-54 codec achieved a similar subjective speech quality to the 13 kbps GSM-FR scheme. The definition of the 6.7 kbps Japanese JDC VSELP codec almost coincided with that of the IS-54 arrangement. This codec development was also followed by a half-rate standardization process, leading to the 3.2 kbps Pitch-Synchroneous Innovation CELP (PSI-CELP) scheme.

The 15-95 Pan-American CDMA system also has its own standardized CELP-based speech codec, which is a variable-rate scheme. It supports bitrates between 1.2 and 14.4 kbps, depending on the prevalent voice activity. The perceived speech quality of these cellular speech codecs contrived mainly during the late 1980s was found to be subjectively similar to each other under the perfect channel conditions of Figure 0.1. Finally, the 5.6 kbps half-rate GSM codec (GSM-HR) also met its specification in terms of achieving a similar speech quality to the 13 kbps original GSM-FR arrangements, though at the cost of quadruple complexity and higher latency.

Recently, the advantages of intelligent multimode speech terminals (IMT), which can reconfigure themselves in a number of different bitrate, quality, and robustness modes, became known in the community. This led to the requirement of designing an appropriate multimode codec, the Advanced Multi- Rate codec referred to as the AMR codec. A range of IMTs are also the subject of this book, as current research on sub-2.4 kbps speech codecs for which auditory masking is more dominant. Lastly, since the wideband codec based on the classic G.722 subband-ADPCM is becoming somewhat obsolete in the light of the exciting new developments in compression, the most recent trend is to consider wideband speech and audio codecs, providing susbtantially enhanced speech quality. As a result of early seminal work on transform-domain or frequency-domain-based compression by Noll and his colleagues, in this field the PictureTel codec (which can be programmed to operate between 10 kbps and 32 kbps and hence is amenable to employment in IMTs), has become the most attractive candidate. The present text portrays this codec in the context of a sophisticated burst-by-burst adaptive wideband turbo-coded Orthogonal Frequency Division Multiplex (OFDM) IMT. This scheme can also transmit high-quality audio signals, behaving essentially as a good waveform codec.

MILESTONES IN SPEECH CODING HISTORY

Over the years a range of excellent monographs and textbooks have been published, characterizing the state-of-the-art and significant milestones. The first major development in the history of speech compression is the invention of the vocoder in 1939. Delta modulation was introduced in 1952 and became well established following Steele's monograph on the topic in 1975 [3]. Pulse Coded Modulation (PCM) was first documented in detail in Cattermole's classic contribution in 1969 [4]. However, in 1967 it was recognized that predictive coding provides advantages over memory-less coding techniques, such as PCM. Predictive techniques were analyzed in depth by Markel and Gray in their 1976 classic treatise [5]. This was followed shortly by the often cited reference [6] by Rabiner and Schafer. In 1979 Lindblom and Ohman also contributed a book on speech communication research [7].

The foundations of auditory theory were laid down as early as 1970 by Tobias [8] but were not fully exploited until the invention of the analysis by synthesis (AbS) codecs, which were heralded by Atal's multi-pulse excited codec in the early 1980s [9]. The waveform coding of speech and video signals was comprehensively documented by Jayant and Noll in their 1984 monograph [10]. During the 1980s, speech codec developments were accelerated by the emergence of mobile radio systems, where spectrum was a scarce resource, potentially doubling the number of subscribers and hence the revenues, if the bitrate could be halved.

The RPE principle, as a relatively low-complexity analysis by synthesis technique, was proposed by Kroon, Deprettere, and Sluyter in 1986 [11]. Further research was conducted by Vary [12, 13] and his colleagues at PKI in Germany and IBM in France, leading to the 13 kbps Pan-European GSM codec. This was the first standardized AbS speech codec, which also employed long-term prediction (LTP), recognizing the important role of pitch determination in efficient speech compression [14, 15]. It was in this period that Atal and Schroeder invented the Code Excited Linear Predictive (CELP) principle [16], leading to perhaps the most productive period in the history of speech coding during the 1980s. Some of these developments were also summarized by, among others, O'Shaughnessy [17], Papamichalis [18], and Deller, Proakis and Hansen [19].

It was also during this era that the importance of speech perception and acoustic phonetics [20] was duly recognized—for example, in the monograph by Lieberman and Blumstein. A range of associated speech quality measures were summarized by Quackenbush, Barnwell, and Clements [21]. Nearly concomitantly, Furui published a book related to speech processing [22]. This period witnessed the appearance of many of the speech codecs seen in Figure 0.1, which found applications in the emerging global mobile radio systems, such as IS-54, JDC. These codecs were typically associated with source-sensitivity matched error protection, for which, for example, Steele, Sundberg, and Wong [23–26] provided early insights on the topic. Further sophisticated solutions were suggested by Hagenauer [27].

During the early 1990s, Atal, Cuperman, and Gersho [28] edited prestigious contributions on speech compression, and Ince [29] contributed a book related to the topic. Anderson and Mohan co-authored a monograph on source and channel coding in 1993 [30]. Most of the recent developments were then consolidated in Kondoz's excellent monograph in 1994 [31] and in the multi-authored contribution edited by Keijn and Paliwal [32] in 1995. The most recent addition to this range of contributions is the second edition of O'Shaughnessy's well-referenced book [19].

PURPOSE AND OUTLINE OF THE BOOK

Against this backdrop (since the publication of Kondoz's monograph in 1994 [31] seven years have elapsed), at the time of writing; this book endeavors to review the recent history of speech compression and communications and to provide the reader with a historical perspective. We begin with a rudimentary introduction to communications aspects, since throughout the book we illustrate the expected performance of the various speech codecs studied in the context of a full wireless transceiver.

The book has four parts. Part I and II cover classic background material, while the bulk of the book comprises research-oriented Parts III and IV, which cover both standardized and proprietary speech codecs and transceivers. Specifically, Part I focuses on classic waveform coding and predictive coding (Chapters 1 and 2). Part II centers on analysis by synthesis-based coding, reviewing the principles in Chapter 3 as well as both narrow and wideband spectral quantization in Chapter 4. RPE and CELP coding are the topic of Chapters 5 and 6, which are followed by a long chapter on the existing forward-adaptive standard CELP codecs in Chapter 7 and on their associated source-sensitivity matched channel coding schemes. Chapter 8 discusses proprietary and standard backward-adaptive CELP codecs, and concludes with a system design example based on a low-delay, multimode wireless transceiver.

The essentially research-oriented Part III is dedicated to a range of standard and proprietary wideband schemes, as well as wireless systems. As an introduction to the scene, the classic G.722 wideband codec is reviewed first, leading to various low-rate wideband codecs. Chapter 9 concludes with a turbo-coded Orthogonal Frequency Division Multiplex (OFDM) wideband audio system design example. The remaining chapters, namely Chapters

10–16 of Part IV, are all dedicated to sub-4 kbps codecs and transceivers. The book is concluded with a brief comparison of a range of various codecs.

This book is limited in terms of its coverage of these aspects simply because of the space limitations. We have nonetheless endeavored to provide the reader with a broad range of applications examples, which are pertinent to a range of typical wireless transmission scenarios.

We hope that the book offers you a range of interesting topics, portraying the current state-of-the-art in the associated enabling technologies. In simple terms, finding a specific solution to a voice communications problem has to be based on a compromise in terms of the inherently contradictory constraints of speech quality, bitrate, delay, robustness against channel errors, and the associated implementational complexity. Analyzing these tradeoffs and proposing a range of attractive solutions to various voice communications problems are basic aims of this book.

Again, it is our hope that the book presents the range of contradictory system design tradeoffs in an unbiased fashion and that you will be able to glean information not only to solve your own particular wireless voice communications problem, but most of all to experience an enjoyable and relatively effortless reading, providing you with intellectual stimulation.

Lajos Hanzo
Clare Somerville
Jason Woodard

Acknowledgments

The book was written by the staff of the Electronics and Computer Science Department at the University of Southampton. We are indebted to our many colleagues who have enhanced our understanding of the subject, in particular to Prof. Emeritus Raymond Steele. These colleagues and valued friends, too numerous to be mentioned, have influenced our views on wireless multimedia communications, and we thank them for the enlightenment gained from our collaborations on various projects, paper, and books. We are grateful to Jan Brecht, Jon Blogh, Steve Braithwaite, Marco Breiling, Marco del Buono, Sheng Chen, Stanley Chia, Byoung Jo Choi, Joseph Cheung, Peter Fortune, Sheyam Domeya, Lim Dongmin, Dirk Didascalou, Stephan Ernst, Eddie Green, David Greenwood, Hee Thong How, Thomas Keller, Ee Lin Kuan, W.H. Lam, C.C. Lee, M.A. Nofal, Xiao Lin, Chee Siong Lee, Tong-Hooi Liew, Matthias Münster, Michael Ng, Vincent Roger-Marchart, Jeff Reeve, Redwan Salami, David Stewart, Jeff Torrance, Spyros Vlahoyiannatos, Stephen Weiss, William Webb, John Williams, Choong Hin Wong, Henry Wong, James Wong, Lie-Liang Yang, Bee-Leong Yeap, Mong-Suan Yee, Kai Yen, Andy Yuen, and many others with whom we enjoyed an association.

We also acknowledge our valuable associations with the Virtual Centre of Excellence in Mobile Communications, especially with its chief executives, Dr. Tony Warwick, Dr. Walter Tuttlebee, Dr. Keith Baughan, and other members of its Executive Committee. Our sincere thanks are also due to the EPSRC, UK; Dr. Joao Da Silva, Dr. Jorge Pereira, Bartholome Arroyo, Bernard Barani, Demosthenes Ikonomou, Fabrizio Sestini and other colleagues from the Commission of the European Communities, Brussels, Belgium; and Andy Wilton, Luis Lopes, and Paul Crichton from Motorola ECID, Swindon, UK, for sponsoring some of our recent research.

We feel particularly indebted to Hee Thong How for his invaluable help with proof-reading parts of the manuscript, to Rita Hanzo, and to Denise Harvey for their skillful assistance in typesetting the manuscript in Latex. Similarly, our sincere thanks are due to Linda Matarazzo, Cathy Faduska and a number of other staff members of the IEEE Press as well as John Wiley and Sons, Inc. for their kind assistance throughout the preparation of the camera-ready manuscript. Finally, our sincere gratitude is due to the numerous authors listed in the Author Index and to those, whose work was not cited due to space limitations for their contributions. Without their help this book would not have materialized.

Lajos Hanzo
Clare Somerville
Jason Woodard

PART I
SPEECH SIGNALS AND
WAVEFORM CODING

1

Speech Signals and Introduction to Speech Coding

1.1 MOTIVATION OF SPEECH COMPRESSION

According to information theory, the minimum bitrate at which the condition of distortionless transmission of any source signal is possible is determined by the entropy of the speech source message. Note, however, that in practical terms the source rate corresponding to the entropy is only asymptotically achievable, for the encoding memory length or delay tends to infinity. Any further compression is associated with information loss or coding distortion. Many practical source compression techniques employ lossy coding, which typically guarantees further bitrate economy at the cost of nearly imperceptible speech, audio, video, and other source representation degradation.

Note that the optimum Shannonian source encoder generates a perfectly uncorrelated source-coded stream, in which all the source redundancy has been removed. Therefore, the encoded source symbols—which in most practical cases are consituted by binary bits—are independent, and each one has the same significance. Having the same significance implies that the corruption of any of the source-encoded symbols results in identical source signal distortion over imperfect channels.

Under these conditions, according to Shannon's fundamental work [33–35], the best protection against transmission errors is achieved if source and channel coding are treated as separate entities. When using a block code of length N channel-coded symbols in order to encode K source symbols with a coding rate of $R = K/N$, the symbol error rate can be rendered arbitrarily low if N tends to infinity and the coding rate to K/N. This condition also implies an infinite coding delay. Based on the above considerations and on the assumption of Additive White Gaussian Noise (AWGN) channels, source and channel coding have historically been separately optimized.

In designing a telecommunications system, one of the most salient parameters is the number of subscribers that can be accommodated by the transmission media utilized. Whether it is a time division multiplex (TDM) or a frequency division multiplex (FDM) system, whether it is analog or digital, the number of subscribers is limited by the channel capacity needed for one speech channel. If the channel capacity demand of the speech channels is halved, the total number of subscribers can be doubled. This gain becomes particularly important in applications like power- and band-limited satellite or mobile radio channels, where the demand for free channels overshadows the inevitable cost constraints imposed by a more complex low-bitrate speech codec. In the framework of the basic limitations of state-of-art very large scale integrated (VLSI) circuit technology, the design of a speech codec is based on an optimum tradeoff between lowest bitrate and highest quality, at the price of lowest complexity, cost, and system delay. Analysis of these contradictory factors pervades all our forthcoming discussions.

1.2 BASIC CHARATERIZATION OF SPEECH SIGNALS

In contrast to deterministic signals, random signals, such as speech, music, video, and other information signals, cannot be described by the help of analytical formulas. They are typically characterized by the help of statistical functions. The power spectral density (PSD), autocorrelation function (ACF), cumulative distribution function (CDF), and probability density function (PDF) are some of the most frequent ones invoked.

Transmitting speech information is one of the fundamental aims of telecommunications, and in this book we concentrate mainly on the efficient encoding of speech signals. The human vocal apparatus has been portrayed in many books dealing with the human anatomy and has also been treated in references dealing with speech processing [5, 17, 22]. Hence, here we dispense with its portrayal and simply note that human speech is generated by emitting sound pressure waves, radiated primarily from the lips, although significant energy emanates through sounds from the nostrils, throat, and the like.

The air compressed by the lungs excites the vocal cords in two typical modes. When generating *voiced sounds*, the vocal cords vibrate and generate a high-energy quasi-periodic speech waveform, while in the case of lower energy *unvoiced sounds*, the vocal cords do not participate in the voice production and the source behaves similarly to a noise generator. In a somewhat simplistic approach, the excitation signal denoted by $E(z)$ is then filtered through the vocal apparatus, which behaves like a spectral shaping filter with a transfer function of $H(z) = 1/A(z)$ that is constituted by the spectral shaping action of the glottis, which is defined as the opening between the vocal folds. Further spectral shaping is carried out by the vocal tract, lip radiation characteristics, and so on. This simplified speech production model is shown in Figure 1.1.

Typical voiced and unvoiced speech waveform segments are shown in Figures 1.2 and 1.3, respectively, along with their corresponding power densities. Clearly, the unvoiced segment appears to have a significantly lower magnitude, which is also reflected by its PSD. Observe in Figure 1.3 that the low-energy, noise-like unvoiced signal has a rather flat PSD, which is similar to that of white noise. In general, the flatter the signal's spectrum, the more unpredictable it becomes, and so it is not amenable to signal compression or redundancy removal.

In contrast, the voiced segment shown in Figure 1.2 is quasi-periodic in the time-domain, and it has an approximately 80-sample periodicity, identified by the positions of the largest time-domain signal peaks, which corresponds to $80 \times 125 \, \mu s = 10 \, ms$.

This interval is referred to as the **pitch period** and it is also often expressed in terms of the **pitch frequency** p, which in this example is $p = 1/(10 \, ms) = 100 \, Hz$. In the case of male speakers, the typical pitch frequency range is between 40 and 120 Hz, whereas for females it can be as high as 300–400 Hz. Observe furthermore that within each pitch period there is a gradually decaying oscillation, which is associated with the excitation and gradually decaying vibration of the vocal cords.

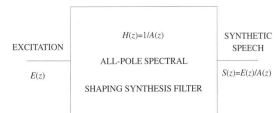

EXCITATION	$H(z)=1/A(z)$	SYNTHETIC
	ALL-POLE SPECTRAL	SPEECH
$E(z)$		$S(z)=E(z)/A(z)$
	SHAPING SYNTHESIS FILTER	

Figure 1.1 Linearly separable speech source model.

A perfectly periodic time-domain signal would have a line spectrum, but since the voiced speech signal is quasi-periodic with a frequency of p (rather than being perfectly periodic) its spectrum in Figure 1.2 exhibits somewhat widened but distinctive spectral needles at frequencies of $n \times p$, rather than being perfectly periodic. As a second phenomenon, we can also observe three, sometimes four, spectral envelope peaks. In our voiced spectrum of Figure 1.2, these **formant frequencies** are observable around 500 Hz, 1500 Hz, and 2700 Hz, and they are the manifestation of the resonances of the vocal tract at these frequencies. In contrast, the unvoiced segment of Figure 1.3 does not have a formant structure; rather it has a more dominant high-pass nature, exhibiting a peak around 2500 Hz. Observe, furthermore, that its energy is much lower than that of the voiced segment of Figure 1.2.

It is equally instructive to study the ACF of voiced and unvoiced segments, which are portrayed on an expanded scale in Figures 1.4 and 1.5, respectively. The voiced ACF shows a set of periodic peaks at displacements of about 20 samples, corresponding to $20 \times 125 \ \mu s = 2.5$ ms, which coincides with the positive quasi-periodic time-domain segments. Following four monotonously decaying peaks, there is a more dominant one around a displacement of 80 samples, which indicates the pitch periodicity. The periodic nature of the ACF can therefore, for example, be exploited to detect and measure the pitch

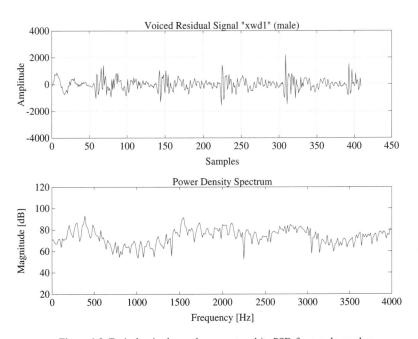

Figure 1.2 Typical voiced speech segment and its PSD for a male speaker.

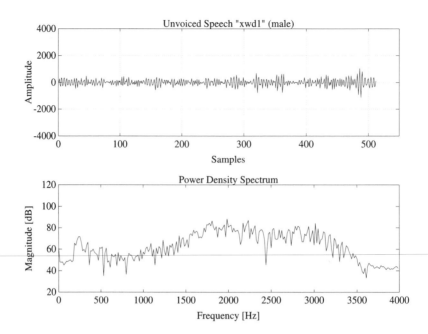

Figure 1.3 Typical unvoiced speech segment and its PSD for a male speaker.

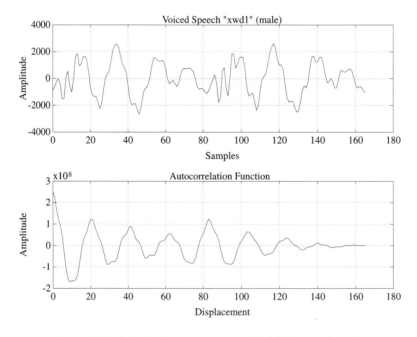

Figure 1.4 Typical voiced speech segment and its ACF for a male speaker.

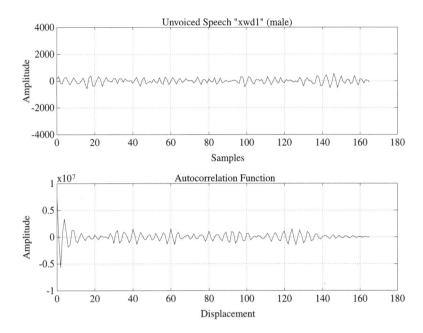

Figure 1.5 Typical unvoiced speech segment and its ACF for a male speaker.

periodicity in a range of applications, such as speech codecs and voice activity detectors. Observe, however, that the first peak at a displacement of 20 samples is about as high as the one near 80. Hence, a reliable pitch detector has to attempt to identify and rank all these peaks in order of prominence, exploiting also the a priori knowledge as to the expected range of pitch frequencies. Recall, too, that, according to the Wiener-Khintshin Theorem, the ACF is the Fourier transform pair of the PSD of Figure 1.2.

By contrast, the unvoiced segment of Figure 1.5 has a much more rapidly decaying ACF, indicating no inherent correlation between adjacent samples and no long-term periodicity. Its sinc-function-like ACF is akin to that of band-limited white noise. The wider ACF of the voiced segment suggests predictability over a time interval of some 3–400 μs. Since the human speech is voiced for about two-thirds of the time, redundancy can be removed from it, using predictive techniques in order to reduce the bitrate required for its transmission.

Having characterized the basic features of speech signals, let us now focus our attention on their digital encoding. Intuitively, it can be expected that the higher the encoder/decoder (codec) complexity, the lower the achievable bitrate and the higher the encoding delay. This is because more redundancy can be removed by considering longer speech segments and employing more sophisticated signal processing techniques.

1.3 CLASSIFICATION OF SPEECH CODECS

Speech coding methods can be broadly categorized as **waveform coding**, **vocoding**, and **hybrid coding**. The principle of these codecs is considered later in this chapter, while the most prominent subclass of hybrid codecs known as analysis-by-synthesis schemes are revisited in detail in Chapter 3 and features throughout the book. Their basic differences become explicit in Figure 1.6, where the speech quality versus bitrate performance of these

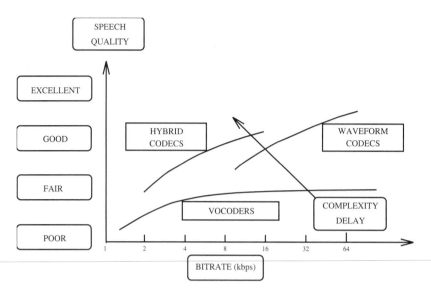

Figure 1.6 Speech quality versus bitrate classification of speech codecs.

codec families is portrayed in qualitative terms. The bitrate is plotted on a logarithmic axis, and the speech quality classes "poor to excellent" broadly correspond to the five-point mean opinion score (MOS) scale values of 2–5 defined by the CCITT, which was recently renamed the International Telecommunications Union (ITU). We will more frequently refer to this diagram and to these codec families during our further discourse in order to allocate various codecs on this plane. Hence, only a rudimentary interpretation is offered here.

1.3.1 Waveform Coding [10]

Waveform codecs have been comprehensively characterized by Jayant and Noll [10]; therefore the spirit of virtually all treatises on the subject follows their approach. Our discussion is no exception.

In general, waveform codecs are designed to be signal independent. They are designed to map the input waveform of the encoder into a facsimile-like replica of it at the output of the decoder. Because of this advantage, they can also encode a secondary type of information such as signaling tones, voice band data, or even music. Because of this signal transparency, their coding efficiency is usually quite modest. The coding efficiency can be improved by exploiting some statistical signal properties, if the codec parameters are optimized for the most likely categories of input signals, while still maintaining good quality for other types of signals as well. The waveform codecs can be further subdivided into time-domain waveform codecs and frequency-domain waveform codecs.

1.3.1.1 Time-Domain Waveform Coding. The most well-known representative of signal-independent time-domain waveform coding is the A-law companded pulse code modulation (PCM) scheme. This coding has been standardized by the CCITT at 64 kbit/s, using non-linear companding characteristics to result in near-constant signal-to-noise ratio

(SNR) over the total input dynamic range. More explicitly, the nonlinear companding compresses large-input samples and expands small ones. Upon quantizing this companded signal, large-input samples will tolerate higher quantization noise than small samples.

Also well-known is the 32 kbit/s adaptive differential PCM (ADPCM) scheme standardized in the ITU Recommendation G.721 (which is the topic of Section 2.7) and the adaptive delta modulation (ADM) arrangement, where usually the most recent signal sample or a linear combination of the last few samples is used to form an estimate of the current one. Then their difference signal, the prediction residual, is computed and encoded with a reduced number of bits, since it has a lower variance than the incoming signal. This estimation process is actually linear prediction with fixed coefficients. However, owing to the nonstationary statistics of speech, a fixed predictor cannot consistently characterize the changing spectral envelope of speech signals. Adaptive predictive coding (APC) schemes utilize two different time-varying predictors to describe speech signals more accurately: a short-term predictor (STP) and a long-term predictor (LTP). We will show that the STP is utilized to model the speech spectral envelope, while the LTP is employed to model the line-spectrum-like fine structure representing the voicing information due to quasi-periodic voiced speech.

All in all, time-domain waveform codecs treat the speech signal to be encoded as a full-band signal and attempt to map it into as close a replica of the input as possible. The difference among various coding schemes is in their degree and way of using prediction to reduce the variance of the signal to be encoded, so as to reduce the number of bits necessary to represent it.

1.3.1.2 Frequency Domain Waveform Coding. In frequency-domain waveform codecs, the input signal undergoes a more or less accurate short-time spectral analysis. The signal is split into a number of sub-bands, and the individual sub-band signals are then encoded by using different numbers of bits in order to obey rate-distortion theory on the basis of their prominance. The various methods differ in their accuracies of spectral resolution and in the bit-allocation principle (fixed, adaptive, semi-adaptive). Two well-known representatives of this class are sub-band coding (SBC) and adaptive transform coding (ATC).

1.3.2 Vocoders

The philosophy of vocoders is based on a priori knowledge of the way the speech signal to be encoded was generated at the signal source by a speaker, which was portrayed in Figure 1.1. The air compressed by the lungs excites the vocal cords in two typical modes. When generating voiced sounds, they vibrate and generate a quasi-periodic speech waveform, while in the case of lower energy unvoiced sounds they do not participate in the voice production and the source behaves similarly to a noise generator. The excitation signal denoted by $E(z)$ in z-domain is then filtered through the vocal apparatus, which behaves like a spectral shaping filter with a transfer function of $H(z) = 1/A(z)$ that is constituted by the spectral shaping action of the glotti, vocal tract, lip radiation characteristics, and so on.

Accordingly, instead of attempting to produce a close replica of the input signal at the output of the decoder, the appropriate set of source parameters is found in order to characterize the input signal sufficiently closely for a given duration of time. First, a decision must be made as to whether the current speech segment to be encoded is voiced or unvoiced. Then the corresponding source parameters must be specified. In the case of voiced sounds, the source parameter is the time between periodic vocal tract excitation pulses, which is often referred to as the pitch p. In the case of unvoiced sounds, the variance or power of the noise-like excitation must be determined. These parameters are quantized and transmitted to the decoder in order to synthesize a replica of the original signal.

The simplest source codec arising from the speech production model is depicted in Figure 1.7. The encoder is a simple speech analyzer, determining the current source parameters. After initial speech segmentation, it computes the linear predictive filter coefficients a_i, $i = 1 \ldots p$, which characterize the spectral shaping transfer function $H(z)$. A voiced/unvoiced decision is carried out, and the corresponding pitch frequency and noise energy parameters are determined. These are then quantized, multiplexed, and transmitted to the speech decoder, which is a speech synthesizer.

The associated speech quality of this type of systems may be predetermined by the adequacy of the source model, rather than by the accuracy of the quantization of these parameters. This means that the speech quality of source codecs cannot simply be enhanced by increasing the accuracy of the quantization, that is, the bitrate, which is evidenced by the saturating MOS curve of Figure 1.6. Their speech quality is fundamentally limited by the fidelity of the model used. The main advantage of the above vocoding techniques is their low bitrate, with the penalty of relatively low, synthetic speech quality. A well-known representative of this class of vocoders is the 2400 bps American Military Standard LPC-10 codec.

In linear predictive coding (LPC), often more complex excitation models are used to describe the voice-generating source. Once the vocal apparatus has been described by the help of its spectral domain transfer function $H(z)$, the central problem of coding is to decide how to find the simplest adequate excitation for high-quality parametric speech representation. Strictly speaking, this separable model represents a gross simplification of the vocal apparatus, but it provides the only practical approach to the problem. Vocoding techniques can also be categorized into frequency-domain and time-domain subclasses. However, frequency-domain vocoders are usually more effective than their time-domain counterparts.

1.3.3 Hybrid Coding

Hybrid coding methods are an attractive tradeoff between waveform coding and source coding, both in terms of speech quality and transmission bitrate, although usually at the price of higher complexity. Every speech coding method, combining waveform and source coding methods in order to improve the speech quality and reduce the bitrate, falls into this broad category. However, adaptive predictive time domain techniques used to describe the human spectral shaping tract, combined with an accurate model of the excitation signal, play the most prominent role in this category. The most important family of hybrid codecs, often referred to as **analysis-by-synthesis** (AbS) codecs, are ubiquitous at the time of writing. Hence, they are

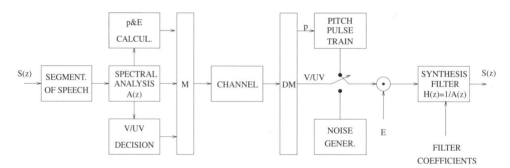

Figure 1.7 Vocoder schematic.

treated in depth in a number of chapters after considering the conceptually simpler category of waveform codecs.

1.4 WAVEFORM OF CODING [10]

1.4.1 Digitization of Speech

The waveform coding of speech and video signals was comprehensively—in fact exhaustively—documented by Jayant and Noll in their classic monograph [10], and hence any treatise on the topic invariably follows a similar approach. This section endeavors to provide a rudimentary overview of waveform coding following the spirit of Jayant and Noll [10]. In general, waveform codecs are designed to be signal independent. They are designed to map the input waveform of the encoder into a facsimile-like replica of it at the output of the decoder. Because of this advantageous property, they can also encode secondary types of information such as signaling tones, voice band data, or even music. Naturally, because of this transparency, their coding efficiency is usually quite modest. The coding efficiency can be improved by exploiting some statistical signal properties, if the codec parameters are optimized for the most likely categories of input signals, while still maintaining good quality for other types of signals.

As noted earlier the waveform codecs can be further subdivided into time-domain waveform codecs and frequency-domain waveform codecs. Let us initially consider the first category. The digitization of analog source signals, such as speech, for example, requires the following steps (see Figure 1.8); the corresponding waveforms are shown in Figure 1.9.

- **Anti-aliasing low-pass filtering** (LPF) is necessary in order to bandlimit the signal to a bandwidth of B before sampling. In case of speech signals, about 1% of the energy resides above 4 kHz and only a negligible proportion above 7 kHz. Hence, commentatory quality speech links, which are also often referred to as wideband speech systems, typically bandlimit the speech signal to 7–8 kHz. Conventional telephone systems usually employ a bandwidth limitation of 0.3–3.4 kHz, which results only in a minor speech degradation, hardly perceptible by the untrained listener.

- The band-limited speech is sampled according to the **Nyquist Theorem**, as seen in Figure 1.8, which requires a minimum sampling frequency of $f_{Nyquist} = 2 \cdot B$. This process introduces time-discrete samples. Due to sampling, the original speech spectrum is replicated at multiples of the sampling frequency. This is why the previous bandlimitation was necessary—in order to prevent aliasing or frequency-domain overlapping of the spectral lobes. If this condition is met, the original analog speech signal can be restored from its samples by passing the samples through a low-pass filter (LPF) with a bandwidth of B. In conventional speech systems, typically a sampling frequency of 8 kHz corresponding to a sampling interval of 125 μs is used.

Figure 1.8 Digitization of analogue speech signals.

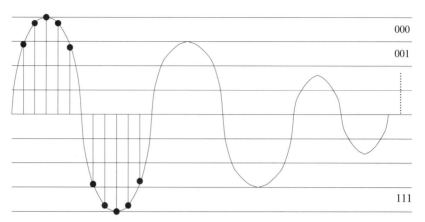

Figure 1.9 Sampled and quantized analog speech signal.

■ Lastly, **amplitude discretization or quantization** must be invoked, according to Figure 1.8, which requires an analog to digital (A/D) converter. The out bits of the quantizer can be converted to a serial bit stream for transmission over digital links.

1.4.2 Quantization Characteristics

Figure 1.9 shows that the original speech signal is contaminated during the quantization process by quantization noise. The severity of contamination is a function of the signal's distribution, the quantizer's resolution, and its transfer characteristic.

The family of **linear quantizers** exhibits a linear transfer function within its dynamic range and saturation above that. They divide the input signal's dynamic range into a number of uniformly or nonuniformly spaced quantization intervals, as seen in Figure 1.10, and assign an R-bit word to each **reconstruction level**, which represents the legitimate output values. In Figure 1.10, according to $R = 3$ there are $2^3 = 8$ reconstruction levels, and a **mid-tread quantizer** is featured, where the quantizer's output is zero, if the input signal is zero. In the case of the **mid-riser quantizer**, the transfer function exhibits a level change at the abscissa value of zero. Note that the quantization error characteristic of the quantizers is also shown in Figure 1.10. As expected when the quantizer characteristic saturates at its maximum output level, the quantization error increases without limit.

The difference between the **uniform and nonuniform quantizer** characteristics in Figure 1.10 is that the uniform quantizer maintains a constant maximum error across its total dynamic range, whereas the nonuniform quantizer employs unequal quantization intervals (quantiles) in order to allow larger granular error, where the input signal is larger. Hence, the nonuniform quantizer exhibits a near-constant signal-to-noise ratio (SNR) across its dynamic range. This may allow us to reduce the number of quantization bits and the required transmission rate, while maintaining perceptually unimpaired speech quality.

In summary, linear quantizers are conceptually and implementationally simple and impose no restrictions on the analog input signal's statistical characteristics such as the probability density function (PDF). Clearly, they do not require a priori knowledge of the input signal. Note, however, that other PDF-dependent quantizers perform better in terms of

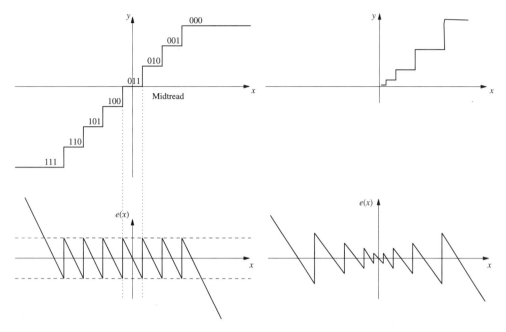

Figure 1.10 Linear quantizers and their quantization errors: left, midtread, right, nonuniform.

overall quantization noise power or signal-to-noise ratio (SNR). These issues will be made more explicit during our further discourse.

1.4.3 Quantization Noise and Rate-Distortion Theory

Observe in Figure 1.10 that the instantaneous **quantization error** $e(x)$ is dependent on the instantaneous input signal level. In other words, $e(x)$ is nonuniform across the quantizer's dynamic range and some amplitudes are represented without quantization error, if they happen to be on a reconstruction level, while others are associated with larger errors. If the input signal's dynamic range exceeds the quantizer's linear range, the quantizer's output voltage saturates at its maximum level and the quantization error may become arbitrarily high. Hence, knowledge of the input signal's statistical distribution is important for minimizing the overall **granular and overload distortion**. The quantized version $\hat{x}(t)$ of the input signal $x(t)$ can be computed as:

$$\hat{x}(t) = x(t) + e(t), \tag{1.1}$$

where $e(t)$ is the quantization error.

If no amplitude discretization is used for a source signal, a sampled analog source has formally an infinite entropy, requiring an infinite transmission rate, which is underpinned by the formal application of Equation 1.2. If the analog speech samples are quantized to R-bit accuracy, there are $q = 2^R$ different legitimate samples, each of which has a probability of occurrence $p_i, i = 1, 2 \ldots q$. It is known from information theory that the R bit/symbol channel capacity requirement can be further reduced using entropy coding to the value of the source's entropy given by:

$$H(x) = -\sum_{i=1}^{q} p_i \cdot \log_2 p_i, \tag{1.2}$$

without inflicting any further coding impairment, if an infinite delay entropy-coding scheme is acceptable. Since this is not the case in interactive speech conversations, we are more interested in quantifying the coding distortion, when using R bits per speech sample.

An important general result of information theory is the **rate-distortion theorem**, which quantifies the minimum required average bitrate R_D in terms of bit/sample in order to represent a random variable (rv) with less than D distortion. Explicitly, for an rv x with variance of σ_x^2 and quantized value $\hat{x}$, the distortion is defined as the mean squared error (mse) expression given by:

$$D = E\{(x - \hat{x})^2\} = E\{e^2(t)\} \tag{1.3}$$

where E represents the expected value.

Observe that if $R_D = 0$ bits are used to quantize the quantity x, then the distortion is given by the signal's variance $D = \sigma_x^2$. If, however, more than zero bits are used, that is, $R_D > 0$, then intuitively one additional bit is needed every time we want to halve the root mean squared (rms) value of D, or quadruple the signal-to-noise ratio of SNR $= \sigma_x^2/D$, which suggests a logarithmic relation between R_D and D. After Shannon and Gallager, we can write for a Gaussian distributed source signal that:

$$R_D = \frac{1}{2}\log_2 \frac{\sigma_x^2}{D} \quad \text{if } D \le \sigma_x^2. \tag{1.4}$$

Upon combining $R_D = 0$ and $R_D > 0$ into one equation, we arrive at:

$$R_D = \begin{cases} \frac{1}{2}\log_2 \sigma_x^2/D & D < \sigma_x^2 \\ 0 & D \ge \sigma_x^2. \end{cases} \tag{1.5}$$

The qualitative or stylized relationship of D versus R_D inferred from Equation 1.5 is shown in Figure 1.11.

In order to quantify the variance of the quantization error, it is reasonable to assume that if the quantization interval q is small and no quantizer overload is incurred, then $e(t)$ is uniformly distributed in the interval $[-q/2, q/2]$. If the quantizer's linear dynamic range is limited to $[\pm V]$, then for a uniform quantizer the quantization interval can be expressed as $q = 2V/2^{R_D}$, where R_D is the number of quantization bits. The quantization error variance can then be computed by squaring the instantaneous error magnitude e and weighting its

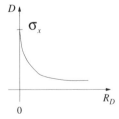

Figure 1.11 Stylized distortion (D) versus coding rate (R_D) curve.

contribution with its probability of occurence expressed by the help of its PDF $p(e) = 1/q$ and finally integrating or averaging it over the range of $[-q/2, q/2]$ as follows:

$$\sigma_e^2 = \int_{-q/2}^{q/2} e^2 p(e)de = \int_{-q/2}^{q/2} e^2 \frac{1}{q} de$$

$$= \frac{1}{q}\left[\frac{e^3}{3}\right]_{-q/2}^{q/2} = \left(\frac{q^3}{8} + \frac{q^3}{8}\right) \cdot \frac{1}{3q} = \frac{q^2}{12}, (1.6)$$

which corresponds to an RMS quantizer noise of $q/\sqrt{12} \approx 0.3q$. In the case of uniform quantizers, we can substitute $q = 2V/2^{R_D}$ into Equation 1.6—where R_D is the number of bits used for encoding—giving the noise variance in the following form:

$$\sigma_q^2 = \frac{q^2}{12} = \frac{1}{12}\left(\frac{2V}{2_D^R}\right)^2 = \frac{1}{3}\frac{V^2}{2^{2R_D}}. \tag{1.7}$$

Similarly, assuming a **uniform signal PDF**, the signal's variance becomes:

$$\sigma_x^2 = \int_{-\infty}^{\infty} x^2 p(x)dx = \int_{-\infty}^{\infty} x^2 \frac{1}{2V} dx = \frac{1}{2V}\left[\frac{x^3}{3}\right]_{-V}^{V} = \frac{1}{6E} \cdot 2V^3 = \frac{E^2}{3}. \tag{1.8}$$

Then the SNR can be computed as:

$$SNR = \frac{\sigma_x^2}{\sigma_q^2} = \frac{V^2}{3} \cdot \frac{2^{2R_D}}{V^2} \cdot 3 = 2^{2R_D}, \tag{1.9}$$

which can be expressed in terms of dB as follows:

$$SNR_{dB} = 10 \cdot \log_{10} 2^{2R} = 20R_D \cdot \log_{10} 2$$
$$SNR_{dB} \approx 6.02 \cdot R_D[dB]. \tag{1.10}$$

This simple result is useful for quick SNR estimates, and it is also intuitively plausible, since every new bit used halves the quantization error and hence doubles the SNR. In practice, the speech PDF is highly nonuniform, and the quantizer's dynamic range cannot be fully exploited, in order to minimize the quantizer characteristic or, synonymously, dynamic range overload error. Hence, Equation 1.10 overestimates the expected SNR.

1.4.4 Nonuniform Quantization for a Known PDF: Companding

If the input signal's PDF is known and can be considered stationary, higher SNR can be achieved by appropriately matched **nonuniform quantization** (NUQ) than in case of uniform quantizers. The input signal's dynamic range is partitioned into nonuniformly spaced segments as we have seen in Figure 1.10, where the quantization intervals are more dense near the origin, in order to quantize the typically high-probability low-magnitude samples more accurately. In contrast, the lower probability signal PDF tails are less accurately quantized. In contrast to uniform quantization, where the maximum error was constant across the quantizer's dynamic range, for nonuniform quantizers the SNR becomes more or less constant across the signal's dynamic range.

It is intuitively advantageous to render the width of the quantization intervals or **quantiles** inversely proportional to the signal PDF, since a larger quantization error is affordable in the case of infrequent signal samples and vice versa. Two different approaches

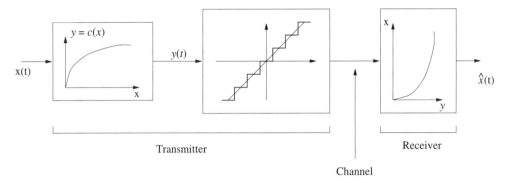

Figure 1.12 Stylised nonuniform quantizer model using companding, when the input signal's PDF is known.

have been proposed, for example, by Jayant and Noll [10] in order to minimize the total quantization distortion in the case of nonuniform signal PDFs.

One system model is shown in Figure 1.12, where the input signal is first compressed using a **nonlinear compander** characteristic and then uniformly quantized. The original signal can be recovered using an expander at the decoder, which exhibits an inverse characteristic with respect to that of the compander. This approach will be considered first, while the design of the minimum mean squared error (mmse) nonuniform quantizer using the Lloyd-Max [36–38] algorithm will be portrayed during our later discussions.

The qualitative effect of nonlinear compression on the signal's PDF is portrayed in Figure 1.13, where it becomes explicit why the compressed PDF can be quantized by a uniform quantizer. Observe that the compander has a more gentle slope, where larger quantization intervals are expected in the uncompressed signal's amplitude range and vice versa. This implies that the compander's slope is proportional to the quantization interval density and inversely proportional to the step-size of any given input signal amplitude.

Following Bennett's approach [39], Jayant and Noll [10] have shown that if the signal's PDF $p(x)$ is a smooth, known function and sufficiently fine quantization is used—implying that $R \geq 6$—then the quantization error variance can be expressed as:

$$\sigma_q^2 \approx \frac{q^2}{12} \int_{-x_{\max}}^{x_{\max}} \frac{p(x)}{|\dot{C}(x)|^2} \, dx, \qquad (1.11)$$

where $\dot{C}(x) = dC(x)/dx$ represents the slope of the compander's characteristic. Where the input signal's PDF $p(x)$ is high, the σ_q^2 contributions are also high due to the high probability

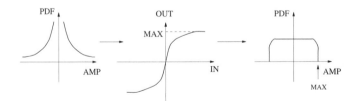

Figure 1.13 Qualitative effect of companding on a known input signal PDF shape.

of occurrence of such signal amplitudes. This effect can be mitigated using a compander exhibiting a high gradient in this interval, since the factor $1/|\dot{C}(x)|^2$ de-weights the error contributions due to the highly peaked PDF near the origin. For an optimum compander characteristic $C(x)$, all quantiles give the same distortion contribution.

Jayant and Noll [10] have also shown that the minimum quantization error variance is achieved by the compander characteristic given by:

$$C(x) = x_{\max} \frac{\int_0^x \sqrt[3]{p(x)}dx}{\int_0^{x_{\max}} \sqrt[3]{p(x)}dx}, \tag{1.12}$$

where the denominator constitutes a normalizing factor. Hence, a simple practical compander design algorithm can be devised by evaluating the signal's histogram in order to estimate the PDF $p(x)$ and by graphically integrating $\sqrt[3]{p(x)}$ according to Equation 1.12 up to the abscissa value x, yielding the companding chracteristic at the ordinate value $C(x)$.

Although this technique minimizes the quantization error variance or maximizes the SNR when there is a known signal PDF, if the input signal's PDF or variance is time-variant, the compander's performance degrades. In many practical scenarios, this is the case; hence it is often advantageous to optimize the compander's characteristic to maximize the SNR independently of the shape of the PDF. Then no compander mismatch penalty is incurred. In order achieve this, the quantization error variance σ_e must be rendered proportional to the value of the input signal $x(t)$ across its dynamic range, implying that large signal samples will have larger quantization error than small samples. This issue is the topic of the next section.

1.4.5 PDF-Independent Quantization using Logarithmic Compression

The input signal's variance is given in the case of an arbitrary PDF $p(x)$ as follows:

$$\sigma_x^2 = \int_{-\infty}^{\infty} x^2 p(x)dx. \tag{1.13}$$

Assuming zero saturation distortion, the SNR can be expressed from Equations 1.11 and 1.13 as follows:

$$SNR = \frac{\sigma_x^2}{\sigma_q^2} = \frac{\int_{-x_{\max}}^{x_{\max}} x^2 p(x)dx}{\frac{q^2}{12} \int_{-x_{\max}}^{x_{\max}} \left(p(x)/|\dot{C}(x)|^2\right)dx}. \tag{1.14}$$

In order to maintain an SNR value that is independent of the signal's PDF $p(x)$, the numerator of Equation 1.14 must be a constant times the denominator, which is equivalent to requiring that:

$$|\dot{C}(x)|^2 \stackrel{!}{=} \left|\frac{K}{x}\right|^2, \tag{1.15}$$

or alternatively that:

$$\dot{C}(x) = K/x \tag{1.16}$$

and hence:

$$C(x) = \int_0^x \frac{K}{z} dz = K \cdot \ln x + A. \tag{1.17}$$

This compander characteristic is shown at the left-hand side of Figure 1.14, and it ensures a constant SNR across the signal's dynamic range, regardless of the shape of the signal's PDF. Intuitively, large signals can have large errors, while small signal must maintain a low distortion, which gives a constant SNR for different input signal levels.

Jayant and Noll also note that the constant A in Equation 1.17 allows for a vertical compander characteristic shift in order to satisfy the boundary condition of matching x_{max} and y_{max}, yielding $y = y_{max}$, when $x = x_{max}$. Explicitly:

$$y_{max} = C(x_{max}) = K \cdot \ln x_{max} + A. \tag{1.18}$$

Upon normalizing Equation 1.17 to y_{max}, we arrive at:

$$\frac{y}{y_{max}} = \frac{C(x)}{y_{max}} = \frac{K \cdot \ln x + A}{K \cdot \ln x_{max} + A}. \tag{1.19}$$

It is convenient to introduce an arbitrary constant B, in order to be able to express A as $A = K \cdot \ln B$, since then Equation 1.19 can be written as:

$$\frac{y}{y_{max}} = \frac{K \cdot \ln x + K \cdot \ln B}{K \cdot \ln x_{max} + K \cdot \ln B} = \frac{\ln xB}{\ln x_{max}B}. \tag{1.20}$$

Equation 1.20 can be further simplified upon rendering its denominator unity by stipulating $x_{max} \cdot B = e^1$, which yields $B = e/x_{max}$. Then Equation 1.20 simplifies to

$$\frac{y}{y_{max}} = \frac{\ln xe/x_{max}}{\ln e} = \ln\left(\frac{e \cdot x}{x_{max}}\right), \tag{1.21}$$

which now gives $y = y_{max}$, when $x = x_{max}$. This logarithmic characteristic, which is shown at the left-hand side of Figure 1.14, must be rendered symmetric with respect to the y-axis, which we achieve upon introducing the signum$(x) = \text{sgn}(x)$ function:

$$\frac{y}{y_{max}} = \frac{C(x)}{y_{max}} = \ln\left(\frac{e \cdot |x|}{x_{max}}\right)\text{sgn}(x). \tag{1.22}$$

This symmetric function is shown in the center of Figure 1.14. However, a further problem is that the logarithmic function is noncontinuous at zero. Thus, around zero amplitude a linear

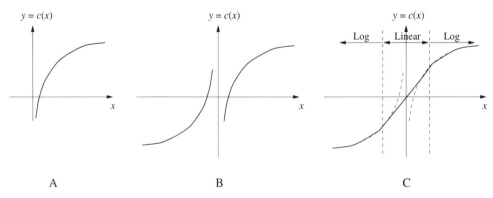

Figure 1.14 Stylized companding characteristic for a near-optimal quantizer.

section is introduced in order to ensure a seamless positive-negative transition in the compression characteristic.

Two practical logarithmic compander characteristics that satisfy the above requirements have emerged. In the United States the μ-law compander was standardized [40–42], while in Europe the A-law compander was proposed [4]. The corresponding stylized logarithmic compander characteristic is depicted at the right-hand side of Figure 1.14.

1.4.5.1 The μ-Law Compander.

This companding characteristic is given by:

$$y = C(x) = y_{\max} \cdot \frac{\ln[1 + \mu \cdot (|x|/x_{\max})]}{\ln(1 + \mu)} \cdot \mathrm{sgn}(x). \tag{1.23}$$

Upon inferring from the $\log(1 + z)$ function that

$$\log(1 + z) \approx z \quad \text{if } z \ll 1, \tag{1.24}$$

in the case of small and large signals, respectively, we have from Equation 1.23 that:

$$y = C(x) = \begin{cases} y_{\max} \cdot \dfrac{\mu \cdot (|x|/x_{\max})}{\ln \mu} & \text{if } \mu \cdot \left(\dfrac{|x|}{x_{\max}}\right) \ll 1 \\[3ex] y_{\max} \cdot \dfrac{\ln[\mu \cdot (|x|/x_{\max})]}{\ln \mu} & \text{if } \mu \cdot \left(\dfrac{|x|}{x_{\max}}\right) \gg 1 \end{cases}, \tag{1.25}$$

which is a linear function of the normalized input signal $x/x_{\max}$ for small signals and a logarithmic function for large signals. The $\mu \cdot |x|/x_{\max} = 1$ value can be considered to be the breakpoint between the small- and large-signal operation, and the $|x| = x_{\max}/\mu$ is the corresponding abscissa value. In order to emphasize the logarithmic nature of the character-istic, μ must be large, which reduces the abscissa value of the beginning of the logarithmic section. The optimum value of μ may be dependent on the quantizer resolution R, and for $R = 8$ the American standard **pulse code modulation** (PCM) speech transmission system recommends $\mu = 255$.

Following the approach proposed by Jayant and Noll [10], the SNR of the μ-law compander can be derived upon substituting $y = C_{\mu}(x)$ from Equation 1.23 into the general SNR formula of Equation 1.14:

$$y = C_{\mu}(x) = y_{\max} \cdot \frac{\ln[1 + \mu(|x|/x_{\max})]}{\ln(1 + \mu)} \cdot \mathrm{sgn}(x) \tag{1.26}$$

$$\dot{C}_{\mu}(x) = \frac{y_{\max}}{\ln(1 + \mu)} \cdot \frac{1}{1 + \mu(|x|/x_{\max})} \cdot \mu\left(\frac{1}{x_{\max}}\right). \tag{1.27}$$

For large-input signals we have $\mu(|x|/x_{\max}) \gg 1$, and hence:

$$\dot{C}_{\mu}(x) \approx \frac{y_{\max}}{\ln \mu} \cdot \frac{1}{x}. \tag{1.28}$$

Upon substituting

$$\frac{1}{\dot{C}_{\mu}(x)} = \frac{\ln \mu}{y_{\max}} \cdot x \tag{1.29}$$

in Equation 1.14 we arrive at:

$$SNR = \frac{\int_{-x_{max}}^{x_{max}} x^2 p(x) dx}{\frac{q^2}{12} \int_{-x_{max}}^{x_{max}} \left(\frac{\ln \mu}{y_{max}}\right)^2 x^2 p(x) dx}$$

$$= \frac{1}{\frac{q^2}{12}\left(\frac{\ln \mu}{y_{max}}\right)^2} = 3\left(\frac{2 y_{max}}{q}\right)^2 \cdot \left(\frac{1}{\ln \mu}\right)^2$$

$$= 3 \cdot 2^{2R} \cdot \left(\frac{1}{\ln \mu}\right)^2. \tag{1.30}$$

Upon exploiting that $2 y_{max}/q = 2^R$ represents the number of quantization levels and expressing the above equation in terms of dB, we get:

$$SNR_{dB}^{\mu} = 6.02 \cdot R + 4.77 - 20 \log_{10}(\ln(1 + \mu)), \tag{1.31}$$

which gives an SNR of about 38 dB in the case of the American standard system using $R = 8$ and $\mu = 255$. Recall that under the assumption of no quantizer characteristic overload and a uniformly distributed input signal, the corresponding SNR estimate would yield $6.02 \cdot 8 \approx 48$ dB. Note, however, that in practical terms this SNR is never achieved, since the input signal does not have a uniform distribution and saturation distortion is also often incurred.

1.4.5.2 The A-law Compander.   Another practical logarithmic compander characteristic is the *A*-**Law Compander** [4], which was standardized by the ITU and is used throughout Europe:

$$y = C(x) = \begin{cases} y_{max} \cdot \dfrac{A(|x|/x_{max})}{1 + \ln A} \cdot \text{sgn}(x); & 0 < \dfrac{|x|}{x_{max}} < \dfrac{1}{A} \\[3mm] y_{max} \cdot \dfrac{1 + \ln[A(|x|/x_{max})]}{1 + \ln A} \cdot \text{sgn}(x); & \dfrac{1}{A} < \dfrac{|x|}{x_{max}} < 1 \end{cases}. \tag{1.32}$$

where $A = 87.56$. Similarly to the μ-law characteristic, it has a linear region near the origin and a logarithmic section above the breakpoint $|x| = x_{max}/A$. Note, however, that in case of $R = 8$ bits $A < \mu$, hence, the A-law characteristic's linear-logarithmic breakpoint is at a higher input value than that of the μ-law characteristic.

Again, substituting:

$$\frac{1}{C_A(x)} = \frac{(1 + \ln A)}{y_{max}} \cdot x \tag{1.33}$$

into Equation 1.14 and exploiting that $2y_{max}/q = 2^R$ represents the number of quantization levels, we have:

$$SNR = \frac{\displaystyle\int_{-x_{max}}^{x_{max}} x^2 p(x)dx}{\displaystyle\frac{q^2}{12}\int_{-x_{max}}^{x_{max}}\left(\frac{(1+\ln A)}{y_{max}}\right)^2 x^2 p(x)dx}$$

$$= \frac{1}{\displaystyle\frac{q^2}{12}\left(\frac{(1+\ln A)}{y_{max}}\right)^2} = 3\left(\frac{2y_{max}}{q}\right)^2 \cdot \left(\frac{1}{(1+\ln A)}\right)^2$$

$$= 3 \cdot 2^{2R} \cdot \left(\frac{1}{(1+\ln A)}\right)^2. \tag{1.34}$$

Upon expressing Equation 1.34 in terms of dB, we arrive at:

$$SNR_{dB}^A = 6.02 \cdot R + 4.77 - 20\log_{10}(1+\ln A)), \tag{1.35}$$

which, similarly to the μ-law compander, gives an SNR of about 38 dB in case of the European standard PCM speech transmission system using $R = 8$ and $A = 87.56$.

Further features of the European A-law standard system are that the characteristic given by Equation 1.32 is implemented in the form of a 16-segment piecewise linear approximation, as seen in Figure 1.15. The segment retaining the lowest gradient of $\frac{1}{4}$ is at the top end of the input signal's dynamic range, which covers half of the positive dynamic range, and it is divided into 16 uniformly spaced quantization intervals. The second segment from the top covers a quarter of the positive dynamic range and doubles the top segment's steepness or gradient to $\frac{1}{2}$, and so on. The bottom segment covers a 64th of the positive dynamic range and has the highest slope of 16 and the finest resolution. The first bit of each $R = 8$-bit PCM codeword represents the sign of the input signal. The next 3 bits specify which segment the input signal belongs to, while the last 4 bits divide a specific segment with 16 uniform-width quantization intervals, as shown in:

$$\underbrace{b_7}_{\substack{\text{sign}\\ \text{(segment)}}} \quad \underbrace{b_6 \quad b_5 \quad b_4}_{\text{segments}} \quad \underbrace{b_3 \quad b_2 \quad b_1 \quad b_0}_{\substack{\text{uniform quant.}\\ \text{in each segment}}}$$

This scheme was standardized by the **International Telegraph and Telephone Consultative Committee (CCITT)** as the G711 Recommendation for the transmission of speech sampled at 8 kHz. Hence the transmission rate becomes $8 \times 8 = 64$ kbit/s (kbps). This results in perceptually unimpaired speech quality, which would require about 12 bits in case of linear quantization.

1.4.6 Optimum Nonuniform Quantization

For nonuniform quantizers the quantization error variance is given by:

$$\sigma_q^2 = E\{|x - x_q|^2\} = \int_{-\infty}^{\infty} e^2(x)p(x)dx, \tag{1.36}$$

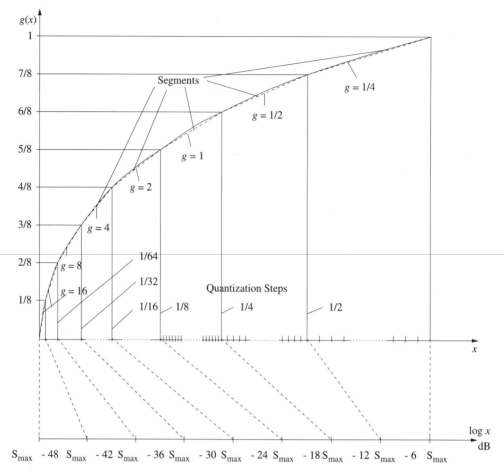

Figure 1.15 Stylized European A-law PCM standard characteristic.

which, again, corresponds to weighting and averaging the quantization error energy over its magnitude range. Assuming an odd-symmetric quantizer transfer function and symmetric PDF $p(x)$, we find that the total quantization distortion power σ_D^2 is as follows:

$$\sigma_D^2 = 2 \int_0^\infty e^2(x)p(x)dx. \tag{1.37}$$

The total distortion can be expressed as the sum of the quantization distortion in the quantizer's linear range, plus the saturation distortion in its nonlinear range, as follows:

$$\sigma_D^2 = \underbrace{2 \int_0^V e^2(x)p(x)dx}_{\sigma_q^2:\ \text{linear region}} + \underbrace{2 \int_V^\infty e^2(x)p(x)dx}_{\sigma_s^2:\ \text{nonlinear region}} \tag{1.38}$$

or more simply as:

$$\sigma_D^2 = \sigma_q^2 + \sigma_s^2. \tag{1.39}$$

In order to emphasize the fact that in case of nonuniform quantization each of the N quantization intervals or quantiles adds a different PDF-weighted contribution to the total quantization distortion, we rewrite the first term of Equation 1.38 as follows:

$$\sigma_q^2 = \sum_{n=1}^{N} \int_{x_n}^{x_{n+1}} e^2(x)p(x)dx$$

$$= \sum_{n=1}^{N} \int_{x_n}^{x_{n+1}} (x - x_q)^2 p(x)dx \tag{1.40}$$

$$= \sum_{n=1}^{N} \int_{x_n}^{x_{n+1}} (x - r_n)^2 p(x)dx, \tag{1.41}$$

where $x_q = r_n$ represents the reconstruction levels.

Given a certain number of quantization bits R and the PDF of the input signal, the optimum Lloyd-Max quantizer, which was independently invented by Lloyd [36, 37] and Max [38], determines the set of optimum quantizer decision levels and the corresponding set of quantization levels.

Jayant and Noll [10] have provided a detailed discussion of two different methods of determining the minimum mean squared error (mmse) solution to the problem. One of the solutions is based on an iterative technique of rearranging the decision thresholds and reconstruction levels, while the other one is an approximate solution valid for fine quantizers using a high number of bits per sample. We first present the general approach to minimizing the mse by determining the set of optimum reconstruction levels r_n, $n = 1 \ldots N$ and the corresponding decision threshold values t_n, $n = 1 \ldots N$.

In general, it is a necessary but not sufficient condition for finding the global minimum of Equation 1.41 for its partial derivatives to become zero. However, if the PDF $p(s)$ is log-concave, that is, the second derivative of its logarithm is negative, then the minimum found is a global one. For the frequently encountered uniform (U), Gaussian (G), and Laplacian (L) PDFs, the log-concave condition is satisfied, but, for example, for Gamma (Γ) PDFs it is not.

Setting the partial derivatives of Equation 1.41 with respect to a specific r_n to zero, only one term in the sum depends on the r_n value considered. Hence we arrive at:

$$\frac{\partial(\sigma_q^2)}{\partial r_n} = 2 \int_{t_n}^{t_{n+1}} (s - r_n) \cdot p(s)ds = 0, \qquad n = 1 \ldots N, \tag{1.42}$$

which leads to

$$\int_{t_n^{opt}}^{t_{n+1}^{opt}} s \cdot p(s)ds = r_n \int_{t_n^{opt}}^{t_{n+1}^{opt}} p(s)ds, \tag{1.43}$$

yielding the optimum reconstruction level r_n^{opt} as follows:

$$r_n^{opt} = \frac{\displaystyle\int_{t_n^{opt}}^{t_{n+1}^{opt}} s \cdot p(s)ds}{\displaystyle\int_{t_n^{opt}}^{t_{n+1}^{opt}} p(s)ds}; \qquad n = 1 \ldots N. \tag{1.44}$$

Note that the above expression depends on the optimum quantization interval thresholds t_n^{opt} and t_{n+1}^{opt}. Furthermore, for an arbitrary nonuniform PDF r_n^{opt} is given by the mean value or the center of gravity of s within the quantization interval n, rather than by $(t_n^{opt} + t_{n+1}^{opt})/2$.

Similarly, when computing $\partial \sigma_q^2 / \partial t_n$, there are only two terms in Equation 1.41, which contain t_n. Therefore, we get:

$$\frac{\partial \sigma_q^2}{\partial t_n} = (t_n - r_{n-1})^2 p(t_n) - (t_n - r_n)^2 p(t_n) = 0 \qquad (1.45)$$

leading to:

$$t_n^2 - 2t_n r_{n-1} + r_{n-1}^2 - t_n^2 + 2t_n r_n - r_n^2 = 0. \qquad (1.46)$$

Hence, the optimum decision threshold is given by:

$$t_n^{opt} = (r_n^{opt} + r_{n-1}^{opt})/2; \qquad n = 2 \ldots N, \qquad t_1^{opt} = -\infty, \qquad t_N^{opt} = \infty \qquad (1.47)$$

which is halfway between the optimum reconstruction levels. Since these nonlinear equations are interdependent they can only be solved by recursive iterations, starting from either a uniform quantizer or from a "hand-crafted" initial nonuniform quantizer design.

Since most practical signals do not obey any analytically describable distribution, the signal's PDF typically has to be inferred from a sufficiently large and characteristic training set. Equations 1.44 and 1.47 will also have to be evaluated numerically for the training set. Below we provide a simple practical algorithm that can be easily implemented by the coding practitioner by the help of the flowchart of Figure 1.16.

Step 1: Input initial parameters such as the number of quantization bits R, maximum number of iteration I, dynamic range minimum t_1 and maximum t_N.

Step 2: Generate the initial set of thresholds $t_1^0 \ldots t_N^0$, where the superscript 0 represents the iteration index, either automatically creating a uniform quantizer between t_1 and t_N according to the required number of bits R, or inputting a "hand-crafted" initial design.

Step 3: While $t < T$, where T is the total number of training samples do:

1. Assign the current training sample s^t, $t = 1 \ldots T$ to the corresponding quantization interval $[t_n^0 \ldots t_{n+1}^0]$ and increment the sample counter $C[n]$, $n = 1 \ldots N$, holding the number of samples assigned to interval n. This corresponds to generating the histogram $p(s)$ of the training set.

2. Evaluate the mse contribution due to assigning s^t to bin[n], that is, $mse^t = (s^t - s_q^t)^2$, and the resultant total accumulated mse, that is, $mse^t = mse^{t-1} + mse^t$.

Step 4: Once all training samples have been assigned to their corresponding quantization bins, that is, the experimental PDF $p(s)$ is evaluated, the center of gravity of each bin is computed by summing the training samples in each bin[n], $n = 1 \ldots N$ and then dividing the sum by the number of training samples $C[n]$ in bin[n]. This corresponds to the evaluation of Equation 1.44, yielding r_n.

Step 5: Rearrange the initial quantization thresholds $t_1^0 \ldots t_N^0$ using Equation 1.47 by placing them halfway between the above computed initial reconstruction levels r_n^0, $n = 1 \ldots N$, where again, the superscript 0 represents the iteration index. This step generates the updated set of quantization thresholds $t_1^1 \ldots t_N^1$.

Step 6: Evaluate the performance of the current quantizer design in terms of

$$SNR = 10 \log_{10} \left[\frac{\sum_{t=1}^{T} (s^t)^2}{mse^t} \right]$$

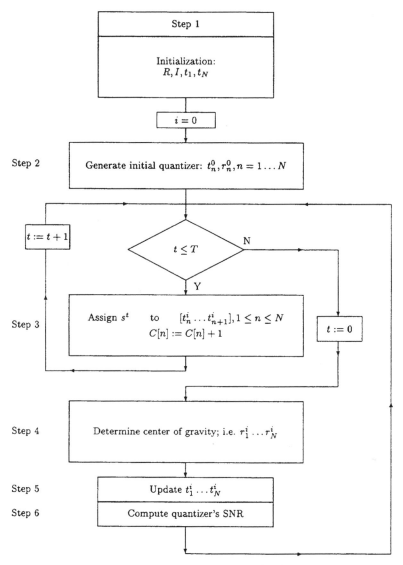

Figure 1.16 Lloyd-Max algorithm flowchart.

Recursion: Repeat **Steps 3–6** by iteratively updating r_n^i, t_n^i for all bins $n = 1 \ldots N$, until the iteration index i reaches its maximum I, while monitoring the quantizer SNR performance improvement given above.

Note that it is important to invoke the algorithm several times, while using a different initial quantizer, in order to ascertain its proper convergence to a global optimum. The inner workings of the algorithm may place the reconstruction levels and thresholds more sparsely, where the PDF $p(s)$ is low and vice versa. If the input signal's statistics obey a U, G, L, or Γ distribution, the Lloyd-Max quantizer's SNR performance can be evaluated using Equations 1.44 and 1.47. Various authors have tabulated the achievable SNR values. Following Max [38], Noll and Zelinski [43] as well as Paez and Glisson [44], both Jayant and Noll [10] as

TABLE 1.1   Maximum Achievable SNR and mse in Case of Zero-mean, Unit-variance Input $[f(R)]$ for Gaussian (**G**) and Laplacian (**L**) PDFs for $R = 1, 2, \ldots 7$ © Prentice Hall, Jayant-Noll [10] 1984, p. 135 and Jain [45] 1989, p. 104

		$R = 1$	$R = 2$	$R = 3$	$R = 4$	$R = 5$	$R = 6$	$R = 7$
G	SNR (dB)	4.40	9.30	14.62	20.22	26.01	31.89	37.81
	$f(R)$	0.3634	0.1175	0.0345	0.0095	0.0025	0.0006	0.0002
L	SNR (dB)	3.01	7.54	12.64	18.13	23.87	29.74	35.69
	$f(R)$	0.5	0.1762	0.0545	0.0154	0.0041	0.0011	0.0003

well as Jain [45] collected these SNR values, which we summarised in Table 1.1 for G and L distributions. Jayant and Noll [10] as well as Jain [45] also tabulated the corresponding t_n and r_n values for a variety of PDFs and R values.

Note in Table 1.1 that apart from the achievable maximum SNR values the associated quantizer mse $f(R)$ is given as a function of the number of quantization bits R. When designing a quantizer for an arbitrary nonunity input variance σ_s^2, the associated quantization thresholds and reconstruction levels must be appropriately scaled by σ_s^2. In case of a large input variance, the reconstruction levels may have to be sparsely spaced in order to cater for the signal's expanded dynamic range. Therefore the reconstruction mse σ_q^2 must also be scaled by σ_s^2, giving:

$$\sigma_q^2 = \sigma_s^2 \cdot f(R).$$

Here we curtail our discussion of **zero-memory quantization** techniques; the interested reader is referred to the excellent in-depth reference [10] by Jayant and Noll for further details. Before we move on to predictive coding techniques, the reader is reminded that in Section 1.2 we observed how both the time- and the frequency-domain features of the speech signal exhibit redundancy. In the next chapter we introduce a simple way of exploiting this redundancy in order to achieve better coding efficiency and reduce the required coding rate from 64 kbps to 32 kbps.

1.5 CHAPTER SUMMARY

In this chapter we provided a rudimentary characterization of voiced and unvoiced speech signals. It was shown that voice speech segments exhibit a quasi-periodic nature and convey significantly more energy than the more noise-like unvoiced segments. Because of their quasi-periodic nature, voiced segments are more predictable; in other words, they are more amenable to compression.

These discussions were followed by a brief introduction to the digitization of speech and to basic waveform coding techniques. The basic principles of logarithmic compression were highlighted, and the optimum nonuniform Max-Lloyd quantization principle was introduced. In the next chapter we introduce the underlying principles of more efficient predictive speech coding techniques.

2

Predictive Coding

2.1 FORWARD PREDICTIVE CODING

In a simplistic but plausible approach one could argue that if the input signal was correlated, the previous sample could be used to predict the present one. If the signal were predictable, the prediction error constituted by the difference of the current sample and the previous one would be significantly smaller on the average than the input signal. This reduces the region of uncertainty in which the signal to be quantized can reside and allows us to use either a reduced number of quantization bits or a better resolution in coding.

Redundancy reduction is achieved by subtracting the signal's predicted value from the current sample to be encoded and hence forming the **prediction error**. We have shown in the previous chapter, how PCM employs an **instantaneous or zero memory quantizer**. Differential Pulse Code Modulation (DPCM) and other linear predictive codecs (LPC) exploit knowledge over the history of the signal and hence reduce its correlation, variance and, ultimately, the bitrate required for its quantization. In a system context this will reduce the bandwidth required for a speech user and thus allow the system to support more users in a given bandwidth.

Recall that **redundancy** exhibits itself both in terms of the power spectral density (PSD) and the autocorrelation function (ACF), as it was demonstrated by Figures 1.2 and 1.4 in the case of voiced speech signals. The flatter the ACF, the more predictable the signal to be encoded and the more efficient its predictive encoding. This redundancy is also exhibited in terms of the nonflat PSD.

Let us now refine this simple predictive approach based on the immediately preceding sample and consider the more general predictive coding schematic shown in Figure 2.1, where the predictor block generates a predicted sample $\tilde{x}(n)$ by some rule to be described at a later stage. This scheme is often referred to as a **forward predictive** arrangement. If the input signal samples are represented by R-bit discrete values and an integer arithmetic is employed, where the quantizer is assumed to be simply a parallel to serial converter, which does not

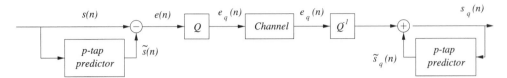

Figure 2.1 Block diagram of a forward predictive codec using p-tap prediction.

introduce any quantization impairment, then $s(n)$, $\tilde{s}(n)$ and $e(n) = e_q(n)$ are all represented by integer values. Since

$$e_q(n) = e(n) = s(n) - \tilde{s}(n), \tag{2.1}$$

we can generate the decoded speech $s_q(n) = s(n)$ by the help of the predictor at the decoder's end of the speech link by simply adding the quantized predicted value $\tilde{s}_q(n)$ to $e_q(n) = e(n)$ as follows:

$$s_q(n) = \tilde{s}_q(n) + e_q(n). \tag{2.2}$$

2.2 DPCM CODEC SCHEMATIC

Recall from the previous chapter that in our forward predictive codec we assumed that no transmission errors occurred. Unfortunately, however the idealistic assumptions of Section 2.1 do not hold in the presence of transmission errors or if the quantizer introduces quantization distortion, which is typically the case, if bitrate economy is an important factor. These problems can be circumvented by the **backward predictive** scheme of Figure 2.2, where the input signal s_n is predicted on the basis of a backward-oriented predictor. The operation of this arrangement is the subject of Section 2.3.

Observe in Figure 2.2 that in contrast to the forward predictive scheme of Figure 2.1 the input signal $s(n)$ is predicted not from the previous values of $s(n - k)$, $k = 1 \ldots p$, but from:

$$s_q(n) = \tilde{s}(n) + e_q(n). \tag{2.3}$$

Since the locally reconstructed signal $s_q(n)$ is contaminated by the quantization noise of $q(n) = e(n) - e_q(n)$ inherent in $e_q(n)$, one could argue that this prediction will be probably a less confident one than that based on $s(n - k)$, $k = 1 \ldots p$, which might affect the coding efficiency of the scheme. Observe, however, that the signal $s_q(n)$ is also available at the decoder, regardless of the accuracy of the quantizer's resolution. Although in case of transmission errors this is not so because of the codec's stabilizing predictive feedback

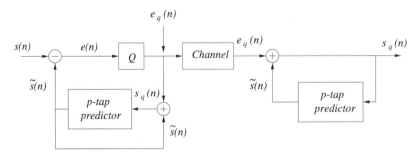

Figure 2.2 Block diagram of a DPCM codec using p-tap prediction.

loop the effect of transmission errors decays, while in case of the forward predictive scheme of Figure 2.1 the transmission errors persist. Observe in Figure 2.2 that the encoder's backward-oriented bottom section is identical to the decoder's schematic. Therefore, it is referred to as the **local decoder**. The local decoder is an important feature of most predictive codecs invoked, in order to be able to mitigate the effects of transmission errors. The output of the local decoder is the **locally reconstructed signal** $s_q(n)$.

The DPCM codec seen in Figure 2.2 is characterized by the following equations:

$$
\begin{aligned}
e(n) &= s(n) - \tilde{s}(n) \\
e_q(n) &= Q[e(n)] \\
s_q(n) &= \tilde{s}(n) + e_q(n).
\end{aligned}
\tag{2.4}
$$

Since the variance of the prediction error $e(n)$ is typically lower than that of the signal $s(n)$, (i.e. $\sigma_e < \sigma_s$), the bitrate required for the quantization of $e(n)$ can be reduced, while maintaining an identical distortion or SNR value.

Following these rudimentary deliberations on redundancy removal using predictive coding, let us now focus our attention on the design of a general p-tap predictor.

2.3 PREDICTOR DESIGN

2.3.1 Problem Formulation

Because of the redundancy inherent in speech, any present sample can be predicted as a linear combination of p past speech samples as follows:

$$
\tilde{s}(n) = \sum_{k=1}^{p} a_k s(n - k),
\tag{2.5}
$$

where p is the predictor order, a_k represents the linear predictive filter coefficients, and $\tilde{s}(n)$ the predicted speech samples. The prediction error, $e(n)$, is then given by

$$
\begin{aligned}
e(n) &= s(n) - \tilde{s}(n) \\
&= s(n) - \sum_{k=1}^{p} a_k s(n - k) \\
&= \sum_{k=0}^{p} a_k s(n - k) \text{ where } a_0 = 1.
\end{aligned}
\tag{2.6}
$$

Upon taking the z-transform of Equation 2.6, we arrive at:

$$
E(z) = S(z) \cdot A(z),
\tag{2.7}
$$

which reflects the **linearly separable speech generation model** of Figure 1.1 in Section 1.2. Observe that:

$$
A(z) = 1 - \sum_{k=1}^{p} a_k z^{-k} = \sum_{k=0}^{p} a_k z^{-k}, \qquad a_0 = 1
\tag{2.8}
$$

can be expressed as:

$$A(z) = 1 - a_1 \cdot z^{-1} - a_2 \cdot z^{-2} - \ldots - a_p \cdot z^{-p}$$
$$= (z - z_i) \ldots (z - z_p), \tag{2.9}$$

which explicitly shows that this polynomial has only zeros but no poles. Thus it is usually referred to as an **all-zero filter**. Expressing the speech signal $S(z)$ in terms of $E(z)$ and $A(z)$ gives:

$$S(z) = \frac{E(z)}{A(z)} = E(z) \cdot H(z), \tag{2.10}$$

suggesting that any combination of $E(z)$ and $H(z) = 1/A(z)$ could adequately model the input signal $S(z)$. However, when the prediction residual $e(n)$ is quantized to $e_q(n)$ in order to achieve bitrate economy, this is not true. We will show that it is an attractive approach to determine the predictor coefficients a_k by minimizing the expected value of the mean squared prediction error of Equation 2.6.

Again, in accordance with our introductory observations in Figure 1.1 of Section 1.2, generating the synthesized speech using Equation 2.10 can also be portrayed as exciting the **all-pole synthesis filter** $H(z) = 1/A(z)$ with the excitation signal $E(z)$. If the predictor removes the redundancy from the speech signal by minimizing the prediction residual, $e(n)$ becomes unpredictable, that is, pseudo-random with an essentially flat spectrum, while $H(z) = 1/A(z)$ models the **spectral envelope of the speech**. Because of the relationship $A(z) = H^{-1}(z)$, the filter $A(z)$ is often referred to as the LPC **inverse filter**.

The expected value (E) of the mean squared prediction error of Equation 2.6 can be written as:

$$E[e^2(n)] = E\left\{ \left[s(n) - \sum_{k=1}^{p} a_k s(n - k) \right]^2 \right\}. \tag{2.11}$$

In order to arrive at the optimum **Linear Predictive Coding** (LPC) coefficients, we compute the partial derivative of Equation 2.11 with respect to all LPC coefficients and set $\partial E / \partial a_i = 0$ for $i = 1 \ldots p$, which yields a set of p equations for the p unknown LPC coefficients a_i as follows:

$$\frac{\partial E[e^2(n)]}{\partial a_i} = -2 \cdot E\left\{ \left[s(n) - \sum_{k=1}^{p} a_k s(n - k) \right] s(n - i) \right\} = 0, \tag{2.12}$$

yielding:

$$E\{s(n)s(n - i)\} = E\left\{ \sum_{k=1}^{p} a_k s(n - k)s(n - i) \right\}. \tag{2.13}$$

Upon exchanging the order of the summation and expected value computation at the right-hand side of Equation 2.13, we arrive at:

$$E\{s(n)s(n - i)\} = \sum_{k=1}^{p} a_k E\{s(n - k)s(n - i)\}, \qquad i = 1, \ldots, p. \tag{2.14}$$

Observe in the above equation that

$$C(i, k) = E\{s(n - i)s(n - k)\}, \tag{2.15}$$

represents the input signal's covariance coefficients, which allows us to rewrite the set of p Equations 2.14 in a terser form as follows [46], [47]:

$$\sum_{k=1}^{p} a_k C(i, k) = C(i, 0), \qquad i = 1, \dots, p. \tag{2.16}$$

2.3.2 Covariance Coefficient Computation

The preceding set of equations is often encountered in various signal processing problems, when minimizing some error term as a function of a set of coefficients. Apart from linear predictive coding, this set of equations is arrived at in optimizing other adaptive filters, such as channel equalizers [48, 49] or in the autoregressive filter representation of error correction block codes [50]. Ideally, the covariance coefficients would have to be determined by evaluating the expected value term in Equation 2.15 over an infinite interval, but this is clearly impractical.

In low-complexity codecs or if the input signal can be considered to possess **stationary statistical properties**, implying that the signal's statistics are time-invariant, the covariance coefficients can be determined using a sufficiently long training sequence. Then the set of p Equations 2.16 can be solved, for example, by **Gauss-Jordan elimination** [51]. A more efficient recursive search algorithm referred to as the **Levison-Durbin algorithm** [6, 47] is highlighted later in this chapter.

In more complex, low-bitrate codecs, the LPC coefficients are determined adaptively for shorter **quasi-stationary** input signal segments in order to improve the efficiency of the predictor, that is, to reduce the prediction error's variance and so to improve the coding efficiency. These time-variant LPC coefficients must be quantized and transmitted to the decoder in order to ensure that the encoder's and decoder's p-tap predictors are identical. This technique, which is often referred to as **forward-adaptive prediction**, implies that at the encoder also the quantized coefficients must be employed, although there the more accurate unquantized coefficients are also available. Another alternative is to invoke the principle of **backward-adaptive prediction**, where the LPC coefficients are not transmitted to the decoder. Instead they are recovered from previous segments of the decoded signal. Again, in order to ensure the identical operation of the local and distant decoders, the encoder also uses previous decoded signal segments rather than unquantized input signal segments in order to determine the LPC coefficients. For the sake of efficient prediction, the delay associated with backward-adaptive prediction must be as low as possible, while the decoded signal quality has to be as high as possible. Hence, this technique is not used in low-bitrate applications, where the typically higher delay and higher coding distortion would reduce the predictor's efficiency. Here we will not analyze the specific advantages and disadvantages of the forward- and backward-adaptive schemes, but during our further discussion we will return to these codec classes and augment their main features by referring to practical standardized coding arrangements belonging to both families.

In spectrally efficient high-quality forward-adaptive predictive codecs, the covariance coefficients $C(i, k)$ of Equation 2.15 are typically computed for intervals during which the signal's statistics can be considered quasi-stationary. A severe limitation, however, is that the quantized coefficients must be transmitted to the decoder: thus, their frequent transmission may result in excessive bitrate contributions. In the case of backward-adaptive arrangements, this bitrate limitation does not exist. Hence, typically higher order predictors can and must be used in order to achieve high prediction gains. The more stringent limiting factor, however, becomes the computational complexity associated with the frequent solution of the high-order set of Equation 2.16. Since the low-delay spectral estimation requirement does not tolerate the too infrequent updating of the LPC coefficients, because the associated coefficients would become obsolete and inaccurate.

2.3.3 Predictor Coefficient Computation

A variety of techniques have been proposed for limiting the range of covariance computation [48, 52], of which the most frequently used ones are the **autocorrelation method** and the **covariance method** [6].

Here we follow the approach proposed by Makhoul [53], Rabiner and Schaefer [6], Haykin [48], and Salami et al. [47], and briefly highlight the **autocorrelation method**, where the prediction error term of Equation 2.11 is now minimized over the finite interval of $0 \le n \le L_a - 1$ rather than $-\infty < n < \infty$. Hence, the covariance coefficients $C(i, k)$ are now computed from the following short-term expected value expression:

$$C(i, k) = \sum_{n=0}^{L_a+p-1} s(n-i)s(n-k), \qquad \begin{matrix} i = 1, \ldots, p, \\ k = 0, \ldots, p. \end{matrix} \tag{2.17}$$

Upon setting $m = n - i$, Equation 2.17 can be expressed as

$$C(i, k) = \sum_{m=0}^{L_a-1-(i-k)} s(m)s(m+i-k), \tag{2.18}$$

which suggests that $C(i, k)$ is the short-time autocorrelation of the input signal $s(m)$ evaluated at a displacement of $(i - k)$, giving:

$$C(i, k) = R(i - k), \tag{2.19}$$

where

$$R(j) = \sum_{n=0}^{L_a-1-j} s(n)s(n+j) = \sum_{n=j}^{L_a-1} s(n)s(n-j), \tag{2.20}$$

and $R(j)$ represents the speech autocorrelation coefficients. Thus, the set of p Equations 2.16 can now be reformulated as:

$$\sum_{k=1}^{p} a_k R(|i - k|) = R(i), \qquad i = 1, \ldots, p. \tag{2.21}$$

Alternatively, Equation 2.21 can be rewritten in a matrix form as:

$$\begin{pmatrix} R(0) & R(1) & R(2) & \ldots & R(p-1) \\ R(1) & R(0) & R(1) & \ldots & R(p-2) \\ R(2) & R(1) & R(0) & \ldots & R(p-3) \\ \vdots & \vdots & \vdots & \ddots & \vdots \\ R(p-1) & R(p-2) & R(p-3) & \ldots & R(0) \end{pmatrix} \cdot \begin{pmatrix} a_1 \\ a_2 \\ a_3 \\ \vdots \\ a_p \end{pmatrix} = \begin{pmatrix} R(1) \\ R(2) \\ R(3) \\ \vdots \\ R(p) \end{pmatrix}. \tag{2.22}$$

This $p \times p$ autocorrelation matrix has a Toeplitz structure in which all the elements along a certain diagonal are identical. Hence, Equation 2.22 can be solved without matrix inversion that would imply a computational complexity cubically related to p. There is a variety of efficient recursive algorithms that have a complexity proportional to the square of p for the solution of Toeplitz-type systems. The most well-known ones are the Berlekamp-

Massey algorithm [50] favored in error correction coding or the recursive Levinson-Durbin algorithm, which can be stated as follows [6, 47, 53]:

$$E(0) = R(0)$$

For $i = 1$ to p do

$$k_i = \left[R(i) - \sum_{j=1}^{i-1} a_j^{(i-1)} R(i-j) \right] / E(i-1) \qquad (2.23)$$

$$a_i^{(i)} = k_i$$

For $j = 1$ to $i-1$ do

$$a_j^{(i)} = a_j^{(i-1)} - k_i a_{i-j}^{(i-1)} \qquad (2.24)$$

$$E(i) = (1 - k_i^2) E(i-1). \qquad (2.25)$$

The final solution after p iterations is given by:

$$a_j = a_j^{(p)} \qquad j = 1, \ldots, p. \qquad (2.26)$$

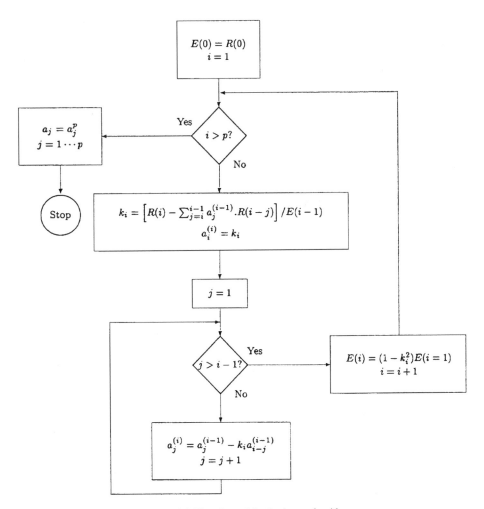

Figure 2.3 Flowchart of the Levinson algorithm.

where $E(i)$ in Equation 2.25 is the prediction error of an ith-order predictor. The flowchart of the Levinson-Durbin algorithm is depicted in Figure 2.3 in order to augment its exposition.

It is beneficial to define the **prediction gain**, which is the ratio of the expected value of the input signal's energy, namely, $R_s(0)$, and that of the prediction error energy $R_e(0)$ expressed in terms of the corresponding autocorrelation coefficients as follows:

$$G = \frac{R_s(0)}{R_e(0)} \qquad (2.27)$$

Note also that the prediction gain is often expressed in terms of dB. Let us now invoke a simple example to augment the above concepts.

EXAMPLE

The long-term one-step autocorrelation coefficient of the input signal was found to be $R_s(1)/R_s(0) = 0.9$. Determine the prediction gain in case of using the optimum one-tap predictor and with the aid of nonoptimum prediction using the previous sample as predicted value. Express these gains in dB.

From Equation 2.11 the prediction error variance can be expressed as:

$$E[e^2(n)] = E[[s(n) - a_1 s(n-1)]^2], \qquad (2.28)$$

yielding:

$$R_e(0) = R_s(0) - 2a_1 R_s(1) + a_1^2 R_s(0), \qquad (2.29)$$

where $R(0)$ and $R(1)$ represent the correlation coefficients at offsets of 0 and 1 sample, respectively. Upon setting the derivative of the above equation with respect to a_1 to zero we get:

$$a_1 = \frac{R_s(1)}{R_s(0)}, \qquad (2.30)$$

which is the **normalized one-step correlation** between adjacent input samples. Lastly, upon substituting the optimum coefficient from Equation 2.30 into Equation 2.29, we arrive at:

$$R_e(0) = R_s(0) - 2\frac{R_s(1)}{R_s(0)} R_s(1) + a^2 R_s(0)$$
$$= R_s(0)(1 - a_1^2). \qquad (2.31)$$

which gives the prediction gain as follows:

$$G = \frac{R_s(0)}{R_e(0)} = 1/(1 - a_1^2). \qquad (2.32)$$

For $a_1 = 0.9$ we have $G = 1/(1 - 0.81) = 5.26$, corresponding to 7.2 dB. When using $a_1 = 1$ in Equation 2.30—which corresponds to using the previous sample to predict the current one—we get $G = 1/0.2 = 5$, which is equivalent to about 7 dB. This result is very similar to the 7.2 dB gain attained, when using the optimum tap, which is due to the high adjacent-sample correlation. For lower correlation values, the prediction gain difference becomes more substantial, eroding to $G < 1$ for uncorrelated signals, where $R_s(1)/R_s(0) < 0.5$.

Returning to the Levinson-Durbin algorithm, we find that the internal variable k_i has a useful physical interpretation, when applying the Levinson-Durbin algorithm to speech signals. Namely, they are referred to as the **reflection coefficients**, and $-1 < k_i < 1$ are defined as

$$k_i = \frac{A_{i+1} - A_i}{A_{i+1} + A_i}, \qquad (2.33)$$

where A_i, $i = 1 \ldots p$ represents the area of an acoustic tube section, assuming that the vocal tract can be modeled by a set of p concatenated tubes of different cross section. The above definition of k_i implies that they physically represent the area ratios of the consecutive sections of the lossless acoustic tube model of the vocal tract [6, 47]. Rabiner and Schafer have shown [6] that the $-1 < k_i < 1$ condition is necessary and sufficient for all the roots of the polynomial $A(z)$ to be inside the unit circle in the z-domain, thereby guaranteeing the stability of the system transfer function $H(z)$. It has been shown that the autocorrelation method always leads to a stable filter $H(z)$. These issues will be revisited in Chapter 4, where the statistical properties of the a_i and k_i parameters will be characterized in terms of their PDFs in Figures 4.1 and 4.2 along with those of a range of other equivalent spectral parameters, which are more amenable to quantization for transmission.

The rectangular windowing of the input signal at the LPC analysis frame edges corresponds in the spectral domain to convolving the signal's spectrum with a sinc-function, which results in the **Gibbs oscillation**. In time-domain the rectangular windowing results in a high prediction error at the beginning and end of the segment, since the signal outside the interval was zero. This undesirable phenomenon can be mitigated by using smooth, tapering windows, such as the time-domain Hamming windowing, which employs the function:

$$w(n) = 0.54 - 0.46\cos(2\pi n/(L_a - 1)), \qquad 0 \leq n \leq L_a - 1, \qquad (2.34)$$

where the Hamming windowing frame length L_a is often longer than the length L of the input signal update frame. The LPC coefficients are typically interpolated between adjacent LPC frames in order to smooth the abrupt signal envelope changes at frame edges.

On the basis of the previously introduced adaptive predictor, which can adjust the predictor coefficients in order to accommodate signal statistics variations, we can modify the DPCM codec schematic of Figure 2.2 to portray these added features, as seen in Figure 2.4.

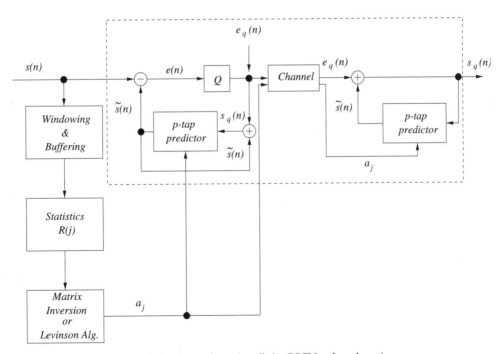

Figure 2.4 Adaptive forward predictive DPCM codec schematic.

Clearly, the filter coefficients a_k, $k = 1 \ldots p$ must be computed, as highlighted earlier in this section, using, for example, the autocorrelation method and the Levinson-Durbin algorithm, before encoding and transmitting them at the cost of an increased bitrate to the decoder. Furthermore, it is necessary to scale the input signal by the help of its variance in order to maintain near-unity input variance and hence achieve best quantization performance. To this effect, a simple but efficient adaptive quantization technique was proposed by Jayant [54], which introduces the notion of memory in the quantization process in order to use the previously quantized sample to control the quantizer's step-size. This method, which can be employed in both PCM and DPCM codecs, is the subject of our next section.

2.4 ADAPTIVE ONE-WORD MEMORY QUANTIZATION [54]

Adaptive one-word-memory quantization is a form of adaptive pulse code modulation (APCM) due to Jayant [54], in which the quantizer's instantaneous step-size is adjusted on the basis of the previous quantized sample in order to minimize the quantization distortion. The schematic of the quantizer is displayed in Figure 2.5. The philosophy behind this scheme is that if the previous quantized sample $s_q(n-1)$ is near the top level of the quantizer characteristic, then action must be taken to increase the quantizer's step-size Δ_n, since in case of correlated samples the forthcoming samples are similar to the current one and so there is a danger of quantizer characteristic overload or saturation. Similarly, if the previous quantized sample $s_q(n-1)$ is near the lowest quantization level, then too high a granular noise is inflicted, since the step-size is too small. This problem can then be mitigated by increasing the step-size. It is plausible, however, that the speed of step-size adaptation is critical, since various source signals have different statistical and spectral domain properties, which result in a different rate of change. Furthermore, the number of quantization bits R is also an important factor in determining the required step-size control parameters. In case of $R = 1$, for example, no magnitude information is available—only the sign of the signal, which precludes employment of this technique. The higher the number of bits R, the finer the step-size control.

Formulating the algorithm displayed in Figure 2.5 more rigorously, the Jayant quantizer [54] adapts its step-size Δ_n at each sampling instant. The quantized outputs $s_q(n)$ of an R-bit quantizer ($R > 1$) is of the form [54]:

$$s_q(n) = Q\{s(n)\}\frac{\Delta_n}{2} \tag{2.35}$$

where $|Q\{s(n)\}| = 1, 3, \ldots, 2^R - 1$; and $\Delta_n > 0$. The step-size Δ_n is given by the previous step-size Δ_{n-1} multiplied by a statistics-dependent, optimized, time-invariant function of the codeword magnitude $|Q\{s_q(n-1)\}|$,

$$\Delta_n = \Delta_{n-1} M(|Q\{s_q(n-1)\}|). \tag{2.36}$$

In practical terms, the step-size Δ_n can vary only over a limited dynamic range, from the minimum step-size $\Delta_{\min}$ to the maximum step-size $\Delta_{\max}$, which is expressed more formally as follows:

$$\Delta_n = \begin{cases} \Delta_{\min}; & \Delta_n < \Delta_{\min} \\ \Delta_{\max}; & \Delta_n > \Delta_{\max} \\ \Delta_n; & \text{otherwise.} \end{cases} \tag{2.37}$$

The multiplier function $M(\cdot)$ determines the rate of adaption for the step-size. For PCM and DPCM-encoded speech and video signals, Jayant tabuluated these multiplier values for a

$$s_q(n) = Q\{s(n)\}.\frac{\triangle n}{2}$$

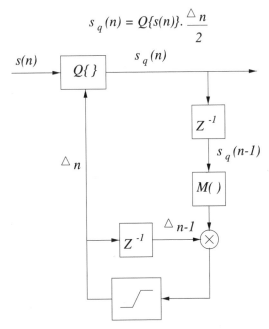

Figure 2.5 Schematic of Jayant's one-word memory quantizer, where the current step-size depends on the previous one, scaled by a multiplier, which is a function of the previous quantized sample.

TABLE 2.1 Jayant-Multipliers for R-bit PCM Quantizers

R	Multiplier $M(\cdot)$
2	0.6, 2.20
3	0.85, 1.00, 1.00, 1.50
	(0.9), 0.95, 1.50, 2.5—for video)
4	0.80, 0.80, 0.80, 0.80, 1.20, 1.60, 2.00, 2.40
5	0.85, 0.85, 0.85, 0.85, 0.85, 0.85, 0.85, 0.85,
	1.20, 1.40, 1.60, 1.80, 2.00, 2.20, 2.40, 2.60

range of quantizer resolutions [54] $R = 2\ldots5$, which are shown in Table 2.1 for PCM codecs. The values in brackets refer to video signals, and similar multipliers apply to DPCM codecs [54] as well. It is also interesting to observe that the step-size increment associated with $M > 1$ is typically more rapid than the corresponding step-size reduction corresponding to $M < 1$, since the onset of speech signals, for example, is more rapid than their decay.

Having highlighted the concept of adaptive one-word-memory quantization, let us now characterize the expected performance of DPCM codecs in contrast to PCM.

2.5 DPCM PERFORMANCE

In this brief performance analysis, we follow the approach proposed by Jain [45] and note that in contrast to PCM, where $s(n)$ is subjected to quantization, in case of DPCM the quantizer operates on the prediction error signal $e(n)$ having a variance of $\sigma_e^2 = E\{[e(n)]^2\}$, while the quantization error variance is assumed to be σ_q^2. Then applying the rate-distortion formula of Equation 1.5, the DPCM coding rate is given by:

$$R_{\mathrm{DPCM}} = \frac{1}{2} \log_2 \frac{\sigma_e^2}{\sigma_q^2} \text{ [bits/pixel]}. \tag{2.38}$$

When compared to PCM and assuming the same quantizer for both PCM and DPCM, we have a constant quantization distortion of σ_q^2, although the quantizer is applied in the former case to quantize the input signal $s(n)$, while in the latter to $e(n)$. Since typically $\sigma_s < \sigma_q$, the coding rate reduction due to using DPCM is yielded as follows [45]:

$$\Delta R = R_{\mathrm{PCM}} - R_{\mathrm{DPCM}} = \frac{1}{2} \log_2 \frac{\sigma_s^2}{\sigma_q^2} - \frac{1}{2} \log_2 \frac{\sigma_e^2}{\sigma_q^2} = \frac{1}{2} \log_2 \frac{\sigma_s^2}{\sigma_e^2} \tag{2.39}$$

giving a coding rate gain of:

$$\Delta R \approx 1.66 \cdot \log_{10} \left(\frac{\sigma_s}{\sigma_e} \right)^2 \text{ [bits/pixel]}. \tag{2.40}$$

For example, if $\sigma_s = 10 \cdot \sigma_e$, then we have $\Delta R = 3.332$, which means that a PCM codec having the same quantization error variance as a DPCM codec would require more than three additional quantization bits per sample, or the DPCM codec ensures in excess of 3 bits/ sample transmission rate saving. In general, the coding rate reduction $\Delta R = (\sigma_s / \sigma_e^2)$ due to DPCM coding depends on the ability to predict the input signal $s(n)$, that is on the intersample correlation. We have seen before that for minimum prediction error variance σ_e^2 the optimum 1-tap predictor coefficient is given by the adjacent-sample correlation.

For the variance of the feedforward prediction error, we have $\sigma_\varepsilon \leq \sigma_e$, since the prediction based on the locally decoded signal contaminated by quantization noise cannot be better than that based on the original signal. This fact does not contradict the previously argued statement that the reconstruction error variance of the DPCM codec is typically lower than that of the feedforward codec. If the number of quantization bits is high, we have $\sigma_\varepsilon \approx \sigma_e$. Hence, **the lower bound** on the DPCM coding rate is given by [45]:

$$R_{\min} = \frac{1}{2} \log_2 \frac{\sigma_\varepsilon^2}{\sigma_q^2} \leq R_{\mathrm{DPCM}}. \tag{2.41}$$

The SNR of the DPCM codec can be written as follows:

$$\mathrm{SNR}_{\mathrm{DPCM}} = 10 \, \log_{10} \frac{\sigma_s^2}{\sigma_q^2} = 10 \, \log_{10} \frac{\sigma_s^2}{\sigma_e^2 \cdot f(R)} \tag{2.42}$$

leading to:

$$\mathrm{SNR}_{\mathrm{DPCM}} \leq 10 \log_{10} \frac{\sigma_s^2}{\sigma_\varepsilon^2 \cdot f(R)} \tag{2.43}$$

where $f(R)$ is the quantizer mean square distortion function for R number of quantization bits in the case of a unit-variance input signal [45]. For an equal number of quantization bits, the SNR improvement of DPCM over PCM is then given by:

$$\Delta SNR = SNR_{DPCM} - SNR_{PCM}$$

$$= 10 \log_{10} \frac{\sigma_s^2}{\sigma_q^2} - 10 \log_{10} \frac{\sigma_d^2}{\sigma_q^2}$$

$$= 10 \log_{10} \frac{\sigma_s^2}{\sigma_d^2} \le 10 \log_{10} \frac{\sigma_s^2}{\sigma_\varepsilon^2}. \tag{2.44}$$

Again, assuming, for example, that $\sigma_s = 10 \cdot \sigma_e$, we have a 20 dB SNR improvement over PCM, while maintaining the same coding rate. In general, the gains achievable will depend on the signal's statistics, as well as on the predictor (P) and quantizer (Q) designs. Usually, Max-Lloyd quantization (MLQ) is used, which is designed to match the prediction error's PDF by allocating more bits, where the PDF is high, and less bits, where the probability of occurence is low. Ideally, the integral of the PDF over each quantization interval is constant.

If the prediction error's PDF matches a Gaussian, Laplacian, or Gamma distribution, the analytic quantizer designs tabulated in the literature [45] can be invoked. Otherwise specially trained Max-Lloyd quantizers must be employed, which can be designed using the training algorithm highlighted in Section 1.4.6. Having considered forward-adaptive predictive coding, in the next section let us now explore some of the features of backward-adaptive predictive schemes.

2.6 BACKWARD-ADAPTIVE PREDICTION

2.6.1 Background

In the preceding sections, we have considered forward-adaptive prediction, where the predictor coefficients must be transmitted to the decoder. Hence, they reserve some of the channel capacity available, and their computation requires buffering a segment of the input signal, over which spectral analysis can take place. These factors limit the affordable rate of predictor updates. In contrast, in backward-adaptive predictive schemes, the predictor coefficients are determined from the previously recovered speech and the frequency of the LPC update is practically limited only by the affordable codec complexity. During our later discourse, we will consider a variety of such backward-adaptive standard and nonstandard codecs, including the CCITT G.721, G.727, G.726, and G.728 codecs.

In this subsection following Jayant's deliberations [10], we introduce a predictor update technique, which is used in a range of standard ADPCM codecs, including the G.721, G.726, and G.727 schemes. The expected value of the mean squared prediction error of Equation 2.6 can also be expressed as:

$$\sigma_e^2 = E[e^2(n)] = E[(s(n) - \tilde{s}(n))^2]$$

$$= E\left[\left(s(n) - \sum_{k-1}^{p} a_k s(n-k)\right)^2\right], \tag{2.45}$$

where σ_e^2 is the variance of the prediction error $e(n)$. This formulation has led us to Equation 2.22, referred to as the Wiener-Hopf equation. Once the matrix was inverted either by Gauss-Jordan elimination or, for example, by the recursive Levinson-Durbin algorithm, the optimum predictor coefficient set becomes known and the actual achievable minimum expected value of the prediction residual energy can be computed. Using a convenient vectorial notation, the predictor coefficient vector and the speech vector used in the prediction process can be expressed as

$$\mathbf{a}^T = [a_{1,2} \ldots a_p]$$
$$\mathbf{s}^T = [s(n-1), s(n-2) \ldots s(n-k)].$$

Upon using this notation, Equation 2.45 can be rewritten as:

$$\sigma_e^2 = E\left[\left(s(n) - \sum_{k=1}^{p} a_k s(n-k)\right)^2\right]$$
$$= E[(s(n) - \mathbf{a}^T \cdot \mathbf{s})(s(n) - \mathbf{s}^T \cdot \mathbf{a})]$$
$$= E[s^2(n) - s(n) \cdot \mathbf{s}^T \cdot \mathbf{a} - \mathbf{a}^T \cdot \mathbf{s} \cdot s(n) + \mathbf{a}^T \cdot \mathbf{s} \cdot \mathbf{s}^T \cdot \mathbf{a}] \tag{2.46}$$

Upon taking the expected value of the individual terms and using the notation of Equations 2.22 and 2.46, we arrive at:

$$\sigma_e^2(\mathbf{a}) = \sigma_s^2 - 2 \cdot \mathbf{a}^T \cdot \mathbf{r} + \mathbf{a}^T \cdot \mathbf{R} \cdot \mathbf{a}, \tag{2.47}$$

where $r = E\{s(n)\mathbf{s}^T\}$ and $R = E\{\mathbf{s}\mathbf{s}^T\}$.

Explicitly, Equation 2.47 quantifies the expected value of the mean squared prediction error as a function of the predictor coefficient vector a and the optimum vector can be computed from Equation 2.22 by matrix inversion or using a recursive solution. However, since the input speech has a time-variant statistical behavior, the optimum coefficient vector is also time-variant. Given an initial vector a^{opt}, it is possible to devise an adaptation algorithm, which seeks to modify the coefficient vector in a sense to reduce $\sigma_e^2(\mathbf{a})$ in Equation 2.47. The gradient of $\sigma_e^2(\mathbf{a})$ with respect to the coefficient vector is an indicator of the required changes in a in order to minimize $\sigma_e^2(\mathbf{a})$. This can be written more formally using Equation 2.47 as:

$$\frac{d\sigma_e^2(\mathbf{a})}{d\mathbf{a}} = 2\mathbf{R} \tag{2.48}$$

$$\Delta\sigma_e^2(\mathbf{a}) \approx 2\mathbf{R} \cdot \Delta\mathbf{a}$$
$$\approx 2\mathbf{R} \cdot (\mathbf{a} - \mathbf{a}^{opt}), \tag{2.49}$$

which demonstrates that the deviation of $\sigma_e^2(\mathbf{a})$ from its minimum value depends on both the speech signal's correlation quantified by $\mathbf{R}$ and the difference between the optimum and current coefficient vector, namely, $(\mathbf{a} - \mathbf{a}^{opt})$. The predictor coefficients can be updated on a sample-by-sample or block-by-block basis using techniques such as the **steepest descent** or Kalman filtering [10] method.

Here we consider a **pole-zero predictor**, which is depicted in Figure 2.6, where $A(z)$ and $B(z)$ represent the all-pole and all-zero filters, respectively. Their transfer functions are given by:

$$A(z) = \sum_{k=1}^{N_f} a_i z^{-k}$$

$$B(z) = \sum_{k=0}^{N_z} b_i z^{-k} \tag{2.50}$$

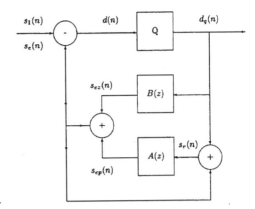

Figure 2.6 Pole-zero predictor schematic.

and a_k, b_k represent the filter coefficients, while N_f and N_z the filter orders. This predictor is studied in more depth in the next section.

2.6.2 Stochastic Model Processes

In order to better understand the behavior of this predictor, we have to embark on a brief discourse concerning stochastic model processes. Our pole-zero predictor belongs to the family of the **autoregressive** (AR), **moving average** (MA) processes, which are jointly referred to as ARMA processes. An all-pole model or autoregressive model is usually derived from the previously introduced predictor formula of Equation 2.5, which is generally presented in the following form:

$$s(n) = \sum_{k=1}^{p} a_k \cdot s(n-k) + e(n) \; \forall_n, \tag{2.51}$$

where $e(n)$ is an uncorrelated, zero-mean, random input sequence with variance σ^2, as one would expect from a reliable predictor, removing all the predictable redundancy. The schematic of an AR process is displayed in Figure 2.7.

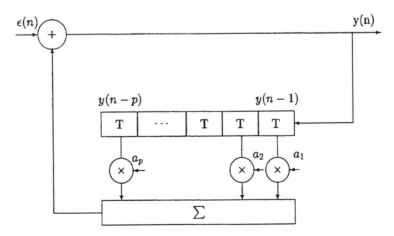

Figure 2.7 Markov model of order p.

From Equation 2.51 the **transfer function of the all-pole AR model** can be derived as follows:

$$e(n) = s(n) - \sum_{k=1}^{p} a_k \cdot s(n-k)$$

$$E(z) = S(z) - \sum_{k=1}^{p} a_k \cdot S(z)z^{-k}$$

$$= S(z)\left[1 - \sum_{k=1}^{p} a_k \cdot z^{-k}\right]$$

leading to:

$$H(z) = \frac{S(z)}{E(z)} = \frac{1}{1 - \sum_{k=1}^{p} a_k \cdot z^{-k}} = \frac{1}{A(z)}. \tag{2.52}$$

As expected, this transfer function exhibits poles at the z values, where $A(z)$ becomes zero.

By contrast, a **MA process** is defined as:

$$s(n) = \sum_{k=0}^{q} b_k e(n-k), \tag{2.53}$$

expressing the random sequence $s(n)$ as a weighted sliding sum of the previous q samples of $e(n)$, where again $e(n)$ is a zero-mean, uncorrelated random sequence with a variance σ^2. The transfer function of a MA model can be expressed as:

$$B(z) = \frac{S(z)}{E(z)} = \sum_{k=0}^{q} b_k z^{-k}, \tag{2.54}$$

which is shown in Figure 2.8.

Upon returning to our pole-zero predictor of Figure 2.6, we have:

$$s_{ep}(n) = \sum_{k=1}^{p} a_k s_r(n-k) \tag{2.55}$$

and

$$s_{ez}(n) = \sum_{k=0}^{q} b_k d_q(n-k) \tag{2.56}$$

and the reconstructed signal $s_r(n)$ of Figure 2.6 can be written as the sum of the quantized prediction residual $d_q(n)$ and the estimated signal $s_e(n)$ as:

$$s_r(n) = s_e(n) + d_q(n),$$

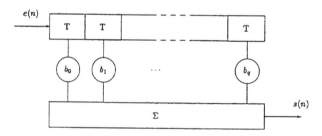

Figure 2.8 MA process generation.

while the estimated signal is the sum of the two predictors, giving:

$$s_e(n) = s_{ez}(n) + s_{ep}(n).$$

Both the all-zero and all-pole predictor coefficients can be updated using the **gradient or steepest descent** algorithm, which modifies each of the pole-zero predictor's coefficients in every iteration by an amount proportional to the error gradient with respect to the specific coefficient concerned, but opposite in terms of its sign, in order to reduce the error variance at each iteration yielding:

$$a_k(i + 1) = a_k(i) - C(i) \cdot \frac{d\sigma_e^2(a_k)}{da_k} \quad k = 1 \dots p$$

$$b_k(i + 1) = b_k(i) - C(i) \frac{d\sigma_e^2(b_k)}{db_k} \quad k = 1 \dots q. \tag{2.57}$$

The coefficients at iteration $(i + 1)$ are derived by subtracting the $C(i)$ scaled gradient of the prediction error variance from the coefficients at iteration i, where $C(i)$ is an adaptation-speed control factor. As expected, the adaptation-speed control factor $C(i)$ has a strong influence on the properties of the predictor adaptation loop. If a larger value is selected, the algorithm achieves a faster convergence at the cost of a higher steady-state tracking error and vice versa. Furthermore, in Equation 2.57 it is possible to use the prediction error itself rather than its longer-term variance. Here we curtail our discussions concerning various practical implementations of the gradient-algorithm based predictor adaptation, and we will revisit this issue in our discourse on the G.721 standard ADPCM codec. Our deliberations concerning backward-adaptive predictive codecs will be further extended at a later stage in the context of the vector-quantized CCITT G.728 low-delay 16 kbps codec in Chapter 8.

Following this rudimentary introduction to backward-adaptive predictive coding, let us now embark on highlighting the details of a specific standardized speech codec—the 32 kbps CCITT G.721 **Adaptive Differential Pulse Code Modulation** (ADPCM) codec. This codec became popular in recent years due to its very low implementational complexity and high speech quality. It has also been adopted by a number of wireless communications standards, such as the British CT2 cordless telephone system [55, 56], the Digital European Cordless Telephone (DECT) system [57, 58], and the Japanese Personal Handy Phone (PHP) system [59].

2.7 THE 32 KBPS G.721 ADPCM CODEC [60]

2.7.1 Functional Description of the G.721 Codec

As mentioned, the 32 kbit/s transmission rate ADPCM codec was specified in the CCITT G.721 Recommendation. The encoder/decoder pair is shown in Figure 2.9, and since it is essentially a waveform codec, apart from speech, it is also capable of transmitting data signals.

As seen in the figure, the A-law or μ-law companded PCM signal is first converted into linear PCM format, since all signal processing steps take place in the linear PCM domain. The input signal's estimate produced by the adaptive predictor is subtracted from the input in order to produce a difference signal having a lower variance. This lower-variance difference signal can then be adaptively quantized with lower noise variance than the original signal, using a 4-bit adaptive quantizer. Assuming a sampling frequency of 8 kHz, an 8-bit PCM sample is represented by a 4-bit ADPCM sample, giving a transmission rate of 32 kbit/s. This

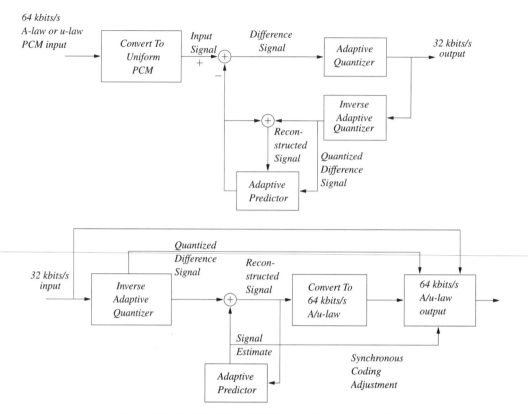

Figure 2.9 Detailed G.721 ADPCM encoder/decoder schematic.

ADPCM stream is transmitted to the decoder. Furthermore, it is locally decoded, using the **inverse adaptive quantizer in G.721 codec**, to deliver the locally reconstructed quantized difference signal. This signal is added to the previous signal estimate in order to yield the locally reconstructed signal. Based on the quantized difference signal and the locally reconstructed signal, the adaptive predictor derives the subsequent signal estimate, and so on.

The ADPCM decoder is constituted by the local decoder part of the encoder, and in addition, it comprises the linear PCM to A-law or μ-law converter. The synchronous coding adjustment block attempts to eliminate the cumulative tandem distortion occurring in subsequent synchronous PCM/ADPCM operations. The further specific implementational details of the G.721 Recommendation will be described with reference to Figure 2.9, where the notation of the G.721 standard [60] has been adopted in order to avoid confusion.

2.7.2 Adaptive Quantizer

A 16-level or $R = 4$-bit adaptive quantizer is used to quantize the prediction error or difference signal $d(k) = s_1(k) - s_e(k)$, which is converted to base 2 logarithmic representation prior to quantization and scaled by the signal $y(k)$, which is output by the quantizer scale factor adaptation block seen in the schematic of Figure 2.9. Note that scaling in the logarithmic domain corresponds to subtracting the scaling factor $y(k)$. The scaled quantizer input/output ranges are given in Table 2.2, where the quantizer output is represented by a 4-bit number $I(k)$, and in its 4-bit binary representation the first bit determines the sign of $d(k)$. The

TABLE 2.2   Scaled Adaptive Quantizer Characteristic, ©CCITT G.721

Scaled Quant. Input Range $\log_2 \lvert d(k) \rvert - y(k)$	$\lvert I(k) \rvert$	Scaled Quant. Output Range $\log_2 \lvert d_q(k) \rvert - y(k)$
3.16–∞	7	3.34
2.78–3.16	6	2.95
2.42–2.78	5	2.59
2.04–2.42	4	2.23
1.58–2.04	3	1.81
0.96–1.58	2	1.29
−0.05–0.96	1	0.53
−∞–0.05	0	−1.05

4-bit sequence $I(k)$ is both transmitted and locally decoded, using the inverse adaptive quantizer reconstruction values at the right hand side of Table 2.2. This delivers the quantized prediction error samples $d_q(k)$. Observe furthermore in Figure 2.9 that $I(k)$ is also input to **the adaptation speed control** and **quantizer scale factor adaptation** blocks, which are considered in more depth below.

2.7.3 G.721 Quantizer Scale Factor Adaptation

The quantizer scale factor adaptation block derives the scaling factor $y(k)$ of Table 2.2. Its inputs are the 4-bit $I(k)$ values and the adaptation speed control parameter $a_l(k)$. The quantizer scaling is rapidly changing for signals characterized by large fluctuations, such as speech signals. By contrast, the scaling is slowly varying for signals resulting in slowly changing difference signals, such as voice band data or signaling tones. The fast scale factor $y_u(k)$ is recursively computed from the previously introduced logarithmic scale factor $y(k)$ in the base 2 logarithmic domain using

$$y_u(k) = (1 - 2^{-5})y(k) + 2^{-5}W[I(k)],$$
$$y_u(k) \approx 0.97y(k) + 0.03 \cdot W[I(k)], \tag{2.58}$$

where $y_u(k)$ is restricted to the range

$$1.06 \leq y_u(k) \leq 10.00.$$

In other words, $y_u(k)$ is a weighted sum of $y(k)$ and $I(k)$, where the dominant part is usually $y(k)$. The leakage factor $(1 - 2^{-5}) \approx 0.971$ allows for the decoder to "forget" the effect of eventual transmission errors. The factor $W(I)$ is specified in the G.721 Recommendation as seen in Table 2.3.

The current value of the slow quantizer scale factor $y_l(k)$ is derived from the fast scale factor $y_u(k)$ and from the slow scaling factor's previous value $y_l(k-1)$, using:

$$y_l(k) = (1 - 2^{-6})y_l(k-1) + 2^{-6}y_u(k),$$
$$y_l(k) \approx 0.984y_l(k-1) + 0.016y_u(k). \tag{2.59}$$

Then, according to the G.721 Recommendation, the fast and slow scale factors are combined to form the scale factor $y(k)$:

$$y(k) = a_l(k)y_u(k-1) + [1 - a_l(k)]y_l(k-1), \tag{2.60}$$

where the **adaptation speed control factor** is constrained to the range $0 \leq a_l \leq 1$, and we have $a_l \approx 1$ for speech signals, whereas $a_l \approx 0$ for data signals. Therefore, for speech signals

TABLE 2.3   Definitioin of the Factor $W(I)$ ©CCITT G.721

$|I|$	7	6	5	4	3	2	1	0
$W(I)$	69.25	21.25	11.50	6.12	3.12	1.69	0.25	−0.75

the fast scaling factor $y_u(k)$ dominates, while for data signals the slow scaling factor $y_l(k)$ prevails.

2.7.4 G.721 Adaptation Speed Control

The computation of the adaptation speed control is based on two measures of the average value of $I(k)$. Namely, d_{ms} describes the relatively short-term average of $I(k)$, while d_{ml} constitutes a relatively long-term average of it, which are defined by the G.721 standard as:

$$d_{ms}(k) = (1 - 2^{-5})d_{ms}(k - 1) + 2^{-5}F[I(k)] \qquad (2.61)$$

and

$$d_{ml}(k) = (1 - 2^{-7})d_{ml}(k - 1) + 2^{-7}F[I(k)], \qquad (2.62)$$

where $F[I(k)]$ is given in the G.721 recommendation as specified by Table 2.4. Explicitly, due to the higher scaling factor of 2^{-5}, the short-term average is more dependent on the current value of $I(k)$ than the long-term average, although both averages more closely resemble their own 2^{-7} scaled previous values due to the near-unity scaling of their preceding values. As a result of the zero-valued weighting function $F[I(k)]$ these averages do not take into account the value of $I(k)$ if it happens to be small.

From the above averages, the variable $a_p(k)$—which will be used in the definition of the adaptation speed control factor—was defined by the G.721 recommendation as follows:

$$\begin{aligned}
a_p(k) &= (1 - 2^{-4})a_p(k - 1) + 2^{-3} \quad \text{if } |d_{ms}(k) - d_{ml}(k)| \geq 2^{-3}d_{ml}(k) \\
a_p(k) &= (1 - 2^{-4})a_p(k - 1) + 2^{-3} \quad \text{if } y(k) < 3 \\
a_p(k) &= (1 - 2^{-4})a_p(k - 1) \qquad\qquad \text{otherwise.}
\end{aligned} \qquad (2.63)$$

More explicitly, the adaption speed control factor $a_p(k)$ is increased and in the long term tends toward the value 2, if the normalized short- and long-term average difference $[d_{ms}(k) - d_{ml}(k)]/d_{ml}(k) \geq 2^{-3}$—that is, if the magnitude of $I(k)$ is changing. Although it is not obvious at first sight, this is because of the factor 2 difference between the positive and negative scaling factors of 2^{-3} and 2^{-4} in Equation 2.63. By contrast, the adaption speed control factor a_p is decreased and tends to zero, if the difference of the above short- and long-term prediction error averages is relatively small—that is, if $I(k)$ is near-constant. This is due to the continuous decrementing action of the 2^{-4} factor in the third line of Equation 2.63. Furthermore, for an idle channel, where no significant scaling is required and the scaling

TABLE 2.4   Definition of the Factor $F[I(k)]$ ©CCITT G.721

$|I(k)|$	7	6	5	4	3	2	1	0
$f[i(K)]$	7	3	1	1	1	0	0	0

factor satisfies $y(k) < 3$, the quantity $a_p(k)$ is increased and also tends to 2, regardless of the value of the above normalized difference.

Finally, the adaptation speed control factor a_l used in Equation 2.60 is derived by limiting a_p according to the G.721 recommendation, as follows:

$$a_l(k) = 1 \qquad \text{if } a_p(k-1) > 1,$$
$$a_l(k) = a_p(k-1) \text{ if } a_p(k-1) \leq 1. \tag{2.64}$$

This limiting operation renders the actual value of $a_p(k)$ irrelevant, as long as it is larger than 1. By keeping $a_l(k)$ constant until $a_p(k)$ falls below 1, this condition postpones the premature start of a fast to slow transition, if the differences in the average value of $I(k)$ were low for only a limited period of a few sampling intervals.

2.7.5 G.721 Adaptive Prediction and Signal Reconstruction

Let us now concentrate on the action of the adaptive predictor of Figure 2.9, which generates the signal estimate $s_e(k)$ from the quantized difference signal $d_q(k)$, as also seen in Figure 2.10. Much of this adaptive predictor design research is due to Jayant. Explicitly, the predictor's transfer function is characterized by six zeros and two poles. This pole-zero predictor models the spectral envelope of a wide variety of input signals efficiently. Note, however, that the previously described Levinson-Durbin algorithm was applicable to the problem of an all-pole model only. The reconstructed signal is given by:

$$s_r(k-i) = s_e(k-i) + d_q(k-i), \tag{2.65}$$

where the signal estimate $s_e(k-i)$ seen in Figure 2.9 is derived from a linear combination of the previous reconstructed samples and that of the previous quantized differences d_q, using:

$$s_e(k) = \sum_{i=1}^{2} a_i(k-1)s_r(k-i) + \sum_{i=1}^{6} b_i(k-1)d_q(k-i), \tag{2.66}$$

where the factors a_i, $i = 1 \ldots 2$ and b_i, $i = 1 \ldots 6$ represent the corresponding predictor coefficients. Both sets of predictor coefficients—$a_i(k)$ and $b_i(k)$—are recursively computed

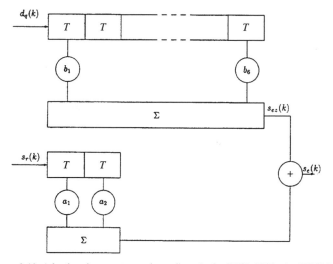

Figure 2.10 Adaptive six-zero, two-pole predictor in the G.721 32 kbp in APPCM codec.

using a somewhat complex gradient algorithm, which we will describe first in analytical terms
following the G.721 Recommendation and then provide a brief verbal interpretation of the
expressions. Explicitly, the first of the second-order predictor coefficients is specified as
follows:

$$a_1(k) = (1 - 2^{-8})a_1(k-1) + (3 \cdot 2^{-8})\text{sgn}[p(k)]\text{sgn}[p(k-1)], \qquad (2.67)$$

where $a_1(k)$ depends strongly on $a_1(k-1)$ as well as on the coincidence of the polarity of two
consecutive samples of the variable $p(k)$, where

$$p(k) = d_q(k) + \sum_{i=1}^{6} b_i(k-1)d_q(k-i)$$

represents the sum of the current quantized prediction error $d_q(k)$ and the estimated or
predicted signal contribution at the output of the sixth-order zero-section of the predictor due
to previous values of $d_q(k)$, while using the coefficients $b_i(k-1)$. Specifically, when updated,
a_1 is increased by the second term of Equation 2.67, if the polarity of two consecutive $p(k)$
values coincides, and decreased otherwise. Note, however, that this adaptation is a slow
process due to the scaling factor of $(3-1) \cdot 2^{-8}$.

A similar adaptation process is applied in order to control the second coefficient of the
predictor's zero section, as follows:

$$a_2(k) = (1 - 2^{-7})a_2(k-1) + 2^{-7}\{\text{sgn}[p(k)]\text{sgn}[p(k-1)]$$
$$- f[a_1(k-1)]\text{sgn}[p(k)]\text{sgn}[p(k-1)]\}, \qquad (2.68)$$

where the function $f(a_1)$ is given by:

$$f(a_1) = \{4a_1 \text{ if } |a_1| \le 1/22 \text{ sgn}(a_1) \text{ if } |a_1| > 1/2 \qquad (2.69)$$

and where $\text{sgn}(0) = +1$. Note that Equation 2.68 is similar to Equation 2.67, but the effect of
the third term governed by the value of a_1 is more dominant, since it is not scaled down by the
factor 2^{-7}. If $|a_1| < 0.5$, then $a_2(k)$ is decreased by the third term, when the adjacent $p(k)$
samples have an identical polarity. If, however, $|a_1| > 0.5$, also the polarity of a_1 enters the
complex interplay of parameters. Lastly, there are two stability constraints, which have to be
satisfied, namely:

$$|a_2(k)| \le 0.75 \wedge |a_1(k)| \le 1 - 2^{-4} - a_2(k). \qquad (2.70)$$

The sixth-order predictor is updated using the following equation:

$$b_i(k) = (1 - 2^{-8})b_i(k-1) + 2^{-7}\text{sgn}[d_q(k)]\text{sgn}[d_q(k-i)] \qquad \text{for } i = 1 \ldots 6, \qquad (2.71)$$

where the predictor coefficients are constrained to the range $-2 \le b_i(k) \le 2$. Observe in
Equation 2.71 that $b_i(k)$ is increased upon updating, if the polarity of the current and previous
quantized prediction error samples $d_q(k)$ and $d_q(k-i)$ $i = 1 \ldots 6$ coincides, and decreased
otherwise. This is because the 2^{-7} scaling factor of the second term outweighs the reduction
caused by the leakage factor of 2^{-8} in the first term.

The ADPCM decoder uses identical functional blocks to those of the encoder, as seen
in Figure 2.9. We point out that when transmitting ADPCM-coded speech, the bit sensitivity
within each 4-bit symbol monotonically decreases from the most significant bit (MSB) toward
the least significant bit (LSB), since the corruption of the MSB inflicts the largest waveform
distortion, while the LSB the smallest. After this rudimentary description of the G.721
ADPCM codec, we first offer a short introduction to speech quality evaluation, which will be

followed by a brief account of two closely related standardized ADPCM-based codecs, namely, the CCITT G.726 and G.727 schemes.

2.8 SUBJECTIVE AND OBJECTIVE SPEECH QUALITY

In order to be able to assess and compare the speech quality of various speech codecs, here we introduce a few speech quality measures, while a more in-depth treatment is offered in Chapter 17. In general, the speech quality of communications systems is difficult to assess and quantify. The most reliable quality evaluation methods are subjectively motivated, such as the **mean opinion score**, (MOS), which uses a five-point scale ranging between one and five. MOS tests facilitate the direct evaluation of arbitrary speech impairments by untrained listeners, but their results depend on the test conditions. Specifically, the selection and ordering of the test material, the language, and listener expectations all influence their outcome. A variety of other subjective measures is discussed in References [61–63], but subjective measures are tedious to derive and difficult to quantify during system development.

By contrast, **objective speech quality** measures do not provide results that could be easily converted into MOS values, but they facilitate quick comparative measurements during research and development. Most objective speech quality measures quantify the distortion between the speech communications system's input and output either in time or in frequency domain. The conventional SNR can be defined as

$$SNR = \frac{\sigma_{in}^2}{\sigma_e^2} = \frac{\sum_n s_{in}^2(n)}{\sum_n [s_{out}(n) - s_{in}(n)]^2}, \tag{2.72}$$

where $s_{in}(n)$ and $s_{out}(n)$ are the sequences of input and output speech samples, while σ_{in}^2 and σ_e^2 are the variances of the input speech and that of the error signal, respectively. A major drawback of the conventional SNR is its inability to give equal weighting to high- and low-energy speech segments, since its value will be dominated by the SNR of the higher-energy voiced speech segments. Therefore, the reconstruction fidelity of voiced speech is given higher priority than that of low-energy unvoiced sounds, when computing the arithmetic mean of the SNR, which can be expressed in dB as $SNR^{dB} = 10 \log_{10} SNR$. Hence, a system optimized for maximum SNR usually is suboptimum in terms of subjective perceptual speech quality.

Some of the ills of speech SNR computation mentioned above can be mitigated by defining the **segmental SNR** (SEGSNR) objective measure as given in

$$SEG - SNR^{dB} = \frac{1}{M} \sum_{m=1}^{M} 10 \log_{10} \frac{\sum_{n=1}^{N} s_{in}^2(n)}{\sum_{n=1}^{N} [s_{out}(n) - s_{in}(n)]^2}, \tag{2.73}$$

where N is the number of speech samples within a segment of typically 15–25 ms, that is, 120–200 samples at a sampling rate of 8 kHz, while M is the number of 15–25 ms segments, over which $SEGSNR^{dB}$ is evaluated. Clearly, the SEGSNR relates the ratio of the **segmental signal energy** to the **segmental noise energy**, computed over 15–25 ms segments and, after expressing this ratio in terms of dB, averages the corresponding values in the logarithmic domain. The advantage of using $SEGSNR^{dB}$ over the conventional SNR is that by averaging the SNR^{dB} values in the logarithmic domain it gives a fairer weighting to low-energy unvoiced segments by computing effectively the geometric mean of the SNR values instead of the arithmetic mean. Hence, the SEGSNR values correlate better with subjective speech quality measures, such as the MOS. Further speech quality measures are discussed in depth in Chapter 17.

2.9 VARIABLE-RATE G.726 AND EMBEDDED G.727 AD-PCM

2.9.1 Motivation

In recent years, two derivatives of the G.721 Recommendation have been approved by the International Telecommunication Union (ITU): the G.726 and G.727 Standards, both of which can operate at rates of 16, 24, 32, and 40 kbps. The corresponding number of bits/sample is 2, 3, 4, and 5. The latter scheme also has the attractive feature that the decoder can operate without prior knowledge of which transmission rate the encoder used. This is particularly useful in packetized speech systems, such as those specified by the ITU G.764 standard, which is referred to as the **packetized voice protocol (PVP)**. Accordingly, congested networks can be relieved by discarding some of the LSBs of the packets at packetization or intermediate nodes.

The schematic of the G.726 codec is identical to that of the previously described G.721 codec, which was shown in Figure 2.9. This is also reflected in the identical structure of the two standard documents. Note, however, that at its various rates different definition tables and constants must be invoked by the G.726 scheme. Hence, Tables 2.2–2.4 are appropriately modified in the G.726 and G.727 standards for the 2, 3, and 5 bits/sample modes of operation, respectively and the reader is referred to the standards for their in-depth study. Here we refrain from discussing the G.726 Recommendation in depth and focus our attention on the G.727 codec in order to be able to describe the principle of **embedded ADPCM coding**.

2.9.2 Embedded G.727 ADPCM Coding

A specific feature of embedded ADCM coding is that the decision levels of the lower-rate codecs constitute a subset of those of the higher-rate codecs. In other words, the embedded codec produces a codeword that consists of **core bits** that cannot be dropped and **enhancement bits** that can be neglected at the decoder. The corresponding codec arrangement is portrayed in Figure 2.11, where the **bit masking block** ensures that the local decoder relies only on a coarser estimate $I_c(k)$ of the quantized prediction error $I(k)$. Explicitly, only the core bits are used to compute the locally reconstructed signal and to control the adaptive predictor, since the enhancement bits may not be available at the decoder. The decoder in Figure 2.11 generates both the lower resolution reconstructed signal used to control the adaptive predictor at both ends and the full-resolution signal based on the assistance of the enhancement bits.

The corresponding ADPCM coding algorithms can be specified by the help of the number of core bits y and enhancement bits x as (x, y). The possible combinations are: (5,2), (4,2), (3,2), (2,2), (5,3), (4,3), (5,4), and (4,4).

EXAMPLE

By referring to a (3,2) embedded scheme, show how the embedded principle can be exploited to drop enhancement bits without unacceptably impairing the speech quality.

The normalized quantization intervals for the 16 kbps and 24 kbps modes are shown in Tables 2.5 and 2.6, respectively.

In these tables, "[" indicates that the endpoint value is included in the range, and "(" or ")" indicates that the endpoint value is excluded from the range. Observe in the first columns of the tables

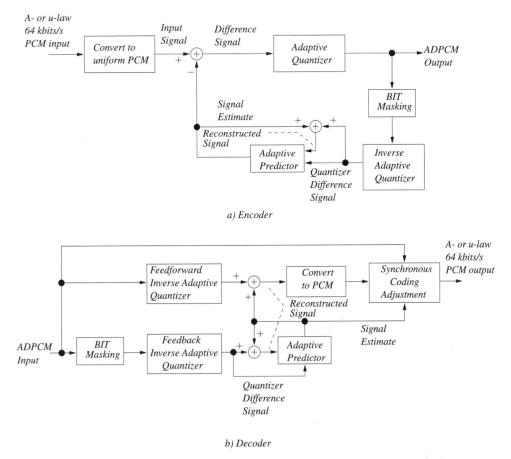

Figure 2.11 G.727 embedded ADPCM codec schematic, where only the core bits are used for prediction, while the enhancement bits, when retained, enhance the reconstructed signal quality.

that in the higher resolution 3-bit mode both the lower and higher quantizer input ranges of the 2-bit mode are split into two further intervals, and in order to differentiate between these intervals, the 24 kbps codec assigns an extra bit to improve the quantizer's resolution. When this enhancement bit is dropped in the network, the decoder will be unable to use it in order to output a reconstruction level, which is in the center of one of the eight quantization intervals of the 24 kbps codec. Instead, it will output a reconstruction level, which is at the center of one of the four quantization intervals of the 16 kbps codec.

TABLE 2.5 Quantizer Normalized Input/Output Characteristic for 16 kbit/s Embedded Operation of the G.727 Codec

Normalized Quantizer Input Range $\log_2 |d(k)| - y(k)$	$|I(k)|$ $|I_c(k)|$	Normalized Quantizer Output $\log_2 |d_q(k)| - y(k)$
$(-\infty, 2.04)$	0	0.91
$[2.04, \infty)$	1	2.85

TABLE 2.6 Quantizer Normalized Input/Output Characteristic for 24 kbit/s Embedded Operation, Where the Decision Thresholds Seen in Table 2.5 Constitute a Subset

Normalized Quantizer Input Range $\log_2 |d(k)| - y(k)$	$|I(k)|$ $|I_c(k)|$	Normalized Quantizer Output $\log_2 |d_q(k)| - y(k)$
$(-\infty, 0.96)$	0	-0.09
$[0.96, 2.04)$	1	1.55
$[2.04, 2.78)$	2	2.40
$[2.78, \infty)$	3	3.09

2.9.3 Performance of the Embedded G.727 ADPCM Codec

In what follows we will characterize the expected performance of the well-established standard G.727 ADPCM speech codec at a range of bitrates. Initially, the efficiency of the adaptive predictor is characterized by the help of the prediction residual and its statistical parameters, the PSD and the ACF, which are portrayed in Figures 2.12–2.15 for the voiced and unvoiced speech segments used earlier in Figures 1.2–1.5.

It becomes clear from the pairwise comparison of Figures 1.2 and 2.12 that in case of voiced speech the prediction residual has a substantially reduced variance, although it still retains some high values in the region of the highest time-domain peaks of the original speech signal. This is because the predictor is attempting to predict an ever-increasing signal on the basis of its past samples, while the signal has actually started to recede. The PSD of the

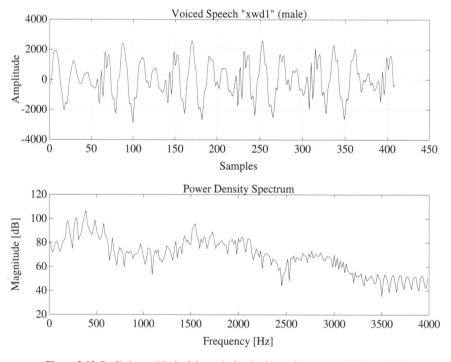

Figure 2.12 Prediction residual of the typical voiced speech segment of Figure 1.2 (top trace) and its PSD for a male speaker (bottom trace).

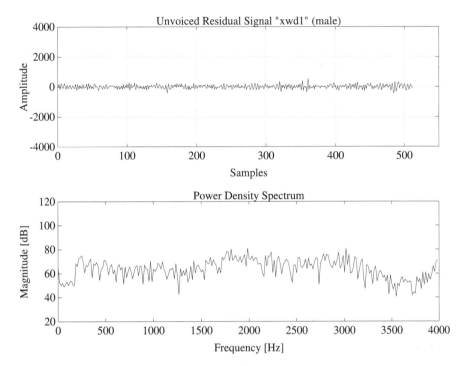

Figure 2.13 Prediction residual of the typical unvoiced speech segment of Figure 1.3 (top trace) and its PSD for a male speaker (bottom trace).

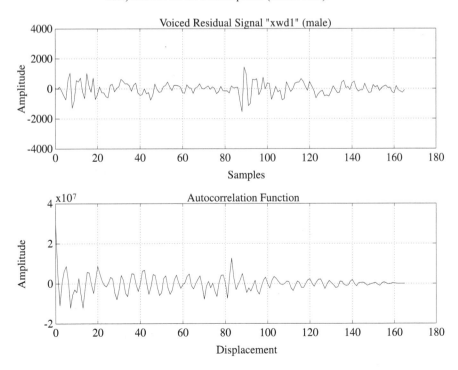

Figure 2.14 Prediction residual or the typical voiced speech segment of Figure 1.4 (top trace) and its ACF for a male speaker (bottom trace).

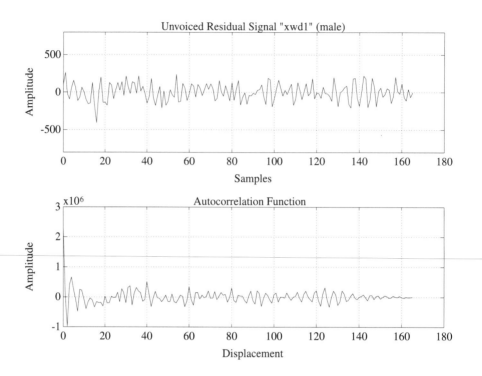

Figure 2.15 Prediction residual of the typical unvoiced speech segment of Figure 1.5 (top trace) and its ACF for a male speaker (bottom trace).

prediction residual is much flatter than that of the original signal, and it becomes similar to that of band-limited white noise.

The unvoiced speech segment of Figure 1.3 and its corresponding residual signal of Figure 2.13 are both noise-like, but the associated PSD functions show that the spectrum of the unvoiced speech segment has also been further flattened. These tendencies are also confirmed by comparing Figures 1.4 and 2.14 in case of the voiced segment using an expanded scale, where the ACF became substantially narrower in case of the voiced residual signal. The unvoiced ACF shown in Figure 1.5 exhibited virtually no correlation. Hence, the prediction residual's ACF is also quite similar, as it was demonstrated by Figure 2.15. As a result, in case of unvoiced speech segments predictive coding does not significantly improve the coding efficiency. However, since human speech is voiced for significantly longer periods of time than it is unvoiced and also due to the typically higher voiced energy, the perceived speech quality is more dependent on that of voice segments, and hence substantial coding efficiency is achieved by predictive codecs.

Figure 2.16 characterizes the codec's performance in the time domain by portraying both a voiced and an unvoiced speech segment, along with the corresponding reconstruction error signal between the original and the reconstructed speech for bitrates of 32, 24, and 16 kbps, that is, using 4, 3, and 2 bits/symbol.

In order to characterize the various operating modes of the G.727 codec in more formal terms, the fluctuations of the segmental signal energy and segmental residual energy are plotted in Figure 2.17 as a function of the frame index for a male speaker using our test file "xwd1" at 32 kbps using 4 core bits and no enhancement bits. Each frame was constituted by 20 ms speech, corresponding to 160 samples at a sampling rate of 8 kHz. Observe in the

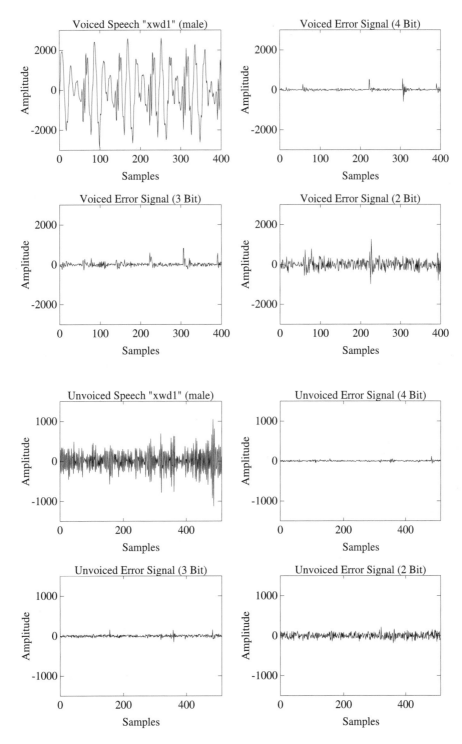

Figure 2.16 Typical voiced and an unvoiced speech segment and the corresponding reconstruction error signal between the original and the reconstructed speech generated by the G.727 codec for the bitrates of 32, 24, and 16 kbps, that is, using 4, 3, and 2 bit/s/symbol.

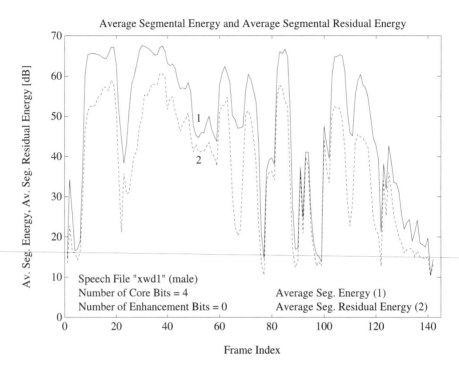

Figure 2.17 Segmental speech energy (1) and segmental residual energy (2) versus frame
index for a male speaker at 32 kbps using 4 core bits and no enhancement
bits.

figure that both functions exhibit a high dynamic range and a bimodal nature, corresponding
to high energy in case of voiced segments and low energy for unvoiced segments. In the
vicinity of very low energy voiced segments, the residual energy is not significantly lower
than the signal's energy, but in case of voiced segments there is typically at least a 10 dB
energy reduction due to the employment of predictive coding, exceeding 30 dB occasionally.

Similar tendencies can be inferred from Figure 2.18, where the fluctuation of the
segmental speech energy, SEGSNR, and prediction gain are displayed as a function of the
frame index for the 40 kbps mode of operation of the G.727 codec, when employing 4 core
bits and one enhancement bit. The SEGSNR fluctuates around 30 dB but occasionally reaches
50 dB, which is associated with perceptually unimpaired, transparent speech quality.

The objective speech quality of a representative subset of the possible modes of
operation of the G.727 codec, namely, that of the (5,4), (4,4), (3,3), and (2,2) modes, is shown
in Figure 2.19 in terms of a set of SEGSNR versus frame index plots for our "xwd1" male
test file. All curves follow the same tendencies and the corresponding average SEGSNR
values can be read from Figure 2.20 for bitrates between 16 and 40 kbps. Observe that the
associated SEGSNR versus bitrate curves are nearly linear and the SEGSNR improvement
due to increasing the bitrate is approximately 15 dB per octave. For example, when doubling
the bitrate from 16 to 32 kbps, the SEGSNR improves from about 15 dB to around 30 dB for a
female speaker and from 13 to around 28 dB for a male speaker.

Before concluding this chapter, in the next section let us briefly consider the
performance estimates provided by the application of rate-distortion theory for the family
of predictive codecs.

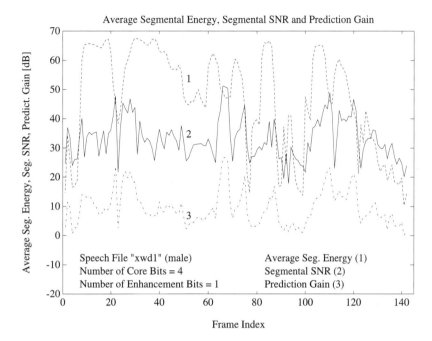

Figure 2.18 Segmental speech energy (1), SEGSNR (2) and prediction gain (3) versus frame index for a male speaker at 40 kbps using 4 core bits and 1 enhancement bit.

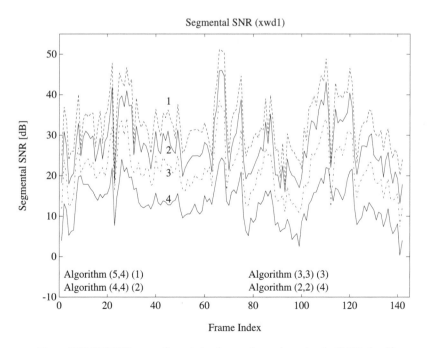

Figure 2.19 SEGSNR versus frame index for a male speaker using the G.727 algorithms (5,4), (1), (4,4) (2), (3,3) (3) and (2,2) (4).

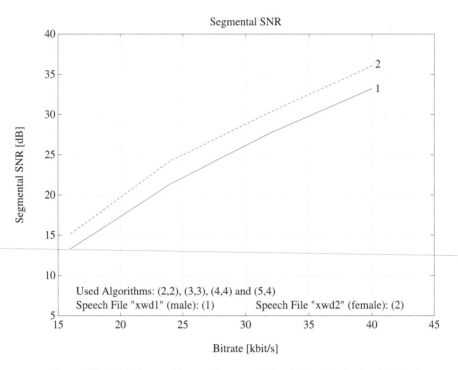

Figure 2.20 SEGSNR verus bitrate performance of the G.727 codec for female (2) and male (1) speakers using the algorithms (2,2), (3,3), (4,4) and (5,4).

2.10 RATE-DISTORTION IN THEORY IN PREDICTIVE CODING

In Section 1.4.3 we applied rate-distortion theory to conventional waveform codecs in order to derive performance estimates. In this section we attempt to provide similar results for the family of predictive codecs. In general, the rate-distortion function is not known in a closed form for arbitrary source distributions and distortion measures. However, for the mean square error (mse) distortion function some results are known, and we consider only this measure. For convenience we repeat Equation 1.5 From Section 1.4.3, stating that for a memoryless Gaussian distributed source x having a variance σ_x^2 the rate-distortion function can be expressed as:

$$R_D = \begin{cases} \frac{1}{2}\log_2 \sigma_x^2/D & 0 \le D \le \sigma_x^2 \\ 0 & D > \sigma_x^2. \end{cases} \tag{2.74}$$

For other memoryless sources, $R(D)$ curves can be calculated numerically, and it can be shown that the rate-distortion function for a Gaussian source upper bounds $R(D)$ for all other sources with the same variance. For example, a memoryless source with the Gamma pdf (which is a close approximation to the long-term pdf of speech signals) can be coded with an SNR of 8.53 dB at the rate of 1 bit/sample [64], compared to an SNR of 6.02 dB for a Gaussian source at the same rate.

For sources exhibiting memory the rate necessary for reproducing the source signal with a given distortion is always less than the rate for a similar source with no memory. This is because sources having memory are predictable; hence, they can be more accurately represented at a given bitrate. Predictability exhibits itself in terms of a non-flat PSD, as it was mentioned before. For a colored, that is, spectrally nonflat Gaussian source having a power spectral density $S(\omega)$, $R(D)$ can be calculated as follows:

$$D(\phi) = \frac{1}{2\pi} \int_{-\pi}^{\pi} \min(\phi, S(\omega)) d\omega$$

$$R(\phi) = \frac{1}{2\pi} \int_{-\pi}^{\pi} \max\left(0, \frac{1}{2}\log_2 \frac{S(\omega)}{\phi}\right) d\omega,$$

(2.75)

since in the low-energy speech spectral bands of Figures 2.21 and 2.22 below the dashed lines the PSD is set to zero.

This implies that the threshold level ϕ is chosen according to the required rate/distortion. In the frequency regions, where the speech PSD dips below ϕ, that is, where $\phi \geq S(\omega)$, no information is transmitted. For these regions—known as the stop-bands—the decoder should set the reconstructed power spectral density to zero in order to minimize the average rate R. Therefore, in the stop-bands the average distortion is equal to the original PSD $S(\omega)$. By contrast, in the frequency regions where $S(\omega) \geq \phi$, known as the pass-bands, the distortion is equal to ϕ and the transmission rate is $\log_2 \sqrt{S(\omega)/\phi}$.

For small distortions, that is, if ϕ is such that $S(\omega) \geq \phi$ for all ω, Equation 2.75 can be simplified to

$$R(D) = \frac{1}{2}\log_2 \frac{\sigma^2 \gamma^2}{D}$$

(2.76)

where σ^2 is the variance and γ^2 is the spectral flatness measure of the source signal. For a memoryless source we have a flat spectrum associated with $\gamma^2 = 1$, and Equation 2.76 is simplified accordingly.

The SNR (in dB) of the reconstructed signal is given by $10 \log_{10}(\sigma^2/D)$. Hence using Equation 2.76, we see that the maximum achievable SNR when coding at a rate of R bits/sample and using a high R value is given by:

$$\begin{aligned} SNR_{max} &= 2R * 10 \log_{10}(2) - 10 \log_{10} \gamma^2 \\ &= T_B + T_P, \end{aligned}$$

(2.77)

where

$$T_B = 2R * 10 \log_{10}(2) \approx 6R$$

(2.78)

and

$$T_P = 10 \log_{10} \frac{1}{\gamma^2}.$$

(2.79)

From Equation 2.79 we see that T_P can be regarded as the best possible gain (in dB) that can be produced by linear prediction of the signal.

As in the memoryless case, for non-Gaussian sources the exact form of $R(D)$ is not known. However, it can be shown that for a source having a given power spectral density, the distortion $R(D)$ will be less than or equal to the rate-distortion function for a Gaussian source with the same PSD.

Rate-distortion theory assumes that the source we are coding is stationary, with a power spectral density known at both the encoder and decoder. Speech, however, is nonstationary and can be considered to be quasi-stationary for only short periods of time, of the order of 20 ms. Furthermore, explicit rate-distortion functions are known only for sources having a

Gaussian distribution, which is not a good model for the long-term pdf of speech signals. Nevertheless, we can use simplified theory in order to give some idea of the optimum performance achievable by a speech codec and as to how such an optimum coder will behave. For example, in [65] the predictions of rate-distortion theory, assuming a Gaussian source, are shown to agree reasonably well with the performance of practical speech codecs.

Equation 2.77 gives the maximum possible *SNR* for a stationary Gaussian source (for small distortions) in terms of the data rate (per sample) and the maximum possible prediction gain T_P of the signal. This gain was taken by O'Neal in [66] to be 21 dB (following suggestions by Atal and Schroeder). Thus, if speech was a stationary Gaussian source, we could write for large rates R

$$\text{SNR}_{\text{max}} \approx 21 + 6R \text{ (dB)}. \qquad (2.80)$$

For lower rates—where $\phi > \min S(\omega)$)—the above equation will not be valid. Hence, we have to use Equation 2.75 in order to calculate the achievable SNR for a given rate. We carried out this experiment using about 7 seconds of speech, sampled at 8 kHz, obtained from two male and two female speakers. The speech signal was split into 256 sample segments, and we used the Fast Fourier Transform (FFT) of the Hamming-windowed samples in order to find the power spectrum $S(\omega)$ for each segment. Then for each segment an iterative procedure was used, invoking Equation 2.75 to find ϕ and hence D for a given rate.

The spectra of two typical segments, one voiced and the other unvoiced, are shown in Figures 2.21 and 2.22. Also shown in these figures in dashed lines are the functions $\min(\phi, S(\omega))$, which give the power spectra of the noise in an optimum encoder. The values of ϕ have been set to give a rate of 1 bit/sample. We found that at this rate the SNRs were about 21 dB for the voiced speech and 14 dB for the unvoiced speech. The voiced segment can be encoded with a lower distortion than the unvoiced segment because of its greater predictability. We found that $T_P = -10 \log_{10} \gamma^2$ was 20 dB for the voiced speech and 15 dB for the unvoiced speech.

Figure 2.23 shows the predicted maximum segmental SNR against the data rate. This was calculated by finding the SNR in decibels for each speech segment as described above and then averaging over the test file's duration. We also calculated the maximum prediction gain T_P in a similar way and found that it was 20.9 dB, agreeing well with the value used in [66]. Notice that for rates above about 1.5 bits/sample the curve in Figure 2.23 becomes approximately a straight line, as predicted by Equation 2.80.

In the discussion above, we have considered each 256 sample (32 ms) segment of speech to be a stationary Gaussian signal. We now discuss how these assumptions are likely to affect our results. First, although the long-term statistics of speech closely match the Gamma pdf, the short-term statistics are approximately Gaussian [67]. Therefore, assuming the 32 ms segments of speech to be Gaussian will probably not affect the validity of our results too gravely. The nonstationarity will have a more significant effect and will result in the true "maximum SNR" function for speech lying somewhere below that drawn in Figure 2.23. Thus, the maximum SNR values we have calculated give an upper bound for the SNR that could be obtained with the aid of a practical speech codec.

We can produce a tighter bound by evaluating how the nonstationary nature of speech will affect our results. The first difference will be that, for a speech codec to obtain a prediction gain close to T_P, it will have to send side information about the current spectrum of the signal to the decoder. The rate necessary for this side information (say $\hat{R}$ bits/sample) will reduce the effective rate R of the codec. The side information necessary to support short-term linear prediction at the current state-of-the-art is about 20 bits per 20 ms or 160 speech samples, requiring on average about $\frac{1}{8}$ bits per sample, and we take this as the necessary rate

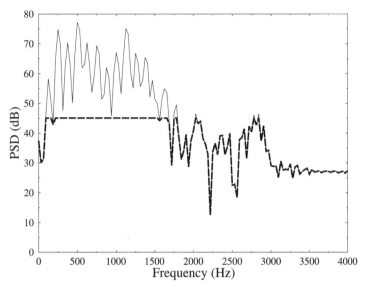

Figure 2.21 Power spectrum density for a segment of voiced speech as well as $\min[\phi, S(\omega)]$ shown using dashed-line.

$\hat{R}$. Second, the prediction gain possible will be reduced below T_P because of the nonstationary nature of speech and also because only limited information about the present correlations in the signal is sent in the side information (that is, the gain will be dependent on $\hat{R}$). For example, for the speech file referred to earlier, the calculated value of T_P is 21 dB, but the gain achieved with the aid of short-term linear prediction (of order 10) is only 17 dB.

At bitrates above about 1.5 bits per sample, Equation 2.80 gives a good approximation to the maximum segmental SNR possible for a speech codec, provided that we take into

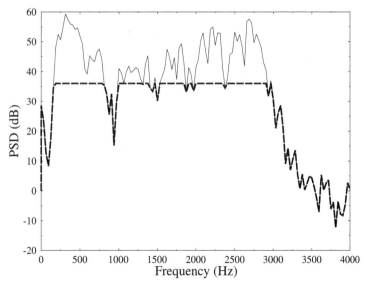

Figure 2.22 Power spectrum density for a segment of unvoiced speech as well as $\min[\phi, S(\omega)]$ shown using dashed line.

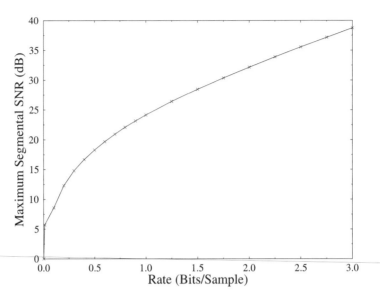

Figure 2.23 Predicted maximum possible segmental SNR.

account the effects mentioned above. For a 16 kbits/s codec, the bitrate is 2 bits per sample. Hence the effective rate R is about 1.875 bits per sample. Thus, using 4 dB as the reduction of the prediction gain T_p, the maximum segmental SNR predicted for a 16 kbits/s speech codec is about 28 dB. At rates below 1 bit per sample, Equation 2.89 is no longer accurate. We therefore have to use Figure 2.23, in order to estimate the maximum segmental SNR of speech codecs at these rates. Furthermore, the effect of the reduction in T_p will be less significant than the 4 dB figure used above because of the fall of the maximum SNR figures below $T_p + 6R$. We take a decrease of about 2 dB to be typical at low rates. These assumptions mean that the effective rate R for a 4.7 kbits/s codec will be about 0.45 bit/sample, giving a maximum segmental SNR of around $17.5 - 2 = 15.5$ dB. Similarly, we predict a maximum possible segmental SNR of about 19 dB at 7.1 kbits/s. It is interesting to compare these figures with those obtained for a range of CELP speech codecs that are to be described later in Chapter 6, which operate at the same rates. It is equally instructive to compare these estimates with the experimentally evaluated results of the G.727 codec in Figure 2.21.

2.11 CHAPTER SUMMARY

In this chapter we initially highlighted the basic principles of forward-predictive as well as DPCM-based coding. This was followed by the design principles of the optimum linear predictor invoking the Levinson-Durbin algorithm. Jayant's adaptive one-word-memory quantizer was then characterized, leading to our discussions on AR, MA, and ARMA processes. The associated predictors were then invoked in the context of the ITU's G.721, G.726 and G.727 standard codes. Finally, the performance of predictive codecs was characterized.

Having considered a range of low-complexity predictive codecs in this chapter, in the forthcoming chapter we will concentrate on the family of lower bit rate, high complexity analysis-by-synthesis (AS) speech codecs. These AB codecs are widely used in most existing mobile radio systems at the time of writing.

PART II
ANALYSIS-BY-SYNTHESIS
CODING

3

Analysis-by-Synthesis Principles

3.1 MOTIVATION

Recall from Section 1.2 that human speech can be adequately described by the help of the linearly separable speech production model of Figure 1.1, where the excitation signal is filtered through a slowly varying spectral shaping system in order to generate the speech signal. In a simple inverse approach, one could view speech production as filtering the excitation $E(z)$ through the spectral shaping system $H(z) = 1/A(z)$ in order to generate the speech signal $S(z)$.

In Chapter 2 we showed in the context of the G.727 codec how a slowly varying two-zero, six-pole predictor can be used to estimate the incoming signal's spectrum, and the corresponding spectral coefficients were determined in Section 2.3.3. It was also demonstrated how this short-term predictor can be rendered adaptive, in order to accommodate changes in the incoming signal's statistics, and it was stated that the predictor coefficients must be transmitted to the decoder in a forward-adaptive predictive codec.

In Sections 2.5 and 2.9.3 we showed how efficient predictive coding was in terms of removing redundancy and reducing the signal's variance. As a result, the prediction residual signal characterized in both time and frequency domains in Section 2.9.3 became nearly unpredictable, which we described by the help of waveform coding techniques using an adaptive quantizer. In the G.727 codec, no predictor coefficients were transmitted, but the number of bits required for the adequate encoding of the near-random prediction residual was quite high, requiring bitrates up to 40 kbps, when using 5 bits/sample in order to maintain a high speech quality.

Although the high-quality encoding of the prediction residual is a sufficient criterion for perceptually high speech quality, it is not a necessary condition. In Section 2.8 we noted that the conventional SNR is not a reliable speech quality measure. In this section we endeavor to improve the bitrate economy, while maintaining perceptually high speech quality.

3.2 ANALYSIS-BY-SYNTHESIS CODEC STRUCTURE

A number of measures can help us achieve the goals incorporated in the analysis-by-synthesis (AbS) codec structure shown in Figure 3.1. In order to improve the coding efficiency, **vector quantization** techniques are invoked, where the synthesis filter is excited by an excitation vector of typically 5 ms or 40 samples length. In addition, a **closed-loop structure** is used. Accordingly, the prediction error between the original input signal and the synthesized speech signal is evaluated for each candidate excitation vector. The specific excitation vector minimizing the **weighted error** rather than the conventional mse is deemed to produce the best synthetic speech quality. Following this rudimentary introduction to the philosophy of AbS codecs, we will elaborate on their salient features during our further discussions.

As seen in Figure 3.1, the slowly varying synthesis filter(s) are excited by the innovation sequences $u(n)$ of the excitation generator in order to produce the synthetic speech $\hat{s}(n)$, which is compared with the input speech $s(n)$ about to be encoded. The prediction error residual $e(n) = s(n) - \hat{s}(n)$ is formed and weighted by the error weighting filter in order to produce the **perceptually weighted error** $e_w(n)$.

Instead of minimizing the usual mean squared error term in an effort to provide best waveform reproduction, these AbS codecs minimize the perceptually weighted error $e_w(n)$. In this way, they actually degrade the waveform representation in favor of better subjective speech quality. The high speech quality of AbS speech codecs is achieved at the cost of relatively high complexity, since the synthetic speech is computed for all legitimate innovation sequences, sometimes several thousand times. A fundamental property of

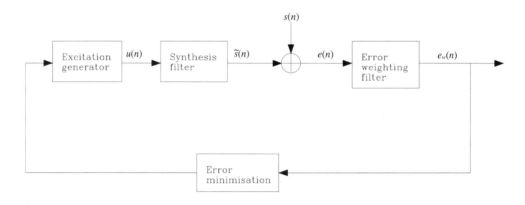

(a) AbS Encoder

(b) AbS Decoder

Figure 3.1 General analysis-by-synthesis codec schematic.

closed-loop AbS codecs is that the prediction residual is encoded by minimizing the perceptually weighted error between the original and reconstructed speech rather than minimizing the mean squared error between the residual and its quantized version as in open-loop structures. The error weighting filter will be derived from the short-term predictor filter, and it is designed to deemphasize the weighted error in the vicinity of formant regions, where the speech signal's spectral prominences sucessfully mask the effects of allowing a higher reconstruction error. This renders the signal-to-noise ratio more or less constant over the speech signal's frequency range, rather than aiming for a near-constant quantized noise PSD.

The **short-term synthesis filter** determined in Section 2.33 is responsible for modeling the spectral envelope of the speech waveform. Its coefficients are computed by minimizing the error of predicting a speech sample from a few (typically 8 to 10), previous samples, where minimization is carried out over a quasi-stationary period of some 20 ms or 160 samples, when the sampling frequency is 8 kHz. The synthesis filter might incorporate an additional **long-term synthesis filter** modeling the fine structure of the speech spectrum, which predicts the long-term periodicity of speech persisting after short-term prediction, reflecting the pitch periodicity of the residual.

As seen in Figure 3.1, the decoder uses an identical structure to that of the encoder for generating the synthetic speech. However, it is considerably less complex since the innovation sequence that minimized the perceptual error is transmitted to the decoder, and it is the only sequence to which the synthesis filter's response is computed.

As detailed in Section 2.3.3, the short-term synthesis filter parameters are determined by minimizing the prediction error over a quasi-stationary interval of about 20 ms outside the optimization loop. The "remainder" of the speech information is carried by the prediction residual, which is not modeled directly. Instead, the best excitation for this short-term synthesis filter is determined by minimizing the weighted error between the input and the synthetic speech. The excitation optimization interval is typically 5 ms, which is a quarter of the 20 ms short-term filter update interval. The 20 ms duration speech frame is therefore typically divided into four subsegments, and the optimum excitation is determined individually for each 5 ms subsegment. The quantized filter parameters and the vector quantized excitation are transmitted to the decoder, where the synthesized speech is generated by filtering the decoded excitation signal through the synthesis filter(s). Let us now consider the effects of choosing different parameters for the short-term synthesis filter.

3.3 THE SHORT-TERM SYNTHESIS FILTER

As mentioned before, the vocal tract can be modeled as a series of uniform lossless acoustic tubes [5, 6]. It can then be shown that for a digital all-pole synthesis filter to approximate the effect of such a model of the vocal tract, its delay should be at least twice the time required for sound waves to travel along the tract. For a vocal tract of length 17 cm voice velocity of 340 m/sec and a sampling rate of 8 kHz, this corresponds to the order p of the filter being at least 8 taps or 8.125 μs = 1 ms. Generally, a few extra taps are added in order to help the filter cope with effects that are not allowed for in the lossless tube model, such as spectral zeros and losses in the vocal tract. We simulated the effect of changing the order p on the prediction gain of the inverse filter $A(z)$. We used about 11 seconds of speech data obtained from two male and two female speakers. The speech was sampled at 8 kHz and split into 20 ms frames. For each frame, the filter coefficients were calculated using the autocorrelation approach applied to the Hamming windowed speech signal, and the prediction gain was calculated and converted into decibels. Here the prediction gain is defined as the energy of the original speech samples $s(n)$ divided by the energy of the prediction error samples $e(n)$. The overall prediction gain was taken as the average of the decibel gains for all the 20 ms frames in the speech file.

The results of our simulations are shown in Figure 3.2. Also shown in this figure is the variation of the segmental SNR of a CELP codec as a function of the order p of its synthesis filter. The filter coefficients were calculated for 20 ms frames as described above and were left unquantized. The excitation parameters for the codec were determined identically to our 7.1 kbits/s CELP codec to be described later in Section 6.2, except no error weighting was used. It can be seen that both the prediction gain of the inverse filter and the segmental SNR of the codec increase, as the order of the synthesis filter is increased. However, in a forward-adaptive system, each synthesis filter coefficient used requires side information to be sent to the decoder. Hence, we wish to keep their number to a minimum. We chose $p = 10$ as a sensible compromise between a high prediction gain and a low bitrate.

The rate required to transmit information about the synthesis filter coefficients also depends on how often this information is updated, that is, on the LPC analysis frame length L. We carried out similar simulations to those described above in order to quantify how the frame length affected the prediction gain of the inverse filter and the segmental SNR of a CELP codec. The order p of the filter was fixed at $p = 10$, and the coefficients were calculated using Hamming windowed speech frames of length L samples. However, the prediction gain and the segmental SNR were calculated using frames 20 ms long to find the gains/SNRs, which were converted into decibels and averaged. This was carried out in order to ensure a fair comparison within our results, which are shown in Figure 3.3. It can be seen that for very

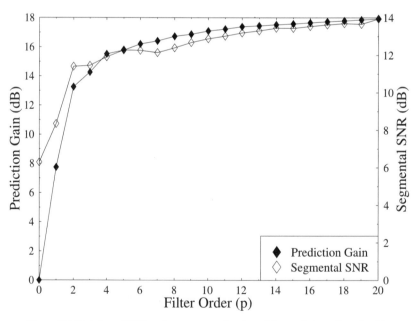

Figure 3.2 Variation of LPC performance as a function of the predictor order p for 20 ms duration speech frames using no error weighting and the 7.1 kbps CELP codec of Section 6.2.

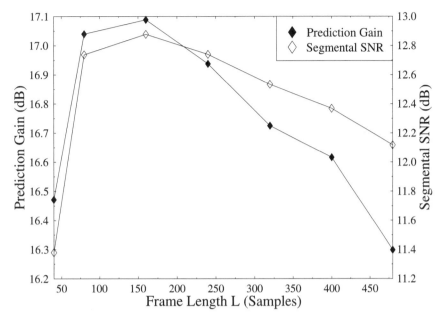

Figure 3.3 Variation of LPC performance versus the analysis frame length L.

short analysis frame lengths, both the prediction gain and the segmental SNR are well below the best values found. This is probably because we have used the autocorrelation method of analysis, and for small values of L we do not have $L \gg p$. Hence, inaccuracies are introduced due to the windowing of the input speech signal. The best values of the prediction gain and the segmental SNR are given for $L = 160$, which corresponds to a 20 ms frame length. For larger frames the performance of the filter gradually decreases due to the nonstationary nature of speech.

The synthesis filter coefficients must be quantized in order to be sent to the decoder. Unfortunately, the filter coefficients themselves are not suitable for quantization because the frequency response of the synthesis filter is very sensitive to changes in them. This means that even a small change in the values of the coefficients when they are quantized can lead to a large change in the spectrum of the synthesis filter. Furthermore, after quantization it is difficult to ensure that a given set of coefficients will produce a stable synthesis filter. Thus, although the autocorrelation approach guarantees a stable filter, this stability could be easily lost through direct quantization of the filter coefficients. Therefore, before quantization, the coefficients are converted into another set of parameters from which they can be recovered, but which are less sensitive to quantization noise and which allow stability to be easily guaranteed. Some schemes use the reflection coefficients, which are related to the lossless acoustic tube model of the vocal tract. These coefficients are calculated as a byproduct of using the Levinson-Durbin algorithm of Figure 2.3 to solve Equation 2.16. Using these coefficients, we can ensure the stability of the synthesis filter by limiting the magnitude of all the coefficients to be less than 1. Typically, the reflection coefficients are transformed, using the inverse-sine transformation or log-area ratios, before quantization. These issues are discussed in the next chapter in detail. Let us now introduce long-term prediction in the AbS codec of Figure 3.1.

3.4 LONG-TERM PREDICTION

3.4.1 Open-Loop Optimization of LTP Parameters

As mentioned earlier, most AbS speech codecs incorporate a long-term predictor (LTP) in order to improve the speech quality and bitrate economy by further reducing the variance of the prediction residual. This improvement can be achieved by predicting and removing the long-term redundancy of the speech signal. Although the short-term predictor (STP) removes the adjacent sample correlation and models the spectral envelope, that is, the formant structure, it still leaves some long-term peaks in the STP residual, since at the onset of quasi-periodic waveform segments of voiced sounds it fails to predict the signal adequately. This is clearly demonstrated by Figure 3.4 for an 800-sample or 100 ms long speech segment. The pitch-related, quasi-periodic LPC prediction error peaks can be efficiently reduced by the LTP, as seen in Figure 3.4c.

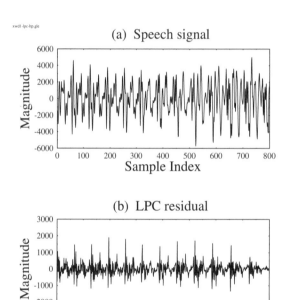

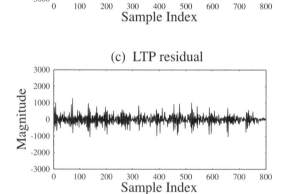

Figure 3.4 Typical 100 ms segment of (a) voiced speech signal, (b) LPC residual and (c) LTP residual.

The operation of the LTP can be explained in a first approximation as subtracting a "pitch-synchronously" positioned or delayed segment of the previous LPC residual from the current segment. If the pitch periodicity is quasi-stationary, that is, near time-invariant, then the properly positioned previous segment will have co-located pitch pulses with the current segment. Hence, after subtracting the previous LPC segment, the pitch-synchronous prediction residual pulses of Figure 3.4b can be eliminated, as evidenced by Figure 3.4c which portrays the LTP residual. We will shortly show that the performance of the above-mentioned simple LTP can be improved if, before subtracting the previous "history," we scale the previous segment by a gain factor G, which can be optimized to minimize the energy of the LTP residual, which will be made explicit in the context of Equation 3.1. Both the LTP delay and the gain factor will have to be transmitted to the decoder in order to be able to reconstruct the LPC residual. Furthermore, at the decoder only the previous reconstructed residual is available, which is based on the transmitted innovation sequence and LTP parameters, such as the delay and gain. The encoder therefore also uses the previously reconstructed LPC residual segments rather than the original ones.

Since periodic signals exhibit a line-spectrum, the quasi-periodic prediction residual's spectrum seen in Figure 2.13 has a periodic fine structure showing peaks and valleys, which is the manifestation of the time-domain long-term periodicity. Thus, the LTP models the spectral fine structure of the speech signal that is similar to the line spectrum of a periodic signal. This pitch-related periodicity is strongly speaker- and gender-dependent. Its typical values are in the range of 100–300 Hz or about 3–10 ms. When employing a LTP, the LTP residual error becomes truly unpredictable. This noise-like process is therefore often modeled by innovation sequences of a zero-mean, unit-variance random Gaussian code book, yielding an extremely efficient vector quantizer. This concept leads to code excited linear predictive coding (CELP), which constitutes the most prominent member of the family of AbS codecs. These codecs are treated in depth in several chapters during our further discourse.

As we have seen in Figure 3.1, the decoder reconstructs the speech signal by passing the specific innovation sequence through the synthesis filter. The best innovation sequence does not necessarily closely resemble the LTP residual. Nor does it guarantee the best waveform match between the original speech and synthetic speech. Rather, it attempts to produce the perceptually best speech quality.

In order to augment our exposition, we now describe the LTP in analytical terms, as follows. When using a 1-tap LTP, the LTP residual $e_L(n)$ is computed as:

$$e_L(n) = r(n) - G_1 r(n - \alpha) \tag{3.1}$$

where $r(n)$ is the STP residual from which its delayed version $r(n - \alpha)$ is subtracted scaled by an optimum gain factor G_1, computed by minimizing the LTP residual error. The z-transform of Equation 3.1 is given by:

$$E_L(z) = R(z)[1 - G_1 z^{-\alpha}], \tag{3.2}$$

which can be rearranged in the following form:

$$R(z) = \frac{E_L(z)}{[1 - G_1 z^{-\alpha}]} = \frac{E_L(z)}{P(z)}, \tag{3.3}$$

where $P(z) = [1 - G_1 z^{-\alpha}]$ is the z-domain transfer function of the LTP.

The total mean squared LTP residual error E_L computed over a segment of N samples can be formulated as follows:

$$E_L = \sum_{n=0}^{N-1} e_L^2(n)$$

$$= \sum_{n=0}^{N-1} [r(n) - G_1 r(n - \alpha)]^2$$

$$= \sum_{n=0}^{N-1} r^2(n) - \sum_{n=0}^{N-1} 2G_1 r(n)r(n - \alpha) + \sum_{n=0}^{N-1} G_1^2 r^2(n - \alpha). \tag{3.4}$$

Setting $\partial E_L / \partial G_1 = 0$ gives

$$\sum_{n=0}^{N-1} -2r(n)r(n - \alpha) + \sum_{n=0}^{N-1} 2G_1 r^2(n - \alpha) = 0, \tag{3.5}$$

yielding the optimum LTP gain factor G_1 in the following form:

$$G_1 = \frac{\sum_{n=0}^{N-1} r(n)r(n - \alpha)}{\sum_{n=0}^{N-1} [r(n - \alpha)]^2}. \tag{3.6}$$

Observe that the gain factor computed can be interpreted as the normalized cross-correlation of $r(n)$, where the normalization factor in the denominator represents the energy of the STP residual segment. If the previous and current segments are identical, they are perfectly correlated, and $G_1 = 1$, yielding $e_L = 0$ in Equation 3.1, which corresponds to perfect long-term prediction. If there is practically no correlation between $r(n)$ and $r(n - \alpha)$, as in case of unvoiced sounds, $G_1 \approx 0$ and no LTP gain is achieved.

In general, upon substituting the optimum LTP gain G_1 back into Equation 3.5, the minimum LTP residual energy is given by:

$$EL, \min = \sum_{n=0}^{N-1} r^2(n) - \frac{\left[\sum_{n=0}^{N-1} r(n)r(n - \alpha) \right]^2}{\sum_{n=0}^{N-1} [r(n - \alpha)]^2}. \tag{3.7}$$

Again, in harmony with our previous argument, minimizing E is equivalent to maximizing the second term in Equation 3.7, which physically represents the normalized correlation between the residual $r(n)$ and its delayed version $r(n - \alpha)$. Hence, the optimum LTP parameters can be determined by computing this term for all possible values of α over its specified range of typically $N = 20 - 147$ samples, when the sampling rate is 8 kHz. The delay α which maximizes the second term is the optimum LTP delay.

The effect of both the STP and LTP becomes explicit by comparing the probability density functions (PDF) of a typical speech signal, as well as that of both the STP residual and the LTP residual, respectively, in Figure 3.5. Note that the speech signal has a long-tailed PDF, while the STP and LTP have substantially reduced the signal's variance. Since the LTP's action is to reduce the relatively low-probability pitch pulses, this effect becomes more explicit from Figure 3.6, where the PDFs were plotted on a logarithmic axis in order to magnify the long low-probability PDF tails. This effect may not appear dramatic, but invoking a LTP typically improves the speech quality sufficiently in order to justify its added complexity.

When using a LTP, our AbS speech codec schematic seen in Figure 3.1 can be re-drawn as portrayed in Figure 3.7. The choice of the appropriate error weighting filter is crucial as regards the codec's performance [46, 68]. Its transfer function is based on findings derived

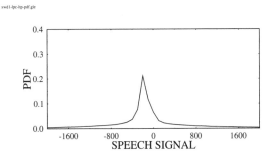

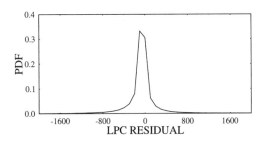

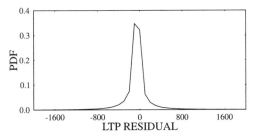

Figure 3.5 PDF of a typical speech signal, (top) LPC residential (middle) and LTP residual (bottom).

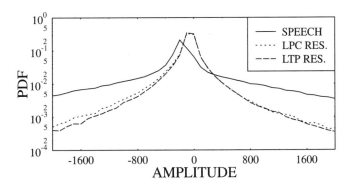

Figure 3.6 Logarithmic PDF of a typical speech signal, (top) LPC residual (middle) and LTP residual (bottom).

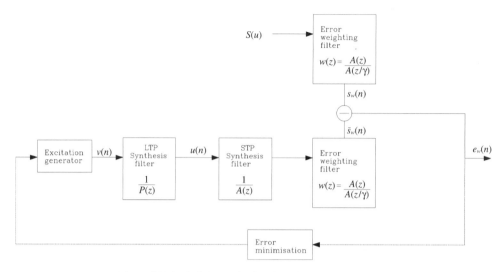

Figure 3.7 Analysis-by-synthesis codec schematic using a LTP.

from the theory of auditory masking. Experience shows that when generating, for example, a sinusoidal signal, often referred to as a single tone due to its single spectral line in the frequency domain, it is capable of masking a high-energy, but spectrally more spread, noise signal residing within the same frequency band. This is due to the human ear's ability to resolve the two signals. Due to the speech signal's spectral prominences in the frequency regions of the formants, this property can be exploited by allowing more quantization noise to be concentrated around them. Clearly, an adaptive quantization noise spectrum shaping filter is required, which de-weights the quantization noise in the formant regions. This allows more quantization noise to reside in these frequency bands than without filtering.

The filter's transfer function may have to depend on the momentary signal spectrum, which is evaluated in the codec in terms of the filter coefficients a_i, describing the polynomial $A(z)$. A convenient choice is to define the error weighting filter's transfer function as follows [68, 69].

$$W'(z) = \frac{A(z)}{A(z/\gamma)} = \frac{1 - \sum_{k=1}^{p} a_k z^{-k}}{1 - \sum_{k=1}^{p} a_k \gamma^k z^{-k}}, \tag{3.8}$$

where the constant γ determines, to what extent the error spectrum is deemphasized in the formant regions. Typical values of γ are in the range of $0.6 \ldots 0.85$. The schematic diagram of Figure 3.7 can also be rearranged in the form shown in Figure 3.8, which is an often favored equivalent configuration.

Recently, other forms of error weighting have been suggested for speech codecs. For example, in the 16 kbits/s G.728 codec [70] the filter

$$W(z) = \frac{A(z/\gamma_1)}{A(z/\gamma_2)} \tag{3.9}$$

is used where $\gamma_1 = 0.9$ and $\gamma_2 = 0.6$. We employed this weighting filter in the context of the low-delay codecs in Chapters 8. In [71] an explicit auditory model is used to take account of the details known about psychoacoustics and auditory masking.

3.4.2 Closed-Loop Optimization of LTP Parameters

According to our previous approach, the LTP parameters were computed from the LPC residual signal using a simple correlation technique, as suggested by Equation 3.7, in a suboptimum two-stage approach often referred to as open-loop optimization. However, Singhal and Atal [72] suggested that a substantially improved speech quality can be attained at the cost of a higher complexity if the LTP parameters are computed inside the AbS loop [72]. This leads to the **adaptive codebook** approach featured in Figure 3.9. This terminology is justified by the fact that the adaptive codebook is regularly replenished using the previous composite excitation patterns $u(n)$ after a delay of one subsegment duration, which will be made more explicit during our forthcoming deliberations following Salami's approach [46, 47].

The composite excitation signal $u(n)$ in Figure 3.9 is given by:

$$u(n) = v(n) + G_1 u(n - \alpha) \tag{3.10}$$

which is the superposition of the appropriately delayed G_1-scaled adaptive codebook entry $u(n - \alpha)$ and the excitation $v(n)$, while $v(n)$ is recomputed for each excitation optimization subsegment. In conventional forward-adaptive AbS codecs, the subsegment length is typically 5–7.5 ms. Thus in an LPC update frame of 10–30 ms there are usually 2–6 excitation optimization subsegments. This provides extra flexibility for the codec to adapt to the changing nature of the speech signal in spite of the LPC parameters and the corresponding spectral envelope being fixed for 10–30 ms due to transmission bitrate constraints. By contrast, in the backward-adaptive codecs, the corresponding intervals can be significantly shorter, since the LPC coefficients are extracted from the previously recovered speech signal rather than being transmitted. This is elaborated on in the context of the ITU low-delay G.728 16 kbps standard codec in Chapter 8.

In forward predictive codecs, the excitation signal $u(n)$ is determined by minimizing the mean squared weighted error (mswe) E_w for a typical duration of a subframe of $N = 40$–60 samples or 5–7.5 ms. Ideally, according to the closed-loop AbS approach, the optimum excitation and the adaptive codebook parameters resulting in the perceptually best synthetic

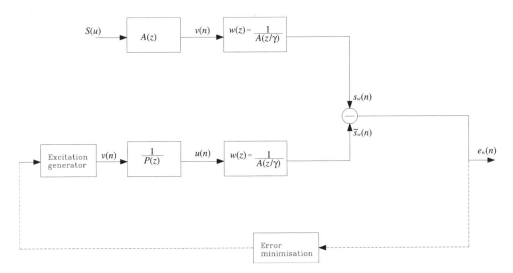

Figure 3.8 Modified analysis-by-synthesis codec schematic with LTP.

speech quality would have to be found jointly—testing each possible combination of the two—in order to minimize E_w. Unfortunately, this would inflict an unacceptable complexity penalty. Hence, usually a suboptimal approach is invoked, where initially the adaptive codebook parameters are computed first, assuming that no excitation is input to the synthesis filter, that is, $v(n) = 0$. This is because the excitation $v(n)$ is not known at this stage, yielding: $u(n) \approx G_1 u(n - \alpha)$.

At this stage, we have not yet specified the set of legitimate excitation patterns, but one might expect that the more attention devoted to designing these sequences and the larger this set, the better the quality of the synthetic speech. During the process of determining the best excitation sequence $u(n)$ from the set of legitimate sequences, which results in the best synthetic speech segment, each excitation segment is filtered through the **weighted synthesis filter** $1/A(z/\gamma)$ of Figure 3.8, which is an infinite impulse response (IIR) system. Hence, following Salami's deliberations [46, 47], the weighted synthetic speech in the current optimization interval can be described as the superposition of the filter's response due to the current excitation sequence and to all previous actual optimum excitation sequences. This memory contribution is not influenced by the current excitation sequence. Hence, it is also often referred to as the IIR filter's **zero input response**. Treating this memory contribution adequately is extremely important in order to ensure that in spite of filtering all the candidate excitations tentatively through this IIR filter, the synthesis filter's output signal becomes a seamless, close replica of the weighted input speech due to the sequence of actual optimum excitation sequences.

Two alternative solutions may be used to treat these memory contributions adequately during the excitation optimization proccess. According to the first technique, once all candidate excitations are tested and the optimum excitation for the current interval is found, the zero input filter response due to the sequence of all concatenated previous optimum excitations can be updated to include the contribution by the current one. This updated memory contribution is then stored in order to be able to add it to the synthesis filter's output due to the set of all candidate excitations during the next optimization interval, before they are compared to the corresponding weighted input speech segment. The disadvantage of this approach is that the zero input response has to be added to all candidate synthetic speech segments, before they are compared to the weighted original speech.

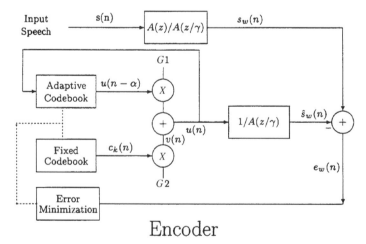

Figure 3.9 Adaptive codebook approach in the context of AbS CELP codecs.

Therefore it is usually more efficient to invoke the second approach and subtract this filter memory contribution from the weighted original speech, before pattern matching, since this operation takes place only once per optimization interval rather than for each candidate excitation pattern. After subtracting the memory contribution of the weighted synthesis filter from the weighted original speech vector, we arrive at the **target vector**, to which all filtered candidate excitation sequences are then compared. However, according to this approach, the IIR filter's memory must be set to zero each time, before a new excitation is tentatively filtered through it, because the effect of the filter memory was taken into account now by modifying the weighted original speech signal. Following the latter approach, the synthetic speech can be described as [46, 47]:

$$\hat{s}_w(n) = \sum_{i=0}^{n} u(i)h_w(n-i) + \hat{s}_0(n), \tag{3.11}$$

where $h_w(n)$ denotes the infinite impulse response of the weighted synthesis filter $W(z) = 1/A(z/\gamma)$, and $\hat{s}_0(n)$ is the zero-input response of the weighted synthesis filter, which is equivalent to the filter's memory contribution due to previous excitations. Hence, the weighted error between the original and synthetic speech can be written as:

$$e_w(n) = x'(n) - \sum_{i=0}^{n} u(i)h_w(n-i), \tag{3.12}$$

where

$$x'(n) = s_w(n) - \hat{s}_0(n) \tag{3.13}$$

represents the weighted input speech after subtracting the memory contribution of the IIR weighted synthesis filter due to previous excitations, and the notation $x'(n)$ is used for later notational convenience.

Having found a solution to treating the IIR filter memory during the excitation optimization process, let us return to finding the closed-loop LTP parameters. Recall that since $v(n)$ is unknown initially, it is set to zero. Thus upon substituting $u(n) \approx Gu(n - \alpha)$ into the weighted error expression of Equation 3.12, we arrive at:

$$\begin{aligned} e_w(n) &= x'(n) - G\sum_{i=0}^{n} u(i-\alpha)h_w(n-i) \\ &= x'(n) - Gu(n-\alpha) * h_w(n) \\ &= x'(n) - Gy_\alpha(n), \end{aligned} \tag{3.14}$$

where we used the short-hand:

$$y_\alpha(n) = u(n-\alpha) * h_w(n) = \sum_{i=0}^{n} u(i-\alpha)h_w(n-i). \tag{3.15}$$

The mean squared weighted error for the excitation optimization subsegment of N samples is given by:

$$E_w = \sum_{n=0}^{N-1} [x'(n) - Gy_\alpha(n)]^2. \tag{3.16}$$

Upon expanding the above equation in analogy to Equations 3.5–3.7 and setting $\partial E_w/\partial G = 0$ leads to [46, 47]:

$$G = \frac{\sum_{n=0}^{N-1} x'(n)y_\alpha(n)}{\sum_{n=0}^{N-1} [y_\alpha(n)]^2}. \tag{3.17}$$

Observe that while the optimum open-loop gain in Equation 3.6 was based on the normalized correlation $r(n)$ of the LPC residual, the closed-loop gain of Equation 3.17 is based on the more elaborate operations summarised in Equation 3.12–3.17.

Since the optimum closed-loop LTP gain is now known, the minimum weighted error is computed by substituting Equation 3.17 into Equation 3.16, which yields:

$$E_w = \sum_{n=0}^{N-1} [x'(n)]^2 - \frac{[\sum_{n=0}^{N-1} x'(n)y_\alpha(n)]^2}{\sum_{n=0}^{N-1} [y_\alpha(n)]^2}. \tag{3.18}$$

The closed-loop LTP delay α is found by maximizing the second term of Equation 3.18, while the optimum LTP gain factor G is determined from Equation 3.17. In conclusion, we note that Salami [46, 47] also proposed a computationally efficient recursive procedure for the successive evaluation of y_α, which is highlighted below.

Salami began his elaborations by noting that the past excitation signal $u(n - \alpha)$ is available only for $n - \alpha < 0$. When $n - \alpha > 0$, the "past excitation" is part of the excitation for the current subframe and so it is not yet known. Thus, for delays less than the subframe length N, only the first α values of $u(n - \alpha)$ are available. We make up the rest of the values by repeating the available pattern—that is taking $u(n - 2\alpha)$ for $\alpha < n < 2\alpha - 1$, and so on, until the range $0 \leq n \leq N - 1$ has been covered.

The computational load required to calculate the convolution $y_\alpha(n)$ for all possible values of the delay α would be large if they were all calculated independently. Fortunately, this can be avoided by calculating the convolution for the lowest value of α and then using an iterative procedure to find $y_\alpha(n)$ for all the other necessary values of α [47]. This iterative procedure is possible because the adaptive codebook codeword for a delay α is merely the codeword for the delay $\alpha - 1$ shifted by one sample, with one new value $u(-\alpha)$ introduced, and one old value $u(N - \alpha)$ discarded. This is true except for delays less than the subframe length N. The iterative procedure becomes slightly more complicated because of the repetition used to construct the codewords.

In summary, we gave a rudimentary introduction to AbS speech coding and showed that the synthetic speech is the output signal of the optimum synthesis filter, when excited by the innovation sequence. Once the STP and LTP analysis and synthesis filters are described by the coefficients a_k, G, and delay α, the central problem of achieving a good compromise in terms of speech quality and bitrate hinges on modeling the prediction residual. A variety of methods for modeling the prediction residual are described in References [46] and [47]. In the next section we will briefly highlight a number of techniques, including the so-called regular pulse excitation (RPE) described in depth in Chapter 5, which is used in the Pan-European mobile radio system known as GSM [73, 74], as well as code-excited linear prediction (CELP), that will be detailed in the context of Chapter 6.

3.5 EXCITATION MODELS

Again, the differences between RPE and CELP codecs arise from the representation of the excitation signal $u(n)$ used. In the multi-pulse excited (MPE) codecs proposed in 1982 by Atal and Remde [9], $u(n)$ is given by a fixed number of nonzero pulses for every frame of speech. The positions of these nonzero pulses within the frame, and their amplitudes, must be determined by the encoder and transmitted to the decoder. In theory, it would be possible to find the very best values for all the pulse positions and amplitudes, but this is not practical due to the excessive complexity it would entail. In practice, some suboptimal method of finding the pulse positions and amplitudes must be used. The positions are usually found one at a time as follows. Initially all the pulses are assumed to have zero amplitude except one. The position and amplitude of this first pulse can then be found by tentatively allocating the pulse to all possible positions and then finding its magnitude in order to minimize the associated perceptually weighted error. Finally, the pulse position and the associated magnitude yielding the lowest weighted error are confirmed. Then, using this information, the position and amplitude of the second pulse can be determined similarly. This procedure continues until all the pulses have been found. Once a pulse position is determined, it is fixed. However, the amplitudes of the previously found pulses can be re-optimized at each stage of the algorithm [75] when a new pulse is allocated. The quality of the reconstructed speech produced by MPE codecs is largely determined by how many nonzero pulses are used in the excitation. However, this is constrained by the bit-rate necessary to transmit information about the pulse positions and amplitudes. Typically, about 4 pulses per 5 ms are used; this leads to good quality reconstructed speech and a bitrate of around 10 kbits/s.

Similarly to the MPE codec, the regular pulse excited (RPE) codec uses a number of nonzero pulses in order to generate the excitation signal $u(n)$. However, in RPE codecs the pulses are regularly spaced with a certain separation. Hence, the encoder only has to determine the position of the first pulse and the amplitude of all the pulses. Therefore, less information has to be transmitted concerning the pulse positions, and hence for a given bitrate the RPE codec can benefit from using many more nonzero pulses than MPE codecs. For example, as will become clear at a bitrate of about 10 kbits/s, around 10 pulses per 5 ms can be used in RPE codecs, compared to 4 pulses for MPE codecs. This allows RPE codecs to give slightly higher quality reconstructed speech than that of the MPE codecs. However, RPE codecs also tend to be more complex. The Pan-European GSM mobile telephone system [74] uses a simplified RPE codec, in conjuction with long-term prediction, operating at 13 kbits/s to provide toll quality speech.

Although MPE and RPE codecs can provide high-quality speech at bitrates around 10 kbits/s and higher, they are unsuitable for significantly lower rates. This is because of the large amount of information that must be transmitted about the excitation pulses' positions and amplitudes. If we attempt to reduce the bitrate by using fewer pulses, or by coarsely quantizing their amplitudes, the reconstructed speech quality deteriorates rapidly. Currently the most commonly used algorithm for producing good quality speech at rates below 10 kbits/s is Code Excited Linear Prediction (CELP). This approach was proposed by Schroeder and Atal in 1985 [16] and differs from MPE and RPE in that the excitation signal is vector-quantized. Explicitly, the excitation is given by an entry from a large vector-quantizer codebook and by a multiplicative gain term invoked in order to control its power. Typically, the codebook index is represented with the aid of about 10 bits (to give a codebook size of 1024 entries), and the codebook gain is coded using about 5 bits. Thus, the bitrate necessary to transmit the excitation information is significantly reduced to around 15 bits compared to the 47 bits used, for example, in the GSM RPE codec.

Originally [16], the codebook used in CELP codecs contained white Gaussian sequences. This was because it was assumed that the long- and short-term predictors would be able to remove nearly all the redundancy from the speech signal in order to produce a random noise-like residual. Furthermore, it was shown that the short-term probability density function (pdf) of this residual was nearly-Gaussian. Schroeder and Atal found that using such a codebook to produce the excitation for long- and short-term synthesis filters could produce high-quality speech. However, each codebook entry had to be passed through the synthesis filters in order to assess how similar the reconstructed speech it produced would be to the original. This implied that the complexity of the original CELP codec was excessive for it to be implemented in real time—it took 125 seconds of Cray-1 CPU time to process 1 second of the speech signal. Since 1985, significant research efforts have been invested in reducing the complexity of CELP codecs—mainly through optimizing the structure of the codebook. Furthermore, significant advances have been made in the design of DSP chips, so that at the time of writing it is relatively easy to implement a real-time CELP codec on a single, low-cost, DSP chip. Several important speech coding standards have been defined based on the CELP principle, for example, the U.S. Department of Defense (DoD) 4.8 kbits/s codec [76], and the ITU's G.728 16 kbits/s low-delay codec [70]. We give a detailed description of CELP codecs in the next chapter.

The CELP coding principle has been very successful in producing communications to toll quality speech at bitrates between 4.8 and 16 kbits/s. The ITU's G.728 standard 16 kbits/s codec produces speech, which is almost indistinguishable from 64 kbits/s log-PCM coded speech, while the DoD's 4.8 kbits/s codec gives good communications-quality speech. Recently, much research has been conducted in the field of codecs operating at rates below 4.8 kbits/s, with the aim of producing a codec at 2.4 or 3.6 kbits/s, while having a speech quality equivalent to that of the 4.8 kbits/s DoD CELP. Here we will briefly describe a few of the approaches that seem promising in contriving such a codec, noting that the last part of this book is dedicated to this codec family.

The original CELP codec's structure can be further improved and used at rates below 4.8 kbits/s by classifying speech segments into a number of classes (for example, voiced, unvoiced, and "transitory" frames) [77]. The different speech segment types are then coded differently with a specially designed encoder for each type. For example, for unvoiced frames the encoder will not use any long-term prediction, whereas for voiced frames such prediction is vital but the fixed codebook may be less important. Such class-dependent codecs have been shown to be capable of producing reasonable quality speech at rates down to 2.4 kbits/s [78]. Multi-Band Excitation (MBE) codecs [79] analyze the speech frequency bands and declare some regions in the frequency domain as voiced and others as unvoiced. For each frame they transmit a pitch period, spectral magnitude, and phase information, as well as voiced/unvoiced decisions for the bands related to the harmonics of the fundamental frequency. Originally, it was shown that such a structure was capable of producing good quality speech at 8 kbits/s. Since then, this rate has been significantly reduced (see, for example [80]). Finally, Kleijn has suggested an approach for coding voiced segments of speech, which he referred to as Prototype Waveform Interpolation (PWI) [81]. This codec operates by sending information about a single pitch cycle every 20 to 30 ms, and by using interpolation between these instances in order to reproduce a smoothly varying quasi-periodic waveform for voiced speech segments using similar principles. Excellent quality reproduced speech can be obtained for voiced speech at rates as low as 3 kbits/s. Such a codec can be combined with a CELP type coding regime for the unvoiced segments, in order to attain good quality speech at rates below 4 kbits/s.

Having highlighted a variety of excitation models used in various previously proposed codecs, in the next section we provide a brief treatise on post-filtering, which has successfully been employed in a range of standard codecs in order to improve their perceptual speech quality.

3.6 ADAPTIVE SHORT-TERM AND LONG-TERM POST-FILTERING

Post-filtering was originally proposed by Jayant and Ramamoorthy [82, 83] for 32 kbps ADPCM coding using the two-pole six-zero synthesis filter of the G.721 codec of Figure 2.11. Later Chen et al. adopted this technique in order to improve the performance of low-rate CELP codecs [84] as well as that of the CCITT G.728 16 kbps low-delay backward-adaptive CELP codec [70, 85]. The basic principle of post-filtering is to further emphasize the spectral peaks of the speech signal, while slightly reducing their bandwidth and attenuating spectral valleys between these prominences. This spectral shaping procedure inevitably alters the waveform shape of the speech signal to a certain extent, which constitutes an undesirable impairment. Nonetheless, the enhancement of the spectral peaks—and in particular the concomitant attenuation of the potentially noise-contaminated low-energy spectral valleys—results in subjective quality improvement. Hence the advantage of reducing the effect of quantization noise in the subjectively important spectral valleys outweighs the waveform distortion penalty inflicted. This is necessary, because despite allocating a reduced amount of quantization noise to the spectral valleys after perceptual error weighting, these low-energy frequency bands remain vulnerable to contamination. This effect can be mitigated by retrospectively attenuating these partially contaminated frequency bands.

During the initial phases of its development, the G.728 codec did not employ adaptive post-filtering, because it was believed that it would result in the accumulation of quantization noise in the speech spectral valleys, when tandeming several codecs. However, tandeming experiments showed that in conjunction with post-filtering the coding noise due to concatenating three asynchronously operated codecs became about 4.7 dB higher than in case no tandeming was used, that is, when using just one codec. Chen et al. concluded [70, 85] that this effect of introducing post-filtering was a consequence of optimizing the extent of post-filtering for maximum noise masking at a concomitant minimum speech distortion for the scenario using no tandeming, that is, when employing a single coding stage. Therefore, upon concatenating up to three asynchronously operated codecs, the amount of post-filtering became exaggerated. These findings prompted a new post-filter design, which was optimized for three stages. As a consequence, the corresponding speech quality over three tandemed codec stages improved by a mean opinion score (MOS) of 0.81 to 3.93.

Modern post-filters [86] operate by emphasizing both the formant and pitch peaks in the frequency-domain representation of speech and simultaneously attenuating the spectral valleys between these peaks. This reduces the audible noise in the reconstructed speech, which persists even after the noise-shaping action of the error weighting filter, since it is in the valleys between the formant and pitch peaks, where the noise energy is most likely to cross the masking threshold and become audible. Therefore, attenuating the speech in these regions reduces the audible noise, and—since our ears are not overly sensitive to the speech intensity in these valleys—only minimal distortion is introduced to the speech signal.

The simplified block diagram of the post-filter arrangement used in the G.728 codec and in our variable-rate codecs proposed in Chapter 8 is shown in Figure 3.10. The components of this schematic are highlighted briefly below. Further specific details concerning the G.728 adaptive post-filter can be found in Section 8.4.6. The long-term post-filter (LTPF) has a transfer function of:

$$H_l(z) = \frac{1}{1+b}(1 + bz^{-p}) \tag{3.19}$$

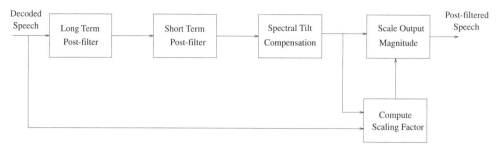

Figure 3.10 The G.728 adaptive post-filter arrangement.

where p is the backward-adapted estimate of the pitch period, which must not be confused with the STP order p. The calculation of the backward-adapted pitch is based on the past encoded speech, as highlighted in Section 8.6, and the coefficient b is given by

$$b = \begin{cases} 0 & \text{if } \beta < 0.6 \\ \lambda\beta & \text{if } 0.6 \le \beta \le 1 \\ \lambda & \text{if } \beta > 1 \end{cases} \tag{3.20}$$

where λ is a parameter, which controls the amount of long-term post-filtering and β is the tap weight of a single-tap long-term predictor having a delay of p, where β is given by

$$\beta = \frac{\sum_{n=-100}^{-1} \hat{s}(n)\hat{s}(n-p)}{\sum_{n=-100}^{-1} \hat{s}^2(n-p)}. \tag{3.21}$$

Note that here β was used instead of the previously introduced conventional forward-adapted LTP delay α. If β is less than 0.6, then the speech is assumed to be unvoiced and b is set to zero, effectively turning off the long-term post-filter.

The short-term post-filter is given by

$$H_s(z) = \frac{1 - \sum_{i=1}^{10} \tilde{a}_i \gamma_1^i z^{-i}}{1 - \sum_{i=1}^{10} \tilde{a}_i \gamma_2^i z^{-i}}, \tag{3.22}$$

where γ_1 and γ_2 are tunable parameters, which control the short-term post-filtering (STPF). Furthermore, $\tilde{a}_i$, $i = 1, 2 \ldots 10$, are the backward-adapted short-term synthesis filter parameters for a filter of order 10, which are derived as a byproduct during calculating the coefficients for the actual 50th-order synthesis filter. Again, this backward-adapted STP calculation process is detailed later in Section 8.4. The all-pole section of $H_s(z)$, which is constituted by its denominator, emphasizes the formants in the reconstructed speech and attenuates the valleys between these formants. However, this filtering operation introduces an undesirable spectral tilt in the post-filtered speech, which leads to a somewhat muffled speech perception. This spectral tilt is partially offset by the all-zero section of $H_s(z)$, namely, by its numerator.

The all-zero section of $H_s(z)$ significantly reduces the muffling effect of the post-filter. However, the post-filtered speech is still slightly muffled; hence, a **spectral tilt compensation** block is used to further reduce this effect. This is a first-order filter with a transfer function of $1 - \mu k_1 z^{-1}$, where μ is a tunable parameter between 0 and 1, and k_1 is the first reflection coefficient calculated from the LPC analysis of the reconstructed speech. During voiced speech, the post-filter introduces a low-pass spectral tilt to the speech, but simultaneously k_1 is close to -1. Hence, the spectral tilt compensation block introduces high-pass filtering in order to offset this spectral tilt. During unvoiced speech, the post-filter tends to introduce a high-pass spectral tilt to the speech, but k_1 becomes positive. Therefore, the spectral tilt compensation block automatically changes to a low-pass filter and again offsets the spectral tilt.

The final section of the post-filter in Figure 3.10 scales the output, so that it has approximately the same power as the original decoded speech. The long-term post-filter has its own gain control because of the factor $1/(1 + b)$ in $H_l(z)$. However, the short-term post-filter tends to amplify the post-filtered speech, when the prediction gain of the short-term filter is high. This leads to the output speech sounding unnatural. The output scaling blocks remove this effect by estimating the average magnitudes of the decoded speech and the output from the spectral tilt compensation block, and determining a scaling factor based on the ratio of these average magnitudes.

The tunable parameters λ, γ_1, γ_2, and μ must be chosen appropriately in order to control the amount of post-filtering used. We want to introduce sufficient post filtering in order to attenuate the audible coding noise as much as possible, without introducing too much distortion to the post-filtered speech. In the G.728 codec, the parameters were chosen to minimize the coding noise after three tandemed codec stages [70], since the ITU allows a maximum of three consecutive tandeming stages. The parameters were set to $\lambda = 0.15$, $\gamma_1 = 0.65$, $\gamma_2 = 0.75$, and $\mu = 0.15$.

In conclusion, post-filtering is important and sophisticated in state-of-the-art codecs. A variety of further solutions will be discussed in Chapter 7 in the context of existing standard codecs. Having considered the basic elements of the analysis-by-synthesis structure, namely, the issues of short- and long-term prediction and various excitation models, in closing of this chapter, an alternative technique of linear predictive AbS coding is presented in the next section, which is referred to as lattice-based short-term prediction. This will also allow us to further familiarize ourselves with the reflection coefficients introduced in the Levinson-Durbin algorithm, as well as with other equivalent ways of describing the speech signal's spectrum and to consider the effect of quantizing these spectral parameters.

3.7 LATTICE-BASED LINEAR PREDICTION

In order to augment our exposition of the linear prediction problem, we note that several authors, including Itakura and Saito [87. 88], Kitawaki [89], Makhoul [52], Rabiner and Schaefer [6], Gordos and Takacs [15] as well as a number of other authors, showed how the key relationship of LPC analysis given by Equation 2.22 can be formulated using the **lattice approach** by combining the correlation computation with an iterative solution for the predictive coefficients.

In order to be able to deduce the linear predictive lattice structure, let us first highlight the analogy between the concept of backward prediction and forward prediction, which rely on a set of symmetric equations. Specifically, let us refer to Figures 3.11 and 3.12, where the current sample $s(n)$ is predicted using the previous p number of samples and coefficients a_i, $i = 1 \ldots p$, and the forward prediction error is given by the usual expression of:

$$e_f(n) = s(n) - \sum_{k=1}^{p} a_k s(n-k) = \sum_{k=0}^{p} a_k s(n-k), \qquad a_0 \equiv 1 \tag{3.23}$$

or in z-domain as:

$$E(z) = A(z) \cdot S(z). \tag{3.24}$$

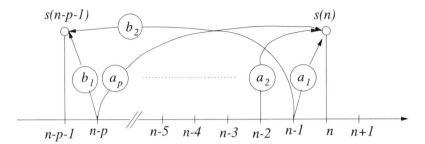

Figure 3.11 Forward and backward prediction of samples in the lattice approach, where
$s(n)$ is forward predicted using a_i, $i = 1 \ldots p$, while $s(n - p - 1)$ is backward
predicted using b_i, $i = 1 \ldots p$.

Similarly, the sample $s(n - p - 1)$ can be predicted in a backward-oriented fashion on the
basis of the samples $s(n - p) \ldots s(n - 1)$, which arrived later than $s(n - p - 1)$, using the
prediction coefficients b_i, $i = 1 \ldots p$. The associated backward prediction error is given by:

$$e_b(n - p - 1) = s(n - p - 1) - \sum_{k=1}^{p} b_k s(n - k). \tag{3.25}$$

It is convenient, however, to relate the backward prediction error to the instant $(n - 1)$,
because this is the time of the latest sample influencing its value. Hence, we rewrite Equation
3.25 as:

$$e_b(n - 1) = s(n - p - 1) - \sum_{k=1}^{p} b_k s(n - k), \tag{3.26}$$

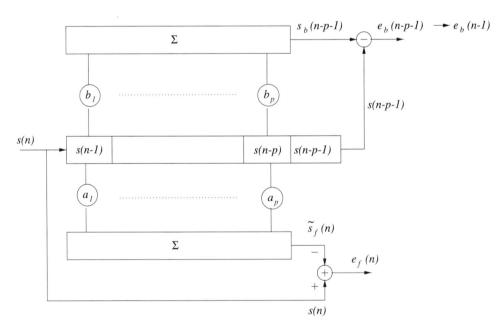

Figure 3.12 Forward and backward prediction scheme using the lattice approach, where
$s(n)$ is forward predicted using a_i, $i = 1 \ldots p$, while $s(n - p - 1)$ is back-
ward-predicted using b_i, $i = 1 \ldots p$.

which allows us to define a causal system, generating $e_b(n - 1)$ on the basis of $s(n - p - 1) \ldots s(n - 1)$. Again, in an analogy to the previously outlined forward predictive approach, the predictor coefficients b_i, $i = 1 \ldots p$ can be determined by minimizing the total squared backward prediction error of

$$E_b = \sum_N e_b^2(n - p - 1)$$

$$= \sum_N [s(n - p - 1) - \sum_{k=1}^{p} b_k s(n - k)]^2. \tag{3.27}$$

Upon expanding Equation 3.27, similarly to our approach previously described by Equation 2.21 in the context of forward prediction, we arrive at:

$$\begin{pmatrix} R(0) & R(1) & \ldots & R(p-1) \\ R(1) & R(0) & \ldots & R(p-2) \\ \vdots & & & \\ R(p-1) & R(p-2) & \ldots & R(0) \end{pmatrix} \begin{pmatrix} b_1 \\ b_2 \\ \vdots \\ b_p \end{pmatrix} = \begin{pmatrix} R(p) \\ R(p-1) \\ \vdots \\ (R(1)) \end{pmatrix}, \tag{3.28}$$

giving a solution of

$$b_i = a_{p+1-i}, \tag{3.29}$$

which, in accordance with Figure 3.11, is symmetric with respect to the forward-oriented predictor.

We can also express this relationship in terms of the corresponding all-zero polynomials $A(z)$ and $B(z)$ upon z-transforming Equation 3.26, yielding [15]:

$$E_b(z)z^{-1} = S(z)z^{-p-1} - S(z)\left[\sum_{k=1}^{p} b_i z^{-k}\right]. \tag{3.30}$$

This allows us to express the backward-oriented all-zero polynomial $B(z)$ as:

$$B(z) = \frac{E_b(z)}{S(z)} = z^{-p} - \sum_{k=1}^{p} b_k z^{-k+1}$$

$$= z^{-p}\left[1 - \sum_{k=1}^{p} b_k z^{-k+1+p}\right] \tag{3.31}$$

and upon exploiting Equation 3.29 we arrive at:

$$B(z) = z^{-p}\left[1 - \sum_{k=1}^{p} a_{p+1-k} z^{-k+1+p}\right]$$

$$= z^{-p}[1 - a_p z^p - a_{p-1} z^{p-1} - \cdots - a_1 z]. \tag{3.32}$$

Lastly, since

$$A(z) = 1 - a_1 z^{-1} - a_2 z^{-2} - \cdots - a_p z^{-p}$$

$$= 1 - a_1 \frac{1}{z} - a_2 \frac{1}{z^2} - \cdots - a_p \frac{1}{z^p} \tag{3.33}$$

and

$$A(z^{-1}) = A\left(\frac{1}{z}\right) = 1 - a_1 z^1 - a_2 z^2 - \cdots - a_p z^p \tag{3.34}$$

we get the plausible relationship of

$$B(z) = z^{-p}A(z^{-1}) \tag{3.35}$$

between the backward and forward-oriented all-zero polynomials. The physical interpretation of Equation 3.35 suggests that the optimum backward prediction polynomial is a close relative of $A(z^{-1})$. The z-domain representation of the backward prediction error is given by:

$$E_b(z) = B(z) \cdot S(z) = z^{-p}A(z^{-1})S(z). \tag{3.36}$$

In order to proceed with the formulation of the lattice-based prediction approach, let us now derive a recursion for the generation of the ith-order all-zero polynomial from the $(i-1)$st-order system, where

$$A^{(i)}(z) = 1 - a_1^{(i)}z^{-1} - a_2^{(i)}z^{-2} - \cdots - a_{i-1}^{(i)}z^{-i+1} - a_i^{(i)}z^{-i}. \tag{3.37}$$

Upon exploiting from the Levinson-Durbin algorithm of Figure 2.3 that for the coefficients of the ith-order system, we have:

$$\begin{aligned}
a_j^{(i)} &= a_j^{(i-1)} - k_i a_{i-j}^{(i-j)} \qquad \text{for } j = 1 \ldots i-1 \\
a_i^{(i)} &= k_i
\end{aligned} \tag{3.38}$$

we arrive at:

$$\begin{aligned}
A^{(i)} &= 1 - (a_1^{(i-1)} - k_i a_{i-1}^{(i-1)})z^{-1} - (a_2^{(i-1)} - k_i a_{i-2}^{(i-1)})z^{-2} - \cdots \\
&\quad - (a_{i-1}^{(i-1)} - k_i a_{i-i+1}^{(i-1)})z^{-i+1} \\
&= 1 + k_i(a_{i-1}^{(i-1)}z^{-1} + a_{i-2}^{(i-1)}z^{-2} + \cdots + a_1^{(i-1)}z^{-i+1} - z^{-i}) \\
&\quad - a_1^{(i-1)}z^{-1} - a_2^{(i-1)}z^{-2} - \cdots - a_{i-1}^{(i-1)}z^{-i+1}.
\end{aligned} \tag{3.39}$$

because of Equation 3.37 we have

$$A^{(i-1)}(z) = 1 - a_1^{(i-1)}z^{-1} - a_2^{(i-1)}z^{-2} - \cdots - a_{i-1}^{(i-1)}z^{-i+1} \tag{3.40}$$

and

$$A^{(i-1)}(z^{-1}) = 1 - a_1^{(i-1)}z - a_2^{(i-1)}z^2 - \cdots - a_{i-1}^{(i-1)}z^{i-1} \tag{3.41}$$

Hence, the required recursion is given by:

$$\begin{aligned}
A^{(i)}(z) &= A^{(i-1)}(z) + k_i(a_{i-1}^{(i-1)}z^{-1} + \cdots + a_1^{(i-1)}z^{-i+1} - z^{-i}) \\
&= A^{(i-1)}(z) + k_i z^{-i}(a_{i-1}^{(i-1)}z^{i-1} + \cdots + a_1^{(i-1)}z^1 - 1) \\
&= A^{(i-1)}(z) - k_i z^{-i} A^{(i-1)}(z^{-1}).
\end{aligned} \tag{3.42}$$

As an example, for $i = 2$ we have:

$$\begin{aligned}
A^{(1)}(z) &= A^{(0)}(z) - k_1 z^{-1} A^{(0)}(z^{-1}) = 1 - k_1 z^{-1} \\
A^{(2)}(z) &= A^{(1)}(z) - k_2 z^{-2} A^{(1)}(z^{-1}); \quad A^{(1)}(z^{-1}) = 1 - k_1 z \\
&= (1 - k_1 z^{-1}) - k_2 z^{-2}(1 - k_1 z) \\
&= 1 - k_1 z^{-1} - k_2 z^{-2} + k_1 k_2 z^{-1} \\
&= 1 - k_1 z^{-1}(1 - k_2) - k_2 z^{-2}.
\end{aligned} \tag{3.43}$$

Observe, however, in both Equation 3.41 and in the above example that the function $A^{(i-1)}(z^{-1})$ represents an unrealizable, noncausal system. Nonetheless, upon using an

$(i − 1)$st-order predictor in Equation 3.35 and invoking Equation 3.42, we can rectify this problem, leading to:

$$A^{(i)}(z) = A^{(i-1)}(z) - k_i z^{-1} z^{(i-1)} A^{(i-1)}(z^{-1})$$
$$= A^{(i-1)}(z) - k_i z^{-1} B^{(i-1)}(z). \tag{3.44}$$

When substituting the recursion of Equation 3.44 into Equation 3.24, the forward-oriented prediction error of the ith-order predictor is yielded as

$$E_f^{(i)}(z) = A^{(i)}(z)S(z)$$
$$= A^{(i-1)}(z)S(z) - k_i z^{-1} B^{(i-1)}(z)S(z) \tag{3.45}$$

Observe in Equation 3.45 that the first term is the forward prediction error of the $(i − 1)$st order predictor, while the second term can be interpreted in an analogous fashion after transforming Equation 3.45 back to the time domain:

$$e_f^{(i)}(n) = e_f^{(i-1)}(n) - k_i e_b^{(i-1)}(n - 1). \tag{3.46}$$

Clearly, this expression generates the forward prediction error of the ith order predictor as a linear combination of the forward and backward prediction errors of the $(i − 1)$st order forward and backward predictors.

In order to arrive at a complete set of recursive formulas, it is also possible to generate the ith order backward prediction error $e_b^{(i)}(n)$ from that of the $(i − 1)$st order forward and backward predictors using the following approach. The ith order backward predictor's prediction error is given in z-domain by:

$$E_b^{(i)}(z) = B^{(i)}(z) \cdot S(z) \tag{3.47}$$

that can be rewritten by the help of Equation 3.35 as:

$$E_b^{(i)}(z) = z^{-i} A^{(i)}(z^{-1}) \cdot S(z) \tag{3.48}$$

which in turn is reformulated using the recursion of Equation 3.42 as:

$$E_b^{(i)}(z) = z^{-i} S(z)[A^{(i-1)}(z^{-1}) - k_i z^i A^{(i-1)}(z)]$$
$$= z^{-i} A^{(i-1)}(z^{-1})S(z) - k_i A^{(i-1)}(z)S(z). \tag{3.49}$$

Exploiting the relationship of Equation 3.35 again and introducing the z-transform of the forward prediction error leads to:

$$E_b^{(i)}(z) = B^{(i)} \cdot S(z)$$
$$= z^{-1} B^{(i-1)}(z)S(z) - k_i A^{(i-1)}(z)S(z)$$
$$= z^{-1} E_b^{(i-1)}(z) - k_i E_f^{(i-1)}(z). \tag{3.50}$$

Finally, after transforming the above equation back to time domain, we arrive at:

$$e_b^{(i)}(n) = e_b^{(i-1)}(n - 1) - k_i e_f^{(i-1)}(n), \tag{3.51}$$

expressing the ith order backward prediction error as a combination of the $(i − 1)$st order forward and backward prediction errors. Furthermore, from the first and second line of Equation 3.50 we can also infer the recursive relationship of

$$B^{(i)}(z) = z^{(-1)} b^{(i-1)}(z) - k_q A^{(i-1)}(z), \tag{3.52}$$

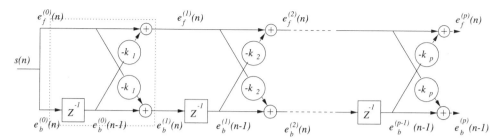

Figure 3.13 Lattice analysis scheme.

producing the optimum ith-order backward-oriented all-zero polynomial from the $(i-1)$st order $B(z)$ and $A(z)$ functions.

The recursions in Equations 3.46 and 3.51 now define the **lattice analysis structure**, delivering both the forward and backward prediction errors from $s(n)$. For the zero-order predictor we have $e_f^{(0)}(n) = e_b^{(0)}(n) = s(n)$, implying that the forward predictor generates $s(n)$ from $s(n)$, while the backward predictor produces $s(n-1)$ from $s(n-1)$. Using Equations 3.46 and 3.51, we now find it easy to confirm that the corresponding schematic obeys the structure of Figure 3.13, which constitutes an alternative implementation of the all-zero analysis filter $A(z)$ without relying on the coefficients a_i, $i = 1 \ldots p$.

The corresponding **synthesis lattice structure** can be readily constructed by adopting an inverse approach in order to generate $s(n)$ from $e_f^{(p)}(n)$. Hence, Equation 3.46 can be rearranged to reflect this approach as follows [15]:

$$e_f^{(i-1)}(n) = e_f^{(i)}(n) + k_i e_b^{(i-1)}(n-1), \tag{3.53}$$

while:

$$
\begin{aligned}
e_b^{(i)}(n) &= e_b^{(i-1)}(n-1) - k_i e_f^{(i-1)}(n) \\
&= e_b^{(i-1)}(n-1) - k_i e_f^{(i-1)}(n) + k_i^2 e_b^{(i-1)}(n-1) - k_i^2 e_b^{(i-1)}(n-1) \\
&= e_b^{(i-1)}(n-1) - k_i[e_f^{(i-1)}(n) - k_i e_b^{(i-1)}(n-1)] - k_i^2 e_b^{(i-1)}(n-1). \tag{3.54}
\end{aligned}
$$

Upon recognizing that the square-bracketed term corresponds to the right-hand side of Equation 3.46, we arrive at:

$$
\begin{aligned}
e_b^{(i)}(n) &= e_b^{(i-1)}(n-1) - k_i e_f^{(i)}(n) - k_i^2 e_b^{(i-1)}(n-1) \\
&\quad - k_i e_f^{(i)}(n) + (1 - k_i^2) e_b^{(i-1)}(n-1). \tag{3.55}
\end{aligned}
$$

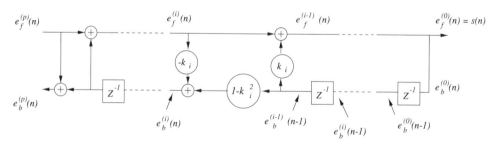

Figure 3.14 Lattice synthesis scheme.

Equations 3.53 and 3.55 are directly realizable, as portrayed in Figure 3.14, which is easily verified by the interested reader. Observe that this circuit contains three multipliers. It is possible to find arithmetically equivalent representations, while requiring two or just one multiplier, which usually requires a higher number of adders [6, 15].

3.8 CHAPTER SUMMARY

In this chapter the AbS structure was introduced, and its building blocks were detailed. The concept of perceptual error weighting was introduced in order to mask the effects of quantization errors in the most vulnerable spectral valleys of the speech signal between the high-energy formant regions. Both open-loop and closed-loop LTPs were analyzed and studied. The latter guarantees a better performance at the cost of a higher complexity. Practical codecs often combine these techniques by invoking an initial coarse open-loop LTP analysis and then a more accurate closed-loop procedure in the vicinity of the pitch value determined by the open-loop search.

These LTP-oriented discussions were followed by a brief discourse on post-filters, which further enhance the perceptual speech quality. Finally, having introduced the reflection coefficients and having studied their characteristics, let us now focus our attention on a range of other spectral coefficients, which are more amenable to quantization. In other words, we are seeking alternative ways of representing the speech signal's spectral envelope, which exhibits a higher robustness against transmission errors inflicted by hostile channels.

4

Speech Spectral Quantization

4.1 LOG-AREA RATIOS

In Section 2.3.3 the filter coefficients a_i, $i = 1 \ldots p$ and their equivalent representations, the reflection coefficients k_i, $i = 1 \ldots p$, were introduced in order to describe the speech signal's spectral envelope. Here we characterize their statistical properties in terms of their experimentally evaluated PDFs, which are portrayed in Figures 4.1 and 4.2, respectively. Experience shows that in case of the a_i coefficients an extremely fine resolution quantization is necessary in order to guarantee the stability of the synthesis filter $H(z) = 1/A(z)$. Clearly, this is undesirable in terms of bitrate.

As seen in Figure 2.3.3, the reflection coefficients have a more limited amplitude range, and the stability of $H(z)$ can be ensured by checking the physically tangible condition of:

$$|k_i| = \left| \frac{A_{i+1} - A_i}{A_{i+1} + A_i} \right| < 1, \tag{4.1}$$

where again, A_i, $i = 1 \ldots p$ represents the area of the ith acoustic tube section modeling the vocal tract [6, 47] and $|k_i| > 1$ would imply a tube cross section of $A_i \leq 0$. If the computed or transmitted value of k_i is outside the unit circle, it can be reduced to below one, which does modify the computed spectrum, but in exchange ensures filter stability. It was shown [90] that for values of $|k_i| \approx 1$ a very fine quantizer resolution must be ensured, requiring a densely spaced Lloyd-Max quantizer, since the filter transfer function $H(z)$ is very sensitive to the quantization errors for values of $|k_i| \approx 1$.

The **log-area ratios** (LAR) defined as:

$$\text{LAR}_i = \log \frac{1 - k_i}{1 + k_i} \tag{4.2}$$

constitute a nonlinear transformation of the reflection coefficients or area ratios and have better quantization properties. Their PDFs are plotted in Figure 4.3. Observe that the range of the LAR coefficients is becoming more limited toward higher order coefficients. This property was exploited, for example, in the Pan-European digital mobile radio system

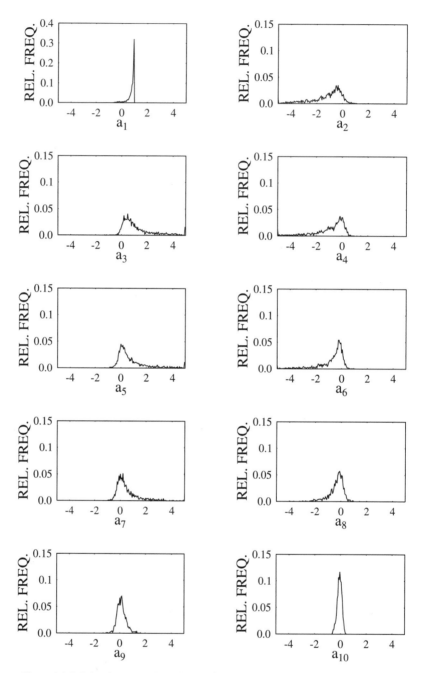

Figure 4.1 Relative frequency plots of the filter coefficients $a_i, i = 1, \ldots 10$ for a typical mixed-gender speech segment.

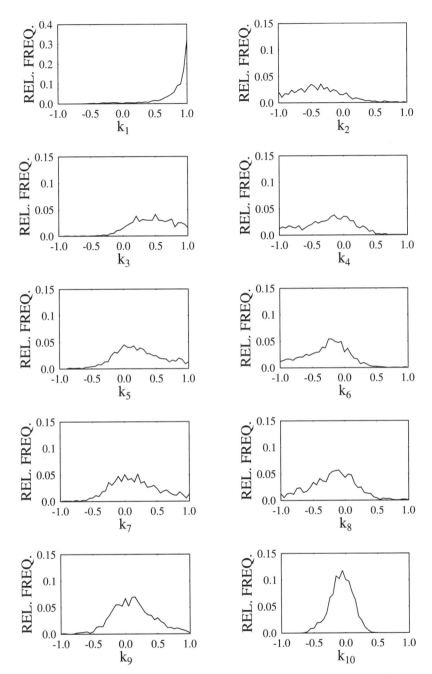

Figure 4.2 Relative frequency plots of the reflection coefficients $k_i, i = 1, \ldots 10$ for a typical mixed-gender speech segment.

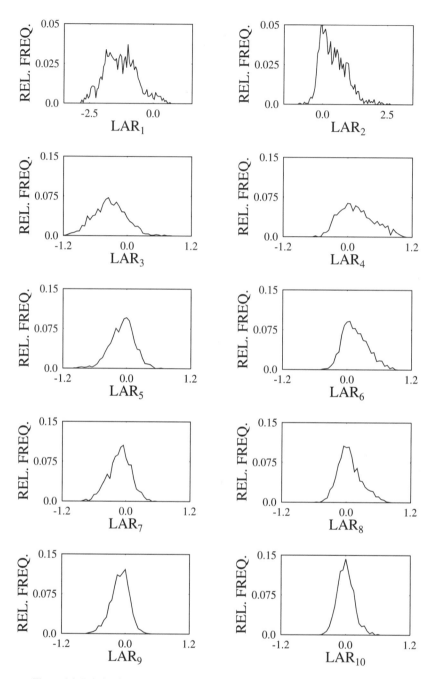

Figure 4.3 Relative frequency plots of the LAR filter coefficients LAR_i, $i = 1, \ldots 10$ for a typical mixed-gender speech segment.

known as GSM [74], where 6, 6, 5, 5, 4, 4, 3, and 3 bits were used to quantise the first eight LAR coefficients, requiring a total of 36 bits per 20 ms LPC analysis frame.

4.2 LINE SPECTRAL FREQUENCIES

4.2.1 Derivation of the Line Spectral Frequencies

Another derivative of the reflection coefficients and the all-zero filter $A(z)$ is the set of **line spectrum frequencies** (LSF) [91, 92], which are often also referred to as **line spectrum pairs** (LSP). In our forthcoming discussion we introduce the LSFs using a detailed mathematical description for the more advanced reader. Then a simple numerical procedure is proposed for their computation, and their statistical properties are contrasted with those of the a_i, k_i and LAR_i parameters.

Recall from Equation 3.42 in Section 3.7, which is repeated here for convenience, that $A^{(i)}(z)$ associated with the ith iteration of the pth order prediction obeys the following recursion:

$$A^{(i)}(z) = A^{(n-i)}(z) - k_i z^{-i} A^{(i-1)}(z^{-1}), \qquad i = 1, \ldots, p, \tag{4.3}$$

where $A^{(0)}(z) = 1$ and the polynomial $A(z^{-1})$ is physically related to the optimum backward-oriented all-zero polynomial $B(z)$ through Equation 3.35. Upon artificially extending the filter order to $i = p + 1$, Equation 4.3 can be formally rewritten as:

$$A^{(p+1)}(z) = A^{(p)}(z) - k^{p+1} z^{-p+1} A^{(p)}(z^{-1}). \tag{4.4}$$

Soong and Juang [93] argued that this extension is legitimate, if no unknown information is exploited, which can be ensured by setting $k_{p+1} = \pm 1$. Then the lattice analysis and synthesis schemes defined by Equations 3.46, 3.51 as well as by Equations 3.53, 3.55 and portrayed in Figures 3.13 and 3.14, respectively, are fully described, since they do not contain unknown quantities. When considering the lattice analysis scheme of Figure 3.13, which generates the prediction residual signal at the output of its $(p + 1)$-st stage, $k_{p+1} = \pm 1$ corresponds to perfect reflection or, in other words, to a complete closure and complete opening of the acoustic tube model at the glottis. From Equation 3.42 according to $k_{p+1} = \pm 1$, at iteration $p + 1$ we can write:

$$A^{(p+1)} = A^{(p)} \pm z^{-}(p+1) A^{(p)}(z^{-1}). \tag{4.5}$$

Specifically, for $k_{p+1} = 1$ the corresponding polynomial defined by Soong and Juang [93] is given by:

$$\begin{aligned} P(z) &= A^{(p+1)}(z) - z^{-(p+1)} A^{(p+1)}(z^{-1}) \\ &= 1 - a_1 z^{-1} - a_2 z^{-2} + \cdots - a_p z^{-p} \\ &\quad - [1 - a_1 z^1 - a_2 z^2 + \cdots - a_p z^p] \ldots z^{-(p+1)}, \end{aligned} \tag{4.6}$$

which is accordingly referred to as the **difference filter**. Similarly, for $k_{p+1} = -1$ we can derive the **sum filter** as follows:

$$Q(z) = A^{(p+1)}(z) + z^{-(p+1)}A^{(p+1)}(z^{-1})$$
$$= 1 - a_1 z^{-1} - a_2 z^{-2} + \cdots - a_p z^{-p}$$
$$+ [1 - a_1 z^1 - a_2 z^2 + \cdots - a_p z^p \ldots z^{-(p+1)}] \tag{4.8}$$

Based on our discussions of forward- and backward-oriented prediction in Section 3.7, and specifically Figure 3.11 and Equation 3.35, the backward-oriented predictor's impulse response can be said to be a time-reversed version of that of the forward-oriented one. In Figure 4.4 a hypothetical all-zero filter impulse response is portrayed, together with its appropriately time-reversed and shifted version and with the impulse responses of the sum

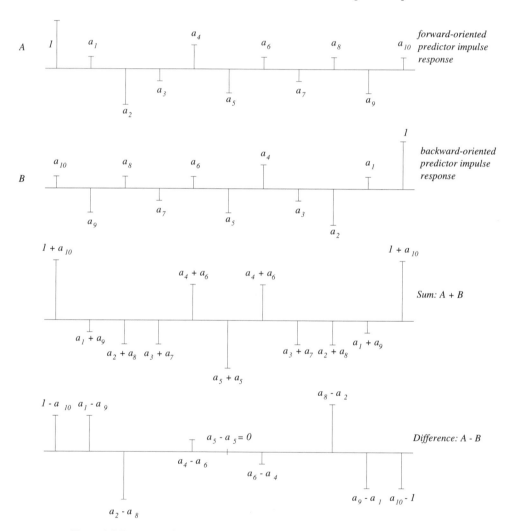

Figure 4.4 From top to bottom: (A) Stylized impulse response of the all-zero filter $A(z)$, (B) The stylised time-reversed shifted impulse response, (Sum: A + B). Stylized impulse response of the sum-filter (Difference: A − B) Stylized impulse response of the difference-filter.

and difference filters. Observe that while the impulse response of the sum filter $Q^{(p+1)}(z)$ is symmetric with respect to its center point, that of the difference filter $P^{(p+1)}(z)$ is antisymmetric or odd-symmetric. From the above two equations, the all-zero analysis filter can then be expressed as:

$$A^p(z) = \frac{1}{2}[P^{(p+1)}(z) + Q^{(p+1)}(z)]. \tag{4.9}$$

This particular formulation is not specific to the linear predictive coding of speech; it is valid for arbitrary finite response filters in general.

From Equation 4.7 we can collect the terms that correspond to the same power of z, or to the same delay in the impulse response of Figure 4.4, which ensues as follows:

$$\begin{aligned}
P^{(p+1)}(z) &= 1 - a_1 z^{-1} + a_p z^{-1} - a_2 z^{-2} + a_{p-1} z^{-2} - \cdots \\
&\quad - a_{p/2} z^{-p/2} + a_{(p/2-1)} z^{-p/2} + a_{p/2} z^{-p/2+1} - a_{(p/2-1)} z^{-p/2+1} + \cdots \\
&\quad + a_2 z^{-p+1} - a_{p-1} z^{-p+1} + \cdots \\
&\quad + a_1 z^{-p} - a_p z^{-p} - z^{-(p+1)} \\
&= 1 + (a_p - a_1) z^{-1} + (a_{p-1} - a_2) z^{-2} + \cdots \\
&\quad + (a_{(p/2-1)} - a_{p/2}) z^{-p/2} - (a_{(p/2-1)} - a_{p/2}) z^{-p/2+1} - \cdots \\
&\quad - (a_{p-1} - a_2) z^{-p+1} - (a_p - a_1) z^{-p} - z^{-(p+1)}. \tag{4.10}
\end{aligned}$$

In harmony with Figure 4.4, Equation 4.10 now explicitly shows the odd symmetry of coefficients which for the first and last terms have an absolute value of one, while for the second and last but one terms $|a_p - a_1|$, and so on. Upon rewriting Equation 4.10 in a more compact form, we arrive at [94]:

$$\begin{aligned}
P^{(p+1)}(z) &= 1 + p_1 z^{-1} + p_2 z^{-2} + \cdots + p_{p/2} z^{-p/2} \\
&\quad - p_{p/2} z^{-p/2+1} - \cdots - p_2 z^{-p+1} - p_1 z^{-p} - z^{-(p+1)}, \tag{4.11}
\end{aligned}$$

where only $p/2$ coefficients are necessary in order to describe $P^{(p+1)}(z)$, and the coefficients are given by:

$$p_1 = (-a_1 + a_p), \qquad p_2 = (-a_2 + a_{p-1}) \ldots p_{p/2} = (-a_{p/2} + a_{p/2-1}). \tag{4.12}$$

Since any odd-symmetric polynomial has a zero at $z = 1$, Equation 4.11 can be rewritten to express this explicitly as [94]:

$$\begin{aligned}
P^{(p+1)}(z) &= (1 - z^{-1})[1 + c_1 z^{-1} + c_2 z^{-2} + \cdots \\
&\quad + c_{p/2-1} z^{-p/2-1} + c_{p/2} z^{-p/2} + \cdots + c_2 z^{-p+2} + c_1 z^{-p+1} + z^{-p}] \\
&= (1 - z^{-1}) \cdot C(z), \tag{4.13}
\end{aligned}$$

where the coefficients $c_1 \ldots c_{p/2}$ can be determined by the help of simple polynomial division. Clearly, the resulting polynomial $C(z)$ now has a total of p coefficients rather than $(p + 1)$, but due to its symmetry only $p/2$ are different. Soong and Juang showed [93] that the roots of such a polynomial occur in complex conjugate pairs on the unit circle, and hence it is sufficient to determine only those on the upper half circle. Explicitly, the roots of $P^{(p+1)}(z)$ are: $1, \pm e^{j\Theta_1}, \pm e^{j\Theta_2}, \ldots, \pm e^{j\Theta_{p/2}}$, which allows us to express $P^{(p+1)}(z)$ as [94]:

$$P^{(p+1)}(z) = (1 - z^{-1}) \prod_{i=1}^{p/2} (1 - e^{j\Theta_i} z^{-1})(1 - e^{-j\Theta_i} z^{-1})$$

$$= (1 - z^{-1}) \prod_{i=1}^{p/2} [1 - z^{-1}(e^{-j\Theta_i} + e^{j\Theta_i}) + z^{-2}]$$

$$= (1 - z^{-1}) \prod_{i=1}^{p/2} [1 - 2z^{-1} \cos 2\pi f_i t_s + z^{-2}], \tag{4.14}$$

where f_i defines the **line spectral frequencies** (LSF) or **line spectral pairs** (LSP), while t_s corresponds to the sampling instants. When using the shorthand

$$d_i = -2 \cos 2\pi f_i t_s \tag{4.15}$$

we arrive at:

$$P^{(p+1)}(z) = (1 - z^{-1}) \prod_{i=1}^{p/2} [1 + d_i z^{-1} + z^{-2}]. \tag{4.16}$$

Following the same approach, a similar expression can be derived for the polynomial $Q^{(p+1)}(z)$, which is given as:

$$Q^{(p+1)}(z) = (1 + z^{-1}) \prod_{i=1}^{p/2} [1 - 2z^{-1} \cos 2\pi f_i t_s + z^{-2}]$$

$$= (1 + z^{-1}) \prod_{i=1}^{p/2} [1 + d_i z^{-1} + z^{-2}]. \tag{4.17}$$

Using Equation 4.9, Kang and Fransen [94] proposed a simple analysis filter implementation on the basis of Equations 4.16 and 4.17. Although this scheme is not widespread in current codec implementations, its portrayal in Figure 4.5 conveniently concludes our previous discourse on the derivation of LSFs. Observe in the figure that it obeys the structure of Equations 4.16 and 4.17, implementing each multiplicative term as a block surrounded by dotted lines.

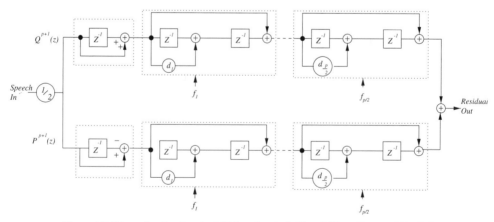

Figure 4.5 Schematic of a pth order LSF-based analysis filter © Kang, Fransen 1985 [94] according to Equation 4.16 and 4.17.

Assuming that the LSFs are known, the coefficients a_i, $i = 1 \ldots p$ can be recovered upon substituting Equations 4.16 and 4.17 in Equation 4.9 and collecting the terms multiplying the appropriate powers of z.

In practical codec implementations, often the lattice-based structures of Figures 3.13 and 3.14 are favored, and the LSFs are computed from the coefficients a_i, $i = 1 \ldots p$ or k_i, $i = 1 \ldots p$ in order to be able to exploit their more attractive quantization properties. More explicitly, in many practical codecs the LSFs are computed by determining the roots of the polynomials $P_{(z)}^{(p+1)}$ and $Q_{(z)}^{(p+1)}$, which are then quantized for transmission to the decoder. At the decoder we have to recover the coefficients a_i. Hence, we next highlight how the predictor coefficients a_i, $i = 1 \ldots p$ can be converted to LSF parameters, and then we summarize the most salient features of LSFs.

4.2.2 Computation of the Line Spectral Frequencies

A number of different techniques have been suggested for the computation of the LSFs [93–97] which have different strengths and weaknesses. Soong and Juang expressed the sum and difference filters $P^{(p+1)}(z)$ and $Q^{(p+1)}(z)$ as follows [93]:

$$P^{(p+1)}(z) = A(z)\left[1 + z^{-(p+1)}\frac{A(z^{-1})}{A(z)}\right] = A(z)[1 + R(z)]$$

$$Q^{(p+1)}(z) = A(z)\left[1 - z^{-(p+1)}\frac{A(z^{-1})}{A(z)}\right] = A(z)[1 - R(z)],$$

(4.18)

where they referred to

$$R(z) = z^{-(p+1)} \cdot \frac{A(z^{-1})}{A(z)}$$

(4.19)

as the **ratio filter**.

Equation 4.19 takes the general form of an **all-pass system**, which has a unity magnitude for all frequencies associated with a phase response. Thus, an all-pass filter is also often referred to as a phase shifter or phase corrector, since it may be invoked to correct the undesirable phase response of the rest of the system. Accordingly, the transfer function of the ratio filter of Equation 4.19 can also be formulated as [93]:

$$R(\omega) = e^{j\phi(\omega)},$$

(4.20)

where $\phi(\omega)$ represents the phase of $R(\omega)$. Equations 4.18 and 4.19 show that in order for $P^{(p+1)}(z)$ and $Q^{(p+1)}(z)$ to disappear, $R(z) = \pm1$ must be maintained, which implies that the zeros of $P^{(p+1)}(z)$ and $Q^{(p+1)}(z)$ must be on the unit circle. Furthermore, the roots are conjugate complex and symmetric to the origin. These facts were already alluded to earlier.

Note that general factorization techniques can be invoked in order to find the roots of $P^{(p+1)}(z)$ as well as $Q^{(p+1)}(z)$. In possession of the roots, we can compute the corresponding LSFs f_i using Equation 4.15. However, upon exploiting our a priori knowledge as regards their locations on the unit circle, more efficient methods can be devised, which is the topic of our forthcoming discussion.

The polynomial $C(z)$ in Equation 4.13 can be rewritten in order to reflect the conjugate complex symmetry of its roots explicitly as:

$$C(z) = z^{p/2}[(z^{p/2} + z^{-p/2}) + c_1(z^{p/2-1} + z^{-(p/2-1)}) + \ldots + c_{p/2}].$$

(4.21)

The equivalent of Equation 4.13 for the polynomial $Q^{(p+1)}(z)$ is:

$$Q^{(p+1)}(z) = (1 + z^{-1})[1 + d_1 z^{-1} + d_2 z^{-2} + \cdots + d_{p/2-1} z^{-p/2-1}$$
$$+ d_{p/2} z^{-p/2} + \cdots + d_2 z^{-p+2} + d_1 z^{-p+1} + z^{-p},$$
$$= (1 + z^{-1}) \cdot D(z), \tag{4.22}$$

yielding the symmetrical formula of

$$D(z) = z^{p/2}[(z^{p/2} + z^{-p/2}) + d_1(z^{p/2-1} + z^{-(p/2-1)}) + \cdots + d_{p/2}]. \tag{4.23}$$

If we now exploit the a priori knowledge that the roots of $C(z)$ and $D(z)$ are on the unit circle, that is, $z = e^{j\omega}$, we can express Equations 4.21 and 4.22 in a real form employing

$$z^{+1} + z^{-1} = e^{j\omega} + e^{-j\omega} = 2\cos\omega$$
$$z^{+2} + z^{-2} = e^{j2\omega} + e^{-j2\omega} = 2\cos 2\omega$$
$$\vdots \tag{4.24}$$
$$z^{+p/2} + z^{-p/2} = e^{j(p/2)\omega} + e^{-j(p/2)\omega} = 2\cos p\omega/2,$$

leading to

$$C(z) = 2e^{j(p/2)\omega}[\cos(p/2)\omega + c_1 \cos(p/2 - 1)\omega + \cdots + c_{p/2-1} \cos\omega + 1/2c_{p/2}] \tag{4.25}$$

$$D(z) = 2e^{j(p/2)\omega}[\cos(p/2)\omega + d_1 \cos(p/2 - 1)\omega + \cdots + d_{p/2-1} \cos\omega + 1/2d_{p/2}]. \tag{4.26}$$

If we can factorize the polynomials $C(z)$ and $D(z)$, then according to Equations 4.13 and 4.22 the roots of the $P^{(p+1)}_{(z)}$ and $Q^{(p+1)}_{(z)}$ have also been found, which determine the LSFs sought.

For the factorization of Equations 4.25 and 4.26, a number of techniques have been proposed. The conceptually most straightforward method is to evaluate the above expressions on a sufficiently fine grid in terms of ω, and observe the abscissa values at which the first expression of Equations 4.25 and 4.26 changes its polarity [93]. Between these positive and negative values there exists a root, which can then be identified more accurately recursively, halving the interval every time, in order to arrive at the required resolution.

The philosophy behind one of the approaches proposed by Kang and Fransen was to calculate the power spectra of $C(z)$ and $D(z)$ in order to be able to locate the frequencies at which these polynomials had local minima. Their alternative proposal was to exploit in Equations 4.18 and 4.19 that when the phase $\phi(\omega)$ of the ratio filter $R(z)$ of Equation 4.19 is a multiple of 2π, we have $Q^{(p+1)}(z) = 0$, since $|R(z)| = 1$. Alternatively, when $\phi(\omega)$ is an odd multiple of π, $P^{(p+1)}(z) = 0$. Hence, the LSFs can be determined by evaluating the phase spectrum $\phi(\omega)$ of the ratio filter $R(z)$ in Equation 4.20. Unfortunately, the above procedures rely on various trigonometric functions of the LSFs, which is an impediment in real-time codecs, since these functions must be pre-stored and thus require memory. Kabal [95] suggested an approach based on expressing $C(z)$ and $D(z)$ in Equations 4.25 and 4.26 in terms of Chebyshev polynomials, which remedy these ills.

4.2.3 Chebyshev—Description of Line Spectral Frequencies

Upon introducing the cosinusoidal frequency transformation of $x = \cos\omega$, for the LSFs, Kabal [95] and Ramachandran noted that Equations 4.25 and 4.26 can be reformulated in terms of the **Chebyshev polynomials**. These polynomials constitute a set of functions that can be generated recursively from lower-order members of the family. This will have implementational advantages. In general, an nth-order Chebyshev polynomial is defined by:

$$T_n(x) = \cos[n \cdot \arccos x] \qquad (4.27)$$

and the recursion generating successive members of the family can be derived by substituting our frequency transformation of $x = \cos \omega$ into Equation 4.27, which yields [98]:

$$T_n(x) = \cos n\omega. \qquad (4.28)$$

Upon formally extending this to $(n - 1)$ and $(n + 1)$, we arrive at:

$$T_{(n+1)}(x) = \cos(n + 1)\omega = \cos n\omega \cos \omega - \sin n\omega \sin \omega \qquad (4.29a)$$

$$T_{(n-1)}(x) = \cos(n - 1)\omega = \cos n\omega \cos \omega + \sin n\omega \sin \omega. \qquad (4.29b)$$

When adding Equations 4.29a and 4.29b and using Equation 4.28, we have $T_{(n+1)}(x) + T_{(n-1)}(x) = 2 \cos n\omega \cos \omega = 2xT_n(x)$, yielding the required recursion as

$$T_{(n+1)}(x) = 2xT_n(x) - T_{(n-1)}(x). \qquad (4.30)$$

From Equation 4.27, for $n = 0$ we have:

$$T_0(x) = 1 \qquad (4.31)$$

$$T_1(x) = x \qquad (4.32)$$

and from Equation 4.30

$$T_2(x) = 2x^2 - 1 \qquad (4.33)$$

$$T_3(x) = 4x^3 - 3x \qquad (4.34)$$

$$T_4(x) = 8x^4 - 8x^2 + 1, \text{ etc.} \qquad (4.35)$$

Upon substituting the corresponding Chebyshev polynomials into Equation 4.25 and 4.26 and neglecting the multiplicative linear-phase term $e^{j(p/2)\omega}$, we arrive at [95]:

$$C'(x) = 2T_{p/2}(x) + 2c_1 T_{p/2-1}(x) + \cdots + 2c_{p/2-1} T_1(x) + c_{p/2} \qquad (4.36a)$$

$$D'(x) = 2T_{p/2}(x) + 2d_1 T_{p/2-1}(x) + \cdots + 2d_{p/2-1} T_1(x) + d_{p/2}. \qquad (4.36b)$$

In order to determine the LSFs from Equations 4.36a and 4.36b, first the roots $x_i = \cos \omega_1$ of $C'(x)$ and $D'(x)$ must be computed, which are then converted to LSFs using $\omega_i = \arccos x_i$. While ω sweeps the range $0 \ldots \pi$ along the positive half of the unit circle, $x = \cos \omega$ takes on values in the range of $[-1, +1]$. This implies that for the roots x_i we have $-1 \leq x_i \leq +1$. At $\omega_i = 0$ we have $x_i = 1$, and the mapping $x = \cos \omega$ ensures that the lowest LSF ω_i is associated with the root x_i closest to unity.

Therefore, Kabal and Ramachandran proposed the following numerical solution for finding the LSF values ω_i at which $C'(x)$ and $D'(x)$ become zero. The principle applied is to a certain extent similar to that suggested by Soong and Juang [93], whereby the intervals in which the sign of the function changes are deemed to contain a single zero. The search is initiated from $x = 1$, since as argued in the previous paragraph, $C'(x)$ has the root closest to unity. Once the region of sign-change is located, the corresponding zero-crossing or change of polarity is identified more accurately by recursively halving of interval.

An attractive property of the Chebyshev polynomials is that, rather than evaluating all independent terms of Equations 4.25 and 4.26 for a high number of abscissa values using, for example, a cosine table, the recursion of Equation 4.30 can be invoked. Hence, during the evaluation of the equivalent set of Equations 4.36a and 4.36b, only two lower order Chebyshev polynomials have to be remembered, as suggested by Equation 4.30.

Upon exploiting the **ordering property of the LSFs** [93], which states that $f_0 < f_1 < f_2 < f_3 \ldots f_p < f_{p+1}$, the search then proceeds to trace the first root of $D'(x)$, commencing from the previously located $C'(x)$ root. This procedure is continued, interchanging the functions $C'(x)$ and $D'(x)$, until all LSFs f_i, $i = 1 \ldots p$ are found. Since $f_0 = 0$ and $f_{p+1} = 0.5$, they are known a priori and so are never transmitted.

The convergence speed of the above procedure is strongly dependent on the choice of the initial evaluation interval δ_1, which has to be sufficiently short in order to avoid having more than one root in an interval over which the polarity of $C'(z)$ and $D'(z)$ changes. Kabal and Ramachandran suggested [95] that $\delta_1 = 0.02$ is an adequate value to use, which implies a resolution of 100 intervals for $-1 \leq x_i \leq +1$.

The refined root-search that invoked interval halving typically required an accuracy of $\delta = 0.0015$, demanding four consecutive interval halving steps. When converting these x-domain root-location ambiguities to ω-domain, the LSF inaccuracy becomes nonlinearly frequency-dependent due to the $\omega_i = \arccos x_i$ conversion. However, several authors, for example, [94], reported that an LSF resolution ambiguity of 10 Hz does not cause any perceptual speech degradation.

In summary, we note that Equations 4.7 and 4.9 defined the odd-symmetric $P^{(p+1)}(z)$ and symmetric $Q^{(p+1)}(z)$ polynomials as the sum and difference polynomials, respectively. They led to the definition of LSFs through Equations 4.16 and 4.17. Assuming that the decoder is informed of the quantized LSFs, Equation 4.9 can be used to reconstruct the all-zero analysis filter $A(z)$. Section 4.2.2 was dedicated to highlighting procedures for the derivation of explicit formulas for the computation of LSFs, while Section 4.2.3 introduced a simple numerical technique for their computation, using a recursive formula for the efficient updating of the associated Chebyshev polynomial coefficients.

In conclusion, the basic properties of LSFs are as follows:

1. The roots of $P(z)$ and $Q(z)$, which are constituted by the LSFs ω_i, $i = 1 \ldots p$ obey the ordering property on the unit circle.
2. The stability of the all-pole synthesis filter $H(z) = 1/A(z)$ is retained upon quantizing the roots of $P(z)$ and $Q(z)$, as long as the ordering property is not violated.
3. The ordering property can also be invoked in order to detect and mitigate the effect of transmission errors in the LSP parameters by reestablishing their right ordering, when transmission errors were made.
4. Experience shows that a concentration of LSFs in a frequency region implies the presence of a spectral peak [99–101].
5. The LSFs evolve smoothly over consecutive frames, as seen in Figure 4.8, which stimulated research in order to further reduce the associated bitrate by exploiting this redundancy using predictive or vector quantization techniques.

Figure 4.6 portrays the PDFs of the LSFs for a tenth-order spectral shaping filter, while the relative frequency histogram of a 35-bit Lloyd-Max quantization scheme is shown in Figure 4.7. Observe that the first three and last two LSFs were quantized using a 3-bit or eight-level Max-Lloyd quantizer, while the other LSFs employed 4-bit or 16-level Max-Lloyd quantization. Accordingly, the latter schemes have a finer resolution or a denser spacing. Furthermore, in the regions of higher relative frequency the Lloyd-Max quantizer allocated the reconstruction levels more closely than in the lower probability intervals. In Figure 4.8 we

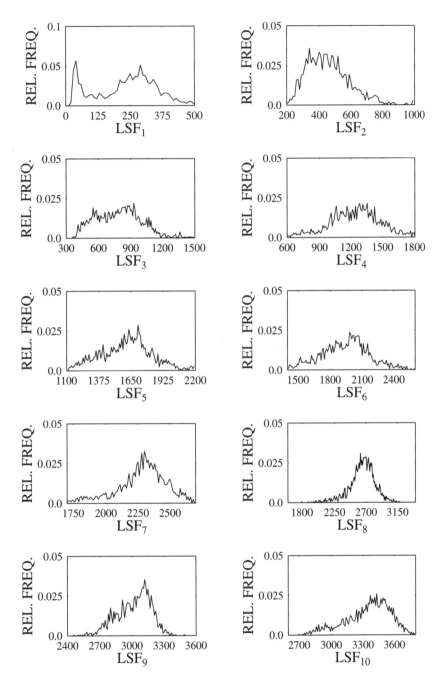

Figure 4.6 Relative frequency plots of the LSF filter coefficients LSF_i, $i = 1, \ldots 10$ for a typical mixed-gender speech segment.

portrayed a typical segment of the evolution of consecutive LSFs for 50 speech frames of 20 ms duration, which corresponds to 1 s speech. Observe that the LSF profiles never cross, and this ordering property is often exploited in error-resilient codecs in order to detect and mitigate the effects of transmission errors, which may have violated this condition. Observe

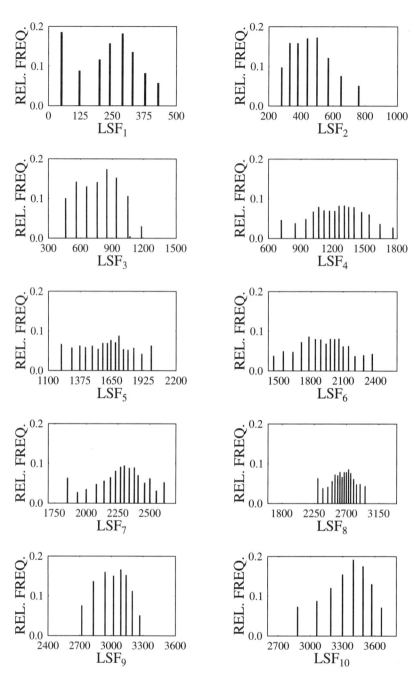

Figure 4.7 Relative frequency plots of the Max-Lloyd quantized LSF filter coefficients
LSF_i, $i = 1, \ldots 10$ for a typical mixed-gender speech segment.

furthermore that the quantized profiles follow closely the pattern of the ideal unquantized functions.

4.3 VECTOR QUANTIZATION OF SPECTRAL PARAMETERS

4.3.1 Background

Vector quantization of various source signals has become popular over the years, and a vast body of research has been incorporated in a range of excellent review papers (e.g., by Makhoul, Roucos, and Gish [67]) and in a monograph by Gersho and Gray [102]. For speech coding with bitrates around 10–16 kbps, the log-area ratios (LAR) or the line spectrum frequencies (LSFs) are usually quantized with 30 to 40 bits per 20 ms LPC update frame. Below 5 kbps encoding rates, either the LPC update frame has to be extended to around 30 ms, or vector quantization of the LPC parameters with at most 25 bits per 20 ms speech frame has to be employed. Conventional vector quantizers (VQ) [67] use trained codebooks, which usually lack robustness over speakers outside the training sequence. Shoham [103] attempted to exploit the similarities among successive spectral envelopes by employing vector predictive coding, where trained codebooks are needed for the predictor and residual vectors. A range of various LPC parameter quantizers have been proposed by Atal and Paliwal, Shoham, Lee, Kondoz and Evans, Yong and Gersho, Ramachandran, Sondhi, Seshadri et al. and Xydeas, respectively in references [92, 104], [103, 105].

A specific low-complexity speaker-adaptive LSF VQ scheme proposed by Lee, Kondoz, and Evans is highlighted next [106], followed by a discussion on a high-complexity vector quantizer arrangement using two consecutive random, stochastic codebooks [107, 108].

4.3.2 Speaker-Adaptive Vector Quantization of LSFs

According to the scheme portrayed in Figure 4.9 proposed by Lee, Kondoz, and Evans [106], the interframe redundancy, which is inherent in consecutive LSF vectors, as evidenced by Figure 4.8, is exploited in order to reduce the number of bits required by scalar quantization. As seen in Figure 4.9, each LSF vector is modeled by a codebook, CB1, containing the previously quantized vectors. Thus the authors refer to this scheme as a **Speaker Adaptive Vector Quantizer** (SAVQ).

Due to the high interframe correlation of the LSFs, this predictive process provides a good estimate of the current frame's LSF vector. Hence the residual error of $E_i > S_i^{K_i}$ from this first stage becomes rather unpredictable. This random prediction error can be quantized using a random Gaussian codebook, namely, CB2. Specifically, the unquantized LSF vector S_i, $i = 1 \ldots p$ is represented by that particular quantized LSF vector $\hat{V}_i$, $i = 1 \ldots p$ from CB1, which minimizes the squared and component-wise accumulated error of:

$$ER = \sum_{i=1}^{p} [S_i - \hat{V}_i]^2. \tag{4.37}$$

Then the prediction error vector $E_i = S_i - \hat{V}_i$, $i = 1 \ldots p$ is quantized with the help of CB2 by minimizing the quantization error term of

$$e = \sum_{i=1}^{p} [G \cdot U_i - E_i]^2. \tag{4.38}$$

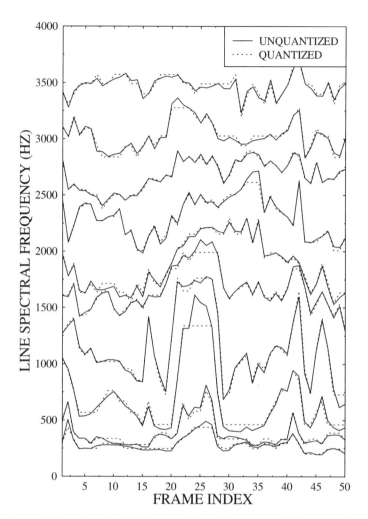

Figure 4.8 Evolution of the LSF filter coefficients LSF_i, $i = 1, \ldots 10$ for a typical 100 ms speech segment using the scalar quantizer of Figure 4.7.

Observe in Figure 4.9 that the codebook gain factor G allows the process to match the power of the codebook entries to that of the LSF prediction residual error. The optimum gain is computed for each entry. In order to find an expression for the gain factor, we set $\partial e / \partial G = 0$, yielding:

$$\sum_{i=1}^{p} 2[G \cdot U_i - E_i] \cdot U_i = 0 \tag{4.39}$$

which gives:

$$G_i = \frac{\sum_{i=1}^{p}[E_i \cdot U_i]}{\sum_{i=1}^{p}[U_i]^2}. \tag{4.40}$$

Observe that this is physically the normalized cross-correlation of the input and output of CB_2. Hence a high gain factor is assigned if E_i and U_i are similar. The effect of using the gain G_i is equivalent to extending the size of the codebook, without increasing the pattern-

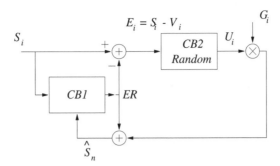

Figure 4.9 LSF vector quantizer schematic.

matching complexity. The codebook indices for CB1 and CB2, along with the quantized gain factor, are transmitted to the decoder, and the encoder also uses the quantized gain in its pattern-matching process. The ordering property must always be checked before an encoded vector is accepted. Notice in Figure 4.9 that the outputs of the two codebooks are superimposed in order to produce the quantized LSF vector, which is then written in CB1 for future use. This scheme has a low complexity, but it has a deficiency in terms of propagating channel errors.

Let us now embark on considering a more complex VQ scheme, which uses stochastic codebook entries and hence requires no training. This has the advantage of exhibiting a similar performance, regardless of the speaker's gender, mother tongue, and so on. This VQ scheme transforms the original stochastic codebook entries in vectors exhibiting similar statistical properties to the original LSFs to be encoded.

4.3.3 Stochastic VQ of LPC Parameters

In this section an academically interesting stochastic VQ scheme is presented for the advanced reader, noting that the practically motivated reader may skip this section and proceed to consider a range of more moderate-complexity LSF quantizers in Section 4.3.4.

4.3.3.1 Background. In Reference [109] a switched-adaptive method was suggested by Yong, Davidson, and Gersho, which exploits the correlation between adjacent LSF vectors in a different fashion. In this section we assume that the reader is familiar with the the the statistical properties of stochastic processes and the Karhunen-Loeve Transform [45], and we propose a stochastic VQ method based on an approach published by Atal [107].

In the original approach, the covariance matrix of the LARs was computed from a buffer containing the previously quantized LAR vectors. Then the covariance matrix of the LARs was decomposed into its eigen vectors and eigen values [51], following a procedure, that is not detailed here [51]. This decomposition was carried out for every new LPC update frame, which is computationally a rather demanding task. Furthermore, the eigen value solution requires an iterative algorithm, for example, the QR algorithm [51], which makes the processing time data dependent. This is undesirable in real-time applications.

According to the approach proposed by Salami et al. [108], an LPC parameter vector, such as the vector of 10 LAR or LSF parameters of an LPC update frame, which possess certain correlation properties, can be quantized using an uncorrelated Gaussian or stochastic codebook by transforming the uncorrelated codebook entries into vectors having correlations similar to those of the LPC parameter vectors. This technique is attractive, since the employment of random or stochastic codebooks ensures speaker-independent performance, which is often a deficiency associated with trained codebooks that may not be robust to speakers outside the training set. In general, a vector x of dimension N having jointly correlated components can be transformed into a vector u exhibiting uncorrelated components using a **orthogonal rotation** by the help of an $N \times N$ matrix $\mathbf{A}$ according to:

$$\mathbf{u} = \mathbf{Ax}. \tag{4.41}$$

Such orthogonal rotations have been used extensively in source coding in order to remove redundancy from the source signal [45]. The effect of orthogonal rotations can be easily made plausible by referring to the **Wiener-Khintchin theorem**, which states that the autocorrelation function (ACF) and power spectral density (PSD) are Fourier transform pairs. A manifestation of this is that the Dirac-delta ACF of AWGN is associated with an infinite bandwidth flat PSD. For example, when correlation is introduced in the uncorrelated AWGN signal by limiting the maximum rate of change at which the source signal can fluctuate using low-pass filtering, the band-limited AWGN has a sinc-function shaped ACF, exhibiting low correlation. In general, the more correlated the signal, the narrower the spectrum. This has been exploited, for example, in the context of **Discrete Cosine Transformation** [45] (DCT) based coding of speech and video signals. More specifically, after discrete cosine transforming the correlated source signal to the frequency domain typically only a small fraction of the signal's spectral coefficients has to be encoded, namely, those that exhibit a high magnitude. By contrast, the remaining low-energy spectral coefficients are neglected without significant loss of energy.

For a correlated source vector $\mathbf{x}$, which, for example, in our case is the vector of 10 LSFs, it was shown that the best decorrelating rotation $\mathbf{A}$ is given by a matrix whose rows are the normalized eigen vectors of $\mathbf{\Gamma_x}$, the covariance matrix of $\mathbf{x}$ [110]. This transformation is usually referred to as the **Karhunen-Loeve Transform** [45] (KLT), and to some extent it can also be applied to non-Gaussian sources [67]. The impediment of the KLT is its high computational complexity, which is due to the fact that the optimum decorrelating rotation matrix $\mathbf{A}$ is dependent on the source signal's correlation properties expressed in terms of $\mathbf{\Gamma_x}$. It can be shown [45] that other time- and data-invariant orthogonal transforms, such as the DCT, have similar decorrelating or energy-compaction properties, while ensuring lower system complexity.

In what follows we describe an inverse approach. Specifically, instead of decorrelating the correlated source vectors in order to achieve better compression, here we use uncorrelated stochastic codebook vectors and impose the required correlation properties in order to be able to model the LPC spectral components adequately.

In general, the covariance matrix $\mathbf{\Gamma_x}$ of a source $\mathbf{x}$ is often used to characterize the source's statistical properties, which can be computed as follows:

$$\mathbf{\Gamma}_x = E[(\mathbf{x} - \bar{\mathbf{x}})(\mathbf{x} - \bar{\mathbf{x}})^T], \tag{4.42}$$

where $E(\bullet)$ denotes the expected value of $\bullet$, the mean value of $\mathbf{x}$ is given by $\bar{\mathbf{x}} = E(\mathbf{x})$, and the superscript T represents matrix transposition.

Before proceeding, we briefly introduce the concept of the previously mentioned **eigen vectors** and **eigen values** [45]. The eigen values γ_k of the matrix $\mathbf{\Gamma}_x$ are defined as the roots of:

$$|\mathbf{\Gamma_x} - \gamma_k \mathbf{I}| = 0, \tag{4.43}$$

where **I** represents the identity matrix having unity diagonal elements, while all other elements are zero. The eigen vectors ϕ_k are defined by all the solutions of the following equation:

$$\Gamma_x \phi_k = \gamma_k \phi_k. \tag{4.44}$$

4.3.3.2 The Stochastic VQ Algorithm.

We now proceed to describe Atal's stochastic VQ algorithm. The covariance matrix Γ_x of the source vector **x** can be decomposed into three matrices according to [51] as follows:

$$\Gamma_x = S \cdot \lambda \cdot S^T, \tag{4.45}$$

where **S** is a matrix whose columns are the normalized eigen vectors of Γ_x and λ is a diagonal matrix whose elements are the eigen values of Γ_x. Equation 4.45 can also be written as:

$$\lambda = S^T \cdot \Gamma_x \cdot S. \tag{4.46}$$

Therefore, the rotated vector **u** in Equation 4.41 is given by

$$u = S^T x. \tag{4.47}$$

Upon exploiting Equation 4.46, the covariance matrix Γ_u of **u** can be formulated as:

$$\begin{aligned} \Gamma_u &= E[(u - \bar{u})(u - \bar{u})^T] \\ &= S^T \cdot \Gamma_x \cdot S = \lambda, \end{aligned} \tag{4.48}$$

which is the diagonal matrix λ, implying that **u** has uncorrelated components. The variances of the components of the uncorrelated vector **u** are the eigen values of Γ_x, and their means are given by:

$$\bar{u} = S^T \bar{x}. \tag{4.49}$$

In order to turn the uncorrelated vector **u** into a vector having unity covariance matrix and zero mean, its mean value $\bar{x}$ is subtracted from it. Then the decorrelating transformation using S^T is carried out, and lastly this quantity is normalized by $\lambda^{-1/2}$ according to:

$$u = \lambda^{-1/2} S^T (x - \bar{x}). \tag{4.50}$$

Hence upon exploiting Equation 4.46 again we have:

$$\begin{aligned} \Gamma_u &= E[(u - \bar{u})(u - \bar{u})^T] \\ &= E[\lambda^{-1/2} S^T (x - \bar{x}) \lambda^{-1/2} S^T (x - \bar{x})^T] \\ &= \lambda^{-1/2} S^T \Gamma_x \lambda^{-1/2} S^T \\ &= \lambda^{-1/2} \lambda \lambda^{-1/2} = I, \end{aligned} \tag{4.51}$$

which explicitly states that the process **u** is uncorrelated, since its covariance matrix is the identity matrix.

Now the stochastic vector quantization method accrues from rearranging Equation 4.50. Specifically, the LPC parameter vector **x** is quantized using the uncorrelated vectors $u^{(k)}$, $k = 1 \dots K$ chosen from a codebook, which contains K number of zero-mean, unit variance Gaussian entries, through the following transformation:

$$\hat{x} = \bar{x} + \beta S \lambda^{1/2} u^{(k)}. \tag{4.52}$$

Equation 4.52 is derived directly from Equation 4.50 with the scalar β introduced in order to allow more flexibility in terms of matching the powers of $\mathbf{x}$ and $\hat{\mathbf{x}}$. The mean squared error between the original and quantized vectors $\mathbf{x}$ and $\hat{\mathbf{x}}$ is given by

$$
\begin{aligned}
E_x &= (\mathbf{x} - \hat{\mathbf{x}})^T (\mathbf{x} - \hat{\mathbf{x}}) \\
&= \|(\mathbf{x} - \hat{\mathbf{x}})^T\|^2 \\
&= \|\mathbf{y} - \beta \lambda^{1/2} \mathbf{u}^{(k)}\|^2,
\end{aligned}
\tag{4.53}
$$

where $\|\bullet\|$ denotes the Euclidean norm of $\bullet$ and

$$
\mathbf{y} = \mathbf{S}^T (\mathbf{x} - \bar{\mathbf{x}}). \tag{4.54}
$$

The optimum codebook gain β is computed by setting $\partial E_x / \partial \beta = 0$. The codebook of K Gaussian vectors $u^{(k)}$, $k = 1 \ldots K$ is exhaustively searched for the index k, which minimizes the error in Equation 4.54. The quantized vector is then computed from Equation 4.52. The long-term covariance matrix $\boldsymbol{\Gamma}_\mathbf{x}$ is precomputed from a large database of LPC vectors. Hence, the decomposition specified in Equation 4.45 is precomputed, saving the effort of decomposing the covariance matrix for every new LPC analysis frame. In fact, no improvement has been achieved when we attempted to update the covariance matrix every LPC analysis frame.

The quality of VQ schemes is typically evaluated in terms of the spectral deviation (SD) metric, which is defined as [62]:

$$
\begin{aligned}
SD &= \frac{1}{2\pi} \int_{-\pi}^{\pi} \left(10 \log |H(\omega)|^2 - 10 \log |\hat{H}(\omega)|^2 \right)^2 d\omega \quad (\text{dB})^2 \\
&= \frac{1}{2\pi} \int_{-\pi}^{\pi} \left(10 \log \frac{|\hat{A}(\omega)|^2}{|A(\omega)|^2} \right)^2 d\omega \quad (\text{dB})^2,
\end{aligned}
\tag{4.55}
$$

and $\hat{H}(z)$ and $\hat{A}(z)$ are the quantized synthesis and analysis filters, respectively. The SD is typically computed for each LPC update frame of 20–30 ms and averaged over a number of speech frames in terms of dB. Although SD = 1 dB is considered as the spectral distortion limen for perceptually transparent coding of the LPC parameters [91], it is also very important to consider its distribution evaluated in terms of its PDF, since the probability of extreme outliers associated with SD values in excess of 2 dB must be very low [92].

Low average spectral deviation values were achieved, when this method was used to quantize the LAR parameters with the aid of 25 bits per LPC update frame. A two-stage VQ approach was adopted in order to reduce the complexity of the error minimization procedure. Exploiting the high correlation between the LSFs in adjacent frames, the method has given better results when the vector $\mathbf{x}$ to be quantized was the difference between the present LSF vector and the previously quantized one. In order to reduce the search complexity from 2^B comparisons, where B is the total number of codebook address bits, a computationally more attractive two-stage approach was employed. Specifically, two codebooks associated with two gain factors were employed, as we have seen in Section 4.3.2 for the speaker-adaptive VQ scheme. For example, when using $B = 20$ bits, initially the first 256-entry codebook was searched in order to find the best entry. Its size was virtually expanded by a factor of 4 using a 2-bit quantized gain factor, which was computed similarly to Equation 4.40. Then the error of this first matching process was further encoded using the second 256-entry codebook and 2-bit quantized gain. In a first approximation, this process reduced the search complexity from an unacceptable 2^{20} comparisons to around 2×2^8.

The performance of this VQ scheme can be further improved by employing a **switched-adaptive vector quantization** approach according to the scheme suggested by Yong, Davidson, and Gersho [109], where a number of fixed covariance matrices are used for different classes of speech. The performance of this approach was characterized by Salami et al. in Reference [111]. We now proceed to consider two recently suggested VQ schemes [92, 104], which have a moderate implementational complexity. Apart from minimizing the average SD, they also limit the probability of high-peak SD values [112].

4.3.4 Robust Vector Quantization Schemes for LSFs

Quite recently, Paliwal and Atal [92] have proposed a moderate complexity 24-bit vector quantization arrangement for the LSFs. They noted that the individual LSFs have a localized effect in terms of spectral distortion in the spectral domain, which facilitates splitting the 10-component LSF vector in shorter vectors, while limiting the spectral distortion spillage from one region to another. They also defined an LSF-based spectral distortion measure. On the basis of the limited distortion spillage to other frequency domains, the most important LSFs were allocated a higher weight during the quantization process and vice versa. In contrast, LARs have a rather widespread effect in the frequency domain.

Specifically, the weighted Euclidean distance measure $d(\mathbf{f}, \hat{\mathbf{f}})$ between the original and quantized LSF vectors was defined as [92]:

$$d(\mathbf{f}, \hat{\mathbf{f}}) = \sum_{i=1}^{10} [c_i w_i (\mathbf{f} - \hat{\mathbf{f}})]^2 \qquad (4.56)$$

where the weighting factor w_i, $i = 1 \dots 10$ is assigned to the ith component of the LSF vector, which is defined as

$$w_i = [|H(f_i)|^2]^r. \qquad (4.57)$$

Specifically, in Equation 4.57 $|H(f_i)|^2$ represents the LPC power spectrum at frequency f_i and the experimentally optimized constant r, allowing Atal and Paliwal to attribute different weights to different LSFs, was 0.15. Lastly, the additional weighting factor c_i was 1.0 for $i \cdots 8$, while a choice of $c_9 = 0.8$ and $c_{10} = 0.4$ allowed the measure to deemphasize high-frequency LSFs.

The VQ complexity is reduced at a concomitant lower performance, if the original 10-component LSF vector is split in smaller vectors. Paliwal and Atal found that a good compromise was to employ a two-way split. An extreme case would be to use 10-way splitting, which is equivalent to scalar quantization. Hence, assuming a total of 24 bits, two 12-bit VQ schemes were employed. Three basic requirements must be satisfied in order to achieve transparent LSF quantization: (1) The average SD is lower than 1 dB, (2) no frames have a SD above 4 dB, and (3) the probability of SD values between 2 and 4 dB is below 2%. Experimental results showed that best overall SD performance in terms of these three criteria was guaranteed, when 5 LSFs were quantized by both 12-bit or 4096-entry codebooks. The LSFs' ordering property can be satisfied by ensuring that only those vectors of the second segment of the codebook are invoked, for which the lowest quantized LSF value within the vector, namely, LSF_6, is higher than the quantized value of the highest frequency component, namely, that of LSF_5 of the first VQ segment. The proposed quantization scheme was shown to have an impressive robustness against channel errors, which was similar to that of scalar arrangements.

In a further attempt to improve the overall LSF quantizer design Ramachandran, Sondhi, Seshadri, and Atal [104] have proposed a hybrid scheme that employs a combination of vector and scalar quantization. The design constraints and objectives were similar to those in Atal's former work reported above, but the weighting factor of Equation 4.57 was modified according to [113]:

$$w_i = \frac{1}{f_i - f_{i-1}} + \frac{1}{f_{i+1} - f_i} \tag{4.58}$$

which attributed higher weights to frequency regions, where the LSFs were grouped closer, indicating a dominant spectral peak. The proposed scheme is memoryless, which improves its robustness against channel errors. Further important design constraints were to reduce the complexity and memory requirements.

The proposed arrangement quantized the differences between consecutive LSFs of the same frame rather than the LSFs themselves [93], since these differences have a lower dynamic range than the LSFs. Initially, an independent vector and a scalar quantizer were designed, both using 29 bits. The authors' conclusion was that the best performance was achieved when each set of 10 LSFs was both scalar and vector-quantized and the specific scheme minimizing the distortion measure was actually used. A further 1-bit flag was then allocated to indicate which scheme was used. A three-way split VQ scheme using (3,3,4) LSF vectors was designed using (10,9,10) bits, respectively. The associated scalar quantizer employed (3,3,3,3,3,3,3,3,3,2) bits for the individual LSFs.

The benefits of using this combined scheme were interpreted by analyzing the quantized vectors. Namely, the two schemes complement each other in that the VQ caters for those LSF sets, where some components are clipped by the scalar arrangement. In contrast, the scalar quantizer can encode the sparse regions of the VQ more efficiently. Lastly, the coding performance can be further improved by employing a codebook adaptation procedure. Specifically, it can be intelligently exploited that due to the LSFs' ordering property, a subset of the second codebook, whose lowest LSF component is lower than the highest one of the first three-component subvectors becomes illegitimate. This fact can be capitalized upon. Rather than restricting the search to that area, the entire codebook can be remapped to the legitimate frequency region, thereby providing a finer quantizer resolution. Specific algorithmic details of this procedure are beyond the scope of our treatment here. The interested reader is referred to [104] for a full description of the associated dynamic programming technique employed.

4.3.5 LSF Vector-Quantizers in Standard Codecs

In recent years, a range of sophisticated, error-resilient, high-quality, low-rate speech codecs emerged, such as, for example, the ITU's 8 kbps G.729 scheme of Section 7.8, the dual-rate G.723.1 scheme of Section 7.12, the 5.6 kbps half-rate GSM codec portrayed in Section 7.7, the enhanced full-rate GSM scheme described in Section 7.10, the 7.4 kbps IS-136 codec arrangement of Section 7.11, or some of the other schemes of Chapter 7. Most of these state-of-the-art codecs employ LSF vector-quantization techniques (detailed in more depth in Chapter 7), but it is beneficial here to put some of the previously detailed principles into practice. Hence below we provide a rudimentary introduction to the split LSF vector quantizer of the recently standardized 7.4 kbps enhanced full-rate IS 136 codec approved in the United States, (see Section 7.11).

The IS-136 scheme requires a bitrate contribution of 26 bits/20 ms for the quantization of the 10 LSFs. The corresponding LSF VQ scheme is shown in Figure 4.10, which is

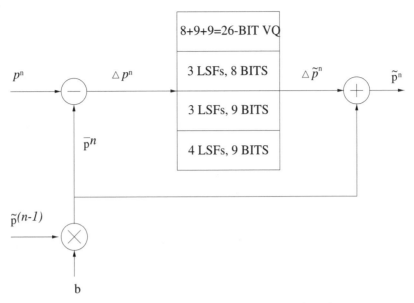

Figure 4.10 The 26-bit IS-136 LSF quantization schematic.

schematically identical to the 24-bit LSF VQ of the G.723.1 dual-rate codec of Section 7.12. Observe in Figures 7.33 and 4.10 that only the codebook sizes are slightly different, since the 7.4 kbps IS-136 codec allocates 26 rather than 24 bits to LSF quantization. As pointed out above, split vector quantization is usually employed, since it reduces the search complexity, although at the cost of some performance degradation. In the IS-136 LSF VQ the first three LSFs are grouped together and vector-quantized using 8 bits, or 256 entries, while the two other groups of LSF quantizers are constituted by three and four LSFs, both employing 9 bits. Observe in Figure 4.10 that the nth unquantized LSF vector p^n is predicted first on the basis of the previous quantized LSF vector $p^{(\tilde{n}-1)}$, after multiplying it with a scaling factor b, which is proportional to the long-term correlation between consecutive LSF vectors. This is often termed a first-order moving-average prediction, since it relies on a simple first-order prediction model. The estimated LSF vector $\bar{p}^n$ is then subtracted from the original unquantized LSF vector in order to generate their difference vector, namely, δp^n, which is split in subvectors of 3, 3, and 4 LSFs and quantized. Finally, the quantized LSF difference vector $\Delta \tilde{p}^n$ is added to the predicted value $\bar{p}^n$, in order to generate the current quantized LSF vector $\tilde{p}^n$. Again, a range of similar LSF VQ schemes can be found in Chapter 7 in the context of other state-of-the-art standard codecs.

4.4 SPECTRAL QUANTIZERS FOR WIDEBAND SPEECH CODING[1]
G. GUIBÉ, H. T. HOW, L. HANZO

4.4.1 Introduction to Wideband Spectral Quantization

In wideband speech codecs, a high number of spectral coefficients—typically 16—has to be quantized in order to represent the spectrum up to frequencies of 7 kHz. However, the

---

[1]This section is based on G. Guibé, H. T. How, and L. Hanzo, submitted to the European Transactions on Telecommunications.

line spectral frequency (LSF) coefficients above 4 kHz are less amenable to vector quantization (VQ) than their low-frequency counterparts.

Table 4.1 summarizes most of the recent approaches to wideband speech spectral quantization found in the literature. The approach employed by Harborg et al. [114] is based on scalar quantization (SQ). However, the resulting bitrate is excessive, requiring 3 or 4 bits for each LSF. Chen et al. [117] as well as Lefebvre et al. [115] utilized low-dimensional split VQ. For instance, a $(2, 2, 2, 2, 2, 3, 3)_{7777777}$ split VQ is invoked in their approach, where only two- or three-dimensional VQs are used, employing 7 bits—that is, 128 codebook entries—per subvector. This reduces the number of bits allocated to the LSF quantization compared to SQ, although the resulting number of bits still remains somewhat high, namely, $7 \cdot 7 = 49$. These approaches are simple, but a large number of bits is required.

Paulus et al. [116] proposed a coding scheme based on sub-band analysis of the speech signal. The speech signal was split into two unequal sub-bands, namely, 0–6 kHz and 6–7 kHz. LPC analysis was only invoked in the lower band, using 14 LSF coefficients quantized with 44 bits per 15 ms. The quantization scheme employed interframe moving-average prediction and split vector quantization. In the 6–7 kHz higher sub-band, only the signal energy was encoded using 12 additional bits. Following a similar approach, Combescure et al. in [119] described a system based on two sub-bands, where the lower band (0–5 kHz) applied a 12th-order LP filter with its coefficients quantized using 33 bits. The upper band (5–7 kHz) uses an eighth-order LP filter encoded with 10 bits, but these coefficients were only transmitted in the higher bitrate mode of the coder, namely, at 24 kbit/s. The lower-band coefficients were quantized using Predictive Multi-Stage Split Vector Quantization (MSVQ). These types of LSF quantizers are not directly amenable to employment in full-band wideband speech codecs. However, the approach using separate coding of the higher- and lower-band LSFs can be helpful in general for LPC quantization.

Finally, Ubale et al. in [118] proposed a scheme using predictive MSVQ of seven stages employing 4 bits each. This method employed a multiple survivor method, where four—rather than one—residual survivors were retained at each pattern-matching stage and were then tested at the next pattern-matching stage. The final decision was taken at the last VQ stage as to which of the split vector combinations gave the lowest quantization error. In addition, the MSVQ was designed by a joint optimization procedure, clearly demonstrating the advantages of using schemes that predictively exploit the knowledge of the signal's past history, in order to improve the coding efficiency.

Having reviewed the background of wideband speech spectral quantization, we now focus on the statistical properties of the wideband speech LSFs, which render it attractive for vector quantization.

TABLE 4.1 Overview of Wideband LPC Quantizers

	Quantization Scheme	No. of Bits per Frame
Harborg et al. [114]	Scalar	60, 70, and 80
Lefebvre et al. [115]	Split VQ	49
Paulus et al. [116]	Predictive VQ	44
Chen et al. [117]	Split-VQ	49
Ubale et al. [118]	Multi-stage VQ	28
Combescure et al. [119]	Multi-stage	33 at 16 kbit/s 43
	Split VQ	43 at 24 kbit/s

4.4.1.1 Statistical Properties of Wideband LSFs.   The employment of the LSF [93, 120, 121] representation for quantization of the LPC parameters is motivated by their statistical properties. Figure 4.11 shows the probability density functions (PDFs) of 16 wideband speech LSFs over the interval of 0–8 kHz. Their different PDFs have to be taken into account in the design of the quantizers.

The essential motivation of vector quantization is the exploitation of the relationship between the LSFs in both the frequency and time-domain. Figure 4.12 shows the time-domain evolution of the wideband (WB) speech LSF traces, demonstrating their strong correlation in consecutive frames in the time domain, which is often referred to as their interframe correlation. Similarly, it demonstrates within each speech frame the ordering property of neighboring LSF values, which is also referred to as intraframe correlation.

Intraframe correlation motivates the employment of vector quantization, since it enables a mapping that matches the multidimensional LSF distribution. We observe at the top of Figure 4.12 that the correlation of the individual LSFs within a given speech frame tends to decrease, as the frequency increases; that is, higher frequency LSFs are more statistically independent of each other, although they still obey the ordering property. This clearly manifests itself, for example, around frame 18 in Figure 4.12. The highest frequency LSFs describe the noisy high-frequency bands of the speech signal, which typically appear to be noise-like. This characteristic will mostly be exploited in the design of memoryless VQ schemes.

Interframe correlation of the LSFs can be exploited by interframe predictive vector quantization schemes having memory, where predictions of the current LSF values are

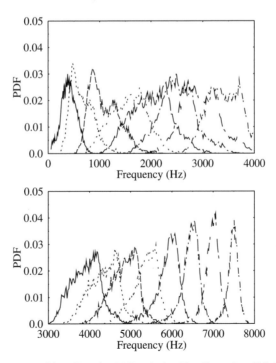

Figure 4.11 PDFs of the LSFs using LPC analysis with a filter order of 16, demonstrating the ordering property of the LSFs.

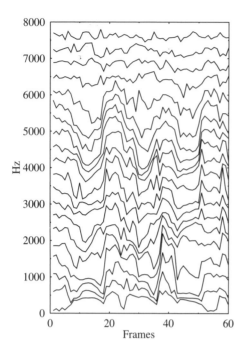

Figure 4.12 Traces of 16 wideband LSFs, demonstrating their inter- and intraframe correlations.

employed, in order to reduce the variance of the vector we want to quantize. Finally, when rapid spectral changes are observed in the LSF traces, affecting both their intra- and interframe correlation, various multimode schemes can be invoked.

4.4.1.2 Speech Codec Specifications. The design of speech codecs is based in general on a tradeoff between the conflicting factors of perceptual speech quality, the required bitrate, the channel error resilience, and the implementational complexity. Wideband speech coding [122] aims to provide a better perceptual quality than narrowband speech codecs. Hence, a fine quantization of the LPC parameters is required.

Listening tests using the scheme depicted in Figure 4.13 indicate that the transparency criterion formulated by Paliwal and Atal [92] in the context of narrowband speech codecs is

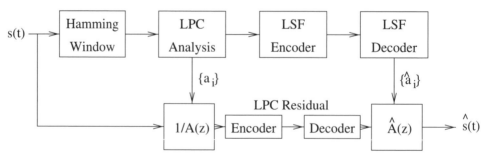

Figure 4.13 Evaluation of the perceptual speech quality after applying LSF vector quantization.

also relevant in wideband scenarios. This criterion uses a spectral distortion (SD) measure given by

$$SD^2 = \frac{1}{f_s} \int_0^{f_s} \left[10 \log_{10}(P(f)) - 10 \log_{10}(\hat{P}(f)) \right]^2 df,$$

where $P(f)$ and $\hat{P}(f)$ are the amplitude spectra of the original and reconstructed signal, respectively. The required criteria are satisfied if an average SD of about 1 dB is maintained and there are only a few "outliers" between SD = 2 and 4 dB, while there are no outliers in excess of SD = 4 dB. In addition, an important issue in speech quality terms is the preservation of the stability of the short term predictor (STP). The STP filter's stability has a dramatic influence on the reconstructed speech quality, which is guaranteed by preserving the ordering property of the LSFs.

Every codec designed for transmission over noisy channels has to exhibit a good robustness against channel errors. The effect of transmission errors is characterized by their immediate effect on both the present speech frame and the forthcoming frames. Complexity reduction is also very important for real-time applications. The codebook storage requirements and codebook search complexity are the main factors to be considered in the field of vector quantization.

4.4.2 Wideband LSF Vector Quantizers

4.4.2.1 Memoryless Vector Quantization. The nearest neighbor vector quantization (NNVQ) scheme [102] theoretically constitutes the optimal memoryless solution for VQ. However, the high number of LSFs—typically 16—required for wideband speech spectral quantization results in a complexity that is not realistic for a real-time implementation, unless the 16-component LSF vector is split into subvectors. As an extreme alternative, low-complexity scalar quantization constitutes the ultimate splitting of the original LSF vector into reduced-dimension subvectors. This method exhibits a low complexity, and a good SD performance can be achieved using 16-entry, or 4-bit, codebooks. Nevertheless, the large number of LSFs required in wideband speech codecs implies a requirement of $4 \cdot 16 = 64$ or $5 \cdot 16 = 80$ bits per 10 ms speech frame. As a result, the contribution of the scalar-quantized LSFs to the codec's bitrate is 6.4 or 8 kbit/s. Slight improvements can be achieved using a nonuniform bit allocation, when more bits are allocated to the perceptually most significant LSFs.

Between these extreme cases, split VQ (SVQ) aims to define a split configuration that minimizes the average SD within a given total complexity. Specifically, split vector quantization operates on subvectors of dimensions that can be vector-quantized within the given constraints of complexity, following the schematic of Figure 4.14.

One of the main issues in split LSF VQ is defining the best possible partitioning of the initial LSF vector into subvectors. Since the high-frequency LSFs typically exhibit a different statistical behaviour from their low frequency counterparts, they have to be encoded separately. For linear predictive filters of order 16, the three highest order LSFs behave differently from the other LSFs, as exemplified by Figure 4.12. Hence, this leads naturally to a (13,3)-split VQ scheme. Figure 4.15 shows the PDF of the SD using a (6,7,3)-split LSF VQ scheme, where the lower frequency 13-component subvector is split into two further six- and seven-component subvectors in order to reduce the implementational complexity. Seven bits (i.e., 128 codebook entries) were used for each subvector. In addition, a (4,4,4,4)-split second

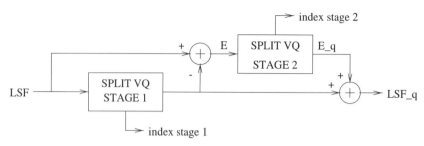

Figure 4.14 Schematic of the multi-stage split VQ.

stage VQ was applied according to Figure 4.14 using five bits (i.e., 32 codebook entries) for each subvector. We refer to this scheme as the $[(6, 7, 3)_{777}; (4, 4, 4, 4)_{5555}]$ 41-bit regime.

The lower intraframe correlation of the higher frequency LSFs imposes a high bitrate requirement on the SVQ in the light of the relatively low energy contained in the corresponding speech band (typically less than 1%). Although split VQ schemes are attractive in terms of complexity and can preserve the LSFs ordering property, they often fail to reach the target SD within a low-bitrate budget.

The introduction of LSF classified vector quantization (CVQ) [102] aims to assign the LSF vectors into classes having a particular statistical behavior, in an effort to improve the coding efficiency.

In Figure 4.16 the LSF vectors are classified into one of m categories $C_1 \cdots C_m$, and then a reduced-size codebook C_m, which reflects the statistical properties of class m is searched in order to find the best matching codebook entry for the unquantized LSF vector. This scheme searches a reduced-size codebook, reducing the matching complexity and the quantization precision in comparison to a VQ using no pre-classification before quantization. In the context of wideband speech LSF quantization, we wish to find a classification of the LSFs, which can provide a more efficient representation of the vector to be quantized, than the previous SVQ. Accordingly, the main issue in classified vector quantization is the design of an accurate classifier. In this context, we briefly investigate the performance of a voiced/unvoiced classifier.

The problem of voicing detection can be solved upon invoking an autocorrelation-based pitch detector [31], exploiting the waveform similarities between the original speech and its pitch-duration shifted version. The highest correlation between these two signals is registered,

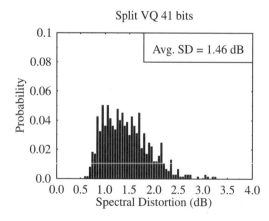

Figure 4.15 PDF of the SD for the 41-bit split VQ scheme using the $[(6, 7, 3)_{777}; (4, 4, 4, 4)_{5555}]$ two-stage regime. (Compare to Figure 4.20 and 4.23.)

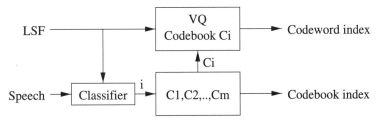

Figure 4.16 Schematic of the classified VQ.

when their displacement corresponds to the pitch. Figure 4.17a shows a low-pass-filtered speech waveform bandlimited to 900 Hz, which was subjected to autocorrelation-based voicing-strength evaluation and thresholding at a normalized cross-correlation of 0.5, in order to generate the binary voiced/unvoiced (V/UV) decisions seen in Figure 4.17b.

Figure 4.18 demonstrates the relevance of this approach, portraying—as an illustrative example—the scatter diagrams of the first two LSFs after classification. For both diagrams, the unoccupied bottom right corner region manifests the dependency between the LSFs due to their ordering property. The first two LSFs of voiced frames at the left of Figure 4.18 are centered around two clusters. One corresponds to the low-frequency LSF 1 occurrences, where LSF 2 appears near constant. The other voiced frame cluster corresponds to frames, where LSF 1 and 2 exhibit similar values, creating a near-linear cluster along the "ordering property border". The unvoiced frames at the right of Figure 4.18 appear more scattered,

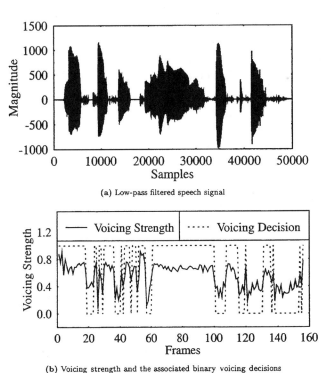

(a) Low-pass filtered speech signal

(b) Voicing strength and the associated binary voicing decisions

Figure 4.17 V/UV speech classification using low-pass filtering of the speech to 900 Hz and autocorrelation-based pitch detection.

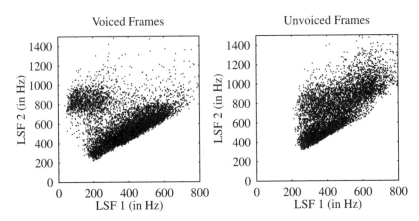

Figure 4.18 Scatter diagrams of the first two LSFs for WB voiced and unvoiced frames.

although they also exhibit an apparent, but less pronounced, clustering along the ordering property border.

Voiced and unvoiced LSFs do not necessarily exhibit a totally different statistical behavior in their clusters along the ordering property border in Figure 4.18. However, the typically more concentrated clusters of the voiced LSF frames can more accurately be vector-quantized, whereas the somewhat more scattered occurrences of the unvoiced frames' LSFs are expected to be less amenable to CVQ. Similar scatter diagrams can also be obtained also for higher frequency LSFs, although the pronounced difference between voiced and unvoiced frames tends to decrease, as the frequency increases. This is directly related to the less pronounced correlation between neighboring LSFs for the higher frequencies of the 8 kHz range.

Although our simulations using this CVQ gave better SD results than the previously discussed split VQ, the overall scheme presents shortcomings. Specifically, if the speech frame classification is carried out before the LSF quantization, classification errors at the voiced/unvoiced speech boundaries increase the average SD, as well as the number of outliers. At the decoder, this method has to rely on the voiced/unvoiced information extracted from the excitation signal in order to reconstruct the LSF coefficients, unless the V/UV mode is explicitly signaled to the decoder. Alternatively, if the V/UV classification is processed after LSF quantization upon selecting the mode having the lower SD, no classification errors occur, although one bit per speech frame is required for transmitting the V/UV mode selection. When using the $[(6, 7, 3)_{777}; (4, 4, 4, 4)_{5555}]$ 41-bit Split LSF VQ for each mode, an average SD of 1.15 dB is obtained upon invoking a mode selection bit, whereas an average SD of 1.35 dB is achieved using the pitch-detection-based classification.

In addition, it is difficult to proceed to a joint optimization of both the voiced and the unvoiced codebooks, since there are regions of the LSF domain where both types of LSFs can be located. The LSF clusters, which are encountered in both modes, are quantized independently by the voiced and the unvoiced codebooks. Hence, the same subdomain of the LSF space is mapped twice by the quantization cells of both modes. This leads to a suboptimal quantization of this area. Let us now consider predictive VQ schemes.

4.4.2.2 Predictive Vector Quantization. In this section, our discussions evolve from memoryless vector quantization to more efficient vector quantization schemes exploiting the time-domain interframe correlation of LSFs. According to this approach, we

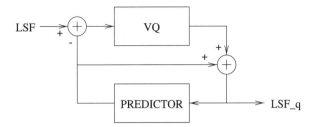

Figure 4.19 Schematic of a predictive vector quantizer (PVQ).

typically quantize a sequence of vectors, where successive vectors may be statistically dependent.

Predictive vector quantization (PVQ) constitutes a vector-based extension of traditional scalar predictive quantization. Its schematic is shown in Figure 4.19. PVQ schemes aim to exploit the correlation between the current vector and its past values in order to reduce the variation range of the signal to be quantized. Provided that there is sufficient correlation between consecutive vectors and the predictor is efficient, the vector components to be quantized are expected to be unpredictable, random noise-like signals, exhibiting a reduced dynamic range. Hence, for a given number of codebook entries, PVQ is expected to give a lower SD than nonpredictive VQ.

Autoregressive (AR) predictors use recursive reconstruction of the LSFs. Thus, they potentially suffer from severe propagation of channel errors over consecutive frames. By contrast, a moving-average (MA) predictor can typically limit the error propagation to a lower number of frames, given by the predictor order. Here, however, we restrict our experiments to first-order AR vector predictors.

Predictive vector quantization does not necessarily preserve the LSFs' ordering property. This may result in instability of the STP filter, deteriorating the perceptual quality. In order to counteract this problem, an LSF rearrangement procedure [123] can be introduced, ensuring a minimum distance of 50 Hz between neighboring LSFs in the frequency domain.

Figure 4.20 shows the PDF of the SD using $(4, 4, 4, 4)_{9999}$ 36-bit split vector quantization of the prediction error, employing a 9-bit codebook per 4-LSF subvector. This

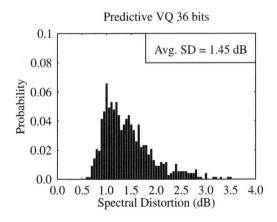

Figure 4.20 PDF of the SD for the 36-bit PVQ scheme. (Compare to Figures 4.15 and 4.23.)

quantizer therefore requires a total of $4 \cdot 9 = 36$ bits per LSF vector. Based on the above experience, we concluded that the 36-bit predictive VQ provides a gain of 5 bits per LSF vector in comparison to our previous 41-bit memoryless SVQ having a similar complexity. Equivalently, predictive VQ generates an average SD gain of approximatively 0.3 dB for a given bitrate. A deficiency of this method is its higher sensitivity to channel error propagation, although this problem can be mitigated by using MA prediction instead of AR prediction. During our investigations, we noted that this scheme was sensitive to unpredictable LSF vectors generated by rapid speech spectral changes, which increase both the average SD and the number of SD outliers beyond $SD = 2$ dB. This problem is addressed in the next section.

4.4.2.3 Multimode Vector Quantization.

Our previous classified vector quantization scheme has primarily endeavored to define V/UV correlation modes. When we observe these voiced/unvoiced speech transitions in the time-domain, they result in the rapid changes of the LSF traces seen in Figure 4.12, for example, around frame 20. Several methods exist for differentiating between these modes. Switched prediction is widely employed [31, 123]. In this section, we investigate the separate encoding of the unpredictable frames due to rapid spectral changes and that of the highly correlated frames. This can be achieved by the combination of a predictive VQ and a fixed memoryless SVQ, referred to as the safety net VQ (SNVQ) scheme [124–126]. In this context, we invoke a full search using both the predictive VQ and the fixed memoryless SVQ schemes for every speech frame, and the better candidate with respect to a mean squared distortion criterion is chosen.

The safety net VQ improves the overall robustness against outliers, which are typically due to input LSF vectors having a low correlation with the previous LSF vectors. In addition, the safety net VQ allows the PVQ to concentrate on the predictable, highly correlated frames. Hence, the variance of the LSF prediction error is reduced, and a higher-resolution LSF prediction error codebook can be designed. The advantage of this method is that when the interframe correlation cannot be successfully exploited in a PVQ scheme, the intraframe correlation is capitalized on instead.

Figure 4.21 shows the structure of the SNVQ scheme. Again, the input LSF vector is quantized using both predictive- and memoryless quantizers. Then both quantized vectors are compared to the input vector in order to select the better quantization scheme. The codebook index selected is transmitted to the decoder, along with a signaling bit that indicates the selected mode. The specific transmitted quantized vector is finally used by the PVQ in order to predict the LSF vector of the next frame.

The performance difference between the memoryless SVQ and predictive VQ sections of the SNVQ suggests the employment of variable bitrate schemes, where the lower

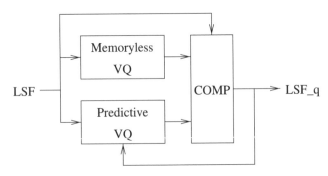

Figure 4.21 Schematic of the safety net vector quantizer (SNVQ) constituted by a memoryless and a predictive VQ.

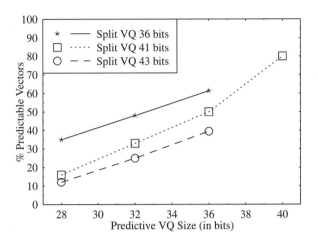

Figure 4.22 Proportion of frames using PVQ in various SNVQ schemes, employing memoryless SVQs of 36, 41, and 43 bits.

performance of the memoryless SVQ can be compensated by using a larger codebook. In our experiments below—as before—a memoryless SVQ 41-bit codebook was used. Hence, the SNVQ is characterized by its average bitrate, depending on the proportion of vectors quantized by the predictive and memoryless VQ, respectively. Eriksson, Linden, and Skoglung [125] argued that the optimum performance is attained, when 50% to 75% of frames invoke the PVQ.

Figure 4.22 shows the proportion of frames quantized using the 28-, 32- and 36-bit PVQs in the context of SNVQ schemes employing 36-, 41-, and 43-bit SVQs. We observe in Figure 4.22 that for a PVQ codebook size of 28 and 32 bits, a relatively low proportion of the LSF vectors was quantized using the PVQ. This indicated that its codebook size was too small, failing to outperform the memoryless 36-, 41-, or 43-bit SVQs. Accordingly, only the 36-bit PVQ was deemed suitable. This figure illustrates that if the predictive VQ exhibits a low performance compared to the memoryless SVQ (i.e., the proportion of its utilization tends to zero), the SNVQ will tend to behave like a simple memoryless SVQ. Alternatively, if the memoryless SVQ exhibits a low performance compared to the PVQ (i.e., the proportion of PVQ LSF vectors tends to 100%), the SNVQ will tend to behave like a PVQ.

The individual PVQ and memoryless SVQ schemes employed so far were designed independently of each other. Hence the resulting scheme is suboptimal. Furthermore, both quantizers were designed without distinction between predictable and unpredictable LSF vectors. Hence, their optimization will aim, on one hand, to have the PVQ focus on predictable frames, which generate LSF prediction errors with a low variation range. On the other hand, the memoryless SVQ codebook is to be matched to the distribution of the unpredictable LSF vectors in the p-dimensional LSF space. In order to obtain an optimal SNVQ, we will proceed as follows:

1. The original training sequence T is passed through our previously used individual suboptimum codebook based SNVQ in order to generate the subtraining sequences T_{PVQ} and T_{SN} of vectors, quantized using either the predictive VQ or the memoryless SVQ, respectively, depending on which generated a lower SD.

TABLE 4.2 Optimization Effects for the [36, 36] and [36, 41] SNVQ Schemes

Scheme	Avg. SD (dB)	Outliers (%)	
		2–4 dB	> 4 dB
[36, 36] SNVQ scheme			
Nonoptimized	1.34	7.19	0.12
Optimized	1.17	2.18	0
[36, 41] SNVQ scheme			
Nonoptimized	1.25	4.5	0.12
Optimized	1.09	0.38	0

2. Then codebooks for both the PVQ and the memoryless SVQ are designed using the subtraining sequences generated above.

Our results to be highlighted with reference to Table 4.2 show that the optimized PVQ results in significant improvements, but only a modest further gain was obtained with the aid of the safety-net approach, invoking the optimized memoryless SVQ. Optimization is the main issue in SNVQ design, requiring the joint design of both parts of the SNVQ. We designed a [36, 36]-bit and a [36, 41]-bit scheme, where the first bracketed number indicates the number of bits assigned to the PVQ, while the second one indicates that of the memoryless SVQ. Again, the performance of these schemes is summarized in Table 4.2. In both cases a SD gain of about 0.15 dB was obtained upon the joint optimization of the component VQs, as seen in Table 4.2. In addition, the number of outliers between 2 and 4 dB was substantially reduced, and all the outliers over 4 dB were removed.

We found that the optimization slightly increased the proportion of frames quantized using the PVQ. For our [36, 36] SNVQ scheme, this proportion increased from 67% to 74%. Similarly, for the [36, 41] SNVQ scheme constituted by the 36-bit PVQ and 41-bit memoryless SVQ, respectively, this proportion increased from 50% to 60%. Hence, in the case of such switched variable-bitrate schemes, the optimization tends to reduce the average SNVQ bitrate, since the PVQ requires fewer bits, than the memoryless SVQ.

Figure 4.23 shows the PDF of the SD for the [36, 41] SNVQ scheme, indicating a significant SD PDF enhancement compared to both the memoryless SVQ and the PVQ. In addition, this system improves the robustness against channel errors, since the propagation of

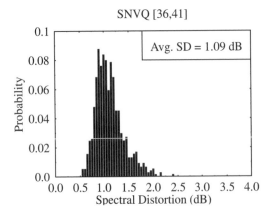

Figure 4.23 PDF of the SD for the [36, 41] bit SNVQ scheme (Compare to Figures 4.15 and 4.20).

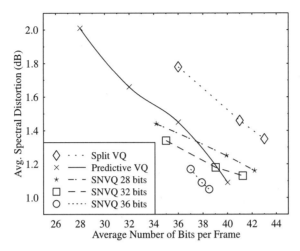

Figure 4.24 Average SD of the various vector quantizers considered in this study.

bit errors was limited due to the low number of consecutive employment of the PVQ. The SNVQ enabled an efficient exploitation of both the interframe correlation and the intraframe correlation of LSF vectors. Its main deficiency is the increased complexity of the codebook search procedure, requiring twice as many comparisons as the memoryless SVQ or the PVQ.

4.4.3 Simulation Results and Subjective Evaluations

Figure 4.24 summarizes the performance of the split memoryless SVQ, the PVQ, and the SNVQ. As observed in the figure, the SD results for the memoryless SVQ are more modest, and in general a better performance was obtained by using the predictive quantization schemes. This figure illustrates a difference of 4 or 5 bits between the memoryless SVQ and the PVQ for the same SD. The three SD curves corresponding to the SNVQ schemes using 28-, 32-, and 36-bit PVQs in conjunction with various associated memoryless SVQ configurations are also shown in Figure 4.24. For the SNVQ using 28- and 32-bit PVQs, the lines crossing the PVQ performance curve drawn using a solid line indicate that at this stage the PVQ starts to attain a better performance than the SNVQ for the equivalent bitrate. Hence, in this scenario there is no benefit from employing SNVQ schemes using 28- and 32-bit PVQs beyond this crossover point. A consistent SD gain in comparison to the PVQ is only ensured for the SNVQ using the 36-bit PVQ. In this case, a 2-bit reduction in the number of required coding bits was obtained. Informal listening tests have shown that the best perceptual performance was obtained by employing the [36, 41] SNVQ scheme.

Table 4.3 details the characteristics of two high-quality quantization schemes. The first configuration utilized a $(4, 4, 4, 4)_{10,10,10,10}$ PVQ scheme employing $4 \cdot 10 = 40$ bits, and the

TABLE 4.3 Transparent Quantization Schemes

Scheme	No. of Bits	Avg. SD (dB)	Outliers (%)	
			2–4 dB	> 4 dB
PVQ	40	1.09	4.24	0
SNVQ	38	1.09	0.38	0

second scheme used a [36, 41] SNVQ arrangement with an average of 38 bits. Although both schemes have a similar average SD, the SNVQ provides a large reduction in the number of SD outliers between 2 and 4 dB, which have a significant effect on the perceptual speech quality. A high speech quality was also obtained for the [36, 36] fixed bitrate SNVQ, as shown in Table 4.2.

4.4.4 Conclusions on Wideband Spectral Quantization

In this section, we have presented a comparative study of various predictive and memoryless vector quantizers. In the context of memoryless vector quantization, a $[(6, 7, 3)_{777}; (4, 4, 4, 4)_{5555}]$ 41-bit multi-stage split vector quantizer was designed. This method enabled a simple implementation. In order to improve the performance of this initial memoryless scheme, we introduced V/UV classification. This approach gave about 0.2 dB SD improvement but increased the complexity. Nonetheless, both of these suboptimum approaches maintained a low computational complexity, as well as a high error resilience.

In the context of 41-bit predictive vector quantization, a SD quality enhancement was achieved compared to memoryless schemes. Alternatively, the number of bits could be reduced to 36, while maintaining a similar average SD. The associated SD PDFs were portrayed in Figures 4.15, 4.20, and 4.23, while their salient features were summarized in Tables 4.2 and 4.3. Unfortunately, the channel error sensitivity increased due to potential error propagation. Lastly, we combined both the memoryless and the predictive approaches in a SNVQ scheme. Even though the SNVQ scheme increased the complexity, it significantly improved the SD performance and mitigated the propagation of channel errors. Our future research considers the design tradeoffs of wideband backward-adaptive speech codecs and transform codecs.

4.5 CHAPTER SUMMARY

This chapter introduced a range of parameters that can be invoked for the error-resilient representation of the speech signal's spectral envelope. Specifically, the LARs and the LSFs were discussed in more detail, and their PDFs were exemplified. These elaborations were followed by the portrayal of a suit of spectral quantizers. VQs were found to be particularly efficient due to the inherent correlation of the LSFs, versus both time and frequency. Finally, a comparative study of various wideband spectral quantizers was provided.

Having treated the issues of spectral representation, in the spirit of the linearly separable speech generation model of Figure 1.1, let us now concentrate on a range of techniques that can be used to represent the prediction residual. However, recall from Section 3.2 and Figure 3.1 that having determined the spectral coefficients of the current segment of speech, the aim in AbS coding is not to find a good waveform replica of the prediction residual but to find a model of it, which results in the **perceptually** best synthetic speech quality. A plethora of techniques have been suggested in the literature, which are based on a range of different design tradeoffs in terms of speech quality, bitrate, implementational complexity, robustness against transmission errors, and so on that will be characterized in the forthcoming chapters.

Let us initially concentrate on the Regular Pulse Excited (RPE) technique in the next chapter, which constitutes an attractive design tradeoff at a bitrate of 13 kbps in terms of low complexity and high speech quality characterized by a mean opinion score (MOS) of about 4. In an international comparative test [127], it outperformed a range of other codecs. Hence, it was selected for the Pan-European mobile radio system known as Global System of Mobile Communications or GSM [73, 74].

5

Regular Pulse Excited Coding

5.1 THEORETICAL BACKGROUND

The schematic of the regular pulse excited (RPE) speech codec is based on the AbS structure of Figure 3.8. The typically $N = 40$ samples or 5 ms duration excitations or innovation sequences $v(n)$ are filtered through the LTP synthesis filter $1/P(z)$, STP synthesis filter $1/A(z)$, and perceptual weighting filter $W'(z) = A(z)/A(z/\gamma)$. There the STP synthesis filter $1/A(z)$ and the numerator of $W'(z)$ cancel, yielding the simplified weighting filter $W(z) = 1/A(z/\gamma)$ again, as seen in Figure 3.8. Hence, the weighted synthetic speech $\tilde{s}_w(n)$ is given by the following convolution:

$$\tilde{s}_w(n) = v(n) * h_p(n) * h_w(n) = v(n) * h_c(n), \qquad (5.1)$$

where $h_p(n)$ and $h_w(n)$ are the impulse responses of the filters $1/P(z)$ and $W(z) = 1/A(z/\gamma)$, respectively, and $h_c(n)$ is that of the cascaded filter complex $1/P(z)$, $W(z)$. Similarly, a 5 ms input speech segment about to be encoded is weighted by the identical perceptual weighting filter. Their difference is computed for each legitimate innovation sequence in order to find the particular one yielding the minimum weighted error, and hence the subjectively best 5 ms duration synthetic speech segment.

Depending on the construction of the innovation vectors, different complexities, bitrates, and speech qualities arise. Historically one of the most important excitation description model is constituted by the multi-pulse excited (MPE) codec invented by Atal and Remde [9, 47]. It was the first AbS codec, yielding good speech quality between 9.6 kbps and 16 kbps at a moderate complexity. Variants of the typically low-bitrate (4.8–8 kbps) CELP codec yield medium quality at a high complexity, while the moderate complexity, medium bitrate (13 kbps) RPE codec used also in the GSM system provides high speech quality (MOS $\approx$ 4.0). Spectrally, it is almost three times more efficient than the previously described 32 kbps ADPCM G.721 ITU codec, while maintaining a higher robustness against channel errors.

In RPE codecs the innovation sequence $v(n)$ holds M number of equidistant excitation samples with amplitudes β_k and positions m_k, yielding a set of legitimate excitation sequences $v(n)$ in the following form:

$$v(n) = \sum_{k=0}^{M-1} \beta_k \delta(n - m_k). \tag{5.2}$$

Since the excitation model $v(n)$ is now given by Equation 5.2, we can embark on determining the optimum excitation parameters β_k, m_k. For LTP delay values longer than the excitation optimization subsegment length, we have $\alpha > N$, implying that the pitch synthesis filter's impulse response $h_p(n)$ is zero inside the current excitation optimization subsegment. Hence, for $n < N$ we have $h_c(n) = h_w(n)$; that is, the composite impulse response is identical to the weighted synthesis filter's impulse response. Hence, if we impose the condition $\alpha > N$ by restricting the LTP delay to values exceeding the excitation frame length N, the LTP synthesis filter will not contribute to the composite synthesis filter's impulse response $h_c(n)$. However, it will have to be considered during the computation of the zero-input response of the combined synthesis filter. Assuming that the LTP synthesis filter is replaced by an adaptive G-scaled codebook $Gu(n - \alpha) = Gc_\alpha$, where the adaptive codebook entry is denoted by c_α, the composite excitation is given by:

$$u(n) = v(n) + Gu(n - \alpha) = v(n) + Gc_\alpha. \tag{5.3}$$

The computation of the LTP parameters α and G was highlighted in Section 3.4.1 and 3.4.2, and the weighted synthetic speech can now be expressed as the convolution of the excitation with the impulse response of the composite synthesis filter, as in:

$$\begin{aligned}
\hat{s}_w(n) &= u(n) * h_w(n) \\
&= v(n) * h_w(n) + Gc_\alpha(n) * h_w(n) + \hat{s}_0(n),
\end{aligned} \tag{5.4}$$

where the convolution is a memoryless process, since the filter memory is treated separately, and $\hat{s}_0(n)$ represents the zero input response of the weighted synthesis filter in the lower branch of Figure 3.8. For details on the analytical description of the zero-input response $\hat{s}_0(n)$, the interested reader is referred to Salami's work [46, 47].

When substituting the excitation model of Equation 5.2 into Equation 5.4, the synthetic speech is yielded in the following form:

$$\begin{aligned}
\hat{s}_w(n) &= \sum_{i=0}^{n} \left(\sum_{k=0}^{M-1} \beta_k \delta(n - m_k) \right) h_w(n - i) + Gc_\alpha(n) * h_w(n) + \hat{s}_0(n), \\
&= \sum_{k=0}^{M-1} \beta_k h_w(n - m_k) + Gy_\alpha(n) + \hat{s}_0(n),
\end{aligned} \tag{5.5}$$

where

$$y_\alpha(n) = c_\alpha(n) * h_w(n)$$

is referred to as the **zero-state response** of the weighted synthesis filter $h_w(n)$, when after resetting its memory to zero, it is excited by the codeword c_α chosen from the adaptive codebook. Now, the weighted error between the original speech and the synthetic speech is given by:

$$e_w(n) = s_w(n) - \hat{s}_w(n),$$

$$= s_w(n) - Gy_\alpha(n) - \hat{s}_0(n) - \sum_{k=0}^{M-1} \beta_k h_w(n - m_k),$$

$$= x(n) - \sum_{k=0}^{M-1} \beta_k h_w(n - m_k), \tag{5.6}$$

where now

$$x(n) = s_w(n) - Gy_\alpha(n) - \hat{s}_0(n), \tag{5.7}$$

implying that not only the zero input response $\hat{s}_0(n)$ of $W(z)$, but also the effect of the scaled adaptive codebook entry $Gy_\alpha(n)$ are subtracted from the weighted original speech in order to generate the **target vector**, to which the candidate synthesis filter responses are then compared in response to the candidate excitation patterns. Explicitly, the target vector $x(n)$ is now computed by updating $x'(n)$ of Equation 3.13, as follows:

$$x(n) = x'(n) - Gy_\alpha(n). \tag{5.8}$$

From Equation 5.6 the total weighted mse (wmse) can now be written as:

$$E_w = \sum_{n=0}^{N-1} e_w^2(n),$$

$$= \sum_{n=0}^{N-1} \left[x(n) - \sum_{k=0}^{M-1} \beta_k h_w(n - m_k) \right]^2. \tag{5.9}$$

Following Kroon's [11] and Salami's deliberations [46, 47], the optimum pulse amplitudes β_k and the pulse positions m_k minimizing the wmse can be determined by setting $\partial E_w / \partial \beta_i = 0$ for $i = 0, \ldots, M - 1$, which yields:

$$\frac{\partial E_w}{\partial \beta_i} = -2 \sum_{n=0}^{N-1} \left[x(n) - \sum_{k=0}^{M-1} \beta_k h_w(n - m_k) \right] h_w(n - m_i) = 0. \tag{5.10}$$

Upon rearranging this formula, we arrive at:

$$\sum_{n=0}^{N-1} x(n) h_w(n - m_i) = \sum_{n=0}^{N-1} \left[\sum_{k=0}^{M-1} \beta_k h_w(n - m_k) \right] h_w(n - m_i). \tag{5.11}$$

Upon exchanging the order of summations at the right-hand side of Equation 5.11, we get:

$$\sum_{k=0}^{M-1} \beta_k \sum_{n=0}^{N-1} h_w(n - m_k) h_w(n - m_i) = \sum_{n=0}^{N-1} x(n) h_w(n - m_i), \qquad i = 0 \ldots M - 1. \tag{5.12}$$

Observe in the above equation that

$$\Phi(m_i, m_k) = \sum_{n=0}^{N-1} h_w(n - m_i) h_w(n - m_k) \tag{5.13}$$

represents the autocorrelation of $h_w(n)$, and

$$\Psi(m_i) = \sum_{n=0}^{N-1} x(n) h_w(n - m_i) = x(n) * h_w(-n) \tag{5.14}$$

the cross-correlation between $x(n)$ and $h_w(n)$, then Equation 5.12 can be simplified to:

$$\sum_{k=0}^{M-1} \beta_k \Phi(m_i, m_k) = \Psi(m_i), \qquad i = 0 \ldots M - 1. \tag{5.15}$$

The above set of M equations can be written in a more explicit matrix form as follows:

$$\begin{pmatrix} \Phi(m_0, m_0) & \Phi(m_0, m_1) & \cdots & \Phi(m_0, m_{M-1}) \\ \Phi(m_1, m_0) & \Phi(m_1, m_1) & \cdots & \Phi(m_1, m_{M-1}) \\ \vdots & \vdots & \ddots & \vdots \\ \Phi(m_{M-1}, m_0) & \Phi(m_{M-1}, m_1) & \cdots & \Phi(m_{M-1}, m_{M-1}) \end{pmatrix} \cdot \begin{pmatrix} \beta_0 \\ \beta_1 \\ \vdots \\ \beta_{M-1} \end{pmatrix} = \begin{pmatrix} \Psi(m_0) \\ \Psi(m_1) \\ \vdots \\ \Psi(m_{M-1}) \end{pmatrix}.$$

$$\tag{5.16}$$

Equation 5.16 represents a set of M equations that should be solved for M pulse positions plus M pulse amplitudes, which is not possible. A computationally attractive, high-quality suboptimum solution was proposed by Kroon, Depretteere, and Sluyter [11] that was also portrayed in Salami's work [46, 47].

According to Kroon et al., the innovation sequence can be derived as a subsampled version of the STP residual. The excitation pulses are d samples apart, and there are d decimated candidate excitation sequences according to the d possible initial grid positions. If a frame of N prediction residual samples is processed, the number of excitation pulses is given by $M = (N)Div(d)$, where Div implies integer division. The legitimate excitation pulse positions are $m[k, i] = k + (i - 1)d, i = 1, 2, \ldots M$, where $k = 0, 1, \ldots (d - 1)$ are the initial grid positions. With the pulse positions fixed, Equation 5.16 is solved d times for each candidate excitation pattern, yielding d sets of M pulse amplitudes. Upon expanding Equation 5.9, we arrive at:

$$E_w = \sum_{n=0}^{N-1} x^2(n) - 2 \sum_{n=0}^{N-1} x(n) \sum_{k=0}^{M-1} \beta_k h_w(n - m_k) + \sum_{n=0}^{N-1} \left[\sum_{k=0}^{M-1} \beta_k h_w(n - m_k) \right]^2,$$

$$= \sum_{n=0}^{N-1} x^2(n) - 2 \sum_{k=0}^{M-1} \beta_k \Psi(m_k) + \sum_{i=0}^{M-1} \sum_{k=0}^{M-1} \beta_i \beta_k \Phi(m_i, m_k). \tag{5.17}$$

The second term of the above expression can be rewritten with the help of Equation 5.15 as follows:

$$\sum_{i=0}^{M-1} \beta_i \Psi(m_i) = \sum_{i=0}^{M-1} \beta_i \sum_{k=0}^{M-1} \beta_k \Phi(m_i, m_k),$$

$$= \sum_{i=0}^{M-1} \sum_{k=0}^{M-1} \beta_i \beta_k \Phi(m_i, m_k), \tag{5.18}$$

which allows us to simplify Equation 5.17 to:

$$E_w = \sum_{n=0}^{N-1} x^2(n) - \sum_{k=0}^{M-1} \beta_k \Psi(m_k), \tag{5.19}$$

where E_w is minimized if the second term of Equation 5.19 is maximized.

Again, the set of M Equations 5.15 or 5.16 contains twice as many unknowns as the number of independent equations; hence, there exists no direct solution to the problem. It would be possible to solve it, assuming a particular legitimate combination of the pulse positions; find the associated optimum excitation pulse amplitudes; and remember the corresponding total wmse E_w from Equation 5.19. This operation could then be continued for all legitimate pulse position combinations, until the optimum one resulting in the minimum E_w term was found. In order to assess the associated computational complexity, we note that the matrix of impulse response autocorrelations can be inverted using Gaussian elimination or employing Cholesky-decomposition [47], which has a complexity proportional to M^3. For the typical values of $N = 40$ and $d = 4$, a total of $M = 10$ equations would have to be solved four times for each 5 ms excitation optimization subsegment. Equation 5.13 and Equation 5.14 would have to be evaluated as well.

The computational complexity incurred in solving Equation 5.16 can be significantly reduced, while maintaining high speech quality. Specifically, substantial algorithmic simplification is achieved at almost imperceptible speech quality degradation assuming that the speech is stationary. This would render the covariance $\Phi(i, j)$ to become $\Phi(|i - j|) = \Phi(k)$. With this assumption the key equation Equation 5.16 is simplified to:

$$
\begin{pmatrix}
\Phi(0) & \Phi(d) & \Phi(2d) & \dots & \Phi[(M-1)d] \\
\Phi(d) & \Phi(0) & \Phi(d) & \dots & \Phi[(M-2)d] \\
\vdots & & & & \\
\Phi[(M-1)d] & & & \dots & \Phi(0)
\end{pmatrix}
\begin{pmatrix}
\beta(k, 1) \\
\beta(k, 2) \\
\vdots \\
\beta(k, M)
\end{pmatrix}
=
\begin{pmatrix}
\Psi[m(k, 1)] \\
\Psi[m(k, 2)] \\
\vdots \\
\Psi[m(k, M)]
\end{pmatrix},
$$
(5.20)

where the correlation matrix Φ becomes a Toeplitz matrix. Hence, Equation 5.20 can again be solved by the help of the Levinson-Durbin algorithm.

Kroon et al. [11] and Salami et al. [46, 47] have reported that $h_w(n)$ is a sharply decaying function. Therefore, its covariance of:

$$
\Phi(i) = \sum_{n=i}^{N-1} h_w(n) h_w(n-i).
$$
(5.21)

decays even faster. This allows us to set all off-diagonal elements in the covariance matrix Φ to zero, resulting in a dramatically reduced complexity when solving Equation 5.16. It can now be written as:

$$
\Phi(0)
\begin{pmatrix}
\beta(k, 1) \\
\beta(k, 2) \\
\vdots \\
\beta(k, M)
\end{pmatrix}
=
\begin{pmatrix}
\Psi[m(k, 1)] \\
\Psi[m(k, 2)] \\
\vdots \\
\Psi[m(k, M)]
\end{pmatrix}.
$$
(5.22)

giving the optimum pulse amplitudes in the form of:

$$
\beta(k, i) = \frac{\Psi[m(k, i)]}{\Phi(0)}.
$$
(5.23)

In order to further simplify the computation of the optimum excitation pulse amplitudes, we briefly return to Equation 5.7. As seen in Figure 3.8, the weighted original speech signal $s_w(n)$ can be expressed as the convolution of the prediction residual $r(n)$ and the weighting filter's response as follows:

$$
s_w(n) = \sum_{i=-\infty}^{n} r(i) h_w(n-i) = \sum_{i=0}^{n} r(i) h_w(n-i) + s_0(n),
$$
(5.24)

where $s_0(n)$ is the zero input response of the filter $W(z)$ in the upper branch of Figure 3.8, processing the orginal speech signal. Then upon substituting $s_w(n)$ from Equation 5.24 into Equation 5.7 we arrive at:

$$\begin{aligned}
x(n) &= r(n) * h_w(n) - Gy_\alpha(n) + s_0(n) - \hat{s}_0(n) \\
&= r(n) * h_w(n) - Gc_\alpha(n) * h_w(n) + s_0(n) - \hat{s}_0(n) \\
&= [r(n) - Gc_\alpha(n)] * h_w(n) + s_0(n) - \hat{s}_0(n) \\
.&= d(n) * h_w(n) + s_0(n) - \hat{s}_0(n),
\end{aligned} \tag{5.25}$$

where the shorthand

$$d(n) = [r(n) - Gc_\alpha] \tag{5.26}$$

was used to denote the LTP residual.

Now, assuming equal memory contributions in the original and synthetic speech paths, since both paths are filtering similar signals, we then have $s_0(n) = \hat{s}_0(n)$. This enables us to compute $\Psi(m_i)$ in Equation 5.14 with the aid of Equation 5.25 as:

$$\Psi(n) = x(n) * h_w(-n) = d(n) * h_w(n) * h_w(-n) = d(n) * \Phi(n). \tag{5.27}$$

Substituting $\Psi(m_i)$ from Equation 5.27 into Equation 5.23 gives the optimum excitation pulse amplitudes, as follows:

$$\beta(k, i) = d[m(k, i)] * \frac{\Phi[m(k, i)]}{\Phi(0)} = d[m(k, i)] * \varphi[m(k, i)]. \tag{5.28}$$

According to Equation 5.28, the derivation of the optimum excitation pulses can be interpreted as filtering the samples of the decimated signal $d[m(k, i)]$ employing a filter described by the help of the impulse response $\varphi[m(k, i)]$. This impulse response was given by Equation 5.21 in the form of the covariance of the weighting filter's impulse response, which is naturally a speech spectrum-dependent, time-variant function akin to the impulse response of a low-pass filter (LPF), which is also often termed a smoother.

Further algorithmic simplifications accrue without significant speech quality degradation, if we derive a time-invariant "compromise smoother" from the long-term averaged weighting filter covariances or employ simply an ideal LPF. For a pulse-spacing or decimation factor of $d = 3$, as in the GSM-standard RPE codec, a cutoff frequency of $f_c = 1.3$ kHz has to be used. For an ideal "rectangular" low-pass FIR-filter of order 11, the symmetric impulse response coefficients are simply derived from the Hamming-windowed sinc-function samples of: $\varphi(0) = \varphi(10) = -0.016256$, $\varphi(1) = \varphi(9) = -0.045649$, $\varphi(2) = \varphi(8) = 0$, $\varphi(3) = \varphi(7) = 0.250793$, $\varphi(4) = \varphi(6) = 0.70079$, $\varphi(5) = 1$.

In conclusion, the simplified RPE codec's operation can be summarized as follows. Initially, the STP coefficients $a_i, i = 1 \ldots p$ are determined, and the STP residual $r(n)$ is computed by filtering the original input speech $s(n)$ through the filter $A(z)$. Then the LTP filter parameters G, α are computed, and the LTP residual $d(n)$ is determined using Equation 5.26, which is then smoothed or low-pass (LP) filtered, before it is decomposed into d number of candidate excitation sequences. In case of $d = 3$ candidate excitation sequences, $d(n)$ is decimated by a factor of $d = 3$. Thus, the LPF's cutoff frequency is $\frac{4}{3} \approx 1.33$ kHz. The specific excitation pulses of each of the d candidate excitation sequences are given by Equation 5.28, which are derived from the smoothed and decimated LTP residual $d(n)$.

The specific candidate excitation sequence minimizing the wmse of Equation 5.19 is then finally selected to generate the synthetic speech by exciting the synthesis filters.

Explicitly, the total wmse $E_w^{(j)}$ for the jth candidate excitation vector is computed using Equation 5.28 as follows:

$$E_w^{(j)} = \sum_{n=0}^{N-1} x^2(n) - \Phi(0) \sum_{k=1}^{M} \beta^2(k, i) = \sum_{n=0}^{N-1} s_w^2(n) - \Phi(0)E(j), \qquad (5.29)$$

where $E(j)$ is the energy of the jth candidate excitation vector. It becomes plausible now that the specific excitation vector having the highest energy, in other words the one maximizing the second term of Equation 5.29, minimizes E_w. This is according to our expectations, since after smoothing the LTP residual was decomposed into d candidate excitations. However, the highest-energy vector is expected to give the best representation of the prediction residual $r(n)$ and hence to generate the closest synthetic speech replica of the original speech segment.

Before we describe the specific implementational details of the standardized GSM speech codec [74], we note that while the original RPE codec as proposed by Kroon et al. [11] was a true AbS codec, the simplified RPE codec deduced above and used by the GSM system is actually an open-loop system. This open-loop codec constitutes an attractive design tradeoff in terms of bitrate, complexity, and speech quality. Salami [46, 47] studied the performance of the RPE codec varying a range of parameters, including the subsegment length N, the number of pulses M per subsegment, and the decimation factor d.

5.2 THE 13 KBPS RPE-LTP GSM SPEECH ENCODER

The selection of the most appropriate speech codec for the GSM system from the set of candidate codecs was based on extensive comparative tests among various operating conditions. The rigorous comparisons published in [127] are interesting and offer deep insights for system designers as regards the pertinent tradeoffs in terms of speech quality, robustness against channel errors, complexity, system delay, and so on. The codecs participating in the final comparative tests were two different sub-band codecs, a multi-pulse excited codec, and the regular pulse excited (RPE) codec, which was finally selected for standardization on the basis of the overall comparison tests. The average mean opinion score (MOS) of the RPE codec on a five-point scale over the various test conditions was found to be four, which is hardly distinguishable from the original uncoded speech among normal operating conditions.

The schematic diagram of the RPE-LTP encoder is shown in Figure 5.1, where the following functional parts can be recognized [73], [12], and [13]: (1) pre-processing, (2) STP analysis filtering, (3) LTP analysis filtering, and (4) RPE computation.

5.2.1 Pre-Processing

Pre-emphasis can be employed to increase the numerical precision in computations by emphasizing the high-frequency, low-power part of the speech spectrum. This can be carried out by the help of a one-pole filter with the transfer function of:

$$H(z) = 1 - c_1 z^{-1}, \qquad (5.30)$$

where $c_1 \approx 0.9$ is a practical value. The pre-emphasized speech $s_p(n)$ is segmented into blocks of 160 samples in a buffer, where they are windowed by a Hamming-window to counteract the spectral domain Gibbs oscillation, caused by truncating the speech signal

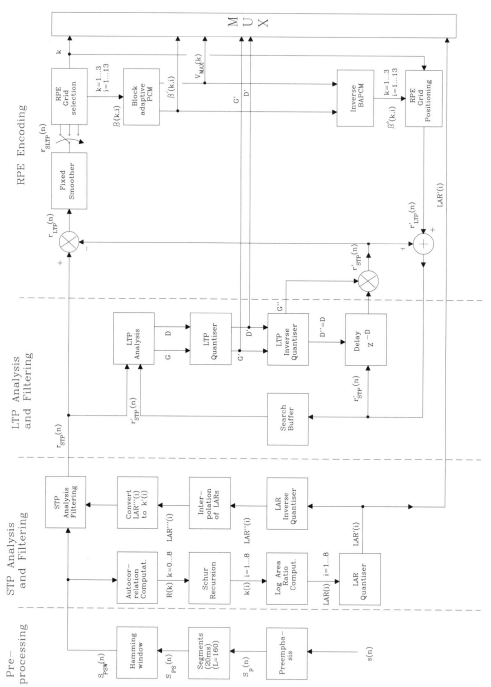

Figure 5.1 Block diagram of the RPE-LTP encoder.

outside the analysis frame. The Hamming-window has a tapering effect toward the edges of a block, while it has no influence in its middle ranges:

$$s_{psw}(n) = s_{ps}(n) \cdot c_2 \cdot \left(0.54 - 0.46 \cos 2\pi \frac{n}{L} \right). \tag{5.31}$$

where $s_{ps}(n)$ represents the pre-emphasized, segmented speech, $s_{psw}(n)$ is its windowed version, and the constant $c_2 = 1.5863$ is determined from the condition that the windowed speech must have the same power as the nonwindowed.

5.2.2 STP Analysis Filtering

For each segment of $L = 160$ samples, nine autocorrelation coefficients $R(k)$ are computed from $s_{psw}(n)$ by:

$$R(k) = \sum_{n=0}^{L-1-k} s_{psw}(k) s_{psw}(n+k) \qquad k = 0 \ldots 8. \tag{5.32}$$

From the speech autocorrelation coefficients, $R(k)$, eight reflection coefficients k_i are computed according to the Schur-recursion [128]. This method is equivalent to the Durbin algorithm used for solving the LPC key equations to derive the reflection coefficients k_i, as well as the STP filter coefficients a_i. However, the Schur-recursion delivers the reflection coefficients k_i only. The reflection coefficients k_i are converted to logarithmic area ratios (LAR(i)) because the logaritmically companded LARs have better quantization properties than the coefficients k_i:

$$\text{LAR}(i) = \log_{10} \left(\frac{1 + k(i)}{1 - k(i)} \right), \tag{5.33}$$

where a piecewise linear approximation with five segments is used to simplify the real-time implementation:

$$\text{LAR}'(i) = \begin{cases} k(i), & \text{if} \quad |k(i)| < 0.675 \\ \text{sign}\,[k(i)][2|k(i)| - 0.675], & \text{if} \quad 0.675 < |k(i)| < 0.95 \\ \text{sign}\,[k(i)][8|k(i)| - 6.375], & \text{if} \quad 0.975 < |k(i)| < 1.0 \end{cases} \tag{5.34}$$

The various LAR(i) $i = 1 \ldots 8$ filter parameters have different dynamic ranges and differently shaped probability density functions (PDFs), as we have seen in Chapter 4. This justifies the allocation of 6, 5, 4, and 3 bits to the first, second, third, and fourth pairs of LARs, respectively. The quantized LAR(i) coefficients LAR$'(i)$ are locally decoded into the set LAR$''(i)$, as well as transmitted to the speech decoder. So as to mitigate the abrupt changes in the nature of the speech signal envelope around the STP analysis frame edges, the LAR parameters are linearly interpolated, and toward the edges of an analysis frame the interpolated LAR$'''(i)$ parameters are used. Now the locally decoded reflection coefficients $k'(i)$ are computed by converting LAR$'''(i)$ back into $k'(i)$, which are used to compute the STP residual $r_{STP}(n)$ in a PARCOR (partial correlation) structure. The PARCOR scheme directly uses the reflection coefficients $k(i)$, in order to compute the STP residual $r_{STP}(n)$. It constitutes the natural analogy to the acoustic tube model of the human speech production.

5.2.3 LTP Analysis Filtering

As we have seen in Chapter 3, the LTP prediction error is minimized by that LTP delay D, which maximizes the cross-correlation between the current residual $r_{STP}(n)$ and its previously received and buffered history at delay D, that is, $r_{STP}(n - D)$. To be more specific, the $L = 160$ samples long STP residual $r_{STP}(n)$ is divided into four $N = 40$ samples long subsegments, and for each of them one LTP is determined by computing the cross-correlation between the presently processed subsegment and a continuously sliding $N = 40$ samples long segment of the previously received 128 samples long STP residual segment $r_{STP}(n)$. The maximum of the correlation is found at a delay D, where the currently processed subsegment is the most similar to its previous history. This is most probably true at the pitch periodicity or at a multiple of the pitch periodicity. Hence, the most redundancy can be extracted from the STP residual, if this highly correlated segment is subtracted from it and multiplied by a gain factor G, which is the normalized cross-correlation found at delay D. Once the LTP filter parameters G and D have been found, they are quantized to give G' and D', where G is quantized only by 2 bits, while to quantize D, 7 bits are sufficient.

The quantized LTP parameters (G', D') are locally decoded into the pair (G'', D'') so as to produce the locally decoded STP residual $r'_{STP}(n)$ for use in the forthcoming subsegments to provide the previous history of the STP residual for the search buffer, as shown in Figure 5.1. Observe that since D is integer, we have $D = D' = D''$. With the LTP parameters just computed, the LTP residual $r_{LTP}(n)$ is calculated as the difference of the STP residual $r_{STP}(n)$ and its estimate $r''_{STP}(n)$, which has been computed by the help of the locally decoded LTP parameters (G'', D) as shown below:

$$r_{LTP}(n) = r_{STP}(n) - r''_{STP}(n) \tag{5.35}$$

$$r''_{STP}(n) = G'' r'_{STP}(n - D). \tag{5.36}$$

Here $r'_{STP}(n - D)$ represents an already known segment of the past history of $r'_{STP}(n)$, stored in the search buffer. Finally, the content of the search buffer is updated by using the locally decoded LTP residual $r'_{LTP}(n)$ and the estimated STP residual $r''_{STP}(n)$ to form $r'_{STP}(n)$, as shown in

$$r'_{STP}(n) = r'_{LTP}(n) + r''_{STP}(n). \tag{5.37}$$

5.2.4 Regular Excitation Pulse Computation

The LTP residual $r_{LTP}(n)$ is weighted with the fixed smoother, which is essentially a gracefully decaying band-limiting low-pass filter with a cutoff frequency of 4 kHz/ $3 = 1.33$ kHz according to a decimation by three about to be employed, as argued in Section 5.1. The impulse response of this filter was also given in Section 5.1. The smoothed LTP residual $r_{SLTP}(n)$ is decomposed into three excitation candidates by actually discarding the 40th sample of each subsegment, since the three candidate sequences can host 39 samples only. Then the energies E1, E2, E3 of the three decimated sequences are computed, and the candidate with the highest energy is chosen to be the best representation of the LTP residual. The excitation pulses are afterward normalized to the highest amplitude $v_{\max}(k)$ in the sequence of the 13 samples, and they are quantized by a 3-bit uniform quantizer, whereas the logarithm of the block maximum $v_{\max}(k)$ is quantized with 6 bits. According to three possible initial grid positions k, 2 bits are needed to encode the initial offset of the grid for each subsegment. The pulse amplitudes $\beta(k, i)$, the grid positions k and the block maxima $v_{\max}(k)$

are locally decoded to give the LTP residual $r'_{LTP}(n)$, where the "missing pulses" in the sequence are filled with zeros.

5.3 THE 13 kbps RPE-LTP GSM SPEECH DECODER

The block diagram of the RPE-LTP decoder is shown in Figure 5.2, which exhibits an inverse structure, consituted by the functional parts of (1) RPE decoding, (2) LTP synthesis filtering, (3) STP synthesis filtering, and (4) post-processing.

RPE decoding: In the decoder the grid position k, the subsegment excitation maxima $v_{max}(k)$, and the excitation pulse amplitudes $\beta'(k, i)$ are inverse quantized, and the actual pulse amplitudes are computed by multiplying the decoded amplitudes with their corresponding block maxima. The LTP residual model $r'_{LTP}(n)$ is recovered by properly positioning the pulse amplitudes $\beta(k, i)$ according to the initial offset k.

LTP synthesis filtering: First the LTP filter parameters (G', D') are inverse quantized to derive the LTP synthesis filter. Then the recovered LTP excitation model $r'_{LTP}(n)$ is used to excite this LTP synthesis filter (G', D') to recover a new subsegment of length $N = 40$ of the estimated STP residual $r'_{STP}(n)$. To do so, the past history of the recovered STP residual $r'_{STP}(n)$ is used, properly delayed by D' samples and multiplied by G' to deliver the estimated STP residual $r''_{STP}(n)$, according to:

$$r''_{STP}(n) = G'.r'_{STP}(n - D'), \qquad (5.38)$$

and then $r''_{STP}(n)$ is used to compute the most recent subsegment of the recovered STP residual, as given in:

$$r'_{STP}(n) = r''_{STP}(n) + r'_{LTP}(n). \qquad (5.39)$$

STP synthesis filtering: In order to compute the synthesized speech $\hat{s}(n)$, the PARCOR synthesis is used, where—similarly to the STP analysis filtering—the reflection coefficients $k(i), i = 1 \ldots 8$ are required. The LAR$'(i)$ parameters are decoded by using the LAR inverse quantizer to give LAR$''(i)$, which are again linearly interpolated toward the analysis frame edges between parameters of the adjacent frames to prevent abrupt changes in the character of the speech spectral envelope. Finally, the interpolated parameter set is transformed back into reflection coefficients, where filter stability is guaranteed, if recovered reflection coefficients, which fell outside the unit circle, are reflected back into it, by taking their reciprocal values. The inverse formula to convert LAR(i) back into $k(i)$ is given by:

$$k(i) = \frac{10^{LAR(i)} - 1}{10^{LAR(i)} + 1}. \qquad (5.40)$$

Post-processing is constituted by the deemphasis, using the inverse of the filter $H(z)$ in Equation 5.30.

The RPE-LTP bit allocation scheme is summarized in Table 5.1 for a period of 20 ms, which is equivalent to the encoding of $L = 160$ samples, while the detailed bit-by-bit allocation is given in the GSM Standard [73].

The 260 bits derived have to be reordered according to their subjective importances before error correction coding, as proposed by GSM, and classified into categories of Class 1a, Class 1b, and Class 2 in descending order of prominence to facilitate a three-level error protection scheme. We note that the true sensitivity order has to be based on subjective tests. Objective bit-sensitivity analysis based on a combination of segmental signal-to-noise ratios

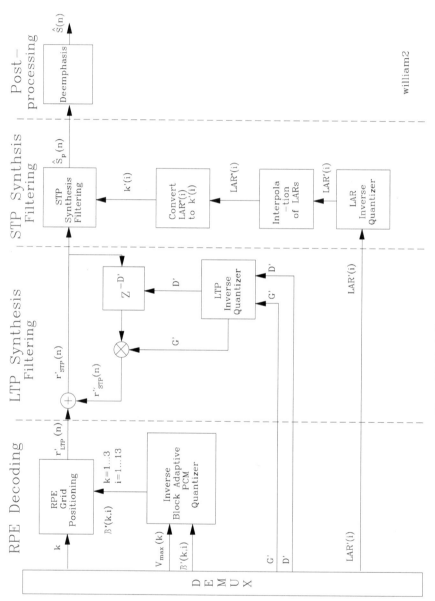

Figure 5.2 Block diagram of the RPE-LTP decoder.

TABLE 5.1 Summary of the RPE-LTP Bit-Allocation Scheme

Parameter to be Encoded	No. of Bits
8 STP LAR coefficients	36
4 LTP Gains G	$4 \times 2 = 8$
4 LTP Delays D	$4 \times 7 = 28$
4 RPE Grid positions	$4 \times 2 = 8$
4 RPE Block maxima	$4 \times 6 = 24$
$4 \times 13 = 52$ amplitudes	$52 \times 3 = 156$
Total number of bits per 20 ms	260
Transmission bitrate	13 kbit/s

and cepstrum distance measures, as defined in Chapter 17, results in a similar significance order [129]. During our experiments, we also designed a modified version of the standard scheme [129], [130], since we found that when using line spectral frequencies (LSF) instead of the standardized logarithmic area ratios (LARs) we obtained a slightly better performance, while encoding them using 36 bits. This is mainly due to their ordering property, as we discussed in Chapter 4, implying that the LSF parameters are monotonically increasing with increasing parameter index. This property allows the detection of channel errors that violate the ordering property, and the speech quality can be improved by LSF extrapolation invoked over consecutive frames. A further difference in this modified RPE-LTP codec was that we used 4 bits to encode the LTP gain instead of the standard 2 bits, which resulted in a improvement in speech quality. Therefore, the total number of bits was 268 per 20 ms frame, and the overall bitrate was increased to 13.4 kbps. The final bit-allocation scheme of the 13.4 kbps RPE-LTP codec is summarised in Table 5.2.

In comparison, the 32 kbps ADPCM waveform codec has a segmental SNR (SSNR) of about 28 dB, while the 13 kbps analysis-by-synthesis (AbS) RPE-LTP codec has a lower SSNR of about 16 dB, associated with similar subjective quality rated as a mean opinion score (MOS) of about 4. This discrepancy in SSNR arises because the RPE-LTP codec utilises perceptual error weighting. The cost of the RPE-LTP codec's significantly lower bitrate and higher robustness compared to ADPCM is its increased complexity and encoding delay.

TABLE 5.2 13.4 kb/s RPE Codec Bit Allocation

Parameter	Bit No.	Bitpos. in Frame
8 LSFs	36	1–36
RPE gridpos.	2	37, 38
Block max.	6	39–44
RPE exc. pulses	$13 \times 3 = 39$	45–83
LTP delay (LTPD)	7	84–90
LTP gain (LTPG)	4	91–94
Per subsegment	58	

Total bitrate: $36 + 4 \times 58 = 268/20$ ms $= 13.4$ kb/s

5.4 BIT SENSITIVITY OF THE 13 kbps GSM RPE-LTP CODEC

The bit allocation of the standard GSM speech codec was summarized in Table 5.1, while the sensitivity of each bit in the 260-bit, 20 ms frame is characterized by Figure 5.3. This figure provides an overview of the SEGSNR degradation inflicted by consistently corrupting one out of the 260 bits of each frame, while keeping the others in the frame intact. This technique masks the effects of error propagation across frame boundaries, since instead of quantifying this potential degradation over a number of consecutive frames, over which it results in observable SEGSNR degradation, the bit concerned is corrupted in each frame. None-theless, because of its simplicity and adequate accuracy, this technique is often used in practice.

Observe in Figure 5.3 that the repetitive structure reflects the periodicity due to the four 5 ms, 40-sample excitation optimization subsegments, while the left-hand-side section corresponds to the 36 LAR coefficients. Focusing more closely on the sensitivity of these LAR bits in the context of both Figure 5.3 and Figure 5.4, the most significant bits (MSB) to least significant bits (LSB) hierarchy is clearly recognized. Furthermore, the higher order LARs, corresponding to the last stages of the acoustic tube-model of the vocal tract are less important and, accordingly, exhibit a lower sensitivity. This is also reflected by the fact that the first, second, third, and fourth pairs of LARs are allocated 6, 5, 4, and 3 bits, respectively. Since the LAR coefficients are re-computed each 20 ms, despite mild error propagation due to LAR interpolation between consecutive frames, their corruption is not as detrimental as that of the excitation pulse block maxima seen in Figure 5.5.

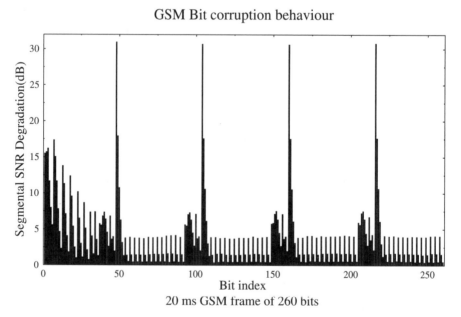

Figure 5.3 Bit sensitivity in the 260-bit, 20 ms RPE-LTP GSM speech frame.

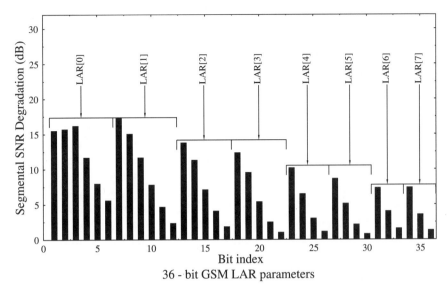

Figure 5.4 Bit sensitivity of the LARs in the 260-bit, 20 ms RPE-LTP GSM speech frame.

By observing Figure 5.5, we note that the MSB-LSB structure of the block maximum bits and normalized excitation pulse magnitude bits is conspicuous. Again, these bits do not result in serious error propagation. This is not true for the long term predictor delay (LTPD) and long-term predictor gain (LTPG) bits, whose sensitivity in practice is more critical, when taking into account the associated error propagation effects.

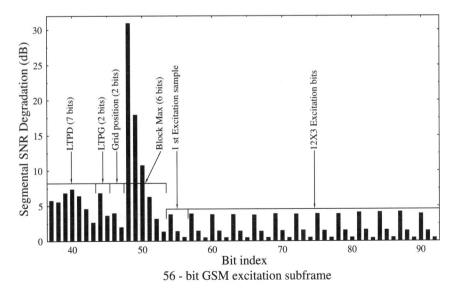

Figure 5.5 Bit sensitivity of the excitation subsegment in the 260-bit, 20 ms RPE-LTP GSM speech frame.

5.5 APPLICATION EXAMPLE: A TOOLBOX-BASED SPEECH TRANSCEIVER [131]

In the comparative study [131], Williams, Hanzo, Steele, and Cheung presented simulation results giving BER, bandwidth occupancy, and an estimate of complexity for 4-bit/symbol 16-Star QAM modems. Their purpose was to characterize the potential of an ambitious multilevel system, that of a 2-bit/symbol $\pi/4$-shifted differential quadrature phase shift keying ($\pi/4$-DQPSK) modems, since they are used in the Pan-American IS-54 [132] and the Japanese JDC [133] systems as well as binary Gaussian minimum shift keying (GMSK) modems. The authors used packet reservation multiple access (PRMA), since it provided substantial improvements over time division multiple access (TDMA) in terms of the number of users supported.

Specifically, in our simulatioins we used gaussian minimum shift keying (GMSK) [134], $\pi/4$-DQPSK, and 16-Star QAM modems [135], combined with the unprotected low-complexity 32 kbps ADPCM codec, as in the digital European cordless telephone (DECT) system, the Japanese Handyphone system (known as PHS), and the British CT2 system. Furthermore, the same modems were also combined with the 13 kbps RPE-LTP GSM codec and a twin-class FEC. Each modem had the option of either a low- or a high-complexity demodulator. The high-complexity demodulator for the GMSK modem was a maximum likelihood sequence estimator based on the Viterbi algorithm [69], whereas the low-complexity one was a frequency discriminator. For the two multilevel modems, either low-complexity noncoherent differential detection or a maximum likelihood correlation receiver (MLH-CR) was invoked. Synchronous transmissions and perfect channel estimation were used in evaluating the relative performances of the systems listed in Table 5.3. Our results represent performance upper bounds, allowing relative performance comparisons under identical circumstances.

The system performances applied to microcellular conditions. The carrier frequency was 2 GHz, the data rate 400 kBd, and the mobile speed 15 m/s. At 400 kBd in microcells, the fading is flat and usually Rician. The best and worst Rician channels are the Gaussian and Rayleigh-fading channels, respectively, and we performed our simulations for these channels to obtain upper and lower bound performances. Our conditions of 2 GHz, 400 kBd, and 15 m/s are arbitrary. They correspond to a fading pattern that can be obtained for a variety of different conditions, for example, at 900 MHz, 271 kBd, and 23 m/s. We compared the performances of the various systems summarised in Table 5.3 when operating according to our standard conditions.

Returning to Table 5.3, the first column shows the system classification letter, the next the modulation used, the third the demodulation scheme employed, the fourth the FEC scheme, and the fifth the speech codec employed. The sixth column gives the estimated relative order of the complexity of the schemes, where the most complex one having a complexity parameter of 12 is the 16-Star QAM, MLH-CR, BCH, RPE-LTP arrangement. All the BCH-coded RPE-LTP schemes have complexity parameters larger than six, while the unprotected ADPCM systems are characterized by values of one to six, depending on the complexity of the modem used. The speech Baud rate and the TDMA user bandwidth are given next.

An arbitrary signaling rate of 400 kBd was chosen for all our experiments, regardless of the number of modulation levels, in order to provide a fair comparison for all the systems under identical propagation conditions. Again, these propagation conditions can be readily converted to arbitrary Bd-rates upon scaling the vehicular speed appropriately. The 400 kBd systems have a total bandwidth of $400/1.35 = 296$ kHz, $2 \cdot 400/1.62 = 494$ kHz, and

TABLE 5.3 System Parameters [131], © IEEE, 1994, Williams, Hanzo, Steele, Cheung

1 Syst.	2 Modulator	3 Detector	4 FEC	5 Speech Codec	6 Complexity Order	7 Baud Rate (KBd)	8 TDMA User Bandw. (kHz)	9 No. of TDMA Users per Carrier	10 No. of PRMA Users per Carrier	11 No. of PRMA Users per slot	12 PRMA User Bandw. (kHz)	13 Min SNR (dB) AWGN	14 Rayleigh
A	GMSK	Viterbi	No.	ADPCM	2	32	23.7	11	18	1.64	14.5	7	∞
B	GMSK	Freq. Discr.	No.	ADPCM	1	32	23.7	11	18	1.64	14.5	21	31
C	π/4-DQPSK	MLH-CR	No.	ADPCM	4	16	19.8	22	42	1.91	10.4	10	28
D	π/4-DQPSK	Differential	No.	ADPCM	3	16	19.8	22	42	1.91	10.4	10	28
E	16-StQAM	MLH-CR	No	ADPCM	6	8	13.3	44	87	1.98	6.7	20	∞
F	16-StQAM	Differential	No	ADPCM	5	8	13.3	44	87	1.98	6.7	21	31
G	GMSK	Viterbi	BCH	RPE-LTP	8	24.8	18.4	12	22	1.83	10.1	1	15
H	GMSK	Freq. Discr.	BCH	RPE-LTP	7	24.8	18.4	12	22	1.83	10.1	8	18
I	π/4-DQPSK	MLH-CR	BCH	RPE-LTP	10	12.4	15.3	24	46	1.92	8	5	20
J	π/4-DQPSK	Differential	BCH	RPE-LTP	9	12.4	15.3	24	46	1.92	8	6	18
K	16-StQAM	MLH-CR	BCH	RPE-LTP	12	6.2	10.3	48	96	2.18	4.7	13	25
L	16-StQAM	Differential	BCH	RPE-LTP	11	6.2	10.3	48	96	2.18	4.7	16	24

GMSK: Gaussian Minimum Shift Keying

π/4-DQPSK: Differential Phase Shift Keying

16-StQAM: 16-level Quadrature Amplitude Modulation

MLH-CR: Maximum Likelihood Correlation Receiver

BCH: Bose-Chaudhuris Hocquenghem FEC Coding

ADPCM: Adaptive Differential Pulse Code Modulation

RPE-LTP: Regular Pulse Excited Speech Codec with Long-Term Prediction

TDMA: Time Division Multiple Access

PRMA: Packet Reservation Multiple Access

$4 \cdot 400/2.4 = 667\,\text{kHz}$, respectively. When computing the user bandwidth requirements we took account of the different bandwidth constraints of GMSK, $\pi/4$-DQPSK, and 16-QAM, assuming an identical Baud rate.

In order to establish the speech performance of systems A-L, we evaluated the segmental signal-to-noise ratio (SEGSNR) and Cepstral Distance (CD), both defined in Chapter 17, versus channel SNR characteristics of these schemes. These experiments yielded 24 curves for AWGN and 24 curves for Rayleigh-fading channels, constituting the best and worst case channels, respectively. Then for the 12 different systems and 2 different channels, we derived the minimum required channel SNR value for near-unimpaired speech quality in terms of both CD and SEGSNR. These values are listed in columns 13 and 14 of Table 5.3.

The bandwidth efficiency gains tabulated are reduced in SIR-limited scenarios due to the less dense frequency reuse of multilevel modems [49]. Nevertheless, multilevel modulation schemes result in higher PRMA gains than their lower level counterparts.

5.6 CHAPTER SUMMARY

This chapter characterized the family of RPE speech codecs. RPE codecs are historically important because they constitute the first AbS codec employed in a public mobile radio system. While the AbS coding principle relies on a closed-loop assisted excitation optimization, the 13 kbps GSM speech codec is, strictly speaking, an open-loop excitation optimization assisted codec, striking a good tradeoff between speech quality and implementational complexity. In this chapter, we also provided some discussions on the bit-sensitivity issues and transmission aspects of a mobile radio system transmitting over fading mobile channels.

Having characterized the family of RPE speech codecs, let us now focus our attention on another prominant class of AbS codecs referred to as code excited linear predictive (CELP) schemes.

6

Forward-Adaptive Code
Excited Linear Prediction

6.1 BACKGROUND

After Schroeder and Atal suggested the basic CELP codec in 1985 [16], it went through a quick evolution and developed into the most prominent speech codec over a wide bitrate range from 4.8–16 kbit/s. The original CELP codec was a **forward-adaptive** predictive scheme, requiring the transmission of spectral envelope and spectral fine-structure information to the decoder. In order to maintain a low bitrate, while using about 36 bits per LPC analysis frame for scalar short-term spectral quantization, the framelength was constrained to be in excess of 20 ms. Then the associated channel capacity requirement becomes 36 bits/20 ms = 1.8 kbps and even upon extending the framelength to 30 ms 1.2 kbps has to be allocated to the LPC coefficients. When using an ingenious split vector-quantization scheme, Salami et al. [123, 136] succeeded in reducing the forward-adaptive frame length and delay to 10 ms, which is an important advance in the state of the art. They used 18 bits/10 ms LPC analysis frame, and the associated coefficients were invoked for the second of its two 5 ms excitation optimization subframes. Those for the first one were inferred by interpolating the LSF parameters of the two adjacent subframes. This scheme was dicussed in Chapter 4 and also discussed in the context of the 8 kbps CCITT G.729 10 ms delay codec in Section 7.8.

The CCITT G.728 Standard codec [85] also employs a CELP-type excitation, but its philosophy has changed substantially from the original CELP concept in many respects. First, constrained by the low-delay requirement of 2 ms, forward-adapted LPC analysis was not a realistic option for the designer team at AT&T. Hence, backward-adaptive prediction was employed, recovering the LPC coefficients from previously decoded speech segments. This was possible at 16 kbps because the decoded speech quality was very good. Hence, the quantization effects did not inflict serious speech degradation, which would have led to precipitated error propagation. In fact, we showed experimentally that the effect of using past decoded speech rather than the current unencoded speech for LPC analysis manifested itself more adversely, due to the inherent time lag and not to quantization effects. Therefore, a frequent LPC filter coefficient update was necessary, which was essentially only restricted by the codec's complexity but did not have any ramifications as regards the bitrate. Specifically, an LPC update interval of 20 samples or 2.5 ms was found to be acceptable in terms of complexity, when using a high-filter order of 50.

A second significant deviation from the original CELP principle is that the choice of the above exceptionally high filter order was justified by a strong preference for avoiding the employment of a LTP. This was justified by the argument that the LTP would have been realistic only in terms of frequent updates without the requirement of added channel capacity—that is, if it was a backward-adaptive LTP sensitive against channel errors due to its long-term memory, reimplanting previous transmission error effects in the adaptive codebook. The presence of a LTP consitutes a particular problem, for example, in packet networks, where the loss of a transmission cell would inflict a long-term speech degradation. The LPC filter order of 50 can remove long-term speech periodicities of up to $50 \times 0.125\,\text{ms} = 6.25\,\text{ms}$, catering for the pitch periodicities of female speakers having pitch frequencies as low as 160 Hz, for whom LTPs have a typically substantial conducive effects.

Third, instead of the original 1024 40-sample, 5 ms random vectors, the G.728 codec uses a smaller 128-entry, 5-sample, 0.625 ms codebook filled with trained rather than stochastic entries. These measures were introduced with reduced implementational complexity, robustness against channel errors, high speech quality, and potential frame loss in transmission networks in mind.

Over the years a number of attractive wideband CELP-based coding schemes have been developed in an attempt to provide improved intelligibility and naturalness using 16 kHz-sampled 7 kHz-bandwidth speech signals. In some proposals, a split-band CELP ocdec was advocated, which allowed greater flexibility in terms of controlling and localizing the frequency-domain effects of quantization noise [137], while maintaining as low a bitrate as 16 kbps. The performance of this codec was similar to the CCITT G.722 standard sub-band-ADPCM codec at 48 kbps [122]. Laflamme et al. [138] and Salami et al. [139] proposed full-band wideband CELP codecs using vast codebooks combined with focused search in order to reduce the associated implementational complexity, while maintaining bitrates of 16 and 9.6 kbps, respectively. Unfortunately, it is unknown whether, for example, either of the above 16 kbps wideband codecs perceptually outperformed the 16 kbps narrowband G.728 codec. Following this brief overview of the advantages and disadvantages of the various CELP codecs, let us dedicate the rest of this chapter to the classic Schroeder-Atal forward-adaptive codec [16]. The tradeoffs and issues are revisited in depth in the forthcoming chapters.

A plethora of computationally efficient approaches has been proposed in order to reduce the excessive complexity of the original CELP codec and ease its real-time implementation. A comprehensive summary of these algorithmic details was published in a range of references by Salami et al. [46, 47], Kondoz [31], and so on. Here we give a rudimentary description of CELP coding.

6.2 THE ORIGINAL CELP APPROACH

In CELP systems, a Gaussian process with slowly varying power spectrum is used to represent the residual signal after short-term and long-term prediction. The speech waveform is generated by filtering Gaussian excitation vectors through the time-varying linear pitch and LPC synthesis filters, as seen in Figure 6.1.

Specifically, in the original CELP codec [16], Gaussian distributed zero-mean, unit-variance random excitation vectors of dimension $N = 40$ were stored in a codebook of 1024 entries, and the optimum excitation sequence was determined by the exhaustive search of the excitation codebook. This scheme essentially followed the schematic of Figure 3.8, where in

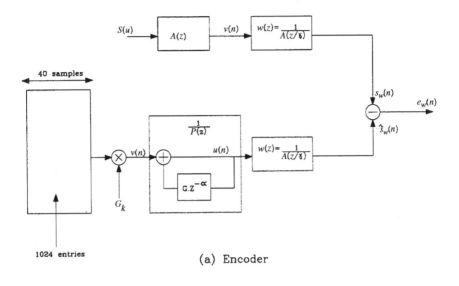

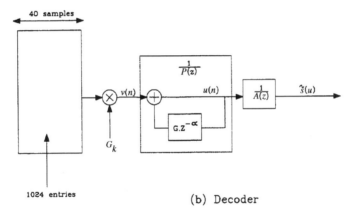

(a) Encoder

(b) Decoder

Figure 6.1 Simplified CELP codec schematic.

contrast to the regularly spaced decimated prediction residual of the RPE codec, the excitation generator was constituted by a fixed stochastic codebook and the long-term predictor's schematic was made explicit.

As argued before, in state-of-art high-quality CELP codecs, the majority of information to be transmitted to the decoder is determined in a closed-loop fashion so that the signal reconstructed by the decoder is perceptually as close as possible to the original speech. Full closed-loop optimization of all the codec parameters is usually not realistic in terms of real-time implementations, but we will attempt to document the range of complexity, speech quality, robustness, and delay tradeoffs. The adaptive codebook-based schematic of CELP codecs is shown in Figure 6.2, which reflects the closed-loop optimized LTP principle explicitly in contrast to Figure 6.1. The adaptive codebook-based schematic differs from the general AbS codec structure shown in Figure 3.8 and from its CELP-oriented interpretation given Figure 6.1. Here the excitation signal $u(n)$ is given by the sum of the outputs from two codebooks. The adaptive codebook is used to model the long-term periodicities present in

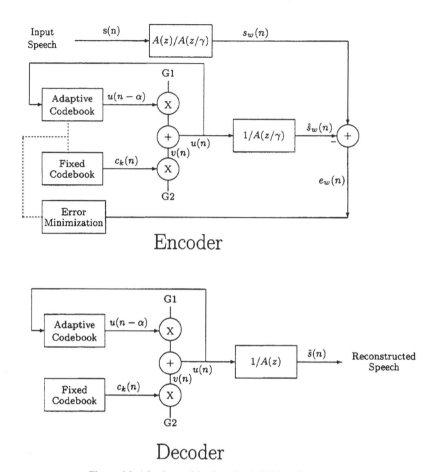

Figure 6.2 Adaptive codebook assisted CELP codec structure.

voiced speech, while the fixed codebook models the random noise-like residual signal that remains after both long- and short-term prediction. Recall that the difference between Figures 3.1 and 3.8 was that the error weighting filter of Figure 3.1 had been moved so that the input speech signal $s(n)$ and the reconstructed speech signal $\hat{s}(n)$ are both separately weighted before their difference is found. This is permissible because of the linear nature of the weighting filter, and it is done because it makes the determination of the codebook parameters less complex. With a synthesis filter of the form $1/A(z)$ and an error weighting filter $A(z)/A(z/\gamma)$, we arrive at the schematic of Figure 6.2. The filter $1/A(z/\gamma)$ in the encoder is referred to as the **weighted synthesis filter**. When fed with an excitation signal, it produces a weighted version $\hat{s}_w(n)$ of the reconstructed speech $\hat{s}(n)$.

In this chapter, we consider only systems in which forward-adaptive filtering is used. For such systems, the input speech is split up into frames for processing, where a frame is of the order of 20 ms long. The frames are usually further divided into subframes, with around four subframes per frame. The short-term synthesis filter coefficients are determined and transmitted once per frame, while the adaptive and fixed codebook parameters are updated once per subframe. The 4.7 kbits/s codec we have simulated has a frame length of 30 ms with four subframes of 7.5 ms each, while our 7.1 kbits/s codec has a frame length of 20 ms with 5 ms long sub-frames.

The encoding procedure generally takes place in three stages. First, the coefficients of the short-term synthesis filter $1/A(z)$ are determined for the frame by minimizing the residual energy obtained when the input speech is passed through the inverse filter $A(z)$. Then for each subframe first the adaptive and then the fixed codebook parameters are calculated using a closed-loop approach. The determination of the synthesis filter was detailed in Section 2.3 and its quantization was addressed in Chapter 4, while the computation of the closed-loop optimized adaptive codebook entry in Section 3.4.2. Let us therefore concentrate our attention in the next section on the fixed codebook-search.

6.3 FIXED CODEBOOK SEARCH

In the final stage of its calculations the coder finds the fixed codebook index and gain which minimize E_w. Following Salami et al. [46, 47] and taking the fixed codebook contribution into account, which was ignored in the last section, we arrive at:

$$
\begin{aligned}
e_w(n) &= s_w(n) - \hat{s}_w(n) \\
&= s_w(n) - \left(\hat{s}_o(n) + G_1 y_\alpha(n)\right) - G_2 c_k(n) * h(n) \\
&= \tilde{x}(n) - G_2 c_k(n) * h(n)
\end{aligned}
\tag{6.1}
$$

where

$$
\tilde{x}(n) = s_w(n) - \hat{s}_o(n) - G_1 y_\alpha(n)
\tag{6.2}
$$

is the target for the fixed codebook search, $c_k(n)$ is the codeword from the fixed codebook, and G_2 is the fixed codebook gain. Thus

$$
\begin{aligned}
E_w &= \frac{1}{N} \sum_{n=0}^{N-1} e_w^2(n) \\
&= \frac{1}{N} \sum_{n=0}^{N-1} \left(\tilde{x}(n) - G_2[c_k(n) * h(n)]\right)^2.
\end{aligned}
\tag{6.3}
$$

Setting $\partial E_w / \partial G_2 = 0$ gives the optimum gain for a given codeword $c_k(n)$ as

$$
\begin{aligned}
G_2 &= \frac{\sum_{n=0}^{N-1} \tilde{x}(n)[c_k(n) * h(n)]}{\sum_{n=0}^{N-1} [c_k(n) * h(n)]^2} \\
&= \frac{\tilde{C}_k}{\xi_k}
\end{aligned}
\tag{6.4}
$$

where

$$
\tilde{C}_k = \sum_{n=0}^{N-1} \tilde{x}(n)[c_k(n) * h(n)]
\tag{6.5}
$$

and

$$
\xi_k = \sum_{n=0}^{N-1} [c_k(n) * h(n)]^2.
\tag{6.6}
$$

Physically, ξ_k is the energy of the filtered codeword, and $\tilde{C}_k$ is the correlation between the target signal $\tilde{x}(n)$ and the filtered codeword. In the search for the fixed codebook parameters, the values of ξ_k and $\tilde{C}_k$ are calculated for every codeword k, and the optimum gain for that codeword is calculated using Equation 6.4. The gain is quantized to give $\hat{G}_2$, which is substituted back into Equation 6.3 to give:

$$E_w = \frac{1}{N} \sum_{n=0}^{N-1} \left(\tilde{x}(n) - \hat{G}_2[c_k(n) * h(n)] \right)^2$$

$$= \frac{1}{N} \left(\sum_{n=0}^{N-1} \tilde{x}^2(n) - 2\hat{G}_2 \sum_{n=0}^{N-1} \tilde{x}(n)[c_k(n) * h(n)] + \hat{G}_2^2 \sum_{n=0}^{N-1} [c_k(n) * h(n)]^2 \right)$$

$$= \frac{1}{N} \left(\sum_{n=0}^{N-1} \tilde{x}^2(n) - 2\hat{G}_2 \tilde{C}_k + \hat{G}_2^2 \xi_k \right). \tag{6.7}$$

The term $T_k = \hat{G}_2(2\tilde{C}_k - \hat{G}_2\xi_k)$ is calculated for every codeword, and the index that maximizes it is chosen. This index along with the quantized gain is then sent to the decoder.

Traditionally, the major part of a CELP coder's complexity comes from calculating the energy ξ_k of the filtered codeword, and the correlation $\tilde{C}_k$ between the target signal $\tilde{x}(n)$ and the filtered codeword for every codebook entry. From Equations 6.5 and 6.6 these are given by:

$$\tilde{C}_k = \sum_{n=0}^{N-1} \tilde{x}(n)[c_k(n) * h(n)]$$

$$= \sum_{n=0}^{N-1} \psi(n)c_k(n) \tag{6.8}$$

and

$$\xi_k = \sum_{n=0}^{N-1} [c_k(n) * h(n)]^2$$

$$= \sum_{i=0}^{N-1} c_k^2(i)\phi(i,i) + 2\sum_{i=0}^{N-2}\sum_{j=i+1}^{N-1} c_k(i)c_k(j)\phi(i,j) \tag{6.9}$$

where

$$\psi(i) = \tilde{x}(i) * h(-i)$$

$$= \sum_{n=i}^{N-1} \tilde{x}(n)h(n-i) \qquad \text{for } i = 0 \ldots N-1 \tag{6.10}$$

and

$$\phi(i,j) = \sum_{n=\max(i,j)}^{N-1} h(n-i)h(n-j) \qquad \text{for } i,j = 0 \ldots N-1. \tag{6.11}$$

The functions $\psi(i)$ and $\phi(i,j)$ can be calculated once per subframe, but then ξ_k and $\tilde{C}_k$ must be calculated for each codeword. This involves a large number of additions and multiplications by the elements of $c_k(n)$. Several schemes, for example, **binary pulse excitation** [111] and **transformed binary pulse excitation** (TBPE) were proposed by Salami [47], and **vector sum excited linear prediction** [140] (VSELP) was proposed by Gerson et al. in order to simplify these calculations. Typically, CELP codecs use codebooks in which most of the entries $c_k(n)$ are zero, which are referred to as **sparse codebooks**, thus greatly reducing the number of additions necessary to find ξ_k and $\tilde{C}_k$. Furthermore, if the nonzero elements of the codebook are equal to $+1$ or -1, then no multiplications are necessary, and $\tilde{C}_k$ and ξ_k can be calculated by a series of additions and subtractions.

6.4 CELP EXCITATION MODELS

6.4.1 Binary-Pulse Excitation

Instead of choosing the excitation pulses in a sparse excitation vector from a Gaussian random process, the pulses can be randomly chosen to be either -1 or 1 without any perceived deterioration in the quality of the CELP reconstructed speech. Using binary-pulse excitation vectors populated by the duo-binary values of -1 or 1, efficiently structured codebooks can be designed, where the codebook structure can be exploited to obtain fast codebook search algorithms [141], [142].

In a further step, we totally eliminate the codebook storage and its corresponding computationally demanding search procedure by utilizing a very simple approach in computing the optimum positions of the duo-binary -1 or 1 excitation pulses. Assuming that M pulses are allocated over the excitation optimization subsegment of N sample positions, we give the excitation vector by:

$$u(n) = \sum_{i=1}^{M} b_i \delta(n - m_i), \qquad \text{for } n = 0 \dots N - 1 \tag{6.12}$$

where b_i represents the duo-binary pulse amplitudes taking values -1 or 1 and m_i are the pulse positions. Having M binary excitation pulses per N-sample excitation vector and assuming that their positions are known is equivalent to a codebook of size 2^M. However, when they can be allocated to any arbitrary positions, the number of position combinations is given by $C_M^N = N!/((N - M)!M!)$. This approach has yielded a performance similar to that of the original CELP system, with the advantage of having a very simple excitation determination procedure characterized by about 10 multiplications per speech sample.

Xydeas et al. [143] and Adoul et al [144] have invoked a useful geometric representation of various CELP excitation codebooks, arguing that regardless of their statistical distribution, they result in similar perceptual speech quality. This is because they represent a vector quantizer, where all codebook entries populate the surface of a unit-radius N-dimensional sphere and 2^N is the maximum possible number of legitimate codebook entries on the sphere's surface. When assuming a sufficiently dense population of the sphere's surface, the subjective speech quality is rather similar for different codebooks, although the coding complexity may vary enormously. In this vector quantizer, each codebook entry constitutes a vector centroid, defining a particular subset of the full codebook. Taking into account the arbitrary positions of the pulses and assuming for the sake of illustration that there are five nonzero pulses, which can take any of the $N = 40$ positions, the total number of combinations in which these can be allocated is $C_5^4 0 = 40!/(35! \cdot 5!) = 658\,008$. Since the total number of possible duo-binary excitations is $2^{40} \approx 1.1 \cdot 10^{12}$, on the average about $2^{40}/658\,008 \approx 1.67 \cdot 10^6$ excitation vectors are possible per actual excitation vectors.

6.4.2 Transformed Binary-Pulse Excitation

6.4.2.1 Excitation Generation. The attraction of Transformed Binary-Pulse Excited (TBPE) codecs when compared to CELP codecs accrues from the fact that the excitation optimization can be achieved in a direct computation step. The sparse Gaussian excitation vector is assumed to take the form of

$$\mathbf{c} = \mathbf{Ab}, \tag{6.13}$$

where the binary vector $\mathbf{b}$ has M elements of ± 1, while the $M \cdot M$ matrix $\mathbf{A}$ represents an orthogonal transformation. Due to the orthogonality of $\mathbf{A}$, the binary excitation pulses of $\mathbf{b}$ are transformed into independent, unit-variance Gaussian components of $\mathbf{c}$. The set of $2M$ binary excitation vectors gives rise to $2M$ Gaussian vectors of the original CELP codec.

Having found the optimum codebook gain G_2 given in Equation 6.4, we can express mean square weighted error expression of Equation 6.3 following Salami et al. as [46, 47]:

$$E_{\min} = \sum_{n=0}^{N-1} x^2(n) - \frac{\left[\sum_{i=0}^{N-1} \psi(i)c_k(i)\right]^2}{\sum_{i=0}^{N-1} c_k^2(i)\phi(i,i) + 2\sum_{i=1}^{N-2}\sum_{j=i+1}^{N-1} c_k(i)c_k(j)\phi(i,j)}, \tag{6.14}$$

and upon using Equations 6.5 and 6.6 the above expression can be simplified as follows:

$$E_{\min} = \sum_{n=0}^{N-1} x^2(n) - \frac{(\tilde{C}_k)^2}{\xi_k}, \tag{6.15}$$

where again, most of the complexity of conventional CELP codecs is due to the computation of the energy ξ_k of the filtered codeword and to the correlation $\tilde{C}_k$ between the target signal $\tilde{x}(n)$ and the filtered codeword, which must be evaluated for all codebook entries.

The direct excitation of the TBPE codec accrues from the matrix representation of Equation 6.14 using Equation 6.13, viz.:

$$E = \mathbf{x}^T\mathbf{x} - \frac{(\mathbf{\Psi}^T\mathbf{Ab})^2}{\mathbf{b}^T\mathbf{A}^T\mathbf{\Phi Ab}} \tag{6.16}$$

The denominator in Equation 6.16 is nearly constant over the entire codebook and hence plays practically no role in the excitation optimization. This is because the autocorrelation matrix $\mathbf{\Phi}$ is strongly diagonal, since the impulse response $h(n)$ decays sharply. Because of the orthogonality of $\mathbf{A}$, we have $\mathbf{A}^T\mathbf{A} = \mathbf{I}$, where $\mathbf{I}$ is the identity matrix, causing the denominator to be constant.

Closer scrutiny of Equation 6.16 reveals that its second term reaches its maximum if the binary vector element $b(i) = -1$, whenever the vector element $\mathbf{\Psi}^T\mathbf{A}$ is negative, and vice-versa, that is, $b(i) = +1$ if $\mathbf{\Psi}^T\mathbf{A}$ is positive. The numerator of Equation 6.16 is then constituted by exclusively positive terms (i.e., it is maximum), and the weighted mean squared error is minimum. The optimum Gaussian excitation is computed from Equation 6.13 in both the encoder and decoder. Only the M-bit index representing the optimum binary excitation vector $\mathbf{b}$ has to be transmitted. The evaluation of the vectors $\mathbf{\Psi}^T\mathbf{A}$ and $\mathbf{c} = \mathbf{Ab}$ requires a mere $2M^2$ number of multiplications/additions, which typically gives five combined operations per output speech sample, a value 400 times lower than the complexity of the equivalent quality CELP codec.

The bit allocation of our TBPE codec is summarized in Table 6.1, while its schematic is portrayed in Figure 6.3. The spectral envelope is represented by 10 line spectrum frequencies (LSFs), which are scalar quantized using 36 bits. The 30 ms long speech frames having 240 samples are divided into four 7.5 ms subsegments having 60 samples. The subsegment excitation vectors $\mathbf{b}$ have 12 transformed duo-binary samples with a pulse-spacing of $D = 5$. The LTP delays (LTPD) are quantized with 7 bits in odd and 5 bits in even indexed subsegments, while the LTP gain (LTPG) is quantized with 3 bits. The excitation gain (EG)

TABLE 6.1 Bit Allocation of 4.8 kBit/s TBPE Codec

Parameter	Bit number
10 LSFs	36
LTPD	$2 \cdot 7 + 2 \cdot 5$
LTPG	$4 \cdot 3$
GP	$4 \cdot 4$
Excitation	$4 \cdot 12$
Total: 14	4

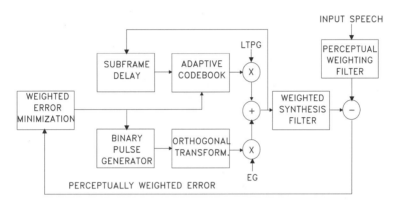

Figure 6.3 Block diagram of the 4.8 kbit/s TBPE codec.

factor is encoded with 4 bits, and the grid position (GP) of candidate excitation sequences by 2 bits. A total of 28 or 26 bits per subsegment is used for quantization, which yields $36 + 2 \cdot 28 + 2 \cdot 26 = 144$ bits/30 ms, resulting in a bitrate of 4.8 kbit/s.

6.4.2.2 Bit-Sensitivity Analysis of the 4.8 Kbps TBPE Speech Codec.   In our bit-sensitivity investigations, we systematically corrupted each bit of a 144-bit TBPE frame and evaluated the SEGSNR and CD degradation. Our results are depicted for the first 63 bits of a TBPE frame in terms of SEGSNR (dB) in Figure 6.4 and in terms of CD (dB) in Figure 6.5. For the sake of completeness, we note that we previously reported our findings on a somewhat more sophisticated sensitivity evaluation technique in reference [145]. According to this procedure, the effects of error propagation across speech frame boundaries due to filter memories were also taken into account by integrating or summing these degradations over all consecutive frames, where the error propagation inflicted measurable SEGSNR and CD reductions. However, for simplicity at this stage we refrain from using this technique, and we demonstrate the principles of source sensitivity-matched error protection using a less complex procedure. We recall from Table 6.1 that the first 36 bits represent the 10 LSFs describing the speech spectral envelope. The SEGSNR degradations shown in Figure 6.4 indicate the most severe waveform distortions for the first 10 bits describing the first 2 to 3 LSFs. The CD degradation, however, was quite severe for all LSFs, particularly for the most significant bits (MSBs) of the individual parameters. This was confirmed by our informal subjective tests. Whenever possible, all LSF bits should be protected against corruption.

The situation is practically reversed for the rest of the 144 bits in the TBPE frame, which represents the LTPD, LTPG, GP, EG, and Excitation parameters for the subsegments. We highlight our findings for the case of the first 27-bit subsegment only, as the other

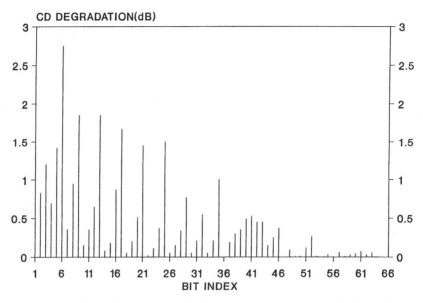

Figure 6.4 Bit sensitivities for the 4.8 kbit/s codec expressed in terms of SEGSNR (dB).

subsegments have identical behaviors. Bits 37–43 represent the LTP delays, and bits 44–47 the LTP gains. Their errors are more significant in terms of SEGSNR than in CD, as demonstrated by Figure 6.5. This is because the LTPD and LTPG parameters describe the spectral fine structure and do not seriously influence the spectral envelope, although they seriously degrade the recovered waveform. As the TBPE codec is a stochastic codec with random excitation patterns, the bits 48–63 assigned to the excitations and their gains are not particularly vulnerable to transmission errors. This is because the redundancy in the signal is removed by the long-term and short-term predictors. Furthermore, the TBPE codec exhibits exceptional inherent excitation robustness, as the influence of a channel error in the excitation diminishes after the orthogonal transformation $\mathbf{c} = \mathbf{Ab}$. In conventional CELP codecs, this is not the case, as a codebook address error causes the decoder to select a different excitation pattern from its codebook, causing considerably more speech degradation than is encountered by the TBPE codec.

In general, most robust performance is achieved if the bit protection is carefully matched to the bit sensitivities, but the SEGSNR and CD sensitivity measures portrayed in Figures 6.4 and 6.5 often contradict. Therefore, we combine the two measures to give a sensitivity figure S, representing the average sensitivity of a particular bit. The bits must first be ordered both according to their SEGSNR and to CD degradations given in Figure 6.4 and 6.5, respectively, to derive their "grade of prominence," with 1 representing the highest and 63 the lowest sensitivity. Observe that the highest CD degradation is caused by bit 6, which is the MSB of the second LSF in the speech frame, while the highest SEGSNR degradation is due to bit 40 in the group of bits 37–43, representing the LTP delay. Furthermore, bit 6 is the seventh in terms of its SEGSNR degradation. Hence, its sensitivity figure is $S = 1 + 7 = 8$, as seen in the first row of Table 6.2. On the other hand, the corruption of bit 40, the most sensitive in terms of SEGSNR, results in a relatively low CD degradation. It does not degrade the spectral envelope representation characterized by the CD, but spoils the pitch periodicity and hence the spectral fine structure. This bit is the nineteenth in terms of its SEGSNR degradation, giving a sensitivity figure contribution of 19 plus 1 due to CD degradation (i.e.,

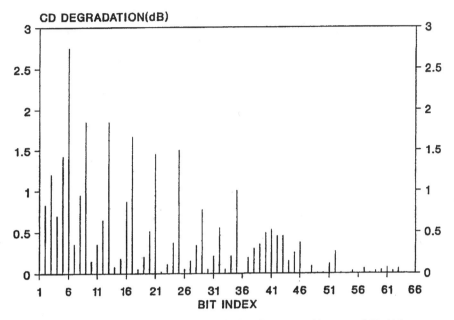

Figure 6.5 Bit sensitivities for the 4.8 kbit/s codec expressed in terms of CD (dB).

the combined sensitivity figure is $S = 20$, as shown by row 6 of Table 6.2. The combined sensitivity figures for all the LSFs and the first 27-bit subsegment are similarly summarised in ascending order in column 3 of Table 6.2, where column 2 represents the bit index in the the first 63-bit segment of the 144-bit TBPE frame.

Having studied the family of TBPE codecs, the next section is dedicated to Vector Sum Excited Linear Prediction [140] (VSELP), which is another successful coding technique. It was standardized not only for the 8 kbps Pan-American dual-mode mobile radio system referred to as IS-54, but also for the 5.6 kbps half-rate Pan-European system GSM [74].

6.4.3 Dual-Rate Algebraic CELP Coding

6.4.3.1 ACELP Codebook Structure. Algebraic Code Excited Linear Predictive (ACELP) codecs have recently conquered the battlefield of speech codec standardization, winning extensive comparative tests aimed at finalizing the 8 kbps CCITT G.729 recommendation. One of the two operating modes of the new CCITT G.723/H.324 codec is also based on ACELP principles, and it is also the most likely candidate for the Pan-European private mobile radio system known as TETRA. The algebraic codebook structure was originally proposed by Adoul et al. in reference [144]. In this section, we briefly introduce the ACELP principle and design a dual-rate ACELP codec, which can be conveniently used in a range of systems. The above-mentioned standard coding schemes will be detailed in Chapter 7.

In the proposed codec, each excitation codeword $c_k(n)$ has only 4 nonzero pulses, which have amplitudes of either $+1$ or -1. In its lower-rate mode, the dual codec allocates these excitation pulses over an excitation optimization subframe of 60 samples or 7.5 ms, while in its higher-rate mode over 40 samples or 5 ms. Also, each nonzero pulse has a limited number of positions within the codeword where it can lie. The amplitudes and possible

TABLE 6.2   Bit-Sensitivity Figures for the 4.8 Kbit/s TBPE Codec

Bit no. in Frame	Bit index in Frame	Sensit. Figure	Bit no. in Frame	Bit index in Frame	Sensit. Figure
1	6	8	36	57	76
2	9	14	37	10	79
3	5	16	38	28	80
4	3	16	39	19	80
5	41	19	40	61	80
6	40	20	41	59	82
7	13	21	42	62	84
8	2	23	43	15	85
9	43	24	44	60	88
10	8	25	45	34	89
11	46	25	46	50	91
12	42	26	47	31	92
13	17	27	48	55	95
14	39	31	49	27	95
15	4	31	50	23	97
16	21	32	51	14	97
17	12	37	52	47	98
18	38	38	53	58	102
19	25	43	54	54	103
20	16	44	55	53	105
21	52	45	56	56	105
22	7	45	57	18	105
23	1	45	58	33	108
24	37	48	59	49	109
25	45	49	60	26	109
26	11	55	61	30	110
27	20	58	62	22	119
28	51	60	63	36	125
29	29	60			
30	35	60			
31	44	63			
32	32	68			
33	48	69			
34	24	71			
35	63	76			

positions within the codeword for each of the four pulses are shown in Table 6.3 for our subframe size 60 4.7 kbits/s codec, and in Table 6.4 for our subframe size 40 7.1 kbits/s codec. In both codecs, each pulse can take up eight positions, and, so the chosen positions can be represented with 3 bits each, giving a total of 12 bits per subframe to represent the codebook index. The gain sign is represented with 1 bit, and its magnitude is quantized with 4 bits using logarithmic quantization. This gives a total of 17 bits per subframe for the fixed codebook information.

The algebraic codebook structure has several advantages: it does not require any codebook storage, since the excitation vectors are generated in real time, and it is robust against channel errors, since a single error corrupts the excitation vector only in one position, leading to a similar excitation vector at the decoder. Most importantly, however, it allows the

TABLE 6.3 Pulse Amplitudes and Positions for the 4.7 kbit/s Codec, © IEEE Adoul et al. [144]

Pulse Number i	Amplitude	Possible Position m_i
0	+1	0,18,16,24,32,40,48,56
1	−1	2,10,18,26,34,42,50,58
2	+1	4,12,20,28,36,44,52
3	−1	6,14,22,30,38,46,54

TABLE 6.4 Pulse Amplitudes and Positions for the 7.1 kbit/s Codec

Pulse Number i	Amplitude	Possible Position m_i
0	+1	1,6,11,16,21,26,31,36
1	−1	2,7,12,17,22,27,32,37
2	+1	3,8,13,18,23,28,33,38
3	−1	4,9,14,19,24,29,34,39

values $\tilde{C}_k$ and ξ_k to be calculated very efficiently. From Equations 6.8 and 6.9, the correlation and energy terms can be computed for the four excitation pulses of Table 6.3:

$$\tilde{C}_k = \psi(m_0) - \psi(m_1) + \psi(m_2) - \psi(m_3) \tag{6.17}$$

and

$$\begin{aligned}
\xi_k = {}& \phi(m_0, m_0) + \phi(m_1, m_1) - 2\phi(m_1, m_0) \\
& + \phi(m_2, m_2) + 2\phi(m_2, m_0) - 2\phi(m_2, m_1) \\
& + \phi(m_3, m_3) - 2\phi(m_3, m_0) + 2\phi(m_3, m_1) - 2\phi(m_3, m_2)
\end{aligned} \tag{6.18}$$

where m_i is the position of the pulse number i. By changing only one pulse position at a time, $\tilde{C}_k$ and ξ_k can be calculated using four nested loops associated with the four excitation pulses used. In the inner loop, $\tilde{C}_k$ is updated with one addition, and ξ_k with three multiplications and four additions. This allows for a very efficient codebook search.

A pair of appropriately extended equations analogous to 6.17 and 6.18 can be written for five and more pulses, leading to a corresponding number of encapsulated search loops, which will be exploited during our discussions on the 8 kbps CCITT G.729 10 ms delay codec in Section 7.8 as well as in Section 9.4. A further major attraction of the ACELP principle is that Salami, Laflamme, Adoul, et al. [136] proposed a computationally efficient focused search technique, which was also advocated by Kataoka, Combescure, Kroon et al [123, 146]. The proposed algorithm invokes a few threshold tests during subsequent search phases upon adding the individual excitation pulses one by one, in order to decide whether a particular subset of vectors characterized by the so far incorporated pulses is likely to lead to the lowest weighted error over the codebook for the subsegment about to be encoded. As we will highlight in Section 9.4, this facilitates a search complexity reduction around a factor of 100 or more without inflicting any significant performance degradation, while searching codebooks of 32,000 entries or even up to 10^6 entries.

In the decoder, the codebook information received from the encoder is used to find an excitation signal $u(n)$. If there are no channel errors, this will be identical to the excitation signal $u(n)$ in the encoder. It is then passed through a synthesis filter $1/A(z)$ to give the reconstructed speech signal $\hat{s}(n)$ as shown in Figure 6.2. The parameters of the synthesis filter are determined from the line spectrum frequencies transmitted from the encoder, using interpolation between adjacent frames.

6.4.3.2 Dual-rate ACELP Bit Allocation.

As mentioned, the excitation signal $u(n)$ is determined for each 5 or 7.5 ms subsegment of a 30 ms speech frame, depending on the targeted output bitrate, and it is described in terms of the following parameters.

- The adaptive codebook delay α that can take any integer value between 20 and 147 and hence is represented using 7 bits.
- The adaptive codebook gain G_1 which is nonuniformly quantized with 3 bits.
- The index of the optimum fixed codebook entry $c_k(n)$, which is represented with 12 bits.
- The fixed codebook gain G_2 which is quantized with a 4-bit logarithmic quantizer and an additional sign bit.

Thus, a total of 27 bits are needed to represent the subsegment excitation signal $u(n)$, and for the low-rate mode we have a total of $(34 + 4 \times 27) = 142$ bits per 30 ms frame, or a rate of about 4.73 kbits/s, while in the high-rate mode the bitrate becomes 142 bits/20 ms = 7.1 kbits/s.

A slightly different higher-rate mode can also be contrived by keeping the 30 ms framelength constant, which may become important in networks operating, for example, on the basis of a fixed 30 ms framelength. In this case, the lower-rate mode's bit allocation remains unchanged, while in the higher-rate mode, six rather than four 5 ms excitation optimization subsegments can be used. Then the number of bits per frame becomes $(34 + 6 \times 27) = 196$, yielding a bitrate of 196 bits/30 ms = 6.54 kbps.

6.4.3.3 Dual-Rate ACELP Codec Performance.

In this chapter, so far we have described in detail the general framework of CELP codecs and considered binary-pulse excitation and transformed binary pulse excitation, vector sum excitation, as well as ACELP codebook structures, which allowed an efficient codebook search. Table 6.6 shows the approximate complexity, in terms of millions of floating point operations per second (MFLOPs), of the various stages of the encoding procedure for our 7.1 kbits/s ACELP codec. Also shown is the complexity for a nonsparse 12-bit conventional CELP codebook search. As can be seen from the table, the fixed codebook search accounts for the majority of the complexity in the encoder, and the algebraic codebook structure gives a huge reduction in this complexity. In total the encoding procedure we have described requires approximately 23 MFLOPs, with most operations being spent on the two codebook searches. The decoder does

TABLE 6.5   Bits Allocated per Frame for the Dual-rate ACELP Codec

Line spectrum frequencies	34
Adaptive codebook delays	28 (4 ∗ 7)
Adaptive codebook gains G_1	12 (4 ∗ 3)
Fixed codebook index k	48 (4 ∗ 12)
Fixed codebook gains G_2	20 (4 ∗ 5)
Total	142

TABLE 6.6   CELP and ACELP Encoder Complexity (MFLOPs)

CELP codebook search	300,000
ACELP codebook search	15
LPC analysis	0.75
Adaptive codebook search	7

not have to do any codebook searches but merely filters the selected excitation through the synthesis filter. As a result it is much less complex and requires only about 0.2 MFLOPs.

The two codecs described here were tested with the speech file described earlier. The 4.7 kbits/s codec produced good communications quality speech with a segmental SNR of 10.5 dB, while the 7.1 kbits/s codec produced speech that was noticeably more transparent and had a segmental SNR of 12.1 dB. An important issue in the design of low-bitrate speech codecs is their robustness to background noise. We tested this aspect of our codec's performance using a speech-correlated noise source called the modulated noise reference unit, as described in [47]. This method was proposed by Law and Seymour in 1962 [147] and was standardized by the CCITT. Figure 6.6 shows how the segmental SNR of our 4.7 kbps codec varies with the signal-to-modulated-noise ratio. It can be seen that the ACELP codec is not seriously affected by the background noise until the signal-to-modulated-noise ratio falls below about 20 dB.

Here we curtail our discourse on the performance of various ACELP codecs, although we will return to the issue of codec robustness in Section 6.6. In the next Section we will revisit the general AbS codec structure in the context of CELP coding in order to identify areas, where the codec performance could be improved at the cost of acceptable implementational complexity.

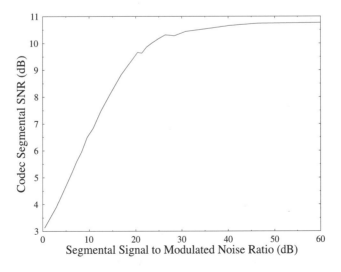

Figure 6.6 Performance of 4.7 kbps ACELP codec for noisy input signals.

6.5 OPTIMIZATION OF THE CELP CODEC PARAMETERS

6.5.1 Introduction

In the previous chapter we discussed the general structure of CELP codecs. This largely closed-loop structure is used in order to produce reconstructed speech, which is as close as possible to the original speech. However, there are two exceptions to an entirely closed-loop approach which are used in most CELP codecs. The first is in the determination of the synthesis filter $H(z)$, which is simply assumed to be the inverse of the short-term linear prediction error filter $A(z)$ which minimizes the energy of the prediction residual. This means that although the excitation signal $u(n)$ is derived, taking into account the form of the synthesis filter, no account is taken of the form of the excitation signal when the synthesis filter parameters are determined. This seems like an obvious deficiency and means, for example, that the synthesis filter may attempt to take account of long-term periodicities, which would be better left to the adaptive codebook.

The second departure from a strict closed-loop approach in most CELP codecs is in determining the codebook parameters. Rather than the adaptive and fixed codebook parameters being determined together to produce an overall minimum in the weighted error signal, the adaptive codebook delay and gain are determined first by assuming that the fixed codebook signal is zero. Then, given the adaptive codebook signal, the fixed codebook parameters are found. This approach is taken in order to reduce the complexity of CELP codecs to a reasonable level. However, it seems obvious that it must lead to some degradation in the reconstructed speech quality.

In this chapter we discuss ways of overcoming the two exceptions to the closed-loop approach, and we attempt to improve the quality of the reconstructed speech from our codecs while maintaining a reasonable level of complexity. We have concentrated our studies on the 4.7 kbits/s forward-adaptive ACELP codec described in the previous chapter, although the techniques described are applicable to other AbS codecs.

6.5.2 Calculation of the Excitation Parameters

In this section we discuss the procedure traditionally used for the adaptive and fixed codebook searches in CELP codecs, and ways in which this procedure can be improved. First, the theory behind a full search procedure is given. Then we describe how the equations derived for a full search reduce to those in Section 6.3 derived for the usual sequential determination of the codebook parameters. In Section 6.5.2.3 we describe the full search procedure, its complexity, and the results it gives. Section 6.5.2.4 describes various suboptimal approaches that can be used, and finally Section 6.5.2.5 describes the quantization of the codebook gains.

6.5.2.1 Full Codebook Search Theory. Consider the weighted error $e_w(n)$ between the weighted input speech and the weighted reconstructed speech. This is given by

$$
\begin{aligned}
e_w(n) &= s_w(n) - \hat{s}_w(n) \\
&= s_w(n) - \hat{s}_o(n) - G_1 y_\alpha(n) - G_2[c_k(n) * h(n)]
\end{aligned}
\tag{6.19}
$$

where the symbols used here have the same meaning as before, throughout Chapter 6. Explicitly, $s_w(n)$ is the weighted input speech, $\hat{s}_o(n)$ is the zero input response of the weighted synthesis filter due to its input in previous subframes, G_1 is the adaptive codebook gain, $y_\alpha(n) = h(n) * u(n - \alpha)$ is the filtered adaptive codebook signal, G_2 is the fixed codebook gain, $c_k(n)$ is the fixed codebook codeword, and $h(n)$ is the impulse response of the weighted synthesis filter.

The search procedure attempts to find the values of the adaptive codebook gain G_1 and delay α and the fixed codebook index k and gain G_2 which minimize the mean square error E_w taken over the subframe length N. This is given by:

$$E_w = \frac{1}{N} \sum_{n=0}^{N-1} e_w^2(n)$$

$$= \frac{1}{N} \sum_{n=0}^{N-1} \left(x(n) - G_1 y_\alpha(n) - G_2[c_k(n) * h(n)]\right)^2$$

$$= \frac{1}{N} \left(\sum_{n=0}^{N-1} x^2(n) + G_1^2 \sum_{n=0}^{N-1} y_\alpha^2(n) + G_2^2 \sum_{n=0}^{N-1} [c_k(n) * h(n)]^2 \right.$$

$$- 2G_1 \sum_{n=0}^{N-1} x(n)y_\alpha(n) - 2G_2 \sum_{n=0}^{N-1} x(n)[c_k(n) * h(n)]$$

$$\left. + 2G_1 G_2 \sum_{n=0}^{N-1} y_\alpha(n)[c_k(n) * h(n)] \right), \tag{6.20}$$

where $x(n) = s_w(n) - \hat{s}_o(n)$ is the target signal for the codebook search, referred to as the LTP target. We can rewrite this formula as

$$E_w = \frac{1}{N} \left(\sum_{n=0}^{N=1} x^2(n) + G_1^2 \xi_\alpha + G_2^2 \xi_k - 2G_1 C_\alpha - 2G_2 C_k + 2G_1 G_2 Y_{\alpha k} \right)$$

$$= \frac{1}{N} \left(\sum_{n=0}^{N-1} x^2(n) - T_{\alpha k} \right) \tag{6.21}$$

where

$$T_{\alpha k} = 2\left(G_1 C_\alpha + G_2 C_k - G_1 G_2 Y_{\alpha k}\right) - G_1^2 \xi_\alpha - G_2^2 \xi_k \tag{6.22}$$

is the term to be maximized by the codebook search. Here

$$\xi_\alpha = \sum_{n=0}^{N-1} y_\alpha^2(n) \tag{6.23}$$

is the energy of the filtered adaptive codebook signal and

$$C_\alpha = \sum_{n=0}^{N-1} x(n)y_\alpha(n) \tag{6.24}$$

is the correlation between the filtered adaptive codebook signal and the codebook target $x(n)$. Similarly, ξ_k is the energy of the filtered fixed codebook signal $[c_k(n) * h(n)]$, and C_k is the correlation between this and the target signal. Finally,

$$Y_{\alpha k} = \sum_{n=0}^{N-1} y_\alpha(n)[c_k(n) * h(n)] \tag{6.25}$$

is the correlation between the filtered signals from the two codebooks. With this notation we intend to emphasize what codebook the variables are dependent on. For example, once the weighted synthesis filter parameters are known, ξ_α depends only on which delay α is chosen for the adaptive codebook, whereas $Y_{\alpha k}$ depends on the indices α and k used for both the adaptive and fixed codebooks.

The codebook search must find the values of the indices α and k, and the gains G_1 and G_2, which maximize $T_{\alpha k}$ and so minimize E_w. For a given pair of indices α and k, we can find the optimum values for G_1 and G_2 by setting the partial derivatives of $T_{\alpha k}$ with respect to G_1 and G_2 to zero. This gives

$$\frac{\partial T_{\alpha k}}{\partial G_1} = 2C_\alpha - 2G_2 Y_{\alpha k} - 2G_1 \xi_\alpha = 0 \tag{6.26}$$

and

$$\frac{\partial T_{\alpha k}}{\partial G_2} = 2C_k - 2G_1 Y_{\alpha k} - 2G_2 \xi_k = 0. \tag{6.27}$$

Solution of these two linear simultaneous equations gives the optimum values of the gains for given codebook indices, as

$$G_1 = \frac{C_\alpha \xi_k - C_k Y_{\alpha k}}{\xi_\alpha \xi_k - Y_{\alpha k}^2} \tag{6.28}$$

and

$$G_2 = \frac{C_k \xi_\alpha - C_\alpha Y_{\alpha k}}{\xi_\alpha \xi_k - Y_{\alpha k}^2}. \tag{6.29}$$

The full search procedure has to find—for every pair of codebook indices α, k—the terms ξ_α ξ_k C_α C_k, and $Y_{\alpha k}$, and use these to calculate the gains G_1 and G_2. These gains can then be quantized and substituted into Equation 6.22 to give $T_{\alpha k}$, which the coder must maximize by the proper choice of α and k.

6.5.2.2 Sequential Search Procedure.
In this section we discuss how the equations derived above relate to those in Section 6.3 for the sequential search procedure, which is usually employed in CELP codecs. In this sequential search, the adaptive codebook parameters are determined first by assuming $G_2 = 0$. Substitution of this into Equation 6.26 gives

$$G_1 = \frac{C_\alpha}{\xi_\alpha} = \frac{\sum_{n=0}^{N-1} x(n) y_\alpha(n)}{\sum_{n=0}^{N-1} y_\alpha^2(n)}. \tag{6.30}$$

If we then substitute the values $G_1 = C_\alpha/\xi_\alpha$ and $G_2 = 0$ into Equation 6.22, the term to be maximized becomes

$$T_{\alpha k} = \frac{C_\alpha^2}{\xi_\alpha} = \frac{\left(\sum_{n=0}^{N-1} x(n) y_\alpha(n) \right)^2}{\sum_{n=0}^{N-1} y_\alpha^2(n)}. \tag{6.31}$$

Once the adaptive codebook parameters have been determined, they are assumed constant during the fixed codebook search. The LTP target $x(n)$ is updated to give the fixed codebook target $\tilde{x}(n)$ where

$$\tilde{x}(n) = x(n) - G_1 y_\alpha(n), \tag{6.32}$$

and for each codebook index k the energy ξ_k and the correlation $\tilde{C}_k$ between $\tilde{x}(n)$ and the filtered codewords are found. The correlation term $\tilde{C}_k$ is given by

$$
\begin{aligned}
\tilde{C}_k &= \sum_{n=0}^{N-1} \tilde{x}(n)[c_k(n) * h(n)] \\
&= \sum_{n=0}^{N-1} (x(n) - G_1 y_\alpha(n))[c_k(n) * h(n)] \\
&= C_k - G_1 Y_{\alpha k}.
\end{aligned}
\tag{6.33}
$$

Substitution of this into Equation 6.27 gives

$$
G_2 = \frac{\tilde{C}_k}{\xi_k}
\tag{6.34}
$$

as in Equation 6.4, and the term to be maximized becomes

$$
\begin{aligned}
T_{\alpha k} &= 2G_1 C_\alpha + 2G_2(C_k - G_1 Y_{\alpha k}) - G_1^2 \xi_\alpha - G_2^2 \xi_K \\
&= 2G_1 C_\alpha - G_1^2 \xi_\alpha + 2G_2 \tilde{C}_k - G_2^2 \xi_k.
\end{aligned}
\tag{6.35}
$$

Now as G_1 and α are fixed, we can ignore the first two terms above and write the expression to be maximized by the fixed codebook search as $G_2(2\tilde{C}_k - G_2\xi_k)$, as in Section 6.3.

6.5.2.3 Full Search Procedure.

Although a full codebook search to find the minimum possible weighted error E_w is not a practical method for use in real speech coders, it does give us an upper bound to the improvements that can be obtained over the sequential search approach.

In order to perform a full search of the two codebooks, the coder must calculate the value of $T_{\alpha k}$ using Equation 6.22 for every possible pair of codebook indices α and k, and select the indices that maximize $T_{\alpha k}$. This means we must calculate ξ_α, and C_α for every adaptive codebook codeword, ξ_k and C_k for every fixed codebook codeword, and $Y_{\alpha k}$ for every pair of codewords. All the necessary values of C_α, ξ_α, C_k, and ξ_k are calculated in the normal sequential search procedure. The extra complexity of the full search comes from calculating $Y_{\alpha k}$ for all values of α and k.

Using a similar approach to that used to find $\tilde{C}_k$ in the normal search, we can write $Y_{\alpha k}$ as

$$
\begin{aligned}
Y_{\alpha k} &= \sum_{n=0}^{N-1} y_\alpha(n)[c_k(n) * h(n)] \\
&= \sum_{n=0}^{N-1} c_k(n)[y_\alpha(n) * h(-n)] \\
&= \sum_{n=0}^{N-1} c_k(n)\Omega_\alpha(n)
\end{aligned}
\tag{6.36}
$$

where $\Omega_\alpha(n)$ is given by

$$
\Omega_\alpha(n) = \sum_{i=n}^{N-1} y_\alpha(i)h(i - n).
\tag{6.37}
$$

Thus, once $\Omega_\alpha(n)$ is known, using the algebraic codebook structure allows $Y_{\alpha k}$ to be calculated using four additions for each fixed codebook index k. Using four nested loops and updating the position of one pulse only in each loop allows us to find $Y_{\alpha k}$ very efficiently. Also because of the nature of the filtered adaptive codebook signal $y_\alpha(n)$, we can find $\Omega_\alpha(n)$ efficiently using an iterative procedure.

We simulated a full search codec in order to evaluate the degradation, inflicted by the sequential approach compared to the ideal full search. We measured the performance of the codec using the conventional segmental SNR and the weighted SNR measures, where the SNR weighting was implemented using the perceptual weighting filter $A(z)/A(z/\gamma)$, averaging the SNR over the entire measurement duration. The delay α of the adaptive codebook was allowed to take any integer value between 20 and 147, and a 12 bit algebraic fixed codebook was used as described in Section 6.3. We found that quantizing the codebook gains with quantizers designed for the normal codec masked the improvements obtained with the full search. Therefore, for all our simulation results reported here and in the next section neither G_1 nor G_2 was quantized. We consider quantization of the gains in Section 6.5.2.5.

We found—for four speech files containing speech from two male and two female speakers—that the full search procedure improved the average segmental SNR of our 4.7 kbits/s ACELP codec from 9.7 dB to 10.8 dB. A similar improvement was seen in the average weighted SNR; it increased from 7.3 dB to 8.2 dB. The reconstructed speech using the full search procedure sounded more full and natural than that obtained using the sequential search procedure.

These gains are obtained only at the expense of a huge increase in the complexity of the codec. Even with the techniques described above to allow the full search to be carried out efficiently, such a codec is almost 60 times more computationally demanding than a codec using the standard approach. Therefore, in the next section we describe some suboptimal approaches to the codebook search, with the aim of keeping the improvement in the reconstructed speech quality we have seen with the full codebook search, but reducing the complexity of the search to a reasonable level.

6.5.2.4 Suboptimal Search Procedures.

The full search procedure described in the previous section allows us to find the best combination of the codebook indices α and k. However, this method is unrealistically complex, and in this section we describe some suboptimal search strategies.

Such a feasible search procedure, which we refer to here as "Method A," is to follow the sequential approach and find G_1 and α by assuming $G_2 = 0$, and then find G_2 and k, while assuming that G_1 and α are fixed. Then—once α and k have been determined—we can use Equations 6.28 and 6.29 in order to jointly optimize the values of the codebook gains. In order to accomplish this, we have to know C_α, ξ_α, C_k, ξ_k, and $Y_{\alpha k}$ for the chosen indices. The values of C_α, ξ_α, and ξ_k are known from the codebook searches, and C_k can be found from $Y_{\alpha k}$ and $\tilde{C}_k$ using Equation 6.33. The main computational requirement for the update of the gains is therefore the calculation of $Y_{\alpha k}$ for the given α and k, and this is relatively undemanding. In fact, updating of the codebook gains given the codebook indices increases the complexity of the codec by about only 2%. Using the same speech-files described earlier, we found this update of the gains increased the average segmental SNR of the codec from 9.7 dB to 10.1 dB, and the average weighted SNR from 7.3 dB to 7.5 dB.

Another possible suboptimal approach to the codebook searches is to find the adaptive codebook delay α using the usual approach (i.e., by assuming $G_2 = 0$), and then use only this value of α during the fixed codebook search in which G_1, G_2, and k are all determined. This is similar to an approach suggested in [149] where a very small (32 entries) fixed codebook was used, and a 1-tap IIR filter was used instead of the adaptive codebook. For our codec we find ξ_k, C_k, and $Y_{\alpha k}$ for every fixed codebook index k using the approach with four nested loops described in Sections 6.3 and 6.5.2.3. The values of ξ_α and C_α are known from the adaptive codebook search. Thus, we can use Equations 6.28 and 6.29 to find G_1 and G_2, and then calculate $T_{\alpha k}$ using Equation 6.22. The value of k which maximizes $T_{\alpha k}$ is chosen as the fixed codebook index. We refer to this joint codebook search procedure as "Method B".

This "Method B"-based search allows the fixed codebook entry to be selected taking full account of the possible variations in the magnitude of the adaptive codebook signal. If we could trust the initial value of α calculated assuming $G_2 = 0$ to be correct, then it would give identical results to the full search procedure. However, it is much less computationally demanding than the full codebook search, and it increases the complexity of the normal codec by only about 30%. In our simulations, we found that it increased the average segmental SNR from 9.7 dB to 10.3 dB. Similarly, the average weighted SNR increased from 7.3 dB to 7.8 dB. Thus, this approach gives significant gains over the normal sequential search, but it still does not match the results of the codec using the full search procedure.

The differences between the results using the full codebook search, and those described above, must be due to differences in the adaptive codebook delay α chosen. We therefore investigated a procedure recalculating or updating this delay, once the fixed codebook index k is known. We refer to this final suboptimal search procedure as "Method C," which operates as follows. The adaptive codebook delay is initially chosen assuming $G_2 = 0$. Then the fixed codebook index is found by calculating G_1, G_2, and $T_{\alpha k}$ for every k, and choosing the index k which maximizes $T_{\alpha k}$ as in the Method B search. Then once k is known, we update the delay α by finding G_1, G_2, and $T_{\alpha k}$ for each possible α, and choosing the delay α which maximizes $T_{\alpha k}$. To do this we need to know ξ_α, C_α, ξ_k, C_k, and $Y_{\alpha k}$ for all values of α and the value of k chosen during the fixed codebook search. As explained previously, ξ_α, C_α, ξ_k, and C_k will all be known already, and so we must calculate $Y_{\alpha k}$ for all possible values of α and a fixed k.

This procedure to update the adaptive codebook delay once the fixed codebook index is known increases the complexity of the codec by about a further 10% relative to the complexity of the normal codec. It improved the average segmental SNR for our four speech-files to 10.6 dB, and the average weighted SNR to 7.8 dB.

The performance of the search procedures we have described in this section, along with the normal and the full search methods, is shown in Table 6.7 in terms of the average segmental and weighted SNRs. Also shown are the complexities of codecs using these search procedures relative to a codec using the normal sequential search. It can be seen that the joint codebook search Method A gives a significant improvement in the codec's performance with very little extra complexity. Furthermore, we can see that Method C—the most complex suboptimal search procedure investigated—increases the codec's complexity by only 40% but gives reconstructed speech, in terms of the segmental SNR at least, very similar to that using the much more complex full search procedure.

The investigations we have reported in this section have ignored the effects of quantization of the codebook gains G_1 and G_2. However, in any real coder we must somehow quantize these gains for transmission to the decoder.

TABLE 6.7 Performance and Complexity of Various Search Procedures

	Segmental SNR	Weighted SNR	Complexity
Sequential Search	9.7	7.3	1
Method A	10.1	7.5	1.02
Method B	10.3	7.8	1.3
Method C	10.6	7.8	1.4
Full search	10.8	8.2	60

6.5.2.5 *Quantization of the Codebook Gains.* In this section we study ways of quantizing the codebook gains G_1 and G_2 to attempt maintaining the improvements due to our various codebook search procedures. This was necessary because we noticed, especially for female speakers, that quantization of the gains had a much more serious effect in the codecs with improved search procedures than for the normal codec. This meant that the improvement that arose from the new search procedures was largely lost when quantization was considered. For example, for one of our speech files containing the sentence "To reach the end he needs much courage" spoken by a woman, the segmental SNR of the normal codec with no quantization was 11.45 dB. With quantization of both gains, this was only slightly reduced to 11.38 dB. The codec using the joint search procedure Method C gave a segmental SNR with no quantization of 12.45 dB. However, with quantization this fell to 11.67 dB, meaning that the increase in the segmental SNR due to the improved search procedure fell from 1 dB without quantization to 0.3 dB with quantization.

There are several possible reasons for this effect. The most obvious is that when the gains are calculated in a different way, their distributions change and so quantizers designed using the old distributions will be less effective. Also, it may just be that the gains calculated with the improved search procedures are more sensitive to quantization than those calculated normally.

Notice that Equation 6.28 gives the optimum value of G_1 only, if G_2 is given by Equation 6.29. When we quantize G_2, the optimum value of G_1 will change. We can find the best value of G_1 by substituting the quantized value of G_2, that is, $\hat{G}_2$, into Equation 6.26. This gives

$$G_1 = \frac{C_\alpha - \hat{G}_2 Y_{\alpha k}}{\xi_\alpha}. \tag{6.38}$$

Similarly, if the adaptive codebook gain has been quantized to give $\hat{G}_1$, then the optimum value of G_2 becomes

$$G_2 = \frac{C_k - \hat{G}_1 Y_{\alpha k}}{\xi_k}. \tag{6.39}$$

We set about improving the quantization of the gains for the codec using our best suboptimal search procedure, namely, Method C. A speech file containing about 11 seconds of speech spoken by two men and two women was used to train our quantizers. None of the speakers, or the sentences spoken, were the same as those used to measure the performance of the codec. Distributions for the two gains were measured using our training data when neither of the gains was quantized. We were then able to train quantizers using the Max-Lloyd algorithm [10].

There is a problem with the adaptive codebook gain G_1 because although most values of G_1 are between $+1.5$ and -1.5, a few values are very high. If we use all these values with the Max-Lloyd algorithm, then the resulting quantizer will have several reconstruction levels that are very high and rarely used. We found that for an eight-level quantizer trained using all the unquantized values of G_1, half the reconstruction levels were greater than 3 or less than -3. Using such a quantizer gives a serious degradation in the segmental SNR of the reconstructed speech. To overcome this problem, the values of G_1 must be cut down to some reasonable range. The DoD [76] codec uses the range -1 to $+2$. Hence, we invoked these values, additionally also experimenting with the range of -1.5 to $+1.5$, which was suggested by the PDF of our own experimental data.

Another problem when using the Max-Lloyd algorithm to design a quantizer for G_1 is that one reconstruction level tends to get allocated very close to zero where the PDF of the gains is low. We overcame this problem by splitting the values of G_1 into positive and negative values and running the Max-Lloyd algorithm separately on each half of the data. Using these techniques, we were able to design quantizers for G_1 which outperformed the quantizer designed for the normal codec.

Our normal codec used a 4-bit logarithmic quantizer for the magnitude of G_2, with the sign being allocated an additional bit. We also used the Max-Lloyd algorithm to design a 5-bit quantizer for G_2 using the distribution derived from our training data.

We conducted our simulations of the codec with G_1 calculated using Equation 6.28, and quantized, and then G_2 calculated using Equation 6.39. Using this technique, we were able to derive distributions for G_2 when G_1 was quantized with various quantizers. Similarly, we were able to find distributions for G_1 when G_2 was quantized with various quantizers. These distributions were then used to train quantizers for G_1 to use in conjunction with those already designed for G_2, and vice versa. We attempted quantizing G_1 first using various different quantizers, and then using the specially trained quantizer for G_2. Similarly, we also attempted quantizing G_2 first and then using various specially trained quantizers for G_1. The best results were obtained when G_2 was calculated first and quantized with the normal logarithmic quantizer, before G_1 was calculated using Equation 6.38 and quantized using a Max-Lloyd quantizer trained with gains cut to the range -1 to $+2$. Such a quantization scheme improved the segmental SNR for the female speech file described earlier from 11.67 dB to 11.97 dB. The improvement was less significant for the two male speech files, but on average using the improved quantization scheme gave a segmental SNR of 10.0 dB and a weighted SNR of 7.5 dB. These figures should be compared to an average segmental SNR of 9.9 dB and an average weighted SNR of 7.4 dB, when using the normal quantizers.

The average segmental SNR and weighted SNR for our four speech files using the codec with the normal search procedure and gain quantizers, and the codec with the improved search procedure (Method C) and quantization, are shown in Table 6.8. It can be seen that on average the improved search procedure and quantization give an increase in the segmental SNR of about half a decibel, and the weighted SNR increases by 0.4 dB. The improvements are similar for both the male and female speech files. In informal listening tests, we found that the reconstructed speech for the improved search procedure again sounded more full and natural than that for the normal search procedure.

Next, we discuss methods of improving the performance of our 4.7 kbits/s forward adaptive ACELP codec by recalculating the synthesis filter parameters after the excitation signal $u(n)$ has been determined. However, in Chapter 8 we will return to joint codebook search procedures and discuss using Method A and Method B described earlier to improve the performance of low-delay backward-adaptive CELP codecs.

6.5.3 Calculation of the Synthesis Filter Parameters

In the previous section, we discussed ways of improving the determination of the codebook parameters that give the excitation signal $u(n)$. At the decoder this excitation signal is passed through the synthesis filter $H(z)$ in order to generate the reconstructed speech $\hat{s}(n)$. As stated before, $H(z)$ is usually assumed to be the inverse of the prediction error filter $A(z)$,

TABLE 6.8 Performance of Search Procedures with Quantization

	Segmental SNR	Weighted SNR
Normal codec	9.5	7.1
Improved search and quant.	10.0	7.5

which minimizes the energy of the prediction residual. It is well known that this is not the ideal way to determine the synthesis filter parameters. For example, when the pitch frequency is close to the frequency of the first formant, which commonly happens for high-pitched speakers, the spectral analysis tends to give spectral envelopes with sharp and narrow resonances [149]. This leads to amplitude booms in the reconstructed speech, which can be annoying.

In this section, we discuss ways of improving the synthesis filter $H(z)$ in order to maximize the SNR of the reconstructed speech. Initially, for simplicity the filter coefficients were not quantized. In these endeavors no overlapping of the LPC analysis frames was implemented, and interpolating the line spectrum frequencies (LSF) of Section 4.2.1 between frames was not used. Discarding of interframe interpolation implies that the filter coefficients for the weighted synthesis filter change only once per frame rather than every subframe. Therefore, the energy of the filtered fixed codebook signals, namely, ξ_k, has to be computed only once per frame. Hence, the complexity of the fixed codebook search is dramatically reduced decreasing overall complexity of the codec by about 40%.

6.5.3.1 Bandwidth Expansion. One well-known and relatively simple way of improving the synthesis filter parameters is to use bandwidth expansion [149]. In this technique the filter coefficients a_k, produced by the autocorrelation or covariance analysis of the input speech, are replaced by $a_k \gamma^k$ where γ is some constant less than 1. This has the effect of expanding the bandwidth of the resonances in the transfer function of the synthesis filter. Therefore, it helps reduce the problems mentioned above which occur when the pitch frequency is close to the first formant frequency.

The constant γ can be expressed as [149]

$$\gamma = \exp(-\sigma \pi T) \tag{6.40}$$

where T is the sampling interval and σ is the bandwidth expansion in Hertz. We attempted using a 15 Hz expansion, which corresponds to $\gamma = 0.9941$, and we found that this improved the segmental SNR of our 4.7 kbits/s codec (with no LSF quantization or interpolation) from 9.90 dB to 10.59 dB. Also, it is reported [150] that such an expansion improves the robustness of a codec to channel errors, and so we used bandwidth expansion in our studies on error sensitivity in Section 6.6. Note that like all the results quoted in this section those above were obtained for a speech file containing one sentence each from two male and two female speakers.

6.5.3.2 Least Squares Techniques. Given an excitation signal $u(n)$ and a set of filter coefficients a_k, $k = 1, 2 \ldots p$, the reconstructed speech signal $\hat{s}(n)$ will be given by

$$\hat{s}(n) = u(n) + \sum_{k=1}^{p} a_k \hat{s}(n - k). \tag{6.41}$$

We wish to minimize E, the energy of the error signal $e(n) = s(n) - \hat{s}(n)$, where $s(n)$ is the original speech signal. E is given by

$$E = \sum_n (s(n) - \hat{s}(n))^2$$

$$= \sum_n \left(s(n) - u(n) - \sum_{k=1}^{p} a_k \hat{s}(n - k) \right)^2$$

$$= \sum_n \left(x(n) - \sum_{k=1}^{p} a_k \hat{s}(n - k) \right)^2 \tag{6.42}$$

where $x(n) = s(n) - u(n)$ is the "target" signal. For a given frame this target is fixed once the excitation signal has been determined. The problem with Equation 6.42 is that E is given in terms of not only the filter coefficients but also the reconstructed speech signal $\hat{s}(n)$, which also depends on the filter coefficients. Therefore, we cannot simply set the partial derivatives $\partial E / \partial a_i$ to zero and obtain a set of p simultaneous linear equations for the optimal set of coefficients.

A feasible approach—which has been used in Multi-Pulse Excited codecs [152, 152]—is to make the approximation

$$\hat{s}(n - k) \approx s(n - k) \tag{6.43}$$

in Equation 6.42, which then gives

$$E \approx \sum_n \left(x(n) - \sum_{k=1}^{p} a_k s(n - k) \right)^2. \tag{6.44}$$

We can then set the partial derivatives $\partial E / \partial a_i$ to zero for $i = 1, 2 \ldots p$ to obtain a set of p simultaneous linear equations as shown in

$$\frac{\partial E}{\partial a_i} = -2 \sum_n \left(x(n) - \sum_{k=1}^{p} a_k s(n - k) \right) s(n - i) = 0 \tag{6.45}$$

so

$$\sum_{k=1}^{p} a_k \sum_n s(n - i)s(n - k) = \sum_n x(n)s(n - i) \tag{6.46}$$

for $i = 1, 2, \ldots, p$. Similarly to our earlier elaborations in this chapter, two different approaches are possible depending on the limits of the summations in Equation 6.46. If we consider $s(n)$ and $u(n)$ to be of infinite duration and minimize the energy of the error signal $e(n)$ from $n = 0$ to $n = L - 1$, where L is the analysis frame length, the summations in Equation 6.46 are from $n = 0$ to $L - 1$, and we have a covariance like approach [69]. Alternatively, we can consider $s(n)$ and $u(n)$ to be nonzero only for $0 \leq n \leq L - 1$, which leads to an autocorrelation-like approach [69] where the simultaneous equations to be solved become

$$\sum_{k=1}^{p} a_k \sum_{n=0}^{L-1-|k-i|} s(n)s(n + |k - i|) = \sum_{n=0}^{L-1-i} s(n)x(n + i). \tag{6.47}$$

We investigated these two approaches, both with and without windowing of $s(n)$ and $u(n)$, in our 4.7 kbits/s codec. We found that the updated filter coefficients were, in terms of the SNR of the reconstructed speech, usually worse than the original coefficients. This is because of the inaccuracy of the approximation in Equation 6.43. To obtain any improvement in the segmental SNR of the reconstructed speech, it was necessary in each frame to find the output of the synthesis filter with the original and updated filter coefficients, and transmit the set of coefficients which gave the best SNR for that frame. Using this technique, we found that the updated filter coefficients were better than the original coefficients in only about 15% of frames, and the segmental SNR of the codec was improved by about 0.25 dB.

These results were rather disappointing. Hence, we attempted to find an improved method of updating the synthesis filter parameters. One possibility comes to light if we write Equation 6.42 in a matrix notation

$$E = |\underline{x} - \hat{\underline{\underline{S}}} \, \underline{a}|^2 \tag{6.48}$$

where

$$\underline{x} = \begin{pmatrix} s(0) - u(0) \\ s(1) - u(1) \\ \vdots \\ s(L-1) - u(L-1) \end{pmatrix} \tag{6.49}$$

$$\underline{\hat{\underline{S}}} = \begin{pmatrix} \hat{s}(-1) & \hat{s}(-2) & \cdots & \hat{s}(-p) \\ \hat{s}(0) & \hat{s}(-1) & \cdots & \hat{s}(-p+1) \\ \vdots & \vdots & \ddots & \vdots \\ \hat{s}(L-2) & \hat{s}(L-3) & \cdots & \hat{s}(L-1-p) \end{pmatrix} \tag{6.50}$$

and

$$\underline{a} = \begin{pmatrix} a_1 \\ a_2 \\ \vdots \\ a_p \end{pmatrix}. \tag{6.51}$$

Note that here we have set the elements of $\underline{x}$ and $\hat{\underline{S}}$, assuming that we are using the covariance-like approach, but similar equations can be written for the autocorrelation approach. We have to attempt to find a set of coefficients $\underline{a}$ such that

$$\hat{\underline{S}}\,\underline{a} \approx \underline{x} \tag{6.52}$$

Similar problems occur in many areas of science and engineering and are solved using least squares (LS) methods [153]. The usual technique is to assume that the "data" matrix $\hat{\underline{S}}$ is known perfectly and that the "observation" vector $\underline{x}$ is known only approximately. Then a set of coefficients $\underline{a}$ are found such that

$$\hat{\underline{S}}\,\underline{a} = \underline{x} + \underline{\Delta x} \tag{6.53}$$

and $|\underline{\Delta x}|^2$ is minimized. One method of solving the LS problem is to use what are called the normal equations

$$\hat{\underline{S}}^T \hat{\underline{S}}\,\underline{a} = \hat{\underline{S}}^T \underline{x}. \tag{6.54}$$

These equations are equivalent to those in Equation 6.46. However, in our problem it is the data matrix $\hat{\underline{S}}$, which is known only approximately, and the observation vector $\underline{x}$, which is known exactly. Therefore, the usual least squares technique will not be ideal for our purposes.

In recent years, a relatively new technique called total least squares (TLS) [154] has been applied to several problems; see, for instance [155]. In this method, errors are assumed to exist in both $\hat{\underline{S}}$ and $\underline{x}$, and we find a set of coefficients $\underline{a}$ such that

$$(\hat{\underline{S}} + \underline{\Delta\hat{S}})\underline{a} = \underline{x} + \underline{\Delta x} \tag{6.55}$$

where $|(\underline{\Delta\hat{S}}||\underline{\Delta x})|_F^2$ is minimized. Here $|\,.\,|_F^2$ denotes the squared Frobenius norm of a matrix, namely, the sum of the squares of the matrix's elements, and $(\underline{\Delta\hat{S}}||\underline{\Delta x})$ is a matrix constructed by adding $\underline{\Delta x}$ to $\hat{\underline{S}}$ as the $p+1$th column of the new matrix.

The solution $\underline{a}$ of the TLS problem can be found using the singular value decomposition of $(\hat{\underline{S}}||\underline{x})$ [154]. We invoked this technique but found that it was not useful, since a high fraction (about 95%) of the sets of filter coefficients it delivered resulted in unstable synthesis filters.

One final least squares method we investigated was the data least squares (DLS) technique [156]. Here all the errors are assumed to lie in the data matrix $\hat{\underline{S}}$, and a set of coefficients are found such that

$$(\hat{\underline{S}} + \underline{\Delta\hat{S}}) \, \underline{a} = \underline{x} \tag{6.56}$$

This is much closer to what we want in our situation, and again the solution can be found using singular value decomposition. However, we found that the filter coefficients produced were very similar to those given by the TLS technique, with again about 95% of the updated synthesis filters being unstable. Unfortunately, then, neither the TLS nor the DLS update is a practical solution for our problem.

6.5.3.3 Optimization via Powell's Method.

Given our input speech signal $s(n)$, the filter's excitation $u(n)$, and the reconstructed speech memory $\hat{s}(-p), \hat{s}(-p+1), \ldots, \hat{s}(-1)$, the error energy E is a function of the p filter coefficients. Thus, we can consider E as a p-dimensional function which we wish to minimize. There are many different methods [153] for the minimisation of multidimensional functions, and we attempted using the direction set, or Powell's, method [153]. This method operates by iteratively carrying out a series of one-dimensional line minimizations, and attempting to find a series of "conjugate" directions for these minimizations, so that the minimum along one direction is not spoiled by subsequent movement along the others. At each iteration a line minimization is carried out along each of p directions, and then the p directions are updated in an effort to obtain the ideal conjugate directions. See [153] for details. The process ends when the decrease in E during a particular iteration is less than some given fractional tolerance. When this happens, it is assumed that we have settled into a minimum, which we hope is the global minimum of E. In our simulations, the line minimizations were carried out using Brent's method [153]. This does a series of evaluations of E for various sets of filter coefficients and hunts down the minimum along a particular direction using either a golden section search or parabolic interpolation.

We invoked Powell's optimization for various values of the fractional tolerance which controls when the process of iterations should end. A good indicator of the complexity of minimization procedures, such as Powell's method, is the number of times the function E to be minimized is evaluated. Every 100 evaluations are approximately as complex as the whole encoding process in our standard ACELP codec. Figure 6.7 shows how the segmental SNR of our 4.7 kbits/s codec with a Powell optimization of the synthesis filter varies with the number of evaluations of E carried out. The best SNR we were able to obtain was 11.85 dB, which was about 2 dB better than the segmental SNR of the codec without interpolation of the LSFs. However, as shown in Table 6.9, this difference is much reduced if we use bandwidth expansion and interpolation of the LSFs in the codec; these methods are much less complex than the Powell's update. The Powell optimization is not a realistic option for a real codec, but it does give us an idea of the AbSolute best performance we can expect from updating the synthesis filter parameters. We see that without LSF quantization, this is only about half a decibel better than a codec with LSF interpolation and bandwidth expansion.

6.5.3.4 Simulated Annealing and the Effects of Quantization.

In any real coder, it is necessary to quantize the synthesis filter parameters for transmission to the decoder. It is not clear whether this need for quantization will make updating the LPC parameters more or less

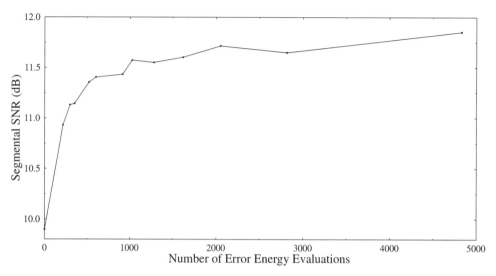

Figure 6.7 Powell optimization performance.

worthwhile. On one hand, the quantization may mask and reduce the improvement due to the update, but on the other hand the updating algorithm can take account of the quantization when it is choosing a set of filter parameters. This may lead to the update having more effect.

We decided to start our investigation of the effects of updating the synthesis filter parameters with quantization of the LSFs by finding an upper limit to the improvement possible. The Powell optimization method was designed to operate on functions of continuous variables, and so it is not suitable when we consider quantization of the LSFs. Instead, we used the technique of simulated annealing [153], which is more suitable for discrete optimization.

Simulated annealing operates—as the termonology suggests—in analogy to the annealing (or slow cooling) of metals. When metals cool slowly from their liquid state, they start in a very disordered and high-energy state and reach equilibrium in an extremely ordered crystalline state. This crystal is the minimum energy state for the system, and simulated annealing similarly allows us to find the global minimum of a complex function with many local minima. The procedure is as follows. The system starts in an initial state, which in our situation is an initial set of quantized LSFs. A temperature-like variable T is defined, and possible changes to the state of the system are randomly generated. For each possible change, the difference ΔE in the error energy between the present state and the possible new state is evaluated. If this is negative, in other words the new state has a lower energy than the old state, then the system always moves to the new state. If, on the other hand,

TABLE 6.9 Performance of Various Synthesis Filter Determination Techniques

	Segmental SNR
Codec with no interpolation or BW expansion	9.90
Codec with least squares optimization	10.13
Codec with LSF interpolation only	10.49
Codec with bandwidth expansion only	10.59
Codec with interpolation and BW expansion	11.29
Codec with Powell optimization	11.85

ΔE is positive, then the new state has higher energy than the old state, but the system may still change to this new state. The probability of this happening is given by the Boltzmann distribution

$$\text{prob} = \exp\left(\frac{-\Delta E}{kT}\right) \qquad (6.57)$$

where k is a constant. The initial temperature is set so that kT is much larger than any ΔE that is likely to be encountered, so that initially most offered moves will be taken. As the optimization proceeds, the "temperature" T is slowly decreased, and the number of moves to states with higher energy reduces. Eventually, kT becomes so small that no moves with positive ΔE are taken, and the system comes to equilibrium in what is hopefully the global minimum of its energy.

The advantage of simulated annealing over other optimization methods is that it should not be deceived by local minima and should slowly make its way toward the global minimum of the function to be minimized. In order to guarantee that this happens, the temperature T must start at a high enough value and is reduced suitably slowly. We followed the suggestions in [153] and reduced T by 10% after every $100p$ offered moves, or every $10p$ accepted moves. The initial temperature was set so that kT was equal to 10 times the highest value of ΔE that was initially encountered. The random changes in the state of the system were generated by randomly choosing an LSF and then moving it up or down by one quantization level, provided that this did not lead to an LSF overlap, as it is necessary to avoid unstable synthesis filters.

We were able to improve the segmental SNR of our 4.7 kbits/s codec with quantization of the LSFs from 9.86 dB to 10.92 dB. Note, furthermore, that we were able to achieve almost the same improvement with a much simpler search technique described below. Rather than choose an LSF at random to modify and accept some changes that increase the error energy as well as all those that reduce the energy, we cycled sequentially through all p LSFs in turn. Each LSF was moved up and down one quantizer level to see if we could reduce the error energy. Any changes that reduced the error energy, but none that increased it, were accepted. This process can be repeated any number of times, with every testing of all p LSFs counting as one iteration. The segmental SNR of our codec against the number of update iterations used is shown in Figure 6.8.

We see that this method of updating the quantized synthesis filter parameters produces a segmental SNR of 10.8 dB after just three iterations. This is almost equal to the improvement produced by simulated annealing of the LSFs, and yet the complexity of the codec is increased by only about 80%. The improvement obtained (about 1 dB) is similar to that quoted in [152] of 10% in multi-pulse codecs at segmental SNRs of around 10 dB. However, the method used in [152] required recalculating the excitation after the update of the synthesis filter parameters, and so approximately doubled the complexity of the codec.

As mentioned in [152], the updating of the synthesis filter not only helps to increase the average segmental SNR, but also to remove the very low minima in SNR that occur for some frames. This effect is shown in Figure 6.9 which shows the variation of SNR for a sequence of 50 frames for 4.7 kbits/s codecs with and without update of the synthesis filter. The update used three iterations of the scheme described above. These low minima that occur can be subjectively annoying, and so it is helpful if they can be partially removed

It is also possible to update the synthesis filter in an attempt to increase the weighted SNR for each frame. We attempted this using the iterative scheme described above, and we found that the improvement in the weighted segmental SNR due to the update saturated after just one iteration. The weighted segmental SNR increased from 7.18 dB to 7.43 dB, and the conventional segmental SNR increased from 9.86 dB to 10.08 dB.

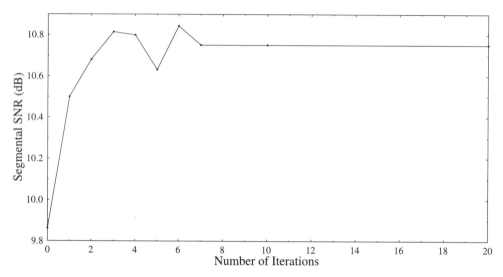

Figure 6.8 Performance of quantized LSF update scheme.

The results described above comparing codecs with updated synthesis filter parameters to a codec with no update are reasonably good. However, as noted earlier for the codecs with no quantization of the LSFs, the results are not so impressive when compared to codecs using the techniques of bandwidth expansion and interpolation of the LSFs. This is shown in Table 6.10. Using both bandwidth expansion and interpolation of the LSFs gives a segmental SNR

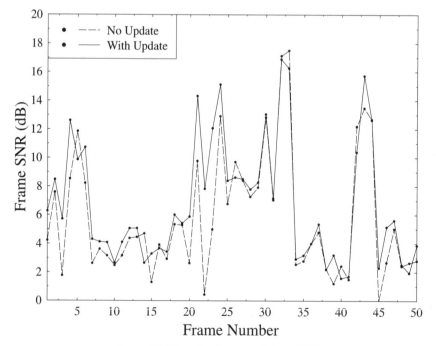

Figure 6.9 Effect of update on variation of SNR.

TABLE 6.10 Performance of Synthesis Filter Techniques with Quantization

	Segmental SNR
Codec with no interpolation or BW expansion	9.86
Codec with bandwidth expansion only	9.89
Codec with LSF interpolation only	10.31
Codec with interpolation and BW expansion	10.76
Codec with iterative update	10.75
Codec with simulated annealing update	10.92

almost identical to that achieved using the iterative update algorithm. Also, the interpolation and bandwidth expansion help remove the very low minima in the SNR in the same way that the update does. Although several papers [152, 157, 158] have appeared reporting reasonable improvements using various methods of update, to our knowledge none of them has considered the effects of LSF interpolation and bandwidth expansion. Our codec with the iterative update of the LSFs is about 10% more complex than the codec with interpolation and bandwidth expansion. However, the LSF interpolation scheme employed increases the delay of the codec by two subframes, or 15 ms. Both interpolation (when used along with bandwidth expansion) and the iterative update scheme give very similar improvements in the performance of the codec. If a 15 ms increase in the delay of the codec is not important, then the LSF interpolation can be invoked. However, our iterative update scheme provides an alternative that gives similar results without increasing the delay of the codec, and it is only slightly more complex.

The research reported here was summarized in [159]. In the next chapter we move on to investigating the error sensitivity of our 4.7 kbits/s ACELP codec.

6.6 THE ERROR SENSITIVITY OF CELP CODECS

6.6.1 Introduction

As we have argued before, CELP codecs can produce good toll quality speech at low bitrates with reasonable complexity. However, almost equally important for a codec which is to be used over a radio channel is its ability to cope with random bit errors between the encoder and decoder. A mobile radio channel is particularly hostile [69] and when there is no line-of-sight path between the receiver and transmitter multipath propagation leads to a channel that can be described by the Rayleigh distribution. Such a channel is not memory-less, and deep fades of −20 dB, or more, are common. Such fades lead to error bursts. Therefore it is necessary to use either interleaving, which attempts to randomize the bit errors, or a channel coder with good burst error correcting abilities. In any case. a channel coder is essential for any speech coder that is to be used over a mobile radio channel at reasonable channel signal-to-noise ratios. However, no channel coder will be able to remove all the bit errors without requiring an unreasonable bandwidth, and so even with channel coding it is important that the speech codec should be as robust as possible to errors.

In this chapter, we describe several methods for improving the bit error sensitivity of our coder. We also show how to measure the error sensitivity of the speech encoder output bits so that the matching channel coder can be carefully designed to give most protection to the bits that are most sensitive. The results of simulations which are reported refer to our 4.7 kbits/s codec. Similar results were found to apply to the 7.1 kbits/s codec.

6.6.2 Improving the Spectral Information Error Sensitivity

It has been noted [160, 161] that the spectral parameters in CELP coders are particularly sensitive to errors. There are many different ways to represent these parameters, but line spectral frequencies (LSFs) [93] offer some definite advantages in terms of error robustness. One advantage is that the spectral sensitivities of the LSFs are localized [92], so that an error in a given LSF produces a change in the resulting spectrum only in the neighborhood of the corrupted LSF. Another advantage is the ordering property of the LSFs. This means that for the synthesis filter to be stable, it is a necessary and sufficient condition that the LSFs from which it was derived are ordered, satisfying the condition $LSF_1 < LSF_2 < LSF_3$, and so on. Therefore, if a set of LSFs are received which are not ordered, the decoder infers that at least one error in the bits must represent these LSFs, and some action must be taken to rectify this error and produce a stable synthesis filter. It is this action which is studied here.

6.6.2.1 LSF Ordering Policies.   There is a high correlation between the LSFs of successive frames. Thus, as reported in [161], occasionally the LSF set for a given frame can be replaced by the set from the previous frame without introducing too much audible distortion. One possible policy for dealing with frames where nonmonotonic LSFs are received is to completely discard the LSFs that were received for that frame and to use those from the previous frame.

A better policy is to attempt replacing those LSFs that have to be replaced, rather than all of them. In [162] when a nonmonotonic set of LSFs is received, the two particular frequencies that cross over are replaced by the corresponding frequencies from the previous frame. Only if the resulting set of LSFs is still not ordered is the whole set replaced.

Several attempts have been made to identify the particular LSF that causes the instability and then to replace only it. In [163] use is made of the long-term statistics of the differences between adjacent LSFs in the same frame. If two frequencies cross over, then an attempt is made to guess which one was corrupted; in general, the guess is correct about 80% of the time. This "hit ratio" can be improved by including a voicing decision. In a frame of voiced speech, the formants are sharper than in unvoiced frames, and so the spacings between adjacent LSFs are generally smaller.

Instead of attempting to guess which LSF from a nonmonotonic set is corrupted, and then replacing this LSF with the corresponding frequency from a previous frame, we attempted to produce a monotonic set of LSFs by inverting various bits in the received bitstream. Initially, we endeavored to determine which set of bits should be examined. For example, if $LSF_i > LSF_{i+1}$, then we know that either LSF_i or LSF_{i+1} has been corrupted. When such a crossover is found, we take the following steps:

1. We check to see if $LSF_i > LSF_{i+2}$. If it is, we assume that LSF_i is corrupted, and we select the bits representing this LSF as those to be examined.

2. We check to see if $LSF_{i-1} > LSF_{i+1}$. If it is, we assume LSF_{i+1} is in error, and we select these bits to be examined.

3. If neither of the checks above indicates whether it is LSF_i or LSF_{i+1} that is corrupted, then the bits representing both these LSFs are selected to be examined.

4. We attempt to correct the LSF crossover by inverting each bit, one at a time, from those to be examined. After each bit inversion, the new value of LSF_i or LSF_{i+1} is

decoded and checked to see if the crossover has been removed, and no new crossovers are introduced. If several possible codes are found, then the one that gives the corrected LSFs as close as possible to their values in the previous frame is chosen.

5. If, as occasionally happens at high bit error rates, no single-bit inversion can be found which corrects the LSF crossover, and introduces no new crossover, then we adopt the policy that is recommended in [46]. First LSF_i, then LSF_{i+1}, then both, and finally the entire LSF set, is replaced by those in the previous frame until a monotonic set is found.

We simulated the effect of the error correction scheme described above over a set of four sentences spoken by different speakers. The predictor coefficients were determined in a 4.7 kbits/s coder using the autocorrelation approach, and a 15 Hz bandwidth expansion was used. The LSFs were nonuniformly quantized with 34 bits. The Cepstral Distance (CD) [61] degradation produced by errors in the bits representing the LSFs is shown in Figure 6.10. The dotted curve represents the effect of the scheme described in [162]. As can be seen, our correction policy gives consistently better results, and a definite subjective improvement was heard in informal listening tests.

Also in [163] a table of "hit ratio" figures is included to indicate how often the correct LSF for replacement was chosen at various bit error rates. The figures for the improved hit ratio which resulted when the voicing decision was used are reproduced in Table 6.11. Also shown in this table is the hit ratio for our scheme, quantifying as to how often the bit that was inverted was part of the codeword for the LSF which had actually been corrupted. As can be

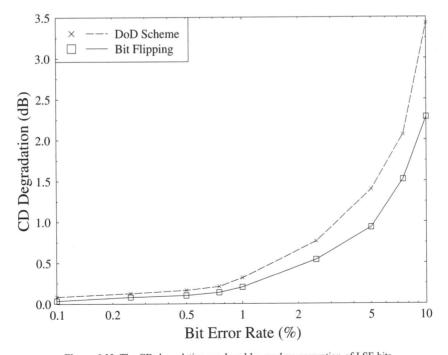

Figure 6.10 The CD degradation produced by random corruption of LSF bits.

TABLE 6.11 Hit Ratios for Various Algorithms

Bit error rate (%)	0.1	1	2	2.5	3	4
Atungsiri's Scheme	100	80	80	82	79	80
Our Scheme	100	88	92	93	93	92
Correct Bit Hit	83	81	80	77	78	78

seen, our scheme performs significantly better than that reported in [163]. In the final row of the table are the figures for how often the correct bit is inverted when a nonmonotonic set of LSFs is received. As can be seen, the bit causing the LSF overlap is corrected about 80% of the time, and when this happens the effect of the bit error is completely removed. As about 30% of corrupted LSF bits produce LSF crossovers, this means that about 25% of all LSF errors can be entirely removed by the decoder.

6.6.2.2 The Effect of FEC on the Spectral Parameters.   Although our scheme can remove the effect of channel errors on the LSF bits about 25% of the time, the reconstructed speech is unacceptably distorted if the bit error rate among the LSF bits is above about 1%. Therefore, some sort of error correction code is necessary if the coder is to be used at higher bit error rates. We found which of the LSF bits were most susceptible to errors by taking one LSF bit at a time and corrupting it 10% of the time. The resulting degradations in the segmental SNR and the Cepstral Distance of the reconstructed speech were noted. The 13 bits that were least sensitive in terms of CD degradation all gave a degradation of less than 0.05 dB when corrupted 10% of the time, and were left unprotected. The remaining 21 bits were protected with a (31,21,2) BCH code which was simulated as follows. If two, or fewer errors were generated in the 31-bit codeword, then they were corrected. If more than two errors were generated, then we assumed that although the BCH code would be unable to correct these errors, it would at least be able to detect that the protected 21 bits might contain errors. Then in the decoding of the speech, if an LSF crossover was found, the decoder attempts to put it right by examining only unprotected bits, unless the BCH code indicates that the 21 protected bits may contain an error.

Thus, the effect of including FEC on some of the LSF bits is not only that the most sensitive bits are completely protected (unless the code fails), but also when an LSF crossover occurs because of an error in one of the less sensitive bits, the bit flipping algorithm is much more likely to select the correct bit to toggle. In fact, we found that for frames where the FEC had not failed, if an LSF crossover occurred it was correctly fixed almost 100% of the time. In informal listening tests, we found that for a bit error rate of 2.5% among the LSF bits the distortions produced were barely noticeable. At 5%, although the distortions were noticeable, the reproduced speech was still of acceptable quality.

Recently, an alternative means of improving the performance of speech and channel codecs, based on similar ideas, was proposed [164]. It uses the ordering property of the LSFs, along with a specific property of multiband excited codecs, to feed back information from the speech decoder to the channel decoder. The speech decoder indicates to the channel decoder if a set of received bits results in an LSF crossover, or is otherwise unlikely to be correct. The channel decoder can then use this information to help it decode the correct information from the received bit stream. Good results, in terms of the error correcting capability of the source-aided channel decoder, are reported.

6.6.2.3 The Effect of Interpolation. In our codec, the usual practice of employing interpolation between the present and the previous set of LSFs is used. This helps minimize sudden sharp changes in the short-term predictor filter coefficients between one frame and the next. However, as can be seen from Figure 6.11, it also leads to increased propagation of the effect of an LSF error from one frame to the next. The upper graph shows the average effect, in terms of degradation of the frame SNR and CD, of an error in one of the LSF bits in the coder with LSF interpolation.

The bit is corrupted in frame 0, and the graph shows how the resultant degradation dies out from one frame to the next. In frame 1 the corrupted set of LSFs is used along with the present set to form the interpolated LSFs. Hence, the effect of the error is almost as serious in the frame following the error as it is in the corrupted frame. After this, the effect of the error quickly disappears.

Because of this error propagation, it might be expected that the error sensitivity of the bits representing the LSFs could be improved by removing the interpolation. However, we found that removing interpolation from the codec reduced its clear channel segmental SNR by about 0.5 dB. At various error rates between 0.1% and 10% the resultant degradations are almost identical to those found in the coder with interpolation. The lower graph in Figure 6.11

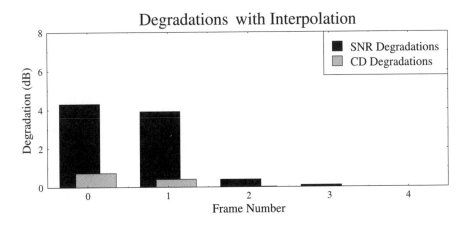

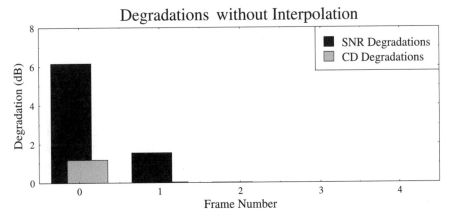

Figure 6.11 The effect of interpolation on error propagation.

shows the effect of an error (on the same LSF bit as was used in the upper graph) in the coder in which interpolation is not used. Although the error propagation is reduced, the degradation in the frame that was corrupted is increased. This is because interpolation helps to smooth out the effect of an LSF error in the corrupted frame.

6.6.3 Improving the Error Sensitivity of the Excitation Parameters

Most of the bits transmitted by a CELP coder are used to represent the excitation for the synthesis filter. In our coder the information that must be sent to the decoder is

1. The fixed codebook index. Twelve bits per subframe are used.
2. The fixed codebook gain. Four bits are used to represent the magnitude, which is logarithmically quantized, and one bit is used to represent the sign.
3. The adaptive codebook delay. The delay can vary between 20 and 147 samples, and so 7 bits per subframe are needed to represent this information.
4. The adaptive codebook gain. Three bits per subframe are used.

6.6.3.1 The Fixed Codebook Index.
The algebraic codebook structure used in our codec is inherently quite robust to channel errors. This is because if one of the codebook index bits is corrupted, the codebook entry selected at the decoder will differ from that used in the encoder only in the position of one of the four nonzero pulses. Hence, the corrupted codebook entry will be similar to the original. This is in contrast to traditional CELP coders which use a nonstructured, randomly filled, codebook. In such codecs when a bit of the index is corrupted, a new codebook address is decoded and the codebook entry used is entirely different to the original. Errors in the codebook index in such coders will therefore be more significant than in ours. Such a codebook is used in [161] where SNR degradations of about 8 dB are recorded when a codebook index bit is corrupted in every frame. In our coder the corresponding degradation is only about 4 dB.

It is generally reported [46, 162] that errors in the fixed codebook index produce reconstructed speech in which the degradations are not perceptually annoying. Therefore, the fixed codebook index is often left unprotected.

6.6.3.2 The Fixed Codebook Gain.
The magnitude of the fixed codebook gain tends to vary quite smoothly from one subframe to the next. Therefore, errors in the codebook gain can be spotted using a smoother to indicate, from the neighbouring gains, what range of values the present codebook gain should lie within. If a codebook gain is found which is not in this range, then it is assumed to be corrupted, and it is replaced with some other gain.

We want a scheme that will spot as many errors in the codebook gain as possible, without introducing too many new errors by replacing gains that were not originally corrupted by the channel. After careful investigation of the effects of bit errors on the fixed codebook gain magnitude, we implemented the following scheme. Every codebook gain quantizer level at the decoder is checked by calculating the mean and standard deviation of its two nearest neighbors. If the standard deviation of these neighbors is less than two quantizer levels, then it is set equal to two. We then check to see if the present level is within 2.25 standard deviations of the mean calculated from its neighbors. If not, it is assumed to be corrupted. When the codebook gain bits are corrupted with an error rate of 2.5%, then this scheme spots almost

90% of the errors in the most significant bit (MSB) of the gain level, while in error-free conditions it falsely spots errors in only about 0.5% of the subframes. This false error spotting produces a small degradation in the decoder performance at a zero bit error rate. However, if some feedback between the channel decoder and the speech decoder is implemented so that the smoother is disabled in error-free conditions, as suggested in [161], then this degradation is removed.

Another important aspect of the smoother is how gains that are thought to be corrupted are replaced. In [161] when a gain magnitude is thought to be in error, it is replaced with the mean of its neighbors' magnitudes. However, we found that a bit flipping scheme, similar to that used to correct LSF crossovers, produced better results. When an error is spotted, the decoder inverts all 4 bits, one at a time, in the received codeword for the gain magnitude. The single-bit inversion that produces a decoded gain level as close as possible to the mean of its neighbors is chosen.

The effect of our smoother on the error sensitivity of the 4 bits per subframe representing the fixed codebook gain magnitude is shown in Table 6.12. This table shows the SNR degradation produced in 4.7 kbits/s codecs with and without smoothing when the bits shown are corrupted in every frame. (The bits are corrupted for one subframe only per frame.) As can be seen, the smoothing improves the error sensitivity of all the bits, especially the MSB in which most of the errors are spotted and corrected by the smoother.

The fixed codebook gain sign shows erratic behavior and is not suitable for smoothing. This bit is among the most sensitive of the coders and should be well protected by the channel codec.

6.6.3.3 Adaptive Codebook Delay.

Seven bits per subframe are used to encode the adaptive codebook delay, and most of these are extremely sensitive to channel errors. An error in one of these bits produces a large degradation not only in the frame in which the error occurred, but also in subsequent frames. Generally, it takes more than 10 frames before the effect of the error dies out.

If the adaptive codebook delay is chosen by the encoder by merely minimizing the weighted mean square error of the reconstructed speech, its behavior will be erratic and not suitable for smoothing. The delay can be forced to take on smooth behavior by modifying the encoder to choose slightly suboptimal delays. This then allows the decoder to use smoothing to minimize the effect of errors. However, there is a noticeable clear channel degradation due to the suboptimal delays chosen by the encoder.

Another approach [161, 165] is to use simulated annealing to assign codewords to delays so that common codewords have good neighbors. This means that when a common codeword is corrupted, the new delay selected is such that the resultant degradation is

TABLE 6.12 SNR Degradations for Fixed Codebook Gain Bits with and without Smoothing

Gain Bit	SNR Deg (dB) without Smoothing	SNR Deg (dB) with Smoothing
LSB	1.4	1.3
Bit 2	3.0	2.8
Bit 3	6.2	4.8
MSB	10.5	2.1

minimized. This approach, along with smoothing, is used in the DoD 4.8 kbits/s standard [76], but as it has been studied extensively already, we have not attempted this.

6.6.3.4 Adaptive Codebook Gain.   The pitch gain is much less smooth than the fixed codebook gain, and it is not suitable for smoothing. However, its error sensitivity can be slightly increased by coding the quantizer level with a Gray code rather than the Natural Binary Code (NBC). The effect off this is shown in Table 6.13, which gives the SNR degradation for the two codes caused by bit errors (at a rate of 10%) in the 3 bits used to represent the gain in one subframe.

6.6.4 Matching Channel Codecs to the Speech Codec

Some bits are much more sensitive to channel errors than others and so should be more heavily protected by the channel coder. However, it is not obvious how the sensitivity of different bits should be measured. One commonly used approach [161] is for a given bit to invert this bit in every frame and measure the segmental SNR degradation that results. The error sensitivity of various bits for our coder measured in this way is shown in Figure 6.12. What information various bits represent is given in Table 6.14. Another similar approach [160] is to measure the degradations in both the SNR and the Cepstral Distance, which result from systematic inversion of a given bit in every frame, and combine these measures to give an overall sensitivity measure.

These approaches do not take adequate account of the different error propagation properties of different bits. This means that if instead of corrupting a bit in every frame it is corrupted randomly with some error probability, then the relative sensitivity of different bits will change. We propose a new measure of error sensitivity. For each bit a graph similar to that in Figure 6.11 is found. In other words, we find the average SNR degradation for a single-bit error in the frame in which the error occurs and in the following frames. The total SNR degradation is then found by adding together the degradations in frames 0,1,2 and so on. This total degradation is equivalent to the average SNR degradation that will be produced by a single error in a given bit. Of course, the effect of a single error on the segmental SNR will be

TABLE 6.13   The Effect of Using a Gray Code for the LTP Gain

Gain Bit	SNR Deg (dB) NBC	SNR Deg (dB) Gray Code
Bit 1	1.9	1.9
Bit 2	3.0	1.7
Bit 3	5.3	4.8

TABLE 6.14   Bit Numbering

Bit Numbers	Represents
1 to 34	LSFs
35 to 41	Adaptive codebook gain (Subframe 1)
42 to 44	Adaptive codebook gain (Subframe 1)
45 to 56	Fixed codebook index (Subframe 1)
57	Fixed codebook gain sign (Subframe 1)
58 to 61	Fixed codebook gain (Subframe 1)

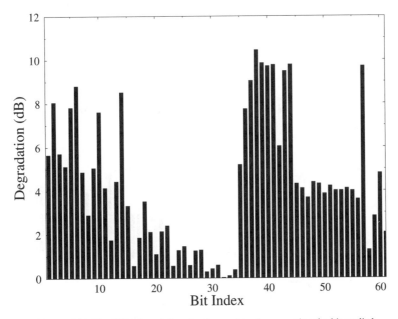

Figure 6.12 The SNR degradation due to consistently corrupting the bit studied.

averaged out over all the frames of the speech file, so that, for example, if a bit with a total SNR degradation of 10 dB is corrupted once in a speech file of 100 frames, then the overall degradation in the segmental SNR will on average be 0.1 dB. The exact degradation depends very much on which frame the bit is corrupted in—Corrupting a given bit in one frame of a speech file can produce a much larger degradation in the segmental SNR for that file than corrupting the same bit in a different frame. This is shown in Figure 6.13 which gives the degradation in the segmental SNR produced by a single-bit corruption, versus the frame in which the corruption takes place, for various different bits.

Figure 6.14 shows, for various bits, the average effect of a bit error in the frame in which the error occurred and in the following frames. The different error propagation properties of different bits can be clearly seen. For example, an error in a bit representing an LSF has a significant effect only in the frame in which the error occurred and in the next two frames. Conversely, an error in a bit representing the LTP delay gives a large degradation in the frame SNR, and this degradation is still significant 10 frames later. Figure 6.15 shows the total SNR degradation for single-bit errors of the various bits. This graph is significantly different from that in Figure 6.12; in particular, the importance of the adaptive codebook delay bits is much clearer because of their memory propagation properties.

Our error sensitivity figure is based on the total SNR degradation and on a similar measure for the total CD degradation. The two sets of degradation figures are combined and are given equal weight by scaling each total SNR degradation by the maximum such degradation, and similarly for the total CD figures. The two sets of scaled degradation figures are then added together to give an overall sensitivity figure between 0 and 2. The higher this figure is, the more sensitive the bit is deemed to be.

Our new scheme was tested as follows. The 12 most sensitive bits were determined using our scheme and the one reported in [160]. These two sets of 12 bits contained 4 in common, which were removed to give two sets of eight bits. The two different sets were

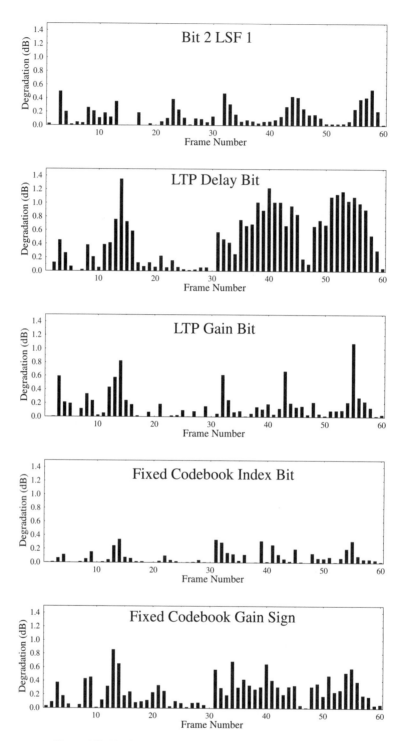

Figure 6.13 The degradation caused by bit errors in different frames.

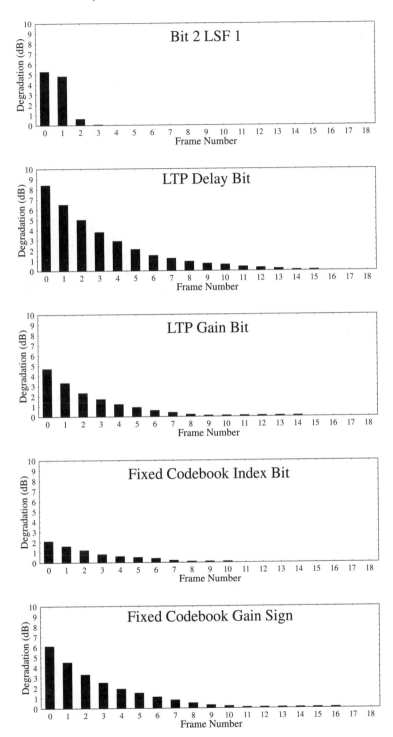

Figure 6.14 The SNR degradation propagation for various bits.

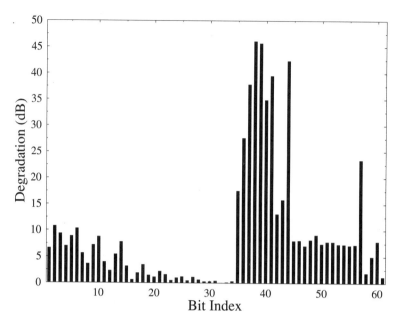

Figure 6.15 Total SNR degradation due to single errors in various bits.

corrupted at a 5% bit error rate for various different speech files. In all cases, we found that both objectively (CD and SNR degradations), and in informal listening tests, the bits our scheme predicted would be most sensitive were much more sensitive than those predicted using the approach in [160].

6.6.5 Error-Resilience Conclusions

In this section we have discussed the error sensitivity of the forward adaptive ACELP codec described earlier in this chapter. We investigated various ways of improving the error sensitivity of the codec and how the sensitivity of different bits could be compared in order to correctly match a channel coder to the speech coder. We have also shown how the degradations produced by errors propagate from one frame to another, and may persist for more than 10 frames, and how the sensitivity of a given bit can vary significantly from frame to frame.

The error sensitivity improvement and evaluation techniques we have described in this chapter were used to match our 4.7 kbits/s speech codec with a set of BCH error correcting codes. The speech and error correction codecs were used in conjunction with 16-level QAM and a Packet Reservation Multiple-Access (PRMA) scheme to simulate a complete multiple-user mobile communication system [145]. Similar studies were also carried out for a 6.5 kbits/s codec, which was similar to our 7.1 kbits/s codec described in Section 6.4.3, except it used six 5 ms subframes to make up a 30 ms frame instead of using four subframes per 20 ms frame. This extension of the frame length of the higher rate codec to be equal to the frame length of the low-rate codec was carried out for reasons of ease of implementation of the PRMA scheme [145, 166], as will become clear in the next section.

6.7 APPLICATION EXAMPLE: A DUAL-MODE 3.1 KBD SPEECH TRANSCEIVER

6.7.1 The Transceiver Scheme

The schematic diagram of the proposed reconfigurable transceiver is portrayed in Figure 6.16. A voice activity detector (VAD) similar to that of the Pan-European GSM system [74] enables or disables the ACELP encoder [167] and queues the active speech frames in the Packet Reservation Multiple-Access (PRMA) [168] slot allocator (SLOT ALLOC) for transmission to the base station (BS). The 4.7 or 6.5 kbits/s (kbps) ACELP coded active speech frames are mapped according to their error sensitivities to n number of protection classes by the Bit Mapper (BIT MAP), as shown in the figure and source sensitivity-matched binary Bose-Chaudhuri-Hocquenghem (BCH) encoded [134] by the BCHE1...BCHEn encoders. The "Map & PRMA Slot Allocator" block converts the binary bit stream to 4- or 6-bit symbols, injects pilot symbols [135] and ramp symbols, and allows the packets to contend for a PRMA slot reservation. After BCH encoding the 4.7 and 6.5 kbps speech bits, they are mapped to 4- or 6-bit symbols, which are modulating a reconfigurable 16- or 64-level Quadrature Amplitude Modulation (QAM) scheme.

We have arranged for the 4.7 kbps/16-QAM and 6.5/64-QAM schemes to have the same signaling rate and bandwidth requirement. Therefore, this transmission scheme can provide higher speech quality, if high channel signal-to-noise ratios (SNR) and signal-to-interference ratios (SIR) prevail. It can be reconfigured under network control to deliver lower but unimpaired speech quality among lower SNR and SIR conditions. Indoors picocellular cordless systems have typically friendly, high-SNR, and high-SIR nondispersive propagation channels, and the partitioning walls also contribute toward attenuating co-channel inter-

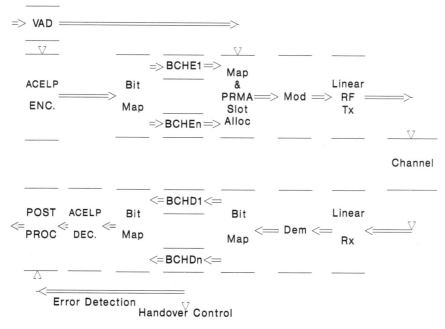

Figure 6.16 Transceiver schematic.

ferences. Furthermore, the PRMA time slots can be classified according to the prevailing interference levels evaluated during idle slots. If sufficiently high SIRs prevail, the higher speech quality mode can be invoked. Otherwise the more robust, lower speech quality mode of operation must be used.

The modulated signal is then transmitted using the linear radio frequency (RF) transmitter (Tx) over the friendly indoors channel, received by the linear receiver (Rx), and demodulated (DEM), and the received speech bits are mapped back to their original bit protection classes by the bit mapper. The n-class BCH decoder BCHD1...BCHDn carries out error correction, before ACELP decoding and post-processing can take place. Observe that the error detection capability of the strongest BCH decoder, which is more reliable than that of its weaker counterparts, can be used to help control handovers to a less interfered PRMA time slot on any other available carrier or to activate speech post-processing in order to conceal the subjective effects of BCH decoding errors.

6.7.2 Reconfigurable Modulation

The choice of the modulation scheme is a critical issue. It has wide-ranging ramifications as regards the system's robustness, bandwidth efficiency, power consumption, whether to use an equalizer, and so on. In reference [131] we have shown that because of Gaussian minimum shift keying (GMSK), $\pi/4$-shifted quaternary phase shift keying ($\pi/4$-DQPSK) and 16-QAM have bandwidth efficiencies of 1.35 bps/Hz, 1.64 bps/Hz, and 2.4 bps/Hz, respectively, 16-QAM achieves the highest PRMA gain. This is explained by the fact that 16-QAM allows us to generate the highest number of time slots among them, given a certain bandwidth. Therefore, the statistical multiplexing gain of PRMA can approach the reciprocal of the voice activity factor. These findings prompted us to opt for multilevel modulation.

In our proposed reconfigurable transceiver, the different source rates of the 4.7 and 6.5 kbits/s ACELP codecs will be equalized using a combination of appropriately designed FEC codecs and 4 bits/symbol or 6 bits/symbol modulators. When the channel signal-to-noise ratio (SNR) and signal-to-interference ratio (SIR) are high, as in friendly indoors picocells, 64-level Quadrature Amplitude Modulation (64-QAM) is used to convey the 196 bits of the 6.5 kbps ACELP codec. In contrast, for worse channel conditions, for example, after a handover to an outdoors microcell, the 142 bits of the lower quality 4.7 kbps codec are delivered by a more robust 16-QAM modem in the same bandwidth as the 64-QAM scheme.

Noncoherent QAM modems [135] are less complex to implement but typically require higher SNR and SIR values than their coherent counterparts. Hence, in our proposed scheme, second-order switched-diversity assisted coherent Pilot Symbol Assisted Modulation (PSAM) using the maximum-minimum-distance square QAM constellation is preferred. For the 16-QAM scheme, it was shown in [135] that it has two independent subchannels exhibiting different integrities, depending on the position of the bits in a 4-bit symbol. On the same note, our 64-QAM modem possesses three different subchannels with different bit error rates. This property naturally lends itself to un-equal error protection, if the source sensitivity-matched integrity requirements are satisfied by the QAM subchannel integrities.

Therefore, we have evaluated the C1 and C2 bit error rate (BER) versus channel signal-to-noise ratio (SNR) performance of our 16-QAM modem using a pilot spacing of $P = 10$ over both the best-case Additive White Gaussian Noise (AWGN) channel and over the worst-case Rayleigh-fading channel with and without second-order diversity. The C1 and C2 BER results are shown in Figure 6.17 for the experimental conditions characterized by a pedestrian speed of 4 mph, propagation frequency of 1.9 GHz, pilot symbol spacing of $P = 10$, and a

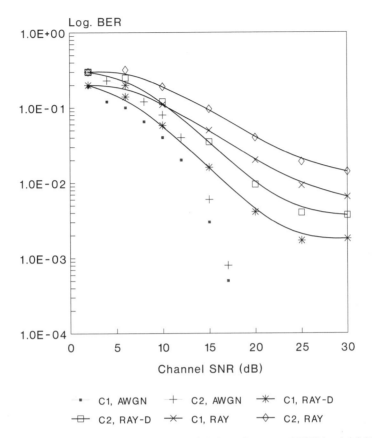

Figure 6.17 C1 and C2 BER versus channel SNR performance of PSAM-assisted 16-QAM using a pilot spacing of $P = 10$ over AWGN and Rayleigh channels at 4 mph, 100 kBd and 1.9 GHz with and without diversity.

signaling rate of 100 kBd. Observe in the figure that over Rayleigh-fading channels (RAY) there is an approximately factor three BER difference between the two subchannels both with and without diversity (D). Because of the violent channel phase fluctuations, our modem was unable to remove the residual BER floor exhibited at higher channel SNR values, although diversity reception reduced its value by nearly an order of magnitude. The diversity receiver operated on the basis of the minimum channel phase shift within a pilot period, since this condition was found to be more effective in terms of reducing the BER than the maximum received power condition. Note that in the case of the chosen 100 kBd signaling rate, the modulated signal will fit in a bandwidth of 200 kHz, when using a 100% excess bandwidth. Since this coincides with the bandwidth of the Pan-European GSM system [74], we will be able to make direct comparisons in terms of the number of users supported. This will allow us to assess the potential benefits of using multimode terminals constituted by third-generation system components in terms of the increased number of users supported.

How this BER difference between the two subchannels can be exploited in order to provide source-matched FEC protection for the 4.7 kbps ACELP codec will be described in the next section. Following a similar approach for the 6.5 kbps/64-QAM scheme leads to the system proposed as a reconfigurable alternative, which is also introduced in the next section.

Suffice to say here that the BER versus channel SNR performance of this more vulnerable but higher speech quality 64-QAM scheme is portrayed under the same propagation conditions as in the case of the 16-QAM modem in Figure 6.18, when using a pilot spacing of $P = 5$. As expected, this diversity and pilot-assisted modem also exhibits a residual BER floor, and there is a characteristic BER difference of about a factor of 2 between the C1 and C2, as well as the C2 and C3 subchannels, respectively. Rather than equalizing these BER differences, we will design an unequal error protection scheme for the speech bits, which capitalizes on this property.

6.7.3 Source-Matched Error Protection

6.7.3.1 Low-Quality 3.1 kBd Mode. In this section we exploit the subchannel integrity differences highlighted in Figures 6.17 and 6.18, and protect these subchannels with source-sensitivity matched binary Bose-Chaudhuri-Hocquenghem (BCH) FEC codecs [134]. Both convolutional [134] and block codes [134] can be successfully employed over bursty mobile channels, and convolutional codes have found favor in systems, such as the Pan-European GSM system [74], where the complexity of soft-decisions is acceptable. Their

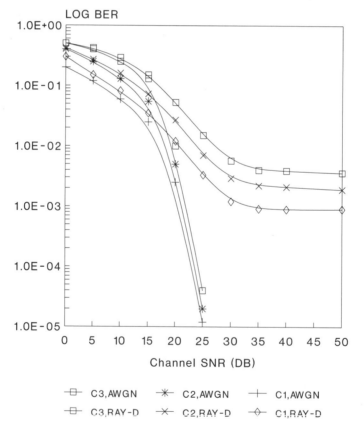

Figure 6.18 C1, C2, and C3 versus channel SNR performance of PSAM-assisted 64-QAM using a pilot spacing of $P = 5$ over AGWN and Rayleigh channels with diversity at 4 mph, 100 kBd and 1.9 GHz.

disadvantage is that they cannot reliably detect decoding errors. Hence, they are typically combined with an external error detecting block code, as in the GSM system. In contrast, powerful block codes have an inherent reliable error detection capability in addition to their error correction capability, which can be exploited to invoke error concealment or to initiate handovers, when the average bit error rate is high, as portrayed in Figure 6.16.

The error sensitivity of the 4.7 kbps ACELP source bits was evaluated in Figures 6.14 and 6.15, but the number of bit protection classes n still remains to be resolved. Intuitively, one would expect that the more closely the FEC protection power is matched to the source sensitivity, the higher the robustness. In order to limit the system's complexity and the variety of candidate schemes, in the case of the 4.7 kbits/s ACELP codec we have experimented with a full-class BCH codec, a twin-class, and a quad-class scheme, while maintaining the same coding rate.

For the full-class system we decided to use the approximately half-rate BCH(127,71,9) codec in both subchannels, which can correct nine errors in each 127-bits block, while encoding 71 primary information bits. The coding rate is $R = 71/127 \approx 0.56$, and the error correction capability is about 7%. Observe that this code curtails BCD decoding error propagation across the speech frame boundaries by encoding each 142-bit speech frame using two BCH(127,71,9) frames, although even a single BCH decoding error will inflict prolonged speech impairments, as portrayed in Figure 6.14.

In order to design the twin-class system, initially we divided the ACELP bits into two sensitivity classes, Class One and Class Two, which are distinct from the C1 and C2 16-QAM subchannels. Both Class One and Two contained 71 bits. Then we evaluated the SEGSNR degradation inflicted by certain fixed-channel BERs maintained in each of the classes using randomly distributed errors, while keeping bits of the other class intact. These experiments suggested that an approximately five times lower BER was required by the more sensitive Class One bits in order to restrict the SEGSNR degradations to similar values of those of the Class Two bits.

Recall from Figure 6.17 that the 16-QAM C1 and C2 subchannel BER ratio was limited to about a factor of 3. Hence, we decided to employ a stronger FEC code to protect the Class One ACELP bits transmitted over the 16-QAM C1 subchannel than for the Class Two speech bits conveyed over the lower integrity C2 16-QAM subchannel, while maintaining the same number of BCH-coded bits in both subchannels. However, the increased number of redundancy bits of stronger BCH codecs requires that a higher number of sensitive ACELP bits are directed to the lower integrity C2 16-QAM subchannel, whose coding power must be concurrently reduced in order to accommodate more source bits. This nonlinear optimization problem can only be solved experimentally, assuming a certain subdivision of the source bits, which would match a given pair of BCH codecs.

Based on our previous findings as regards the C1 and C2 16-QAM BERs and taking account of the practical FEC correcting power limitations, we decided to increase the C1-C2 16-QAM subchannel BER ratio from about 3 by about a factor of 2 so that the Class One ACELP bits were guaranteed a BER advantage of about a factor of 6 over the more robust Class Two bits. After some experimentation, we found that the BCH(127,57,11) and BCH(127,85,6) codes employed in the C1 and C2 16-QAM subchannels provided the required integrity. The SEGSNR degradation caused by a certain fixed BER assuming randomly distributed errors is portrayed in Figure 6.19 for both the full-class and the above twin-class system, where the number of ACELP bits in the protection classes One and Two is 57 and 85, respectively. Note that the coding rate of this system is the same as that of the full-class scheme and each 142-bit ACELP frame is encoded by two BCH codewords. This again yields $2 \cdot 127 = 254$ encoded bits and curtails BCH decoding error propagation across speech segments, although the speech codec's memory will still be corrupted and hence will prolong speech impairments. The FEC-coded bitrate became ≈8.5 kbps.

The BER versus channel SNR performance of our twin-class C1, BCH(127,57,11)-protected and C2, BCH(127,85,6)-protected diversity-assisted 16-QAM modem is shown in Figure 6.20 along with the curves C1, Ray-D and C2, Ray-D characteristic of the diversity-assisted no-FEC Rayleigh-fading scenarios, which are repeated here from Figure 6.17 for ease of reference. Observe that between the SNR values of 15–20 dB there is about an order of magnitude BER difference between the FEC-coded subchannels, as required by the 4.7 kbps speech codec.

With the incentive of perfectly matching the FEC coding power and the number of bits in the distinct protection classes to the ACELP source-sensitivity requirements, we also designed a quad-class system, while maintaining the same coding rate. We used the BCH(63,24,7), BCH(63,30,6), BCH(63,36,5), and BCH(63,51,2) codes and transmitted the most sensitive bits over the C1 16-QAM subchannel using the two strongest codes and relegated the rest of them to the C2 subchannel, protected by the two weaker codes.

The PRMA control header [168] was for all three schemes allocated a BCH(63,24,7) code, hence, the total PRMA frame length became 317 bits, representing 30 ms speech and yielding a bitrate of ≈10.57 kbps. The 317 bits give 80 16-QAM symbols and 9 pilot symbols as well as $2 + 2 = 4$ ramp symbols, resulting in a PRMA frame length of 93 symbols per 30 ms slot. Hence, the signaling rate becomes 3.1 kBd. Using a PRMA bandwidth of 200 kHz, similarly to the Pan-European GSM system [74] and a filtering excess bandwidth of 100% allowed us to accommodate 100 kBd/3.1 kBd ≈ 32 PRMA slots.

6.7.3.2 High-Quality 3.1 kBd Mode. Following the approach proposed in the previous subsection, we designed a triple-class source-matched protection scheme for the 6.5 kbps ACELP codec. The C1, C2, and C3 64-QAM subchannel performance was

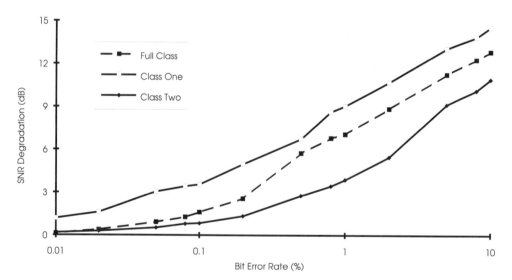

Figure 6.19 SEGSNR degradation versus bit error rate for the 4.7 kbps ACELP codec when mapping 71 ACELP bits to both Classes One and Two in the full-class system and 57 as well as 85 bits to Classes One and Two in the twin-class scheme, respectively.

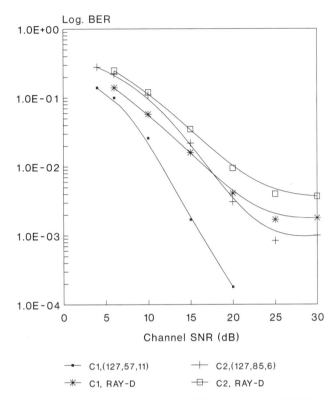

Figure 6.20 C1 and C2 BER versus channel SNR performance of PSAM-assisted
16QAM using a pilot spacing of $P = 10$ over Rayleigh channels at 4 mph,
100 kBd, and 1.9 GHz with diversity and FEC coding.

characterized by Figure 6.18, when using second-order switched-diversity and pilot-symbol-assisted coherent square-constellation 64-QAM [49] among our previously stipulated propagation conditions, with a pilot-spacing of $P = 5$. The BER ratio of the C1, C2, and C3 subchannels was about $1 : 2 : 4$.

The SEGSNR degradation versus channel BER performance of the 6.5 kbits/s higher-quality mode is portrayed in Figure 6.21, when using randomly distributed bit errors and assigning 49, 63, and 84 bits to the three sensitivity classes. For reference we have also included the sensitivity curve for the full-class codec. As we have seen for the lower-quality 16-QAM mode of operation, the modem subchannel BER differences had to be further emphasized using stronger FEC codes for the transmission of the more vulnerable speech bits.

The appropriate source-sensitivity-matched codes for the C1, C2, and C3 subchannels were found to be the shortened 13-error correcting BCH13 = BCH(126,49,13), the 10-error correcting BCH10 = BCH(126,63,10), and the 6-error correcting BCH6 = BCH(126,84,6) codes, while the packet header was again allocated a BCH(63,24,7) code. The corresponding BER versus channel SNR curves are presented for our standard propagation conditions in Figure 6.22, where the nonprotected diversity-assisted Rayleigh BER curves are also repeated for convenience. These codes allowed us to satisfy both the integrity and the bit-packing requirements, while curtailing bit error propagation across speech frame boundaries.

The total number of BCH-coded bits becomes $3 \times 126 + 63 = 441/30$ rms, yielding a bitrate of 14.7 kbps. The resulting 74 64-QAM symbols are amalgamated with 15 pilot and 4

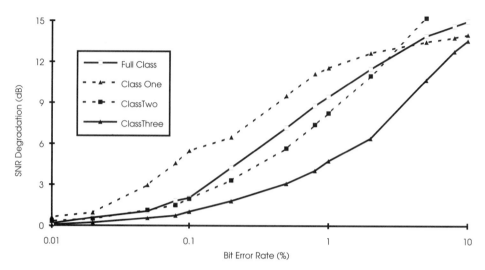

Figure 6.21 SEGSNR degradation versus bit error rate for the 6.5 kbps ACELP codec when using either the full-class scheme or mapping 49, 63, and 84 ACELP bits to Classes One, Two, and Three in the triple-class scheme, respectively.

ramp symbols, giving 93 symbols/30 ms, which is equivalent to a signaling rate of 3.1 kBd, as in the case of the low-quality mode of operation. Again, 32 PRMA slots can be created, as for the low-quality system, accommodating more than 50 speech users in a bandwidth of 200 kHz and yielding a speech user bandwidth of about 4 kHz, while maintaining a packet dropping probability of about 1%.

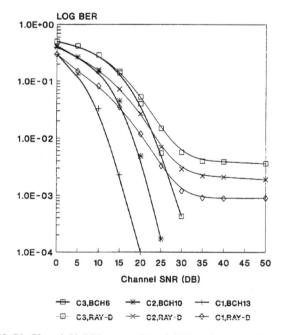

Figure 6.22 C1, C2, and C3 BER versus channel SNR performance of PSAM-assisted 64QAM using a pilot spacing of $P = 5$ over Rayleigh channels at 4 mph, 100 kBd, and 1.9 GHz with diversity and FEC coding.

6.7.4 Voice Activity Detection and Packet Reservation Multiple Access

In the modulation section, we have noted that multilevel modulation conveniently increases the number of time slots, which in turn results in higher PRMA statistical multiplexing gain than in the case of binary modulation. The operation of the voice activity detector (VAD) [74] has a profound effect on the overall subjective speech quality. The fundamental design problem is that on one hand the VAD must respond to an active speech spurt almost instantaneously in order to queue the active speech packet for transmission to the BS and hence minimize front-end speech spurt clipping. On the other hand, it has to have a low false triggering rate even in the presence of high-level acoustic background noise. This imposes a taxing design problem, since the input signal's statistics must be observed for some length of time in order to differentiate reliably between speech and noise. In our GSM-like VAD [74], a combination of signal power, stationarity, and spectral envelope-based decisions is carried out before speech is deemed to be present. In order to prevent prematurely curtailing active spurts during low-energy voiced sounds, a hangover switch-off delay of one speech frame length or 30 ms was imposed. The GSM VAD was designed and extensively tested by an international expert body. For further details, the interested reader is referred to Reference [74].

PRMA was designed to convey speech signals on a flexible demand basis via Time Division Multiple Access (TDMA) systems [168]. In our system a (VAD) similar to that of the GSM system [74] queues the active speech spurts to contend for an up-link TDMA time slot for transmission to the BS. Inactive users' TDMA time slots are offered by the BS to other users, who become active and are allowed to contend for the unused time slots with a given permission probability P_{perm}. In order to prevent colliding users from consistently colliding in their further attempts to attain a time-slot reservation, we have $P_{perm} < 1$. If several users attempt to transmit their packets in a previously free slot, they collide and none of them will attain a reservation. In contrast, if the BS receives a packet from a single user, or succeeds in decoding an uncorrupted packet despite a simultaneous transmission attempt, then a reservation is granted. When the system is heavily loaded, the collision probability is increased. Hence, a speech packet might have to keep contending in vain, until its life span expires due to the imminence of a new speech packet's arrival after 30 ms. In this case, the speech packet must be dropped, but the packet dropping probability must be kept below 1%. Since packet dropping is typically encountered at the beginning of a new speech spurt, its subjective effects are perceptually insignificant.

Our transceiver used a signaling rate of 100 kBd in order for the modulated signal to fit in a 200 kHz GSM channel slot, when using a QAM excess bandwidth of 100%. The number of time slots created became TRUNC(100 kBd/3.1 kBd) = 32, where TRUNC represents truncation to the nearest integer, while the slot duration was 30 ms/32 = 0.9375 ms. One of the PRMA users was transmitting speech signals recorded during a telephone conversation, while all the other users generated negative exponentially distributed speech spurts and speech gaps with mean durations of 1 and 1.35 s. These PRMA parameters are summarized in Table 6.15.

In conventional Time Division Multiple-Access (TDMA) systems, the reception quality degrades due to speech impairments caused by call blocking, handover failures, and corrupted speech frames due to noise, as well as co- and adjacent-channel interference. In PRMA systems, calls are not blocked due to the lack of an idle timeslot. Instead, the number of contending users is increased by one, slightly inconveniencing all other users, but the packet dropping probability is increased only gracefully. Handovers are performed in the form of

TABLE 6.15   Summary of PRMA Parameters

PRMA Parameters	
Channel rate	100 kBd
Source rate	3.1 kBd
Frame duration	30 ms
No. of slots	32
Slot duration	0.9375 ms
Header length	63 bits
Maximum packet delay	30 ms
Permission probability	0.2

contention for an uninterfered idle time slot provided by the specific BS offering the highest signal quality among the potential target BSs.

If the link degrades before the next active spurt is due for transmission, the subsequent contention phase is likely to establish a link with another BS. Hence, this process will have a favorable effect on the channel's quality, effectively simulating a diversity system having independent fading channels and limiting the time spent by the MS in deep fades, thereby avoiding channels with high noise or interference.

This attractive PRMA feature can be capitalized upon in order to train the channel segregation scheme proposed in Reference [169]. Accordingly, each BS evaluates and ranks the quality of its idle physical channels constituted by the unused time slots on a frame-by-frame basis and identifies a certain number of slots, N, with the highest quality (i.e., lowest noise and interference). The slot status is broadcast by the BS to the portable stations (PSs), and top-grade slots are contended for using the less robust, high speech quality 64-QAM mode of operation, while lower quality slots attract contention using the lower speech quality, more robust 16-QAM mode of operation. Lastly, the lowest quality idle slots currently impaired by noise and interference can be temporarily disabled. When using this algorithm, the BS is likely to receive a signal benefiting from high SNR and SIR values, minimizing the probability of packet corruption due to interference and noise. However, due to disabling the lowest-SNR and -SIR slots, the probability of packet dropping due to collision is increased, reducing the number of users supported. When a successful, uncontended reservation takes place using the high speech quality 64-QAM mode, the BS promotes the highest quality second-grade time slot to the set of top-grade slots, unless its quality is unacceptably low. Similarly, the best temporarily disabled slot can be promoted to the second-grade set in order to minimize the collision probability, if its quality is adequate for 16-QAM transmissions.

With the system elements described we now focus our attention on the the the performance of the re-configurable transceiver proposed.

6.7.5 3.1 kBd System Performance

The number of speech users supported by the 32-slot PRMA system becomes explicit from Figure 6.23, where the packet dropping probability versus number of users is displayed. Observe that more than 55 users can be served with a dropping probability below 1%. The effect of various packet dropping probabilities on the objective speech SEGSNR quality measure is portrayed in Figure 6.24 for both the 4.7 kbps and the 6.5 kbps mode of operation. This figure implies that packet dropping due to PRMA collisions is more detrimental in case of the higher quality 6.5 kbps codec, since it has an originally higher SEGSNR. In order to

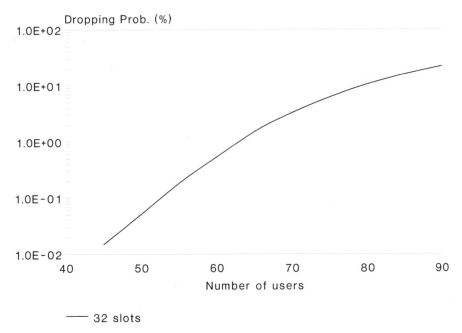

Figure 6.23 Packet dropping probability versus number of users for 32-slot PRMA.

restrict the subjective effects of PRMA-imposed packet dropping, according to Figure 6.23 the number of users must be below 60. However, in generating Figure 6.24, packets were dropped on a random basis. The same 1% dropping probability associated with initial clipping only imposes much less subjective annoyance or speech quality penalty than intra-spurt packet loss would. As a comparative basis, it is worth noting that the 8 kbps CCITT/ITU ACELP candidate codec's target was to inflict less than 0.5 mean opinion score (MOS) degradation in case of a speech frame error rate of 3%.

The overall SEGSNR versus channel SNR performance of the proposed speech transceiver is displayed in Figure 6.25 for the various systems studied, where no packets were dropped, as in a TDMA system supporting 32 subscribers. Observe that the source-sensitivity-matched twin-class and quad-class 4.7 kbps ACELP-based 16-QAM systems have a virtually identical performance, suggesting that using two appropriately matched protection classes provides adequate system performance, while maintaining a lower complexity than the quad-class scheme. The full-class 4.7 kbps/16-QAM system was outperformed by both source-matched schemes by about 4 dB in terms of channel SNR, the latter systems requiring an SNR in excess of about 15 dB for nearly unimpaired speech quality over our pedestrian Rayleigh-fading channel. When the channel SNR was in excess of about 25 dB, the 6.5 kbps/ 64-QAM system outperformed the 4.7/16-QAM scheme in terms of both objective and subjective speech quality. When the proportion of corrupted speech frames due to channel-induced impairments and due to random packet dropping as in Figure 6.25 was identical, similar objective and subjective speech degradations were experienced. Furthermore, at around a 25 dB channel SNR, where the 16-QAM and 64-QAM SEGSNR curves cross each other in Figure 6.25, it is preferable to use the inherently lower quality but unimpaired mode of operation.

When supporting more than 32 users, as in our PRMA-assisted system, speech quality degradation is experienced due to packet corruption caused by channel impairments and

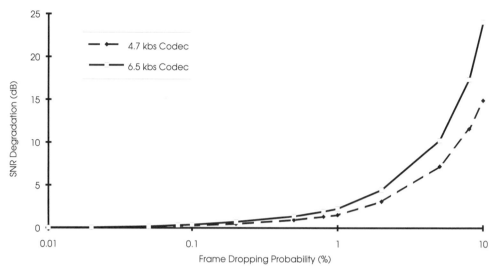

Figure 6.24 Speech SEGSNR degradation versus packet dropping probability for the 4.7 and 6.5 kbit/s ACELP codecs.

packet dropping caused by collisions. These impairments yield different subjective perceptual degradation, which we will attempt to compare in terms of the objective SEGSNR degradation. Quantifying these speech imperfections in relative terms in contrast to each other will allow system designers to adequately split the tolerable overall speech degradation between packet dropping and packet corruption. The corresponding SEGSNR versus channel SNR curves for the twin-class 4.7 kbps/16-QAM and the triple-class 6.5 kbps/64-QAM

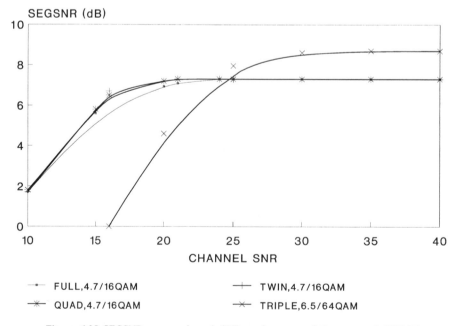

Figure 6.25 SEGSNR versus channel SNR performance of the proposed 100 kBd transceiver using 32-slot TDMA.

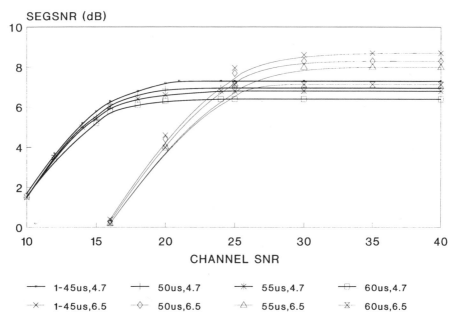

Figure 6.26 SEGSNR versus channel SNR performance of the reconfigurable 100 kBd transceiver using 32-slot PRMA for different number of conversations.

operational modes are shown in Figure 6.26 for various numbers of users between 1 and 60. Observe that the rate of change of the SEGSNR curves is more dramatic due to packet corruption caused by low-SNR channel conditions than due to increasing the number of users.

As long as the number of users does not significantly exceed 50, the subjective effects of PRMA packet dropping show an even more benign speech quality penalty than that suggested by the objective SEGSNR degradation because frames are typically dropped at the beginning of a speech spurt due to a failed contention.

6.7.6 3.1 kBd System Summary

In conclusion, our reconfigurable transceiver has a single-user rate of 3.1 kBd and can accommodate 32 PRMA slots at a PRMA rate of 100 kBd in a bandwidth of 200 kHz. The number of users supported is in excess of 50, and the minimum channel SNR for the lower speech quality mode is about 15 dB, while for the higher quality mode it is about 25 dB. The number of time slots can be further increased to 42, when opting for a modulation access bandwidth of 50%, accommodating a signaling rate of 133 kBd within the 200 kHz system bandwidth. This will inflict a slight bit error rate penalty but will pay dividends in terms of increasing the number of PRMA users by about 20. The parameters of the proposed transceiver are summarized in Table 6.16. In order to minimize packet corruption due to interference, the employment of a time-slot quality ranking algorithm is essential for invoking the appropriate mode of operation. When serving 50 users, the effective user bandwidth becomes 4 kHz, which guarantees the convenience of wireless digital speech communication in a bandwidth similar to conventional analog telephone channels.

Future research in the field of speech coding and modulation is being targeted at creating a more finely graded set of reconfigurable subsystems in terms of speech quality,

TABLE 6.16   Transceiver Parameters

Parameter	Low/High-Quality Mode
Speech codec	4.7/6.5 kbps ACELP
FEC	Twin-/Triple-class binary BCH
FEC-coded rate	8.5/12.6 kbps
Modulation	Square 16-QAM/64-QAM
Demodulation	Coherent diversity PSAM
Equalizer	No
User's signaling rate	3.1 kBd
VAD	GSM-like [74]
Multiple acess	32-slot PRMA
Speech frame length	30 ms
Slot length	0.9375 ms
Channel rate	100 kBd
System bandwidth	200 kHz
No. of users	>50
Equiv. user bandwidth	4 kHz
Min. channel SNR	15/25 dB

transmission rate, and robustness. These new subsystems will enable us to match the mode of operation more closely with the prevailing channel quality. Further algorithmic research is required in order to define specific control algorithms to accommodate various operating conditions, in particular in the area of appropriate time-slot classification algorithms to invoke the best matching mode of operation and find the best compromise between packet dropping due to collision and packet corruption due to channel impairments.

In the next section we invoke a similar reconfigurable transceiver, but we employ a modem-mode dependent number of PRMA slots for conveying the speech information.

6.8 MULTI-SLOT PRMA TRANSCEIVER [170]

6.8.1 Background and Motivation

In another study by Williams, Hanzo, and Steele [170], packet reservation multiple-access (PRMA)-assisted adaptive modulation using 1, 2, and 4 bit/symbol transmissions was proposed as an alternative to dynamic channel allocation (DCA) in order to maximize the number of users supported in a traffic cell. The cell was divided into three concentric rings. In the central high signal-to-noise ratio (SNR) region, 16-level star quadrature amplitude modulation (16-StQAM) was used, in the first ring differential quaternary phase shift keying (DQPSK) was invoked, while in the outer ring differential phase shift keying (DPSK) was utilized. In our diversity-assisted modems a channel SNR of about 7, 10, and 20 dB, respectively, was required in order to maintain a bit error ratio (BER) of about 1%, which can then be rendered error-free by the binary BCH error correction codes used. Our previously designed 4.7 kbps algebraic code excited linear predictive (ACELP) speech codec of Section 6.4.3 was assumed, protected by the quad-class source-sensitivity matched BCH coding scheme of Section 6.7.3, yielding a total bitrate of 8.4 kbps. A GSM-like voice activity detector [74] (VAD) controls the PRMA-assisted adaptive system, which ensures a capacity improvement of a factor of 1.78 over PRMA-aided binary schemes.

The dynamic channel allocation (DCA) and (PRMA) techniques potentially allow large increases in capacity over a fixed channel allocation (FCA) time division multiple access

(TDMA) system. Although both DCA and PRMA can offer a significant system capacity improvement, their capacity advantages typically cannot be jointly exploited, since the rapid variation of slot occupancy resulting from the employment of PRMA limits the validity of interference measurements, which are essential for the reliable operation of the DCA algorithm. One alternative to tackle this problem is to have mixed fixed and dynamic frequency reuse patterns, although this approach reduces the number of slots per carrier for the PRMA scheme, thus decreasing its efficiency.

Diversity-assisted adaptive modulation can be used as an alternative to DCA. The cells must be frequency planned as in a FCA system using a binary modulation scheme. When adaptive modulation is employed, the throughput is increased by permitting high-level modulation schemes to be used by the mobiles roaming near the center of the cell, which therefore will require a lower number of PRMA slots to deliver a fixed number of channel-encoded speech bits to the base station (BS). In contrast, mobile stations (MS) near the fringes of the cell will have to use binary modulation in order to cope with the prevailing lower signal-to-noise ratio (SNR) and hence will occupy more PRMA slots for the same number of speech bits. Specifically, our adaptive system uses three modulation schemes, namely, binary differential phase shift keying (DPSK) transmitting 1 bit per symbol at the cell boundary, quaternary differential phase shift keying (DQPSK), transmitting 2 bits per symbol at medium distances from the BS, and 16-level Star Quadrature Amplitude Modulation [49] (16-StQAM), which carries 4 bits per symbol close to the center of the cell.

6.8.2 PRMA-Assisted Multi-Slot Adaptive Modulation

Standard PRMA schemes [168] have been discussed, for example, in [135]. However, in the proposed PRMA-assisted adaptive modulation scheme, mobile stations (MS) can reserve more than one slot in order to deliver up to four bursts per speech frame, when DPSK is invoked toward the cell edges. When a free slot appears in the frame, each mobile that requires a new reservation contends for it based on a permission probability, P_p. If the slot is granted to a 16-StQAM user, that slot is reserved in the normal way. If the slot is granted to a DQPSK user, then the next available free slot is also reserved for that user. Lastly, if the slot is granted to a DPSK user, then the next three free slots must also be reserved for this particular user. In this way, users that require more than one slot are not disadvantaged by forcing them to contend for each slot individually. If, however, fewer than three slots are available, DQPSK or 16-StQAM users still may be able to exploit the remaining slots.

Again, we found that the difference in signal-to-noise ratio (SNR) required for the different diversity-assisted modulation schemes in order to maintain similar bit error ratios (BER) was approximately 3 dB between DPSK and DQPSK, and 12dB between DPSK and StQAM, when transmitting over Rayleigh-fading channels in our GSM- and DECT-type systems. The BER curves for these modulation schemes in narrowband Rayleigh channels with second-order diversity, a propagation frequency of 2 GHz and a vehicular speed of 15 m/s are shown in Figures 6.27 and 6.28 in case of the GSM- and DECT-type systems, respectively.

Thus, using an inverse fourth power pathloss law, DPSK was invoked between radii $0.84R$ and the cell boundary, R, which is one-quarter of the cell area. StQAM was used between the cell center and $0.5R$, which is a further quarter of the cell area and DQPSK in the remaining area, which constitutes half of the total cell area. Accordingly, considering the

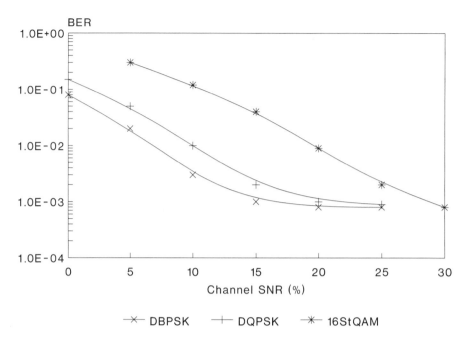

Figure 6.27 BER performance of our modulation schemes in Rayleigh fading with second order diversity at a symbol rate of 133 kBd, carrier frequency 2 GHz, and mobile velocity of 15 m/s © IEE, Williams, Hanzo, Steele, 1995, [170].

number of slots needed by the various modulation schemes invoked and assuming a uniform traffic density, we can calculate the expected number of required slots per call as

$$E(n) = \frac{1}{4} \cdot 4 + \frac{1}{2} \cdot 2 + \frac{1}{4} \cdot 1 = 2.25 \text{ slots.}$$

Since a binary user would require four slots, this implies a capacity improvement of a factor of $4/2.25 \approx 1.78$.

6.8.3 Adaptive GSM-Like Schemes

The basic systems features are summarised in Table 6.17, where all modulation schemes assumed an excess bandwidth of 50%, resulting in a symbol rate, which is two-thirds of the total bandwidth. The 8.4 kbps channel-coded rate after accommodating the packet header allowed us to create 48 or 416 slots per 30 ms frame in the GSM-like and DECT-like systems, respectively, as shown in the table. Specifically, when using the 133.33 kBd GSM-like adaptive PRMA schemes, we can create 48 slots per 30 ms speech frame, which is equivalent to 12 slots for a binary-only system, since four slots are required for the transmission of a 30 ms speech packet. When the quaternary system is used, 24 pairs of slots can be created. Note that when fixed channel allocation is used, the adaptive scheme and the binary-only scheme can use the same cluster size. A quaternary only system requires a 3dB greater SIR than the binary scheme. According to Lee [171], we have:

$$\frac{D}{R} = \sqrt{3K} \tag{6.58}$$

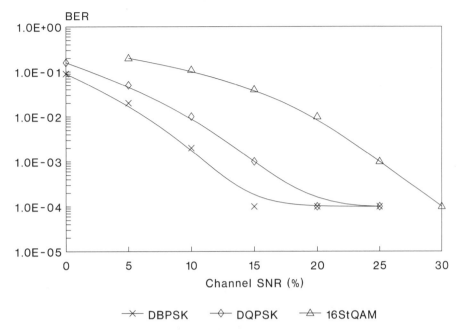

Figure 6.28 BER performance of our modulation schems in Rayleigh fading with second-order diversity at a symbol rate of 1152 kBd, carrier frequency 2 GHz, and mobile velocity of 15 m/s © IEE, Williams, Hanzo, Steele, 1995, [170].

where D is the distance to the closest interferer, R is the cell radius, and K is the cluster size. The prevailing signal-to-interference ratio (SIR) can be expressed as

$$\text{SIR} \approx \left(\frac{D}{R}\right)^{\gamma} \qquad (6.59)$$

where γ is the pathloss exponent and hence

$$K = \frac{1}{3}(\text{SIR})^{2/\gamma}. \qquad (6.60)$$

In this study we have used a pathloss exponent of $\gamma = 4$. Therefore increasing the SIR by 3dB requires that the cluster size be increased by a factor of $\sqrt{2}$. The packet dropping versus number of users performance of the 12-slot binary scheme is shown in Figure 6.29 together with the 24-slot quaternary and the 48-slot adaptive scheme. For all schemes, their associated optimum permission probability was used, which allowed us to support the highest number of users, assuming a packet dropping probability of 1%. We found that a maximum of 19 simultaneous calls can be supported at a packet dropping probability of 1%, when using the

TABLE 6.17 Parameters of the GSM Like and DECT Like Adaptive Modulation PRMA Systems

Parameter	GSM	DECT	Unit
Channel bandwidth	200	1728	kHz
Symbol rate	133	1152	kBd
Bursts per frame	48	416	

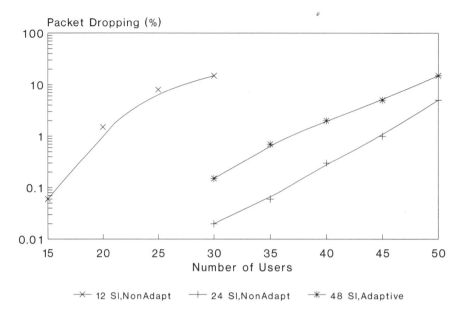

Figure 6.29 Packet dropping performance of the GSM-like PRMA schemes © IEE, Williams, Hanzo, Steele, 1995, [170].

binary scheme with a PRMA permission probability of 0.5. In contrast, the 24-slot quaternary scheme can support 44 simultaneous calls when using a permission probability of 0.4. Lastly, our 48-slot adaptive scheme can accommodate 36 simultaneous calls with a permission probability of 0.5. The capacity improvements attainable by the proposed GSM-like scheme are presented in Table 6.18.

6.8.4 Adaptive DECT-Like Schemes

In our DECT-like schemes, we have INT{1152 kBd/2.77 kBd} = 416 slots per frame for the adaptive PRMA system. This is equivalent to 104 slots for a binary-only system and 216 slots for a quaternary-only system. Note that when fixed channel allocation is used, the adaptive scheme and the binary-only scheme can use the same cluster size. Again, a quaternary-only system requires a 3 dB greater SIR than the binary scheme, and so the cluster size should be increased by a factor of $\sqrt{2}$.

The packet dropping versus number of users performance of the 104-slot binary scheme is portrayed in Figure 6.30 when using a permission probability of 0.1. Observe from the figure that the binary scheme can support up to 220 simultaneous calls at a packet dropping

TABLE 6.18 Improvements in Capacity Possible with Adaptive Modulation PRMA with 48 Slots © IEE, Williams, Hanzo, Steele, 1995, [170]

System	Slots	P_p	Simult. Calls	Normalized by Cluster Size K	Improvement over Binary with PRMA	Improvement over Binary without PRMA
DBPSK	12	0.5	19	19	—	58%
DQPSK	24	0.4	44	31.1	64%	159%
Adaptive	48	0.5	36	36	89%	200%

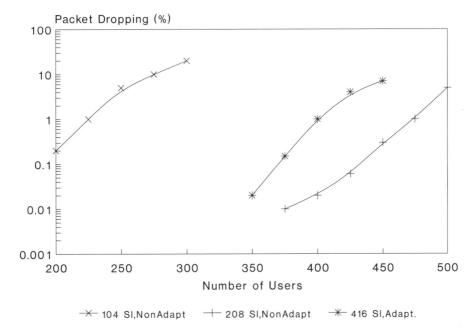

Figure 6.30 Packet dropping performance of the DECT-like PRMA schemes © IEE,
Williams, Hanzo, Steele, 1995, [170].

probability of 1%. When opting for the 208-slot quaternary scheme, the packet dropping
versus number of users performance curve reveals that this system can accommodate 470
simultaneous calls with a permission probability of 0.1. Finally, the packet dropping
performance of the 416-slot adaptive scheme suggests that the number of supported
simultaneous conversations is about 400, when opting for a permission probability of 0.1.
% % The achievable capacity improvements for our DECT-like system are displayed in Table
6.19.

6.8.5 Summary of Adaptive Multi-Slot PRMA

In conclusion, adaptive modulation with PRMA gives the expected three- to four-fold
capacity increase over the binary scheme without PRMA. Generally, the greater the number of
slots, the greater the advantage of PRMA over non-PRMA systems, since the statistical
multiplexing gain approaches the reciprocal of the speech activity ratio. Furthermore, PRMA-
assisted adaptive modulation achieves an additional 80% capacity increase over PRMA-
assisted binary modulation. The speech performance of our adaptive system evaluated in

TABLE 6.19 Achievable Capacity Improvements for the Adaptive Modulation
PRMA with 416 Slots © IEE, Williams, Hanzo, Steele, 1995, [170]

System	Slots	P_p	Simult. Calls	Normalized by Cluster Size K	Improvement over Binary with PRMA	Improvement over Binary without PRMA
DBPSK	104	0.1	220	200	—	112%
DQPSK	208	0.1	470	332	51%	219%
Adaptive	416	0.1	400	400	82%	285%

terms of segmental signal-to-noise ratio (SEGSNR) and cepstral distance (CD) is unimpaired by channel effects for SNR values in excess of about 8, 10, and 20 dB, when using diversity-assisted DPSK, DQPSK, and 16-StQAM, respectively, although in dispersive environments a reduced performance is expected.

6.9 CHAPTER SUMMARY

This chapter highlighted the design principles of forward-adaptive CELP codecs and reviewed various CELP excitation models proposed over the years. The philosophy of ACELP codecs was detailed in more depth, since these codecs have been successful at various rates and have found their way into various standardized codecs.

The sensitivity of these schemes against transmission errors was also analyzed, and a new sensitivity measure was proposed, which was capable of quantifying the effects of error propagation inflicted upon the codec. Finally, a PRMA-assisted dual-rate system design example was offered, which was capable of operating at two different speech coding rates, while maintaining a constant system bandwidth. This was achieved by adjusting the number of bits per symbol conveyed by the transceiver as a function of the channel quality experienced.

Having introduced the concept of CELP codecs, in the next chapter we attempt to provide a review of most of the forward-adaptive CELP-based standard speech codecs that emerged during the past few years.

7

Standard Forward-Adaptive CELP Codecs

7.1 BACKGROUND

With the rapid development of DSP technology on one hand and the advent of recent speech compression advances on the other, the late 1980s and 1990s have witnessed the emergence of a whole host of new speech coding standards. Some of these standards are summarized in this chapter in order to put our earlier theoretical elaborations into practice. Considerable improvements have taken place in terms of both speech quality and robustness against channel errors, partially rendered affordable by more capable DSPs. The ITU's 8 kbps G.729 codec, for example, maintains a similar speech quality to that of the 32 kbps G.726 ADPCM codec, which is equivalent to wireline quality, while maintaining a high robustness against transmission errors. More explicitly, over the years the speech quality of 64 kbps standard PCM codecs has been maintained by the various codecs, which gradually reduced this rate to 8 kbps, at the cost of ever increasing implementational complexity. At the time of writing, researchers are attempting to further halve the 8 kbps rate of the G.729 codec to 4 kbps, an initiative referred to as the ITU 4 codec development. Further important factors are that modern codecs tolerate both background noise, such as engine noise in cars, and tandeming in mobile-to-mobile connections. Since the standard codecs are reviewed here in chronological order, this chapter also constitutes a historical portrayal of advances in the field. The objective and subjective performance of most existing standard codecs is compared in Chapter 17. Let us commence our discourse considering the first CELP-based standard codec, namely, the U.S. Department of Defense (DoD) 4.8 kbps codec in the next section.

7.2 THE US DoD FS-1016 4.8 kbits/s CELP CODEC [76]

7.2.1 Introduction

In 1984 the United States Department of Defense (US DoD) launched a program to develop a third-generation secure telephone unit in order to supplement the 2.4 kbits/s LPC-10 vocoder. The vocoder produced speech *almost* as intelligible as natural speech [172], but it sounded synthetic and lacked any speaker recognizability. In 1988, a survey of 4.8 kbits/s codecs was conducted, and a CELP codec [162], jointly developed by the DoD and AT&T Bell Laboratories, was selected. This codec, which was later enhanced and standardized as Federal Standard 1016 (FS-1016) [76], was very advanced for its time and outperformed all US government standard codecs at rates below 16 kbits/s [173].

The FS-1016 codec uses a standard CELP structure, with both a fixed and an adaptive codebook producing the excitation to an all-pole synthesis filter. A frame length of 30 ms is used, and each frame is split into four 7.5 ms subframes. The filter coefficients for a tenth-order all-pole synthesis filter are determined for each frame using forward-adaptive LPC analysis, and are then converted to LSFs and scalar quantized with 34 bits. The excitation to this filter is coded every subframe, using a 512-entry ternary-valued overlapping fixed codebook, and a 256-entry adaptive codebook with fractional delays. Both codebooks are searched by the encoder using a closed-loop search to minimize the weighted squared error between the original and the reconstructed speech, and the codebook gains are scalar quantized with 5 bits each. In odd subframes the adaptive codebook index is coded with 8 bits, but, to reduce the complexity and the bitrate of the codec, in even subframes this delay is differentially encoded with 6 bits. One bit per frame is used for synchronization, and 4 bits per frame are used to provide simple forward error correction for the most sensitive bits transmitted by the codec. Finally, 1 bit per frame is allocated for future expansion of the codec, as suggested by Bishnu Atal. This bit is intended to ensure that the standard does not become obsolete as technology advances. It could be used, for example, to indicate that some as yet unknown improved decoding technique should be used. The bit allocation for the codec is summarized in Table 7.1.

At the decoder the received bit stream is used to give filter coefficients for the synthesis filter and to select codebook entries from the adaptive and fixed codebooks to excite this filter and produce the reconstructed speech. Adaptive post-filtering can then be applied to this reconstructed speech to improve its perceptual quality.

An interesting aspect of the FS-1016 standard is that it allows some flexibility in both encoders and decoders that comply with the standard. For example, the encoder can search only a subset of the fixed or adaptive codebook in order to reduce its complexity. Also, the post-filter recommended at the decoder is optional. However, we now describe in more detail the blocks that would be required for full implementation of the standard.

TABLE 7.1   Bit-Allocation Scheme of the FS-1016 Codec

Parameter	Per Subframe	Total per Frame
LPC coefficients	–	34
Adaptive codebook delay	8 or 6	28
Fixed codebook index	9	36
Adaptive codebook gain	5	20
Fixed codebook gain	5	20
Forward error correction	–	4
Synchronization	–	1
Expansion Bit	–	1
Total	–	144

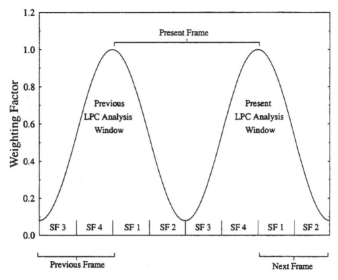

Figure 7.1 LPC analysis window used in FS-1016.

7.2.2 LPC Analysis and Quantization

LPC analysis is carried out for every 30 ms frame at the encoder to derive filter coefficients for use in the synthesis and weighting filters. A 30 ms Hamming window covering the last two subframes of the current frame and the first two subframes of the next frame is used, as shown in Figure 7.1. Autocorrelation coefficients are found from the windowed speech signal, which can then be used to calculate a set of 10 filter coefficients, a_i. A bandwidth expansion of 15 Hz is applied to the filter by replacing the original coefficients a_i with $a_i\gamma^i$, where $\gamma = 0.994$. This bandwidth expansion improves the reconstructed speech quality of the codec and also aids the quantization of the coefficients.

The expanded filter coefficients are converted to line spectral frequencies (LSFs) and scalar quantized with 34 bits. Interpolation between the quantized LSFs from the previous frame and those from the present frame is then used to give a set of LSFs for each subframe. The interpolation coefficients used are shown in Table 7.2. The interpolated LSFs for each subframe are converted back to give the filter coefficients to be used in that subframe.

A simple weighting filter of the form

$$W(z) = \frac{A(z)}{A(z/\gamma)},\tag{7.1}$$

TABLE 7.2   LSF Interpolation Used in the FS-1016 Codec

Subframe	Contribution of LSFs from Previous Frame	Contribution of LSFs from Present Frame
1	7/8	1/8
2	5/8	3/8
3	3/8	5/8
4	1/8	7/8

where $\gamma = 0.8$, is used at the encoder. It is then the squared weighted error which is minimized by the adaptive and fixed codebook searches.

7.2.3 The Adaptive Codebook

A 256-entry adaptive codebook is used in the FS-1016 codec to model the long-term periodicities present in voiced speech. The adaptive codebook delay ranges between 20 and 147, and noninteger as well as integer delays are used. Different delay resolutions are used for different delay ranges as shown in Table 7.3. These resolutions were chosen to give the highest resolutions for typical female speakers, where the improvements in reconstructed speech quality given by noninteger pitch resolution are especially significant [174]. Adaptive codebook codewords for noninteger delays are formed using interpolation with Hamming windowed sinc functions. Interpolating functions at least 8 points long are recommended for the codebook search, and 40 points long for the synthesis of the selected adaptive codebook codeword.

The entire adaptive codebook is searched, and an index is coded with 8 bits in the first and third subframe, whereas in the second and fourth subframes the delay is delta encoded, relative to the previous subframe's delay, using only 6 bits. This was found to reduce the bitrate and the complexity of the encoder while causing no perceivable loss in the codec's reconstructed speech quality. Submultiples of the delay value which gives the minimum weighted squared error between the original and the weighted speech are checked and favored if they give a match to the original speech, which is almost as good. This results in a smoothly varying pitch contour, which is important for the delta coding of the speech in odd subframes. It also enables the receiver to use a smoother to check for channel errors in the received adaptive codebook index.

Once the adaptive codebook delay has been chosen, the corresponding gain term is calculated and scalar quantized using 5-bit nonuniform quantization between -1 and $+2$. This gives an adaptive codebook signal that is filtered through the synthesis and weighting filters and subtracted from the target for the adaptive codebook search to give the target signal for the fixed codebook search.

7.2.4 The Fixed Codebook

The fixed codebook in the FS-1016 codec contains 512 sparse, ternary-valued, over-lapped codewords. The codewords are overlapped by -2 so that each codeword contains all but two samples of the previous codeword plus two new samples. This overlapping dramatically reduces the storage necessary for the fixed codebook as only $N + 2(L - 1) = 1082$, rather than $LN = 30720$, elements need to be stored at the encoder and decoder. Here $L = 512$ is the number of entries in the codebook, and $N = 60$ is the

TABLE 7.3 Delay Resolutions Used in the FS-1016 Codec

Delay Range	Resolution
20–25 2/3	1/3 sample
26–33 3/4	1/4 sample
34–79 2/3	1/3 sample
80–147	1 sample

dimension of each entry. The overlapped nature of the codebook also allows fast calculation of the energy and correlation terms that must be calculated for each codebook entry to allow the fixed codebook search to be carried out, and it is reported in [175] to give performance equivalent to that of a nonoverlapped codebook.

The 1082 codebook entries are derived using a zero-mean unit-variance white Gaussian sequence. This sequence is center-clipped at 1.2, and all values that are greater than 1.2, or less than -1.2, are set equal to $+1$ or -1. This gives a ternary-valued codebook that is approximately 77% sparse and whose nonzero elements are either $+1$ or -1. The sparse ternary-valued nature of the codebook gives a further reduction in the storage necessary for the codebook and further simplifies the codebook search procedure.

A novel feature of the FS-1016 codec is in the calculation of the fixed codebook gain. This gain is nonuniformly quantized with 5 bits, and initially is calculated and quantized for each codebook entry as in most CELP codecs. As reported in [173], this joint optimization of the codebook index and quantized gain are subjectively similar to searching twice as large a fixed codebook without joint optimization. However, once the fixed codebook gain and index have been determined, the fixed codebook gain is adaptively attenuated or amplified depending on the efficiency of the adaptive codebook. This is similar to Shoham's constrained excitation idea [176] and attenuates the stochastic element of the excitation during voiced segments of speech. This reduces roughness heard during sustained voiced segments of speech and significantly improves the subjective quality of the reconstructed speech. During unvoiced segments of speech, the stochastic element of the excitation signal is increased, which provides a more subjectively pleasing match between the reconstructed and input speech.

The efficiency of the adaptive codebook is measured using the normalized cross-correlation R between the target signals for the fixed and adaptive codebook searches. This is given by

$$R = \frac{\sum_{n=0}^{N-1} x(n)y(n)}{\sum_{n=0}^{N-1} x^2(n)}, \tag{7.2}$$

where $x(n)$ is the target signal for the adaptive codebook search and $y(n)$ is the target signal for the fixed codebook search. The quantized codebook gain $\hat{G}_2$ is then modified to $\tilde{G}_2$ depending on the value of R as follows:

$$\tilde{G}_2 = \begin{cases} 0.2\hat{G}_2, & |R| < 0.04 \\ 1.4\hat{G}_2\sqrt{|R|}, & |R| > 0.81 \\ \hat{G}_2\sqrt{|R|}, & \text{otherwise.} \end{cases} \tag{7.3}$$

This modification of the stochastic excitation component has a negligible effect on the complexity of the codec, but as stated earlier gives a significant improvement in the subjective quality of the codec.

Once the fixed and adaptive codebook indices and gains have been found at the encoder, locally reconstructed speech can be calculated at the encoder and used to update the filter memories. Also, indices representing the fixed and adaptive codebook signals are coded and sent to the decoder, allowing it to find the reconstructed speech. The decoder also incorporates a simple post-filter to further improve the subjective quality of the reconstructed speech, and error detection and concealment techniques to improve the robustness of the codec to channel errors.

7.2.5 Error Concealment Techniques

The FS-1016 codec uses several techniques to improve its performance over noisy channels. A Hamming (15,11,1) forward error correcting (FEC) code is used to protect the 11 most sensitive bits of each frame. This, together with careful assignment of binary indices to codebook indices and the use of adaptive smoothers at the decoder, yields a codec that is reasonably resilient to channel errors.

The power of the Hamming code is concentrated on the adaptive codebook information because of the sensitivity of this information to channel errors as described in Section 6.6. The 3 most significant bits of the index representing the two absolute adaptive codebook delays are protected by the FEC. The two absolute delays are heavily protected in this way, whereas the two delta-coded delays are not protected at all because of the importance of correctly decoding the absolute delays in order for the delta-coded delays to be received correctly. Also, the most significant bit representing the adaptive codebook gain for each subframe is protected. This gives a total of 10 bits to be protected per frame, and the final protected bit is the "Bishnu" expansion bit described earlier.

Along with the FEC code, the indices of the adaptive codebook delay are assigned using simulated annealing to minimize the effect of a single-bit error in the 8 bits representing each absolute adaptive codebook delay. Adaptive smoothers, which are disabled when the decoding of the (15,11,1) Hamming code indicates error-free conditions, operate on both the fixed and adaptive codebook gains, as well as the adaptive codebook index. Finally, when an error in the 34 bits representing the quantized LSFs causes adjacent LSFs to overlap, this error can be detected by the decoder and action can be taken to mitigate it. When overlapping LSFs are detected at the decoder, an attempt is made to correct them by repeating the two corresponding LSFs from the previous frame. If this does not result in a monotonic set of LSFs, then the entire set of 10 LSFs is replaced with the set from the previous frame. The combination of the measures described above allows the FS-1016 to provide reasonable speech quality at bit error rates as high as 1%.

7.2.6 Decoder Post-Filtering

A traditional short-term pole–zero filter with adaptive spectral tilt compensation, as suggested by Chen [84], is recommended for use at the FS-1016 decoder. Cautious application of the post-filter is suggested, especially in situations where the codec is used in tandem or in noisy conditions. When the codec is used in noisy environments the post-filter may enhance the noise because it is based on the LPC coefficients. Also, post-filtering can be detrimental when the codec is used several times in tandem. In such circumstances it is suggested that all the post-filters are disabled except for that operating at the final decoder.

7.2.7 Conclusion

The FS-1016 standard provided the first use of the CELP principle in a standard codec and allowed reconstructed speech of communications quality for the first time at a bitrate as low as 4.8 kbits/s. It also showed reasonable resilience to channel errors, and it operated well in the presence of acoustic background noise. Use of a ternary-valued overlapped fixed codebook meant that the codec could be implemented in real time using readily available DSP chips. Also, the standard was flexible enough to allow only segments of either the fixed or adaptive codebook to be searched at the encoder. Thus, it allowed the complexity of the codec to be reduced if lower reconstructed speech quality was acceptable. More recently, a dynamic partial search scheme has been proposed [177] for the fixed codebook which reduces the codebook search complexity significantly without degrading the reconstructed speech quality.

In closing, we note that this scheme, denoted by FS1016, is compared to various existing standard codecs in Figure 17.4 of Chapter 17.

7.3 THE 7.95 kbps PAN-AMERICAN SPEECH CODEC [132] KNOWN AS IS-54 DAMPS CODEC

This section gives a rudimentary overview of the operation of the 7.95 kbps Pan-American advanced mobile phone system's (DAMPS) [132] speech codec. Similarly to the half-rate Pan-European system, known as GSM (detailed in Section 7.7), the DAMPS' speech codec is also based on the vector sum excited linear predictive (VSELP) principle proposed by Gerson and Jasiuk [178, 179]. Its schematic, portrayed in Figure 7.2, is also similar to the half-rate GSM codec's schematic seen in Figure 7.12 in that the fixed codebook entry is a linear combination of two scaled vectors. This allows for a high grade of flexibility in terms of the excitation vector shape, as it is also argued in more depth in Section 7.7. This scheme is compared to a range of existing standard codecs in Figure 17.4 of Chapter 17.

The codec's bit-allocation scheme is shown in Table 7.4, while the reasons for using the specified number of bits are detailed during our later discussions. Similarly to other medium-rate speech codecs, 38 spectral quantization bits per 20 ms are allocated for the reflection coefficients, corresponding to a 1.9 kbps bitrate contribution. The specific number of bits used for the individual reflection coefficients is 6, 5, 5, 4, 4, 3, 3, 3, 3, and 2, starting from the first one. Similarly to the half-rate GSM codec, the fixed point lattice technique (FLAT) [180, 181] was proposed for the IS-54 standard by Gerson. The lattice-based prediction algorithms were the subject of Section 3.7, while for the implementational details the interested reader is

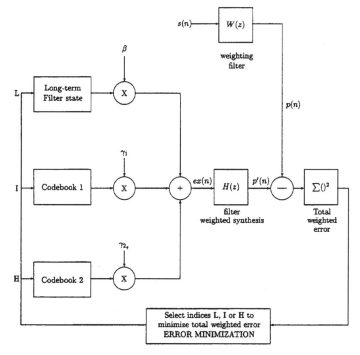

Figure 7.2 The 7.95 kbps DAMPS VSELP encoder's schematic © 1992, [132].

TABLE 7.4   7.95 kbps IS-54 VSELP Codec Bit Allocation © 1992, [132]

Parameter	Bit/Subfr.	Bit/Frame
Reflection coeff.		38
Frame energy $R(0)$		5
Pitchlag L	7	$4 \times 7 = 28$
CB-entries I, H	$7 + 7$	$4 \times 14 = 56$
Gains, β, γ_1, γ_2	8	32
Total		159/20 ms

referred to the Standard [132]. Suffice to say here that the technique computes the reflection coefficients iteratively, always producing the optimum jth order predictor at each stage of the iteration, which can be quantized, before the next coefficient is determined. Each forthcoming predictor stage can therefore compensate for the quantization error of the optimum reflection coefficient determined at the previous stage.

Five bits are used to quantize the energy of the speech frame. As we will see, the same procedure was used for the half-rate VSELP GSM codec's bit-allocation scheme in Table 7.10, which was also partially designed by Gerson and Jaisuk [178,179]. The gains, β, γ_1, γ_2 in Figure 7.2 are quantized using a total of 8 bits/5 ms subframe, contributing 32 bits/20 ms frame. The adaptive codebook index or pitch-lag is represented by 7 bits/5 ms subframe, corresponding to 128 possible pitch values. For reasons of robustness against channel errors, no differential coding of the pitch-lag was used. Lastly, both fixed codebook entries are represented by a 128-entry, 7-bit codebook. When combining all possible codebook entries and gain factors, the total number of legitimate excitation patterns per subsegment becomes $128 \times 128 \times 128 \times 256 = 536\,870\,912$, while maintaining an acceptable computational complexity. This is achieved using a suboptimum solution, whereby the three codebooks are searched through consecutively, identifying the best entry of the adaptive codebook first and then the fixed entries. The decoder's operation is characterized by Figure 7.3, which is essentially constituted by the synthesizer section of the encoder, extended by the spectral post-filter.

In order to provide source-sensitivity matched error protection for the speech bits, similarly to most mobile radio speech transmission schemes, the 159 speech bits are divided in a number of protection classes. The most sensitive 12 bits are assigned a 7-bit cyclic redundancy checking (CRC) pattern, which is used by the decoder for invoking bad frame masking, as shown in Figure 7.4. This could be due to channel errors, or to the fast associated control channel stealing a speech frame for conveying a very urgent control message, such as a handover request. In this case, the speech frame is obliterated, and at the decoder it has to be replaced by a repeated speech segment. However, this simple post-processing can only mitigate the frame loss for periods below 100 ms or five consecutive frames.

As portrayed in Figure 7.4, the 159 speech bits are subdivided into 77 Class One bits and 82 Class Three bits. The more important class-1 bits are half-rate convolutionally encoded, while the remaining 82 bits are transmitted unprotected. This implies that the Class Two bits are always more prone to error. The convolutional encoder processes $77 + 7 + 5 = 89$ bits, where the 5 tailing bits are required by the constraint-length five code to flush its buffer before the transmission of the next speech frame. This allows us to curtail the propagation of transmission errors across frame boundaries, which would otherwise result in prolonged speech degradation due to the decoder's deviation from the

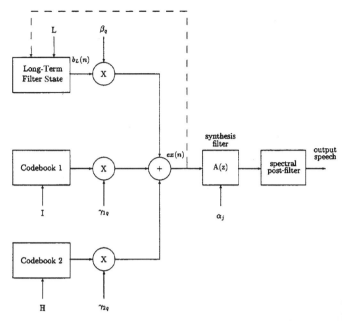

Figure 7.3 The 7.95 kbps DAMPS VSELP decoder's schematic © 1992, [132].

error-free trellis path. There will, however, still be error propagation through the codec's adaptive codebook. The $2 \times 89 = 178$ protected Class-One bits and the 82 Class-Two bits are then ciphered for the sake of confidentiality, and 260 bits/20 ms are transmitted to the decoder. The resulting error-protected bitrate is the same as the unprotected full-rate RPE-coded GSM rate. Lastly, interleaving over two consecutive speech frames takes place in order to disperse bursty channel errors, which tend to overload the error correction capability of the channel

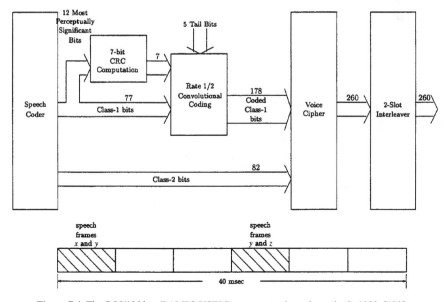

Figure 7.4 The 7.95/13 kbps DAMPS VSELP error protection schematic © 1992, [132].

decoder. As displayed in Figure 7.4, there are three time slots per transmission frame in IS-54, and the channel-coded Class-One bits are dispersed over two consecutive 20 ms speech frames. Hence, each transmission packet contains bits from two consecutive speech frames, namely, frame x, y and y, z, respectively.

Again, this scheme is compared to a range of existing standard codecs in Figure 17.4 of Chapter 17 in subjective speech quality terms.

7.4 THE 6.7 kbps JAPANESE DIGITAL CELLULAR SYSTEM'S SPEECH CODEC [133]

Similarly to the 7.95 kbps IS-54 Pan-American codec of Section 7.3, the Japanese Digital Cellular (JDC) system's 6.7 kbps speech codec [133] is also based on the VSELP excitation optimization principle introduced by Gerson and Jasiuk [178, 179]. The schematic of the JDC codec is also quite similar to that of the IS-54 arrangements shown in Figures 7.2 and 7.3, apart from the fact that in the JDC system only one codebook is used. This naturally restricts the number of different excitation vectors and so results in a somewhat lower bitrate and speech quality. The corresponding bitrate allocation is summarized in Table 7.5, while the associated subjective speech quality of this scheme is compared to a range of existing standard codecs in Figure 17.4 of Chapter 17.

As seen in Table 7.5, a total of 36 bits are used for spectral quantization, where the specific number of bits used for the individual reflection coefficients is 5, 5, 4, 4, 4, 3, 3, 3, 3, and 2. Similarly to the 5.6 kbps half-rate GSM codec and the 7.95 kbps IS-54 codec, the fixed point lattice technique (FLAT) [180, 181] was proposed by Gerson. Recall that lattice-based prediction algorithms were detailed in Section 3.7, and the implementational aspects can be found in the Standard [133]. We remind the reader here that the reflection coefficients are determined iteratively, generating the optimum jth order lattice-based predictor at each stage of the iteration, which can be quantized, before the next coefficient is determined. Therefore, as argued for both the half-rate GSM codec and the IS-54 scheme, the effect of the quantization errors of the reflection coefficients of each predictor stage can be taken into account during the computation of the next reflection coefficient.

Similarly to the IS-54 codec characterized by Table 7.4 and to the half-rate GSM scheme of Table 7.10, 5 bits are used to quantize the energy of the speech frame. The adaptive and fixed codebook gains β, γ_1 of Figure 7.2 are jointly vector-quantized using a total of 7 bits/5 ms subframe, requiring 28 bits/20 ms frame. The adaptive codebook index or pitch-lag is represented by a 7 bits/5 ms subframe, encoding 128 possible pitch values. Despite the low-bitrate constraint, for error-resilience reasons no differential coding of the pitch-lag was employed. Lastly, the 512-entry fixed codebook address is encoded using a 9 bits/5 ms

TABLE 7.5   6.7 kbps JDC Codec Bit Allocation
© [133]

Parameter	Bit/Subfr.	Bit/Frame
Reflection coeff.		36
Frame energy $R(0)$		5
Pitchlag L	7	$4 \times 7 = 28$
CB-entries I	9	$4 \times 9 = 36$
Gains, β, γ_1	7	$4 \times 7 = 28$
Soft-interp. bit		1
Total		134/20 ms

subframe. When combining all the codebook entries and gain factors, the total number of excitations per subsegment becomes $128 \times 128 \times 512 = 8\,388\,608$, which is substantially lower than the corresponding number of $128 \times 128 \times 128 \times 256 = 536\,870\,912$ used by the higher-rate IS-54 codec. Naturally, for complexity reasons a full search is impractical; hence, a suboptimum solution is to search through the three codebooks consecutively. First, the best entry of the adaptive codebook is found and then the fixed entry.

The soft-interpolation flag of Table 7.5 is used to signal for the decoder which of two specific sets of filter coefficients was used by the encoder, where the first one is generated without interpolation and the second with interpolation. Explicitly, this bit is used to inform the decoder whether the current frame's prediction residual energy was lower with or without interpolating the direct form LPC coefficients. Again, we will see during our further discourse in the context of Table 7.10 that this technique was also employed in the half-rate GSM codec. For details of the codebook construction and other algorithmic issues, the interested reader is referred to the Recommendation [133]. Let us now consider the associated channel coding aspects.

The JDC system's channel coding scheme seen in Figure 7.5 exhibits a similar structure to that of the IS-54 arrangement portrayed in Figure 7.4. The 134/20 ms speech bits are divided into two main protection classes. The perceptually most sensitive 44 bits are assigned a 7-bit cyclic redundancy checking (CRC) segment, which is again invoked by the decoder for activating bad frame masking. As displayed in Figure 7.5, the 134 speech bits are grouped in 75 Class-One bits and 59 Class-Two bits. The perceptually more significant Class-One bits are 9/17-rate convolutionally encoded, while the 59 Class-Two bits remain unprotected. This implies that the Class-Two bits are always more prone to errors. The $75 + 7 + 5 = 87$ input bits are convolutionally encoded, where the 5 tailing bits are necessitated by the constraint-length five code to clear its buffer, before the next speech frame is transmitted. As a result, both the encoder and decoder begin their operation from an identical, known state, which is beneficial in error-resilience terms.

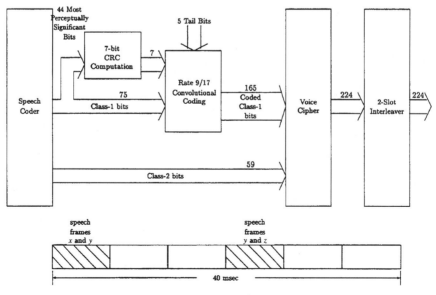

Figure 7.5 The 6.7/11.4 kbps JDC VSELP error protection schematic © [133].

The $(9/17) \times 87 = 164.33$ encoded bits are represented naturally by 165 protected Class-One bits, and the 59 Class-Two bits are then ciphered for the sake of confidentiality and 224 bits/20 ms are transmitted to the decoder. The resulting error-protected bitrate of 11.2 kbps is very close to that of the 11.4 kbps half-rate GSM rate. The latter, however, has a lower speech rate of 5.6 kbps and yet exhibits an improved speech quality. Interleaving is carried out over two consecutive speech frames or 40 ms in order to randomize the bursty channel error statistics and hence to improve the scheme's error resilience. Again, similarly to the IS-54 system's 3-slot per channel structure seen in Figure 7.4, there are 3 time slots per transmission frame in the JDC system and the channel-coded Class-One bits are dispersed over two consecutive 20 ms speech frames. Each transmission packet hosts bits from two consecutive speech frames, namely, frame x, y and y, z, respectively.

Because of the advances in speech compression technology, recently it became realistic to further reduce the bitrate of the first VSELP-based codecs, such as the 7.95 kbps IS-54 and the 6.7 kbps JDC codecs, which led to the development of the 5.6 kbps half-rate GSM codec of Section 7.7. The subjective speech quality of the previously discussed IS54 and JDC coding arrangements is compared to a range of existing standard codecs in Figure 17.4 of Chapter 17.

7.5 THE QUALCOMM VARIABLE-RATE CELP CODEC [182]

7.5.1 Introduction

Among the several different digital cellular mobile phone systems in use around the world, most, including the European GSM, the Japanese JDC, and the American IS-54 standards, use Time Division Multiple Access (TDMA) to allow groups of several users to share the same frequency. However, in July 1993 the American Telecommunications Industry Association (TIA) gave approval to a Code Division Multiple Access (CDMA) system known as IS-95, which is designed to offer an alternative digital system to IS-54. This system was designed by Qualcomm and claims to offer large increases in capacity over IS-54 [182]. Qualcomm designed a speech codec, known as Qualcomm CELP (QCELP) [182], for use in IS-95, and this speech codec is described here. A more detailed description can be found in the IS-95 standard [183].

QCELP is a variable-rate CELP codec that operates at one of four data rates for every 20 ms frame. Which data rate the codec uses is determined by the encoder, depending on the input signal. The four possible data rates are 8, 4, 2 kbits/s, and 800 bits/s. These different rates are known as full-rate, $\frac{1}{2}$ rate, $\frac{1}{4}$ rate, and $\frac{1}{8}$ rate. The speech encoder tries to determine the nature of the input signal and codes active speech frames at full rate and background noise and silence at one of the lower rates. Testing has shown that for a typical conversation the QCELP codec operates at an average bitrate of under 4 kbits/s but provides speech quality equivalent to the 8 kbits/s VSELP codec used in IS-54. As CDMA is used in IS-95, this reduction in the average bitrate of the speech codec is easily exploited to improve the capacity of the system.

In the next section we give an overview of the coding used in the QCELP codec, and the bit allocation used at its various rates. Then we describe how the encoder determines at which rate to code a given speech frame. Finally we give details of the various components used in the QCELP codec.

7.5.2 Codec Schematic and Bit Allocation

A schematic of the QCELP codec is shown in Figure 7.6. For all the data rates except the $\frac{1}{8}$ rate, the codec uses a relatively standard CELP codec structure with a fixed codebook, a pitch filter, and a short-term synthesis filter. At the $\frac{1}{8}$ rate, the codec structure is modified so as to code background noise more efficiently, which is what the $\frac{1}{8}$ rate is used for. No pitch filter is used, and instead of an entry chosen by analysis-by-synthesis (AbS) techniques from a fixed codebook, a gain-scaled pseudorandom series from a random noise generator is used as the excitation to the synthesis filter. At the decoder, post-filtering is used to improve the perceptual quality of the reconstructed speech.

The bit allocation for the various data rates is shown in Table 7.6. For each rate this table shows how many bits are used to code the LPC, pitch, and fixed codebook parameters, and how many times per frame these parameters are determined and coded. At all the rates, the LPC coefficients are determined and transmitted once per 20 ms frame, but at lower rates fewer bits are used for their quantization. Both the pitch filter (when used) and the fixed codebook parameters are coded with 10 bits at all rates, but at different rates these parameters are coded more or less frequently.

7.5.3 Codec Rate Selection

For most frames at the encoder, the QCELP codec decides which of its four data rates to use by comparing the energy of its input over the 20 ms frame to an estimate of the background noise energy. This estimate of the background noise energy is updated in each frame, depending on whether the current input frame has a lower or higher energy than the estimate. If the estimate is higher than the current input energy, then the estimate is reset to the input energy. If, on the other hand, the estimate is lower than the input energy, then it is slightly increased. This means that when no speech is present, the estimate of the background noise energy follows the input energy. When speech is present, the estimate slowly increases, but the fluctuations inherent in the input speech energy cause it to be frequently reset. An

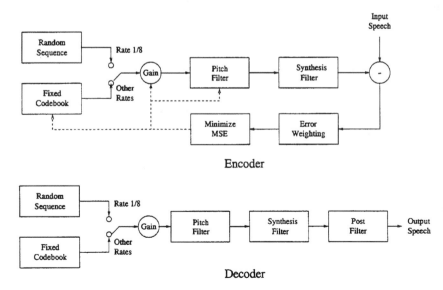

Figure 7.6 QCELP codec.

TABLE 7.6   Bit-Allocation Scheme of the QCELP Codec

Bit Rate	8 kbits/s	4 kbits/s	2 kbits/s	800 bits/s
LPC	40 bits once	20 bits once	10 bits once	10 bits once
Pitch	10 bits 4 times	10 bits twice	10 bits once	0 bits
Code Book	10 bits 8 times	10 bits 4 times	10 bits twice	6 bits once
Total	160 bits per frame	80 bits per frame	40 bits per frame	16 bits per frame

example of the variations in the input speech energy and the background noise estimate can be found in [182].

To select which of the four data rates to use, the encoder uses a set of three thresholds that "float" above the running estimate of the background noise energy. If the energy of the input signal is higher than all three thresholds, then the encoder selects the full rate. Otherwise one of the lower rates is selected.

This comparison of the input signal's energy to three floating thresholds is how the encoder data rate is selected for most frames. However, the encoder can also be instructed to generate a blank packet to allow for "blank and burst" transmission of signaling information. Also, the encoder can be instructed not to code at the full rate for certain given frames. This allows the network to reduce the average data rate of its existing users and hence increase its capacity to accommodate extra users. This means that the CDMA system has a "soft capacity." When the number of users is greater than the usual system capacity, extra users can be accommodated by slightly decreasing each user's codec data rate and hence voice quality.

In the following sections we describe the formant and pitch filters used in the QCELP codec, the excitation used for these filters, the post-filtering used at the decoder to improve the perceptual quality of the reconstructed speech and finally the error protection and concealment techniques used.

7.5.4 LPC Analysis and Quantization

A tenth-order LPC synthesis filter is used in the QCELP codec. The filter coefficients are determined from the input speech using the autocorrelation method. A 160-sample Hamming window, centered between the 139th and 140th sample of the current 160-sample frame is used to calculate autocorrelation values. These are then converted to filter coefficients using the Levinson-Durbin algorithm, and a 15-Hz bandwidth expansion is applied before the coefficients are converted to line spectral pairs (LSPs). The 10 LSPs are quantized using a scalar predictive quantizer as shown in Figure 7.7.

This predictive quantizer operates as follows. Initially, each LSP has an offset or a "bias" value subtracted. These bias values are the values of the LSPs when the input speech has flat spectrum. The bias used for LSP_i is given by

$$\text{Bias} = \frac{0.5i}{p+1}(0.0454545\ldots)i \quad 1 \le i \le 10, \tag{7.4}$$

where $p = 10$ is the order of the filter. The bias offset LSPs then have a predicted value for the offset LSP subtracted, and the difference between the actual offset LSP and the predicted offset LSP is quantized with either 4, 2, or 1 bits, depending on the codec rate. The predicted values for the offset LSPs are simply given by the value of the quantized offset LSP for the

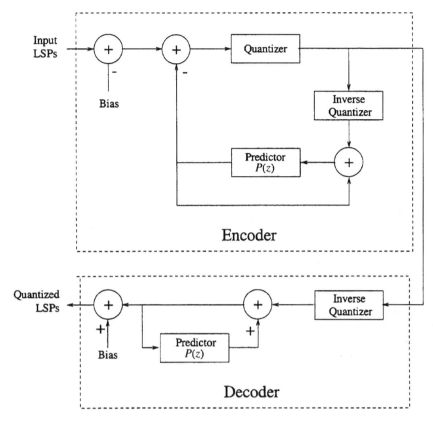

Figure 7.7 LSP quantization.

previous frame multiplied by 29/32, implying that the predictor has a transfer function $P(z)$ given by

$$P(z) = 0.90625z^{-1}. \tag{7.5}$$

At the decoder, and also in the local decoder in the encoder, the LSPs are reconstructed by inverse quantizing the transmitted quantizer index, adding first the predicted offset LSP and then the bias value to give the quantized LSPs. The stability of the synthesis filter is ensured by forcing the LSPs to be ordered. The decoder also ensures that the frequencies are separated by at least 80 Hz in order to avoid unusually large peaks in the frequency response of the synthesis filter. For rates below the full rate, only 1 or 2 bits are used to quantize each LSP. The quantization is therefore very noisy. In order to remove some of this quantization noise for codec rates below the full rate, the LSPs are low-pass filtered. The extent of this filtering is set depending on the current rate of the codec. Also, if 10 or more consecutive-rate $\frac{1}{4}$ or $\frac{1}{8}$ frames are received, or if a frame erasure occurs, then the extent of the filtering is dramatically increased. Finally, before converting the LSPs back to filter coefficients, LSP interpolation between the quantized LSPs for the current and the previous frame is used to determine LSPs for each pitch and fixed codebook subframe. At different rates the QCELP codec has different numbers of subframes per 20 ms frame, and so the interpolation used depends on the rate of the codec.

Having determined the synthesis filter coefficients to use, the encoder next calculates the pitch filter coefficient and delay.

7.5.5 The Pitch Filter

The QCELP codec uses a pitch filter of the form

$$\frac{1}{1 - bz^{-L}} \tag{7.6}$$

to represent the long-term periodicities present in voiced speech, where the pitch gain b and lag L are determined once per pitch subframe by AbS techniques in the encoder and transmitted to the decoder. The pitch subframes are 5 ms, 10 ms, and 20 ms long for full, $\frac{1}{2}$, and $\frac{1}{4}$ rate frames, respectively. In $\frac{1}{8}$ rate frames, the pitch filter is not used. The pitch lag L is represented with 7 bits and can take integer values between 17 and 143. This means that it can take only 127 different values rather than the 128 distinct values that can be represented with 7 bits. The 128th value ($L = 16$) is used to denote a pitch gain b of 0. The gain b has nine possible values, uniformly spaced in the range $0 \le b \le 2$ with steps of 0.25. Three bits are used to code the chosen value of b when this value is greater than 0, and $b = 0$ is coded by setting $L = 16$.

The pitch delay and gain L and b are determined using AbS techniques to minimize the weighted error between the original speech and the synthesized speech. When these parameters are determined, the output from the fixed codebook is unknown and so is assumed to be zero. A weighting filter of the form

$$W(z) = \frac{\hat{A}(z)}{\hat{A}(z/\gamma)} \tag{7.7}$$

is used to determine both the pitch filter and fixed codebook parameters. Here $\hat{A}(z)$ is the inverse synthesis filter using the quantized interpolated filter coefficients, and γ is a constant (0.8). In the AbS search for the best pitch lag L and gain b, each possible value of L and b is tested to see which pair of values minimizes the weighted error between the synthesized and the original speech. This is in contrast to the more usual approach in determining the pitch parameters where the best delay is chosen using AbS techniques, but the quantization of the gain is done outside the AbS loop.

Another difference between the representation of the voicing information in QCELP and in most CELP type codecs is that a pitch filter is used instead of an adaptive codebook. Using a pitch filter gives the same excitation signal to the synthesis filter as an adaptive codebook except for pitch delays L shorter than the pitch subframe. To recap, using the notation of Section 3.4, the output from an adaptive codebook is given by $G_1 * u(n - \alpha)$, where $u(n)$ is the excitation to the synthesis filter, G_1 is the codebook gain, and α is the delay. When the adaptive codebook parameters G_1 and α, which are equivalent to b and L for the pitch filter in the QCELP codec, are determined, the excitation signal $u(n)$ is known only for the previous subframes. Hence, $u(n - \alpha)$ cannot be determined for $n \ge \alpha$. This problem is most often overcome by repeating the available excitation signal in the adaptive codebook—in other words, by using $u(n - \alpha)$ for $0 \le n < \alpha$, $u(n - 2\alpha)$ for $\alpha \le n < 2\alpha$, and so on.

In the QCELP pitch filter, an alternative approach is taken. In the AbS determination of the pitch lag L and gain b for lags L shorter than the subframe length, the available past excitation $u(n - L)$ is extended for $n \ge L$ using the delayed "formant residual" as an estimate. This formant residual is given by the original speech signal $s(n)$ filtered through the inverse synthesis filter $\hat{A}(z)$. This estimate is used only when determining L and b—for the determination of the fixed codebook parameters and in the decoder when generating the synthesized speech this estimate is not needed.

Once the pitch filter parameters L and b have been determined, the fixed codebook parameters are found.

7.5.6 The Fixed Codebook

For all the coding rates except the $\frac{1}{8}$ rate, a fixed codebook searched using AbS techniques is used to provide the excitation to the pitch and synthesis filters. This codebook is described in this section. For the $\frac{1}{8}$ rate, the codec uses a pseudorandom sequence for the excitation.

For the full, $\frac{1}{2}$, and $\frac{1}{4}$ coding rates, the fixed codebook is searched every 2.5, 5, or 10 ms. This means that there are two fixed codebook subframes for every pitch subframe. A 7-bit Gaussian vector codebook is used, together with a 3-bit gain codebook. This gives a total of 10 bits per subframe to represent the fixed codebook information. In order to reduce the complexity of the codebook search, a circular recursive codebook with 128 entries $c(0), c(1), \ldots c(127)$ is used. The kth codebook entry $c_k(n)$ is then given by $c([n - k]_{\mathrm{mod}128})$. This means the $k + 1$th codebook entry is equal to the kth entry shifted by one place, with one new sample added at $c_{k+1}(0)$ and one sample dropped at $c_{k+1}(Lc - 1)$, where L_c is the fixed codebook subframe length (20, 40, or 80 samples depending on the coding rate). The recursive nature of the codebook then allows the convolutions of $c_k(n)$ with the impulse response $h(n)$ of the weighted synthesis filter, which are carried out during the AbS search of the codebook, to be calculated recursively. This significantly reduces the complexity of the codebook search. To further simplify this search, the 128 entries of the recursive codebook are center clipped so that approximately 80% of them are zero.

Like the AbS search for the pitch filter parameters, the fixed-codebook parameters are determined by searching for both the best codebook entry k and the best quantized gain G within the AbS loop. This codebook gain is quantized with 3 bits: one for its sign and two for its magnitude. The magnitude is quantized in the log domain using a scalar predictive quantizer similar to that used for the LSPs and shown in Figure 7.7. However, the gain quantizer does not subtract a bias value before the quantization, and it uses a second-order nonlinear predictor function rather than the simple first-order predictor $P(z)$ given in Equation 7.5 used in the LSP quantizer.

7.5.7 Rate $\frac{1}{8}$ Filter Excitation

At the $\frac{1}{8}$ rate, the excitation to the synthesis filter (the pitch filter is not used at this rate) is modified to allow the codec to encode background noise more efficiently. Instead of using the recursive center clipped Gaussian codebook used at the higher rates and described in the previous section, a pseudorandom number generator is used to give the filter excitation. This excitation is scaled by a gain, which is always positive, but its magnitude is quantized with 2 bits in the same way as the codebook gain is quantized for the higher rate frames. For rate $\frac{1}{8}$ packets, only the gain that is then used is the average of the gain magnitude for the previous frame (or subframe if the previous frame was at a higher rate than $\frac{1}{8}$) and the quantized gain for the present frame. This effectively low-pass filters the gain and prevents burstiness in the level of the background noise. The gain is also interpolated during the length of the frame to give a smooth variation in the level of the reconstructed background noise.

To ensure that the encoder and decoder use the same pseudorandom sequence, and hence keep the memory of their filters identical, the random number generators in both the encoder and decoder use the transmitted 16-bit packet as their seed. As well as the 10 bits used to represent the LSPs for the frame and the 2 bits used to quantize the gain for the frame, the encoder adds 4 pseudo-random bits to ensure that the 16-bit packet that is transmitted is random.

7.5.8 Decoder Post-Filtering

At the decoder a post-filter similar to those described for other codecs is used. The post-filter has a transfer function $PF(z)$ given by

$$PF(z) = B(z)\frac{\hat{A}(z/\alpha)}{\hat{A}(z/\beta)}, \tag{7.8}$$

where $\hat{A}(z)$ is the inverse synthesis filter and α and β are constants, equal to 0.5 and 0.8, respectively. $B(z)$ is a spectral tilt compensation filter, and the filter $\hat{A}(z/\alpha)/\hat{A}(z/\beta)$ gives a short-term post-filter that emphasizes the formant peaks in the reconstructed speech and attenuates the valleys between these peaks. This renders the coding noise less audible and so improves the perceptual quality of the reconstructed speech. However, this filter also introduces a spectral tilt to the reconstructed speech which can result in the speech sounding muffled. This effect is corrected using the spectral tilt compensation filter $B(z)$, which is given by

$$B(z) = \frac{1 - gz^{-1}}{1 + gz^{-1}}, \tag{7.9}$$

where g is determined based on the average of the 10 interpolated LSPs.

Finally, gain compensation is applied at the output of the post-filter to ensure that the energy of its input and output are roughly equal. A scaling factor given by the square root of the ratio of the energies of the input and output of $PF(z)$ is calculated and filtered with a first-order IIR filter before being used to scale the output from $PF(z)$ to give the decoded speech.

7.5.9 Error Protection and Concealment Techniques

For full-rate frames, the 18 most perceptually sensitive bits (the most significant bits from the 10 LSPs and the 8 fixed codebook gain magnitudes) are protected with 11 parity bits from a (29,18) cyclic code. This code allows the decoder to provide error detection and correction for these 18 most sensitive bits. The decoder is also able to deal with packets that are declared "erased." Such packets occur when the decoder has been unable to satisfactorily determine the coding rate or when the decoder determines that a full-rate frame was sent, but the (29,18) cyclic code protecting the 18 most sensitive bits of the frame is overloaded. When such erasures occur, the decoder takes the following steps

1. The LSPs are decayed toward their white noise "bias" values.
2. The previous pitch-lag L is used.
3. The pitch gain b is decayed toward zero.
4. A random codebook index is chosen.
5. The fixed-codebook gain is decayed toward zero.

Decaying these parameters toward their background levels helps prevent annoying squeaks or whistles that can occur in the reconstructed speech when bit errors arise. It is stated in [182] that when operating under typical conditions in a CDMA system the quality of the reconstructed speech in the QCELP codec is very close to that achieved over an error-free channel.

7.5.10 Conclusion

In this section we have described the techniques used in the QCELP variable-rate speech codec. This codec produces speech quality equivalent to that of the 7.95 kbits/s IS-54 VSELP codec of Section 7.3, but at an average bitrate of less than 4 kbits/s. This reduction in the average bitrate of the codec is exploited by the CDMA system used in IS-95 to almost double the user capacity of the system. This codec marks the end of the first-generation CELP-based codecs. CELP schemes and their relatives, such as VSELP codecs, constituted an important milestone in the history of speech compression, reaching bitrates as low as 4.8 kbps in the DoD codec.

For rates below 4.8 kbps however, further advances were required. These advances were fueled by two factors. First, the ever increasing demand for accommodating more speech users in the allocated bandwidth of existing mobile radio systems led to the development of a half-rate coding standard, doubling the number of users supported. This trend is hallmarked by the 3.6 kbps half-rate Japanese codec of the next section, as well as by the 5.6 kbps half-rate Pan-European GSM standard of Section 7.7. The second trend was the arrival of a range of enhanced full-rate codecs, such as that of the Pan-American Qualcomm system and the enhanced version of the IS-54 system's speech codec, referred to as the IS-136 standard arrangement, which is the subject of Section 7.11. The Pan-European GSM system was also endowed by a new enhanced full-rate scheme, which is discussed in Section 7.7. As the first representative of this new generation of CELP-based schemes in the next Section, let us now consider the 3.6 kbps half-rate Japanese codec.

7.6 JAPANESE HALF-RATE SPEECH CODEC [133]

7.6.1 Introduction

The Japanese full-rate speech codec [133] was developed by Ohya, Suda, and Miki [184,185], and in this section we follow their approach in describing the codec's main features, which employs vector sum excited linear prediction (VSELP) based coding at 6.7 kb/s. This speech-coded rate is increased by 4.5 kb/s of error protection, giving a total channel-coded rate of 11.2 kb/s. Thus, the half-rate speech codec is expected to operate at a rate beneath 4 kb/s. In order to sufficiently reduce the codec's bitrate, the excitation vector length passed to the synthesis filter has to be increased to about 8 to 10 ms. Hence, the excitation vector may contain several pitch period cycles. The random nature of the conventional CELP codevectors, described in Chapter 6, poorly represents these excitations. Thus, employing purely random excitations seriously degrades the performance of any CELP-based speech codec operating at rates beneath 4 kb/s. The Japanese half-rate speech codec [133] overcomes these excitation vector problems by employing the pitch synchronous innovation code-excited linear prediction (PSI-CELP) principle [186]. This PSI-CELP codec operates at a rate of 3.6 kb/s, with an additional 2.0 kb/s allocated for error protection, producing a channel-coded rate of 5.6 kb/s. This codec is described in detail next.

7.6.2 Codec Schematic and Bit Allocation

The Japanese half-rate speech codec operates on the basis of 40 ms speech frames, processing an input speech bandwidth of 0.3 kHz to 3.4 kHz, and a sampling frequency of 8 kHz. The encoder and decoder schematics are given in Figures 7.8 and 7.9, respectively. Four 10 ms subframes are used within each 40 ms speech frame. Initially, the power of each subframe is vector-quantized using a total of 6 bits/frame. The 10 LPC coefficients are then vector-quantized in the LSP domain for the second and fourth subframe, while the parameters of the first and third subframes are not explicitly transmitted. They are regenerated at the decoder with the aid of interpolation between the transmitted subframes' parameters. In the second and fourth subframes, a moving-average predictor and two-stage vector quantization are employed for each subframe, using a total of 30 bits/frame for LSP quantization.

Similarly to traditional CELP codecs [76], the PSI-CELP speech codec uses two excitation vectors in order to represent the excitation for each subframe. The first excitation vector is generated either by an adaptive codebook using fractional delays or by a fixed random codebook. More explicitly, if the input excitation has insufficient periodicity in order to warrant the employment of the adaptive codebook, which would be typically used in voiced segments, then the first excitation vector is generated by a fixed random codebook. A total of 8 bits/subframe are used for encoding the first excitation vector, as detailed in Section 7.6.6.

The second excitation vector is generated by the superposition of two fixed codebooks (see Section 7.6.7). As regards the construction of the second excitation vector, if the first excitation vector is selected from the fixed random codebook, then the two subcodebooks operate as in a typical CELP system. However, if the first excitation vector is generated by the adaptive codebook, then the second excitation vector is constituted by a sequence having the length of the pitch determined by the adaptive codebook. This sequence is selected from each

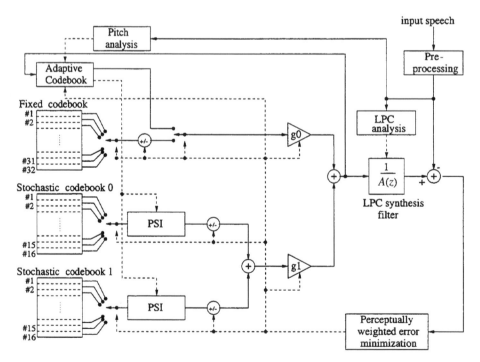

Figure 7.8 The Japanese half-rate speech encoder's schematic.

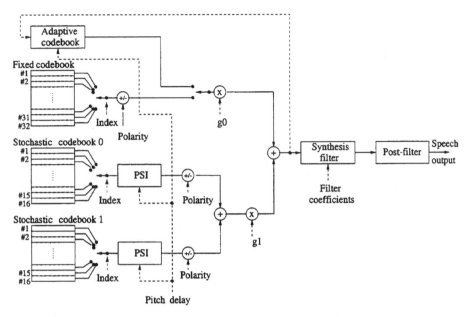

Figure 7.9 The Japanese half-rate speech decoder's schematic.

subcodebook. This sequence is then repeated in order to represent the subcodebook's contribution to the second excitation vector. This unique feature of the PSI-CELP codec enhances the periodicity of voiced speech. This principle is often used in speech codecs operating at rates beneath 4 kb/s. For each excitation optimization subframe, the gains of the excitation vectors are vector-quantized using a 7-bit codebook. The entries constituting the two excitation vectors and the codebook gains are optimized for each subframe using an analysis-by-synthesis search in order to minimize the weighted error between the original and reconstructed speech.

At the decoder the transmitted parameters are decoded in order to produce the filter coefficients for the synthesis filter and to select the codebook entries for the excitation vectors and gain codebooks. Adaptive post-filtering [86] is used to improve the perceptual quality of the reconstructed speech. The bit-allocation scheme of the Japanese half-rate speech codec is summarized in Table 7.7. We now describe in more detail the various blocks shown in Figures 7.8 and 7.9.

TABLE 7.7 Bit Allocation for the 40 ms Duration LPC Frame of the Japanese Half-rate Speech Codec, which is constituted by four 10 ms duration subframes

Parameter	Bits/Frame
LSP parameters	30
Power	6
Excitation vector 1	8×4
Excitation vector 2	12×4
Gain vector	7×4
Total/40 ms	144 (3.6 kb/s)

7.6.3 Encoder Pre-Processing

The input speech signal is bandlimited to the frequency range of 0.3 kHz to 3.4 kHz. Subsequently, the power of each 10 ms subframe is computed, transformed to the logarithmic domain, and stored in a four-dimensional vector. Since the energy levels of the consecutive 1-ms speech subsegments are similar, vector quantization of these values results in coding economy. A total of 6 bits per 40 ms speech frame was found to be adequate for their quantization, which corresponds to a codebook size of 64 entries, as seen in Table 7.7.

7.6.4 LPC Analysis and Quantization

For the Japanese half-rate speech codec, tenth-order LPC analysis is performed, and the LPC coefficients are transformed to the LSP domain, as highlighted in Chapter 4. The LPC coefficients are calculated twice for every 40 ms duration speech frame, namely, for the second and fourth 10 ms subframes. By contrast, for the first subframe the LPC coefficients are calculated from the average of the LSPs of the fourth subframe in the previous speech frame and from those of the second subframe in the current 40 ms speech frame. Concerning the third subframe, the LPC coefficients are determined from the average of the LSPs in the second and fourth subframes of the current speech frame. For the second and fourth subframes, the window employed during the LPC analysis is a nonsymmetric window of 35.8 ms, which is calculated from the impulse response of an autoregressive (AR) filter. This nonsymmetric window ensures that no future samples are required for the LPC analysis.

The LSP vector quantization process allocates a total of 30 bits per 40 ms speech frame, as shown in Tables 7.7 and 7.8. The LSP vector to be quantized is initially estimated by moving-average (MA) prediction. This moving-average prediction exploits the high correlation between the LSPs in adjacent frames, while ensuring that channel errors only propagate to a fixed number of frames. For details of the specific MA predictor, the interested reader is referred to [133]. Suffice to say here that a 1-bit flag is used by the codec in order to differentiate between two different MA predictor coefficients, as seen in Table 7.8.

In each of the 10 ms subframes, the LSP vector estimates produced by the moving-average predictor are quantized using a two-stage vector quantization (VQ) process. For the second 10 ms subframe, the first VQ stage involves a 7-bit or 128-entry vector quantizer. The difference between the LSP vector to be quantized and the best-matching entry of the above-mentioned 7-bit codebook is calculated, and their difference is passed to the second VQ stage,

TABLE 7.8 Bit Allocation of the Japanese Half-rate Speech Codec's LSP Quantization Scheme

Parameter	Bits/Frame
MA coefficients	1
2nd subframe	
1st stage	7
2nd stage	3
4th subframe	
1st stage	7
2nd stage (low freq)	6
2nd stage (high freq)	6
Total/20 ms	30

which employs 3-bit vector quantization. The associated number of LSP coding bits is also shown in Table 7.8. In the fourth subframe, the first-stage quantization is also based on a 7-bit vector quantizer. However, in the second-stage vector quantization, the 10-dimensional LSP difference vector created by subtracting the 7-bit codebook-entry from the unquantized LSP vector is split into a five-dimensional low-frequency LSP vector and a five-dimension high-frequency LSP vector. Both the lower- and higher-frequency vectors are vector-quantized using 6 bits. Again, Table 7.8 summarizes the bit allocation of the LSP quantization process.

To elaborate a little further, initially eight candidate vectors are selected for the first stage of the fourth subframe's vector quantization process. By contrast, four LSP candidate vectors are selected for the second VQ stage of the fourth subframe. To proceed further, from these candidate vectors four candidates are selected for the fourth subframe's vector quantization process. Concerning the second subframe, for every candidate LSP vector chosen for the fourth subframe, eight further candidate vectors are selected in each vector quantization stage. The final process is to select the best combination of candidate LSP vectors for both the second and fourth subframes. This VQ process achieves a good tradeoff between complexity and LSP vector quantization accuracy.

7.6.5 The Weighting Filter

The weighting filter used by the Japanese half-rate speech codec is based on the unquantized LPC filter coefficients α_i. The transfer function of the weighting filter is comprised of a spectral weighting filter $W_f(z)$ and a pitch weighting filter $W_p(z)$, which is formulated as:

$$W(z) = W_f(z) \cdot W_p(z) \approx W_f'(z) \cdot W_p(z) \tag{7.10}$$

where the individual filter transfer functions are given by:

$$W_f(z) = \frac{1 + \sum_{i=1}^{p} \alpha_i \gamma_1^i z^{-i}}{1 + \sum_{i=1}^{p} \alpha_i \gamma_2^i z^{-i}} \quad (0 \le \gamma_2 \le \gamma_1 \le 1) \tag{7.11}$$

$$W_f'(z) = \sum_{i=0}^{m} \alpha_i z^{-i} \tag{7.12}$$

$$W_p(z) = 1 + \epsilon_1 \sum_{i=-1}^{1} \beta_i z^{-L-i} \tag{7.13}$$

and $p = 10$ and $m = 11$, $\gamma_1 = 0.9$ and $\gamma_2 = 0.4$, and $\epsilon_1 = 0.4$. The calculation of the filter $W_f(z)$ is computationally complex; hence, it is approximated by the FIR filter $W_f'(z)$.

7.6.6 Excitation Vector 1

For the Japanese half-rate speech codec, the structure of the first excitation vector is portrayed at the top left corner of Figure 7.10. Specifically, the first excitation vector contains the adaptive codebook entry used in the traditional CELP codecs described in Chapter 6, together with an entry from a fixed random codebook. This first excitation vector is encoded with a total of 8 bits, resulting in 256 different possible excitations. The adaptive codebook is constituted by 192 entries, while the fixed random codebook has 32 entries, each of which can be multiplied by ±1—again, yielding a total of 256 excitation patterns.

For the adaptive codebook, noninteger pitch delays are used in the range of $L_{min} = 16$ and $L_{min} = 97$, invoking the closed-loop search described in Section 3.4.2. The fixed codebook-based section of the first excitation vector was designed to improve the representa-

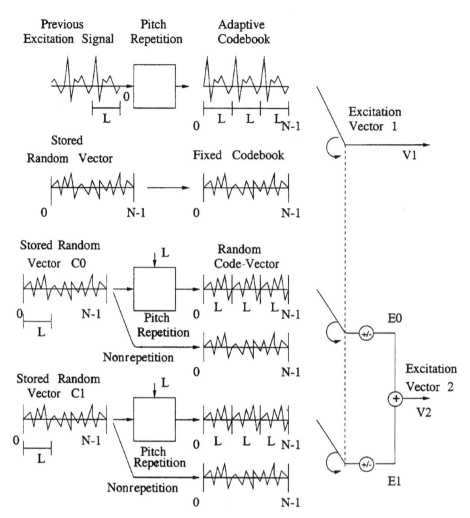

Figure 7.10 Excitation formulation for the Japanese half-rate speech codec.

tion quality of the uncorrelated portions of speech, namely, that of silence, unvoiced, and transient segments. This choice of codebooks follows the technique often used in low-bitrate speech codecs, where voiced and unvoiced speech, as described in Section 1.2, are encoded separately.

The selection between the adaptive codebook and the fixed random codebook is based on a perceptually weighted distortion metric given by:

$$D = \| W(X^* - Y)\|^2 \tag{7.14}$$

where W is a matrix constituted by the impulse response of the perceptual error weighting filter, X^* is the input speech vector containing the current speech subframe, and Y is the corresponding vector containing the synthesized speech.

Initially for the adaptive codebook, six pitch delay candidates are selected, using an open-loop technique. These six candidates are eventually reduced to two candidates following an analysis-by-synthesis-based closed-loop search. An additional two candidate excitations are selected from the fixed codebook. Finally, from these four candidate excitations the best two candidate excitations are selected for the first excitation vector.

7.6.7 Excitation Vector 2

The structure of the second excitation vector is portrayed in the lower portion of Figure 7.10. Each subcodebook is assigned 6 bits, which includes 1 bit for their polarity. The outputs of the two subcodebooks are then combined in order to create the second excitation vector.

During the determination of the second excitation vector, initially the result of optimizing the first excitation is examined. If the fixed codebook was selected, then the two 10 ms duration excitation vectors output by the 6-bit subcodebooks are combined in order to produce the second excitation vector. However, if the adaptive codebook was selected for generating the first excitation vector, then the length of the excitation vectors generated by the subcodebooks is limited to the duration of the associated pitch delay, as seen at the bottom of Figure 7.10. These pitch-length fixed-codebook vector segments are then repeated a number of times until the subframe is filled. Thus, when encountering voiced speech segments, pitch synchronous excitation vectors are produced by each subcodebook. Hence, for uncorrelated portions of speech, such as unvoiced speech segments, the combined excitation vector will contain only random signals. By contrast, for predictable portions of speech, such as voiced speech, the combined excitation vector will contain two pitch synchronous vectors. The employment of the adaptive codebook and fixed random codebooks follows the philosophy of conventional CELP codecs, while the separate encoding of predictable and unpredictable portions of speech reflects the principles of a vocoder. Thus, the PSI-CELP Japanese half-rate speech codec constitutes a hybrid of traditional CELP codecs and traditional vocoders.

7.6.8 Channel Coding

The Japanese half-rate speech codec was designed to cope with a maximum of 3% burst error rate inflicted by the transmission channel. It has been found that the power parameter, the first excitation vector, some of the LSP parameters, and the most significant gain signaling bit are particularly sensitive to noisy channels.

Figure 7.11 shows the channel coding scheme protecting the different bits generated by the encoder, while Table 7.9 summarizes the associated bit allocation of the channel coding

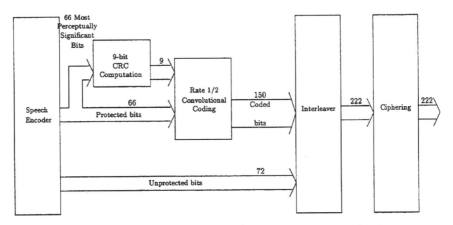

Figure 7.11 Channel coding employed by the Japanese half-rate speech codec.

TABLE 7.9   Bit Allocation for the Channel Coding Scheme Used in the Japanese Half-rate Speech Codec

Parameter	Protection Bits/Frame	Source Bits/Frame
LSP parameters	21	30
Power	6	6
Excitation vector 1	32	32
Excitation vector 2	0	48
Gain vector	4	28
Total/40 ms	63	144

arrangement. Specifically, the 66 most sensitive bits are protected bits, while the remaining 72 bits are left unprotected. Initially, the protected bits are assigned a 9-bit cyclic redundancy check (CRC) code for error detection. Subsequently, the protected bits and the output of the CRC code are passed to a half-rate convolutional code having a memory of 7. Finally, interleaving is performed to randomize the effect of channel error bursts.

7.6.9 Decoder Post-Processing

The initial process at the decoder is to check whether the transmitted frame has been lost due to channel errors. If a lost frame has occurred, then a parameter recovery process is activated, with the lost parameters interpolated from the previous frames' corresponding parameters that have been successfully received. The power is attenuated depending on the current and past frame error events. For a lost speech-coded frame, the LPC coefficients, the first excitation vector, and the gain parameters are replaced by the previous correctly received values. However, in a lost frame, for the second excitation vector, no form of error recovery was undertaken. Following the determination of the parameters, the reconstructed speech is formed and subsequently passed to a post-filter [86].

The post-filter $F(z)$ used in the Japanese half-rate speech decoder is adaptive, and it was designed to improve the perceptual quality of the reconstructed speech. The post-filter's transfer function is described by:

$$F(z) = F_f(z) \cdot F_p(z) \cdot F_h(z) \tag{7.15}$$

where the individual parameters are given by

$$F_f(z) = \frac{1 + \sum_{i=1}^{p} \alpha_{qi}\gamma_3^i z^{-i}}{1 + \sum_{i=1}^{p} \alpha_{qi}\gamma_4^i z^{-i}} \quad (0 \le \gamma_3 \le \gamma_4 \le 1) \tag{7.16}$$

$$F_p(z) = \frac{1}{1 + \epsilon_2 \sum_{i=-}^{1} v_i z^{-L-i}} \tag{7.17}$$

$$F_h(z) = 1 - \eta z^{-1} \tag{7.18}$$

and the individual parameters are given by $p = 10$, $\gamma_3 = 0.5$, $\gamma_4 = 0.8$, $\epsilon_2 = 0.7$, and $\eta = 0.4$.

Specifically, the filter $F_f(z)$ is based on the LPC filter coefficients and was designed to augment the spectral domain formants in the speech spectrum. Hence, it is effectively a short-term post-filter. By contrast, the filter $F_p(z)$ is a 3-tap pitch comb-filter designed to enhance the pitch harmonics in the speech spectrum. It therefore constitutes a long-term post-filter. The third filter described by $F_h(z)$ is a single-tap differential high-pass filter, designed to combat the muffling effect of the long-term and short-term post-filters. If a speech frame loss occurs at the decoder, then the post-filter parameters are changed to $\epsilon_2 = 0.4$ and $\eta = 0.0$, reducing the effect of long-term post-filtering and removing the high-pass filter. Following the post-filter, automatic gain control is employed in order to restore the original speech energy level.

As in the context of the other speech codecs considered, this coding arrangement will be compared in subjective speech quality terms to a range of existing standard codecs in Figure 17.4 of Chapter 17, where it is denoted by JDC/2. In the next section we will consider the 5.6 kbps half-rate GSM codec, which is based on a refined VSELP codec, a principle that was used in the 7.95 kbps IS-54 and the 6.7 kbps JDC codecs.

7.7 THE HALF-RATE GSM SPEECH CODEC

7.7.1 Half-rate GSM Codec Outline and Bit Allocation

In what follows we briefly highlight the techniques proposed by Gerson, Jasiuk, Mueller, Nowack, and Winter [187], which led to the definition of the half-rate GSM standard codec employing a 5.6 kbps vector sum excited linear predictive (VSELP) codec [178, 179]. The codec's schematic is shown in Figure 7.12, where two different block diagrams characterize its operation in four different operational modes. In Mode 0 the codec obeys the schematic portrayed at the top of Figure 7.12, while in the remaining three modes, Mode 1, 2, and 3, it is configured, as seen at the bottom of Figure 7.12. The analysis synthesis filter's coefficients are determined every 20 ms, and this interval is divided in four 5 ms excitation optimization subsegments, corresponding to 160 and 40 samples, respectively, when using a sampling frequency of 8 kHz. In our forthcoming discussion we focus our attention on the above-mentioned different operating modes and the corresponding schematics.

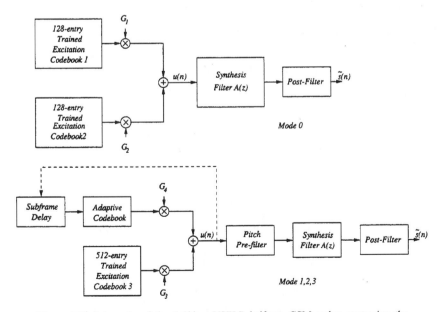

Figure 7.12 Schematic of the 5.6 kbps VSELP half-rate GSM codec, portraying the unvoiced Mode 0 and the voiced Modes 1, 2, and 3.

TABLE 7.10 Bit-allocation Scheme of the 5.6 kbps
VSELP Half-rate GSM Codec

Parameter	Bits/frame
LPC coefficients	28
LPC interpolation flag	1
Excitation mode	2
Mode 0:	
Codebook 1 index	$4 \times 7 = 28$
Codebook 2 index	$4 \times 7 = 28$
Modes 1, 2, 3	
LTPD (subframe 1)	8
Δ LTPD (subframes 2, 3, 4)	$3 \times 4 = 12$
Codebook 3 index	$4 \times 9 = 36$
Frame energy E_F	5
Excitation gain-related quantity $[E_s E_1]$	$4 \times 5 = 20$
Total no. of bits	112/20 ms
Bitrate	5.6 kbps

The codec's bit-allocation scheme is summarized in Table 7.10 for the synthesis modes of 0–3. The speech spectral envelope is encoded by allocating 28 bits/20 ms synthesis frame for the vector quantization of the reflection coefficients. A soft interpolation bit is used to inform the decoder as to whether the current frame's prediction residual energy was lower with or without interpolating the direct form LPC coefficients.

As mentioned before, there are four different synthesis modes, corresponding to different excitation modes, implying the presence of different grades of voicing in the speech signal. As seen in the table, 2 bits/frame are used for excitation mode selection. The decisions as to what amount of voicing is present, and hence which excitation mode has to be used, are based on the Long-Term Predictor (LTP) gain, which is typically high for highly correlated voiced segments and low for noise-like, uncorrelated unvoiced segments.

In the unvoiced Mode 0, the schematic at the top of Figure 7.12 is used, where the speech is synthesized by superimposing the G_1- and G_2-scaled outputs of two 128-entry trained codebooks in order to generate the excitation signal, which is then filtered through the synthesis filter $A(z)$ and the spectral post-filter. Accordingly, both Excitation Codebook 1 and 2 have a 7-bit address in each of the four subsegments, as shown in Table 7.10.

In Modes 1–3, where the input speech exhibits some grade of voicing, the schematic at the bottom of Figure 7.12 is used. The excitation is now generated by superimposing the G_3-scaled 512-entry trained codebook's output onto that of the G_4-scaled adaptive codebook. The fixed codebook in these modes requires a 9-bit address, yielding a total of $4 \times 9 = 36$ coding bits for the 20 ms frame, as seen in Table 7.10. The adaptive codebook delay or long-term predictor delay (LTPD) is encoded in the first subsegment using 8 bits, allowing for 256 integer and noninteger delay positions. In consecutive subframes, the LTPD is encoded differentially with respect to the previous subframe's delay, which we indicated as ΔLTPD in Table 7.10. The 4 encoding bits allow for a maximum difference of $[-8, +7]$ positions with respect to the previous LTPD value. The legitimate LTPD values are listed in Table 7.11.

Observe in the table that for low LTPD values a finer resolution is used, and the highest resolution is assigned for the range $23 \ldots (34 + \frac{5}{6})$, corresponding to a pitch lag of between $2.875 \ldots 4.35$ ms or pitch frequency of 230–348 Hz.

Returning to Table 7.10, the overall frame energy is encoded with 5 bits, which allows spanning a dynamic range of 64 dB, when using a stepsize of 2 dB and 32 steps. The

TABLE 7.11 Legitimate Noninteger LTPD Values and LTP Resolution in the 5.6 kbps VSELP Half-rate GSM Codec

LTPD range	Resolution
$21\ldots(22+\frac{2}{3})$	$\frac{1}{3}$
$23\ldots(34+\frac{5}{6})$	$\frac{1}{6}$
$35\ldots(49+\frac{2}{3})$	$\frac{1}{3}$
$50\ldots(89+\frac{1}{2})$	$\frac{1}{2}$
$90\ldots142$	1

excitation gains G_1–G_4 are not directly encoded. Instead, the energy of each subframe E_s is expressed, normalized by the frame energy E_F, which is then jointly vector-quantized with another parameter about to be introduced. Specifically, it was found advantageous to express the relative contribution E_1 of the first excitation component constituted by Codebook 1 at the top of Figure 7.12 in Mode 0 and by the adaptive codebook at the bottom of Figure 7.12 in Modes 1–3 to the overall excitation. Clearly, this relative contribution must be limited to the range of $0\ldots1$. Then the parameter pair $[E_s,\ E_1]$ is vector-quantized using 5 bits/5 ms subsegment, which allowed for 32 possible combinations. Accordingly, Table 7.10 assigns a total of 20 bits/20 ms frame for the encoding of this gain-related information.

7.7.2 Spectral Quantization in the Half-Rate GSM Codec

According to Table 7.10, the codec employs 28-bit vector quantization (VQ) of the reflection coefficients, where the best set is deemed to be the one that minimizes the prediction residual energy. A reduced-complexity version of the Fixed Point Lattice Technique (FLAT) [180, 181] was proposed for the standard, which will be briefly highlighted below.

It would be impractical to use a 2^{28}-entry codebook for both search-complexity and storage-capacity reasons, whence a suboptimum three-way split-vector implementation was proposed by Gerson [180], where the reflection coefficients $k_1 - k_3$, $k_4 - k_6$ and $k_7 - k_{10}$ are stored in separate codebooks. The number of quantization or codebook address bits is $Q_1 = 11$, $Q_2 = 9$ and $Q_3 = 8$ bits, respectively. A particularly attractive property of the reflection coefficient-based lattice-type predictors is that in case of the split-vector quantizers the choice of the current acoustic tube model segment's reflection coefficient quantizer can partially compensate for the quantization effects of the preceding tube section quantizer.

To elaborate on these issues, Gerson, Jasiuk et al. [187] introduced the ingenious concept of pre-quantization, where in each of the three split codebooks a pre-quantizer using $P_1 = 6$, $P_2 = 5$, and $P_3 = 4$ bits is invoked. Each vector of the pre-quantizer is associated with a set of vectors in the actual quantizer. For example, each of the $P_1 = 6$ bit quantizer entries is associated with $n_1 = 2^{Q_1}/2^{P_1} = 2^{11}/2^6 = 2^5 = 32$ vectors in the first actual VQ codebook, and so on. In order to reduce the overall complexity, the prediction residual error is computed for each of the pre-quantizer vectors at a given acoustic tube-model segment, and the four vectors resulting in the four lowest error energy values are earmarked. These four vectors are then used as pointers to identify four sets of vectors, which are associated with the earmarked pre-quantizer vectors. The four sets of actual quantized vectors are then exhaustively searched in order to find the set, which minimizes the prediction residual energy.

This technique results in a substantial complexity reduction. Specifically, instead of searching the $2^{Q_1} = 2^{11} = 2048$-entry codebook storing the reflection coefficients $(k_1 - k_3)$, initially the $2^{P_1} = 2^6 = 64$-entry pre-quantizer codebook is searched to find the best four "pointers," around each of which then the prediction residual is evaluated 32 times, requiring its computation 128 times. For simplicity, assuming an identical evaluation complexity for both steps, the complexity of the full search was reduced by a factor of $2048/(64 + 128) \approx 10.67$. The corresponding factors for the $(k_4 - k_6)$ and $(k_7 - k_{10})$ codebooks are $2^9/(32 + 64) \approx 5.3$ and $2^8/(16 + 64) = 3.2$, respectively.

The reflection coefficients themselves have been reported to have a high spectral sensitivity in the vicinity of the unit circle, when $k_i \approx 1$. This may result in a large speech spectrum variation due to the quantization of the reflection coefficients. Hence, a very fine Max-Lloyd quantizer would be required for their quantization in this domain, instead of uniform quantization. Therefore, two widely used nonlinear transformations have been proposed for circumventing this problem, namely, the log-area ratios (LAR) and the inverse sine transformation $S_i = \sin^{-1}(k_i)$, which are more amenable to uniform quantization. The GSM half-rate codec uses the latter, employing an efficient 8-bit representation for the codebook entries, which were generated by uniformly sampling their inverse-sine representations. Let us now briefly consider the error protection strategy used.

7.7.3 Error Protection

The error control strategy used is based on the schematic of Figure 7.13, which is quite similar in terms of its philosophy to that of other mobile radio systems, such as, for example, the full-rate or the enhanced full-rate GSM schemes or the IS-54 system, portrayed in Figure 7.4. The 112 bits/20 ms are divided into 95 more sensitive Class-One bits and 17 more robust Class-Two bits. The most sensitive 22 Class-One bits are assigned a 3-bit cyclic redundancy checking (CRC) pattern, which is then invoked by the decoder for initiating bad frame masking. Bad frames may be encountered due to channel errors or to fast associated control channel messages replacing a speech frame, for example, in order to signal an urgent handover request. In this case, the speech frame is wiped out by this fast associated control channel message, and at the decoder it has to be replaced by a post-processed speech segment.

As displayed in Figure 7.13, the 17 robust Class-Two bits are unprotected, while the 95 Class-One bits are $\frac{1}{3}$-rate, constraint-length 7 convolutionally encoded. Here we note that the

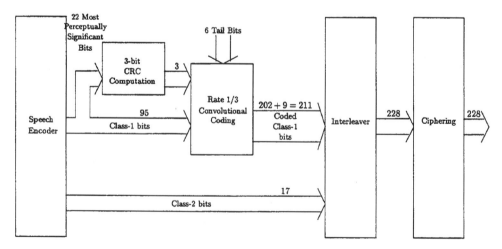

Figure 7.13 The 5.6/11.4 kbps GSM half-rate error protection schematic.

definition of constraint length in this case includes the current input bit of the encoder, plus the six shift-register stages. Six tailing bits are therefore necessary for flushing the encoder's shift registers after each transmission burst in order to prevent error propagation across transmission frame boundaries. We note, however, that a punctured code was employed, where the effective coding rate becomes $\frac{1}{2}$ due to puncturing. More explicitly, puncturing implies obliterating some of the encoded bits. The 95 Class-One bits and the 6 tailing bits yield 101 bits, which generate 202 punctured convolutionally coded bits, while the 3 CRC bits are $\frac{1}{3}$-rate coded, yielding a total of 211 bits. After concatenating the 17 unprotected bits, the total rate becomes 228 bits/20 ms = 11.4 kbps, which is exactly half that of the full rate and enhance full-rate systems.

Having highlighted the basic features of the 5.6 kbps half-rate GSM codec, we now address some of the issues specific to the 8 kbps ITU G.729 scheme in the next Section.

7.8 THE 8 kbits/s G.729 Codec [123]

7.8.1 Introduction

In 1990 the CCITT invited candidate codecs for a low-delay 8 kbits/s speech coding standard. Requirements regarding speech quality, robustness to channel errors, and frame length were specified. However, no candidate codec submitted by the July 1991 deadline satisfied all the requirements, and so in November 1991 the frame-length requirement was relaxed from the original 5 ms to 16 ms. In November 1992 two candidate codecs were submitted. One, from NTT in Japan, used Conjugate Structure CELP (CS-CELP) with a frame length of 13 ms. The other was designed by France Telecom and the University of Sherbrooke in Canada, and used Algebraic CELP (ACELP) with a frame length of 12 ms. It was decided that considering potential applications a 10 ms frame length would be preferable, and so both groups agreed to reduce the frame length of their codecs to 10 ms. Aspects of both the CS-CELP codec [188] and the ACELP codec [136, 189] were used in the final standardized codec, which uses Conjugate Structure Algebraic CELP (CS-ACELP), and provides toll quality speech at 8 kbits/s with a 10 ms frame length. This codec is described in detail below.

7.8.2 Codec Schematic and Bit Allocation

Schematics of the G.729 encoder and decoder are shown in Figures 7.14 and 7.15. It can be seen that the structure of this codec is similar to that of other forward-adaptive codecs described earlier. Forward adaption is used to determine the synthesis filter parameters once per 10 ms frame. These filter coefficients are then converted to LSFs and quantized with 18 bits using predictive two-stage vector quantization. Each 10 ms frame is split into two 5 ms subframes, and the excitation for the synthesis filter is determined for each subframe. The long-term correlations in the speech are modeled using an adaptive codebook with fractional delays, using 8 bits to represent the delay in the first subframe and 5 bits to differentially encode the delay in the second subframe. Also, to improve the robustness of the codec to channel errors, the six most significant bits of the adaptive codebook index in the first subframe have a parity bit added. This allows most errors in these bits to be detected at the decoder, and when such errors are detected, an error concealment procedure is applied. A 17-bit algebraic codebook with a focused search procedure is used as the fixed codebook. Finally, the adaptive and fixed-codebook gains are vector-quantized with 7 bits using a two-stage conjugate structured codebook, with fourth-order moving-average prediction applied to the

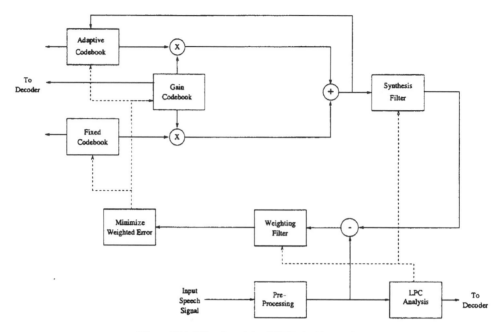

Figure 7.14 8 kbps low-delay CCITT G.729 encoder.

fixed-codebook gain to aid the efficiency of the quantizer. The entries from the fixed, adaptive, and gain codebooks are chosen every subframe using an analysis-by-synthesis search to minimize the weighted error between the original and the reconstructed speech.

At the decoder the transmitted parameters are used to give the filter coefficients for the synthesis filter and to select entries from the fixed, adaptive, and gain codebooks to represent the excitation to this filter. The reconstructed speech is then post-processed to improve its perceptual quality. The bit allocation of the G.729 codec is summarized in Table 7.12. We now describe in more detail the various blocks shown in the G.729 encoder and decoder.

7.8.3 Encoder Pre-Processing

Simple pre-processing is applied to the input speech signal in the G.729 encoder. The input signal is assumed to be a 16-bit linear PCM signal and is initially divided by a factor of 2 to reduce the possibility of overflows in fixed-point implementations of the codec. The

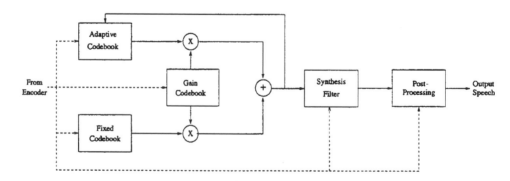

Figure 7.15 8 kbps low-delay CCITT G.729 decoder.

TABLE 7.12   Bit-Allocation Scheme of the G.729 Codec

Parameter	Sub-Fr. 1	Sub-Fr. 2	Total/frame
LPC: LSPQ1-LSPQ4			$1 + 7 + 5 + 5 = 18$
Pitch delay: PD1, PD2	8	5	13
Parity for pitch delay: PPD	1	0	1
Fixed codebook ind.: FC1, FC2	$3 \times 3 + 4 = 13$	$3 \times 3 + 4 = 13$	26
Sign of fixed codebook: SFC1, SFC2	4	4	8
Codebook gains (Stage 1): GC1A, GC2A	3	3	6
Codebook gains (Stage 2): GC1B, GC2B	4	4	8
Total			80

signal is also high-pass filtered using a second-order pole-zero filter with a cutoff frequency of 140 Hz. This serves as a precaution against undesired low-frequency components in the input signal. The pre-processed speech signal acts as the input to the speech encoder and is referred to as the input speech in our descriptions below.

7.8.4 LPC Analysis and Quantization

LPC analysis is carried out in the G.729 encoder to derive filter coefficients to be used by the tenth-order synthesis and weighting filters. The coefficients for these filters are calculated at the encoder for every 10 ms frame, using the autocorrelation method, with a 30 ms asymmetric window. This window is shown in Figure 7.16, where the sample indices 0, 1 ... 79 correspond to the present 10 ms frame. The window consists of half a Hamming window for 25 ms and a quarter of a cosine cycle for the final 5 ms of the window. It can be seen that, although the frame length of the codec is 10 ms, a 5 ms lookahead is used, which increases the total delay of the codec by 5 ms.

The windowed speech signal is used to compute 11 autocorrelation coefficients $R(k)$, $k = 0, 1 \ldots 10$. These autocorrelations are then slightly modified as follows. $R(0)$ is given a lower bound of 1.0 to avoid arithmetic problems with low-level-input signals. A 60 Hz

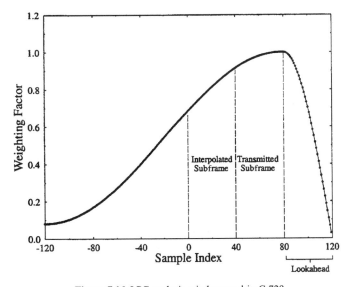

Figure 7.16 LPC analysis window used in G.729.

bandwidth expansion is applied to the filters by multiplying the autocorrelation coefficients $R(k)$ by $\exp[-\frac{1}{2}(2\pi f_o k/f_s)^2]$ for $k = 1, 2 \ldots 10$. Here $f_o = 60\,\mathrm{Hz}$, and f_s is the sampling frequency of $8000\,\mathrm{Hz}$. Finally, $R(0)$ is multiplied by a white noise correction factor of 1.001.

These modified autocorrelation coefficients are used to calculate the filter coefficients a_i, $i = 1, 2 \ldots 10$, using the Levinson-Durbin algorithm. Then the filter coefficients are converted into line spectral frequencies (LSFs) before quantization and interpolation. The synthesis filter coefficients to be used are derived from the quantized set of LSFs. Interpolation is used on these LSFs so that in the first subframe the LSFs used are the average of the quantized LSFs from the present and the previous frames, whereas in the second subframe the quantized LSFs from the present frame are used.

The simplified block diagram of the 18-bit predictive two-stage LSF vector quantizer used in G.729 is shown in Figure 7.17, which will be elaborated on below. Here it is exploited that due to the inherent correlation between consecutive sets of LSF vectors, the previous quantized LSS vector provides a good estimate of the current vector to be quantized, resulting in a lower-variance quantity to be quantized. Hence the number of LSF quantization bits required by the prediction error is reduced. The switched fourth-order moving-average (MA) predictor seen in the figure is constituted by a pair of predictors, which is not explicitly shown in the figure. Both of these predictors are tentatively invoked, and the one minimizing the LSF prediction error of Figure 7.17 is actually employed in the prediction, which is signaled to the decoder using the 1-bit flag at its output. This MA predictor is used to predict the set of LSFs for the current frame on the basis of the previous quantized LSFs, and then the LSF prediction error between the resultant LSF vector prediction and the actual set of zero-mean LSFs is quantized using a two-stage vector quantizer. According to the set of 10 LSFs, in the first stage a 10-dimensional, 7-bit, 128-entry codebook is used in order to crudely estimate the LSF vector and to derive the Stage 1 LSF prediction error of Figure 7.17. The Stage 1 LSF prediction error is then modeled by invoking the Stage 2 LSF vector quantizer, which attempts to match the five-dimensional split LSF vectors using the 5-bit or 32-entry codebooks. Together with the 1-bit flag at the output of the MA predictor of Figure 7.17 that is used to specify which of the pair of LSF predictors implicit in this block should be employed, this gives a total of $7 + 5 + 5 + 1 = 18$ bits per frame for the quantization of the LSFs.

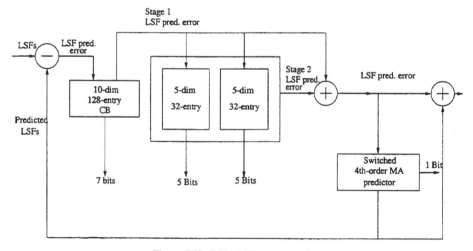

Figure 7.17 G.729 LSF vectorquantizer.

To elaborate further on the inner working of the G.729 LSF vector-quantizer, for each speech frame both possible LSF MA predictors give a set of predicted LSFs, yielding two sets of LSF prediction errors, which must be vector-quantized. Then for each set of LSF prediction errors, the following procedure is carried out. Initially the first-stage, 7-bit, 10-dimensional codebook is searched to find the codebook entry, which gives the closest match to the set of zero-mean LSF prediction errors. This closeness of match is measured using the simple squared-error measure. Then the difference between the codebook entry selected from the first codebook and the set of prediction errors to be quantized is itself quantized in the second stage of the vector quantizer. The second stage is a 10-bit quantizer, but in order to reduce the complexity of the quantizer, it is split into two. One 5-bit, five-dimension quantizer is used to code the first 5 LSFs, and the other 5-bit quantizer codes the final 5 LSFs. Entries from the two codebooks are chosen to minimize the weighted squared error E_{lsf}, where

$$E_{lsf} = \sum_{i=1}^{i=10} W_i (\omega_i - \hat{\omega}_i)^2 \qquad (7.19)$$

and ω_i are the set of input LSFs, $\hat{\omega}_i$ are the quantized LSFs, and W_i are a set of weighting coefficients derived from the input LSFs. Thus, for each of the two sets of predictors, a 7-bit index for the first-stage quantizer is chosen to minimize the squared quantization error, and then two 5-bit indices are chosen for the second-stage quantizer to minimize the weighted squared error E_{lsf}. The LSF predictor, which gives the lowest weighted squared error E_{lsf}, is chosen as the predictor to be used, and 1 bit is sent to the decoder to indicate which predictor to use. The stability of the synthesis and weighting filters is guaranteed by ensuring that the quantized LSFs are ordered and that adjacent LSFs are separated by at least a given minimum distance.

7.8.5 The Weighting Filter

The weighting filter used in the G.729 encoder is based on the unquantized filter coefficients a_i derived from the LPC analysis described above. The transfer function of the weighting filter is given by

$$\begin{aligned} W(z) &= \frac{A(z/\gamma_1)}{A(z/\gamma_2)} \\ &= \frac{1 + \sum_{i=1}^{10} \gamma_1^i a_i z^{-i}}{1 + \sum_{i=1}^{10} \gamma_2^i a_i z^{-i}} \end{aligned} \qquad (7.20)$$

where γ_1 and γ_2 control the amount of weighting. This amount of weighting is made adaptive in order to improve the performance of the codec for input signals with a flat frequency response, by adapting γ_1 and γ_2 based on the spectral shape of the input signal. This adaption is done for every 10 ms frame, but interpolation is used in the first subframe to smooth the adaption process. The adaption is based on log-area ratio (LAR) coefficients obtained as a byproduct of the LPC analysis carried out on the input speech. The LARs of a second-order filter are used to characterize the input speech as either flat or tilted, and the values of γ_1 and γ_2 are adjusted depending on this classification. We note here that in the G.729 standard the Levinson-Durbin algorithm delivers a set of LPC coefficients, which have the opposite sign in comparison to the G.723.1 standard, for example. This is why there is a positive sign in front of the summation in the weighting filter of Equation 7.52, while the G.723.1 weighting filter, for example, has a negative sign in the weighting filter.

7.8.6 The Adaptive Codebook

For each subframe an adaptive codebook index must be chosen which minimizes the weighted error between the input speech and the reconstructed speech. In analysis-by-synthesis codecs, such as G.729, the best adaptive codebook index is determined using a closed-loop search as described in Section 3.4.2. However, in order to reduce the complexity of this closed-loop search, in the G.729 encoder the search range is limited to around a candidate delay T_{op} which is obtained by an open-loop pitch analysis on the input weighted speech $s_w(n)$. This open-loop pitch analysis is carried out over the 10 ms frame and attempts to maximize the autocorrelation $R_w(k)$ of the weighted input speech. This correlation is given by

$$R_w(k) = \sum_{n=0}^{79} s_w(n)s_w(n-k),\qquad(7.21)$$

and its maximum is found in the following three ranges: 20–39, 40–79, and 80–143. These three maxima in $R_w(k)$ are normalized, and the open-loop pitch value T_{op} is selected from among the three values of k which give a local maxima by favoring the delays with values in the lower ranges. This division of the delay range into sections, and favoring the delays in the lower sections, is done in order to avoid choosing pitch multiples as the open-loop pitch T_{op}.

Once the open-loop pitch has been determined, the closed-loop search for the adaptive codebook index T_1 in the first subframe is limited to the six samples around T_{op}. This index is coded with 8 bits and takes fractional values with resolution $\frac{1}{3}$ in the range $19\frac{1}{3}$–$84\frac{2}{3}$, and integer values only in the range 85–143. In the second subframe, the closed-loop search for the adaptive codebook index T_2 is limited to delays around the delay T_1 chosen in the first subframe. A codebook index with resolution $\frac{1}{3}$ is selected in the range between $\left(\text{int}(T_1) - 5\frac{2}{3}\right)$ and $(\text{int}(T_1) + 4\frac{2}{3})$, where $\text{int}(T_1)$ is the integer part of T_1. This index T_2 is coded with 5 bits.

The closed-loop pitch search for T_1 and T_2 is achieved by maximizing the term

$$\chi_\alpha = \frac{\sum_{n=0}^{39} x(n)y_\alpha(n)}{\sum_{n=0}^{39} y_\alpha(n)y_\alpha(n)}\qquad(7.22)$$

where $x(n)$ is the target for the filtered adaptive codebook signal and $y_\alpha(n)$ is the past filtered excitation at delay α. The fractional pitch search is carried out, when necessary, using interpolated values of χ_α. This interpolation is done using an FIR filter based on a Hamming windowed $\sin(x)/x$ function.

Once the adaptive codebook index T_1 or T_2 for the subframe has been determined, the resulting output from the adaptive codebook must be calculated at both the encoder and decoder. This adaptive codebook signal is the delayed past excitation signal, but if a fractional delay has been selected, then interpolation must be carried out on this past excitation signal. Again, a FIR filter based on a Hamming windowed $\sin(x)/x$ function is used for this interpolation.

At the encoder once the adaptive codebook signal has been determined, the fixed codebook is searched, again using a closed-loop search designed to minimize the weighted error between the reconstructed and input speech signals. The structure of this fixed codebook, and the techniques used to search it, are described below.

7.8.7 The Fixed Algebraic Codebook

G.729 uses a 17-bit fixed codebook. Using traditional random codebooks, the closed-loop search of such a large codebook would be extremely complex and render the use of such a codebook in a real-time speech codec unrealistic. However, in G.729 an algebraic codebook is used, with only four nonzero pulses per subframe. This allows the codebook to be searched efficiently using a series of four nested loops. Also, a focused search is used to further simplify the determination of the codebook parameters. These measures mean that the huge 17-bit codebook can be searched with reasonable complexity and thus used in the G.729 codec, which is intended for real-time operation on a single DSP.

The structure of the algebraic codebook used in G.729 is shown in Table 7.13. Each codeword contains only four nonzero pulses, each of which has its amplitude fixed to either $+1$ or -1 and coded with 1 bit. The first three nonzero pulses have eight possible positions and have their positions coded with 3 bits each. The final pulse has 16 possible positions, and its position is coded with 4 bits. Thus, a total of 17 bits are used to represent the fixed codebook index. The fixed codebook signal is then given by

$$c_k(n) = s_0\delta(n - m_0) + s_1\delta(n - m_1) + s_2\delta(n - m_2) + s_3\delta(n - m_3), \quad (7.23)$$

where s_i is the sign and m_i the position of pulse i.

A special feature of the codebook used in G.729 is that for pitch delays less than 40 the codebook signal $c_k(n)$ is modified according to

$$c_k(n) = \begin{cases} c_k(n) & n = 0, \ldots, T - 1 \\ c_k(n) + \beta c_k(n - t) & n = T, \ldots, 39 \end{cases} \quad (7.24)$$

where T is the integer part of the pitch delay used in the current subframe, and the value of β is based on the quantized pitch gain of the previous subframe. This modification is incorporated into the codebook search by modifying the impulse response $h(n)$ of the synthesis and weighting filters used in the codebook search. It is equivalent to including an adaptive pre-filter in the codebook, and it enhances the harmonic components in the reconstructed speech and improves the performance of the codec.

The fixed codebook search is carried out as follows. The target signal $\tilde{x}(n)$ for the filtered fixed codebook signal is given by the target signal $x(n)$ from the pitch search with the filtered adaptive codebook contribution subtracted, that is,

$$\tilde{x}(n) = x(n) - G_1 y_\alpha(n) \quad (7.25)$$

where G_1 is the unquantized pitch gain, given by

$$G_1 = \frac{\sum_{n=0}^{39} x(n)y_\alpha(n)}{\sum_{n=0}^{39} y_\alpha^2(n)} \quad \text{bounded by } 0 \le G_1 \le 1.2 \quad (7.26)$$

and $y_\alpha(n)$ is the filtered adaptive codebook signal. As explained in Chapter 6, the best codebook vector is then found by determining which vector k maximizes the term $T_k = C_k^2/\xi_k$. Here C_k is the correlation between the filtered fixed codebook signal and the target signal $\tilde{x}(n)$, and ξ_k is the energy of the filtered fixed codebook signal. As there are only

TABLE 7.13 Pulse Amplitudes and Positions for the G.729 Codec

Pulse Number i	Amplitude	Possible Positions m_i
0	±1	0,5,10,15,20,25,30,35
1	±1	1,6,11,16,21,26,31,36
2	±1	2,7,12,17,22,27,32,37
3	±1	3,8,13,18,23,28,33,38
		4,9,14,19,24,29,34,39

four nonzero pulses per codeword, with positions m_i and amplitudes s_i, these terms can be written as

$$
\begin{aligned}
C_k &= \sum_{n=0}^{39} \tilde{x}(n)[c_k(n)*h(n)] \\
&= \sum_{n=0}^{39} \psi(n)c_k(n) \\
&= \sum_{i=0}^{3} s_i\psi(m_i)
\end{aligned}
\tag{7.27}
$$

where

$$
\begin{aligned}
\psi(i) &= \tilde{x}(i)*h(-i) \\
&= \sum_{n=i}^{39} \tilde{x}(n)h(n-i) \quad \text{For } i = 0\cdots 39,
\end{aligned}
\tag{7.28}
$$

and

$$
\begin{aligned}
\xi_k &= \sum_{n=0}^{39} [c_k(n)*h(n)]^2 \\
&= \sum_{i=0}^{39} c_k^2(i)\phi(i,i) + 2\sum_{i=0}^{38}\sum_{j=i+1}^{39} c_k(i)c_k(j)\phi(i,j) \\
&= \sum_{i=0}^{3} \phi(m_i,m_i) + 2\sum_{i=0}^{2}\sum_{j=i+1}^{3} s_i s_j\phi(m_i,m_j)
\end{aligned}
\tag{7.29}
$$

where

$$
\phi(i,j) = \sum_{n=max(i,j)}^{39} h(n-i)h(n-j) \quad \text{For } i,j = 0\cdots 39.
\tag{7.30}
$$

The functions $\psi(i)$ and $\phi(i,j)$ can be calculated once per subframe, but then ξ_k and C_k must be calculated for each codeword. To simplify the search procedure for a given set of pulse positions m_i, the signs of the four pulses s_i are set equal to the signs of $\psi(m_i)$ at the pulse positions. This means that the correlation term C_k will be maximized for the given set of pulse positions and is given by

$$
C_k = |\psi(m_0)| + |\psi(m_1)| + |\psi(m_2)| + |\psi(m_3)|.
\tag{7.31}
$$

It also allows the calculation of the energy term ξ_k to be simplified by modifying $\phi(i,j)$. The sign information is included in $\phi(i,j)$ by modifying it to $\tilde{\phi}(i,j)$ as follows

$$
\tilde{\phi}(i,j) = |\psi(i)||\psi(j)|\phi(i,j).
\tag{7.32}
$$

Also, the diagonal elements in ϕ are scaled so as to remove the factor of 2 in Equation 7.29, that is,

$$
\tilde{\phi}(i,i) = \frac{1}{2}\phi(i,j),
\tag{7.33}
$$

so that the energy term ξ_k that must be calculated for every codeword k is simplified to

$$
\xi_k/2 = \sum_{i=0}^{3} \tilde{\phi}(m_i,m_i) + \sum_{i=0}^{2}\sum_{j=i+1}^{3} \tilde{\phi}(m_i,m_j),
\tag{7.34}
$$

which is significantly less complex to calculate than the expression in Equation 7.29.

The codebook search is further simplified using a focused search procedure. As usual in algebraic CELP codecs, a series of four nested loops are used to test the value of T_k for each set of pulse positions. However, in G.729 the final loop, which is the largest because of the 16 possible positions of the fourth pulse, is entered only if the correlation C_k due to the first three pulses exceeds a certain threshold Thr_3. This threshold is precomputed before the codebook search commences for each subframe, and is set to

$$Thr_3 = av_3 + 0.4 * (max_3 - av_3) \qquad (7.35)$$

where av_3 is the average correlation due to the first three pulses, and max_3 is the maximum value of the correlation due to the first three pulses. Also, the maximum number of times the final loop can be entered is set to 180 per frame, to give a definite upper bound to the complexity of the codebook search.

Using the methods described above, at most 180×16 codebook entries per frame are tested to see if they maximize T_k. This is only about 1% of the total number of tests of 2×2^{17} per frame that would be necessary if all possible pulse positions and signs were tested. However, the performance of the codec using such focused search procedures is reported [138] to be close to that which would be achieved using the much more complex full search.

Once the adaptive and fixed codebook indices have been determined by the decoder, the two codebook gains are vector-quantized with 7 bits.

7.8.8 Quantization of the Gains

The two codebook gains in G.729 are quantized using a predictive, two stage, conjugate structured vector quantizer. Fourth-order moving-average prediction, based on the energies (in the logarithmic domain) of the previous gain scaled fixed codebook signals, is used to find a predicted fixed codebook gain $\tilde{G}_2$. The optimum gain G_2 is then given by

$$G_2 = \gamma \tilde{G}_2 \qquad (7.36)$$

where γ is a correction factor that is quantized along with the adaptive codebook gain G_1.

The quantized values of G_1 and γ are chosen from a two-stage codebook. The first stage consists of a 3-bit two-dimensional codebook, while the second stage is a 4-bit two-dimensional codebook. Thus, the quantized values $\hat{G}_1$ and $\hat{G}_2$ of the adaptive and fixed codebook gains are given by

$$\hat{G}_1 = G_1 CB_1(k_1) + G_1 CB_2(k_2) \qquad (7.37)$$

and

$$\hat{G}_2 = \tilde{G}_2(G_2 CB_1(k_1) + G_2 CB_2(k_2)) \qquad (7.38)$$

where k_1 and k_2 are the chosen indices from the two codebooks, $G_1 CB_1$ and $G_2 CB_1$ are the entries from the first-stage codebook, and $G_1 CB_2$ and $G_2 CB_2$ are the entries from the second codebook.

The indices k_1 and k_2 from the two codebooks must be chosen so as to minimize the weighted squared error between the input and the reconstructed speech. As explained previously in Section 6.5.2.1, this is equivalent to maximizing $T_{\alpha k}$, where

$$T_{\alpha k} = 2\left(\hat{G}_1 C_\alpha + \hat{G}_2 C_k - \hat{G}_1 \hat{G}_2 Y_{\alpha k}\right) - \hat{G}_1^2 \xi_\alpha - \hat{G}_2^2 \xi_k. \qquad (7.39)$$

The definitions and interpretations of the terms C_α, ξ_α, C_k, ξ_k, and $Y_{\alpha k}$ were given in Section 6.5. Hence, suffice to say here that they are all fixed, once the adaptive and fixed codebook indices have been selected. The gain vector quantization must therefore simply select codebook indices that give values of $\hat{G}_1$ and $\hat{G}_2$ which maximize $T_{\alpha k}$ above.

The conjugate structure of the codebooks simplifies this search procedure as follows. The two codebooks are arranged so that in general the first codebook contains entries in which the elements corresponding to $\hat{G}_2$ are larger than those corresponding to $\hat{G}_1$. Similarly, in the second codebook the elements corresponding to $\hat{G}_1$ are generally larger than those corresponding to $\hat{G}_2$. When the codebooks are searched, a pre-selection process is applied to simplify the search. The optimum value of G_2, derived from Equation 7.39, is used to select the 4 from 8 codebook entries of the first codebook whose values of $\hat{G}_2$ are closest to the optimum. Similarly, the optimum value of G_1 is used to select 8 from the 16 values of the second codebook whose values of $\hat{G}_1$ are closest to the optimum. Then an exhaustive search of the $4 \times 8 = 32$ possible codebook index combinations is carried out, and the indices k_1 and k_2 which maximize $T_{\alpha k}$ are chosen. The quantized gains $\hat{G}_1$ and $\hat{G}_2$ are then given by Equations 7.37 and 7.38.

We have described how the encoder finds codebook indices from an 18-bit LSF vector quantizer, a 17-bit algebraic codebook, the adaptive codebook, and a 7-bit vector gain quantizer. These indices are transmitted to the decoder which uses them to determine the coefficients for the synthesis filter, the excitation signal for this filter, and hence the reconstructed speech signal $\hat{s}(n)$. The decoder also applies post-filtering to the speech to improve its perceptual quality, and it uses error concealment techniques to improve the robustness of the codec to channel errors. The post-processing and error concealment techniques used at the decoder are described below.

7.8.9 Decoder Post-Processing

After the reconstructed speech signal $\hat{s}(n)$ is calculated at the decoder, post-processing is applied. The reconstructed speech is passed through an adaptive post-filter to improve its perceptual quality. It is then high-pass filtered using a second-order pole-zero filter with a cutoff frequency of 100 Hz, and finally the filtered signal is multiplied by a factor of 2 to restore the input signal level.

The adaptive post-filtering used in G.729 is similar to the post-filtering used in G.728. It improves the perceptual quality of the decoded speech [86] by emphasizing the formant and pitch peaks in the speech and attenuating the valleys between these peaks. This reduces the audible noise in the reconstructed speech because, even with the noise shaping of the error weighting filter, it is in the valleys between the formant and pitch peaks that the noise energy is most likely to cross the masking threshold and become audible. Therefore, attenuating the speech in these regions reduces the audible noise, and because our ears are not very sensitive to the speech intensity in these valleys, only minimal distortion is introduced to the speech signal.

A block diagram of the post-filter used in G.729 is shown in Figure 7.18. It consists of both a long- and a short-term post-filter, together with a spectral tilt compensation filter and adaptive gain scaling. We describe here the blocks shown in Figure 7.18, but for more information on the ideas behind post-filtering, see the excellent paper by Chen and Gersho [86] as well as Section 8.4.6.

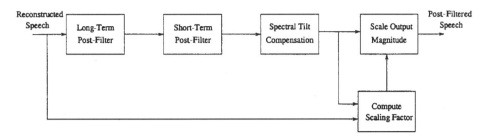

Figure 7.18 The G.729 adaptive post-filter.

The long-term post-filter has a transfer function of:

$$H_p(z) = \frac{1}{1 + \gamma_p g_l}(1 + \gamma_p g_l z^{-T}) \tag{7.40}$$

where γ_p is a constant that controls the amount of post-filtering and is set to 0.5, T is the pitch delay, and g_l is the gain coefficient. Both T and g_l are derived from the residual signal $\hat{r}(n)$ obtained by filtering the reconstructed speech $\hat{s}(n)$ through $\hat{A}(z/\gamma_n)$, which is the numerator of the short-term post-filter. The delay T is calculated with resolution $\frac{1}{8}$, using a two-pass procedure, by searching for the maximum of the correlation of $\hat{r}(n)$ around the integer part of the delay T_1 in the first subframe. Once the delay T is found, the gain g_l is calculated, again using the residual signal $\hat{r}(n)$. This gain term is bounded by $0 \le g_l \le 1.0$ and is set to zero in order to disable the long-term post-filtering if the long-term prediction gain is less than 3 dB.

After the long-term post-filtering, a short-term post-filter $H_f(z)$ is used. This filter has a transfer function of:

$$\begin{aligned} H_f(z) &= \frac{1}{g_f}\frac{\hat{A}(z/\gamma_n)}{\hat{A}(z/\gamma_d)} \\ &= \frac{1}{g_f}\frac{1 + \sum_{i=1}^{10} \gamma_n \hat{a}_i z^{-i}}{1 + \sum_{i=1}^{10} \gamma_d \hat{a}_i z^{-i}}, \end{aligned} \tag{7.41}$$

where $\hat{A}(z)$ is the quantized inverse synthesis filter, $\gamma_n = 0.55$ and $\gamma_d = 0.7$ are constants that control the amount of short-term post-filtering, and g_f is a gain term given by

$$g_f = \sum_{n=0}^{19} |h_f(n)|, \tag{7.42}$$

where $h_f(n)$ is the impulse response of the filter $\hat{A}(z/\gamma_n)/\hat{A}(z/\gamma_d)$. Because the short-term post-filter numerator $\hat{A}(z/\gamma_n)$ is used to calculate the residual signal $\hat{r}(n)$ used in determining the long-term post-filter parameters, to reduce the complexity of the post-filter the all-zero section of the short term post-filter $\hat{A}(z/\gamma_n)$ is in fact used before the long-term post-filter. It is then the residual signal $\hat{r}(n)$ that is passed through the long-term post-filter and the all-pole section $1/(g_f \hat{A}(z/\gamma_d))$ only of the short-term post-filter. However, this moving of the all-zero section of the short-term post-filter does not affect the transfer function of the overall post-filter but merely reduces the complexity of the post-filtering.

After the short-term post-filtering, tilt compensation is used to compensate for the spectral tilt introduced by $H_f(z)$. The tilt compensation filter $H_t(z)$ is a first-order all-zero filter with a transfer function of:

$$H_t(z) = \frac{1}{1 - |\gamma_t k_1|}(1 + \gamma_t k_1 z^{-1}), \tag{7.43}$$

where k_{1} is the first reflection coefficient derived from the impulse response $h_f(n)$ of $\hat{A}(z/\gamma_n)/\hat{A}(z/\gamma_d)$ and γ_t is set to 0.9 if k_1 is negative and 0.2 if k_1 is positive.

The final block in the post-filter is adaptive gain control, used to compensate for energy differences between the reconstructed speech signal $\hat{s}(n)$ and the post-filtered signal $S_f(n)$. For each subframe a gain factor G is calculated according to

$$G = \frac{\sum_{n=0}^{39} |\hat{s}(n)|}{\sum_{n=0}^{39} |sp(n)|}. \tag{7.44}$$

Then each post-filtered sample $s_f(n)$ is scaled by a factor $g(n)$ that is updated sample by sample according to

$$g(n) = 0.9875g(n-1) + 0.125G. \tag{7.45}$$

This results in a smoothly varying gain scaling factor $g(n)$. The signal $g(n) \cdot S_f(n)$ is then high-pass filtered and multiplied by a factor of 2, as explained earlier, to give the output speech from the decoder.

7.8.10 G.729 Error Concealment Techniques

An important part of any speech codec used over channels subject to errors is its resilience to these errors. Two measures are used in G.729 to help improve its error resilience: a parity bit is used to protect the adaptive codebook index T_1 in the first subframe, and a frame erasure concealment procedure is used to improve the decoder performance when frame erasures occur in the received bit stream.

As noted in Section 6.6, the bits representing the adaptive codebook index in CELP codecs are extremely sensitive to channel errors. This is particularly true of the delay T_1 from the first subframe in G.729 because the second subframe delay T_2 is calculated and coded relative to T_1. Therefore, a parity bit is added at the encoder to the six most significant bits representing T_1. When a single error occurs in one of the six most significant bits of T_1, or in the parity bit itself, this error is detected by the decoder based on the parity information transmitted by the encoder. When such an error is detected, the decoded value of T_1 is considered to be incorrect and is replaced by the integer part of the delay T_2 from the previous subframe. This helps reduce the impact of errors among the bits representing T_1.

The G.729 decoder also employs a frame erasure concealment technique. The method of detecting which frames have been erased is not specified but depends on the application in which the codec is used. However, when the decoder is told that a frame of 80 bits has not been received correctly, because, for example, a packet of information has been dropped by the transmission system, it employs techniques to reconstruct the current frame based on previously received information. Both the synthesis filter and the excitation to this filter must be derived. In addition, the memory of the LSF and the fixed codebook gain predictors must be updated.

The coefficients of synthesis filter for an erased frame are simply set equal to those from the last good frame. Also, the LSF predictor, which uses the output of the two stage 17-bit vector quantizer as its input, has its memory updated. This is done using the set of quantized LSFs from the last good frame to derive an output from the vector quantizer, which would have led to this set of LSFs in the current frame. This derived codebook output is then used to update the memory of the LSF predictor.

In an erased frame, the two codebook gains $\hat{G}_1$ and $\hat{G}_2$ are given by attenuated versions of the gains used in the previous subframe. $\hat{G}_2$ is attenuated by a factor of 0.98 each subframe, and $\hat{G}_1$ is bounded by $\hat{G}_1 < 0.9$ and is attenuated by a factor of 0.9 each subframe. The adaptive codebook delay is based on the integer part of the delay in the last good subframe. This delay is then used in any following erased frames, but to avoid excessive periodicity the delay is increased by 1 for each subframe (but bounded by 143). The fixed codebook index is randomly generated. The excitation signal to use in erased frames is then determined based on whether the frame is considered to be periodic or nonperiodic. This decision is made based on the long-term prediction gains derived when calculating the long-term post-filter coefficients g_l in the previous frame. If this long-term prediction gain in either of the previous two subframes is greater than 3 dB, then the present frame is considered to be periodic. Otherwise the frame is classified as nonperiodic. In periodic frames the excitation signal to be used is taken entirely from the adaptive codebook. In other words, the fixed codebook contribution is set to zero, whereas in nonperiodic frames it is taken entirely from the fixed codebook. These methods allow the G.729 decoder to cope well with frame erasures in the received bit stream.

7.8.11 G.729 Bit Sensitivity

The bit sensitivity of the G.729 scheme was characterized using the previously introduced "consistent corruption technique" and was plotted in terms of SEGSNR degradation versus bit index in Figure 7.19. Observe in the figure that while the line spectrum pair predictor choice flag-bit LSPQ1 of Table 7.12 and the higher-order second-stage 5-bit, 32-entry VQ address bits LSPQ4 appear quite robust to transmission errors, the vulnerability of the first-stage 10-dimensional, 128-entry address bits LSPQ2 is quite pronounced. This is as expected, since corruption of any of these bits will affect all 10 LSFs. The sensitivity of the low-frequency second-stage 5 bits LSPQ3 is lower than that of the 7 LSPQ2 bits but higher than that of the high-frequency bits LSPQ4. Clearly, these findings are in harmony with our expectations.

The pitch delay (PD) parameters PD1 and PD2 are both quite sensitive, in particular the 8 bits of PD1, since due to the differential encoding of PD2 any errors in PD1 automatically corrupt PD2 as well. Similar comments apply to the jointly vector-quantized fixed and adaptive codebook gains, where the more important 3-bit first-stage indices GC1A and GC2A of both subsegments exhibit a very pronounced error sensitivity, while the second-stage VQ address bits GC1B and GC2B have a somewhat more mitigated sensitivity. There also appears

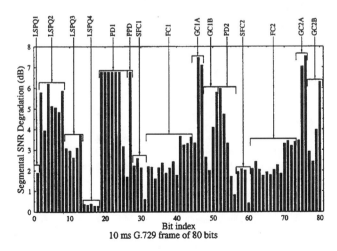

Figure 7.19 Bit sensitivity of the forward-adaptive 8 kbps ACELP G.729 speech frame.

to be a more robust category of bits, which is constituted mainly by the set of 13 fixed codebook index bits FC1 and FC2 and their corresponding sign bits, namely, bits SFC1 and SFC2. This excitation robustness is an attractive property of ACELP codecs, where corrupting one of the pulses does not vitally affect the shape of the excitation vector and the synthesized speech quality.

7.8.12 Turbo-coded Orthogonal Frequency Division Multiplex Transmission of G.729 Encoded Speech [190] J.P Woodard, T. Keller, L. Hanzo

7.8.12.1 Background.   In this section we study the performance of a range of parallel concatenated or turbo codecs (TC) in conjunction with various interleavers, which profoundly affect the TC performance over wideband Orthogonal Frequency Division Multiplexing (OFDM) systems, which were highlighted in [135]. Because of their diversity effect, wideband propagation channels provide similar gains for OFDM modems to those of equalized narrowband channels [49], resulting in substantial coding gains, when combined with turbo coding.

In the proposed system, the source and channel-coded bits are transmitted using a wideband Orthogonal Frequency Division Multiplexing (OFDM) system in the framework of the Mode-I FRAMES proposals [191]. We illustrate the benefits of using OFDM with channel coding to alleviate some of the problems associated with wideband fading channels. Furthermore, we discuss how OFDM can be used in conjunction with the G.729 speech codec and half-rate channel coding in order to utilize one speech/data FRAMES subburst. Finally, some of the issues and problems associated with using turbo-coded OFDM in speech transmission systems are considered using the system characterized in Table 7.14. As shown in Figure 7.25, a channel SNR of 6 dB appears sufficiently high under the stipulated system conditions for near-unimpaired speech transmission.

TABLE 7.14 Turbo-Coded OFDM System for Speech Transmission—System Parameters © [190], Woodard, Keller, Hanzo, 1997

System Parameters	
Carrier frequency	2 GHz
Sampling rate	1.3 Mhz
Channel	
Impulse response	COST207 BU
Normalized Doppler frequency	$6.7664 \cdot 10^{-5}$
OFDM	
Number of subcarriers	64
Cyclic extension	24 samples
Data subcarriers	43
Pilot subcarriers	21
Modulation scheme	Coherent QPSK
Turbo Channel Coding	
Constraint length	3
Generator polynomials	7, 5
Interleaver length	169
Decoding algorithm	MAP
Number of iterations	8

7.8.12.2 System Overview. The system model employed in this study is depicted in Figure 7.20. At the transmitter, a G.729 speech coder generates data packets of 80 bits per 10 ms from a speech file, and this speech data is encoded by a half–rate channel encoder. The encoded bits are modulated by a quadrature phase shift keying (QPSK) modulator, and the resulting signals are transmitted using an OFDM modem to the receiver. During transmission, the signal is corrupted in the frequency-selective, time-varying channel, and white Gaussian noise is added at the receiver's input stage. At the receiver, the OFDM signal is demultiplexed and demodulated, and the resulting bits are passed to the channel decoder. The received bits are decoded by the G.729 decoder, and the segmental signal-to-noise ratio (SEGSNR) degradation of the recovered speech is evaluated.

7.8.12.3 Turbo Channel Encoding. Turbo coding is a novel form of channel coding that reportedly produces excellent results [192,193]. The information sequence is encoded twice, with an interleaver between the two encoders serving to make the two encoded data sequences approximately statistically independent of each other. In our simulations we have used half-rate recursive systematic convolutional (RSC) encoders, but turbo coding is also possible with other constituent codes [194]. Each constituent RSC encoder produces a systematic output, which is equivalent to the original information sequence, as well as a stream of parity information. The two parity sequences are then punctured before being transmitted along with the original information sequence to the decoder. This puncturing of the parity information allows a wide range of coding rates to be realized. We have chosen to use the commonly adopted scheme of sending alternative parity bits from each encoder. Along with the original data sequence, this results in an overall coding rate of $\frac{1}{2}$.

The original, near-Shannonian performance results for turbo codes were achieved using a very long block length L of 65,536 bits. It is well known that the performance of the codes decreases as the frame length L decreases, but that good performance is still achievable with relatively short frame lengths. It is also well known that the design of the interleaver used within the turbo coder has a vital influence on its performance. For long frame lengths random interleavers are used, but for shorter frame lengths of 100 or 200 bits, such as in a speech transmission system, Jung and Naßhan [195] reported that block interleavers should be used. However, Jung and Naßhan used a 12×16 block interleaver in their work, while we have found that block interleavers with an odd number of rows and columns significantly outperform those with an even number of rows or columns. This is because, as Barbulescu and Pietrobon [196] note, with an odd number of rows and columns the odd and even data bits are kept separate. When alternate puncturing from each constituent encoder is used, as it most often is, this ensures that for each information bit one and only one parity bit is

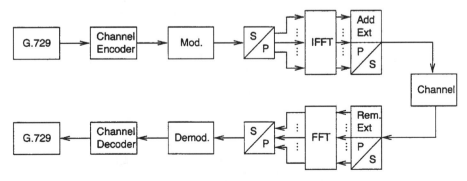

Figure 7.20 Schematic model of the G.729 OFDM system.

transmitted. This "odd-even" separation improves the performance of the turbo code [196], especially for short frame length systems in our experience.

As mentioned above, the G.729 speech codec provides 80 coded speech bits per 10 ms frame. All our simulations have used two constraint-length, three RSC constituent encoders, with generator polynomials expressed in octal form as 7 and 5. The Maximum A Posteriori (MAP) [197] algorithm has been used with 8 decoding iterations. For each 10 ms G.729 frame to be turbo encoded separately, we need to convey 80 information bits, plus 2 bits for trellis termination, where the number of trellis-terminating bits required to flush the encoder's shift-register corresponds to the number of shift-register stages in the encoder. This gives a required interleaver length of 82, which is very close to the interleaver length of 81 given by a 9×9 square interleaver.

For BER comparisons we have simulated systems using both a square $L = 81$ interleaver, which can transport 79 data bits per turbo-coded frame, and a system with an $L = 82$ interleaver. Because of the known benefit of using block interleavers for short frame transmission systems [195], we generated this length-82 interleaver by merely copying the elements of a square 81 interleaver and leaving the final additional element in the $L = 82$ interleaver un-interleaved. We have also simulated a system with $L = 169$, using a 13×13 square interleaver. As described later, this turbo encoder is used to code the 160 bits from two 10 ms G.729 frames. Finally, in order to characterize the near-optimum performance that can be achieved with turbo coding, we have simulated a system using a random interleaver with $L = 10{,}000$. Naturally, such an encoder could not be used for speech systems because of the delay it would introduce, but it may be useful for video or data transmission. Let us now consider the frame structure of the proposed system.

7.8.12.4 OFDM in the FRAMES Speech/Data Sub-Burst.

The emerging UMTS standard will have to accommodate a wide range of user profiles and data rates. The Advanced Communications Technologies and Services (ACTS) program's FRAMES project [191] aims to propose such a system, incorporating a wide variety of possible system parameters. For these experiments, the FRAMES Mode 1 Speech/Data sub-burst was chosen, offering sufficient data bandwidth for half-rate coded speech transmission. Figure 7.21 shows the timing of the frame and the chosen time slot, where the frame and the Speech/Data sub-burst durations are 4.615 µs and 72.1 µs, respectively, and the channel symbol rate is 1.3 MHz. Originally, the FRAMES proposal specified Offset-QPSK as the modulation scheme in these slots, leading to a channel bitrate of 2.6 Mbits/s.

The FRAMES Speech/Data sub-burst offers a convenient environment for 64 subcarrier OFDM transmission, as is demonstrated in Figure 7.21. The 64 data samples of the OFDM symbol are preceded by a 24-sample cyclic extension, which allows operation in wideband

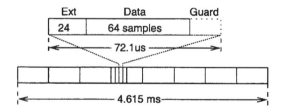

Figure 7.21 ACTS FRAMES Mode 1 frame and Speech/Data sub-burst. The subburst has been modified to hold a 64-subcarrier OFDM signal and a 24-samples cyclic extension. The symbol rate and the guard time duration have not been altered.

channels with an impulse response length of up to 24 samples or 18.5 μs without interburst interference. Let us now consider our wideband channel model in the next Section.

7.8.12.5 Channel Model.

All experiments were conducted utilizing the COST207 bad urban (BU) compliant impulse response [198]. The continuous COST207 BU impulse response was discretized to a seven-path model exhibiting a delay spread of 2.45 μs and a maximum delay of 7.7 μs, as seen in Figure 7.22. Each of the paths constituting the impulse response was faded independently, employing a Rayleigh-fading channel. The carrier frequency and the vehicular velocity were set to 2 GHz and 50 km/h, respectively, which leads to a Doppler frequency of 92.6 Hz for the Rayleigh channel. The normalized Doppler frequency is therefore $6.7664 \cdot 10^{-5}$.

The magnitude of the resulting time- and frequency-variant channel transfer function for a duration of 200 frames or 0.923 seconds is shown in Figure 7.23. Although the transfer function exhibits considerable variations in the frequency domain, the average received subcarrier energy per OFDM symbol, indicated by the bold line, shows little fluctuation. This relative stability of the OFDM symbol energy, over a period of time substantially longer than the inverse of the Doppler frequency, is an effect of the inherent multipath diversity. This leads to a more even distribution of errors, which enables the channel codec to work more efficiently.

7.8.12.6 Turbo-Coded G.729 OFDM Parameters.

Since the end-to-end delay of speech transmission should be less than 100 ms, the speech frame length should ideally not exceed 20 to 30 ms. The performance of a turbo decoder, on the other hand, improves with an increasing number of coded bits per block. As a compromise, a 20 ms speech block size was chosen. The G.729 speech codec produces 80 data bits per 10 ms input speech, resulting in a total of 160 data bits per speech block. We will demonstrate that the performance of turbo codes is very dependent on the internal interleaver's algorithm and latency. For short block lengths, as stated earlier, square interleavers with an odd number of rows and columns exhibit

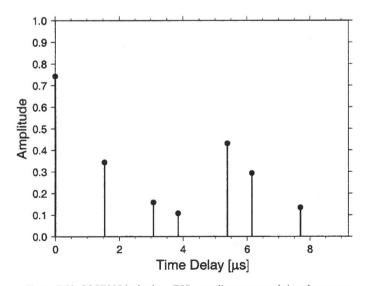

Figure 7.22 COST207 bad urban (BU) compliant seven-path impulse response.

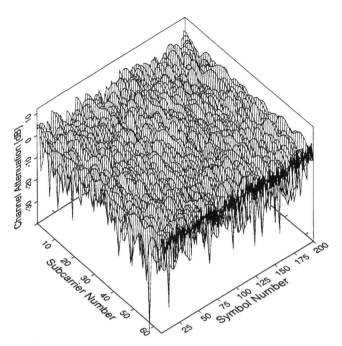

Figure 7.23 Amplitude plot of the frequency and time-varying channel impulse response for 200 OFDM symbols © [190], Woodard, Keller, Hanzo, 1997.

the best performance. The smallest square interleaver holding 160 input bits is $13 \times 13 = 169$ bits long, allowing for the transmission of 160 data, two termination, and seven unused padding bits.

The 169 uncoded data bits produce 338 coded output bits, which are transmitted using OFDM in one time slot over four consecutive frames. Employing QPSK as the modulation scheme for the OFDM subcarriers, only 43 of the 64 subcarriers in each OFDM symbol are employed for data transmission. The 21 remaining subcarriers are used for Pilot-Symbol Assisted Modulation (PSAM), allowing coherent detection of the symbols at the receiver. This PSAM was not simulated, but instead perfect channel estimation was used at the receiver. This means that both the demodulator and the turbo decoder operated with perfect estimates of both the fading amplitude and the noise variance. Having described the system, let us now focus our attention on the results.

7.8.12.7 Turbo-Coded G.729 OFDM Performance.
Figure 7.24 shows the BER performance of our system for the various turbo encoder/interleaver combinations described earlier, as well as for a constraint-length three convolutional code for comparison. It can be seen that the $L = 10,000$ turbo code gives an extremely impressive performance even in the Rayleigh-fading channel. Both the $L = 81$ and $L = 169$ turbo-decoded systems give performances significantly better than the convolutional coded system, showing that turbo codes can be useful in speech transmission schemes. However, disappointingly, the $L = 82$ system performs much worse than the $L = 81$ system, illustrating the importance of choosing a good interleaver for use with turbo encoders.

The $L = 169$ turbo-coded system was used to transmit G.729 coded speech. The segmental SNR degradation relative to the performance of G.729 over a perfect channel, against channel SNR for both this and the convolutional coded system, is shown in Figure

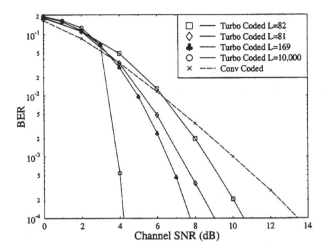

Figure 7.24 The effect of frame length on BER © [190], Woodard, Keller, Hanzo, 1997.

7.25. It can be seen that the turbo-coded system gives a gain of about 3 dB in channel SNR over the convolutional coded system in the segmental SNR degradation region of less than 1 dB, which corresponds to near-unimpaired speech quality. We note, however that this is achieved at the cost of an increased decoding complexity due to the eight decoding iterations employed.

7.8.12.8 Turbo-Coded G.729 OFDM Summary.
In conclusion, the attractive G.729 speech codec can be advantageously combined with turbo coding and OFDM transmission using the system illustrated in Table 7.14. Due to the multipath diversity of wideband channels, the OFDM modem performance is quite impressive. Furthermore, the error distribution is less bursty than over narrowband channels; hence, the channel codec is less

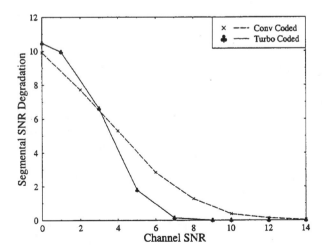

Figure 7.25 The segmental SNR degradation with convolutional and turbo encoding © [190], Woodard, Keller, Hanzo, 1997.

frequently overloaded by channel errors. In Figure 7.25, a channel SNR of 6 dB appears sufficiently high under the stipulated system conditions for near-unimpaired speech transmission.

7.8.13 G.729 Summary

In this section we have described the ITU's G.729 scheme. This codec operates at 8 kbits/s with a 10 ms frame length and gives output speech of quality equivalent to the 32 kbits/s ADPCM G.726 codec in error-free conditions. In the presence of channel errors, the G.729 codec significantly outperforms G.726. As described above, various techniques are used to reduce the complexity of the codec, and implementations on single fixed-point DSPs are available already. This codec looks set to become very widely used in many applications. In the previous section we also provided an application example for the G.729 codec.

Similarly to the other coding schemes considered, the G.729 arrangement is compared in subjective speech quality terms to the family of existing standard codecs in Figure 17.4 of Chapter 17. In the next section we briefly highlight the techniques used in the reduced-complexity G.729A codec.

7.9 THE REDUCED-COMPLEXITY G.729 ANNEX-A CODEC

7.9.1 Introduction

In this section we describe the recently adopted ITU-T Recommendation G.729 Annex A (also known as G.729A) [199,200]. This codec is a modification of the standard 8 kbits/s G.729 codec with significantly reduced complexity and only a slight degradation in performance.

G.729A grew from the interest in digital simultaneous voice and data (DSVD) applications in 1995. Although several standard low-bitrate speech codecs existed or were being finalized, at the time it was felt that, in order for speech coding and modem algorithms to be integrated on the same processor, a lower complexity speech codec was needed. A limit of 10 MIPS was set on the complexity of the codec, which was required to give speech quality as good as G.726 at 32 kbits/s in most conditions, and to operate at a bitrate of 11.4 kbits/s or lower. In the summer of 1995 five candidate codecs were submitted for subjective testing, including one from the University of Sherbrooke in Canada, which was based on G.729. This University of Sherbrooke codec had the advantage of being bit-stream interoperable with G.729. In other words, speech encoded using G.729 could be decoded with the new algorithm, and vice versa. This was considered important by the study group of the ITU-T, and so the Sherbrooke codec was chosen to be used in DSVD applications. Rather than forming a new recommendation, it was decided to make the reduced complexity version of G.729 a new Annex to the original G.729 recommendation. Hence, G.729 Annex A was formed.

G.729A operates at 8 kbits/s and gives speech quality equivalent to G.729 and G.726 at 32 kbits/s in most conditions, with only a small degradation in performance over G.729 in the case of three tandems and in the presence of background noise. It is approximately 50% less complex than G.729 and has been implemented on a fixed-point DSP (the Texas Instruments TMS320C50) using only 12 MIPS for full-duplex operation (compared to 22 MIPS for G.729) [199]. Most of the codec is identical to G.729, with changes made to the following aspects of the codec to reduce complexity:

1. The perceptual weighting filter.
2. The open-loop search for the pitch delay.
3. The closed-loop pitch search.
4. The algebraic codebook search.
5. The decoder post processing.

7.9.2 The Perceptual Weighting Filter

In G.729, as described in Section 7.8.5, an error weighting filter of the form $W(z) = A(z/\gamma_1)/A(z/\gamma_2)$ is used. The unquantized LPC filter coefficients a_i are used to form $A(z)$, and both γ_1 and γ_2 are adapted. In G.729A a more traditional error weighting filter $W(z)$, where

$$W(z) = \frac{\hat{A}(z)}{\hat{A}(z/\gamma)}, \tag{7.46}$$

is used. The quantized LPC filter coefficients $\hat{a}_i$, identical to those used in the synthesis filters in the encoder and decoder, are used to form $\hat{A}(z)$ so that the concatenation of the synthesis filter $1/\hat{A}(z)$ and the weighting filter $W(z)$ becomes $W(z)/\hat{A}(z) = 1/\hat{A}(z/\gamma)$. This simplifies the filtering operations involved in the speech encoding. Also, the weighting factor γ used in G.729A is constant (0.75), and so the adaption procedures for γ_1 and γ_2 used in G.729 are not needed.

7.9.3 The Open-Loop Pitch Search

In both G.729 and G.729A, the adaptive codebook search is simplified by first finding an open-loop pitch-delay value T_{op} for each 10 ms frame and then doing a closed-loop search around T_{op} in each 5 ms subframe to find the optimum adaptive codebook delay. The open-loop pitch search attempts to maximize the autocorrelation $R_w(k)$ of the weighted input speech $s_w(n)$ in three ranges of k: 20–39, 40–79, and 80–143. In G.729A, the calculation of the autocorrelation function $R_w(k)$ is simplified by using only even samples of $s_w(n)$, so that $R_w(k)$ is given by

$$R_w(k) = \sum_{n=0}^{39} s_w(2n) * s_w(2n - k), \tag{7.47}$$

rather than $\sum_{n=0}^{79} s_w(n) * s_w(n - k)$ as in G.729 (see Equation 7.21). The search for the best open-loop pitch is also further simplified by initially only testing even values of k in the third range ($80 \le k \le 143$) and then testing the two odd values of k around the chosen even value. This almost halves the number of calculations of $R_w(k)$ that must be carried out in the third range.

7.9.4 The Closed-Loop Pitch Search

In G.729A the closed-loop search for the best adaptive codebook indices T_1 and T_2 for the two subframes is also simplified. In the G.729 codec in the first subframe, χ_α, as given in Equation 7.22, is calculated for values around T_{op} to find the value of α which maximizes χ_α. This value of α is chosen as the adaptive codebook index T_1 for the first subframe. Similarly, in the second subframe values of χ_α around $int(T_1)$ are calculated to find the index T_2. In

G.729A these search operations are simplified by considering only the numerator of Equation 7.22 giving χ_α; in other words, instead of χ_α the term $\tilde{\chi}_\alpha$ is maximized, where

$$\tilde{\chi}_\alpha = \sum_{n=0}^{39} x(n)y_\alpha(n)$$

$$= \sum_{n=0}^{39} x_b(n)u(n-\alpha). \qquad (7.48)$$

Here $x(n)$ is the target for the filtered adaptive codebook signal, $u(n-\alpha)$ is the past excitation signal, $y_\alpha(n)$ is the filtered version of $u(n-\alpha)$, and $x_b(n)$ is the backward filtered target signal.

This change to the closed-loop adaptive codebook search results in some degradation compared to G.729. The chosen adaptive codebook delay sometimes differs by $\frac{1}{3}$ from that chosen in G.729. However, calculating $\tilde{\chi}_\alpha$ rather than χ_α approximately halves the complexity of the closed-loop pitch search [200].

7.9.5 The Algebraic Codebook Search

Both G.729 and G.729A employ a huge 17-bit algebraic codebook. Exhaustively searching such a large codebook would be unrealistic for a real-time speech codec. The codebook search is simplified using an algebraic structure; each codebook entry consists of only four nonzero pulses. Each pulse can be either $+1$ or -1, and has its sign encoded with 1 bit per pulse. The possible positions of the pulses are shown in Table 7.13 and are encoded with a total of 13 bits for the four pulses. Both codecs predetermine the sign of the four pulses depending on the sign of the backward-filtered target signal $\psi(m_i)$ at the pulse positions m_i. This leaves effectively a 13-bit codebook to be searched—still too large to be realistic for a real-time codec.

In G.729 the codebook search is further simplified using a series of four nested loops and a focused search procedure as described in Section 7.8.7. This results in a maximum of 2880 codebook entries being tested per frame—only about 1% of the total number of tests that would be necessary to exhaustively search the entire 17-bit codebook (if the signs were not predetermined). In G.729A, the algebraic codebook search is further simplified using a depth-first tree search. In this approach only 320 codebook entries are tested per subframe (i.e., 640 per frame), reducing the number of tested codebook entries by a factor of 4.5 relative to G.729.

The depth-first tree search used in G.729A is responsible for about 50% of the reduction in complexity of G.729A relative to G.729. Using this technique, the algebraic codebook search consumes about 3 MIPS, as opposed to about 8.5 MIPs for the search technique used in G.729. The simpler codebook search technique gives only a slight degradation in the codec's performance—about a 0.2 dB drop in the SNR [199, 200].

7.9.6 The Decoder Post-Processing

In both G.729 and G.729A, at the decoder the reconstructed speech signal is passed through an adaptive post-filter to improve its perceptual quality. This post-filter consists of both short- and long-term filters, spectral tilt compensation, and gain scaling. The post-filtering operations used in G.729 are described in Section 7.8.9 and use about 2.5 MIPS. In G.729A several changes are implemented to simplify the post-filtering. The main change is in the adaptive long-term post-filtering. In G.729 the delay T of the long-term post-filter in each subframe is calculated with $\frac{1}{8}$ sample resolution using a two-stage search around the integer part of the transmitted adaptive codebook delay T_1 for the first subframe. In G.729A the delay T is always an integer and is computed by searching the range $T_{cl} - 3$ to $T_{cl} + 3$, where T_{cl} is the integer part of the transmitted adaptive codebook delay for the current subframe. This, along with several other minor modifications, reduces the complexity of the decoder post-filtering to about 1 MIPS.

7.9.7 Conclusions

In this section we have described the G.729A 8 kbits/s codec. This codec is very similar to, and is bit-stream compatible with, the G.729 codec described in Section 7.8. However, due to several complexity reducing modifications, it has a full-duplex complexity of only about 12 MIPS, compared to about 22 MIPS for G.729. This reduction in complexity is achieved at the expense of a small degradation in the performance of the codec in the case of three tandems and in the presence of background noise. The codec was originally conceived for use in DSVD applications, but it is suitable for use in many other applications as well. Indeed, because of its bit-stream compatibility, it can be used as a direct replacement for G.729, when complexity reduction is necessary.

In the next two sections, we consider a pair of ACELP-based codecs that essentially also originated from the University of Sherbrooke speech compression team. Hence, many of the attractive features of the G.729 scheme can be recognized in different incarnations. We note furthermore that these schemes were contrived, in order to improve the perceived speech quality of the existing GSM and IS-54 systems, in an attempt to render the wireless service quality similar to that of wireline-based systems. Let us commence this excursion by considering the enhanced full-rate GSM codec.

7.10 THE 12.2 kbps ENHANCED FULL-RATE GSM SPEECH CODEC [201, 202]

7.10.1 Enhanced Full-rate GSM Codec Outline

This section presents a brief account of the operation of the 12.2 kbps enhanced full-rate GSM speech codec, which will replace the 13 kbps RPE speech codec. This scheme was standardized by the European Telecommunications Standardization Institute (ETSI) in 1996. Here we follow the approach of Salami, Laflamme, Adoul et al. [201, 202] and the interested reader is referred to [202] for a more in-depth discussion. The codec employs the successful ACELP excitation model invented in 1987 by Adoul et al. at Sherbrooke University [144], which was detailed in Section 6.3. The enhanced full-rate GSM scheme uses a bitrate of 10.6 kbps for channel coding, resulting in a channel-coded rate of 22.8 kbps, similarly to the 13 kbps RPE GSM codec, which was the topic of Chapter and was characterized by the schematics of Figures 5.1 and 5.2.

The enhanced full-rate GSM (EFR-GSM) encoder schematic is portrayed in Figure 7.26, while that of the decoder is displayed in Figure 7.29, both of which are detailed here. Similarly to the RPE GSM encoder of Figure 5.1, the input speech is initially preemphasized using a high-pass filter in order to augment the low-energy, high-frequency components, before the speech signal is processed. Observe in Figure 7.26 that, as usual, the spectral quantization is carried out on a frame-by-frame basis, while the excitation optimization is on a subsegment-by-subsegment basis, although we note at this early stage that the spectral quantization is quite original.

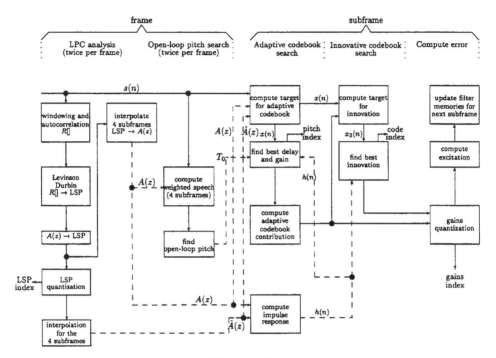

Figure 7.26 Enhanced full-rate 12.2 kbps GSM encoder schematic.

The codec's bit-allocation scheme is summarized in Table 7.15, while the rationale behind using the specified number of bits is detailed during our forthcoming discourse. The 38 LSF quantization bits per 20 ms constitute a 1.9 kbps bitrate contribution, which is typical for medium-rate codecs, although the quantization scheme to be highlighted below is unconventional. The fixed ACELP codebook (CB) gains are quantized using 5 bits/subframe, while the fixed ACELP codes are represented by 35 bits per subframe, which again, is justified with reference to Table 7.15. The adaptive codebook index, corresponding to the pitch-lag, is represented by 9 bits, catering for 512 possible positions in the first and third subframes using a very fine oversampling by a factor of 6 in the low-delay region. In the second and fourth subframes, the pitch-lag is differentially encoded with respect to the odd subframes, again employing an oversampling by six in the low-delay domain.

We note that historically the DoD codec was the first scheme to invoke the differential coding of the pitch-lag and oversampling in the low-lag pitch domain. These measures became fairly widely employed in state-of-the-art codecs, despite the inherent error sensitivity of differential coding. The high-resolution pitch-lag coding of low values is important

TABLE 7.15   12.2 kbps Enhanced Full-Rate GSM Codec Bit Allocation

Parameter	1. & 3. Subfr.	2. & 4. Subfr.	No. of Bits	Total (kbps)
Two LSF sets			38	1.9
Fixed CB gain	5	5	$4 \cdot 5 = 20$	1
ACELP code	35	35	$4 \cdot 35 = 140$	7
Adaptive CB index	9	6	$2 \cdot 9 + 2 \cdot 6 = 30$	1.5
Adaptive CB gain	4	4	16	0.8
Total			244/20 ms	12.2

because it is beneficial to ensure a more-or-less constant relative pitch resolution rather than a constant absolute resolution, as in the case of uniformly applied $125\,\mu s$ sample-spaced pitch encoding. Lastly, the pitch-gains are encoded using 4 bits per subframe.

7.10.2 Enhanced Full-Rate GSM Encoder

7.10.2.1 Spectral Quantization and Windowing in the Enhanced Full-Rate GSM Codec.
Let us initially consider the spectral quantization employed in the EFR-GSM codec, where tenth-order LPC analysis is invoked twice for each 20 ms speech frame, upon using two different 30 ms-duration asymmetric windows. In contrast to the 8 kbps ITU G.729 ACELP codec's window function shown in Figure 7.16, where a 5 ms or 40-sample lookahead was used, the EFR-GSM codec employs no "future speech samples," or—synonymously—no lookahead in the filter coefficient computation, and both asymmetric window functions act on the same set of 240 speech samples, corresponding to the 30 ms analysis interval. Whereas in the 10 ms-framelength G.729 codec an additional 5 ms lookahead delay was deemed acceptable in exchange for a smoother speech spectral envelope evolution, in the 20 ms-framelength EFR-GSM scheme, this was deemed unacceptable. This implies that there is a 10 ms or 80-sample "look-back" interval in the window functions.

Before specifying the shape of the window functions, let us state the rationale behind using two LSF sets, which are used for the second and fourth subframes, respectively. Accordingly, the peak of the first window $w_1(n)$ of Figure 7.27 is concentrated near the center of the second subframe, while that of the second window $w_2(n)$ is near the center of the fourth subframe. Hence, the latter has to exhibit a rapidly decaying slope, given that no lookahead is employed. For the first and third subframes, the LSFs are interpolated on the basis of the surrounding subframes. Specifically, the first window $w_1(n)$ is constituted by two Hamming-window segments of different sizes, as in:

$$
w_1(n) = \begin{cases} 0.54 - 0.46 \cdot \cos\dfrac{\pi n}{L_1 - 1}, & n = 0, \ldots L_1 - 1 \\[2mm] 0.54 - 0.46 \cdot \cos\dfrac{\pi(n - L_1)}{L_2 - 1}, & n = L_1, \ldots L_1 + L_2 - 1 \end{cases}, \tag{7.49}
$$

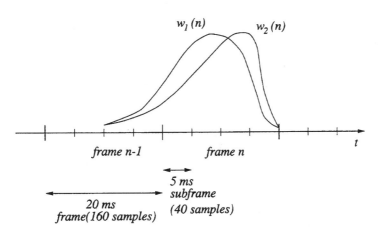

Figure 7.27 Stylized enhanced full-rate GSM window functions.

where the parameters $L_1 = 160$ and $L_2 = 80$ were standardized. Although this window is asymmetric, it is gently decaying toward both ends of the current 20 ms frame, as seen in Figure 7.27. By contrast, since the center of gravity of the second window is close to the beginning of the frame, it has to be tapered more abruptly, which is facilitated by using a short raised-cosine segment, as in:

$$w_2(n) = \begin{cases} 0.54 - 0.46 \cdot \cos\dfrac{\pi n}{L_1 - 1}, & n = 0, \ldots L_1 - 1 \\[4mm] \cos\dfrac{2\pi(n - L_1)}{4L_2 - 1}, & n = L_1, \ldots L_1 + L_2 - 1 \end{cases}, \qquad (7.50)$$

where the parameters $L_1 = 232$ and $L_2 = 8$ were employed.

As seen in Figure 7.26, the autocorrelation coefficients are computed from the windowed speech, and the Levinson-Durbin algorithm is employed in order to derive both the reflection and the linear predictive coefficients, which describe the speech spectral envelope by the help of the $A(z)$ polynomial. Further details of Figure 7.26 concerning, for example, the pitch-lag search and excitation optimization will be unraveled during our later discussions. The LPC coefficients are then converted to LSFs and quantized using the split matrix quantizer (SMQ) of Figure 7.28, which is considered next.

First, the long-term mean of both LSF vectors is removed, yielding the zero-mean LSF vectors p_1^n and p_2^n for frame n, corresponding to the two windows in Figure 7.27. Then both LSF sets of frame n are predicted from the previous quantized LSF set $\tilde{p}_2^{n-1}$, taking into account their long-term correlation of 0.65, as portrayed in Figure 7.28. Both LSF difference vectors are then input to the split matrix quantizer. Specifically, the LSFs of both vectors are paired, as suggested by Figure 7.28, creating a 2×2 submatrix from the first two LSFs of both LSF vectors and quantizing them by searching through a 7-bit, 128-entry codebook. Similarly, the third and fourth LSFs of both LSF vectors are paired and quantized using the 8-bit, 256-entry codebook of Figure 7.28, and so on. Observe that the most important LSFs corresponding to the medium-frequency range are quantized using a larger codebook than those toward the lower and higher frequencies. Finally, after finding the best-matching codebook entries for all 2×2 submatrices, the previous subtracted predicted values are added to them in order to produce both quantized LSF vectors, namely, $\tilde{p}_1^n$ and $\tilde{p}_2^n$, respectively.

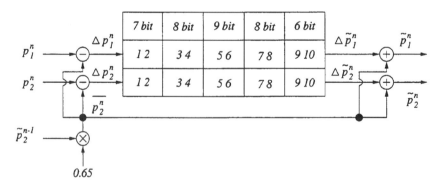

Figure 7.28 The 38-bit split matrix LSF quantization of the sets generated using windows $w_1(n)$ and $w_2(n)$ of Figure 7.27 in the 12.2 kbps enhanced full-rate GSM codec.

7.10.2.2 Adaptive Codebook Search. A combined open- and closed-loop pitch analysis is used, which is similar to that employed in the G.729 codec discussed in Section 7.8. Salami et al. [201] summarized the procedure as follows:

■ As seen in Figure 7.26, based on the weighted speech an open-loop pitch search is carried out twice per 20 ms frame or once every two subframes, favoring low-pitch values in order to avoid pitch doubling. In this search integer sample-based search is used, and the open-loop lag T_o is identified.

■ Then a closed-loop search for integer pitch values is conducted on a subframe basis. This is restricted to the range $[T_o \pm 3]$ in the first and third subframes, in order to maintain a low search complexity. As to the second and fourth subframes, the closed-loop search is concentrated around the pitch values of the previous subframe, in the range of $[-5 \ldots + 4]$.

■ Finally, fractional pitch delays are also tested around the best closed-loop lag value in the second and fourth subframes, although only for the pitch delays below 95 in the first and third subframes, corresponding to pitch frequencies in excess of about 84 Hz.

■ Having determined the optimum pitch-lag, the adaptive codebook entry is uniquely identified, while its gain is restricted to the range of $[0 \ldots 1.2]$ and quantized using 4 bits, as seen in Table 7.15.

Let us now consider the optimization of the fixed codebook in the next subsection.

7.10.2.3 Fixed Codebook Search. Only a rudimentary overview of the principles of ACELP coding [144] is given here because details were presented in Section 6.3. As shown in Table 7.15, 35 bits per subsegment are allocated to the ACELP code. The 5 ms, 40-sample excitation vector hosts 10 nonzero excitation pulses, each of which can take the values ± 1. Salami et al. [201] subdivided the 40-sample subframe into five tracks, each comprising two excitation pulses. The two pulses in each track are allowed to be co-located, potentially resulting in pulse amplitudes of ± 2. The standardized pulse positions are summarized in Table 7.16. Since there are eight legitimate positions for each excitation pulse, 3 bits are necessary for signaling each pulse position. Given that there are 10 excitation pulses, a total of 30 bits are required for their transmission. Furthermore, the sign of the first pulse of each of the five tracks is encoded using 1 bit, yielding a total of 35 bits per subsegment. The sign of the second pulse is inherently determined by the order of the pulse positions, an issue that is elaborated on in References [201,202]. The 3-bit pulse positions were also Gray-coded, implying that adjacent pulse positions are different only in the 1-bit position. Hence, a bit error results in the closest possible excitation pulse position to the one that was transmitted.

TABLE 7.16 12.2 kbps Enhanced Full-Rate GSM Codec's ACELP Pulse Allocation © IEEE, Salami et al., 1997, [201]

Track	Pulses	Positions
1	p_0, p_1	0, 5, 10, 15, 20, 25, 30, 35
2	p_2, p_3	1, 6, 11, 16, 21, 26, 31, 36
3	p_4, p_5	2, 7, 12, 17, 22, 27, 32, 37
4	p_6, p_7	3, 8, 13, 18, 23, 28, 33, 38
5	p_8, p_9	4, 9, 14, 19, 24, 29, 34, 39

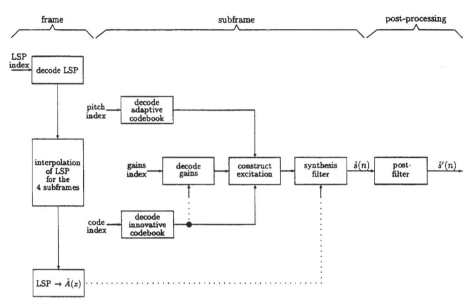

Figure 7.29 Enhanced full-rate GSM decoder schematic.

This ACELP codebook is then invoked in order to generate the 20 ms synthetic speech frame, which is compared to the original speech segment in order to identify the best excitation vector.

At the decoder portrayed in Figure 7.29, the received codec parameters are recovered and the synthetic speech is reconstructed. Specifically, the decoded LSF parameters are interpolated for the individual subframes. Both the fixed and adaptive codebook vectors are regenerated, and with the aid of the corresponding gain factors the excitation signal is synthesized. The excitation is then filtered through the synthesis filter and the post-filter in order to generate the synthetic speech.

Following the above brief description of the enhanced full-rate GSM codec, we now consider another enhanced full-rate codec in the next section, namely, that of the IS-54 system, which was standardized as the IS-136 scheme.

7.11 THE ENHANCED FULL-RATE 7.4 kbps IS-136 SPEECH CODEC [203, 204]

7.11.1 IS-136 Codec Outline

In this section we provide a rudimentary introduction to the operation of the 7.4 kbps IS-136 speech codec, which is a successor of the 7.95 kbps IS-54 DAMPS speech codec [132]. This scheme was standardized in the IS-641 recommendation [205] as part of the enhanced IS-136 standard in the United States [203]. This new scheme was the result of a collaboration between Nokia and Sherbrook University, and here we follow the approach of Honkanen, Vainio, Jarvinen, Haavisto, Salami, Laflamme, and Adoul [204]. The interested reader is referred to [205] for a more detailed overview. Similarly to a number of other standard schemes, the codec employs the ACELP excitation model contrived in 1987 by Adoul et al. at Sherbrook University [144], which was described in depth in Section 6.3. The original IS-54 VSELP codec was discussed in Section 7.3. This scheme is, however, more similar to the enhanced full-rate ACELP GSM codec. In fact, the schematic of these schemes is quite similar. Thus, here we do not duplicate the corresponding block diagrams, rather, we simply refer the reader to Figures 7.26 and 7.29, which are briefly highlighted. We note, however, that the spectral quantization, windowing, and interpolation regime is, for example, radically different from that of the enhanced full-rate GSM codec, since a more stringent bitrate constraint has been imposed. Further differences are inevitable in terms of the number of bits allocated to the various codec parameters.

As, for example, in the full-rate and enhanced full-rate GSM encoders, the input speech is initially preemphasized using a high-pass filter in order to boost the low-energy, high-frequency components and hence mitigate the associated number representation problems. As seen in Figure 7.26 for the enhanced GSM codec, the spectral quantization is carried out on a frame-by-frame basis, while the excitation optimization is on a subsegment-by-subsegment basis. The codec's bit-allocation scheme is presented in Table 7.17. Below we provide some rudimentary justification for the specific parameter quantization schemes used.

7.11.2 IS-136 Bit-Allocation Scheme

In comparison to the schematically similar ACELP enhanced full-rate GSM codec, there is a reduced bitrate contribution by the 10 LSFs due to using only one set of LSFs per 20 ms, as opposed to two. For each 20 ms speech frame a 30 ms-duration asymmetric windows is applied. Whereas for the enhanced full-rate GSM codec, for example, no window-lookahead was used, in the IS-136 codec a 5 ms or 40-sample lookahead was employed, similarly to the 8 kbps ITU G.729 ACELP codec. Bitrate savings are achieved by vector quantization (VQ), requiring a total of 26 bits/20 ms, which constitutes a 1.3 kbps bitrate contribution. The corresponding LSF VQ scheme is shown in Figure 7.30, which is very similar to the corresponding arrangement of the G.723.1 dual-rate codec of Section 7.12. By comparing Figures 7.33 and 7.30, it becomes clear that essentially only the codebook sizes are slightly different, since the 7.4 kbps IS-136 scheme invests a total of 26 rather than 24 bits in spectral quantization—due to its slightly less stringent bitrate budget. Explicitly, split vector-quantization is used for reasons of complexity reduction, where the first three LSFs are grouped together and vector-quantized using 8 bits, or 256 entries, while the two other groups of LSF quantizers are constituted by 3 and 4 LSFs, employing 9 and 9 bits, respectively. As seen in Figure 7.30, the nth unquantized LSF vector p^n is predicted first on the basis of the previous quantized LSF vector $p^{(\tilde{n}-1)}$, after multiplying it with a scaling factor b, which is proportional to the long-term correlation between adjacent LSF vectors. The predicted LSF vector $\bar{p}^n$ is then subtracted from the original unquantized LSF vector in order to generate their difference vector, namely, δp^n, which is split in subvectors of 3, 3, and 4 LSFs and quantized. Finally, the quantized LSF difference vector $\Delta \tilde{p}^n$ is added to the predicted value $\bar{p}^n$ in order to generate the current quantized LSF vector $\tilde{p}^n$.

TABLE 7.17 The 7.4 kbps Enhanced Full-Rate IS-136 Codec's Bit Allocation

Parameter	1. & 3. Subfr.	2. & 4. Subfr.	No. of Bits	Bitrate (kbps)
10 LSFs			$8 + 9 + 9 = 26$	1.3
Gain VQ	7	7	28	1.4
ACELP code	17	17	68	3.4
Pitch-lag	8	5	26	1.3
Total			148/20 ms	7.4

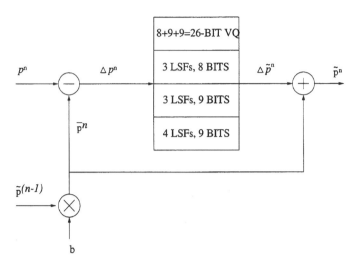

Figure 7.30 The 26-bit IS-136 LSF quantization schematic.

The fixed ACELP codebook (CB) gains and adaptive CB gains are jointly vector quantized using 7 bits/5 ms subframe, hence contributing 1.4 kbps to the total bitrate. The fixed ACELP codes are assigned 17 bits per subframe, which will be justified in the context of Table 7.18. As seen in the table, the adaptive codebook index or pitch-lag is encoded using 8 bits, corresponding to 256 positions in the first and third subframes. In the second and fourth subframes, the pitch-lag is differentially encoded by 5 bits with respect to the corresponding lags in subframes 1 and 3, allowing 32 possible positions. Similarly to the G.729 and to the enhanced full-rate GSM codecs, Salami et al. employed a combination of open- and closed-loop search for the pitch-lag in the sample index range of [19 1/3–143], implying that, as in most modern codecs, in the low-delay range oversampling is used. Specifically, the 1/3-sample based search is carried out over the interval [19 1/3–85]. This guarantees a similar relative resolution to the high-delay-lag range, where integer-sampling is sufficiently accurate. The open-loop search is explicitly shown in the schematic of Figure 7.26, which is found from the perceptually weighted input speech.

In order to ensure a smooth evolution of the pitch-lag and hence also to aid the operation of the differential pitch-lag coding in even subframes, the open-loop pitch-lag is determined once per 10 ms—in other words in every other subframe. This implies giving preference to low pitch-lag values and hence preventing opting for pitch-harmonics, rather

TABLE 7.18   7.4 kbps Enhanced Full-Rate IS-136 Codec's ACELP Pulse Allocation ©IEEE, Honkanen et al., 1997, [204]

Track	Pulses	Positions
1	p_0	0, 5, 10, 15, 20, 25, 30, 35
2	p_1	1, 6, 11, 16, 21, 26, 31, 36
3	p_2	2, 7, 12, 17, 22, 27, 32, 37
4	p_3	3, 8, 13, 18, 23, 28, 33, 38
5		4, 9, 14, 19, 24, 29, 34, 39

than for the true pitch values. This initial pitch-lag search is then followed by a subframe-based closed-loop pitch search in the range of $[\pm3]$ around the open-loop values for subframes 1 and 3. Finally, the pitch-lag of the even-indexed subframes is found by restricting the closed-loop search to the range $[-5\ldots+4]$ around the previous odd-indexed subframe. Again, these measures ensure the well-behaved evolution of the pitch-lag over time.

7.11.3 Fixed Codebook Search

As noted before, the principles of ACELP coding proposed by Adoul et al. [144] were outlined in Section 6.3, and the fixed codebook search of this scheme is akin to that of the enhanced full-rate GSM codec of Section 7.10. The ACELP codebook of Table 7.18 is also similar to that of Table 7.15. However, instead of allocating two pulses per excitation track in each 40 ms subsegment, due to the lower bitrate constraint of 7.4 kbps here only one pulse per excitation track is employed. The corresponding bitrate contribution was reduced from 35 bits per 40-sample subsegment to 17 bits.

Explicitly, the 5 ms, 40-sample excitation vector hosts four nonzero excitation pulses, each of which can take the values ±1. Salami et al. [201, 204] subdivided the 40-sample subframe into five tracks. Each of the first three tracks hosts an excitation pulse, while tracks 4 and 5 share a pulse. Since there are eight possible positions for each pulse in the first three tracks, their encoding requires 9 bits, while the encoding of the fourth pulse necessitates 4 bits, yielding a total of 13 bits per subsegment for the pulse positions, while another bit is used to encode the sign of the bit. Hence a total of 17 bits/5 ms subsegment are required for the ACELP code. Honkanen et al. employed focused search strategies similar to those first proposed by Salami et al. for the G.729 codec [136]. According to the 13-bit ACELP code, a total of 8192 entries per subsegment have to be tested for identifying the optimum one, which was reduced to about 9% of the total range at the cost of low perceptual penalty. The decoder's operations are also similar to those of the enhanced full-rate GSM decoder, which was portrayed in Figure 7.29. Let us now consider some of the channel coding aspects of the IS-136 codec.

7.11.4 IS-136 Channel Coding

Source-sensitivity matched error protection is provided for the IS-136 codec by dividing the speech bits in two protection classes, Class-One and Class-Two. The codec itself is more robust against channel errors than the original 7.95 kbps IS-54 scheme, and due to the reduced speech-rate more robust channel coding can be assigned. The schematic of the channel coding and mapping scheme is very similar to that of the IS-54 arrangement of Figure 7.4; only the number of associated bits has changed, as seen in Figure 7.31. The most sensitive 48 bits are allocated a 7-bit cyclic redundancy checking (CRC) pattern, which can assist the decoder for activating bad frame masking.

As indicated by Figure 7.31, the 148 speech bits are classified as 96 Class-One bits and 52 Class-Two bits. The more error-prone Class-One bits are half-rate, constraint-length five convolutionally encoded, while the remaining 52 bits are transmitted unprotected. The convolutional encoder processes $96 + 7 + 5 = 108$ bits, where the 5 tailing bits are again necessitated by the constraint-length five code to flush its buffer before the transmission of the next speech frame. This prevents the propagation of channel errors across speech frame boundaries, which would otherwise result in prolonged speech degradation due to the convolutional decoder's deviation from the error-free trellis path. The $2 \times 108 = 216$ protected Class-One bits have to undergo light puncturing, since only 260 channel-coded bits can be transmitted in the current IS-136 transmission burst structure. Figure 7.31 shows

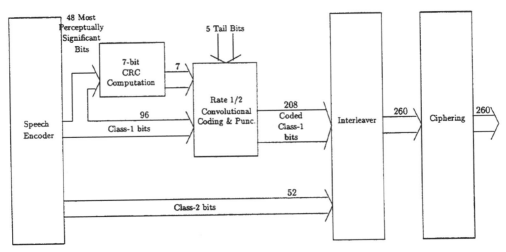

Figure 7.31 The 7.4/13 kbps IS-136 ACELP error protection schematic ©1996, [203].

that a total of 208 bits are generated after puncturing, which are then amalgamated with the 52 unprotected bits to yield the required 260 channel coded speech bits. These are then interleaved and ciphered before they are transmitted over the channel.

Having considered the family of recent enhanced full-rate codecs, which were based on the ACELP principle, we now focus our attention on another ITU scheme, namely the dual-rate G.723 codec.

7.12 THE ITU G.723.1 DUAL-RATE CODEC [206]

7.12.1 Introduction

The ITU G.723.1 dual-rate codec was contrived to form part of the H.324 multimedia compression and transmission standard, which also includes the well-known H.263 video codec. Initially, this speech codec was referred to as G.723, but since there exists an older ADPCM-based G.723 standard, this scheme was renamed G.723.1 in order to avoid confusion. The G.723.1 encoding and decoding processes are based on linear prediction carried out for 30 ms or 240-sample speech segments with a lookahead of 7.5 ms, giving a total delay of 37.5 ms. Analysis-by-synthesis excitation optimization is used on the basis of four 60-sample subsegments. The G.723.1 scheme is a dual-rate speech codec that employs algebraic code excited linear prediction at 5.3 kbit/s, a technique also adopted by the 8 kbit/s ITU G.729 codec. For its 6.3 kbit/s mode of operation multi-pulse maximum likelihood quantization (MP-MLQ) excitation is utilized. The codec's bit allocation is shown in Table 7.20 in its 5.3 kbps mode of operation, while that of the 6.3 kbps mode is portrayed in Table 7.21, both of which will be elaborated on at a later stage. This dual-rate principle has been demonstrated to be a useful system design option for intelligent multimode transceivers [145], which facilitate a transceiver reconfiguration at each speech frame boundary in order to provide, for example, a more robust but lower speech quality mode of operation or a higher speech quality and higher speech-rate associated with weaker error correction. The G.723.1 codec is also amenable to voice-activity controlled discontinuous transmission and comfort noise injection during untransmitted passive speech spurts. A further feature of this scheme is that it was designed to require a relatively low implementational complexity.

7.12.2 G.723.1 Encoding Principle

The schematic of the G.723.1 encoder is shown in Figure 7.32. Similarly to other ITU speech codecs, the G.723.1 scheme bandlimits the speech signal to the conventional 300–3400 Hz telephone band, samples it at 8 kHz, and then converts it to 16-bit linear PCM for further processing. Hence, this scheme actually constitutes a transcoder. Before further processing the speech signal is high-pass-filtered, which removes any residual DC-offset, and excitation optimization is carried out on the basis of 7.5 ms or 60-sample segments.

As seen in Figure 7.32, the original speech signal $y(n)$ is segmented to yield the 240-sample speech $s(n)$ before it is subjected to LPC analysis. The LPC filter order is 10, and, similarly to most forward-adaptive LPC-based scheme, its coefficients are determined from the original speech signal. A 180-sample duration Hamming-window is centered on each subsegment, and 11 autocorrelation coefficients are determined before invoking the Levinson-Durbin algorithm, in order to compute four LPC sets per subsegment. These coefficients are then used in the formant-based perceptual weighting filter. The LPC-coefficients used in the last of the four excitation optimization subsegments are quantized using a technique referred to as Predictive Split Vector Quantization (PSVQ). The LPC coefficients are transformed to LSP format before quantization in the LSP-quantizer block of Figure 7.32. Observe furthermore in the figure that, as in most state-of-the-art codecs, the LSP parameters are locally decoded and interpolated across subsegments.

The conventional formant-based perceptual error weighting filter of Figure 7.32 is followed by a harmonic noise-shaping filter, which, as suggested by its name, relies on the harmonic pitch estimate delivered by the corresponding block of Figure 7.32. For a detailed

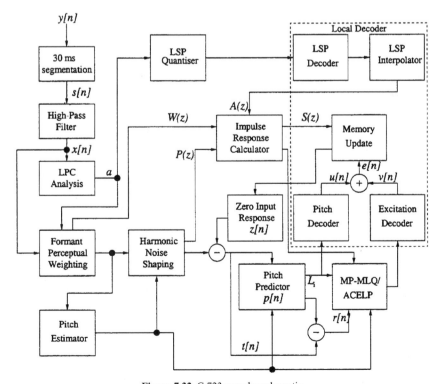

Figure 7.32 G.723 encoder schematic.

discourse on the mathematical description of this filter, the interested reader is referred to the G.723.1 standard [206]. In order to maintain a low complexity, this pitch-lag estimate L_{ol} is derived from the formant-weighted speech signal initially in an open-loop search in the range of 18 to 142 samples on the basis of two consecutive subsegments. This open-loop pitch estimate can then be used for a more accurate closed-loop analysis-by-synthesis search in a limited range, which takes place in the corresponding "Pitch Predictor" block of Figure 7.32, operating on the weighted speech signal following the formant-based and harmonic-based filtering blocks and the deduction of the filters' zero-input response. As in all other analysis-by-synthesis codecs, the zero input response of the synthesis filter corresponds to its memory from the previous excitation optimization cycle.

An unusually high, fifth-order pitch predictor is employed. For the first and third subframes, the pitch-lag is refined around the open-loop estimate within the range of ± 1 and its value is transmitted using 7 bits. For the second and fourth subframes, the pitch delay is differentially encoded using 2 bits, allowing a deviation in the range of $[-1 \ldots + 2]$. The pitch predictor gains are vector-quantized employing a 170-entry codebook for the 5.3 kbps mode of operation and an additional 85-entry codebook for the 6.3 kbps mode. The 85-entry codebook is activated for quantizing open-loop pitch gains, when the associated pitch-lag is below 58, while the 170-entry codebook is dedicated to quantizing the pitch gains related to high pitch-delay scenarios. The effect of the refined pitch-predictor can then be deducted from the speech signal, and, depending on the required bitrate, the resultant residual signal is consecutively subjected to either MP-MLQ or ACELP excitation optimization in the MP-LPQ/ACELP block of Figure 7.32. As usual, the local decoder decodes the pitch-, excitation-as well LSP parameters in order to ensure that the encoder and decoder rely on the same set of parameters in reconstructing the speech. The Impulse Response Calculator block determines the response of the combined closed-loop synthesis filter, constituted by the formant-based perceptual weighting filter and the harmonic noise-shaping filter.

7.12.3 Vector-Quantization of the LSPs

The previously mentioned PSVQ LSP-quantization scheme is depicted in Figure 7.33. Initially, the long-term average $\mathbf{P}_{mean}$ of the unquantized LSP parameters is subtracted from the current set of unquantized LSP in order to arrive at the set $\mathbf{P}_n$, although this subtraction step is not shown in the figure. Because of the inherent correlation between consecutive LSPs, the previous quantized LSP vector $\tilde{\mathbf{P}}_{n-1}$ provides a good estimate of the current vector to be quantized. Their long-term adjacent-vector correlation was found to be $\frac{12}{32}$, which is used here in a simple first-order predictor to produce an estimate $\bar{\mathbf{P}}_n$ of the current vector $\mathbf{P}_n$ that has to be quantized. As portrayed in the figure, the difference ΔP_n of the estimate and the original vector is computed, which is now likely to exhibit a more uncorrelated behavior. This vector could be scalar quantized, but better performance is achieved using vector quantization at the cost of higher complexity. However, the vector quantization of a 10-dimensional vector may become excessively high, if a low quantization distortion has to be maintained, requiring a large trained codebook. A good compromise is to use the split vector quantization principle advocated by the G.729 codec, for example. The G.723.1 scheme employs a three-way split LSP VQ, constituted by three subvectors, which have dimensions of 3, 3, and 4, respectively. Having found the best-matching codebook entry for the three subvectors, the quantized LSP vector is determined by adding the predicted value $\bar{\mathbf{P}}_n$ to the codebook entry and super-imposing the previously subtracted mean value, as portrayed in the figure.

As mentioned before, the LSPs are then interpolated across the subframes for maintaining a seamless spectral envelope evolution. If $\mathbf{P}_{\mathbf{quant}}^{\mathbf{n}}$ is the quantized LPC vector

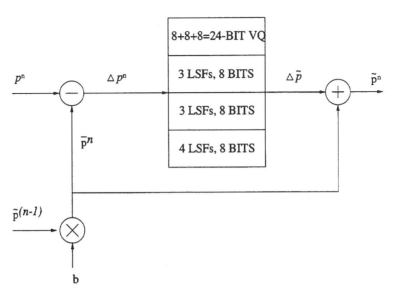

Figure 7.33 The 24-bit G.723.1 LSF quantization schematic.

in the present frame and $\mathbf{P}_{quant}^{n-1}$ is the quantized LPC vector from the past frame, then the interpolated LSP vector for the four subframes is given by:

$$\mathbf{P}_{quant,1}^{n} = 0.75\mathbf{P}_{quant}^{n-1} + 0.25\mathbf{P}_{quant}^{n}$$

$$\mathbf{P}_{quant,2}^{n} = 0.5\mathbf{P}_{quant}^{n-1} + 0.5\mathbf{P}_{quant}^{n}$$

$$\mathbf{P}_{quant,3}^{n} = 0.25\mathbf{P}_{quant}^{n-1} + 0.75\mathbf{P}_{quant}^{n}$$

$$\mathbf{P}_{quant,4}^{n} = 0.75\mathbf{P}_{quant}^{n}$$

7.12.4 Formant-Based Weighting Filter

The weighting filter employed in the G.723.1 encoder is similar to that of the G.729 scheme, and it is based on the unquantized LPC filter coefficients a_i, which are updated for each subsegment on the basis of a 180-sample Hamming-windowed speech segment. The transfer function of the weighting filter is given by:

$$W_j(z) = \frac{A_j(z/\gamma_1)}{A_j(z/\gamma_2)}$$

$$= \frac{1 - \sum_{i=1}^{10} \gamma_1^i a_{ij} z^{-i}}{1 - \sum_{i=1}^{10} \gamma_2^i a_{ij} z^{-i}} \quad 0 \leq j \leq 3 \tag{7.51}$$

where $\gamma_1 = 0.9$ and $\gamma_2 = 0.5$ determine the amount of spectral weighting. In the G.729 codec the amount of weighting (i.e., the factors γ_1 and γ_2) were adaptively controlled in order to improve the performance of the codec for input signals with a flat frequency response. We note furthermore that in the G.729 standard, for example, the Levinson-Durbin algorithm delivers a set of LPC coefficients, which have the opposite sign in comparison to the G.723.1 LPC coefficients. Hence, we have a negative sign before the summation in the weighting filter of Equation 7.52, while the G.729 weighting filter contains a positive sign in the weighting filter.

7.12.5 The 6.3 kbps High-Rate G.723.1 Excitation

The target vector modeling is carried out in the MP-MLQ/ACELP block of Figure 7.32 using the following convolution:

$$r'(n) = \sum_{j=0}^{n} h(j) \cdot v(n-j) \ 0 \le n \le 59, \tag{7.52}$$

where $v(n)$ is the excitation vector and $h(n)$ is the impulse response of the combined formant-based perceptual filter and harmonic noise filter. The excitation vector has the form

$$v(n) = G \cdot \sum_{m=0}^{M-1} \alpha_m \cdot \delta(n - n_m) \ 0 \le n \le 59, \tag{7.53}$$

where G is the excitation gain factor, allowing the excitation's energy to fluctuate and hence to cater for speech segments exhibiting different energy, α_m, $m = 0 \ldots M$ represents the sign of the Dirac-delta excitation pulses, while n_m $m = 0 \ldots M$ denote the positions of the excitation pulses. The number of excitation pulses in the 6.3 kbps mode is $M = 6$ in even subframes and 5 in odd ones. The pulse positions are restricted either to be all odd or even, which is encoded using a grid-position bit, as seen in the bit-allocation scheme of Table 7.20. Given that the odd or even positions are preselected by the grid-position bit, there are 30 possible pulse locations, and the six excitation pulses of the even subframes hence can take

$$\binom{30}{6} = 593775$$

different positions. Similarly, in the odd subframes

$$\binom{30}{5} = 142506$$

position combinations can be encountered. Since $2^{20} = 1048576$ and $2^{18} = 262144$, the required number of bits using the enumerative coding technique is 20 and 18 in the even and odd subframes, respectively.

It is also noted in the recommendation, however, that the resultant bitrate can be further reduced if the pulse positions are not separately represented for the individual subframes. This is plausible, since we noted above that $2^{20} = 1048576$ and $2^{18} = 262144$, allowing us to represent by nearly a factor of two more than the possible number of 593775 and 142506 excitation pulse positions of the two modes. Hence, it was recommended that the first four MSBs from each subframe pulse position index be combined and that 13 bits be employed to encode these 16 bits. This reduces the total number of bits from 192 to 189/30 ms, yielding a bitrate of 6.3 kbps. This bitrate reduction is not reflected in the bit-allocation scheme of Table 7.20 for the sake of simplicity. However, in the bit-sensitivity plot of Figure 7.35, this bitrate reduction becomes explicit, portraying the 13-bit pulse position MSBs (POS MSB) and the 16-, 14-, 16- and 14-bit position indices.

The excitation is found using the classic approach—in other words by minimizing the mse between the target vector and the candidate excitation vectors over the set of legitimate excitation patterns, where the error term concerned is formulated as follows:

$$e(n) = r(n) - r'(n)$$
$$= r(n) - G \cdot \sum_{m=0}^{M-1} \alpha_m \cdot h(n - n_m) \ 0 \le n \le 59. \tag{7.54}$$

Minimizing the mean squared error term of:

$$E = \sum_{0}^{59} e^2(n) \tag{7.55}$$

for all the previously stipulated legitimate excitation patterns leads to the following optimum excitation gain expression:

$$G_{max} = \frac{max|d(j)|_{j=0...59}}{\sum_{n=0}^{59} h^2(n)}, \tag{7.56}$$

where we have:

$$d(j) = \sum_{n=j}^{59} r(n) \cdot h(n - j) \ 0 \le n \le 59. \tag{7.57}$$

The optimum excitation gain G_{max} is then logarithmically scalar-quantized using 24 quantization steps, which are spaced by 3.2 dB. Taking the logarithm of a quantity, which exhibits a highly nonuniform PDF, compresses the large values to be quantized and expands the range of lower values, hence rendering the PDF typically more uniform and more amenable to uniform quantization on the resulting logarithmic scale. In order to further improve the speech quality, the optimum quantized gain G_{max} is tentatively reduced by one 3.2 dB step and increased by two such steps, and the excitation pulses are reoptimized to find the best combination of these parameters, resulting in the minimum mse. Finally, these parameters are encoded and transmitted to the decoder.

7.12.6 The 5.3 kbps Low-Rate G.723.1 Excitation

ACELP codecs have been discussed in depth earlier both in general terms and in the context of the 8 kbps G.729 codec. Hence here we refrain from detailing the excitation optimization procedure. Suffice to say here that a 17-bit ACELP codebook is used in the 5.3 kbps mode, where the innovation vector is constituted by at most four nonzero pulses, which can have the signs and positions summarized in Table 7.19.

As mentioned before, the pulses can occupy either even or odd positions in the subframe, which is ensured by testing the mse associated with the set of pulses shifted by one position with respect to that indicated in Table 7.19. This is signaled to the decoder using the grid-position bit. Observe furthermore from the table that the bracketed pulse positions are actually outside the subframe limits, and hence they are not used. According to the three legitimate positions of the excitation pulses, their position is signaled to the decoder using 3 bits, while their sign is transmitted using a fourth bit. Hence, for the four excitation pulses 16 bits are required, totaling 17 with the additional grid-position bit. The excitation optimization is structured in four nested loops, according to identifying the best position for each of the four excitation pulses.

TABLE 7.19 G.723.1 ACELP Excitation Pulses in the 5.3 kbps Mode

Sign	Positions
±1	0, 8, 16, 24, 32, 40, 48, 56
±1	2, 10, 18, 26, 34, 42, 50, 58
±1	4, 12, 20, 28, 36, 44, 52, (60)
±1	6, 14, 22, 30, 38, 46, 54, (62)

The computational complexity of the codec is further reduced by applying a focused search strategy, similarly to the G.729 8 kbps ACELP codec. Explicitly, before entering the last of the four nested loops, a thresholding operation is invoked, in order to test whether it is sufficiently promising to continue the search in terms of synthesized speech quality. This loop is then searched only if the thresholding condition is met. Furthermore, the maximum number of entering this loop is also fixed for the sake of setting a maximum for the search complexity, which is an important aspect of real-time implementations. Specifically, the last loop is entered maximum 150 times per subsegment. Before the excitation optimization for the next subsegment begins, the memory of the concatenated synthesis filter, formant-based perceptual weighting filter, and harmonic noise filter has to be updated both at the encoder and decoder. This operation is carried out by filtering the optimum excitation through this cascaded filter complex at both the encoder and decoder and storing it, until it is invoked as the zero input response or filter memory during the optimization of the next subsegment excitation.

7.12.7 G.723.1 Bit Allocation

The bit-allocation schemes of the two modes of operation are summarized in Tables 7.20 and 7.21. Since the innovation sequences of the two modes are different, the encoding of the excitation pulses is different, but the remaining parameters are encoded identically. Hence, the bitrate reduction accrues from the lower number of excitation quantization bits in the 5.3 kbps ACELP mode. Explicitly, instead of the $76 + 22 = 98$ bits/30 ms ≈ 3.27 kbps pulse position and pulse sign excitation bitrate contribution of the 6.3 kbps scheme, the 5.3 kbps codec requires $48 + 16 = 64$ bits/30 ms ≈ 2.13 kbps, resulting in a bitrate reduction of about 1.1 kbps.

In order to elaborate further on gain quantization, we note that 12 gain quantization bits are allocated for encoding the 24 3.2 dB-spaced excitation gain levels and the 5-tap pitch

TABLE 7.20 Bit-Allocation Scheme of the 5.3 kbps Mode of the G.723.1 Codec

Parameter	Sub-Fr.1	Sub-Fr.2	Sub-Fr.3	Sub-Fr.4	Total/30 ms
LPC Indices					$3 \cdot 8 = 24$
Adaptive codebook lag: ACL0–ACL3	7	2	7	2	18
Excitation and pitch gains combined: GAIN0–GAIN3	12	12	12	12	48
Pulse positions: POS0–POS3	12	12	12	12	48
Pulse signs: PSIG0–PSIG3	4	4	4	4	16
Grid index: GRID0–GRID3	1	1	1	1	4
Total					158/30 ms

TABLE 7.21 Bit-Allocation Scheme of the 6.3 kbps Mode of the G.723.1 Codec

Parameter	Sub-Fr.1	Sub-Fr.2	Sub-Fr.3	Sub-Fr.4	Total/30 ms
LPC Indices					$3 \cdot 8 = 24$
Adaptive codebook lag: ACL0–ACL3	7	2	7	2	18
Excitation and pitch gains combined: GAIN0–GAIN3	12	12	12	12	48
Pulse positions: POS0–POS3	20	18	20	18	76
Pulse signs: PSIG0–PSIG3	6	5	6	5	22
Grid index: GRID0–GRID3	1	1	1	1	4
Total					192/30 ms

predictor gains, using 170 levels for the latter. The associated quantization schemes are identical in both operational modes. There is a total of $170 \times 24 = 4080$ possible combinations of the excitation and pitch gains, which is less than 4096 and hence can be jointly encoded using 12 bits. However, this is not a robust quantization, since a single-bit error in the 12-bit index will affect all gains. Alternatively, it would have been possible to employ 8 bits for the pitch gains, implementing a finer, 256-level quantizer and 5 bits for the excitation gain, again ensuring a somewhat finer 32-level quantization scheme. Quantizing the two gains separately would have required a total of 13 bits, requiring only one additional bit, while ensuring a higher error resilience. A similar combinatorial coding was also used for the excitation pulse locations.

There is a further fine detail concerning the 6.3 kbps gain quantization.[1] Namely, if the pitch-lag in the first or third subframe is less than 58, then the number of levels for pitch gains is 85 instead of 170 for two consecutive subframes. This will save one gain quantization bit, since $85 \times 24 = 2040 < 2048$, and hence 11 bits will suffice. This saved bit is used to signal whether pitch sharpening is invoked in the excitation code. If pitch sharpening is employed, then an excitation pulse is replaced with a series of excitation pulses, separated by the pitch period within the limits of the subframe boundary, as will be illustrated using the following example. Let us assume that there are three excitation pulses at the pulse positions of 0, 20, and 30 and that the pitch period is 25, corresponding to $25 \times 125\,\mu s = 3.125\,\text{ms}$ or to a pitch frequency of 320 Hz.

When invoking pitch sharpening, the above three pulses are replaced by the following pitch-spaced pulses: 0, 25, 50; 20, 45; 30, 55, while using the previously determined excitation gains. We note, however, that the pitch-spaced pulses are not extended beyond the subframe boundary at position 60. Hence, there are still three excitation gains but six excitation pulses in this case. Both the original and the pitch-sharpened excitation configurations will be tentatively invoked, and the one that minimizes the error criterion will be selected. The 1-bit flag saved in the modified gain quantization will be used to indicate the presence or absence of pitch sharpening.

Since the quantization and encoding philosophies of the various remaining parameters were discussed during our earlier discourse, here we refrain from detailing these bit-allocation tables further. The encoded bit stream is transmitted to the decoder, where the individual parameters are reconstructed. They are employed in reconstructing the synthesized speech signal similarly to the local decoder, which was briefly summarized during our description of

---

[1] The authors are grateful to Redwan Salami for this private communication.

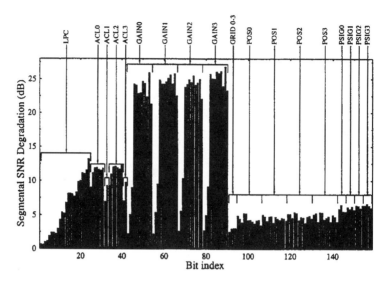

Figure 7.34 Bit sensitivity of the 5.3 kbps GH.723 speech frame.

the encoder schematic. With the main G.723.1 algorithms known, let us now consider the error sensitivity of the codec.

7.12.8 G.723.1 Error Sensitivity

The bit-allocation scheme of the G.723.1 codec is summarized in Tables 7.21 and 7.20, respectively. In this section knowledge of these tables is assumed, and only a brief summary of the associated bit-sensitivity issues is offered with reference to Figures 7.34 and 7.35.

Considering the sensitivity of the 5.3 kbps codec first, the first 24 bits of the frame illustrated in Figure 7.34 represent the LPC vector quantizer address bits, which exhibit a gradually increasing SEGSNR degradation for the more significant address bits. This

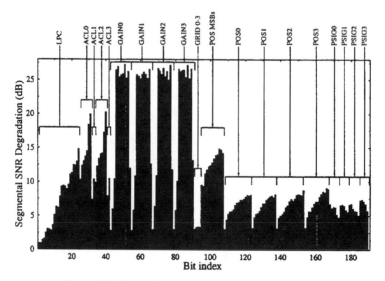

Figure 7.35 Bit sensitivity of the 6.3 kbps G.723 speech frame.

suggests that the codebook is structured by allocating codebook entries representing spectral envelopes similar to each other in each other's vicinity, since corrupting the least significant bits (LSBs) of the codebook address results in small SEGSNR degradations. By contrast, when corrupting the most significant bits (MSBs), a high SEGSNR degradation is experienced. The nondifferentially encoded 7-bit adaptive codebook lag bits exhibit more or less uniform bit sensitivities, while the differentially coded 2-bit lags are less sensitive. This is intuitively expected, since the corruption of the nondifferential values also corrupts the differential values. The highest sensitivity is exhibited by the jointly quantized 12-bit gain indices. As expected, similar observations can also be made with respect to the corresponding bits of the 6.3 kbps codec, as seen in Figure 7.35.

Focusing our attention on excitation pulse parameters, we see that the 5.3 kbps ACELP excitation bits of Figure 7.34 shows a rather flat sensitivity as a function of the bit index. This is in harmony with our expectations for ACELP codecs, since corrupting one pulse position does not dramatically change the whole of the excitation vector. This is also valid for the excitation pulse sign, although the associated SEGSNR degradation is slightly higher than that due to the position bits.

The 6.3 kbps excitation encoding relies on the enumerative technique of the multi-pulse excitation. The sensitivity of the excitation grid bits is relatively low, while that of the jointly encoded four subsegment excitation pulse position MSBs is significantly higher. The observed staircase effect again suggests that similar excitation pulse positions are encoded using similar indices. Hence, if one of the index LSBs is corrupted, its SEGSNR effects are more mitigated than in the case of some of the MSBs. The excitation pulse signs again exhibit similar sensitivities to the pulse position index bits.

7.13 CHAPTER SUMMARY

In Sections 7.2–7.12 we have described many CELP-related forward-adaptive standard speech codecs. We remind the reader of the stylized Figure 1.6, which was refined by Cox [1,2], portraying the associated formally evaluated subjective quality in Figure 17.4 of Chapter 17.

Since the 1990s, many standard speech codecs have emerged, each of which characterized a certain state-of-the-art. The ultimate aim has been to find the best quality, complexity, delay, and robustness tradeoff at the given stage of development. The standard codecs considered can be broadly divided into two classes: mobile radio speech codecs and ITU schemes. As seen in Figure 17.4, the G.-series ITU codecs always endeavored to maintain a subjective speech quality in excess of a MOS of 4, while reducing the bitrate by a factor of 2, which was only possible at the cost of increasing the implementational complexity. By contrast, the mobile radio codecs often had to accept a compromise associated with a lower speech quality, compromised, for example, by the complexity constraints imposed by limited battery-consumption.

The first standard CELP-based was the 4.8 kbps DoD scheme of Section 7.2, where differential pitch-lag encoding and oversampled lag representation was first invoked in an attempt to minimize the prediction residual and to ensure near-constant relative lag-representation error across the entire legitimate lag range. This scheme invoked scalar quantization of the LPC coefficients.

This arrangement was followed by the family of VSELP codecs, such as the 7.95 kbps IS-54 and the 6.7 kbps JDC schemes of Sections 7.3 and 7.4, which are closely related to each other. However, whereas the IS-54 codec possessed two fixed codebooks, the JDC codec had only one. Hence, the former had a significantly higher combination of fixed codebook vectors and a higher associated complexity and speech quality. Similarly to the DoD scheme, both of these codecs employed scalar quantization of the LPC coefficients, which required in excess of 30 bits per LPC frame. The variable-rate Qualcomm codec of Section 7.5 also represented a similar stage of development to these codecs, using scalar quantization of the LPC coefficients.

The half-rate codec family was spawned by the 3.6 kbps Japanese PSI-CELP scheme of Section 7.6, which was the first codec to invoke pitch-synchronous excitation. Another half-rate codec is the 5.6 kbps GSM coding arrangement of Section 7.7, which followed similar VSELP coding principles of the IS-54 and JDC codecs. However, in order to operate at such a low rate, it employed four different excitation modes. In the unvoiced mode, the combination of two 7-bit excitation codebooks was employed, but no long-term prediction was utilized. In voiced speech segments, long-term prediction was invoked, and a larger single codebook of 512 entries was used. Furthermore, the reflection coefficients were quantized using a three-way split vector quantizer, reducing the number of LPC quantization bits below 30 per LPC frame. An advantageous feature of the reflection-coefficient-based lattice predictors was that the quantization error of the previous quantization stages was partially taken into account by the later quantizer stages. The half-rate GSM codec also employed a multi-resolution long-term predictor.

Perhaps the most prominent speech codec to date is the G.729 scheme of Section 7.8 and its reduced-complexity version described in Section 7.9 employing ACELP principles. For the sake of maintaining a low delay, it employs asymmetric windowing with a short, 40 sample or 5 ms lookahead and differentially encoded pitch-lag. A sophisticated 18-bit LSF vector quantizer is employed, along with a refined perceptual error weighting filter. Both short-term and long-term post-filtering as well as spectral tilt compensation are invoked at the output of the decoder in order to improve the perceived speech quality. Strong consideration was given to the error concealment aspects in order to tolerate high error rates and frame loss rates. Similar techniques dominated the design of the enhanced full-rate GSM codec of Section 7.10 and that of the enhanced Pan-American codec referred to as the IS-136 scheme. The G.723.1 dual-rate codec of Section 7.12 also used ACELP coding in one of its operational modes, while multi-pulse excited techniques were utilized in its other mode.

Following the above brief chronological overview of the speech codec standardization scene, let us now consider backward-adaptive codecs in the next chapter.

8

Backward-Adaptive Code Excited Linear Prediction

8.1 INTRODUCTION

In the previous chapter a range of medium- to high-delay forward-adaptive CELP codecs have been described, which constituted different trade-offs in terms of speech quality, bitrate, delay and implementational complexity. In this chapter our work moves on to low-delay, backward-adaptive codecs.

This chapter discusses why the delay of a speech codec is an important parameter, methods of achieving low-delay coding, and the problems with these methods. Much of the material presented centers around the recently standardized 16 kbits/s G728 Low-Delay CELP codec [70, 85] and the associated algorithmic issues are described. We also discuss our attempts to extend the G728 codec in order to propose a low-delay, programmable bitrate codec operating between 8 and 16 kbits/s. The chapter also focuses on the potential speech quality improvements that can be achieved in such a codec by adding a long-term predictor (LTP), albeit at the cost of increased error sensitivity due to error propagation effects introduced by the backward-adaptive LTP. These error propagation effects can be mitigated at system level, for example, by introducing reliable error control mechanisms, such as automatic repeat request (ARQ), an issue to be discussed in a system context at a later stage. We also discuss the means of training the codebooks used in our variable rate codec to optimize its performance and an alternative variable-rate codec, which has a constant vector size. Finally we describe the post-filtering used to improve the perceptual quality of our codecs.

8.2 MOTIVATION AND BACKGROUND

The delay of a speech codec can be an important parameter for several reasons. In the public switched telephone network, 4- to 2-wire conversions lead to echoes, which are subjectively annoying if the echoes are sufficiently delayed. Experience shows that the 57.5 ms speech coding and interleaving delay of the Pan-European GSM system already introduces an undesirable echoing effect. This value can be considered as the maximum tolerable margin in toll-quality communications. Even if echo cancelers are used, a high-delay speech codec makes the echo cancellation more difficult. Therefore, if a codec is to be connected to the telephone network, it is desirable that its delay should be as low as possible. If the speech codec used has a lower delay, then other elements of the system, such as bit interleavers, will have more flexibility and should be able to improve the overall quality of the system.

The one-way **coding delay** of a speech codec is defined as the time from when a sample arrives at the input of the encoder to when the corresponding sample is produced at the output of the decoder, assuming the bit stream from the encoder is fed directly to the decoder. This one-way delay is typically made up of three main components [70]. The first is the **algorithmic buffering delay** of the codec—the encoder operates on frames of speech and must buffer a frame-length's worth of speech samples before it can start encoding. The second component is the **processing delay**—speech codecs typically operate in just real time, and so it takes almost one frame length in time to process the buffered samples. Finally, there is the bit **transmission delay**—if the encoder is linked to the decoder by a channel with capacity equal to the bitrate of the codec, then there will be a further time delay equal to the codec's frame length while the decoder waits to receive all the bits representing the current frame.

The overall one-way delay of the codec is equal to about three times the frame length of the codec. However, this delay can be reduced by careful implementation of the codec. For example, if a faster processor is used, the processing delay can be reduced. Also, it may not be necessary to wait until the whole speech frame has been processed before we can start sending bits to the decoder. Finally, a faster communications channel, for example, in a time division multiplexed system, can dramatically reduce the bit transmission delay. Other factors may also result in an increase in the total delay. For example, the one subframe lookahead used to aid the interpolation of the LSFs in our ACELP codecs described earlier will increase the overall delay by one subframe. Nonetheless, typically the one-way coding delay of a speech codec is assumed to be about 2.5 to 3 times the frame length of the codec.

The most effective way to produce a low-delay speech codec is to use as short a frame length as possible. Traditional CELP codecs have a frame length of 20 to 30 ms, leading to a total coding delay of at least 50 ms. Such a long frame length is necessary because of the forward adaption of the short-term synthesis filter coefficients. As explained in Chapter 6, a frame of speech is buffered, LPC analysis is performed, and the resulting filter coefficients are quantized and transmitted to the decoder. As we reduce the frame length, the filter coefficients must be sent more often to the decoder, and so more and more of the available bitrate is taken up by LPC information. Although efficient speech windowing and LSF quantization schemes have allowed the frame length to be reduced to 10 ms (with a 5 ms lookahead) in a candidate codec [136] for the CCITT 8 kbits/s standard, a frame length of between 20 and 30 ms is more typical. If we want to produce a codec with delay of the order of 2 ms, which was the objective for the CCITT 16 kbits/s codec [85], we cannot use forward adaption of the synthesis filter coefficients.

The alternative is to use backward-adaptive LPC analysis. This means that rather than window and analyze present and future speech samples in order to derive the filter coefficients, we analyze previous quantized and locally decoded signals to derive the coefficients. These past quantized signals are available at both the encoder and decoder, and so no side information about the LPC coefficients needs to be transmitted. This allows us to update the filter coefficients as frequently as we like, with the only penalty being a possible increase in the complexity of the codec. Thus, we can dramatically reduce the codec's frame length and delay.

As already explained, backward-adaptive LPC analysis has the advantages of allowing us to dramatically reduce the delay of our codec and removing the information about the filter coefficients that must be transmitted. This side information is usually about 25% of the bitrate of a codec, and so it is very helpful if it can be removed. However, backward adaption has the disadvantage that it produces filter coefficients that are typically degraded in comparison to those used in forward-adaptive codecs. The degradation in the coefficients has two sources [207]:

1. **Noise Feedback**. In a backward-adaptive system, the filter coefficients are derived from a quantized signal, and so there will be a feedback of quantization noise into the LPC analysis which will degrade the performance of the coefficients produced.

2. **Time Mismatch**. In a forward-adaptive system, the filter coefficients for the current frame are derived from the input speech signal for the current frame. In a backward-adaptive system, we have only signals available from previous frames to use, and so there is a time mismatch between the current frame and the coefficients we use for that frame.

The effects of noise feedback increase dramatically as the bitrate of the codec is reduced, which means that traditionally backward adaption has only been used in high-bitrate, high-quality codecs. Recently, however, as researchers have attempted to reduce the delay of speech codecs, backward-adaptive LPC analysis has been used at bitrates as low as 4.8 kbits/s [208].

Clearly, the major design challenge associated with the ITU G728 codec was due to the complexity of its specifications, which are summarized in Table 8.1. Although many speech codecs can produce good speech quality at 16 kbps, at such a low rate most previous codecs have inflicted significantly higher delays than the targeted 2 ms. This is because in order to achieve such a low rate in linear predictive coding, the update interval of the LPC coefficients must be around 20–30 ms. As argued earlier in the case of scalar LPC parameter coding typically 36 bits/20 ms = 1.8 kbps channel capacity is required for their encoding. Hence, in the case of a 2 ms delay forward predictive coding is not a realistic alternative. We have also seen in Section 2.9 that low-complexity, low-delay ADPCM coding at 16 kbps is possible, which would satisfy the first two criteria of Table 8.1, but the last three requirements are not satisfied.

Chen, Cox, Lin, Jayant, and Melchner have contributed a major development to the state-of-art of speech coding [70], which satisfied all the design specifications and was standardized by the ITU [85]. In this section we follow their discussions in References [70] and [85, pp. 625–627] in order to describe the operation of their proposed backward-adaptive codec. The ITU's call for proposals stimulated a great deal of research, and a variety of candidate codecs were proposed, which typically satisfied some but not all requirements of

TABLE 8.1 G 728 Codec Specifications

Parameter	Specification
Bitrate	16 kbps
One-way delay	<2 ms
Speech quality at $BER = 0$	<4 QDU for one codec
	<14 QDU for three tandems
Speech quality at $BER = 10^{-3}$ and 10^{-2}	Better than that of G721 32 kbps ADPCM
Additional requirement	Pass DTMF and CCITT No. 5, 6, and 7 signaling

Table 8.1. Nonetheless, a range of endeavors, including those of References [82, 209], have contributed to the standardization process.

CELP coding emerged as the best candidate, which relied on backward prediction using a high-order (50) filter; where the coefficients did not have to be transmitted, they were extracted from the past decoded speech. Due to the high-order short-term predictor (STP), there was no need to include an error-sensitive long-term predictor (LTP). The importance of adaptive post-filtering was underlined by Jayant and Ramamoorthy in [82, 83], where the quality of 16 kbps ADPCM-coded speech was reportedly improved, which was confirmed by Chen and Gersho [84].

The delay and high speech quality criteria were achieved by using a short STP-update interval of 20 samples or $20 \cdot 125 \ \mu s = 2.5$ ms and an excitation vector length of 5 samples or $5 \cdot 125 \ \mu s = 0.625$ ms. The speech quality was improved using a trained rather than a stochastic codebook, which was "virtually" extended by a factor of 8 using a 3-bit codebook gain factor. Lastly, a further novel element of the codec is the employment of backward-adaptive gain scaling [210, 211]. In the next section we describe the 16 kbits/s G728 low-delay CELP codec, particularly the ways it differs from the ACELP codecs we have used previously. We also attempt to quantify the effects of both noise feedback and time mismatch on the backward-adaptive LPC analysis used in this codec.

8.3 BACKWARD-ADAPTIVE G728 CODEC SCHEMATIC [70, 85]

The G728 encoder and decoder schematics are portrayed in Figures 8.1 and 8.2, respectively. The input speech segments are compared with the synthetic speech segments as in any AbS codec, and the error signal is perceptually weighted, before the specific codebook entry associated with the lowest error is found in an exhaustive search procedure. For the G728 codec a vector size of five samples corresponding to $5 \cdot 125 \ \mu s = 0.625$ ms was found appropriate in order to curtail the overall speech delay to 2 ms.

Having fixed the length of the excitation vectors, let us now consider the size of the excitation codebook. Clearly, the larger the codebook size, the better the speech quality, but the higher the computational complexity and the bitrate. An inherent advantage of backward-

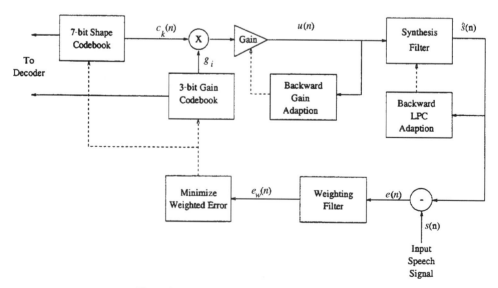

Figure 8.1 16 kbps low-delay CCITT G728 encoder.

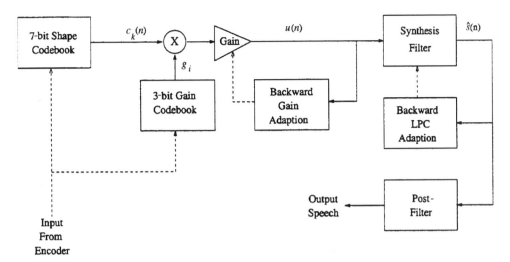

Figure 8.2 16 kbps low-delay CCITT G728 decoder.

adaptive prediction is that the LPC coefficients are not transmitted; hence, a high-order filter can be used, and we can dispense with using an LTP. Therefore, a design alternative is to allocate all bits transmitted to the codebook indices. Assuming a transmission rate of 16 kbps and an 8 kHz sampling rate, we are limited to a coding rate of 2 bits/sample or 10 bits/5 samples. Logically, the maximum possible codebook size is then $2^{10} = 1024$ entries. Recall that in the case of forward predictive codecs the codebook gain was typically quantized using 4 to 5 bits, which allowed a degree of flexibility in terms of excitation envelope fluctuation. In this codec, it is unacceptable to dedicate such a high proportion of the bitrate budget to the gain quantization. Chen and Gersho [211] noted that this slowly fluctuating gain information is implicitly available and hence predictable on the basis of previously scaled excitation segments. This prompted them to contrive a **backward-adaptive gain predictor**, which infers the required current scaling factor from its past values using predictive techniques. The actual design of this gain predictor will be highlighted at a later stage. This allowed the total of 10 bits to be allocated to the codebook index, although the codebook finally was trained as a 128-entry scheme in order to reduce the search complexity by a factor of 8, and the remaining 3 bits were allocated to quantize another multiplicative gain factor. This two-stage approach is suboptimum in terms of coding performance, since it replaces eight independent codebook vectors by eight identically shaped, different magnitude excitation vectors. Nonetheless, the advantage of the eightfold reduced complexity outweighted the significance of a slight speech degradation.

As mentioned before, Chen et al. decided to opt for a fiftieth-order backward-adaptive STP filter in order to achieve the highest possible prediction gain and to be able to dispense with LTP filtering, without having to transmit any LPC coefficients. However, the complexity of the Levinson–Durbin algorithm used to compute the LPC coefficients is proportional to the square of the filter order $p = 50$, which constitutes a high complexity. This is particularly so if the LPC coefficients are updated for each five-sample speech vector. In order to compromise, an update interval of 20 samples or 2.5 ms was deemed to be appropriate. This implies that the LPC parameters are kept constant for the duration of four excitation vectors, which is justifiable since the speech spectral envelope does not vary erratically.

A further ramification of extending the LPC update interval is that the time lag between the speech segment to be encoded and the spectral envelope estimation is increased. This is a disadvantage of backward-adaptive predictive systems because in these schemes the current speech frame is used for the speech spectral estimation. On the same note, backward-adaptive arrangements have to infer the LPC coefficients from the past decoded speech, which is prone to quantization effects. For high-rate, high-quality coding, this is not a significant problem, but it is aggravated by error propagation effects, inflicting future impairments in future LPC coefficients. Hence, at low bitrates, below 8 kbps, backward-adaptive schemes found only limited favor in the past. These effects can be readily quantified using the unquantized original delayed speech signal and the quantized but not delayed speech signal to evaluate the codec's performance. Woodard [159] found that these factors degraded the codec's SEGSNR performance by about 0.2 dB due to quantization noise feedback and by about 0.7 dB due to the time mismatch, yielding a total of 0.9 dB SEGSNR degradation. At lower rates and higher delays, these degradations become more dominant. Let us now concentrate our attention on specific algorithmic issues of the codec schematics given in Figures 8.1 and 8.2.

8.4 BACKWARD-ADAPTIVE G728 CODING ALGORITHM [70, 85]

8.4.1 G728 Error Weighting

In contrast to the more conventional **error weighting filter** introduced in Equation 3.8, the G728 codec employs the filter [84]:

$$W(z) = \frac{1 - A(z/\gamma_1)}{1 - A(z/\gamma_2)} = \frac{1 - \sum_{i=1}^{10} a_i \gamma_1^i z^{-k}}{1 - \sum_{i=1}^{10} a_i \gamma_2^i z^{-k}} \qquad (8.1)$$

where $\gamma_1 = 0.9$ and $\gamma_2 = 0.6$, and the filter is based on a tenth order LPC analysis carried out using the unquantized input speech. This was necessary to prevent the introduction of spectral distortions due to quantization noise. Since the error weighting filter is only used at the encoder, where the original speech signal is available, this error weighting procedure does not constitute any problem at all. The choice of the $\gamma_1 = 0.9$ and $\gamma_2 = 0.6$ parameters was motivated by the requirement of optimizing the tandemized performance for three asynchronous coding operations. Explicitly, listening tests proved that the pair $\gamma_1 = 0.9$ and $\gamma_2 = 0.4$ gave a better single-coding performance, but for three tandemed codec $\gamma_2 = 0.6$ was found to exhibit a superior performance. The coefficients of this weighting filter are computed from the windowed input speech, and the particular choice of the window function is highlighted in the next section.

8.4.2 G728 Windowing

The choice of the windowing function plays an important role in capturing the time-variant statistics of the input speech, which in turn influences the subsequent spectral analysis. In contrast to more conventional Hamming windowing, Chen et al. [70] proposed to use a hybrid window, which is constituted by an exponentially decaying long-term past history section and a nonrecursive section, as is depicted in Figure 8.3.

Let us assume that the LPC analysis frame size is $L = 20$ samples, which hosts the samples $s(m)$, $s(m + 1) \ldots s(m + L - 1)$, as portrayed in Figure 8.3. The N-sample window section immediately preceding the current LPC frame of L samples is then termed the nonrecursive portion, since it is described mathematically by the help of a sinusoid nonrecursive function of $w(n) = -\sin[c(n - m)]$, where the sample index n is limited to the previous N samples $(m - N \leq n \leq (m - 1))$. In contrast, the recursive section of the

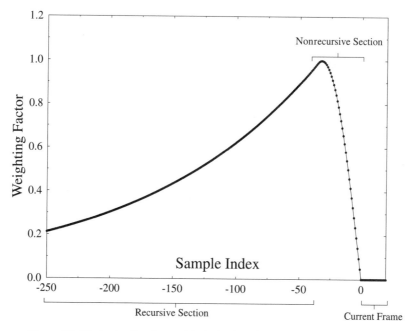

Figure 8.3 Windowing function used in the backward adaption of the synthesis filter.

window function weights the input speech samples preceding $(m - N)$, as suggested by Figure 8.3, using a simple negative exponential function given by

$$w(n) = b \cdot \alpha^{-[n-(m-N-1)]} \text{ if } n \leq (m - N - 1), \tag{8.2}$$

where $0 < b, \alpha < 1$. Evaluating Equation 8.2 for sample index values at the left of $n = (m - N)$ in Figure 8.3 yields weighting factors of b, $b \cdot \alpha$, $b \cdot \alpha^2 \ldots$. In summary, the hybrid window function can be written as:

$$w_m(n) = \begin{cases} f_m(n) = b \cdot \alpha^{-[n-(m-N-1)]} & \text{if } n \leq (m - N - 1) \\ g_m(n) = -\sin[c(n - m)] & \text{if } (m - N) \leq n \leq (m - 1) \\ 0 & \text{if } n \geq m \end{cases} \tag{8.3}$$

It is important to maintain a seamless transition between the recursive and nonrecursive section of the window function in order to avoid introducing spectral sidelobes, which would be incurred in case of a noncontinuous derivative at $n = (m - N)$ [212], where the two sections are joined.

Cheng et al. also specify in the Recommendation [85] how this recursive windowing process can be exploited to calculate the required autocorrelation coefficients, using the windowed speech signal given by:

$$s_m(n) = s(n) \cdot w_m(n) \tag{8.4}$$

where the subscript m indicates the beginning of the current L-sample window in Figure 8.3.

For an Mth-order LPC analysis at instant m, the autocorrelation coefficients $R_m(i)$ $i = 0, 1, 2 \ldots M$ are required by the Levinson–Durbin algorithm, where

$$R_m(i) = \sum_{n=-\infty}^{m-1} s_m(n) \cdot s_m(n-i)$$

$$= \sum_{n=-\infty}^{m-N-1} s_m(n) \cdot s_m(n-i) + \sum_{n=m-N}^{m-1} s_m(n) \cdot s_m(n-i). \tag{8.5}$$

Upon taking into account Equations 8.3 and 8.4 in Equation 8.5, the first term of Equation 8.5 can be written as follows:

$$r_m(i) = \sum_{n=-\infty}^{m-N-1} s(n) \cdot s(n-i) \cdot f_m(n) \cdot f_m(n-i), \tag{8.6}$$

which constitutes the recursive component of $R_m(i)$, since it is computed from the recursively weighted speech segment. The second term of Equation 8.5 relates to the section given by $(m - N) \leq n \leq (m - 1)$ in Figure 8.3, which is the nonrecursive section. The N-component sum of the second term is computed for each new N-sample speech segment, while the recursive component can be calculated recursively following the procedure proposed by Chen et al. [70, 85].

Assuming that $r_m(i)$ is known for the current frame, we proceed to the frame commencing at sample position $(m + L)$, which corresponds to the next frame in Figure 8.3, and we express $r_{m+L}(i)$ in analogy with Equation 8.5 as follows:

$$r_{m+L}(i) = \sum_{n=-\infty}^{m-1} s_m(n) \cdot s_m(n-i)$$

$$= \sum_{n=-\infty}^{m-N-1} s_m(n) \cdot s_m(n-i) + \sum_{n=m-N}^{m-1} s_m(n) \cdot s_m(n-i)$$

$$= \sum_{n=-\infty}^{m-N-1} s(n) \cdot f_m(n) \cdot \alpha^L \cdot s(n-i) f_m(n-i) \alpha^L \tag{8.7}$$

$$+ \sum_{n=m-N}^{m+L-N-1} s_{m+L}(n) \cdot s_{m+L}(n-i)$$

$$= L^{2L} r_m(i) + \sum_{n=,-N}^{m+L-N-1} s_{m+L}(n) \cdot s_{m+L}(n-i)$$

This expression is the required recursion, which facilitates the computation of $r_{m+L}(i)$ on the basis of $r_m(i)$. Finally, the total autocorrelation coefficient $R_{m+L}(i)$ is generated by the help of Equation 8.5. When applying the general hybrid windowing process to the LPC analysis associated with the error weighting, the following parameters are used: $M = 10$, $L = 20$, $N = 30$, $\alpha = (\frac{1}{2}) 40 \approx 0.983$, yielding $\alpha^{2L} = \alpha^{40} = \frac{1}{2}$. Then the Levinson and Durbin algorithm is invoked in the usual manner, as described by Equation 2.25 and Figure 2.3.

The performance of the synthesis filter, in terms of its prediction gain and segmental SNR of the G728 codec using this filter, is shown against the filter order p in Figure 8.4 for a single sentence spoken by a female. Also shown in Table 8.2 is the increase in performance obtained when p is increased above 10, which is the value most commonly used in AbS codecs. There is a significant performance gain due to increasing the order from 10 to 50, but little additional gain is achieved as p is further increased.

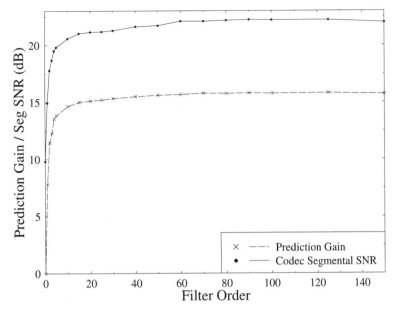

Figure 8.4 Performance of the synthesis filter in a G728-like codec.

We also tested the degradations in the synthesis filter's performance at $p = 50$ due to use of backward adaption. This was done as follows. To measure the effect of quantization noise feedback, we updated the synthesis filter parameters exactly as in G728 except we used the previous speech samples rather than the previous reconstructed speech samples. To measure the overall effect of backward adaption, we updated the synthesis filter using both past and present speech samples. The improvements obtained, in terms of the segmental SNR of the codec and the filter's prediction gain, are shown in Table 8.3. We see that due to the high SNR of the G728 codec, noise feedback has relatively little effect on the performance of the synthesis filter. The time-mismatch gives a more significant degradation in the codec's performance. Note, however, that the forward-adaptive figures given in Table 8.3 could not be obtained in reality because they do not include any effects of the LPC quantization that must be used in a real forward-adaptive system.

Having familiarized ourselves with the hybrid windowing process in general terms, we note that this process is invoked during three different stages of the G728 codec's operation. The next scheme, where it is employed using a different set of parameters is the codebook gain adaption arrangement, which will be elaborated on in the forthcoming section.

TABLE 8.2 Relative Performance of the Synthesis Filter as p Is Increased

Filter Order p	A Prediction Gain (dB)	Δ Seg-SNR (dB)
10	0.0	0.0
25	+0.68	+0.70
50	+1.05	+1.21
75	+1.12	+1.41
100	+1.11	+1.46
150	+1.10	+1.42

TABLE 8.3   Effects of Backward Adaption of the Synthesis Filter

	Δ Prediction Gain (dB)	Δ Seg-SNR (dB)
No noise feedback	+0.50	+0.18
No time mismatch	+0.74	+0.73
Use forward adaption	+1.24	+0.91

8.4.3 Codebook Gain Adaption

Let us describe the codebook vector scaling process at iteration n by the help of:

$$e(n) = \delta(n) \cdot y(n), \tag{8.8}$$

where $y(n)$ represents one of the 1029 five-sample codebook vectors, $\delta(n)$ the scaling gain factor, and $l(n)$ the scaled excitation vector. The associated root-mean-squared (RMS) values are denoted by $\delta_e(n)$ and $\delta_y(n)$, respectively. As regards the RMS values we also have:

$$\delta_e(n) = \delta(n) \cdot \delta_y(n) \tag{8.9}$$

or in logarithmic domain:

$$\log[\delta_e(n)] = \log[\delta(n)] + \log[\delta y(n)].$$

The philosophy of the gain prediction scheme is to exploit the correlation between the current required value of $\delta(u)$ and its past history, which is a consequence of the slowly varying speech envelope. Chen and his colleagues suggested employing a tenth-order predictor operation on the sequence $\log[\delta_e(n-1)]$, $\log[\delta_e(n-2)] \ldots \log[\delta_e(n-10)]$ in order to predict $\log[\delta(n)]$. This can be written more formally as:

$$\log[\delta(n)] = \sum_{i=1}^{10} pi \log[\delta_e(n-i)], \tag{8.10}$$

where the coefficients $pi \ i = 1 \ldots 10$ are the predictor coefficients.

When using a tenth-order predictor relying on 10 gain estimates derived for five speech samples each, the memory of this scheme is 50 samples, which is identical to that of the STP. This predictor therefore analyzes the same time interval as the STP and assists in modeling any latent residual pitch periodicity. The excitation gain is predicted for each speech vector n from the 10 previous gain values on the basis of the current set of predictor coefficients $pii = 1 \ldots 10$. These coefficients are then updated using conventional LPC analysis every fourth five-sample speech vector, or every 20 samples.

The schematic of the gain prediction scheme is depicted in Figure 8.5, where the gain-scaled excitation vector $e(n)$ is buffered and the logarithm of its RMS value is computed in order to express it in terms of dB. At this stage the average excitation gain of voiced speech, namely, an offset of 32 dB is subtracted in order to remove the bias of the process, before hybrid windowing and LPC analysis takes place.

The **bandwidth expansion** module modifies the predictor coefficients $\hat{\alpha}_i$ computed according to

$$\alpha_i = \left(\frac{29}{32}\right)^i \hat{\alpha} = (0.90625)^i \hat{\alpha}_i i = 1 \ldots 10 \tag{8.11}$$

This process is equivalent in z-domain to moving all the poles of the corresponding synthesis filter toward the origin according to the factor $\left(\frac{29}{32}\right)$. Poles outside the unit circle

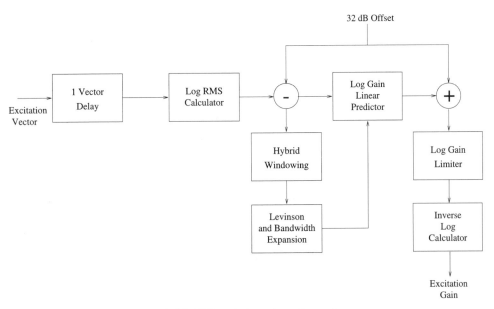

Figure 8.5 G728 excitation gain predictor scheme.

imply instability, while those inside but close to the unit circle are associated with narrow but high spectral prominences. Moving these poles further away from the unit circle expands their bandwidth and mitigates the associated spectral peaks. If the encoder and decoder are misaligned, for example, because the decoder selected the wrong codebook vector due to channel errors, both the speech synthesis filter and the gain prediction scheme will be "deceived." The above bandwidth expansion process assists in reducing the error sensitivity of the predictive coefficients by artificially modifying them at both the encoder and decoder using a near-unity leakage factor.

Returning to Figure 8.5, we see that finally the modified predictor coefficients of Equation 8.11 are employed to predict the required logarithmic gain $\log[\sigma(n)]$. Before the gain factor is used in the current frame, its 32 dB offset must be restored, while its extreme values are limited to the range of 0–60 dB and finally $\sigma(n)$ is restored from the logarithmic domain. The linear gain factor is limited accordingly to the range 1–1000.

The efficiency of the backward gain adaption can be seen from Figure 8.6. This shows the PDFs, on a log scale for clarity, of the excitation vector's optimum gain both with and without gain adaption. Here the optimum vector gain is defined as

$$\sqrt{\frac{1}{vs}\sum_{n=0}^{vs} g^2 c_k^2(n)} \tag{8.12}$$

where g is the unquantized gain chosen in the codebook search. For a fair comparison, both PDFs were normalized to have a mean of 1. It can be seen that gain adaption produces a PDF that peaks around 1 and has a shorter tail and a reduced variance. This makes the quantization of the excitation vectors significantly easier. Shown in Figure 8.7 are the PDFs of the optimum unquantized codebook gain g and its quantized value, when backward gain adaption is used. It can be seen that most of the codebook gain values have a magnitude less than or close to 1, but it is still necessary to allocate two gain quantizer levels for the infrequently used high-magnitude gain values.

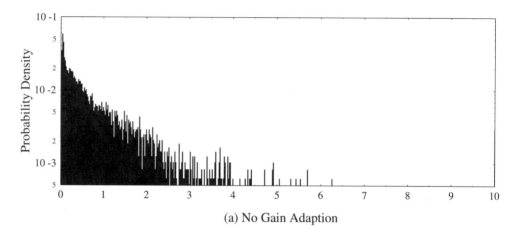

(a) No Gain Adaption

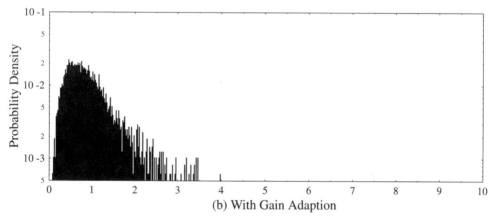

(b) With Gain Adaption

Figure 8.6 PDFs of the normalized codebook gains with and without backward gain adaption.

By training a split $\frac{7}{3}$ bit shape/gain codebook, as described in Section 8.7, for G728-like codecs both with and without gain adaption we found that the gain adaption increased the segmental SNR of the codec by 2.7 dB and the weighted segmental SNR by 1.5 dB. These are significant improvements, especially when it is considered that the gain adaption increases the encoder complexity by only about 3%.

8.4.4 G728 Codebook Search

The standard recognized technique of finding the optimum excitation in CELP codecs is to generate the **target vector** for each input speech vector to be encoded and match the filtered candidate excitation sequences to the target. During synthesizing the speech signal for each codebook vector, the excitation vectors are filtered through the concatenated LPC synthesis filter and the error weighting filter, which are described by the impulse response $h(n)$ as seen in Figure 8.1. Since this filter complex is an infinite impulse response (IIR) system, upon exciting it with a new codebook entry its output signal will be the superposition of the response due to the current entry plus the response due to all previous entries. We note that the latter contribution is not influenced by the current input vector. Hence, this **filter memory contribution** plays no role in identifying the best codebook vector for the current

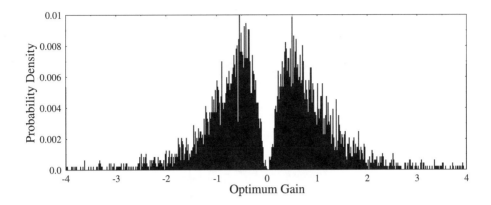

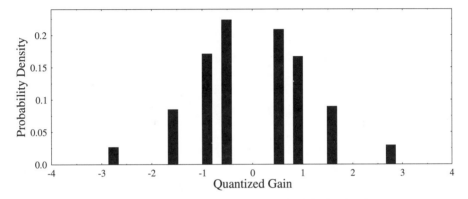

Figure 8.7 PDFs of the optimum and quantized codebook gain values.

five-sample frame. Therefore, the filter memory contribution due to previous inputs has to be buffered, before a new excitation is input and subtracted from the current input speech frame, in order to generate the target vector $x(n)$. All filtered codebook entries are compared to the target vector in order to find the best innovation sequence resulting in the best synthetic speech segment. A preferred alternative to subtracting the filter memory from the input speech in generating the target vector is to set the filter memory to zero before a new codebook vector is fed into it. Since the backward-adaptive gain $\sigma(n)$ is known at frame n, before the codebook search commences, the normalized target vector $x(n) = x(n)/\sigma(n)$ can be used during the optimization process.

Let us follow the notation used in the G728 Recommendation and denote the codebook vectors by y_j, $j = 1 \ldots 128$ and the associated gain factor by g_i, $i = 1 \ldots 8$. Then the filtered and gain-scaled codebook vectors are given by the convolution:

$$\hat{x}_{ij} = \sigma(n) \cdot g_i[h(n) * y_j], \tag{8.13}$$

where again $\sigma(n)$ represents the codebook gain determined by the backward-adaptive gain recovery scheme of Figure 8.5. By the help of the lower triangle convolution matrix of:

$$\mathbf{H} = \begin{bmatrix} h_0 & 0 & 0 & 0 & 0 \\ h_1 & h_0 & 0 & 0 & 0 \\ h_2 & h_1 & h_0 & 0 & 0 \\ h_3 & h_2 & h_1 & h_0 & 0 \\ h_4 & h_3 & h_2 & h_1 & h_0 \end{bmatrix} \tag{8.14}$$

Equation 8.13 can be expressed in a terser form as follows:

$$\hat{x}_{ij} = \mathbf{H}\sigma(n)g_i y_j. \tag{8.15}$$

The best innovation sequence is deemed to be the one that minimizes the following mse distortion expression:

$$D = \|x(n) - \hat{x}_{ij}\|^2 = \sigma^2(n)\|\hat{x}(n) - g_i \cdot \mathbf{H}y_j\|^2 \tag{8.16}$$

where again $\hat{x}(n) = x(n)/\sigma(n)$ is the normalized target vector. Upon expanding the above term, we arrive at:

$$D = \sigma^2(n)\left[\|\hat{x}(n)\|^2 - 2g_i\hat{x}^T\mathbf{H}y_j + g_i^2\|\mathbf{H}g_j\|^2\right] \tag{8.17}$$

Since the normalized target vector energy $\|\hat{x}(n)\|^2$ and the codebook gain $\sigma(n)$ are constant for the duration of scanning the codebook, minimizing D in Equation 4.36 is equivalent to:

$$\hat{D} = -2g_i \cdot p^T(n) \cdot y_j + g_i^2 E_j, \tag{8.18}$$

where the shorthand of $p(n) = \mathbf{H}^T \cdot \hat{x}(n)$ and $E_j = \|\mathbf{H}y_j\|^2$ was employed. Notice that E_j represents the energy of the filtered codebook entry y_j. Since the filter coefficients are only updated every 20 samples, E_j $j = 1 \ldots 128$ is computed once per LPC update frame.

The optimum codebook entry can now be found by identifying the best g_i, $i = 1 \ldots 8$. A computationally more efficient technique is to compute the optimum gain factor for each entry and then quantize it to the closest prestored value. Further specific details of the codebook search procedure are given in [70, 85], while the codebook training algorithm was detailed in [213].

In the original CELP codec proposed by Schroeder and Atal, a stochastic codebook populated by zero-mean unit-variance Gaussian vector was used. The G728 codec uses a 128-entry trained codebook.

In a conceptually simplistic but suboptimum approach, the codebook could be trained by simply generating the prediction residual using a stochastic codebook and then employing the **pairwise nearest neighbor** or the **pruning method** [102] to cluster the excitation vectors in order to arrive at a trained codebook. However, upon using this trained codebook, the prediction residual vectors generated during the codec's future operation may now be different, necessitating the retraining of the codebook recursively a number of times. This is particularly true in case of backward-adaptive gain recovery because the gain factor will be dependent on the codebook entries, which in turn again will depend on the gain values. According to Chen [213], the codec performance is dramatically reduced if no closed-loop training is invoked. The robustness against channel errors was substantially improved following the proposals by De Marca and Jayant [214], as well as Zeger and Gersho [215], using pseudo-Gray coding of the codebook indices, which ensured that in case of a single channel error the corresponding codebook entry was similar to the original one.

8.4.5 G728 Excitation Vector Quantization

At 16 kbits/s there are 10 bits that can be used to represent every five sample vectors, and as the LPC analysis is backward adaptive these bits are used entirely to code the excitation signal $u(n)$ that is fed to the synthesis filter. The five sample excitation sequences are vector-quantized using a 10-bit split shape-gain codebook. Seven bits are used to represent the vector shapes, and the remaining 3 bits are used to quantize the vector gains.

This splitting of the 10-bit vector quantizer reduces the complexity of the closed-loop codebook search. To measure the degradations introduced by this splitting, we trained codebooks for a $\frac{7}{3}$ bit shape/gain split vector quantizer, and a pure 10-bit vector quantizer. We found that the 10-bit vector quantizer gave no significant improvement in either the segmental SNR of the segmental weighted SNR of the codec. It also increased the complexity of the codebook search by about 550% and the overall codec complexity by about 300%. This splitting of the vector quantizer is therefore a very efficient way to reduce the complexity of the encoder.

The closed-loop codebook search is carried out as follows. For each vector, the search procedure finds values of the gain quantizer index i and the shape codebook index k which minimize the squared weighed error E_w for that vector. E_w is given by

$$E_w = \sum_{n=0}^{vs-1} \left(s_w(n) - \hat{s}_o(n) - \hat{\sigma} g_i h(n) * c_k(n) \right)^2 \tag{8.19}$$

where $s_w(n)$ is the weighted input speech, $\hat{s}_o(n)$ is the zero input response of the synthesis and weighting filters, $\hat{\sigma}$ is the predicted vector gain, $h(n)$ is the impulse response of the concatenated synthesis and weighting filters, and g_i and $c_k(n)$ are the entries from the gain and shape codebooks. This equation can be expanded to give:

$$E_w(n) = \hat{\sigma}^2 \sum_{n=0}^{vs-1} \left(x(n) - g_i[h(n) * c_k(n)] \right)^2 \tag{8.20}$$

$$= \hat{\sigma}^2 \sum_{n=0}^{vs-1} x^2(n) + \hat{\sigma}^2 g_i^2 \sum_{n=0}^{vs-1} [h(n) * c_k(n)]^2$$

$$- 2\hat{\sigma}^2 g_i \sum_{n=0}^{vs-1} x(n)[h(n) * c_k(n)] \tag{8.21}$$

$$= \hat{\sigma}^2 \sum_{n=0}^{vs-1} x^2(n) + \hat{\sigma}^2 \left(g_i^2 \xi_k - 2g_i C_k \right)$$

where $x(n) = (s_w(n) - \hat{s}_o(n))/\hat{\sigma}$ is the codebook search target,

$$C_k = \sum_{n=0}^{vs-1} x(n)[h(n) * c_k(n)] \tag{8.22}$$

is the correlation between this target and the filtered codeword $h(n) * c_k(n)$, and

$$\xi_k = \sum_{n=0}^{vs-1} [h(n) * c_k(n)]^2 \tag{8.23}$$

is the energy of the filtered codeword $h(n)*c_k(n)$. Note that this is almost identical to the form of the term in Equation 6.7 which must be minimized in the fixed codebook search in our ACELP codecs.

In the G728 codec, the synthesis and weighting filters are changed only once every four vectors. Hence, ξ_k must be calculated for the 128 codebook entries only once every four vectors. The correlation term C_k can be rewritten as

$$C_k = \sum_{n=0}^{vs-1} x(n)[h(n)*c_k(n)]$$

$$= \sum_{n=0}^{vs-1} c_k(n)\psi(n) \tag{8.24}$$

where

$$\psi(n) = \sum_{i=n}^{vs-1} x(i)h(i-n) \tag{8.25}$$

is the reverse convolution between $h(n)$ and $x(n)$. This means that we need to carry out only one convolution operation for each vector to find $\psi(n)$. Then we can find C_k for each codebook entry k with a relatively simple series of multiply-add operations.

The codebook search finds the codebook entries $i = 1$–8 and $k = 11$–28 which minimize E_w for the vector. This is equivalent to minimizing

$$D_{ik} = g_i^2 \xi_k - 2g_i C_k. \tag{8.26}$$

For each codebook entry k, C_k is calculated, and then the best quantized gain value g_i is found. The values g_i^2 and $2g_i$ are precomputed and stored for the eight quantized gains, and these values along with ξ_k and C_k are used to find D_{ik}. The codebook index k which minimizes this, together with the corresponding gain quantizer level i, are sent to the decoder. These indices are also used in the encoder to produce the excitation and reconstructed speech signals used to update the gain predictor and the synthesis filter.

The decoder's schematic was portrayed in Figure 8.2, which carries out the inverse operations of the encoder seen in the figure. Without delving into specific algorithmic details of the decoder's functions in the next section, we briefly describe the operation of the post-filter at its output stage.

Post-filtering was originally proposed by Jayant and Ramamoorthy [82, 83] in the context of ADPCM coding using the two-pole six-zero synthesis filter of the G721/codec of Figure 2.11 to improve the perceptual speech quality.

8.4.6 G728 Adaptive Post-Filtering

Since post-filtering was shown to improve the perceptual speech quality in the G721 ADPCM codec, Chen et al. have also adopted this technique in order to improve the performance of CELP codecs [84]. The basic philosophy of post-filtering is to augment spectral prominences, while slightly reducing their bandwidth and attenuating spectral valleys between them. This procedure naturally alters the waveform shape to a certain extent, which constitutes an impairment. But its perceptual advantage in terms of reducing the effect of quantization noise outweighs the former disadvantage.

Early versions of the G728 codec did not employ adaptive post-filtering in order to prevent the accumulation of speech distortion during tandeming several codecs. However, without post-filtering, the coding noise due to concatenating three asynchronously operated codecs became about 4.7 dB higher than in the case of one codec. Chen et al. found that this was due to optimizing the extent of post-filtering for maximum noise masking at a concomitant minimum speech distortion, while using a single coding stage. Hence, the amount of post-filtering became excessive in case of tandeming. This then led to a design, which was optimized for three concatenated coding operations. The corresponding speech quality improved by a mean opinion score (MOS) point of 0.81 to 3.93.

8.4.6.1 Adaptive Long-Term Post-Filtering. The schematic of the G728 adaptive post-filter is shown in Figure 8.8. The **long-term post-filter** is a comb filter that enhances the spectral needles in the vicinity of the upper harmonics of the pitch frequency. Although the G728 codec dispenses with using LTP or pitch predictor for reasons of error resilience,

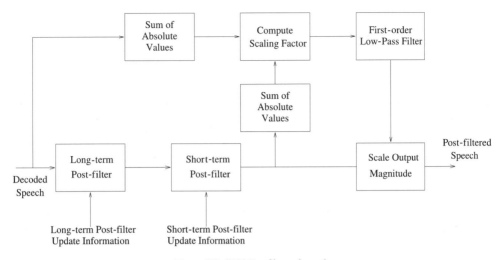

Figure 8.8 G728 Postfilter schematic.

the pitch information is recovered in the codec using a pitch detector to be described at a later stage. Assuming that the true pitch periodicity p is known, the long-term post-filter can be described by the help of the transfer function:

$$H_l = g_l(1 + bz^{-p}),\qquad(8.27)$$

where the coefficients g_l, b, and p are updated during the third five-sample speech segment of each four-segment, or 2.5 ms duration LPC update frame, as suggested by Figure 8.8.

The **postfilter adapter** schematic is displayed in Figure 8.9. A tenth-order LPC inverse filter and the **pitch detector** act in unison to extract the pitch periodicity p. Chen et al. also proposed a possible implementation for the pitch detector. The tenth order LPC inverse filter of

$$\tilde{A}(z) = 1 - \sum_{i=1}^{10} \tilde{a}_i z^{-i}\qquad(8.28)$$

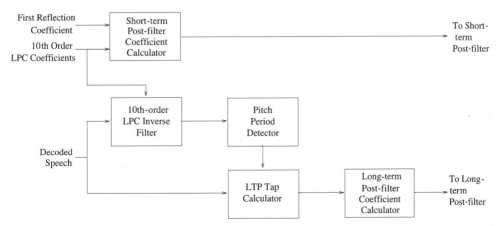

Figure 8.9 Postfilter adapter schematic.

employs the filter coefficients $\tilde{a}_i$, $i = 1 \ldots 10$ computed from the synthetic speech in order to generate prediction residual $d(k)$. This signal is fed to the pitch detector of Figure 8.9, which buffers a 240-sample history of $r(k)$. It would now be possible to determine the pitch periodicity using the straightforward evaluation of Equation 3.7 for all possible delays in the search scope, which was stipulated in the G78 codec to be $[20 \ldots 140]$, employing a summation limit of $N = 100$. However, the associated complexity would be unacceptably high.

Therefore the Recommendation suggests bandlimiting filter $d(k)$ using a third-order elliptic low-pass filter to a bandwidth of 1 kHz and then decimate it by a factor of 4, allowing a substantial complexity reduction. The second term of Equation 3.7 is maximized over the search scope of $\alpha = [20, 21 \ldots 140]$, but in the decimated domain this corresponds to the range $[5, 6, \ldots 35]$. Now Equation 3.7 only has to be evaluated for 31 different delays, and the $\log \alpha_1$ maximizing the second term of Equation 3.7 is inferred as an initial estimate of the true pitch periodicity p. This estimate can then be refined to derive a better estimate α_3 by maximizing the above-mentioned second term of Equation 3.7 over the undecimated $r(k)$ signal within the log range of $[\alpha_1 \pm 3]$. In order to extract the true pitch periodicity, it has to be established whether the refined estimate α_2 is not a multiple of the true pitch. This can be ascertained by evaluating the second term of Equation 3.7 in the range $[\alpha_3 \pm 6]$, where α_3 is the pitch determined during the previous 20-sample LPC update frame. Due to this frequent pitch-picking update at the beginning of each talk-spurt, the scheme will be able to establish the true pitch-lag, since the true pitch-lag is always longer than 20 samples or 2.5 ms and hence no multiple-length lag values will be detected. This will allow the codec to recursively check in the absence of channel error, whether the current pitch-lag is within a range of ± 6 samples or 1.5 ms of the previous one, namely, α_3. If this is not the case, the lag $(\alpha_3 - 6) < \alpha_4 < (\alpha_3 + 6)$ is also found, for which the second term of Equation 3.7 is maximum.

Now a decision must be taken as to whether α_4 or α_2 constitutes the true pitch-lag. This can be established by ranking them on the basis of their associated gain terms $G = \beta$ given by Equation 3.6, which is physically the normalized cross-correlation of the residual segments at delays 0 and α, respectively. The higher this correlation, the more likely that α represents the true pitch-lag. In possession of the optimum LTP lag α and gain β, Chen et al. defined the LT post-filter coefficients b and g_e in Equation 8.27 as

$$b = \begin{cases} 0 & if & \beta < 0.6 \\ 0.15\beta & if & 0.6 \leq \beta \leq 1 \\ 0.15 & if & \beta = 1 \end{cases} \tag{8.29}$$

$$g_e = \frac{1}{1+b}, \tag{8.30}$$

where the factor 0.15 is an experimentally determined constant controlling the weighting of the LT post-filter. If the LTP gain of Equation 3.6 is close to unity, the signal $v(k)$ is almost perfectly periodic. If, however, $\beta < 0.6$, the signal is unvoiced, exhibiting almost no periodicity. Hence, the spectrum has no quasi-periodic fine structure. Therefore, according to $b = 0$, no long-term post-filtering is employed, since $H_l(z) = 1$ represents an all-pass filter. Lastly, in the range of $0.6 \leq \beta \leq 1$ we have $b = 0.5\beta$; that is, β controls the extent of long-term post-filtering, allowing a higher degree of weighting in case of highly correlated $r(k)$ and speech signals.

Having described the adaptive long-term post-filtering, let us now turn our attention to details of the **short-term (ST) post-filtering**.

8.4.6.2 G728 Adaptive Short-Term Post-Filtering. The adaptive ST post-filter standardized in the G728 Recommendation is constituted by a tenth-order pole-zero filter concatenated with a first-order single-zero filter, as in:

$$H_s(z) = \frac{1 - \sum_{i=1}^{10} \overline{b}_i z^{-i}}{1 - \sum_{i=1}^{10} \overline{a}_i z^{-i}} [H\mu z^{-1}], \tag{8.31}$$

where the filter coefficients are specified as follows:

$$\begin{aligned}
\overline{b}_i &= \tilde{a}_i (0.65)^i & i &= 1, 2, \ldots 10 \\
\overline{a}_i &= \tilde{a}_i (0.75)^i & i &= 1, 2, \ldots 10 \\
\mu &= 0.15.k_i
\end{aligned} \tag{8.32}$$

The coefficients $\tilde{a}_i$, $i = 1 \ldots 10$ are obtained in the usual fashion as byproducts of the fiftieth order LPC analysis at iteration $i = 10$, while k_1 represents the first reflection coefficient in the Levison-Durbin algorithm of Figure 2.3. Observe in Equation 8.32 that the coefficients $\overline{a}_i$ and $\overline{b}_i$ are derived from the progressively attenuated $\tilde{a}_i$ coefficients. The pole-zero section of this filter emphasizes the formant structure of the speech signal while attenuating the frequency regions between formants. The single-zero section has a high-pass characteristic and was included in order to compensate for the low-pass nature or spectral delay of the pole-zero section.

Returning to Figure 8.8, observe that the output signal of the adaptive post-filter is scaled in order for its input and output signals to have the same power. The sum of the post-filter's input and output samples is computed, and the required scaling factor is calculated and low-pass filtered in order to smooth its fluctuation, before the output scaling takes place.

Here we conclude our discussions on the standard G728 16 kbps codec with a brief performance analysis, before we embark on contriving a range of programmable-rate 8–16 kbps codecs.

8.4.7 Complexity and Performance of the G728 Codec

In the previous subsections we have described the operation of the codec. The associated implementational complexities of the various sections of the G728 codec are shown in Table 8.4 in terms of millions of arithmetic operations (mostly multiplies and adds) per second. The weighting filter and codebook search operations are carried out only by the encoder, which requires a total of about 12.4 million operations per second. The post-filtering is carried out only by the decoder, which requires about 8.7 million operations per second. The full-duplex codec requires about 21 million operations per second.

TABLE 8.4 Millions of Operations per Second Required by G728 Codec

Synthesis filter	5.1
Backward gain adaption	0.4
Weighting filter	0.9
Codebook search	6.0
Postfiltering	3.2
Total encoder complexity	12.4
Total decoder complexity	8.7

We found that the codec gave an average segmental SNR of 20.1 dB and an average weighted segmental SNR of 16.3 dB. The reconstructed speech was difficult to distinguish from the original, with no obvious degradations. In the next section we discuss our attempts to modify the G728 codec in order to produce a variable bitrate 8–16 kbits/s codec, which gives a graceful degradation in speech quality as the bitrate is reduced. Such a programmable-rate codec is useful in intelligent systems, where the transceiver may be reconfigured under network control in order to invoke a higher or lower speech quality mode of operation or to assign more channel capacity to error correction coding in various traffic loading or wave propagation scenarios.

8.5 REDUCED-RATE G728-LIKE CODEC: VARIABLE-LENGTH EXCITATION VECTOR

Having detailed the G728 codec in the previous section, we now describe our work in reducing the bitrate of this codec and producing an 8–16 kbits/s variable rate low-delay codec. The G728 codec uses 10 bits to represent each five sample vector. To reduce the bitrate of this codec, we must either reduce the number of bits used for each vector or increase the number of speech samples per vector. If we were to keep the vector size fixed at five samples, then in an 8 kbits/s codec we would have only 5 bits to represent both the excitation shape and gain. Without special codebook training, this leads to a codec with unacceptable performance. Therefore, initially we concentrated on reducing the bitrate of the codec by increasing the vector size. In Section 8.8 we discuss the alternative approach of keeping the vector size constant and reducing the size of the codebooks used.

In this section, at all bitrates we use a split $\frac{7}{3}$ bit shape/gain vector quantizer for the excitation signal $u(n)$. The codec rate is varied by changing the vector size vs used, from $vs = 5$ for the 16 bits/s codec to $vs = 10$ for the 8 kbits/s codec. For all the codecs, we used the same 3-bit gain quantizer as in G728, and for the various shape codebooks we used randomly generated Gaussian codebooks with the same variance as the G728 shape codebook. Random codebooks with a Gaussian PDF were used for simplicity and because historically such codebooks give a relatively good performance [16]. We found that replacing the trained shape codebook in the G728 codec with a Gaussian codebook reduced the segmental SNR of the codec by 1.7 dB and the segmental weighted SNR by 2 dB. However, these losses in performance are recovered in Section 8.7 when we consider closed-loop training of our codebooks.

In the G728 codec the synthesis filter, weighting filter, and gain predictor are all updated every four vectors. With a vector size of 5, this means the filters are updated every 20 samples or 2.5 ms. Generally, the more frequently the filters are updated, the better the codec will perform, and we found this to be true for our codec. However, updating the filter coefficients more frequently significantly increases the complexity of the codec. Therefore, we decided to keep the period between filter updates as close as possible to 20 samples as the bitrate of our codec is reduced by increasing the vector size. This means reducing the number of vectors between filter updates as the vector size is increased. For example, at 8 kbits/s the vector size is 10, and we updated the filters every two vectors, which again corresponds to 2.5 ms.

The segmental SNR of our codec against its bitrate as the vector size is increased from 5 to 10 is shown in Figure 8.10. Also shown in this figure is the segmental prediction gain of the synthesis filter at the various bitrates. This figure shows that the segmental SNR of our codec decreases smoothly as its bitrate is reduced, falling by about 0.8 dB for every 1 kbits/s drop in the bitrate.

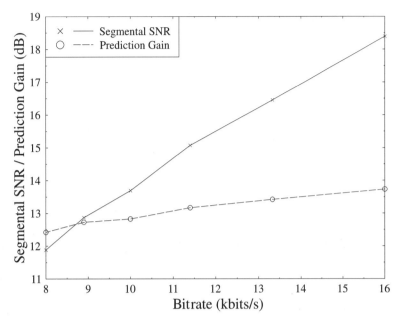

Figure 8.10 Performance of the reduced-rate G728-like codec I with variable-length excitation vectors.

As explained in the previous section, an important part of the codec is the backward-adaptive synthesis filter. Figure 8.10 shows that the prediction gain of this filter falls by only 1.3 dB as the bitrate of the codec is reduced from 16 to 8 kbits/s. This suggests that the backward-adaptive synthesis filtering copes well with the reduction in bitrate from 16 to 8 kbits/s. We also carried out tests at 16 and 8 kbits/s, similar to those used for Table 8.3, to establish how the performance of the filter would be improved if we were able to eliminate the effects of using backward adaption (i.e., the noise feedback and time mismatch). The results are shown in Tables 8.5 and 8.6 for the 16 kbits/s codec (using the Gaussian codebook rather than the trained G728 codebook used for Table 8.3) and the 8 kbits/s codec. As expected, the effects of noise feedback are more significant at 8 than 16 kbits/s, but the overall effects on the codec's segmental SNR of using backward adaption are similar at both rates.

It has been suggested [207] that high-order backward-adaptive linear prediction is inappropriate at bitrates as low as 8 kbits/s. However we found that this was not the case for our codec and that increasing the filter order from 10 to 50 gave almost the same increase in the codec performance at 8 kbits/s as at 16 kbits/s. This is shown in Table 8.7.

Another important part of the G728 codec is the backward gain adaption. Figure 8.6 shows how at 16 kbits/s this backward adaption makes the optimum codebook gains cluster around 1, and hence become easier to quantize. We found that the same was true at 8 kbits/s.

TABLE 8.5 Effects of Backward Adaption of the Synthesis Filter at 16 kbits/s

	Δ Prediction Gain (dB)	Δ Segmental SNR (dB)
No noise feedback	+0.74	+0.42
No time mismatch	+0.85	+0.83
Use forward adaption	+1.59	+1.25

TABLE 8.6   Effects of Backward Adaption of the Synthesis Filter at 8 kbits/s

	Δ Prediction Gain (dB)	Δ Segmental SNR (dB)
No noise feedback	+2.04	+0.75
No time mismatch	+0.85	+0.53
Use forward adaption	+2.89	+1.28

TABLE 8.7   Relative Performance of the Synthesis Filter as p is Increased at 8 and 16 kbits/s

	Δ Prediction Gain (dB)	Δ Segmental SNR (dB)
8 kbits/s $p = 10$	0.0	0.0
8 bits/s $p = 50$	+0.88	+1.00
16 kbits/s $p = 10$	0.0	0.0
16 kbits/s $p = 50$	+1.03	+1.04

To quantify the performance of the gain prediction, we defined the following signal to noise ratio

$$SNR_{\text{gain}} = \frac{\sum \sigma_o^2}{\sum (\sigma_o - \hat{\sigma})^2}. \tag{8.33}$$

Here σ_o is the optimum excitation gain given by

$$\sigma_o = \sqrt{\frac{1}{vs} \sum_{n=0}^{vs} \left(\hat{\sigma} g c_k(n)\right)^2} \tag{8.34}$$

where g is the unquantized gain chosen by the codebook search and $\hat{\sigma}$ is the predicted gain value. We found that this gain prediction SNR was on average 5.3 dB for the 16 kbits/s codec and 6.1 dB for the 8 kbits/s codec. Thus, the gain prediction is even more effective at 8 kbits/s than at 16 kbits/s.

In the next section we discuss the addition of long-term prediction to our variable-rate codec.

8.6 THE EFFECTS OF LONG-TERM PREDICTION

In this section we describe the improvements in our variable-rate codec that can be obtained by adding backward-adaptive long-term prediction (LTP). This work began when we found that significant long-term correlations remained in the synthesis filter's prediction residual, even when the pitch period was lower than the order of this filter. This can be seen from Figure 8.11, which shows the prediction residual for a segment of voiced female speech with a pitch period of about 45 samples. It can be seen that the residual has clear long-term redundancies, which could be exploited by a long-term prediction filter.

In a forward-adaptive system, the short-term synthesis filter coefficients are determined by minimizing the energy of the residual signal found by filtering the original speech through the inverse synthesis filter. Similarly for open-loop LTP, we minimize the energy of the long-term residual signal which is found by filtering the short-term residual through the inverse long-term predictor. If $r(n)$ is the short-term residual signal, then for a 1-tap long-term

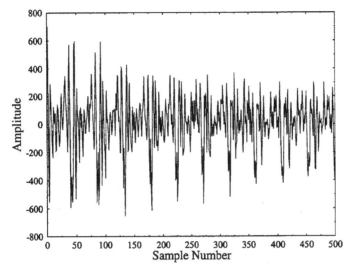

Figure 8.11 Short-term synthesis-filter prediction residual in G728.

predictor we want to determine the delay L and gain β which minimize the long-term residual energy E_{LT} given by

$$E_{LT} = \sum_n \left(r(n) - \beta r(n - L)\right)^2. \tag{8.35}$$

The best delay L is found by calculating

$$X = \frac{\left(\sum_n r(n)r(n - L)\right)^2}{\sum_n r^2(n - L)} \tag{8.36}$$

for all possible delays and choosing the value of L which maximizes X. The best long-term gain β is then given by

$$\beta = \frac{\sum_n r(n)r(n - L)}{\sum_n r^2(n - L)}. \tag{8.37}$$

In a backward-adaptive system, the original speech signal $s(n)$ is not available. Instead, we use the past reconstructed speech signal $\hat{s}(n)$ to find the short-term synthesis filter coefficients. These coefficients can then be used to filter $\hat{s}(n)$ through the inverse filter to find the "reconstructed residual" signal $\hat{r}(n)$. This residual signal can then be used in Equations 8.36 and 8.37 to find the LTP delay and gain. Alternatively, we can use the past excitation signal $u(n)$ in Equations 8.36 and 8.37. This approach is slightly simpler than using the reconstructed residual signal because the inverse filtering of $\hat{s}(n)$ to find $\hat{r}(n)$ is not necessary. We found in our codec that the two approaches gave almost identical results.

Initially, we used a 1-tap LTP in our codec. The best delay L was found by maximizing

$$X = \frac{\left(\sum_{n=-100}^{-1} u(n)u(n - L)\right)^2}{\sum_{n=-100}^{-1} u^2(n - L)} \tag{8.38}$$

over the range of delays 20 to 140 every frame. The LTP gain β was updated every vector by solving

$$\beta = \frac{\sum_{n=-100}^{-1} u(n)u(n-L)}{\sum_{n=-100}^{-1} u^2(n-L)}. \qquad (8.39)$$

We found that this backward-adaptive LTP improved the average segmental SNR of our codec by 0.6 dB at 16 kbits/s and 0.1 dB at 8 kbits/s. However, the calculation of X as given in Equation 8.38 for 120 different delays every frame dramatically increases the complexity of the codec. The denumerator $\sum u^2(n-L)$ for delay L need not be calculated independently but instead can be simply updated from the equivalent expression for delay $L - 1$. Even so, if the frame size is 20 samples, then to calculate X for all delays increases both the encoder and the decoder complexity by almost 10 million arithmetic operations per second, which is unacceptable.

Fortunately, the G728 post-filter requires an estimate of the pitch period of the current frame. This is found by filtering the reconstructed speech signal through a tenth-order short-term prediction filter to find a reconstructed residual-like signal. This signal is then low-pass-filtered with a cutoff frequency of 1 kHz and 4 : 1 decimated, which dramatically reduces the complexity of the pitch determination. The maximum value of the autocorrelation function of the decimated residual signal is then found to give an estimated τ_d of the pitch period. A more accurate estimate τ_p is then found by maximizing the autocorrelation function of the undecimated residual between $\tau_d - 3$ and $\tau_d + 3$. This lag could be a multiple of the true pitch period. To guard against this possibility, the autocorrelation function is also maximized between $\tau_o - 6$ and $\tau_o + 6$, where τ_o is the pitch period from the previous frame. Finally, the pitch estimator chooses between τ_p and the best lag around τ_o by comparing the optimal tap weights β for these two delays.

This pitch estimation procedure requires only about 2.6 million arithmetic operations per second and is carried out at the decoder as part of the post-filtering operations anyway. Using this method to find a LTP delay has no effect on the decoder complexity and increases the encoder complexity by only 2.6 million arithmetic operations per second. We also found that not only was this method of calculating the LTP delay much simpler than finding the maximum value of X from Equation 8.38 for all delays between 20 and 140, but it also gave better results. This was due to the removal of pitch doubling and tripling by the checking of pitch values around that used in the previous frame. The average segmental SNR and segmental weighted SNR for our codec at 16 kbits/s both with and without 1-tap LTP using the pitch estimate from the post-filter is shown in Table 8.8. Similar figures for the codec at 8 kbits/s are given in Table 8.9. We found that when LTP was used, there was very little gain in having a filter order any higher than 20. Therefore, the figures in Tables 8.8 and 8.9 have a short-term filter order of 20 when LTP is used.

Tables 8.8 and 8.9 also give the performance of our codec at 16 and 8 kbits/s when we use multi-tap LTP. As the LTP is backward adaptive, we can use as many taps in the filter as

TABLE 8.8   Performance of LTP at 16 kbits/s

	Segmental SNR (dB)	Segmental Weighted SNR (dB)
No LTP	18.43	14.30
1-Tap LTP	19.08	14.85
3-Tap LTP	19.39	15.21
5-Tap LTP	19.31	15.12

TABLE 8.9 Performance of LTP at 8 kbits/s

	Segmental SNR (dB)	Segmental Weighted SNR (dB)
No LTP	11.86	8.34
1-Tap LTP	12.33	8.64
3-Tap LTP	12.74	9.02
5-Tap LTP	12.49	8.81

we like, with the only penalty being a slight increase in complexity. Once the delay is known, for a $(2p+1)$th-order predictor the filter coefficients b_{-p}, b_{-p+1}, ..., b_0, ..., b_p are given by $(2p+1)$st-order solving the following set of simultaneous equations

$$\sum_{j=-p}^{j=p} b_j \sum_{n=-100}^{n=-1} u(n-L-j)u(n-L-i) = \sum_{n=-100}^{-1} u(n)u(n-L-i) \qquad (8.40)$$

for $i = -p, -p+1, \ldots, p$. The LTP synthesis filter $H_{LTP}(z)$ is then given by

$$H_{LTP}(z) = \frac{1}{1 - b_{-p}z^{-L+p} - \cdots - b_0 z^{-L} - \cdots b_p z^{-L-p}}. \qquad (8.41)$$

Tables 8.8 and 8.9 show that at both 16 and 8 kbits/s the best performance is given by a 3-tap filter, which improves the segmental SNR at both bitrates by almost 1 dB. Also, because when LTP is used the short-term synthesis filter order was reduced to 20, the complexity of the codecs is not significantly increased by the use of a long-term prediction filter.

It was possible to slightly increase the performance of the codec with LTP by modifying the signal $u(n)$ used to find the filter coefficients in Equation 8.40. This modification involves simply repeating the previous vector's excitation signal once. Hence, instead of using the signal $u(-1)$, $u(-2)$, ..., $u(-100)$ to find the LTP coefficients, we use $u(-1)$, $u(-2)$, ..., $u(-vs)$, $u(-1)$, $u(-2)$, ..., $u(-100+vs)$. This single repetition of the previous vector's excitation in calculating the LTP coefficients increased both the segmental and weighted SNR of our codec at 16 kbits/s by about 0.25 dB. It also improved the codec performance at 8 kbits/s, although by only about 0.1 dB. The improvements that this repetition brings in the codec's performance seem to be due to the backward-adaptive nature of the LTP. No such improvement is seen when a similar repetition is used in a forward-adaptive system.

Figure 8.12 depicts the variation in the codec's segmental SNR as the bitrate is reduced from 16 to 8 kbits/s. The codec uses 3-tap LTP with the repetition scheme described above and a short-term synthesis filter of order 20. Also shown in this figure is the equivalent variation in segmental SNR for the codec without LTP, repeated here from Figure 8.10. The addition of long-term prediction to the codec gives a uniform improvement in its segmental SNR of about 1 dB from 8 to 16 kbits/s. The effectiveness of the LTP can also be seen from Figure 8.13, which shows the long-term prediction residual in the 16 kbits/s codec for the same segment of speech as was used for the short-term prediction residual in Figure 8.11. It is clear that the long-term correlations have been significantly reduced. It should be noted, however, that the addition of backward-adapted long-term prediction to the codec will degrade its performance over noisy channels. This aspect of our codec's performance is the subject of ongoing work [216].

Finally, we tested the degradations in the performance of the long-term prediction due to use of backward adaption. To measure the effect of quantization noise feedback, we used past values of the original speech signal rather than the reconstructed speech signal to find the LTP delay and coefficients. To measure the overall effect of backward adaption as opposed to

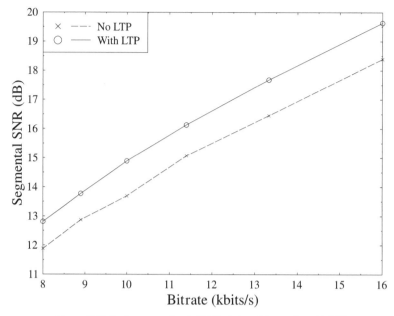

Figure 8.12 Performance of a 8–16 kbits/s low delay codec with LTP.

open-loop forward adaption, we used both past and present speech samples to find the LTP delay and coefficients. The improvements obtained in terms of the segmental SNR and the segmental weighted SNR are shown in Table 8.10 for the codec at 16 kbits/s and in Table 8.11 for the codec at 8 kbits/s. It can be seen that the use of backward adaption degrades the codec's performance by just under 1 dB at 16 kbits/s and just over 1 dB at 8 kbits/s. At both

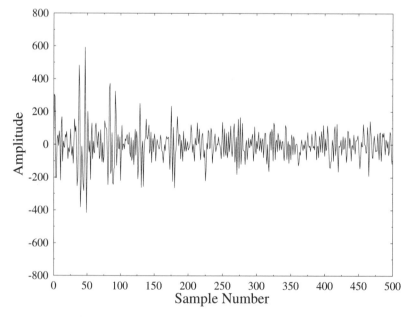

Figure 8.13 Long-term filter prediction residual at 16 kbits/s.

TABLE 8.10   Effects of Backward Adaption of the LTP at 16 kbits/s

	Δ Segmental Weighted SNR (dB)	Δ Segmental SNR (dB)
No noise feedback	−0.03	+0.01
No time mismatch	+0.87	+0.85
Use forward adaption	+0.84	+0.86

TABLE 8.11   Effects of Backward Adaption of the LTP at 8 kbits/s

	Δ Segmental Weighted SNR (dB)	Δ Segmental SNR (dB)
No noise feedback	−0.18	+0.02
No time mismatch	+1.17	+1.17
Use forward adaption	+0.99	+1.19

bitrates, noise feedback has very little effect, with most of the degradation coming from the time mismatch inherent in backward adaption.

8.7 CLOSED-LOOP CODEBOOK TRAINING

In this section we describe the training of the shape and gain codebooks used in our codec at its various bitrates. In Sections 8.5 and 8.6 Gaussian shape codebooks were used, together with the G728 gain codebook. These codebooks were used for simplicity and in order to provide a fair comparison between the different coding techniques.

Because of the backward-adaptive nature of the gain and synthesis filter and LTP adaption used in our codec, it is not sufficient to generate a training sequence for the codebooks and use the Lloyd algorithm [217] to design the codebooks. This is because the codebook entries required from the shape and gain codebooks depend very much on the effectiveness of the gain adaption and the LTP and synthesis filters used. However, because these are backward adapted, they depend on the codebook entries that have been selected in the past. Therefore, the effective training sequence needed changes as the codebooks are trained. It is reported in [210], for example, that in a gain-adaptive vector quantization scheme unless the codebook is properly designed, taking into account the gain adaption, the performance is worse than simple nonadaptive vector quantization.

We used a closed-loop codebook design algorithm similar to that described in [213]. A long speech file consisting of four sentences spoken by two males and two females is used for the training. Both the sentences spoken and the speakers are different from those used for the performance figures quoted in this chapter. The training process commences with an initial shape and gain codebook and codes the training speech as usual. The total weighted error E_k from all the vectors that used the codebook entry $c_k(n)$ is then given by

$$E_k = \sum_{m \in N_k} \left(\hat{\sigma}_m^2 \sum_{n=0}^{vs-1} \left(x_m(n) - g_m[h_m(n) * c_k(n)] \right)^2 \right) \tag{8.42}$$

where N_k is the set of vectors that use $c_k(n)$, $\hat{\sigma}_m$ is the backward-adapted gain for vector m, g_m is the gain codebook entry selected for vector m, and $h_m(n)$ is the impulse response of the concatenated weighting filter and the backward-adapted synthesis filter used in vector m.

Finally, $x_m(n)$ is the codebook target for vector m, which with $(2p+1)$th-order LTP is given $(2p+1)$st-order by

$$x_m(n) = \frac{s_{wm}(n) - \hat{s}_{om}(n) - \sum_{j=-p}^{j=p} b_{jm} u_m(n - L_m - j)}{\hat{\sigma}_m}. \tag{8.43}$$

Here $s_{wm}(n)$ is the weighted input speech in vector m, $\hat{s}_{om}(n)$ is the zero input response of the weighting and synthesis filters, $u_m(n)$ is the previous excitation, and L_m and b_{jm} are the backward-adapted LTP delay and coefficients in vector m.

Equation 8.42 giving E_k can be expanded to yield:

$$
\begin{aligned}
E_k &= \sum_{m \in N_k} \left(\hat{\sigma}_m^2 \sum_{n=0}^{vs-1} \left(x_m(n) - g_m[h_m(n) * c_k(n)] \right)^2 \right) \\
&= \sum_{m \in N_k} \left(\hat{\sigma}_m^2 \sum_{n=0}^{vs-1} x_m^2(n) + \hat{\sigma}_m^2 g_m^2 \sum_{n=0}^{vs-1} [h_m(n) * c_k(n)]^2 - 2\hat{\sigma}_m^2 g_m \sum_{n=0}^{vs-1} x_m(n)[h_m(n) * c_k(n)] \right) \\
&= \sum_{m \in N_k} \left(\hat{\sigma}_m^2 \sum_{n=0}^{vs-1} x_m^2(n) + \hat{\sigma}_m^2 g_m^2 \sum_{n=0}^{vs-1} [h_m(n) * c_k(n)]^2 - 2\hat{\sigma}_m^2 g_m \sum_{n=0}^{vs-1} P_m(n) c_k(n) \right),
\end{aligned}
\tag{8.44}
$$

where $p_m(j)$ is the reverse convolution between $h_m(n)$ and the target $x_m(n)$. This expression can be partially differentiated with respect to element $n = j$ of the codebook entry $c_k(n)$ to give

$$\frac{\partial E_k}{\partial c_k(j)} = \sum_{m \in N_k} \left(2\hat{\sigma}_m^2 g_m^2 \sum_{n=0}^{vs-1} c_k(n) H(n,j) - 2\hat{\sigma}_m^2 g_m P_m(j) \right) \tag{8.45}$$

where $H_m(n,j)$ is the autocorrelation of the delayed impulse response $h_m(n)$ and is given by

$$H_m(i,j) = \sum_{n=0}^{vs-1} h_m(n-i) h_m(n-j). \tag{8.46}$$

Setting these partial derivatives to zero gives the optimum codebook entry $c_k^*(n)$ for the cluster of vectors N_k as the solution of the set of simultaneous equations

$$\sum_{m \in N_k} \left(\hat{\sigma}_m^2 g_m^2 \sum_{n=0}^{vs-1} c_k^*(n) H_m(n,j) \right) = \sum_{m \in N_k} \left(\hat{\sigma}_m^2 g_m P_m(j) \right) \quad \text{for } j = 0, 1, \ldots, vs - 1. \tag{8.47}$$

A similar expression for the total weighted error E_i from all the vectors that use the gain codebook entry g_i is

$$
\begin{aligned}
E_i &= \sum_{m \in N_i} \left(\hat{\sigma}_m^2 \sum_{n=0}^{vs-1} \left(x_m(n) - g_i[h_m(n) * c_m(n)] \right)^2 \right) \\
&= \sum_{m \in N_i} \left(\hat{\sigma}_m^2 \sum_{n=0}^{vs-1} x_m^2(n) + g_i^2 \hat{\sigma}_m^2 \sum_{n=0}^{vs-1} [h_m(n) * c_m(n)]^2 - 2g_i \hat{\sigma}_m^2 \sum_{n=0}^{vs-1} x_m(n)[h_m(n) * c_m(n)] \right)
\end{aligned}
$$

where N_i is the set of vectors that use the gain codebook entry g_i, and $c_m(n)$ is the shape codebook entry used by the mth vector. Differentiating this expression with respect to g_i gives:

$$
\frac{\partial E_i}{\partial g_i} = \sum_{m \in N_i} \left(2g_i \hat{\sigma}_m^2 \sum_{n=0}^{vs-1} [h_m(n) * c_m(n)]^2 \right.
$$
$$
\left. -2\hat{\sigma}_m^2 \sum_{n=0}^{vs-1} x_m(n)[h_m(n)*c_m(n)] \right)
\tag{8.49}
$$

and setting this partial derivative to zero gives the optimum gain codebook entry g_i^* for the cluster of vectors N_i as:

$$
g_i^* = \frac{\sum_{m \in N_i} \left(\hat{\sigma}_m^2 \sum_{n=0}^{vs-1} x_m(n)[h_m(n)*c_m(n)] \right)}{\sum_{m \in N_i} \left(\hat{\sigma}_m^2 \sum_{n=0}^{vs-1} [c_m(n)*h_m(n)]^2 \right)}.
\tag{8.50}
$$

The summations in Equations 8.48 and 8.50 over all the vectors that use $c_k(n)$ or g_i are carried out for all 128 shape codebook entries and all 8 gain codebook entries as the coding of the training speech takes place. At the end of the coding, the shape and gain codebooks are updated using Equations 8.48 and 8.50, and then the codec starts coding the training speech again with the new codebooks. This closed-loop codebook training procedure is summarized as follows:

1. Start with an initial gain and shape codebook.
2. Code the training sequence using the given codebooks. Accumulate the summations in Equations 8.48 and 8.50.
3. Calculate the total weighted error of the coded speech. If this distortion is less than the minimum distortion so far, keep a record of the codebooks used as the best codebooks so far.
4. Calculate new shape and gain codebooks using Equations 8.48 and 8.50.
5. Return to step 2.

Each entire coding of the training speech file counts as one iteration, and Figure 8.14 shows the variation in the total weighted error energy E, and the codec's segmental SNR, as the training progresses for the 16 kbits/s codebooks. From this figure it can be seen that this closed-loop training sequence does not give a monotonic decrease in the total weighted error from one iteration to the next. This is because of the changing of the codebook target $x_m(n)$, as well as the other backward-adapted parameters, from one iteration to the next. However, it is clear from Figure 8.14 that the training gives a significant improvement in the codec's performance. Because of the nonmonotonic decrease in the total weighted error energy, it is necessary during the codebook training to keep a record of the lowest error energy achieved so far and the corresponding codebooks. If a certain number of iterations passes without this minimum energy being improved, then the codebook training can be terminated. As seen in Figure 8.14, we get close to the minimum within about 20 iterations.

An important aspect of vector quantizer training is the initial codebook used. In Figure 8.14 we used the G728 gain codebook and the Gaussian shape codebook as the initial codebooks. We also tried using other codebooks such as the G728 fixed codebook and Gaussian codebooks with different variances as the initial codebooks. Although these gave very different starting values of the total weighted error E and took different numbers of

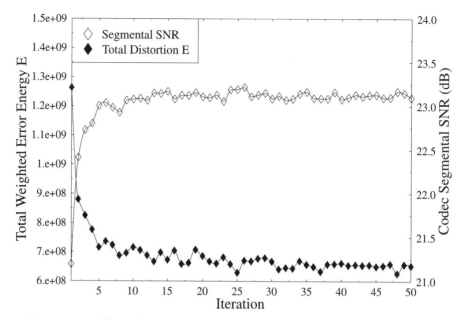

Figure 8.14 Codec's performance as the codebooks are trained.

iterations to give their optimum codebooks, they all resulted in codebooks that gave very similar performances. Therefore, we concluded that the G728 gain codebook and the Gaussian shape codebook are suitable for use as the initial codebooks.

We trained different shape and gain codebooks for use by our codec at all of its bitrates between 8 and 16 kbits/s. The average segmental SNR given by the codec using these codebooks is shown in Figure 8.15 for the four speech sentences that were not part of the training sequence. Also shown in this figure for comparison is the curve from Figure 8.12 for the corresponding codec with the untrained codebooks. It can be seen that the codebook training gives an improvement of about 1.5 to 2 dB across the codec's range of bitrates.

It can be seen from Figure 8.14 that a decrease in the total weighted error energy E does not necessarily correspond to an increase in the codec's segmental SNR. This is also true for the codec's segmental weighted SNR, and it is because the distortion D calculated takes no account of the different signal energies in different vectors. We tried altering the codebook training algorithm to take account of this, hoping that it would result in codebooks that gave lower segmental SNRs. However, the codebooks trained with this modified algorithm gave similar performances to those trained by minimizing E.

We also attempted training different codebooks at each bitrate for voiced and unvoiced speech. The voicing decision can be made backward adaptive based on the correlations in the previous reconstructed speech. A voiced/unvoiced decision like this is made in the G728 post-filter to determine whether to apply pitch post-filtering. We found, however, that although an accurate determination of the voicing of the speech could be made in a backward-adaptive manner, no significant improvement in the codec's performance could be achieved by using separately trained voiced and unvoiced codebooks. This agrees with the results in [207] when fully backward-adaptive LTP is used.

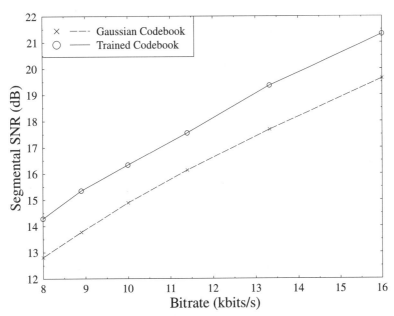

Figure 8.15 Performance of the 8–16 kbits/s codec with trained codebooks.

8.8 REDUCED-RATE G728-LIKE CODEC: CONSTANT-LENGTH EXCITATION VECTOR

In the previous sections we discussed a variable-rate codec based on G728, which varied its bitrate by changing the number of samples in each vector. The excitation for each vector was coded with 10 bits. In this section we describe the alternative approach of keeping the vector size constant and varying the number of bits used to code the excitation. The bitrate of the codec is varied between 8 and 16 kbits/s with a constant vector size of five samples by using between 5 and 10 bits to code the excitation signal for each vector. We used a structure for the codec identical to that described earlier, with backward gain adaption for the excitation and backward-adapted short- and long-term synthesis filters. With 10, 9, or 8 bits to code the excitation, we used a split vector quantizer, similar to that used in G728, with a 7-bit shape codebook and a 3, 2, or 1-bit gain codebook. For the lower bitrates we used a single 7, 6, or 5-bit vector quantizer to code the excitation. Codebooks were trained for the various bitrates using the closed loop codebook training technique described in Section 8.7.

The segmental SNR of this variable-rate codec is shown in Figure 8.16. Also shown in this graph is the segmental SNR of the codec with a variable vector size, copied here from Figure 8.15 for comparison. At 16 kbits/s the two codecs are, of course, identical, but at lower rates the constant vector size codec performs worse than the variable vector size codec. The difference between the two approaches increases as the bitrate decreases, and at 8 kbits/s the segmental SNR of the constant vector size codec is about 1.75 dB lower than that of the variable vector size codec.

Although the constant vector size codec gives lower reconstructed speech quality, it does have certain advantages. The most obvious is that it has a constant delay equal to that of G728 (i.e., less than 2 ms). Also, the complexity of its encoder, especially at low bitrates, is lower than that of the variable vector size codec. This is because of the smaller codebooks used. At 8 kbits/s the codebook search procedure has only to examine 32 codebook entries.

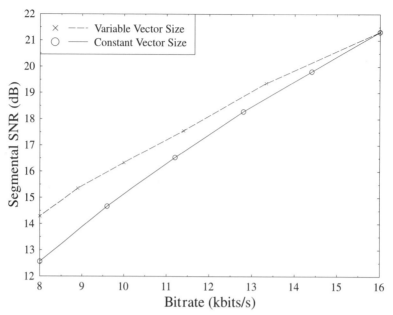

Figure 8.16 Performance of the reduced-rate G728-like codec II with constant-length excitation vectors.

Therefore, for some applications this codec may be more suitable than the higher speech quality variable vector size codec.

In this chapter so far we have described the G728 16 kbps low-delay codec and investigated a variable-rate low-delay codec, which is compatible with the 16 kbits/s G728 codec at its highest bitrate and exhibits a graceful degradation in speech quality down to 8 kbits/s. The bitrate can be reduced while the buffering delay is kept constant at five samples (0.625 ms). Alternatively, better speech quality is achieved if the buffering delay is increased gradually to 10 samples as the bitrate is reduced down to 8 kbits/s.

8.9 PROGRAMMABLE-RATE 8–4 kbps LOW-DELAY CELP CODECS

8.9.1 Motivation

Having discussed low-delay 16–8 kbits/s programmable-rate coding in the previous section, in this section we consider methods of improving the performance of the proposed 8 kbits/s backward-adaptive predictive codec while maintaining as low a delay and complexity as possible. Our proposed 8 kbits/s codec developed in Sections 8.5 and 8.8 uses a 3-bit gain codebook and a 7-bit shape codebook with backward adaption of both the long- and short-term synthesis filters, and gives an average segmental SNR of 14.29 dB. In this section we also describe the effect of increasing the size of the gain and shape codebooks in this codec while keeping a vector length of 10 samples. Next we consider the improvements that can be achieved, again while maintaining a vector length of 10 samples, by using forward adaption of the short- and long-term synthesis filters. Then we show the performance of three codecs, based on those developed in the earlier sections, operating at bitrates between 8 and 4 kbits/s. Finally, as an interesting benchmarker, we describe a codec, with a vector size of 40 samples, based on the algebraic codebook structure we described in Section 6.4.3. The performance of this codec is compared to the previously introduced low-delay codecs from Section 8.9.5 and the higher delay forward-adaptive predictive ACELP codec described in Section 6.4.3.

8.9.2 8–4 kbps Codec Improvements Due to Increasing Codebook Sizes

In this section we use the same structure for the codec as before but increase the size of the shape and gain codebooks. This codec structure is shown in Figure 8.17, and we refer to it as "Scheme One." We used 3-tap backward-adapted LTP and a vector length of 10 samples with a 7-bit shape codebook, and we varied the size of the gain codebook from 3 to 4 and 5 bits. Then in our next experiments we used a 3-bit gain codebook and trained 8- and 9-bit shape codebooks. Finally, we attempted to increase the size of both the shape and gain codebooks by 1 bit. In each case the new codebooks were closed-loop trained using the technique described in Section 8.7.

The segmental SNRs of this Scheme One codec with various size shape and gain codebooks is shown in Table 8.12. It can be seen that adding 1 bit to either the gain or the shape codebook increases the segmental SNR of the codec by about 1 dB. Adding two extra bits to the shape codebook, or 1 bit each to both codebooks, increases the segmental SNR by almost 2 dB.

8.9.3 8–4 kbps Codecs—Forward Adaption of the Short-Term Synthesis Filter

In this section we consider the improvements that can be achieved in the vector size 10 codec by using forward adaption of the short-term synthesis filter. In Table 8.6 we examined the effects of backward adaption of the synthesis filter at 8 kbits/s. However, these figures gave the improvements that can be achieved by eliminating the noise feedback and time mismatch that are inherent in backward adaption when using the same recursive windowing function and update rate as the G728 codec. In this section we consider the improvements that could be achieved by significantly altering the structure used for determining the synthesis filter parameters.

The codec structure used is shown in Figure 8.18, and we refer to it as "Scheme Two." Its only difference from our previously developed 8 kbits/s backward-adaptive codec is that we replaced the recursive windowing function shown in Figure 8.3 with an asymmetric analysis window which was used in a candidate codec for the CCITT 8 kbits/s standard [136, 189]. This window, which is shown in Figure 8.19, is made up of half a Hamming window and a quarter of a cosine function cycle. The windowing scheme uses a frame length of 10 ms (or 80 samples), with a 5 ms lookahead. The 10 ms frame consists of two subframes, and a line spectral frequency (LSF) interpolation scheme similar to that described in Section 6.4.3 is used.

We implemented this method of deriving the LPC coefficients in our codec. The vector length was kept constant at 10 samples, but instead of the synthesis filter parameters being updated every 20 samples, as in the Scheme One codec, they were updated every 40 samples using either the interpolated or transmitted LSFs. In the candidate 8 kbits/s CCITT codec [136] a filter order of 10 is used, and the 10 LSFs are quantized with 19 bits using differential split vector quantization. However, for simplicity and in order to see the best performance gain possible for our codec by using forward adaption of the short-term synthesis filter, we used the 10 unquantized LSFs to derive the filter coefficients. A new 3-bit gain codebook and

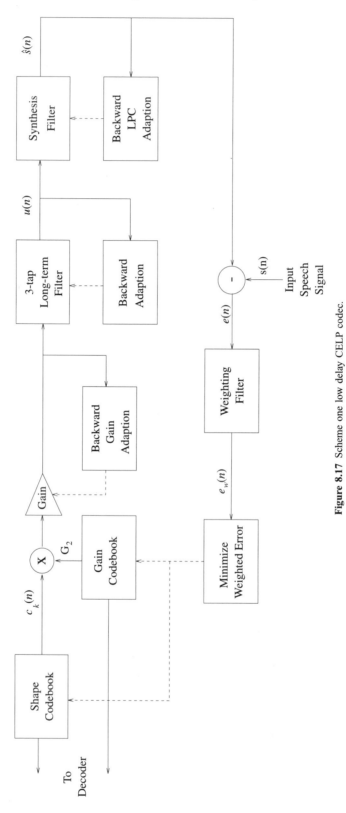

Figure 8.17 Scheme one low delay CELP codec.

TABLE 8.12 Performance of the Scheme One Codec with Various Size Gain and Shape Codebooks

Gain Codebook Bits	Shape Codebook Bits	Segmental SNR (dB)
3	7	14.29
4	7	15.24
5	7	15.62
3	8	15.33
3	9	16.12
4	8	16.01

7-bit shape codebook were derived for this codec using the codebook training technique described in Section 8.7. We found that this forward adaption increased the segmental SNR of the codec by only 0.8 dB. Even this rather small improvement would, of course, be reduced by the quantization of the LSFs. Using a 19-bit quantization scheme to transmit a new set of LSFs every 80 sample frame would mean using on average about 2.4 bits per 10 sample vector.

Traditionally, codecs employing forward-adaptive LPC are more resilient to channel errors than those using backward-adaptive LPC. However, a big disadvantage of using such a forward-adaptive LPC scheme is that it would increase the delay of the codec by almost an order of magnitude. Instead of a vector length of 10 samples, we would need to buffer a frame of 80 speech samples, plus a 40-sample lookahead, to calculate the LPC information. This would increase the overall delay of the codec from under 4 ms to about 35 ms.

8.9.4 Forward Adaption of the Long-Term Predictor

8.9.4.1 Initial Experiments. In this section we consider the gains in our codec performance which can be achieved using forward adaption of the long term predictor (LTP) gain. Although forward adaption of the LTP parameters would improve the codec's robustness to channel errors, we did not consider forward adaption of the LTP delay because to transmit this delay from the encoder to the decoder would require around 7 extra bits per vector. However, we expected to be able to improve the performance of the codec, at the cost of significantly fewer extra bits, by using forward adaption of the LTP gain.

Previously, we employed a 3-tap LTP with backward-adapted values for the delay and filter coefficients. Initially, we replaced this LTP scheme with an adaptive codebook arrangement, where the delay was still backward adapted, but the gain was calculated as in forward-adaptive CELP codecs, which was detailed in Section 6.5. This calculation assumes that the fixed codebook signal, which is not known until after the LTP parameters are calculated, is zero. The "optimum" adaptive codebook gain G_1, which minimizes the weighted error between the original and reconstructed speech, is then given by:

$$G_1 = \frac{\sum_{n=0}^{vs-1} x(n)y_\alpha(n)}{\sum_{n=0}^{vs-1} y_\alpha^2(n)}. \tag{8.51}$$

Here $x(n) = s_w(n) - \hat{s}_o(n)$ is the target for the adaptive codebook search, $s_w(n)$ is the weighted speech signal, $\hat{s}_o(n)$ is the zero input response of the weighted synthesis filter, and

$$y_\alpha(n) = \sum_{i=0}^{n} u(i - \alpha)h(n - i) \tag{8.52}$$

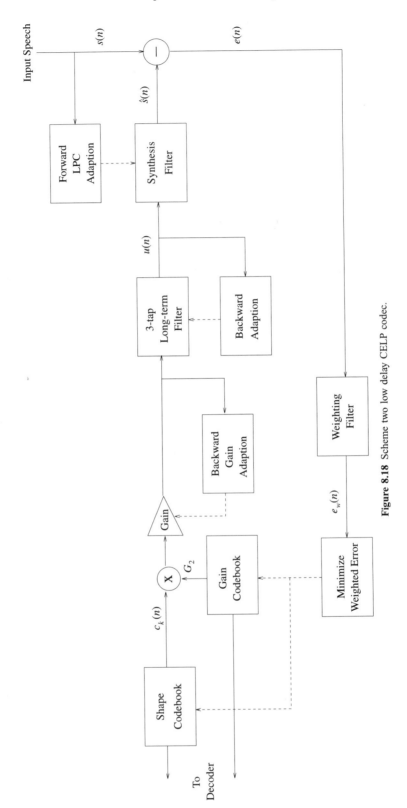

Figure 8.18 Scheme two low delay CELP codec.

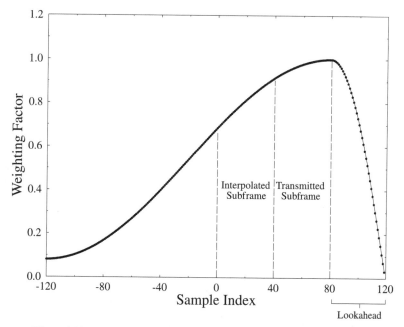

Figure 8.19 LPC windowing function used in the CCITT 8 kbits/s candidate codec.

is the convolution of the adaptive codebook signal $u(n - \alpha)$ with the impulse response $h(n)$ of the weighted synthesis filter, where α is the backward-adapted LTP delay.

Again, we trained new $\frac{7}{3}$-bit shape/gain fixed codebooks, and we used the unquantized LTP gain G_1 as given by Equation 8.51. However, we found that this arrangement improved the segmental SNR of our codec by only 0.1 dB over the codec with 3-tap backward-adapted LTP. Therefore, we decided to invoke some of the joint adaptive and fixed codebook optimization schemes described in Section 6.5.2.4. These joint optimization schemes are described below.

The simplest optimization scheme—Method A from Section 6.5.2.4—involves calculating the adaptive and fixed codebook gains and indices as usual and then updating the two gains for the given codebook indices k and α using Equations 6.28 and 6.29, which are repeated here for convenience:

$$G_1 = \frac{C_\alpha \xi_k - C_k Y_{\alpha k}}{\xi_\alpha \xi_k - Y_{\alpha k}^2} \tag{8.53}$$

$$G_2 = \frac{C_k \xi_\alpha - C_\alpha Y_{\alpha k}}{\xi_\alpha \xi_k - Y_{\alpha k}^2}. \tag{8.54}$$

Here G_1 is the LTP gain, G_2 is the fixed codebook gain,

$$\xi_\alpha = \sum_{n=0}^{vs-1} y_\alpha^2(n) \tag{8.55}$$

is the energy of the filtered adaptive codebook signal, and

$$C_\alpha = \sum_{n=0}^{vs-1} x(n) y_\alpha(n) \tag{8.56}$$

is the correlation between the filtered adaptive codebook signal and the codebook target $x(n)$. Similarly, ξ_k is the energy of the filtered fixed codebook signal $[c_k(n)*h(n)]$, and C_k is the correlation between this and the target signal. Finally,

$$Y_{\alpha k} = \sum_{n=0}^{vs-1} y_\alpha(n)[c_k(n) * h(n)] \qquad (8.57)$$

is the correlation between the filtered signals from the two codebooks.

We studied the performance of this gain update scheme in our vector length 10 codec. A 7-bit fixed shape codebook was trained, but the LTP and fixed codebook gains were not quantized. We found that the gain update improved the segmental SNR of our codec by 1.2 dB over the codec with backward-adapted 3-tap LTP and no fixed codebook gain quantization. This is a much more significant improvement than that reported in Section 6.5.2.4 for our 4.7 kbits/s ACELP codec because of the much higher update rate for the gains used in our present codec. In our low-delay codec, the two gains are calculated for every 10-sample vector, whereas in the 4.7 kbits/s ACELP codec used in Section 6.5 the two gains are updated only every 60-sample subframe.

Encouraged by these results, we also invoked the second suboptimal joint codebook search procedure described in Section 6.5.2.4. In this search procedure, the adaptive codebook delay α is determined first by backward adaption in our present codec, and then for each fixed codebook index k the optimum LTP, and fixed codebook gains G_1 and G_2 are determined using Equations 8.53 and 8.54. The index k which maximizes $T_{\alpha k}$, given in Equation 8.58, will minimize the weighted error between the reconstructed and original speech for the present vector, and is transmitted to the decoder. This codebook search procedure was referred to as Method B in Section 6.5.2.4.

$$T_{\alpha k} = 2\big(G_1 C_\alpha + \hat{\sigma} G_2 C_k - \hat{\sigma} G_1 G_2 Y_{\alpha k}\big) - G_1^2 \xi_\alpha - \hat{\sigma}^2 G_2^2 \xi_k \qquad (8.58)$$

We trained a new 7-bit fixed shape codebook for this joint codebook search algorithm, and the two gains G_1 and G_2 were left unquantized. We found that this scheme gave an additional improvement in the performance of the codec so that its segmental SNR was now 2.7 dB higher than the codec with backward-adapted 3-tap LTP and no fixed gain quantization. Again, this is a much more significant improvement than that which we found for our 4.7 kbits/s ACELP codec.

8.9.4.2 Quantization of Jointly Optimized Gains.
The improvements for our vector size 10 codec when we use an adaptive codebook arrangement with joint calculation of the LTP and fixed codebook gains, and no quantization of either gain, are quite promising. Next we considered the quantization of the two gains, G_1 and G_2. In order to minimize the number of bits used, we decided to use a vector quantizer for the two gains. A block diagram of the coding scheme used is shown in Figure 8.20. We refer to this arrangement as "Scheme Three."

This Scheme Three codec with forward-adaptive LTP was tested with 4, 5, 6, and 7-bit vector quantizers for the fixed and adaptive codebook gains and a 7-bit shape codebook. The vector quantizers were trained as follows. For a given vector quantizer level i, the total weighted energy E_i for speech vectors using this level will be

$$E_i = \sum_{m \in N_i} \left(\sum_{n=0}^{vs-1} \big(x_m(n) - G_{1i} y_{\alpha m}(n) - G_{2i} \hat{\sigma}_m [h_m(n) * c_m(n)]\big)^2 \right). \qquad (8.59)$$

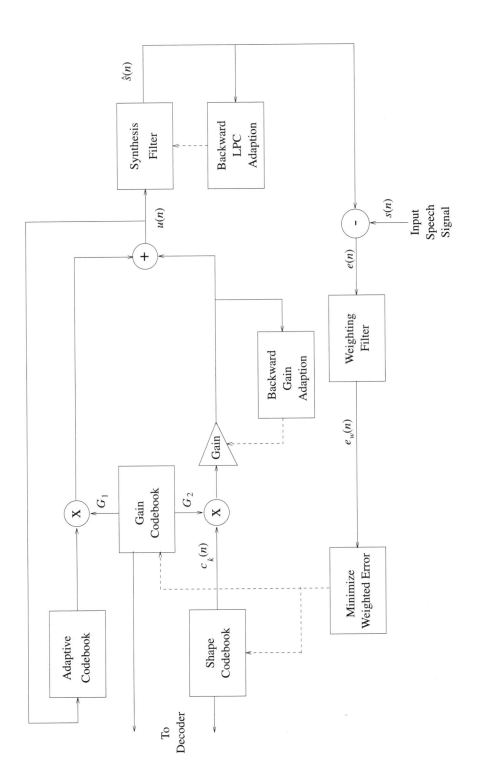

Figure 8.20 Scheme three low delay CELP codec.

Here $x_m(n)$, $y_{\alpha m}(n)$, and $h_m(n)$ are the signals $x(n)$, $y_\alpha(n)$, and $h(n)$ in the mth vector, $\hat{\sigma}_m$ is the value of the backward-adapted gain $\hat{\sigma}$ in the mth vector, $c_m(n)$ is the fixed codebook entry $c_k(n)$ used in the mth vector, G_{1i} and G_{2i} are the values of the two gains in the ith entry of the joint vector quantizer, and N_i is the set of speech vectors that use the ith entry of the vector quantizer. As before, vs is the vector size used in the codec, which in our present experiments is 10.

Expanding Equation 8.59 gives:

$$E_i = \sum_{m \in N_i} (X_m + G_{1i}^2 \xi_{\alpha m} + G_{2i}^2 \hat{\sigma}_m^2 \xi_{km} - 2G_{1i}C_{\alpha m}$$
$$- 2\hat{\sigma}_m G_{2i} C_{km} + 2\hat{\sigma}_m G_{1i} G_{2i} Y_{\alpha km}) \tag{8.60}$$

where $X_m = \sum_{n=0}^{vs-1} x_m^2(n)$ is the energy of the target signal $x_m(n)$, and $\xi_{\alpha m}$, ξ_{km}, $C_{\alpha m}$, C_{km}, and $Y_{\alpha km}$ are the values in the mth vector of ξ_α, ξ_k, C_α, C_k, and $Y_{\alpha k}$ defined earlier.

Differentiating Equation 8.61 with respect to G_{1i} and setting the result to zero gives

$$\frac{\partial E_i}{\partial G_{1i}} = \sum_{m \in N_i} (2G_{1i}\xi_{\alpha m} - 2C_{\alpha m} + 2\hat{\sigma}_m G_{2i} Y_{\alpha km}) = 0 \tag{8.61}$$

or

$$G_{1i} \sum_{m \in N_i} \xi_{\alpha m} + G_{2i} \sum_{m \in N_i} \hat{\sigma}_m Y_{\alpha m} = \sum_{m \in N_i} C_{\alpha m}. \tag{8.62}$$

Similarly, differentiating with respect to G_{2i} and setting the result to zero gives:

$$G_{1i} \sum_{m \in N_i} \hat{\sigma}_m Y_{\alpha km} + G_{2i} \sum_{m \in N_i} \hat{\sigma}_m^2 \xi_{km} = \sum_{m \in N_i} \hat{\sigma}_m C_{km}. \tag{8.63}$$

Solving these two simultaneous equations gives the optimum values of G_{1i} and G_{2i} for the cluster of vectors N_i as:

$$G_{1i} = \frac{\left(\sum_{m \in N_i} C_{\alpha m}\right)\left(\sum_{m \in N_i} \hat{\sigma}_m^2 \xi_{km}\right) - \left(\sum_{m \in N_i} \hat{\sigma}_m C_{km}\right)\left(\sum_{m \in N_i} \hat{\sigma}_m Y_{\alpha km}\right)}{\left(\sum_{m \in N_i} \xi_{\alpha m}\right)\left(\sum_{m \in N_i} \hat{\sigma}_m^2 \xi_{km}\right) - \left(\sum_{m \in N_i} \hat{\sigma}_m Y_{\alpha km}\right)^2} \tag{8.64}$$

and

$$G_{2i} = \frac{\left(\sum_{m \in N_i} \hat{\sigma}_m C_{km}\right)\left(\sum_{m \in N_i} \xi_{\alpha m}\right) - \left(\sum_{m \in N_i} C_{\alpha m}\right)\left(\sum_{m \in N_i} \hat{\sigma}_m Y_{\alpha km}\right)}{\left(\sum_{m \in N_i} \xi_{\alpha m}\right)\left(\sum_{m \in N_i} \hat{\sigma}_m^2 \xi_{km}\right) - \left(\sum_{m \in N_i} \hat{\sigma}_m Y_{\alpha km}\right)^2}. \tag{8.65}$$

Using Equations 8.64 and 8.65, we performed a closed-loop training of the vector quantizer gain codebook along with the fixed shape codebook similarly to the training of the shape and single gain codebooks described in Section 8.7. However, we found a similar problem to that which we encountered when training scalar codebooks for G_1 and G_2 in Section 6.5.2.5. Specifically, although almost all values of G_1 have magnitudes less than 2, a few values have very high magnitudes. This leads to a few levels in the trained vector quantizers having very high values and being very rarely used. Following an in-depth investigation into this phenomenon, we solved the problem by excluding all vectors for which the magnitude of G_1 was greater than 2, or the magnitude of G_2 was greater than 5, from the training sequence. This approach solved the problems of the trained gain codebooks having some very high and very rarely used levels.

We trained vector quantizers for the two gains using 4, 5, 6, and 7 bits. The values of the 4-bit trained vector quantizer for G_1 and G_2 are shown in Figure 8.21. It can be seen that

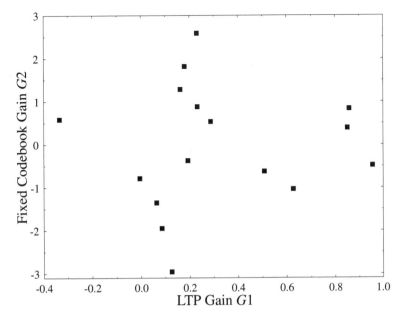

Figure 8.21 Values of G_1 and G_2 in the 4 bit gain quantizer.

when G_1 is close to zero, the values of G_2 have a wide range of values between -3 and $+3$, but when the speech is voiced and G_1 is high, the fixed codebook contribution to the excitation is less significant, and the quantized values of G_2 are closer to zero.

Our trained joint gain codebooks are searched as follows. For each fixed codebook entry k, the optimum gain codebook entry is found by tentatively invoking each pair of gain values in Equation 8.58 in order to test which level maximizes $T_{\alpha k}$ and hence minimizes the weighted error energy. The segmental SNR of our Scheme Three codec with a trained 7-bit shape codebook and trained 4, 5, 6, and 7-bit joint $G1/G_2$ vector quantizers is shown in Table 8.13. The segmental SNRs in this table should be compared with the value of 14.29 dB obtained for the Scheme One codec with a 3-bit scalar quantizer for G_2 and 3-tap backward-adapted LTP.

Table 8.13 shows that the joint G_1/G_2 gain codebooks give a steady increase in the performance of the codec as the size of the gain codebook is increased. In the next section we describe the use of backward-adaptive voiced/unvoiced switched codebooks to further improve the performance of our codec.

TABLE 8.13 Performance of the Scheme Three Codecs

Gain Codebook Bits	Segmental SNR (dB)
4 bits	14.81
5 bits	15.71
6 bits	16.54
7 bits	17.08

8.9.4.3 8–4 kbps Codecs—Voiced/Unvoiced Codebooks.   In Section 8.7 we discussed using different codebooks for voiced and unvoiced segments of speech and using a backward-adaptive voicing decision to select which codebooks to use. However, we found that in the case of a codec with fully backward-adaptive LTP no significant improvement in the codec's performance was achieved by using switched codebook excitation. In this section we discuss using a similar switching arrangement in conjunction with our Scheme Three codec described above.

The backward-adaptive voiced/unvoiced switching is based on the voiced/unvoiced switching used in the postfilter employed in the G728 codec [85]. In our codec, the switch uses the normalized autocorrelation value of the past reconstructed speech signal $\hat{s}(n)$ at the delay α which is used by the adaptive codebook. This normalized autocorrelation value β_α is given by

$$\beta_\alpha = \frac{\sum_{n=-100}^{-1} \hat{s}(n)\hat{s}(n-\alpha)}{\sum_{n=-100}^{-1} \hat{s}^2(n-\alpha)}, \tag{8.66}$$

and when it is greater than a set threshold, the speech is classified as voiced. Otherwise the speech is classified as unvoiced. In our codec, as in the G728 post-filter, the threshold is set to 0.6.

Figure 8.22 shows a segment of the original speech and the normalized autocorrelation value β_α calculated from the reconstructed speech of our 8 kbits/s codec. To aid the clarity of this graph, the values of β_α have been limited to lie between 0.05 and 0.95. It can be seen that the condition $\beta_\alpha > 0.6$ gives a good indication of whether the speech is voiced or unvoiced.

The backward-adaptive voicing decision was incorporated into our Scheme Three codec shown in Figure 8.20 to produce a new coding arrangement which we referred to as "Scheme Four." Shape and joint gain codebooks were trained as described earlier for both the voiced and unvoiced modes of operation in a vector length 10 codec. The quantized values of G_1 and G_2 in both the 4-bit voiced and unvoiced codebooks are shown in Figure 8.23. It can

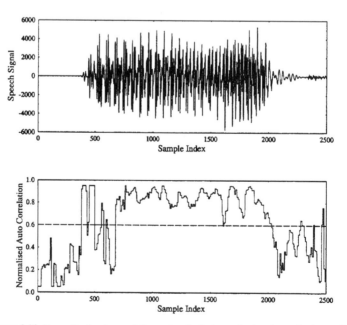

Figure 8.22 Normalized autocorrelation value β_α during voiced and unvoiced speech.

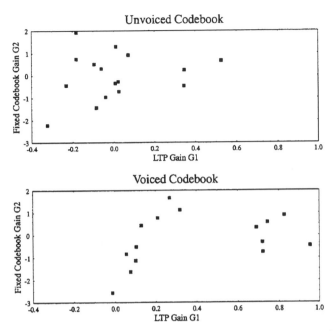

Figure 8.23 Values of G_1 and G_2 in the 4 bit voiced and unvoiced gain quantizers.

be seen that similarly to Figure 8.21 when G_1 is high the range of values of G_2 is more limited than when G_1 is close to zero. Furthermore, as expected, the voiced codebook has a group of quantizer levels with G_1 close to 1, whereas the values of the LTP gain in the unvoiced codebook are closer to 0.

The results we achieved with 7-bit shape codebooks and joint gain codebooks of various sizes are shown in Table 8.14. Comparing this table to Table 8.13 reveals that the voiced/unvoiced switching gives an improvement in the codec's performance of about 0.25 dB for the 4- and the 5-bit gain quantizers, and a smaller improvement for the 6- and 7-bit gain quantizers.

8.9.5 Low-Delay Codecs at 4–8 kbits/s

The improvements that can be achieved in our vector size 10 codec by increasing the size of the shape and gain codebooks, and by using forward adaption of the short-term predictor coefficients and the long-term predictor gain, are summarized in Table 8.15. This table shows the various gains in the codec's segmental SNR against the number of extra bits used to represent each 10-sample vector.

TABLE 8.14 Performance of the Scheme Four Codecs

Gain Codebook Bits	Segmental SNR (dB)
4 bits	15.03
5 bits	15.92
6 bits	16.56
7 bits	17.12

TABLE 8.15 Improvements Obtained Using Schemes One to Four

	Synthesis Filter	Long Term Predictor	Shape C.B.	Gain C.B.	Extra Bits	Δ Seg. SNR
Scheme One	Backward adapted $p = 20$	3-Tap backward adapted	7 bits	3 bits	0	0 dB
			7 bits	4 bits	1	+0.95 dB
			8 bits	3 bits	1	+1.04 dB
			7 bits	5 bits	2	+1.33 dB
			8 bits	4 bits	2	+1.72 dB
			9 bits	3 bits	2	+1.83 dB
Scheme Two	Forward adapted $p = 10$	3-Tap backward adapted	7 bits	3 bits	≈2.4	≤ +0.82 dB
Scheme Three	Backward adapted $p = 20$	Forward adapted	7 bits	4 bits	1	+0.52 dB
			7 bits	5 bits	2	+1.42 dB
			7 bits	6 bits	3	+2.25 dB
			7 bits	7 bits	4	+2.79 dB
Scheme Four	Backward adapted $p = 20$	Switched forward adapted	7 bits	4 bits	1	+0.74 dB
			7 bits	5 bits	2	+1.63 dB
			7 bits	6 bits	3	+2.27 dB
			7 bits	7 bits	4	+2.83 dB

In this table the Scheme One codec (see Section 8.9.2) is the vector size 10 codec, with a 3-tap backward-adapted LTP and a 20-tap backward-adapted short-term predictor. The table shows the gains in the segmental SNR of the codec that are achieved by adding one or two extra bits to the shape or the scalar gain codebooks.

The Scheme Two codec (see Section 8.9.3) also uses 3-tap backward-adapted LTP but employs forward adaption to determine the short-term synthesis filter coefficients. Using these coefficients without quantization gives an improvement in the codec's segmental SNR of 0.82 dB, which would be reduced if quantization were applied. In Reference [136], where forward adaption is used for the LPC parameters, 19 bits are used to quantize a set of LSFs for every 80-sample frame. This quantization scheme would require us to use about 2.4 extra bits per 10-sample vector.

The Scheme Three codec (see Section 8.9.4) uses backward adaption to determine the short-term predictor coefficients and the long-term predictor delay. However, forward adaption is used to find the LTP gain, which is jointly determined along with the fixed codebook index and gain. The LTP gain and the fixed codebook gain are jointly vector-quantized using 4, 5, 6, or 7-bit quantizers, which implies using between 1 and 4 extra bits per 10-sample vector.

Finally, the Scheme Four codec uses the same coding strategy as the Scheme Three codec, but it also implements a backward-adapted switch between specially trained shape and vector gain codebooks for the voiced and unvoiced segments of speech.

It is clear from Table 8.15 that, for our vector size 10 codec, using extra bits to allow forward adaption of the synthesis filter parameters is the least efficient way of using these extra bits. If we use two extra bits, the largest gain in the codec's segmental SNR is given if we simply use the Scheme One codec and increase the size of the shape codebook by 2 bits. This gain is almost matched if we allocate one extra bit to both the shape and gain codebooks in the Scheme One codec. This would increase the codebook search complexity less dramatically than allocating both extra bits to the shape codebook.

In order to give a fair comparison between the different coding schemes at bitrates between 4 and 8 kbits/s, we tested the Scheme One, Scheme Three, and Scheme Four codecs using 8-bit shape codebooks, 4-bit gain codebooks, and vector sizes of 12, 15, 18, and 24 samples. This gave three different codecs at 8, 6.4, 5.3, and 4 kbits/s. Note that as the vector size of the codecs increases, their complexity also increases. Methods of reducing this complexity are possible [218] but have not been studied in our work. The segmental SNRs of our three 4–8 kbits/s codecs against their bitrates are shown in Figure 8.24.

Several observations can be made from this graph. At 8 kbits/s, as expected from the results in Table 8.15, the Scheme One codec gives the best quality reconstructed speech, with a segmental SNR of 14.55 dB. However, as the vector size is increased and hence the bitrate is reduced, it is the Scheme One codec whose performance is most adversely affected. At 6.4 kbits/s and 5.3 kbits/s, all three codecs give very similar segmental SNRs, but at 4 kbits/s the Scheme One codec is worse than the other codecs, which use forward adaption of the LTP gain. This indicates that, although the 3-tap backward-adapted LTP is very effective at 8 kbits/s and above, it is less effective as the bitrate is reduced. Furthermore, the backward-adaptive LTP scheme is more prone to channel error propagation.

Similarly, as indicated in Table 8.15, the backward-adaptive switching between specially trained voiced and unvoiced gain and shape codebooks improves the performance of our Scheme Four codec at 8 kbits/s so that it gives a higher segmental SNR than the Scheme Three codec. However, as the bitrate is reduced, the gain due to this codebook switching is eroded, and at 4 kbits/s the Scheme Four codec gives a lower segmental SNR than the Scheme Three codec. This is due to inaccuracies in the backward-adaptive voicing decisions at the lower bitrates. Figure 8.25 shows the same segment of speech as was shown in Figure 8.22 and the normalized autocorrelation value β_α calculated from the reconstructed speech of our Scheme Four codec at 4 kbits/s. The condition $\beta_\alpha > 0.6$ no longer gives a good indication of the voicing of the speech. Again for clarity of display, the values of β_α have been limited to between 0.05 and 0.95 in this figure.

In listening tests we found that all three codecs gave near toll quality speech at 8 kbits/s, with differences between the codecs being difficult to distinguish. However, at 4 kbits/s the

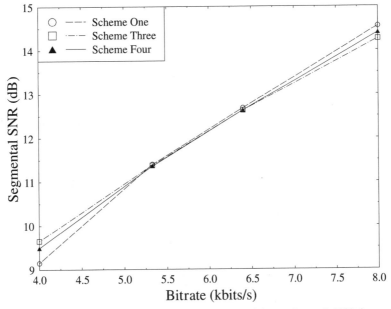

Figure 8.24 Performance of schemes one three and four codecs at 4–8 kbits/s.

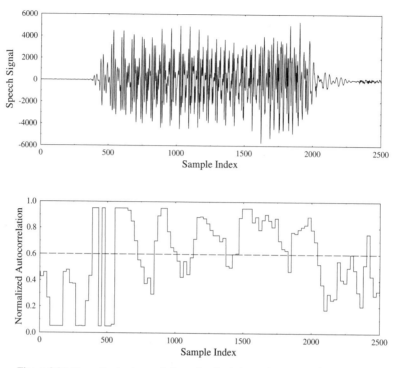

Figure 8.25 Normalized autocorrelation value β_α during voiced and unvoiced speech.

Scheme Two codec clearly sounded better than the Scheme One codec and gave reconstructed speech of communications quality.

8.9.6 Low-Delay ACELP Codec

We implemented a low-delay version of our algebraic CELP (ACELP) codec, which was described in Section 6.4.3, and we developed a series of low-delay codecs with a frame size of 40 samples or 5 ms. Hence, the total delay was about 15 ms, and there were various bitrates between 5 and 6.2 kbits/s. All of these codecs use backward adaption with the recursive windowing function described in Section 8.4.2 in order to determine the coefficients for the synthesis filter, which has an order of $p = 20$. Furthermore, they employ the same weighting filter, which was described in Section 8.4.1, as our other low-delay codecs. However, apart from this they have a structure similar to the codecs described in Section 6.4.3. An adaptive codebook is used to represent the long-term periodicities of the speech, with possible delays taking all integer values between 20 and 147 and being represented using 7 bits. As described in Section 6.4.3, the best delay is calculated once per 40-sample vector within the analysis-by-synthesis loop at the encoder and then transmitted to the decoder.

Initially, we used the 12-bit ACELP fixed codebook structure shown in Table 6.4, which is repeated in Table 8.16. Each 40-sample vector has a fixed codebook signal given by 4 nonzero pulses of amplitude $+1$ or -1, whose possible positions are shown in Table 8.16. Each pulse position is encoded with 3 bits, giving a 12-bit codebook. As explained in Section 6.3, the pulse positions can be found using a series of four nested loops, leading to a very efficient codebook search algorithm [69, 138].

TABLE 8.16 Pulse Amplitudes and Positions for the 12-bit ACELP Codebook

Pulse Number i	Amplitude	Possible Position m_i
0	+1	1,6,11,16,21,26,31,36
1	−1	2,7,12,17,22,27,32,37
2	+1	3,8,13,18,23,28,33,38
3	−1	4,9,14,19,24,29,34,39

In our first low-delay ACELP codec, which we refer to as Codec A, we used the same 3- and 5-bit scalar quantizers as were used in the codecs in Section 6.4.3 to quantize the adaptive and fixed codebook gains G_1 and G_2. This meant that 12 bits were required to represent the fixed codebook index, 7 bits for the adaptive codebook index, and 8 bits to quantize the two codebook gains. This gave a total of 27 bits to represent each 40-sample vector, giving a bitrate for this codec of 5.4 kbits/s. We found that this codec gave an average segmental SNR of 10.20 dB, which should be compared to the average segmental SNRs for the same speech files of 9.83 dB, 11.13 dB and 11.42 dB for our 4.7 kbits/s, and 6.5 kbits/s and 7.1 kbits/s forward-adaptive ACELP codecs described in Section 6.4.3. All of these codecs have a similar level of complexity, but the backward-adaptive 5.4 kbits/s ACELP codec has a frame size of only 5 ms, compared to the frame sizes of 20 or 30 ms for the forward-adaptive systems. Furthermore, Figure 8.26 shows that, upon interpolating the segmental SNRs between the three forward-adaptive ACELP codecs, the backward-adaptive ACELP codec at 5.4 kbits/s gives a very similar level of performance to the forward-adaptive codecs. In this figure we have marked the segmental SNRs of the three forward-adaptive ACELP codecs with circles, and the segmental SNR of our low-delay ACELP codec at 5.4 kbits/s with a diamond.

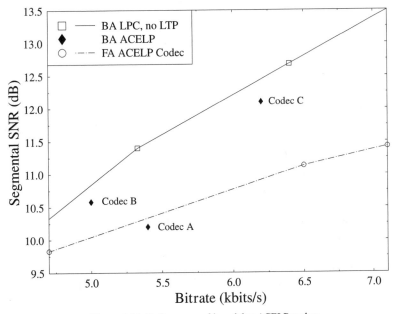

Figure 8.26 Performance of low delay ACELP codecs.

Also marked with diamonds are the segmental SNRs and bitrates of other backward-adaptive ACELP codecs which will be described later. For comparison, the performance of the Scheme One low-delay codec, described in Section 8.9.5 and copied from Figure 8.24, is also shown.

It can be seen from Figure 8.26 that, although the 5.4 kbits/s low-delay backward-adaptive ACELP codec gives a similar performance in terms of segmental SNR to the higher delay forward-adaptive ACELP codecs, it performs significantly worse than the Scheme One codec of Table 8.15, which uses a shorter vector size and a trained shape codebook. We therefore attempted to improve the performance of our low-delay ACELP codec by introducing vector quantization and joint determination of the two codebook gains G_1 and G_2. Note that similar vector quantization and joint determination of these gains was used in the Scheme Three and Scheme Four codecs described in Section 8.9.5. We also reintroduced the backward adaption of the fixed codebook gain G_2—as known from the schematic of the G.728 decoder seen in Figure 8.2, which was used in our other low-delay codecs as detailed in Section 8.4.3. We replaced the 3- and 5-bit scalar quantizers for G_1 and G_2 with a 6-bit joint vector quantizer for these gains, which resulted in a total of 25 bits being used to represent each 40-sample vector and therefore gave us a 5 kbits/s codec. We refer to this as Codec B. The joint 6-bit vector quantizer for the gains was trained as described in Section 8.9.4.2. A joint codebook search procedure was used so that for each fixed codebook index k the joint gain codebook was searched to find the gain codebook index which minimized the weighted error for that fixed codebook index. The best shape and gain codebook indices are therefore determined together. This codebook search procedure results in a large increase in the complexity of the codec but also significantly increases the performance of the codec.

We found that our 5 kbits/s Codec B, using joint vector quantization of G_1 and G_2 and backward adaption of G_2, gave an average segmental SNR of 10.58 dB. This is higher than the segmental SNR of the codec with scalar gain quantization (i.e., Codec A), despite Codec B having a lower bitrate. The performance of this Codec B is marked with a diamond in Figure 8.26, which shows that it falls between the segmental SNRs of the ACELP codecs with scalar gain quantization and the Scheme One codecs.

Next we replaced the 12-bit algebraic codebook detailed in Table 8.16 with the 17-bit algebraic codebook used in the G.729 ACELP codec described in Section 7.8. Also, the 6-bit vector quantization of the two gains was replaced with 7-bit vector quantization. This gave a 6.2 kbits/s codec, referred to as Codec C, which is similar to the G.729 codec. The main difference between G.729 and our Codec C is that G.729 uses forward adaption to determine the LPC coefficients, whereas Codec B uses backward adaption. This implies that it does not transmit the 18 bits per 10 ms that G.729 uses to represent the LPC parameters, and hence it operates at a bitrate 1.8 kbits/s lower. Also its buffering delay is halved to only 5 ms.

We found that this Codec C gave reconstructed speech with a segmental SNR of 12.1 dB, as shown in Figure 8.26. Our G.729-like codec gives a better segmental SNR than the forward-adaptive ACELP codecs described earlier. This is because of the more advanced 17-bit codebook, together with the joint determination and vector quantization of the fixed and the adaptive codebook gains, used in the backward-adaptive ACELP codec. It is also clear from Figure 8.26 that Codec C gives a similar performance to the backward-adaptive variable rate codecs with trained codebooks. Subjectively, we found that Codec C gave speech of good communications quality, but significantly lower than that of the toll quality produced by the forward-adaptive G729. Even so, this codec may be preferred to G.729 in situations where a lower bitrate and delay are required, and the lower speech quality can be accepted.

The characteristics of our low-delay ACELP codecs are summarized in Table 8.17. In the next section we discuss error sensitivity issues relating to the low-delay codecs described in this chapter.

TABLE 8.17   Performance and Structure of Low-Delay ACELP Codecs

	Algebraic Codebook	Gain Quantization	Bitrate (kbits/s)	Segmental SNR
Codec A	12 bit	3 + 5 bit scalar	5.4	10.2 dB
Codec B	12 bit	6 bit vector	5	10.6 dB
Codec C	17 bit	7 bit vector	6.2	12.1 dB

8.10 BACKWARD-ADAPTIVE ERROR SENSITIVITY ISSUES

Traditionally, one serious disadvantage of using backward adaption of the synthesis filter is that it is more sensitive to channel errors than forward adaption. In this section, we consider the error sensitivity of the 16 kbits/s G728 codec and discuss the error sensitivity of the 4–8 kbits/s low-delay codecs, as well as the means of improving this error sensitivity. Finally, we investigate the error sensitivity of our low-delay ACELP codec and compare this to the error sensitivity of a traditional forward-adaptive ACELP codec.

8.10.1 The Error Sensitivity of the G728 Codec

For each five-sample speech vector, the G728 codec produces a 3-bit gain codebook index, and an 8-bit shape codebook index. Figure 8.27 shows the sensitivity of these 10 bits to channel errors. For each bit, the error sensitivities were measured by corrupting the given bit only with a 10% bit error rate (BER). This approach was taken, rather than the more usual method of corrupting the given bit in every frame, in order to account for the possible different error propagation properties of different bits [145]. Bits 1 and 2 in Figure 8.27 represent the magnitude of the excitation gain, bit 3 represents the sign of this gain, and the remaining bits are used to code the index of the shape codebook entry chosen to represent the

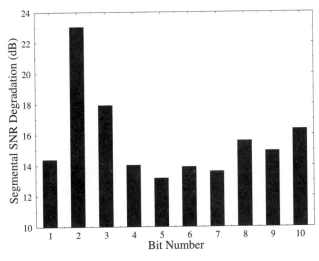

Figure 8.27 Degradation in G728 segmental SNR caused by 10% BER in given bits.

excitation. It can be seen from this figure that not all 10 bits are equally sensitive to channel errors. Notice, for example, that bit 2, representing the most significant bit of the excitation gain's magnitude, is particularly sensitive.

This unequal error sensitivity can also be seen from Figure 8.28, which shows the segmental SNR of the G728 codec for channel BERs between 0.001% and 1%. The solid line shows the performance of the codec when the errors are equally distributed among all 10 bits, whereas the dashed lines show the performance when the errors are confined to the 5 most sensitive bits (the Class One bits) or the 5 least sensitive bits (the Class Two bits). The 10 bits were arranged into these two groups based on the results shown in Figure 8.27—bits 2, 3, 8, 9, and 10 formed Class One, and the other 5 bits formed Class Two. It can be seen that the Class One bits are about two or three times more sensitive than the Class Two bits. Therefore, when the G728 codec is employed in an error-prone transmission scheme, for example, in a mobile radio transmission system, the error resilience of the system will be improved if unequal error protection is employed [216]. The use of unequal error protection for speech codecs is discussed in detail later in this chapter.

8.10.2 The Error Sensitivity of Our 4–8 kbits/s Low-Delay Codecs

We now consider the error sensitivity of some of our 4–8 kbits/s codecs which were described in Section 8.9.5. It is well known that codecs using backward adaption for both the LTP delay and gain are very sensitive to bit errors; this is why LTP was not used in G728 [70]. Thus, as expected, we found that the Scheme One codec gave a very poor performance when subjected to even a relatively low bit error rate (BER). Unfortunately, we also found similar results for the Scheme Three and Scheme Four codecs, which, although they used backward adaption for the LTP delay, used forward adaption for the LTP gain. We therefore decided that none of these codecs is suitable for use over error-prone channels. However, the Scheme One

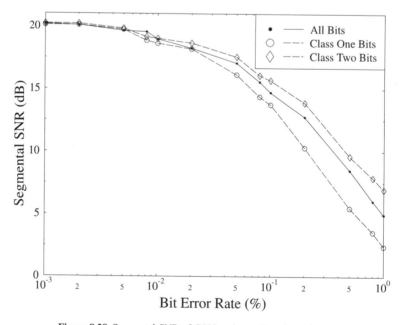

Figure 8.28 Segmental SNR of G728 codec against channel BER.

codec can be easily modified by removing its entirely backward-adapted 3-tap LTP and increasing the order of its short-term filter to 50 as in G728, to make it less sensitive to channel errors. Although this impairs the performance of the codec, as can be seen from Figure 8.29 the resulting degradation in the codec's segmental SNR is not too serious, especially at low bitrates. Therefore, in this section we detail the error sensitivity of the Scheme One codec with its LTP removed, and we describe the means of making this codec less sensitive to channel errors. For simplicity, only the error sensitivity of the codec operating with a frame length of 15 samples and a bitrate of 6.4 kbits/s are detailed here. However, similar results also apply at the other bitrates.

At 6.4 kbits/s our codec transmits only 12 bits per 15-sample frame from the encoder to the decoder. Of these 12 bits, 8 are used to represent the index of the shape codebook, and the remaining 4 bits are used to represent the index of the gain codebook entry used. The error resilience of these bits can be significantly improved by careful assignment of codebook indices to the various codebook entries. Ideally, each codebook entry would be assigned an index so that corruption of any of the bits representing this index would result in another entry being selected in the decoder's codebook, which is in some way "close" to the intended codebook entry. If this ideal can be achieved, then the effects of errors in the bits representing the codebook indices will be minimized.

Consider first the 8-bit shape codebook. Initially, the 256 available codebook indices are effectively randomly distributed among the codebook entries. We seek to rearrange these codebook indices so that when the index representing a codebook entry is corrupted, the new index will represent a codebook entry that is "close" to the original entry. In our work we chose to measure this "closeness" by the squared error between the original and corrupted codebook entries. We considered only the effects of single-bit errors among the 8 codebook bits because at reasonable bit error rates (BERs) the probability of two or more errors occurring in 8 bits will be small. Thus, for each codebook entry the "closeness" produced by a certain arrangement of codebook entries is given by the sum of the squared errors between

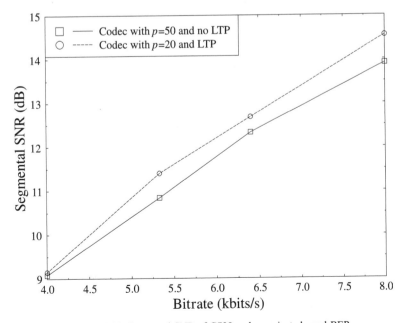

Figure 8.29 Segmental SNR of G728 codec against channel BER.

the original codebook entry and the eight corrupted entries that would be produced by inverting each of the 8 bits representing the entry's index. The overall "cost" of a given arrangement of codebook indices is then given by the closeness for each codebook entry, weighted by the probability of that codebook entry being used. Thus, the cost we seek to minimize is given by

$$\text{Cost} = \sum_{j=0}^{255} P(j) \left[\sum_{i=1}^{8} \left(\sum_{n=1}^{15} (c_j(n) - c_j^i(n))^2 \right) \right] \tag{8.67}$$

where $P(j)$ is the probability of the jth codebook entry being used, $c_j(n)$, $n = 1 \cdots 15$, is the jth codebook entry, and $c_j^i(n)$ is the entry that will be received if the index j is transmitted but the ith bit of this index is corrupted.

The problem of choosing the best arrangement of the 256 codebook indices among the codebook entries is similar to the famous traveling salesman problem. In this problem, the salesman must visit each of N cities and must choose the order in which he visits the cities so as to minimize the total distance he travels. As N becomes large, it becomes impractical to solve this problem using an exhaustive search of all possible orders in which he could visit the cities—the complexity of such a search is proportional to $N!$ Instead, a nonexhaustive search must be used, which we hope will find the best order possible in which to visit the N cities.

The minimization method of simulated annealing has been successfully applied to this problem [153] and has also been used by other researchers as a method of improving the error resilience of quantizers [219]. Simulated annealing works, as its name suggests, in analogy to the annealing (or slow cooling) of metals. When metals cool slowly from their liquid state, they start in a very disordered and high-energy state and reach equilibrium in an extremely ordered crystalline state. This crystal is the minimum-energy state for the system, and simulated annealing similarly allows us to find the global minimum of a complex function with many local minima. The procedure works as follows. The system starts in an initial state, which in our situation is an initial assignment of the 256 codebook indices to the codebook entries. A temperature-like variable T is defined, and possible changes to the state of the system are randomly generated. For each possible change, the difference $\Delta Cost$ in the cost between the present state and the possible new state is evaluated. If this is negative, that is, the new state has a lower cost than the old state, then the system always moves to the new state. If, on the other hand, $\Delta Cost$ is positive, then the new state has a higher cost than the old state, but the system may still change to this new state. The probability of this happening is given by the Boltzmann distribution

$$\text{prob} = \exp\left(\frac{-\Delta Cost}{kT} \right) \tag{8.68}$$

where k is a constant. The initial temperature is set so that kT is much larger than any $\Delta Cost$ that is likely to be encountered, so that initially most offered moves will be taken. As the optimization proceeds, the "temperature" T is slowly decreased, and the number of moves to states with higher costs decreases. Eventually, kT becomes so small that no moves with positive $\Delta Cost$ are taken, and the system comes to equilibrium in what is hopefully the global minimum of its cost.

The advantage of simulated annealing over other optimization methods is that it should not be deceived by local minima and should slowly make its way toward the global minimum of the function to be minimized. In order to make this likely to happen, it is important to ensure that the temperature T starts at a high enough value and is reduced suitably slowly. We followed the suggestions in [153] and reduced T by 10% after every $100\,N$ offered moves, or every $10\,N$ accepted moves, where N is the number of codebook entries (256). The initial temperature was set so that kT was equal to 10 times the highest value of $\Delta Cost$ that was initially encountered. The random changes in the state of the system were generated by randomly choosing two codebook entries and swapping the indices of these two entries.

The effectiveness of the simulated annealing method in reducing the cost given in Equation 8.67 is shown in Figure 8.30. This graph shows the cost of the present arrangement of codebook indices against the number of arrangements of codebook indices that have been tried by the minimization process. The initial randomly assigned arrangement of indices to codebook entries gives a cost of 1915 in the sense of Equation 8.67. As can be seen in Figure 8.30, initially the temperature T is high, and so many index assignments that have a higher cost than this are accepted. However slowly as the number of attempted configurations increases the temperature T decreases, and so fewer rearrangements that increase the cost of the present arrangement are accepted. Thus, as can be seen in Figure 8.30, the cost of the present arrangement slowly falls, and the curve narrows as the temperature increases and fewer rearrangements that increase the cost of the present arrangement are accepted. The cost of the final arrangement of codebook indices to codebook entries is 1077 in the sense of Equation 8.67, which corresponds to about a 44% reduction in cost.

The effectiveness of this rearrangement of codebook indices in increasing the resilience of the codec to errors in the bit stream between its encoder and decoder can be seen in Figure 8.31. This graph shows the variation in the segmental SNR of our 6.4 kbits/s low-delay codec with the bit error rate (BER) between its encoder and decoder. The solid line shows the performance of the codec with the original codebook index assignment, and the lower dashed line shows the performance when the shape codebook indices are rearranged. At BERs of between 0.1% and 1%, the codec with the rearranged codebook indices has a segmental SNR about 0.5 to 1 dB higher than the original codec.

Apart from the 8-shape codebook bits that the codec transmits from its encoder to the decoder, the only other information that is explicitly transmitted are the 4 bits representing the

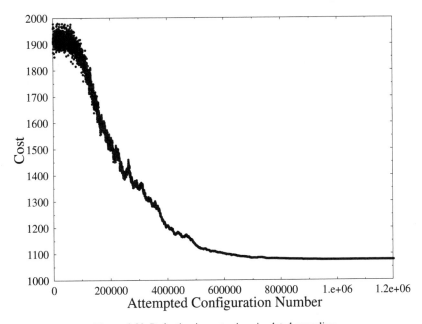

Figure 8.30 Reduction in cost using simulated annealing.

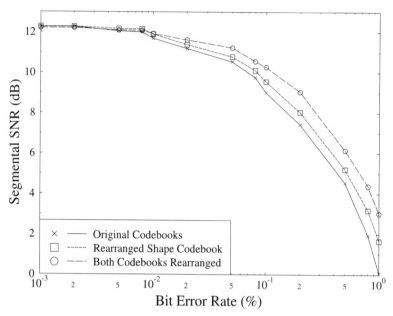

Figure 8.31 The error sensitivity of our low delay 6.4 kbits/s codec.

gain codebook entry selected. Initially, indices were assigned to the 16 gain codebook entries using the simple Natural Binary Code (NBC). However, because the gain codebook levels do not have an equiprobable distribution, this simple assignment can be improved upon in a similar way to that described for the shape codebook. Again, we defined a cost function that was to be minimized. This cost function was similar to that given in Equation 8.67 except that the gain codebook is scalar, whereas the shape codebook has a vector dimension of 15. No summation over n is needed in the cost function for the gain codebook index arrangement. We used simulated annealing again to reduce the cost function over that given using a NBC, and we found that we were able to reduce the cost by over 60%. The effect of this rearrangement of the gain codebook indices is shown by the upper curve in Figure 8.31. The figure gives the performance of the Scheme One codec, with LTP removed, with both the gain and shape codebooks rearranged. It can be seen that the rearrangement of the gain codebook indices gives a further improvement in the error resilience of the codec. Moreover, the codec with both the shape and gain codebooks rearranged has a segmental SNR more than 1 dB higher than the original codec at BERs around 0.1%.

8.10.3 The Error Sensitivity of Our Low-Delay ACELP Codec

The segmental SNR of our 6.2 kbits/s low-delay ACELP codec described in Section 8.9.6 is shown in Figure 8.32. Also shown in this figure are the error sensitivities of our 6.4 kbits/s Scheme One codec with no LTP and a traditional 6.5 kbits/s forward-adaptive ACELP codec. As noted above, at 0% BER the two backward-adaptive codecs give similar segmental SNRs, but the forward-adaptive codec gives a segmental SNR of about 1 dB lower. However in subjective listening tests, the better spectral match provided by the forward-adaptive codec, which is not adequately reflected in the segmental SNR distortion measure, results in it providing better speech quality than the two backward-adaptive codecs. As the BER is increased, the backward-adaptive ACELP is the most adversely affected, but

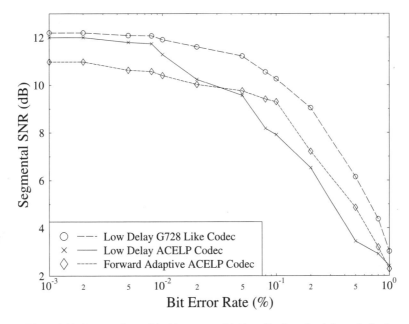

Figure 8.32 A comparison of the bit error sensitivities of backward and forward adaptive codecs.

surprisingly, the other backward-adaptive codec is almost as robust to channel errors as the forward-adaptive ACELP codec. Both of these codecs give a graceful degradation in their reconstructed speech quality at BERs up to about 0.1%, but they provide impaired reconstructed speech for BERs much above this.

Let us now in the next section provide an application scenario for employing the previously designed G.728-like 8–16 kbps speech codecs and evaluate the performance of the transceiver proposed.

8.11 A LOW-DELAY MULTIMODE SPEECH TRANSCEIVER

8.11.1 Background

The intelligent, adaptively reconfigurable wireless systems of the near future require programmable source codecs in order to optimally configure the transceiver to adapt to time-variant channel and traffic conditions. Hence, we designed a flexible transceiver for the previously portrayed programmable 8–16 kbits/s low-delay speech codec, which is compatible with the G728 16 kbits/s ITU codec at its top rate and offers a graceful tradeoff between speech quality and bitrate in the range 8–16 kbits/s. Source-matched Bose-Chaudhuri-Hocquenghem (BCH) codecs combined with unequal protection pilot-assisted 4- and 16-level quadrature amplitude modulation (4QAM, 16QAM) are employed in order to transmit both the 8 and the 16 kbits/s coded speech bits at a signaling rate of 10.4 kBd. In a bandwidth of 1728 kHz, which is used by the Digital European Cordless Telephone (DECT) system, 55 duplex or 110 simplex time slots can be created. We will show that good toll quality speech is delivered in an equivalent user bandwidth of 15.71 kHz, if the channel signal-to-noise ratio (SNR) and signal-to-interference ratio (SIR) are in excess of about 18 and 26 dB for the lower and higher speech quality 4QAM and 16QAM modes, respectively.

8.11.2 8–16 kbps Codec Performance

The segmental SNR versus bitrate performance of our 8–16 kbits/s codec was shown in Figure 8.16. The unequal bit error sensitivity of the codec becomes explicit in Figure 8.28, showing the segmental SNR of the G728 codec for channel BERs between 0.001% and 1%. The 10 bits were arranged into these two groups based on the results shown in Figure 8.27— bits 2, 3, 8, 9, and 10 formed Class One, and the other 5 bits formed Class Two. The Class One bits are about two or three times more sensitive than the Class Two bits and therefore should be more strongly protected by the error correction and modulation schemes. For robustness reasons, we have refrained from using a LTP.

We also investigated the error sensitivity of the 8 kbits/s mode of our low-delay codec. LTP was not invoked, but the codec with a vector size of 10 was used because, as was seen earlier, it gave a segmental SNR almost 2 dB higher than the 8 kbits/s mode of the codec with a constant vector size of 5. As discussed in Section 8.7, the vector codebook entries for our codecs were trained as described in [213]. However, the 7-bit indices used to represent the 128 codebook entries are effectively randomly assigned. This assignment of indices to codebook entries does not affect the performance of the codec in error-free conditions, but it is known that the robustness of vector quantizers to transmission errors can be improved by the careful allocation of indices to codebook entries [214]. This can be seen from Figure 8.33, which shows the segmental SNR of the 8 kbits/s codec for BERs between 0.001% and 1%. The solid line shows the performance of the codec using the codebook with the original index assignment, whereas the dashed line shows the performance of the codec when the index assignment was modified to improve the robustness of the codebook. A simple, nonoptimum algorithm was used to perform the index assignment, and it is probable that the codec's robustness could be further improved by using a more effective minimization algorithm such as simulated annealing. Also, as in the G728 codec, a natural binary code was used to

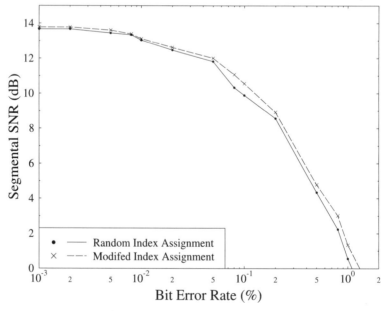

Figure 8.33 Segmental SNR of 8 kbits/s codec against channel BER for original and rearranged codebooks.

represent the eight quantized levels of the excitation gain. It is likely that the use, for example, of a Gray code to represent the eight gain levels could also improve the codec's robustness.

The sensitivity of the 10 bits used to represent each 10-speech sample vector in our 8 kbits/s codec is shown in Figure 8.34. Again, bits 1, 2, and 3 are used to represent the excitation gain, and the other 7 bits represent the index of the codebook entry chosen to code the excitation shape. As in the case of the G728 codec, the unequal error resilience of different bits can be clearly seen. Note in particular how the least significant of the 3 bits representing the excitation gain is much less sensitive than the 7 bits representing the codebook index, but that the two most sensitive gain bits are more sensitive than the codebook index bits.

Figure 8.35 shows the segmental SNR of the 8 kbits/s codec for BERs between 0.001% and 1%. Again, the solid line shows the performance of the codec when the errors are equally distributed among all 10 bits, whereas the dashed lines show the performance when the errors are confined to the 5 most sensitive Class One bits or the 5 least sensitive Class Two bits. The need for the more sensitive bits to be more protected by the FEC and modulation schemes is again apparent. These schemes, and how they are used to provide the required unequal error protection, are discussed in the next section.

8.11.3 Transmission Issues

8.11.3.1 Higher-Quality Mode.   Based on the bit-sensitivity analysis presented in the previous section, we designed a sensitivity-matched transceiver scheme for both the higher and lower quality speech coding modes. Our basic design criterion was to generate an identical signaling rate in both modes in order to facilitate the transmission of speech within the same bandwidth, while providing higher robustness at a concomitant lower speech quality, if the channel conditions degrade.

Specifically, in the more vulnerable, higher-quality mode, 16-level pilot symbol assisted quadrature amplitude modulation (16-PSAQAM) [135] was used for the transmission of speech encoded at 16 kbps. In the more robust, lower-quality mode, the 8 kbps encoded

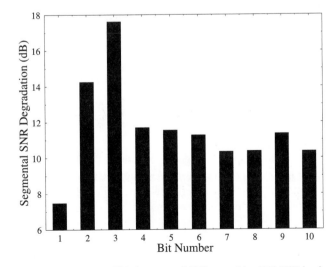

Figure 8.34 Degradation in 8 kbits/s segmental SNR caused by 10% BER in given bits.

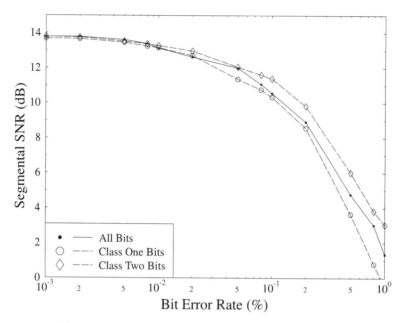

Figure 8.35 Segmental SNR of 8 kbits/s codec against channel BER.

speech is transmitted using 4-PSAQAM at the same signaling rate. In our former work [145], we have found that typically it is sufficient to use a twin-class unequal protection scheme rather than more complex multiclass arrangements. We have also shown [49] that the maximum minimum distance square 16QAM constellation exhibits two different-integrity subchannels, namely, the better quality C1 and lower quality C2 subchannels, where the bit error rate (BER) difference is about a factor 2 in our operating signal-to-noise ratio (SNR) range. This was also argued in [135].

Hence, we would require a forward error correction (FEC) code of twice the correction capability for achieving a similar overall performance of both subchannels over Gaussian channels, where the errors have a typically random rather than bursty distribution. Over bursty Rayleigh channels, an even stronger FEC code would be required in order to balance the differences between the two subchannels. After some experimentation, we opted for the binary Bose-Chaudhuri-Hocquenghem BCH(127,92,5) and BCH(124,68,9) codes [134] for the protection of the 16 kbps encoded speech bits. The weaker code was used in the lower BER C1 subchannel and the stronger in the higher BER C2 16QAM subchannel. Upon evaluating the BERs of the coded subchannels over Rayleigh channels, which are not presented here due to lack of space, we found that a ratio of two in terms of coded BER was maintained.

Since the 16 kbps speech codec generated a 160 bits/10 ms frame, the 92 most vulnerable speech bits were directed to the better BCH(127,92,5) C1 16QAM subchannel, while the remaining 68 bits to the other subchannel. Since the C1 and C2 subchannels have an identical capacity, after adding some padding bits 128 bits of each subchannel were converted to 32 4-bit symbols. A control header of 30 bits was BCH(63,30,6) encoded, which was transmitted employing the more robust 4QAM mode of operation using 32 2-bit symbols. Finally, two ramp symbols were concatenated at both ends of the transmitted frame, which also incorporated four uniformly spaced pilot symbols. A total of 104 symbols/10 ms therefore represented 10 ms speech, yielding a signaling rate of 10.4 kBd. When using a bandwidth of 1728 kHz, as in the digital European cordless telephone (DECT) system and an excess bandwidth of 50%, the multi-user signaling rate becomes 1152 kBd. Hence, a total of INT[1152/104] = 110 time slots can be created, which allows us to support 55 duplex conversations in time division duplex (TDD) mode. The time-slot duration becomes 10 ms/(110 slots) ≈ 90.091 μs.

8.11.3.2 Lower-Quality Mode. In the lower-quality 8 kbps mode of operation, 80 bits/10 ms are generated by the speech codecs, but the 4QAM scheme does not have two different integrity subchannels. Here we opted for the BCH(63,36,5) and BCH(62,44,3) codes in order to provide the required integrity subchannels for the speech codec. Again, after some padding the 64-bit coded subchannels are transmitted using 2-bit/symbol 4QAM, yielding 64 symbols. After incorporating the same 32-symbol header block, 4 ramp and 4 pilot symbols, as for the higher-quality mode, we arrive at a transmission burst of 104 symbols/10 ms, yielding an identical signaling rate of 10.4 kBd.

8.11.4 Speech Transceiver Performance

The SEGSNR versus channel SNR performance of the proposed multimode transceiver is portrayed in Figure 8.36 for both 10.4 kBd modes of operation. Our channel conditions were based on the DECT-like propagation frequency of 1.9 GHz, signaling rate of 1152 kBd and pedestrian speed of 1 m/s = 3.6 km/h, which yielded a normalized Doppler frequency of 6.3 Hz/1152 kBd ≈ 5.5 · 10^{-3}. Observe in the figure that unimpaired speech quality was experienced for channel SNRs in excess of about 26 and 18 dBs in the less and more robust

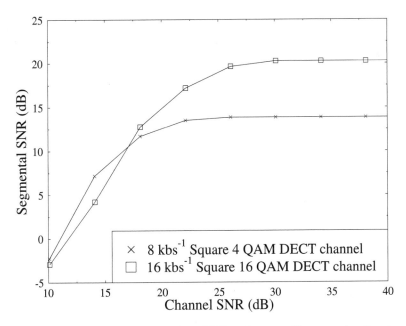

Figure 8.36 Segmental SNR versus channel SNR performance of the proposed multimode transceiver.

modes, respectively. When the channel SNR degrades substantially below 22 dB, it is more advantageous to switch to the inherently lower quality but more robust and essentially error-free speech mode, demonstrating the advantages of the multimode concept. The effective single-user simplex bandwidth is 1728 kHz/110 slots ≈ 15.71 kHz, while maintaining a total transmitter delay of 10 ms. Our current research is targeted at increasing the number of users supported using Packet Reservation Multiple Access.

8.12 CHAPTER SUMMARY

In this chapter we highlighted the operation of the CCITT G728 16 kbps standard codec and proposed a range of low-delay coding schemes operating between 16–8 and 8–4 kbits/s. While in the higher bitrate range entirely backward-adaptive predictive arrangements were used, in the lower range codecs the employment of both forward and backward adaption of the long-term filter has been considered. But all the codecs use backward adaption of the short-term synthesis filter and so have frame sizes of at most 5 ms. Both relatively small, trained, shape codebooks and large algebraic codebooks were used. We found that the resulting codecs offered a range of reconstructed speech qualities between communications quality at 4 kbits/s to near-toll quality at 8 kbits/s. Lastly, an application example was given, demonstrating the practical applicability of the codecs portrayed. Let us now concentrate our attention on high-quality wideband speech compression in the next chapter.

PART III
WIDEBAND CODING
AND TRANSMISSION

9

Wideband Speech Coding

9.1 SUB-BAND-ADPCM WIDEBAND CODING AT 64 kbps [220]

9.1.1 Introduction and Specifications

In our previous chapters, we assumed that the speech signal was band-limited to 0.3–3.4 kHz and sampled according to the Nyquist principle at 8 kHz. This filtering process, however, removes some of the energy of the speech signal, which amounts to about 1%. This does not significantly reduce the perceived quality of reproduction, but nonetheless, in many applications, such as in commentatory quality channels, a better quality is desirable. Therefore, the CCITT standardized a wideband codec, referred to as G722, which filters the signal to 50–7000 Hz, before sampling at 16 kHz takes place. We commence our discourse on wideband speech coding by considering the specifications and algorithmic details of the G722 sub-band-split adaptive differential pulse code modulated (SB-ADPCM) speech codec. In our discussion, we follow the G722 Recommendation and Maitre's deliberations. The range of requirements to be satisfied by the standardized G722 codec encompassed the following specifications [220]:

1. A speech quality better than that of 128 kbps PCM was sought, and the encoding quality of music signals was not considered of highest priority.

2. There was no consideration given to the transmission of voice band data or in-band signaling.

3. A total of four tandemed sections, including digital transcoding to and from linear PCM, were considered a realistic requirement.

4. The codec was required to have no significant quality degradation at a BER of 10^{-4} and have a better performance at 10^{-3} than 128 kbps linear PCM.

5. The total delay was specified to be less than 4 ms.

6. There was a need to accommodate a data channel at the cost of a reduced speech quality, which was satisfied by defining the following three modes of operation:

TABLE 9.1   G722 Codec Specification

Mode	Speech Rate (kbps)	Data Rate (kbps)
1	64	0
2	56	8
3	48	16

Mode 1—Speech only at 64 kbps, Mode 2—56 kbps speech plus 8 kbps data, and Mode 3—48 kbps speech plus 16 kbps data, which are also summarized in Table 9.1. Two candidate codecs emerged: a full-band ADPCM codec and a sub-band-ADPCM (SB-ADPCM) scheme. Comparative tests have shown that the latter significantly outperformed the full-band codec at the lower rates of 48 and 56 kbps, which justified its standardization.

9.1.2 G722 Codec Outline

The basic codec schematic is shown in Figure 9.1, where the full-band input signal $x(n)$ is split in two sub-band signals: the higher-band component $X_h(n)$ and the lower-band component $X_L(n)$. The band-splitting operation is carried out by the aliasing-free **quadrature mirror filter** (QMF), which is characterized in analytical terms in the next section.

The QMF stage is constituted by two linear-phase finite impulse response (FIR) filters, whose impulse responses are symmetric. These filters split the 0–8000 Hz frequency band to 0–4000 Hz and 4000–8000 Hz, which now can be sampled at 8 kHz due to halving their bandwidths.

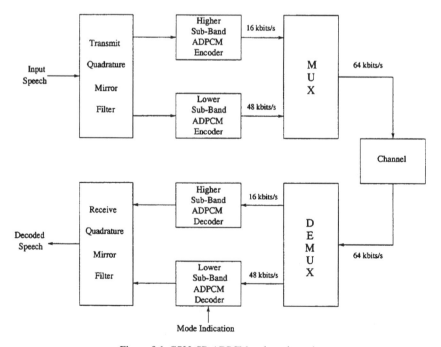

Figure 9.1 G722 SB-ADPCM codec schematic.

The 0–4000 Hz lower band retains a significantly higher proportion of the signal energy than the higher band. Furthermore, it is more important subjectively than the 4000–8000 Hz higher band. Hence, it is encoded using 6 bits/sample ADPCM coding, at 8 ksamp·6 bits/sample = 48 kbps in Mode 1. The lower-significance 4–8 kHz band is encoded using 2 bits/sample, that is, at 16 kbps. The resulting signals are denoted by I_L and I_M, which are then multiplexed for transmission over the digital channel. It is important to note that the 6-bit lower-band ADPCM quantizer could generate 64 reconstruction levels, but only 60 levels are actually generated.

This is explained as follows. As mentioned before, the codec can drop its rate to 48 kbps, in which case the lower-band ADPCM codec transmits at 32 kbps or 4 bits/sample. In certain systems, the all-zero codeword's transmission must be avoided in order to refrain from generating long strings of zeros, which may result in synchronization problems. Thus, in Mode 3 only 15 levels are used for quantization. Similarly, adding a further bit to the quantizer output generates 30 levels, and including two additional bits allows us to differentiate among 60 levels.

The principle of **embedded ADPCM** coding has been detailed with reference to the CCITT G727 codec. Similar principles are followed in the G722 codec in order to support the synchronous operation of the encoder and decoder in the event of one or two bits are dropped from the transmitted sequence for the sake of supporting 8 or 16 kbps data transmission.

The two least significant bits (LSBs) of the lower-band signal I_L are punctured in the predictive feedback loop to produce the truncated representation $I_{Lt}(n)$ in Figure 9.2. The coarsely quantized prediction residual $d_{Lt}(n)$ and the reconstructed, truncated-precision lower-band signal $r_{Lt}(n)$ are input to the pole-zero predictor, which was featured in the G727 codec in order to produce the predicted signal $s_L(n)$. Using 4 bits rather than 6 bits in the encoder's prediction loop allows the decoder's prediction loop to be synchronized with that of the encoder even in the event, when data bits are transmitted along with 48 kbps-coded speech. The operation of the higher-band ADPCM encoder depicted in Figure 9.3 is very similar to that of the lower-band scheme, except that it uses 2 bits/sample, 15-level quantization without deleting any bits from the predictor loop. (For further details on the embedded ADPCM codec, the interested reader is referred to Section 2.9.2.)

The schematic of the SB-ADPCM decoder shown in Figure 9.1 follows the inverse structure of the encoder. After demultiplexing the higher and lower sub-band bits, these sequences are decoded using the decoders of Figures 9.4 and 9.5, respectively. The 16 kbps higher band signal I_H is input to the decoder of Figure 9.4, and its reconstructed signal is r_H. The operation of the adaptive quantizer and that of the adaptive predictor is identical to those of the encoder, and these are described during our further discussions.

The lower-band's schematic is somewhat more complex because of the embedded tri-modal operation. The received 48 kbps bit stream is processed according to the Mode Indication signal, and the corresponding decoded signal is assigned into the low-band decoded residual d_L. In the highest quality Mode 1, the 60-level inverse adaptive quantizer is used, while in Mode 2 the LSB delivering the 8 kbps data-signal is deleted from each received 6-bit sample and the remaining 5 bits are input to the 30-level inverse quantizer. Lastly, in Mode 3 the 2LSBs conveying the 16 kbps data signal are removed from the signal, before invoking the 15-level inverse quantizer for their decoding. Observe at the bottom of Figure 9.5 that the truncated low-band signal I_{Lt} is also input to the quantizer adaptation block. The lower-band quantizer step-size ΔL is then used by all three too, increase adaptive quantizers. Note, that due to the embedded operation the adaptive predictor uses the truncated 4-bit resolution decoded residual d_{Lt} and the resulting truncated reconstructed signal $v_{Lt} = d_{Lt} + s_L$, where S_L is the estimate of the low-band input signal x_L seen in Figure 9.2.

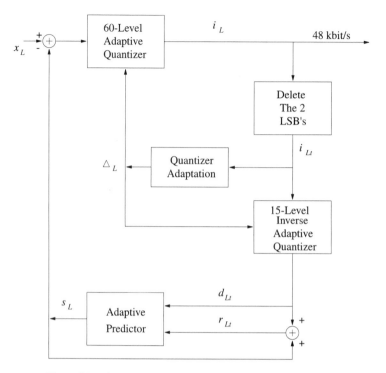

Figure 9.2 Schematic of the lower-band G722 SB-ADPCM encoder.

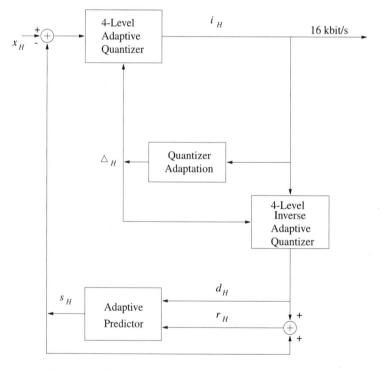

Figure 9.3 Schematic of the higher-band G722 SB-ADPCM encoder.

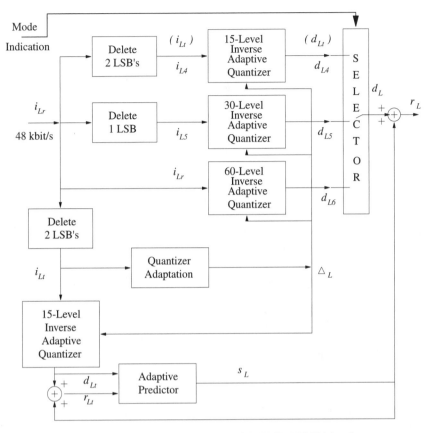

Figure 9.4 Schematic of the lower-band G722 SB-ADPCM decoder.

In the absence of transmission errors, s_L produced by the Adaptive Predictor of Figure 9.4 is identical to that of Figure 9.2 and the transmission of data in Modes 2 and 3 does not affect this estimate. Finally, the low-band reconstructed signal r_L is generated by adding the estimate s_L to the decoded residual d_L to yield $r_L = s_L + d_L$.

In our deliberations so far we have not considered the operation of the adaptive predictors, quantizer adaption processes and QMF-based bandpass splitting. In the next sections we concentrate on these issues.

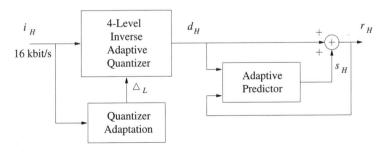

Figure 9.5 Schematic of the lower-band G722 SB-ADPCM decoder.

9.1.3 Principles of Sub-Band Coding

Let us briefly introduce the operation of sub-band codecs (SBC) [221, 222], where the speech signal is initially split into a number of sub-bands, which are separately encoded. The main attraction of sub-band codecs is that they allow an arbitrary bit allocation to be applied to each sub-band according to their perceptual importance, thereby confining the corresponding quantization noise to the sub-bands concerned. Then output bits generated by the sub-band encoders are multiplexed and transmitted to the receiver, where after demultiplexing and decoding each sub-band signal the original full-band signal is reconstructed by combining the individual sub-band components.

The success of this technique hinges on the design of appropriate band-splitting analysis and synthesis filters, which do not interfere with each other in their transition bands, that is, avoid the introduction of the **aliasing distortion** induced by sub-band overlapping due to an insufficiently high sampling frequency, (i.e., sub-sampling). If, on the other hand, the sampling frequency is too high, or for some other reason the filterbank employed generates a spectral gap, again, the speech quality suffers. In simplistic approach this would imply employing filters having a zero-width transition band, associated with an infinite-steepness cutoff slope. This would require an infinite filter order, which is impractical. As a practical alternative, an ingenious band-splitting structure referred to as quadrature mirror filter (QMF) was proposed by Esteban and Galand [223], which will be detailed at a later stage. QMFs have a finite filter order and remove aliasing effects by cancellation in the overlapping transition bands.

9.1.4 Quadrature Mirror Filtering [223] [47]

9.1.4.1 Analysis Filtering. As noted, quadrature mirror filters were introduced by Esteban and Galand [223], while Johnston [224] designed a range of QMFs for a variety of applications. The principle of QMF analysis/synthesis filtering can be highlighted following their deliberations and considering the twin-channel scheme portrayed in Figure 9.6, where the sub-band signals are initially unquantized for the sake of simplicity. The corresponding spectral-domain operations can be viewed in Figure 9.7.

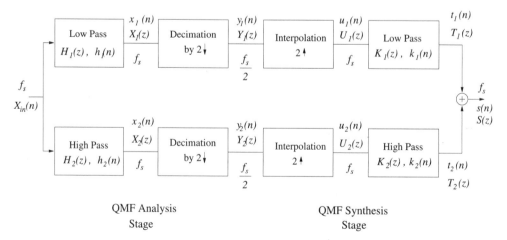

Figure 9.6 QMF analysis/synthesis arrangement.

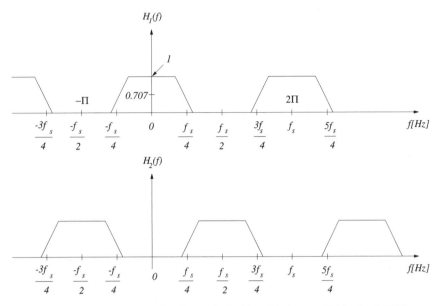

Figure 9.7 Stylized spectral domain transfer function of the lower and higher-band QMFs.

If most of the energy of the speech signal is confined to the frequency $f_s/2$, it can be band-limited to this range and sampled at $f_s = 1/T = \omega_s/2\pi$, to produce the QMF's input signal $x_{in}(n)$, which is input to the QMF analysis filter of Figure 9.6. As seen in the figure, this signal is filtered by the low-pass filter $H_1(z)$ and the high-pass filter $H_2(z)$ in order to yield the low-band signal $x_1(n)$ and the high-band signal $x_2(n)$, respectively. Since the energy of $x_1(n)$ and $x_2(n)$ is now confined to half of the original bandwidths of $x(n)$, the sampling rate of the sub-bands can be halved by discarding every second sample to produce the **decimated signals** $y_1(n)$ and $y_2(n)$.

In the sub-band synthesis stage of Figure 9.6, the decimated signals $y_1(n)$ and $y_2(n)$ are **interpolated** by inserting a zero-valued sample between adjacent samples in order to generate the up-sampled sequences $u_1(n)$ and $u_2(n)$. These are then filtered using the z-domain transfer functions $K_1(z)$ and $K_2(z)$ in order to produce the discrete-time sequences $t_1(n)$ and $t_2(n)$, which now again have a sampling frequency of f_s. The filtering operation reintroduced nonzero samples in the positions of the previously injected zero in the process of interpolation. Finally, the $t_1(n)$ and $t_2(n)$ are superimposed over each other, delivering the recovered speech $s(n)$.

Esteban and Galand [223] have shown that if the low-pass (LP) filter transfer functions $H_1(z)$, $K_1(z)$ and their high-pass (HP) counterparts $H_2(z)$, $K_2(z)$ satisfy certain conditions, perfect signal reconstruction is possible, provided the sub-band signals are unquantized. Let us assume that the transfer functions obey the following constraint:

$$|H_1(e^{j\omega T})| = |H_2(e^{j((\omega_s/2-\omega)T)})| \tag{9.1}$$

where ω is the angular frequency, $2\pi = \omega_s$ and the imposed constraint implies a mirror-symmetric magnitude response around $f_s/4$, where the 3 dB-down frequency responses, corresponding to $|H(\omega)| = 0.5$ cross at $f_s/4$. This can be readily verified by the following argument referring to Figure 9.7. Observe that $H_1(\omega)$ is equal to the $\omega_s/2 = \pi$-shifted version of the mirror image $H_2(-\omega)$, which becomes explicit by shifting $H_2(-\omega)$ to the right by

$\omega_s/2 = \pi$ at the bottom of the figure. It can also be verified in the figure that by shifting $H_1(\omega)$ to the left by $\omega_s/2 = \pi$ the following relationship holds:

$$|H_2(e^{j\omega T})| = |H_1(e^{-j((\omega_s/2)-\omega)T})| \tag{9.2}$$

Upon exploiting that:

$$e^{-j((\omega_s/2)-\omega)T} = e^{-j(\pi-\omega T)}$$
$$= \cos(\pi - \omega T) - j\sin(\pi - \omega T)$$
$$= -\cos(\omega T) - j\sin(\omega T)$$
$$= -e^{j\omega T} \tag{9.3}$$

Equation 9.2 can also be written as:

$$|H_2(e^{j\omega T})| = |H_1(-e^{j\omega T})|, \tag{9.4}$$

and upon taking into account that $z = e^{j\omega T}$, in z-domain we have $H_1(z) = H_2(-z)$. Following a similar argument, it can also be easily shown that the corresponding HP filters $K_1(z)$ and $K_2(z)$ also satisfy Equation 9.4.

Let us now show how the original full-band signal can be reproduced using the required filters. The z-transform of the LP-filtered signal $x_1(n)$ can be expressed as:

$$X_1(z) = H_1(z)X(z) \tag{9.5}$$

or alternatively as:

$$X_1(z) = a_0 + a_1 z^{-1} + a_2 z^{-2} + a_3 z^{-3} + a_4 z^{-4} + \cdots \tag{9.6}$$

where a_i, $i = 1, 2, \ldots$, are the z-transform coefficients. Upon decimating $x_1(n)$, we arrive at $y_1(n)$, which can be written in z-domain as:

$$Y_1(z) = a_0 + a_2 z^{-1} + a_4 z^{-2} + \cdots, \tag{9.7}$$

where every other sample has been discarded and the previous even samples now become adjacent samples, which corresponds to halving the sampling rate. Equation 9.8 can also be decomposed to the following expression:

$$Y_1(z) = \frac{1}{2}\left[a_0 + a_1 z^{-1/2} + a_2(z^{-1/2})^2 + a_3(z^{-1/2})^3 + a_4(z^{-1/2})^4 + \cdots\right] \tag{9.8}$$

$$+ \frac{1}{2}\left[a_0 + a_1(-z^{-1/2}) + a_2(-z^{-1/2})^2 + a_3(-z^{-1/2})^3 \cdots\right]$$

$$= \frac{1}{2}\left[X_1(z^{1/2}) + X_1(-z^{1/2})\right], \tag{9.9}$$

which represents the decimation operation in z-domain.

9.1.4.2 Synthesis Filtering.
The original full-band signal is reconstructed by interpolating both the low-band and high-band signals, filtering them and adding them, as shown in Figure 9.6. Considering the low-band signal again, $y_1(n)$ is interpolated to give $u_1(n)$, whereby the injected new samples are assigned zero magnitude, yielding:

$$U_1(z) = a_0 + 0 \cdot z^{-1} + a_2 z^{-2} + 0 \cdot z^{-3} + a_4 z^{-4} + \cdots$$
$$= Y_1(z^2). \tag{9.10}$$

From Figure 9.6 the reconstructed low-band signal is given by:

$$T_1(z) = K_1(z)U_1(z). \tag{9.11}$$

When using Equations 9.5 to 9.11, we arrive at:

$$
\begin{aligned}
T_1(z) &= K_1(z)U_1(z) \\
&= K_1(z)Y_1(z^2) \\
&= K_1(z)\frac{1}{2}\big[X_1(z) + X_1(-z)\big] \\
&= \frac{1}{2}K_1(z)\big[H_1(z)X(z) + H_1(-z)X(-z)\big].
\end{aligned} \tag{9.12}
$$

Following similar arguments in the lower branch of Figure 9.6 as regards the high-band signal, we arrive at:

$$T_2(z) = \frac{1}{2}K_2(z)\big[H_2(z)X(z) + H_2(-z)X(-z)\big]. \tag{9.13}$$

Upon adding the low-band and high-band signals, we arrive at the reconstructed signal:

$$
\begin{aligned}
S(z) &= T_1(z) + T_2(z) \\
&= \frac{1}{2}K_1(z)\big[H_1(z)X(z) + H_1(-z)X(-z)\big] \\
&\quad + \frac{1}{2}K_2(z)\big[H_2(z)X(z) + H_2(-z)X(-z)\big].
\end{aligned}
$$

This formula can be rearranged in order to reflect the partial system responses due to $X(z)$ and $X(-z)$:

$$
\begin{aligned}
S(z) &= \frac{1}{2}\big[H_1(z)K_1(z) + H_2(z)K_2(z)\big]X(z) \\
&\quad + \frac{1}{2}\big[H_1(-z)K_1(z) + H_2(-z)K_2(z)\big]X(-z),
\end{aligned} \tag{9.14}
$$

where the second term reflects the aliasing effects due to decimation-induced spectral overlap around $f_s/4$, which can be eliminated following Esteban and Galand [223], if we satisfy for the following constraints:

$$K_1(z) = H_1(z) \tag{9.15}$$

$$K_2(z) = -H_1(-z) \tag{9.16}$$

and invoke Equation 9.4, satisfying the following relationship:

$$H_2(z) = H_1(-z). \tag{9.17}$$

Upon satisfying these conditions, Equation 9.14 can be written as:

$$
\begin{aligned}
S(z) &= \frac{1}{2}\big[H_1(z)H_1(z) - H_1(-z)H_1(-z)\big]X(z) \\
&\quad + \frac{1}{2}\big[H_1(-z)H_1(z) - H_1(z)H_1(-z)\big]X(-z),
\end{aligned}
$$

simplifying the aliasing-free reconstructed signal's expression to:

$$S(z) = \frac{1}{2}\big[H_1^2(z) - H_1^2(-z)\big]X(z). \tag{9.18}$$

If we exploit that $z = e^{j\omega T}$, we arrive at:

$$S(e^{j\omega T}) = \frac{1}{2}\big[H_1^2(e^{j\omega T}) - H_1^2(-e^{j\omega T})\big]X(e^{j\omega T})$$

and from Equation 9.3 by symmetry we have:

$$-e^{-j\omega T} = e^{j((\omega_s/2)+\omega)T}, \tag{9.19}$$

leading to:

$$S(e^{j\omega T}) = \frac{1}{2}\big[H_1^2(e^{j\omega T}) - H_1^2(e^{j((\omega_s/2)+\omega)T})\big]X(e^{j\omega T}). \tag{9.20}$$

9.1.4.3 Practical QMF Design Constraints.

Having considered the analysis/synthesis filtering, we find that the elimination of aliasing becomes more explicit in this subsection. Let us now examine how the imposed filter design constraints can be satisfied. Esteban and Galand [223] have proposed an elegant solution in the case of finite impulse response (FIR) filters, having a z-domain transfer function given by

$$H_1(z) = \sum_{n=0}^{N-1} h_1(n)z^{-n}, \tag{9.21}$$

where N is the FIR filter order. Since $H_2(z)$ is the mirror-symmetric replica of $H_1(z)$, below we show that its impulse response can be derived by inverting every other tap of the filter impulse response $h_1(n)$. Explicitly, from Equation 9.17 we have

$$\begin{aligned}
H_2(z) &= H_1(-z) \\
&= \sum_{n=0}^{N-1} h_1(n)(-z)^{-n} \\
&= \sum_{n=0}^{N-1} h_1(n)(-1)^{-n}z^{-n} \\
&= \sum_{n=0}^{N-1} h_1(n)(-1)^{n}z^{-n}, \tag{9.22}
\end{aligned}$$

which obeys the above stated symmetry relationship between the low-band and high-band impulse responses.

According to Esteban and Galand, the low-band transfer function $H_1(z)$, which is a symmetric FIR filter, can be expressed by its magnitude response $H_1(\omega)$ and a linear phase term, corresponding to the filter-delay $(N-1)$, as follows:

$$H_1(e^{j\omega T}) = H_1(\omega)e^{-j(N-1)\pi(\omega/\omega_s)}. \tag{9.23}$$

Upon substituting this linear-phase expression in the reconstructed signal's expression in Equation 9.20 and taking into account that $2\pi/\omega_s = 2\pi/(2\pi f_s) = T$, we arrive at:

$$S\left(e^{j\omega T}\right) = \frac{1}{2}\left[H_1^2(\omega)e^{-j2(N-1)\pi(\omega/\omega_s)} - H_1^2\left(\omega + \frac{\omega_s}{2}\right)e^{-j2(N-1)\pi((\omega/\omega_s)+\frac{1}{2})}\right]X\left(e^{j\omega T}\right)$$

$$S\left(e^{j\omega T}\right) = \frac{1}{2}\left[H_1^2(\omega) - H_1^2\left(\omega + \frac{\omega_s}{2}\right)e^{-j(N-1)\pi}\right]e^{-j(N-1)2\pi(\omega/\omega_s)}X\left(e^{j\omega T}\right). \tag{9.24}$$

As to whether the aliasing can be perfectly removed, we have to consider two different cases, depending on whether the filter order N is even or odd.

1. **The filter-order N is even**

   In this case we have:

$$e^{-j(N-1)\pi} = -1, \tag{9.25}$$

   since the expression is evaluated at odd multiples of π on the unit circle. Hence, the reconstructed signal's expression in Equation 9.24 can be formulated as:

$$S\left(e^{j\omega T}\right) = \frac{1}{2}\left[H_1^2(\omega) + H_1^2\left(\omega + \frac{\omega_s}{2}\right)\right]e^{-j(N-1)\omega T}X\left(e^{j\omega T}\right). \tag{9.26}$$

   In order to satisfy the condition of a perfect all-pass system, we have:

$$H_1^2(\omega) + H_1^2\left(\omega + \frac{\omega_s}{2}\right) = 1 \tag{9.27}$$

   yielding:

$$S\left(e^{j\omega T}\right) = \frac{1}{2}e^{-j(N-1)\omega T}X\left(e^{j\omega T}\right), \tag{9.28}$$

   which can be written in the time domain as:

$$s(n) = \frac{1}{2}x(n - N + 1). \tag{9.29}$$

   In conclusion, if the FIR QMF filter-order N is even, the reconstructed signal is an $N - 1$-sample delayed and 1/2-scaled replica of the input speech, implying that all aliasing components have been removed.

2. **The filter-order N is odd**

   For an odd filter-order N we have:

$$e^{-j(N-1)\pi} = 1, \tag{9.30}$$

   since the exponential term is evaluated now at even multiples of π; hence, the reconstructed signal's expression is now formulated as:

$$S\left(e^{j\omega T}\right) = \frac{1}{2}\left[H_1^2(\omega) - H_1^2\left(\omega + \frac{\omega_s}{2}\right)\right]e^{-j(N-1)\omega T}X\left(e^{j\omega T}\right). \tag{9.31}$$

   Observe that due to the symmetry of $H_1(\omega)$ we have $H_1(\omega) = H_1(-\omega)$, and so the square-bracketed term becomes zero at $\omega = -\omega_s/4$. Therefore, the reconstructed signal $S\left(e^{j\omega T}\right)$ is now different from the transmitted signal. As a consequence, perfect-reconstruction QMFs have to use even filter orders.

In conclusion, the conditions for perfect reconstruction QMFs are summarized in Table 9.2. Johnston [224] has proposed a range of perceptually optimized real QMF filter designs, which

TABLE 9.2   Conditions for Perfect Reconstruction QMF

$H_1(z)$ is a symmetric FIR filter of even order:

$$h_1(n) = h_1(N - 1 - n), \ n = 0 \dots (N - 1)$$

$H_2(z)$ is an antisymmetric FIR filter of even order:

$$h_2(n) = -h_2(N - 1 - n), \ n = 0 \dots (N/2) - 1$$

Mirror symmetry:

$$H_2(z) = H_1(-z)$$
$$h_2(n) = (-1)^n h_1(n), n = 0 \dots (N - 1)$$
$$K_1(z) = H_1(z)$$
$$K_2(z) = -H_2(z)$$

All-pass criterion:

$$H_1^2(\omega) + H_1^2\left(\omega + \frac{\omega_s}{2}\right) = 1$$

process real-time signals. Various complex quadrature mirror filters (CQMF) potentially halving the associated computational complexity have been suggested by Nussbaumer [225] and Galand [226].

Let us now apply these results to the G722 codec. The G722 analysis QMF stage is shown in Figure 9.8, where a joint tapped delay-line is used by the low-pass and high-pass stages. The filter coefficients are tabulated in Table 9.3, and the symmetry of the LP impulse response becomes explicit in the table. The antisymmetric HP impulse response of Table 9.2 is implemented using a single (-1) multiplier in Figure 9.8, which is an attractive implementation suggested by Maitre. The input speech is clocked into the shift register at a rate of 16 kHz, and decimation is implemented by outputting the split-band signals x_L and x_M at 8 kHz.

The structure of the QMF synthesis stage is shown in Figure 9.6, which also obeys the conditions summarized in Table 9.2, requiring $K_1(z) = H_1(z)$ in the lower band and $K_2(z) = -H_2(z)$ in the higher band.

Following Maitre's approach [220], the above operations can be summarized as follows:

$$x_1(j) = x_A(j) + x_B(j)$$
$$x_2(j) = x_A(j) - x_B(j), \tag{9.32}$$

TABLE 9.3   Transmit and Receive QMF Coefficient Values

h0	,	h23	.366211E-03
h1	,	h22	−.134277E-02
h2	,	h21	−.134277E-02
h3	,	h20	.646973E-02
h4	,	h19	.146484E-02
h5	,	h18	−.190430E-01
h6	,	h17	.390625E-02
h7	,	h16	.441895E-01
h8	,	h15	−.256348E-01
h9	,	h14	−.982666E-01
h10	,	h13	.116089E+00
h11	,	h12	.473145E+00

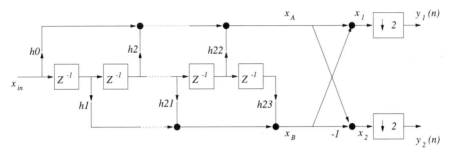

Figure 9.8 G722 QMF analysis stage.

where we have:

$$x_A(j) = \sum_{i=0}^{11} h(2i)x_{in}(j-2i)$$

$$x_B(j) = \sum_{i=0}^{11} h(2i+1)x_{in}(j-2i-1). \tag{9.33}$$

Substituting Equation 9.33 into Equation 9.32 yields

$$x_1(j) = \sum_{i=0}^{11} h(2i)x_i(j-2i) + \sum_{i=0}^{11} h(2i+1)\cdot x_{in}(j-2i-1)$$

$$= \sum_{i=0}^{23} h(i)x_{in}(j-i)$$

$$x_2(j) = \sum_{i=0}^{23} (-1)^i h(i)\cdot x_{in}(j-i) \tag{9.34}$$

In z-domain we have:

$$X_1(z) = \sum_{i=0}^{23} h(i)z^{-1}X_{in}(z) = H_1(z)x_{in}(z), \tag{9.35}$$

where

$$H_1(z) = \sum_{i=0}^{23} h(i)z^{-1} \tag{9.36}$$

and

$$X_2(z) = \sum_{i=0}^{23} (-1)^i h(i)z^{-1} = H_2(z)\cdot X_{in}(z), \tag{9.37}$$

where

$$H_2(z) = \sum_{i=0}^{23} (-1)^i h(i)z^{-1}. \tag{9.38}$$

Recall from Equation 9.9 that the decimated signals can be written as:

$$Y_1(z) = \frac{1}{2}\left[X_1(z^{1/2}) + X_1(-z^{1/2})\right]$$

$$Y_2(z) = \frac{1}{2}\left[X_2(z^{1/2} + X_2(-z^{1/2})\right] \tag{9.39}$$

and in the case of no transmission errors we get:

$$U_1(z) = Y_1(z^2) = \frac{1}{2}\left[X_1(z) + X_1(-z)\right]$$

$$U_2(z) = Y_2(z^2) = \frac{1}{2}\left[X_2(z) + X_2(-z)\right]. \tag{9.40}$$

Upon substituting Equations 9.35 and 9.37 into Equation 9.40, we arrive at:

$$U_1(z) = \frac{1}{2}\left[H_1(z) \cdot X_{in}(z) + H_1(-z) \cdot X_{in}(-Z)\right]$$

$$U_2(z) = \frac{1}{2}\left[H_2(z) \cdot X_{in}(z) + H_2(-z) \cdot X_{in}(-z)\right]. \tag{9.41}$$

As seen in Figure 9.6, the original speech is reconstructed as:

$$S(z) = T_1(z) + T_2(z)$$
$$= K_1(z) \cdot U_1(z) + K_2(z) \cdot U_2(z) \tag{9.42}$$

and taking into account Equation 9.41 and the condition $K_1(z) = H_1(z)$ and $K_2(z) = -H_2(z)$ in Table 9.2 we have:

$$S(z) = H_1(z)U_1(z) - H_2(z)U_2(z). \tag{9.43}$$

The equivalent expression in time domain using Equations 9.36 and 9.38 is:

$$s(j) = 2\left[\sum_{i=0}^{23} h(i)u_1(j-i) - \sum i = 0^{23}(-1)^i h(i)u_2(j^{-i})\right]$$

$$= 2\sum_{i=0}^{23} h(i)\left[u_1(j-i) - (-1)^i u_2(j-i)\right]$$

$$= 2\sum_{i=0}^{11} h(2i)\left[u_1(j-2i) - u_2(j-2i)\right]$$

$$+ 2\sum_{i=0}^{11} h(2i+1)\left[u_1(j-2i-1) + u_2(j-2i-1)\right]$$

$$= 2\sum_{i=0}^{11} h(2i)x_3(i) + 2\sum_{i=0}^{11} h(2i+1)x_4(i), \tag{9.44}$$

where

$$x_3(i) = u_1(j-2i) - u_2(j-2i)$$

$$x_4(i) = u_1(j-2i-1) + u_2(j-2i-1). \tag{9.45}$$

In summary, these operations justify the simplified QMF analysis/synthesis operations portrayed in Figures 9.8 and 9.9, which now necessitate two filters only. Having considered the QMF stages, let us now concentrate on the operation of the adaptive quantizers.

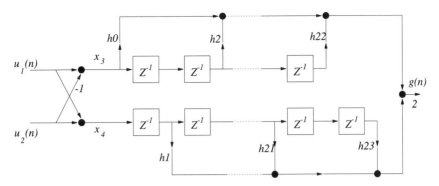

Figure 9.9 G722 QMF synthesis stage.

9.1.5 G722 Adaptive Quantization and Prediction

Let us now consider the quantization of the low-band and high-band prediction error signals $e_L(n)$ and $e_H(n)$, respectively, which are generated by subtracting the corresponding estimates $s_L(n)$ and $s_H(n)$ from the respective low-band and high-band QMF outputs:

$$e_L(n) = x_L(n) - s_L(n)$$
$$e_H(n) = x_H(n) - s_H(n).$$

Then, as mentioned before, in the low-band 60-, 30-, or 15-level quantization is used to avoid transmitting long strings of the all-zero codeword, while the high-band is quantized with 2 bits/sample. For the sake of brevity we do not detail the quantization tables containing the output codewords and decision levels for both sub-bands. For details on these issues, the interested reader is referred to the Recommendation G722.

In the embedded codec, the truncated 4-bit low-band codeword I_{Lt} is converted to the truncated locally decoded low-band difference d_{Lt} using the 4-bit low-band inverse quantizer Q_{L4}^{-1} of Figure 9.10 and Table 9.4 and scaling it by $\Delta_L(u)$ as follows:

$$d_{Lt}(n) = Q_{L4}^{-1}[I_{Lt}(n)] \cdot \Delta_L(n) \cdot \text{sgn}[I_{Lt}(n)],$$

where the $\text{sgn}[I_{Lt}(n)]$ function indicates the sign of the low-band prediction error $e_L(n)$. The untruncated high-band difference signal is regenerated similarly using an analogous formula.

The quantizer scaling factors are updated in the logarithmic domain in order to maintain a high dynamic range, and then they are converted to the linear domain using a lookup table. The logarithmic scaling factors $\nabla_L(n)$ and $\nabla_H(n)$ are computed using the following recursive relationships:

$$\nabla_L(n) = \frac{127}{128} \cdot \nabla_L(n-1) + W_L[I_{Lt}(n-1)]$$

$$\nabla_H(n) = \frac{127}{128} \cdot \nabla_H(u-1) + W_H[H(n-1)],$$

where W_L and W_H are the **logarithmic scaling factors** and the low-band factor W_L is given in Table 9.4 for the sake of illustration. This expression is similar to the corresponding scaling factor update formula of Equation 2.4 used in the previously detailed G721 32 kbps ADPCM codec, although the logarithmic scalers of Tables 2.3 and 9.4 are different. On the same note, the G722 codec uses a **leakage factor** of $p = (127/128)$, while the G721 scheme used $\beta = \frac{31}{32}$, implying a somewhat higher innate robustness or tolerable bit error rate in case of the

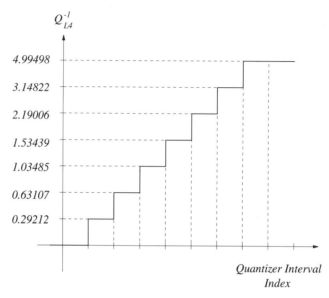

Figure 9.10 The 4-bit low-band inverse quantizer Q_{L4}^{-1}.

latter. The logarithmic scaling factors are limited to the range: $0 \leq \nabla_L(n) \leq g$, $0 \leq \nabla_H(n) \leq 11$, and they are converted to the linear domain using an approximation to the inverse of the $\log_2(\cdot)$ function, namely:

$$\Delta_L(n) = 2^{(\nabla_L(n)+2)} \cdot \Delta_{\min}$$

$$\Delta_H(n) = 2^{\nabla_H(n)} \cdot \Delta_{\min}$$

where $\Delta_{\min}$ was set to half the step-size of the 14-bit A/D converter used, which minimized the codecs' idle noise.

The **adaptive predictor** used in the G722 codec is identical to the G721 two-pole, six-zero ARMA arrangement, detailed in Section 2.7. Specifically, Equations 2.66–2.71 have to be used in order to periodically update the predictors using a simplified gradient algorithm.

In closing, we note that the low-band decoder can make use of the mode information, which can be inferred by the data extraction unit preceding the G722 decoder. The G722 codec can operate without the availability of this side-information, albeit at the cost of some performance degradation.

TABLE 9.4 Lower-band Inverse Quantizer (Q_{L4}^{-1}) and Logarithmic Scale-factor (W_L) Characteristic

Quant. Int Index	Q_{L4}^{-1}	W_L
1	0	−0.02930
2	0.29212	−0.01465
3	0.63107	0.02832
4	1.03485	0.08398
5	1/53439	0.16309
6	2.19006	0.26270
7	3.14822	0.58496
8	4.99498	1.48535

9.1.6 G722 Coding Performance

Since the standardization of the G722 codec compression technology has made substantial advances, it has not yet led to lower-rate wideband speech coding standards.

9.2 WIDEBAND TRANSFORM CODING AT 32 kbps [227]

9.2.1 Background

In this section we briefly consider a scheme proposed by Quackenbush, which processes 7 kHz bandwidth speech sampled at 16 kHz [227]. This codec achieves a compression of 8 : 1, when compared to the near-transparent quality 16-bit PCM input signal and hence transmits at 2 bits/sample or 32 kbps. In his contribution Quackenbush adopted the transform-coding approach proposed by Johnston [228] for audio signals and reduced the bitrate required.

9.2.2 Transform-Coding Algorithm

The codec's schematic is shown in Figure 9.11, which processes 240-sample blocks, corresponding to 15 ms at a sampling rate of 16 kHz, and concatenates 16 samples from the previous block, yielding an overall block length of 256 samples. This block is then windowed, smoothing the first 16 samples of the block at both ends. This 256-sample real signal is then transformed to 128 complex coefficients using the fast fourier transform (FFT), where the coefficients are quantized for transmission.

The codec distributes the quantization noise in the spectral domain so that its perceptual effects are minimized by adjusting the signal-to-quantization noise ratio appropriately across the frequency band. This process could also be conveniently carried out using a QMF stage to split the frequency band into the required width sub-bands, as we have seen in the case of the G722 codec. Quackenbush [227] here followed Scharf's suggestions [8] in order to determine the tolerable noise threshold T_i for the frequency band i, where i is the **critical band** index. Quackenbush opted for a fixed, rather than dynamically adjusted, critical band energy evaluation, determining the energy C_i for band i from the long-term analysis of speech. According to Scharf's suggestion, the noise threshold can be adjusted to

$$T_i = 14.5 + i[\text{dB}]$$

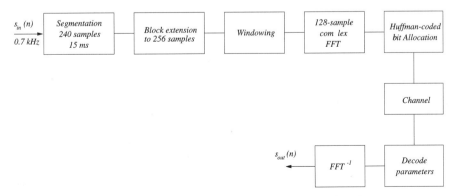

Figure 9.11 Transform-coded wideband speech codec schematic.

below the signal energy C_i, while inflicting negligible perceptual distortion. This simple masking model allows us to determine the required bit allocation as a function of frequency, which is carried out dynamically, using an iterative procedure.

In each frame, 26 bits have a time-invariant assignment, allocating 16 bits to the lowest-frequency FFT bin, 4 bits to indicate the number of iterations during the bit-allocation process, which can be accordingly 16, 2 bits for the selection of one of four Huffman codebooks, and 4 bits for frame-synchronization. Therefore, assuming 2 bits/sample coding, there are $480 - 76 = 454$ bits for dynamic spectral-domain coding of the FT coefficients. The bit-allocation scheme is summarized in Table 9.5.

Following an initial tentative bit allocation, the iterative bit allocation is activated. The FFT spectrum is subdivided into 16 frequency bands, and "sub-bands" $k = 1 \dots 11$ are assigned 6 FFT spectral lines, while "sub-bands" $k = 12 \dots 16$ are allocated 12 spectral lines. The above $11.6 + 5.12 = 126$ spectral lines are encoded by the iterative technique to be highlighted, line 0 has a fixed allocation of 16 bits and line 127 is not encoded. Then the maximum spectral magnitude M_k of each "sub-band" k is found and quantized logarithmically, yielding

$$m_k = \{\log_2 M_k\} \qquad k = 1 \dots 16,$$

where $\{\cdot\}$ indicates the smallest integer greater or equal to $\cdot$ and an, m_k-bit, quantizer is needed in "sub-band" k.

Once the "sub-band" spectral maxima and the spectral lines are encoded, the bitrate economy can be further improved using Huffman coding. Two sets of codebooks are used for both the maxima and the spectral lines. Quackenbush argues that this technique does not dramatically improve the average bitrate overall, but it reduces the peak rate.

Huffman coding has been treated in a range of classic books, such as, for example, Reference [10] by Jayant and Noll. Huffman coding is a simple practical technique, which arranges the messages to be encoded in descending order and assigns a variable-length code to them on the basis of their probability of occurrence. Specifically, more frequent messages are encoded using a low number of bits, while infrequent ones can be transmitted using longer codes. Overall, if the source-message probabilities vary over a wider range, the average bitrate is significantly reduced at the cost of some coding complexity and delay.

Returning to the process of Huffman coding the spectral maxima, Quackenbush trained a separate codebook for the 16 "sub-bands". For the spectral lines, a more complex scheme was used, where 1, 2, or 3 complex spectral lines were concatenated into a single word before invoking a specific Huffman coding table. Choice of the Huffman coding table was governed by the value of m_k; hence no side-information had to be sent to the decoder, since m_k was transmitted for all sub-bands. The choice of Huffman codebooks is summarized in Table 9.6.

TABLE 9.5   Bit-allocation Table for 32 kbps Wideband Transform Codec

Parameter	No of bits/15 ms
Lowest freq. FFT bin	16
No of bit alloc iterations	4
Selection of Huffman codebook	2
Frame synchr.	2
FFT coefficients	454
Total	480 bits/15 ms = 32 kbps

TABLE 9.6 Dependence of Huffman Coding Scheme on Sub-band Maxima, ©
IEEE, Quackenbush, 1991

Condition	No. of Quant. Levels	Codebook	Complex vector-length
$15 < |m_k|$	1771	5	1 (real)
$7 < |m_k| \leq 15$	31	4	1
$3 < |m_k| \leq 7$	15	3	1
$1 < |m_k| \leq 3$	7	2	2
$0 < |m_k| \leq 1$	3	1	3

Specifically, one of five Huffman codebooks is invoked in each of the sub-bands, depending on $|m_k|$, and all spectral lines belonging to this band are encoded by the same book. If $|m_k| = 0$, no bits are allocated to the given band. Furthermore, observe in the table that if $|m_k| > 15$, then the real and imaginary spectral lines are encoded separately by Codebook 5, which is similar to Codebook 4, but the encoded values are limited to $[7 \dots 14]$.

Quackenbush used the following iterative codebook design approach. Initially, a set of simple linear scalar quantizers was used in order to generate a histogram of the quantities to be encoded, such as the sub-band maxima and the spectral line, and the generated bitrate was estimated. Then Huffman codebooks were generated on the basis of these tentative histograms for both the sub-band maxima and spectral lines. These codebooks were then used in a subsequent session to estimate the bitrate again, and lastly a new set of histograms was generated again. These iterations can be repeated a number of times in order to arrive at a near-optimum set of Huffman codebooks.

As mentioned before, there are two Huffman codebooks for both the sub-band maxima and the spectral lines. First, m_k is encoded tentatively invoking both codebooks for each sub-band maxima, and the one resulting in a lower bitrate is selected. This side-information flag is also signaled to the decoder. Then depending on $|m_k|$, the corresponding set of twin codebooks of Table 9.6 is used, checking the generated number of bits due to both of the twin codebooks. The number of bits generated is also stored.

Quackenbush also suggested an iterative bitrate control mechanism for maintaining a rate of 32 kbps or 480 bits/15 ms. If after the first encoding pass the bitrate is not between 1.9 and 2 bits/sample, corresponding to $456 \dots 480$ bits per frame, then a new coding cycle ensues. An integer scaling factor m was introduced to control the number of quantizer levels used, and on each iteration the quantized spectral lines were scaled by a factor of $2^{(1/m)}$, until the number of bits generated was between 456 and 480. If the number of bits generated is too low, the number of quantization levels is increased and vice versa.

The value of the scaling factor m is determined as follows. If the number of bits produced by the initial spectral line quantization is below 456, m is set to 1, otherwise to -1. This would imply a spectral-line quantizer up- or down-scaling by a factor of 2 or $1/2$ respectively. During the subsequent iterations it is observed whether the direction of scaling is maintained; if so, the preceding value of m is retained. If, however, the direction of scaling has to change, since due to a previous correction step now the number of bits generated deviates from the target in the opposite direction, the value of m is doubled and the sign of it is toggled, before the next iteration takes place.

As an example, let us assume that $m = 1$ was used in the last iteration, and as an effect of scaling by $2^{1/m} = 2$ the number of bits became too high. Now it would not improve the bitrate iteration to set $m = -1$ again, since that would reduce the bitrate exactly to that value, which activated the choice of $m = 1$, requiring a bitrate increase. Hence, m is doubled to $m = 2$, and its sign is toggled, leading to a scaling by $2^{(1/m)} = 2^{-1/2} = 1/\sqrt{2} = \sqrt{2} \approx 1.41$.

This sequence of iterative scaling operations is encoded using 4 bits, allowing for a selection of 16 possible consecutive scaling protocols to take place. The iterative bit-allocation process is concluded, when either the number of bits falls between 456 and 480, or the maximum allowed number of iterations took place. The total bitrate can be maintained at 32 kbps, but in some cases the total bitrate budget may not be fully exploited.

9.3 SUB-BAND-SPLIT WIDEBAND CELP CODECS

9.3.1 Background

Since CELP codecs are so successful in coding narrowband speech signals, they have also been employed in wideband coding. A simple and realistic alternative is to invoke the previously described CCITT G728 16 kbps low-delay narrowband codec and to operate it at an increased sampling rate of 16 kHz rather than at 8 kHz, which would result in a bitrate of 32 kbps. However, better results can be achieved if the codec is designed specifically for wideband applications, since the efficient encoding of high frequencies present in the 4–7 kHz band requires special attention. Ordentlich and Shoham proposed a low-delay Celp-based 32 kbps wideband codec [229], which achieved a similar speech quality to the G722 64 kbps codec at a concomitant higher complexity.

A philosophy similar to that of the backwardly adaptive G728 16 kbps codec was proposed by Ordentlich and Shoham, and two codec versions were tested, one with and one without LTP, although their preferred codec refrained from using LTP. The backward-adaptive LPC filter had an order of 32, which was significantly lower than the filter order of 50 used in the G728 codec. Recall that the G728 filter order of 50 was able to cater for long-term periodicities of up to 6.25 ms, corresponding to pitch frequencies down to 160 Hz without a LTP, allowing better reconstruction for female speakers. The filter order of 32 at a sampling frequency of 16 kHz cannot cater for long-term periodicities.

In contrast to the G728 codebook of 128 entries here, 1024 entries were used to model the 5-sample excitations. Let us now examine how the bitrate can be further reduced using split-band coding.

9.3.2 Sub-Band-Based Wideband CELP Coding

9.3.2.1 Motivation. One problem associated with full-band coding of wideband speech is the codec's inability to treat the less predictable high-frequency, low-energy speech band, which was tackled by the G722 codec using split-band coding. Although this band is important for maintaining an improved intelligibility and naturalness, it only contains a small fraction of the speech energy; therefore, its bitrate contribution has to be limited appropriately. In a contribution by Black, Kondoz, and Evans [137], the backward-adaptive principle was retained for the sake of low delay, but it was combined with a split-band approach. This is motivated mainly by the fact that in a full-band CELP codec the excitation is typically chosen on the basis of providing good low-frequency regeneration, since the majority of the energy resides in that band. Hence, the lower-energy high-frequency region may not be treated adequately in full-band CELP codecs, unless appropriate measures are taken, such as choosing vast codebooks, which then require sophisticated measures in order to mitigate their complexity.

It is well understood that in backward-adaptive narrowband codecs, such as the G728 scheme, an LPC frame update rate of 2 ms is sufficiently frequent in order to achieve similar LPC prediction gains to forward-adaptive arrangements. However, when using a similar

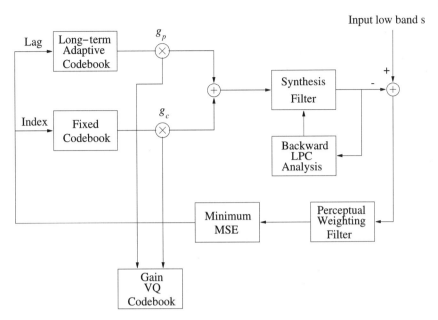

Figure 9.12 Low-band encoder in 16 kbps subband-CELP wideband Codec © Black et al. [137]

update rate in wideband coding, the high-frequency spectrum above 4 kHz was reported to have been distorted [137] due to the dominantly low-frequency-matched synthesis process. Black, Kondoz, and Evans have found that the backward-adaptive LPC spectrum, or the spectral envelope, often exhibited a higher energy toward high frequencies than the forward-adaptive spectrum, which was again attributable to the predominantly low-frequency matching. All in all, the sub-band approach has a number of advantages for wideband coding, allowing the codec to restrict quantization error spillage from one band to the other. Hence Black et al. [137] favored this technique. The previously described G722 standard QMF band-splitting scheme was used, and the proposed low-band and high-band schemes are depicted in Figures 9.12 and 9.13, both of which now operate independently at an 8 kHz sampling rate.

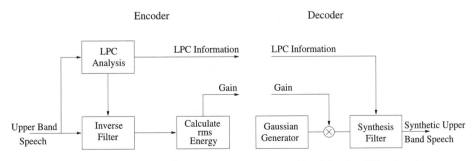

Figure 9.13 High-band encoder/decoder in 16 kbps subband-CELP wideband codec © Black et al. [137].

9.3.2.2 Low-Band Coding. The low-band was encoded by a backward-adaptive CELP codec using a tenth-order LPC filter updated over 148 kHz sampled samples, or 1.75 ms. This narrowband LPC analysis was free from the high-band (> 4 kHz) spectral envelope distortion problem of backward-adaptive wideband codecs. For the preferred innovation sequence length of 14 samples, the authors argued that it was necessary to incorporate a forward-adaptive LTP in order to counteract the potentially damaging error feedback effect of the backward-adaptive LPC analysis. A conventional perceptual weighting filter was employed, and noninteger LTP delays were incorporated. Specifically, a resolution of $1/3 \cdot 1/8$ kHz $\approx 41.67 \mu$s was used between LTP delays of $19\frac{1}{3}$ and $84\frac{2}{3}$, while in the range $85 \ldots 143$ no oversampling was utilized. The LTP delay was represented by 8 bits. Black et al. first initiated a closed-loop synthesis for all integer delays. If the delay found fell in the high-resolution region, a range of fractional delays surrounding the identified integer position was also tested, which was found to improve the codec performance for female speakers. In contrast to the G728 trained 128-entry codebook, a 256-entry fixed stochastic codebook was used in the low-band, which contained overlapping entries.

9.3.2.3 High-Band Coding. The upper-band typically contains a less structured, noise-like signal, which has a slowly varying dynamic range. Black et al. here proposed to use a sixth-order forward-adaptive predictor updated over a 56-sample interval, which is quadrupled in comparison to the low-band. Backward-adaptive prediction would be unsuitable for this less accurately quantized band, which would precipitate the effect of quantization errors in future segments. This crude LPC analysis did not attempt to give a waveform-matching representation of the upper-band signal; it merely endeavored to model its spectral envelope. Therefore, the decoder regenerated the high-band signal by exciting the LPC synthesis filter using a scaled random zero-mean, unit-variance excitation pattern. The magnitude of this vector was determined by the encoder upon inverse-filtering the high-band signal using the sixth-order LPC filter and calculating the energy of the residual over 56 samples. Listening tests confirmed that the excitation gain was typically too high, in particular for voiced sounds. Hence this gain factor was scaled by 0.5, which was then quantized and transmitted to the decoder. Let us now consider the parameter quantization schemes proposed by Black et al.

9.3.2.4 Bit-Allocation Scheme. In the backward-predictive low-band, no LPC spectral information is transmitted, and so all the bits are assigned to the frequently updated fixed codebook and adaptive codebook parameters. The fixed codebook gain can be predicted by a technique proposed by Soheili [230] which was referred to as backward average mean smoothing, where the current gain is predicted by the average of the preceding three quantized gains. This predicted gain G_p was then used to normalize the current stochastic codebook gain determined during the excitation optimization. This normalized fixed codebook gain was jointly vector-quantized with the LTP gain in a closed-loop optimization process. Similar schemes were discussed in Section 6.5.2.5. Black et al. used the Linde-Buzo-Gray (LBG) [213, 217] clustering algorithm for training the joint gain codebook.

The six high-band LPC coefficients were transformed to LSFs and vector-quantized with a total of 12 bits, while the random excitation vector gains were quantized with 4 bits. The overall bit-allocation scheme is portrayed in Table 9.7.

Informal listening tests showed that the codec had a similar performance to the G722 scheme at 48 kbps.

TABLE 9.7 Bit allocation of 16 kbps SP-CELP Wideband Codec © Black et al. [137]

Parameter	Bits	Update [ms]	Bitrate [bps]
Low-band			
LTP delay	8	1.75	4571.4
Codebook index	8	1.75	4571.4
Gain VQ	8	1.75	4571.4
High-band			
LSFs	12	7	1714.4
Gain	4	7	571.4
Total			16000

A range of further attractive wideband schemes was proposed by the prestigious speech coding group at Sherbrook University, who have contributed a plethora of successful narrowband and wideband ACELP codecs to the state of art.

9.4 FULLBAND WIDEBAND ACELP CODING

9.4.1 Wideband ACELP Excitation [138]

One difficulty associated with wideband CELP coding without band-splitting is that upon doubling the sampling rate and hence the bitrate, while maintaining the same relative bitrate contribution for all parameters, as in narrowband schemes, the codec's complexity may become excessively high. For example, assuming a forward-adaptive codec and a 10-bit codebook for narrowband coding, the corresponding 20-bit wideband codec would be unrealizably complex, requiring the generation of the synthetic speech for $2^{20} = 1\,048\,576$ codebook entries. Hence, suboptimum approaches, such as multi-stage codebooks or split-band coding must be used.

In Reference [138] Laflamme, Adoul Salami, et al. argued, however, that ACELP codecs are amenable to wideband coding, when employing a **focused codebook search strategy** using a number of encapsulated search loops. This technique facilitates searching only a fraction of a vast codebook, while achieving a similar performance to that of a full search. Without repeating the algorithmic details, this technique was also proposed by the authors for the CCITT G729 8 kbps low-delay codec using a 15-bit ACELP codebook and five encapsulated loops (described earlier in Section 7.8).

As one would expect, according to the 16 kHz sampling frequency, the authors doubled the length of the excitation vectors to 80 samples, corresponding to a 5 ms excitation optimization subframe. A codebook size of 2^{20} was proposed, which can be realistically invoked with the proviso of using the focused search strategy, and the excitation pulse magnitudes were fixed to 1, −1, 1, −1, 1 implying that 5 pulses per excitation vector were used. Assuming that each pulse can occupy 16 legitimate interlaced positions, the 5 pulses are encoded by a total of $4 \cdot 5 = 20$ bits, yielding a 20-bit codebook. The codebook structure can be described more explicitly as [138]:

$$c(n) = \sum_{i=0}^{4} b_i \delta(u - m_i) n = 0 \dots 79,$$

where

$$b_i = \begin{cases} +1 & \text{for } i = \text{even} \\ -1 & \text{for } i = \text{odd} \end{cases} \tag{9.46}$$

are the excitation pulse amplitudes and m_i the legitimate pulse locations given by

$$m_i^{(j)} = i + 5j, \; i = 0 \ldots 4, \; j = 0 \ldots 16. \tag{9.47}$$

As mentioned before, Adoul et al. [144] and Xydeas et al. [143] offered a plausible geometric interpretation of different CELP codebooks by allocating the zero-mean unit-variance codebook vector to the surface of a unit-radius sphere. They invoked this useful "visual aid" in supporting the perceptual equivalence of different excitation models, populating the surface of the N-dimensional hyper sphere by randomly or uniformly spaced excitation vectors, where N represents the length of the excitation patterns used. Since due to Equation 9.47 any of the 80 excitation pulse positions can host a pulse, Laflamme et al. [138] noted that for $N = 80$ and 5 pulses per vector the synthesized number of 5 ms audio segments becomes $C_5^{80} = 80!/75!5! \approx 24.04 \cdot 10^6$ out of the potentially possible $2^{80} \approx 1.21 \cdot 10^{24}$ segments that would be generated by the full search of a fully populated, that is, nonsparsed 80-pulse binary excitation codebook. The proposed ACELP codebook ensures sufficiently dense coverage of the excitation vector space, while reducing the number of search operations by a factor of $\sim 5 \cdot 10^{16}$.

The ACELP codebook search is inherently structured, which alleviates its real-time implementation by referring to Equations 6.8 and 6.9. In Section 6.4.3, it was argued that updating $\tilde{C}_k$ and ξ_k for the testing of a new excitation vector becomes very efficient, if always only one pulse position is updated upon cycling through the legitimate set of excitation vectors. In general, when there are p legitimate pulse positions for each pulse, p nested loops can be created for this recursive search technique. Nonetheless, even this efficient update technique is excessively complex for a codebook of 2^{20} entries. Hence Laflamme et al. [138] suggested a focused search strategy similar to that which the Sherbrooke Laboratory, CNET, and NTT proposed for the G729 codec. The G729 codec's ACELP search strategy was hence adopted in their 16 kbps wideband codec.

The philosophy behind this focused search technique is to quantify the chances of each particular excitation subset to contain the optimum excitation vector. Upon testing the incremental effect of each newly included excitation pulse from the set of p pulses of a vector as regards the overall weighted error of this specific excitation vector, it becomes possible to quantify the chances of this vector leading to the minimum error over the whole codebook without actually adding all p pulses. Specifically, after adding, say, 3 to 4 pulses, the weighted error can be tested against an experimentally optimized threshold inferred from the statistical evaluation of the weighted error of the best vectors after entering 3–4 rather than 5 nested loops.

In order to be more explicit, the error term of Equations 6.14 and 6.15 must be minimized over the codebook by maximizing its second term, given by [47, 139]:

$$\tau_k = \frac{(\tilde{C}_k)^2}{\xi_k}.$$

The higher the ratio $\tilde{C}_k$ in Equation 6.15, the lower the wmse, which facilitates the focused search. Equations 6.17 and 6.18, which were valid for $b_i = +1, -1, +1, -1, p = 4$

pulses and 4 nested loops per excitation vector, can be reformulated to reflect the situation $b_i = +1, -1, +1, -1, +1, p = 5$ pulses and 5 loops as follows:

$$\tilde{C}_k = 4(m_0) - 4(m_1) + 4(m_2) - 4(m_3) + 4(m_5) \tag{9.47}$$

$$\begin{aligned}
\xi_k = {} & \phi(m_0, \ m_0) \\
& + \phi(m_{11}m_1) - 2\phi(m_1 m_0) \\
& + \phi(m_{21}m_2) + 2\phi(m_{21}m_0) - 2\phi(m_{21}m_1) \\
& + \phi(m_3, \ m_3) - 2\phi(m_3, \ m_0) + 2\phi(m_{31}m_1) - 2\phi(m_{31}m_2) \\
& + \phi(m_{41}m_4) + 2\phi(m_{41}m_0) - 2\phi(m_{41}m_1) + 2\phi(m_41m_2) - 2\phi(m_{41}m_3).
\end{aligned} \tag{9.48}$$

Recall that physically C_k is the cross-correlation between the target vector $\mathbf{X}$ and the filtered excitation vector $\mathbf{HC_k}$, while ξ_k is the energy of the filtered codeword $\mathbf{HC_k}$.

Laflamme et al. [138] evaluated the ratio $\tau_k = (\tilde{C}_k)^2/\xi_k$ in Equation 6.15 at every stage, when considering the cumulative effect of including one pulse at a time out of the p legitimate pulses, which allowed the authors to derive a set of thresholds for the consecutive search stages. The proportion of the total set of legitimate excitation vectors over which the search is carried out can be controlled by eliminating particular vector subsets from further search, if they fail to produce "promising" $\tau_k = (\tilde{C}_k)^2/\xi_k$ ratios after adding 3 to 4 excitation pulses. The higher this statistically optimized threshold at stages 3–4, the higher the proportion of eliminated vectors and the lower the search complexity. This naturally increases the chances of occasionally prematurely eliminating certain excitation subsets. In most cases, however, the second best vector will still be retained, and the reward of reduced complexity far outweighs the inflicted slight performance penalty.

Laflamme et al. [138] quantified the associated wideband speech quality degradation, which is shown in Table 9.8 for various proportions of the codebook, which the authors adjusted using two statistically optimized thresholds at stages 3 and 4. Observe in the table that upon searching a mere 0.05% of the $2^{20} \approx 10^6$ entry codebook, a relatively low SNR degradation of 0.4 dB was inflicted, while reducing the search complexity by a factor of 2000. This would correspond to the full search of a 512-entry, 9-bit address codebook, while maintaining a SNR in excess of 21 dB. In general, the number of threshold controlled search operations varies on a frame-by-frame basis, which results in a time-variant implementational complexity. In order not to hamper real-time implementations, the search must be curtailed, once a predefined maximum number of operations is reached.

TABLE 9.8 Wideband ACELP Speech Degradation Versus Fraction of Codebook Searched © IEEE Laflamme et al. [138]

Search Complexity (%)	SNR dB
100	22.2
4	22.14
1.6	22.0
0.2	22.05
0.15	21.83
0.05	21.8
0.03	21.5

Although Laflamme et al. [138] outlined a tentative bit-allocation scheme in their treatise, they have moved on to propose slightly different wideband ACELP codecs [231].

9.4.2 Backward-Adaptive 32 kbps Wideband ACELP [231]

In [231] the Sherbrooke team attempted to contrive a backward-adaptive predictive CELP codec, which used a conventional backward-adaptive CELP schematic, except for the fact that two excitation generators were employed. Both excitation generators had a separate gain factor and were constituted by a modified ACELP-type codebook, where each binary pulse could take an arbitrary sign. Hence, the excitation vector retained a higher flexibility in terms of pulse amplitude, and the codebook search required $2d$ nested loops in order to arrive at the optimum excitation.

In their backward-adaptive ACELP codec, the authors used a 32nd-order LPC filter and a 3-tap pitch predictor, both of which were updated every 2 ms, corresponding to 32 samples at a sampling rate of 16 kHz. They used a windowing function similar to that of the 16 kbps G728 codec of Chapter 8. The input speech was also preemphasized. The excitation frame length was 16 samples or 1 ms, hosting 4 pulses per excitation vector; each had four legitimate locations encoded by 2 bits and magnitudes of ± 1. Therefore, each pulse required a total of 3 bits, and each vector holding 4 pulses needed 12 encoding bits. The two codebook gains were assigned 4 bits each. Therefore, the two codebooks were allocated a total of $2 \times (12 + 4) = 32$ bits, 1 ms, yielding a bitrate of 32 kbps, while maintaining a delay of 1 ms. The codecs' bit-allocation scheme is summarized in Table 9.9. Sanchez-Calle et al. noted that the achieved speech quality was similar to that maintained by the previously described 16 kbps scheme of Reference [138], but the ability of the low-delay 32 kbps backward-adaptive scheme to encode music rather than speech was superior. The achieved SEGSNR of the 32 kbps codec was in the range of 20–22 dB for wideband speech signals.

9.4.3 Forward-Adaptive 9.6 kbps Wideband ACELP [139]

In a further contribution, Salami, Laflamme, and Adoul [139] returned to the forward-adaptive ACELP philosophy, while using the previously described dual-codebook ACELP structure. A range of innovative techniques was proposed in order to mitigate the codec's complexity, escalating due to the doubled sampling rate. Specifically, the real-time wideband codec has half the time, namely, $1/(16 \text{ kHz}) = 62.5 \text{ μs}$ to process twice as many samples in comparison to conventional narrowband codecs, assuming a certain fixed analysis interval duration.

Here we restrict ourselves to a portrayal of the proposed bit-allocation scheme, which is summarized in Table 9.10. The LPC update frame length was 30 ms, and a filter order of 16

TABLE 9.9 Bit allocation of 32 kbps Backward-Adaptive Full-band Wideband ACELP Codec © IEEE Sanchez-Calle et al., 1992 [231]

Parameter	No of bits/1 ms	Bitrate (kbps)
Codebook Index 1	$4 \cdot (2 + 1) = 12$	12
Codebook Index 2	$4 \cdot (2+) = 12$	12
Codebook Gain 1	4	8
Codebook Gain 2	4	8
Total	32	32

TABLE 9.10 Bit-allocation Scheme of Wideband Forward-adaptive Full-band ACELP Codec © IEEE Salami, 1992 [139]

Parameter	Update (ms)	No. of Bits	Bitrate (kbps)
LPC filter	30	54	1.8
LTP delay	6	8	1.33
LTP gain	6	4	0.67
Codebook ind. 1	6	12	2
Codebook ind. 2	6	13	2.17
Codebook gain 1	6	6	1
Codebook gain 2	6	3	0.5
Padding bits	30	4	0.13
Total	30	$54 + (5 \cdot 46) + 4 = 288$	9.6 kbps

was used, quantizing the LSFs with a total of 54 bits using a channel capacity of $54/30$ ms $= 1.8$ kbps. There were five 6 ms excitation optimization subsegments, constituted by 96 62.5 µs-spaced samples. The pitch-delay was restricted to the range $40 \ldots 295$. Accordingly, 8 bits were used for its encoding. The LTP gain was quantized with 4 bits.

The two codebooks in this scheme are different from each other. The first one contains 4 conventional $+1, -1, +1, -1$ interlaced pulses per 6 ms, or 96 sample excitation optimization vectors, where the pulse positions m_i were defined as [139]:

$$m_i^{(j)} = 3i + 12j, \ i = 0 \ldots 3, \ j = 0 \ldots 7.$$

As can be inferred from the above equation, there are eight possible positions for each interlaced pulse. Hence a total of $4.3 = 12$ bits per 6 ms excitation vector are needed for their encoding. The associated bitrate contribution is 12 bits/6 ms $= 2$ kbps. In order to maintain a near-constant implementational complexity, the fourth encapsulated loop is entered at most 64 times, and a maximum total of 512 excitation vectors are used out of the possible $2^{12} = 4096$ sequences. The first codebook gain was encoded using 6 bits/6 ms, yielding a bitrate contribution of 1 kbps.

After taking into account the contribution of the ACELP codebook, the second codebook has to model an essentially random process. Hence, this codebook has a simple structure, populated with regularly spaced binary pulses. In order to incorporate some flexibility in this codebook, the excitation pulses ± 1 were spaced at positions $k + 9 \cdot n$, where the initial grid position k can take the values $k = 0, 1, 2,$ and 3 and $n = 0, 1, 2 \ldots 10$. Hence, there are four possible initial grid positions, and a total of 11 pulses are allocated with a spacing of 9. Therefore, the second codebook requires a total of $(11 + 2) = 13$ coding bits per 96-bit subsegment, yielding a bitrate contribution of 2.17 kbps. The second codebook gain was quantized relative to the first gain, using 3 bits per subsegment, requiring a channel capacity of 0.5 kbps. Finally, 4 padding bits per 30 ms LPC update frame were used, giving a total of $54 + (5 \cdot 46) + 4 = 288$ bits per 30 ms, corresponding to a bitrate of 9.6 kbps. This codec was reported to have an SNR of around 17 dB, while in perceptual terms a higher rate, 14 kbps version of it was formally found equivalent to the G722 SB-ADPCM codec operated at 56 kbps. Recall that the 16 kbps SB-CELP codec proposed by Black et al. [137], which was described earlier in this chapter, was deemed to have a similar performance to the 48 kbps G722 operating mode.

Roy and Kabal [232] made a comparative study of a conventional single- and dual-codebook CELP codec for wideband speech coding. In harmony with Salami et al. [139], they also found that the dual-codebook arrangement was more natural-sounding, although their findings were based mainly on experiments carried out without quantization of the filter parameters.

9.5 A TURBO-CODED BURST-BY-BURST ADAPTIVE WIDEBAND SPEECH TRANSCEIVER[1]
T. KELLER, M. MÜNSTER, L. HANZO

9.5.1 Background and Motivation

Burst-by-burst adaptive quadrature amplitude modulation (AQAM) transceivers [135] have recently generated substantial research interest within the wireless communications community [233–241]. The transceiver reconfigures itself on a burst-by-burst basis, depending on the instantaneous perceived wireless channel quality. More explicitly, the associated channel quality of the next transmission burst is estimated, and the specific modulation mode, which is expected to achieve the required performance target, is then selected for the transmission of the current burst. In other words, modulation schemes of different robustness and of different data throughput are invoked. In the event of expected error burst due to a low expected instantaneous channel quality, the transmitter can also be temporarily disabled, while the data is delayed and buffered, until the channel quality improves, provided that the associated delay is not excessive for the service supported. Because of this feature, the distribution of channel errors becomes typically less bursty than in conjunction with nonadaptive modems. This is an attractive feature in conjunction with channel codecs, resulting in potentially increased coding gains [242–245]. Furthermore, the soft-decision channel codec metrics can also be invoked in estimating the instantaneous channel quality. Recently, block turbo-coded AQAM transceivers have also been proposed for dispersive wideband channels in conjunction with conventional decision feedback equalisers [242–245] (DFE), where the mean squared error (MSE) at the DFE's output was used as the channel quality metric, controlling the choice of modes. An alternative neural-network radial basis function (RBF) DFE-based AQAM modem design was proposed in [246], where the RBF DFE provided the channel quality estimates for the modem mode switching regime.

Further recent work on combining various conventional channel coding schemes with adaptive modulation has been reported by Matsuoka et al. [247], Lau et al. [248], and Goldsmith et al. [249]. For data transmission systems, which do not necessarily require a low transmission delay, variable-throughput adaptive schemes can be devised, which operate efficiently in conjunction with powerful error correction codecs, such as long block-length turbo codes [192, 193]. By contrast, fixed-rate burst-by-burst adaptive systems, which sacrifice a guaranteed bit error rate (BER) performance for the sake of maintaining a fixed data throughput, are more amenable to employment in the context of low-delay interactive speech and video communications systems. The above burst-by-burst adaptive principles can also be extended to adaptive orthogonal frequency division multiplexing (AOFDM) schemes [250] and to adaptive joint-detection based code division multiple access (ACDMA) arrangements [251].

OFDM was first proposed by Chang in his 1966 paper [252] and revived by Cimini's often cited paper [253], but it was not developed to its full potential until the 1990s, when a whole host of contributions appeared, for example, in [254]. Other developments were due to Rohling et al. [255] at the University of Hamburg, Huber et al. at Erlangen University [256], Meyr et al. [257, 258], at Aachen University, Jones, Wilkinson and Barton in the UK [259,

---

[1] This section is based on T. Keller, M. Münster, and L. Hanzo, *A Turbo-coded Burst-by-burst Adaptive Wideband Speech Transceiver*; IEEE JSAC Wireless Series, Nov. 2000 pp. 2363–2372

TABLE 9.11 Basic Wideband Codec Features

	Coding Algorithm	Bitrate (kbps)	Encoding Delay (ms)	Bit Allocation
G722	SB-ADPCM 0–4 kHz: 4–6 bit ADPCM, 4–8 kHz: 2-bit ADPCM	64	1.5	Table 9.1
Quackenbush [227]	Adaptive 256-FFT	32	16	
Laflamme et al. [138]	Full-band Fwd-adapt. ACELP	16	15	Not avail.
Sanchez-Calle [231]	Full-band Bwd-adapt. ACELP	32	1	Table 9.9
Salami et al. [139]	Full-band Fwd-adapt. ACELP	9.6–14	30	Table 9.10
Black et al. [137]	Split-band 0–4 kHz: 13.7 kbps backw-adapt. CELP, 4–8 kHz: 2.3 kbps Vocoder	16	7	Table 9.7

260], and di Benedetto and Mandarini at the University of Rome [261], in order to name just a few of the key contributors without completeness. Further significant advances over more benign, slowly varying dispersive Gaussian fixed links are due to Chow, Cioffi, and Bingham [262] from the United States, where OFDM became the dominant solution for asymmetric digital subscriber loop (ADSL) applications, potentially up to a bitrate of 54 Mbps. In Europe OFDM has been favored for both digital audio broadcasting (DAB) and Digital Video Broadcasting [263, 264] (DVB), as well as for high-rate wireless asynchronous transfer mode (WATM) systems due to its ability to combat the effects of highly dispersive channels [265]. The notion of adaptive bit allocation in the context of OFDM was proposed as early as 1989 by Kalet [266] and was further developed by Chow et al. [262] and refined for duplex wireless links, for example, in [250]. Lastly, an OFDM-based narrowband speech system was proposed in [190]. The co-channel interference sensitivity of OFDM can be mitigated with the aid of adaptive beam-forming [267, 268].

Against this backcloth, in this section we propose a burst-by-burst adaptive 7-KHz bandwidth audio transceiver scheme, based on turbo-coded multimode constant throughput OFDM. The rationale behind proposing this system was that nonadaptive OFDM was also a contender for the Pan-European universal mobile telecommunications system (UMTS). Hence, it was beneficial to explore the potential of a substantially enhanced turbo-coded fixed-rate AQAM wideband audio arrangement. First, OFDM provides a powerful framework for exploiting both the time- and frequency-domain channel properties by adapting the bit allocation to subcarriers, as we will demonstrate. Second, OFDM is amenable to powerful soft-decision-based turbo coding [190, 269, 270]. Third, although our adaptive transceiver requires a programmable-rate speech or audio codec, to date only a limited number of such codecs have been proposed in the literature. Specific examples are the lower-quality 4 kHz bandwidth (i.e., narrowband) advanced multirate (AMR) speech codec, which was designed for UMTS, and the higher-quality 7 kHz bandwidth G.722.1 codec, which can be programmed to operate between 10 kbps and 32 kbps.

We will explore the design tradeoffs and show that the AOFDM bitrate can be adaptively controlled in an effort to find the best compromise in terms of loading the AOFDM subcarriers more heavily in an effort to increase the available throughput bitrate for maintaining a higher speech coding rate and higher speech quality, while also maintaining a high robustness against transmission errors. A further tradeoff is that, although the more heavily loaded, higher-throughput AOFDM modem is more vulnerable against transmission errors due to using more corrupted subcarriers, the longer turbo interleaving improves the turbo codecs' performance.

The proposed AOFDM system is constituted by two adaptation loops: an inner constant throughput transmission regime and an outer switching control regime. Jointly they maintain the required target bitrate of the system, while employing a set of distinct operating modes. This system was contrived in order to highlight the system design aspects of joint burst-by-burst adaptive modulation, channel coding, and source coding. This system design section is structured as follows. Subsection 9.5.2 provides a brief system overview, listing also our experimental conditions, which is followed by Subsection 9.5.4, detailing the philosophy of our constant throughput burst-by-burst adaptive OFDM modem. Subsection 9.5.6 investigates the multimode modem adaptation regime proposed, leading to a discussion on the adaptive audio source codec employed in Subsection 9.5.8. Our system performance results are summarized in Subsection 9.5.10 along with our future research endeavors.

9.5.2 System Overview

The structure of the proposed adaptive OFDM transceiver is depicted schematically in Figure 9.14. The top half of the diagram is the transmitter chain, which consists of the source and channel coders, a channel interleaver to decorrelate the channel's frequency-domain fading, an adaptive modulator, a multiplexer adding signaling information to the transmitted data, and an inverse fast fourier transform/radio frequency (IFFT/RF) OFDM stage. The receiver, seen in the lower half of the graph, consists of an RF/FFT OFDM receiver, a demultiplexer extracting the signaling information, an adaptive demodulator, a de-interleaver/channel decoder, and the source decoder. The parameter adaptation linking the receiver- and transmitter-chain consists of a channel-quality estimator and the mode selection, as well as the modulation adaptation blocks.

The open-loop control structure of the adaptation algorithms can be observed in the figure, where the receiver's operation is controlled by the signaling information that is contained in the received OFDM symbol, while the channel quality information estimated by the receiver is employed in order to determine the parameter set to be employed by the transmitter. The two distinct adaptation loops distinguished by dotted and dashed lines are the inner and outer adaptation regimes, respectively. The outer adaptation loop controls the overall throughput of the system, so that a fixed–delay decoding of the received data packets becomes possible. This controls the packet size of the channel codec, the block length of the channel encoder and interleaver, as well as the target throughput of the inner adaptation loop. The operation of the adaptive modulator, controlled by the inner loop, is transparent to the rest of the system. The operation of the adaptation loops is described in more detail below.

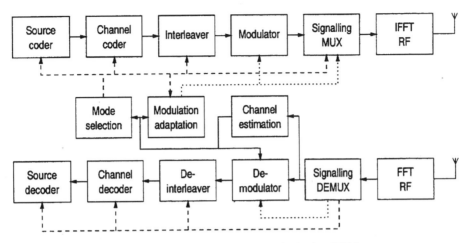

Figure 9.14 Schematic model of the multimode adaptive OFDM system.

9.5.3 System Parameters

The transmission parameters have been adopted from the TDD-mode of the Pan-European UMTS system [271], having a carrier frequency of 1.9 GHz and a TDD-frame and time-slot duration of 4.615 ms and 122 μs, respectively. The sampling rate is assumed to be 3.78 MHz, leading to a 1024-subcarrier OFDM symbol, having a cyclic extension of 64 samples in each time slot. In order to assist in the spectral shaping of the OFDM signal, there are a total of 96 virtual subcarriers at the bandwidth boundaries. Table 9.12 gives an overview of the transmission parameters employed for this system.

The 7-kHz bandwidth G.722.1 audio codec [272] designed by the PictureTel company has been chosen for this system because of its good audio quality, robustness to packet dropping, and adjustable bitrate, which will be discussed in more depth below.

The channel encoder/interleaver combination is a convolutional constituent coding-based turbo encoder [192, 193] employing block interleavers with a subsequent pseudorandom channel interleaver. The constituent recursive systematic convolutional (RSC) encoders are of constraint length 3, with octal generator polynomials of (7,5), and eight iterations are performed at the decoder, utilizing the maximum aposteriory (MAP) algorithm [197] and the log-likelihood ratio soft inputs provided by the demodulator.

The channel model consists of a four-path COST 207 Typical Urban impulse response [273], where each impulse is subjected to independent Rayleigh fading having a normalized Doppler frequency of $2.25 \cdot 10^{-6}$, corresponding to a pedestrian scenario with a walking speed of 3 mph. The unfaded channel impulse response and the amplitude of the correspond-

TABLE 9.12 OFDM System Parameters of Adaptive System

OFDM FFT length	1024
Active subcarriers	928
Guard interval length	64 samples
Sampling rate	3.78 MHz
TDD frame duration	4.615 ms
TDD slot duration	122 μs

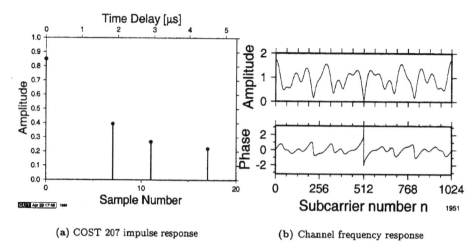

(a) COST 207 impulse response   **(b)** Channel frequency response

Figure 9.15 Channel model: (a) COST 207 impulse response; (b) unfaded frequency
domain channel transfer function $H(n)$. The grey shaded area represents the
virtual subcarriers.

ing frequency domain channel transfer function are shown in Figure 9.15. The grey shaded
areas in Figure 9.15b represent the virtual subcarriers.

9.5.4 Constant Throughput Adaptive Modulation

The constant throughput adaptive OFDM algorithm attempts to allocate the required
number of bits for transmission to the specific OFDM subcarriers exhibiting a low BER due
to their unattenuated spectral envelope in Figure 9.15b, while the use of high BER subcarriers
is minimized. We assume an open-loop adaptive system, basing the decision on the next
transmit OFDM symbol's modulation scheme allocation on the channel estimation gained at
the reception of the most recent OFDM symbol by the local station. Sub-band adaptive
modulation [250], where the modulation scheme is adapted not on a subcarrier-by-subcarrier
basis, but for blocks of adjacent subcarriers, is employed in order to simplify the adaptive
OFDM modem mode signaling requirements.

If the impulse response of the channel $h(t, \tau)$ varies slowly compared to the OFDM
symbol duration, then the Fourier transform of the impulse response during the OFDM
symbol exists, and the data symbols transmitted in the subcarriers $n \in [0, \ldots, N]$ are exposed
to the frequency-domain fading determined by the instantaneous channel transfer function
$H(t, n \cdot \Delta f) = H_n$.

The allocation of bits to subcarriers is based on the estimated frequency-domain
channel transfer function $\hat{H}_n$. On the basis of this and the overall signal-to-noise ratio (SNR)
γ, the local SNR in each subcarrier n can be calculated as: $\gamma_n = \gamma/|\hat{H}_n|^2$. The predicted BER
$p_e(\gamma_n, m)$ in each subcarrier n and each of the possible modulation schemes $m \in [0, \ldots, M]$
can now be computed and summed over the N_j subcarriers in sub-band j in order to yield the
expected number of bit errors for each sub-band and for each modulation scheme, which is
given by:

$$e(j, m) = \sum_i p_e(\gamma_i, m)$$

for all subcarrier indices i in sub-band j. In our case, four modulation schemes are employed for $m = 0, \ldots, 3$, which are "no transmission," BPSK, QPSK, and 16QAM, respectively. Clearly, $e(j, 0) = 0$, and the other bit error probabilities can be evaluated using the Gaussian Q-function [274]. The number of bits transmitted in sub-band j, when using modulation scheme m, is denoted by $b(j, m)$.

The bit-allocation regime operates iteratively, allocating bits to subcarriers by choosing the specific subcarriers for transmitting the next bit to be assigned for transmission, which increases the system's BER by the smallest amount. In other words, the bits to be transmitted are allocated consecutively, commencing by assigning bits to the highest channel quality subcarriers, gradually involving the lower channel quality carriers.

More explicitly, for each sub-band, a state variable s_j is initialized to 0, and then the sub-band index j, for which the differential BER increment $(e_{s_j+1} - e_{s_j})/(b_{s_j+1} - b_{s_j})$ due to assigning the next bit to be transmitted is the lowest is found. The state variable s_j is incremented from 0, if it is not yet set to the index of the highest order modulation mode, that is, to 16QAM. This search for the lowest BER "cost" or BER penalty, when allocating additional bits, is repeated until the total number of bits allocated to the current OFDM symbol is equal to or higher than the target number of bits to be transmitted. Clearly, the higher the target number of bits to be transmitted by each OFDM symbol, the higher the BER, since gradually lower and lower channel quality subcarriers have to be involved.

The transmitter modulates the subcarriers using the specific modulation schemes indexed by the state variables s_j, eventually padding the data with dummy bits, in order to maintain the required constant data throughput. The specific modulation schemes chosen for the different sub-bands have to be signaled to the receiver for demodulation. Alternatively, blind sub-band modem mode detection algorithms can be employed at the receiver [270]. For the scope of these investigations, we assume 32 sub-bands of 32 subcarriers in each 1024-subcarrier OFDM symbol. Perfect channel estimation and sub-band modem mode signaling were assumed.

9.5.5 Adaptive Wideband Transceiver Performance

Figure 9.16 shows an example of the fixed throughput adaptive modulation scheme's performance under the channel conditions characterized above, for a block length of 578 coded bits. As a comparison, a fixed BPSK modem transmitting the same number of bits in the same channel, employing 578 out of 1024 subcarriers, is depicted. The number of bits per OFDM symbol is based on a 200-bit useful data throughput, which corresponds to a 10 kbps data rate, padded with 89 bits which can contain a checksum for error detection and high-level signaling information, as well as half-rate channel coding.

The BER plotted in the figure is the hard decision bit error rate at the receiver before channel decoding. It can be seen that the adaptive modulation yields a significantly improved performance, which is also reflected in the frame error rate (FER). This FER is the probability of a decoded block containing errors, in which case it is unusable for the source decoder and hence it is dropped. This error event can be detected by using the checksum of the data symbol.

The modulation scheme allocation for the 578 data bit adaptive modem for an average channel SNR of 5 dB is given in Figure 9.17a for 100 consecutive OFDM symbols. The unused sub–bands with indices 15 and 16 contain the virtual carriers, and therefore do not transmit any data. It can be seen that the adaptation algorithm allocates data to the better quality subcarriers on a symbol-by-symbol basis, while keeping the total number of bits per OFDM symbol constant. As a comparison, Figure 9.17b shows the equivalent overview of the

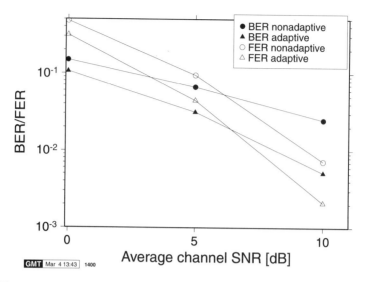

Figure 9.16 FER and uncoded BER for fixed throughput adaptive and nonadaptive
modulation in the fading time dispersive channel of Figure 9.15 for a
block length of 578 coded bits per 1024 subcarrier for the system of Table
9.12.

modulation schemes employed for the fixed bitrate of 1458 bits per OFDM symbol. In order
to hit the throughput target, hardly any sub-bands are in "no transmission" mode, and overall
higher order modulation schemes have to be employed.

Figure 9.18 shows the subcarrier SNR for the first transmitted frame over the channel of
Figure 9.15 for a long-term SNR of 5 dB. The subcarrier SNR experienced by the modem
varies greatly both across the overall OFDM bandwidth and within the sub-bands, delineated
by the dotted vertical lines. The different-shade grey markers at the bottom of the graph
indicate the modem mode employed for each sub-band, and the circular markers indicate the
expected BER, averaged over the subcarriers of each sub-band. Figure 9.18a gives the modem
mode allocation and BER for the 10 kbps mode, corresponding to the first column of Figure
9.17a, while Figure 9.18b depicts the same information for the 32 kbps mode, which
corresponds to the first column of Figure 9.17b.

9.5.6 Multimode Transceiver Adaptation

While the fixed throughput adaptive algorithm copes well with the frequency-domain
fading of the channel, there is also a medium-term time-domain variation of the overall
channel capacity. Hence—in addition to the previously proposed fixed-rate frequency-domain
bit-allocation scheme—in this section we propose the employment of a time-variant bitrate
scheme in order to gauge its additional performance potential benchmarked against the fixed-
rate schemes. We will then also contrive appropriate matching audio transceivers at a later
stage. However, our experience demonstrated that it was an arduous task to employ powerful
block-based turbo channel coding schemes in conjunction with variable throughput adaptive
schemes for real-time applications, such as voice or videotelephony. Nonetheless, a multimode adaptive system can be designed that allows us to switch between a set of different
source and channel coders as well as transmission parameters, depending on the overall
instantaneous channel quality. We have investigated the employment of the estimated overall

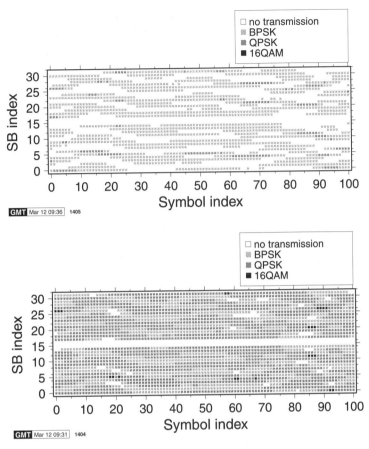

Figure 9.17 Overview of modulation scheme allocation for the 578-bit fixed throughput adaptive modem over the fading time-dispersive channel of Figure 9.15 at 5 dB average channel SNR.

BER at the output of the receiver, which is the sum of all the $e(j, s_j)$ sub-band BER contributions after modem mode adaptation. On the basis of this expected input error rate of the channel decoder, the probability of a frame error (FER) must be estimated and compared with the expected FER of the other modem modes. Then, the mode having the lowest FER is selected and the source coder, the channel coder, and the adaptive modem are set up accordingly.

We have defined four different operating modes, which correspond to the uncoded audio data rates of 10, 16, 24, and 32 kbps at the source encoder's output. With half-rate channel coding and allowing for checksum and signaling overheads, the number of transmitted coded bits per OFDM symbol is 578, 722, 1058, and 1458 for the four source-coded modes, respectively.

9.5.7 Transceiver Mode Switching

Figure 9.19 shows the observed FER for all four modes versus the uncoded BER that was predicted at the transmitter during the channel estimation and modem mode adaptation. The predicted BER was discretized into intervals of 1%, and the FER was averaged over these

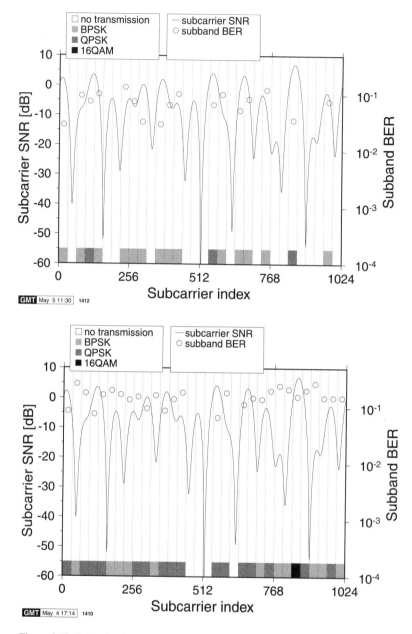

Figure 9.18 Subcarrier SNR versus subcarrier index for the first transmitted frame in the channel of Figure 9.15 for a long-term SNR of 5 dB, with selected modem mode and average estimated sub-band BER for the 32 sub-bands. The two sub-bands around carrier 512 are virtual carriers. (a) 10 kbps mode; (b) 32 kbps mode.

intervals. It can be seen that for estimated BER values below 5% no frame errors were observed for any of the modes. For higher estimated BER values, the higher throughput modes exhibited a lower FER than the lower throughput modes, which was consistent with the turbo coder's performance increase for longer block lengths. A FER of 1% was observed

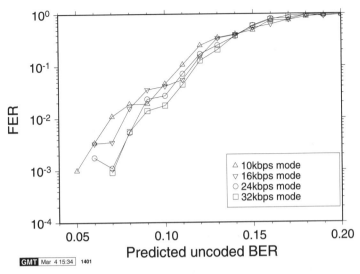

Figure 9.19 Frame error rate versus the predicted uncoded BER for 10 kbps, 16 kbps, 24 kbps and 32 kbps modes.

for a 7% predicted input error rate for the 10 kbps mode, while BERs of 8% to 9% were allowed for the longer blocks.

In this study we assumed the best-case scenario of using the measured FER statistics of Figure 9.19 for the mode-switching algorithm. In this case, the FER corresponding to the predicted overall BER values for the different modes are compared, and the mode with the lowest FER is chosen for transmission. The mode-switching sequence for the first 500 OFDM symbols at 5 dB channel SNR is depicted in Figure 9.20. It can be seen that in this segment of the sequence 32 kbps transmission is the most frequently employed mode, followed by the

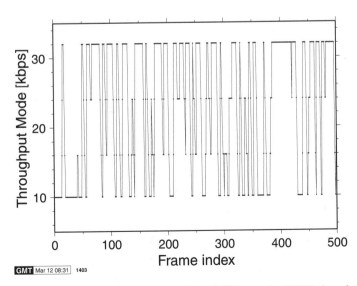

Figure 9.20 Mode switching pattern at 5 dB channel SNR over the WATM channel of Figure 9.15b.

TABLE 9.13 FER and Relative Frequency of Different Bitrates in the Fixed Bitrate and in the Burst-by-Burst Switching Schemes (successfully transmitted frames) for an SNR of 5 dB

Scheme	FER[%]	Rel.Fr.: [%] 10 kbps	Rel.Fr.: [%] 16 kbps	Rel.Fr.: [%] 24 kbps	Rel.Fr. [%] 32 kbps
Fixed-10 kbps	4.45	95.55	0.0	0.0	0.0
Fixed-16 kbps	5.58	0.0	94.42	0.0	0.0
Fixed-24 kbps	10.28	0.0	0.0	89.72	0.0
Fixed-32 kbps	18.65	0.0	0.0	0.0	81.35
Switch-I	4.44	21.87	13.90	11.59	48.20
Switch-II	5.58	0.0	34.63	11.59	48.20

10 kbps mode. The intermediate modes are mostly transitory, as the improving or deteriorating channel conditions render switches between the 10 kbps and 32 kbps modes necessary. This behavior is consistent with Table 9.13, for the "Switch-I" scheme, which will be discussed in depth during our forthcoming discourse. Let us now briefly consider the 7 kHz bandwidth audio codec, which can be reconfigured in a range of different quality and bitrate modems and hence can exploit the time-variant bitrate of the AOFDM modem.

9.5.8 The Wideband G.722.1 Codec

9.5.8.1 Audio Codec Overview. In recent years, speech coding research has focused on coding 7 kHz bandwidth rather than 3.4 kHz bandwidth speech signals, in an effort to increase the perceived speech quality [116, 119]. The challenge in this context has been the encoding of the speech components above 3.4 kHz, which on average account for less than 1% of the speech energy; yet they substantially influence the perceived speech quality. A plausible approach is to separate these two bands, which allows the designer to independently control the number of bits allocated to them. A more refined approach is to invoke frequency-domain coding techniques, such as transform coding [119, 272], which allows a more intelligent, finely grained distribution of the available coding bits to the most important audio signal frequencies. Furthermore, the bitrate can be adaptively controlled in an effort to find the best compromise in terms of loading the AOFDM subcarriers more heavily and to increase the available bitrate for maintaining a higher speech coding rate and higher speech quality, while maintaining a high robustness against transmission errors.

The current 64 kbps G.722 ITU standard wideband speech codec [122] is becoming antiquated, and the PictureTel Transform Codec (PTC) was selected for the new ITU-T G.722.1 wideband audio coding standard [272]. It is based on the Modulated Lapped Transform (MLT) [275], followed by a quantization stage using a perceptually motivated psychoacoustic quantization model and Huffman coding for encoding the residual frequency domain coefficients.

At its input, the G.722.1 expects frames of 320 Pulse Code Modulated (PCM) audio samples, obtained by sampling an audio signal at a frequency of 16 kHz with a quantizer resolution of 14, 15, or 16 bit. Furthermore, the input samples are assumed to contain frequency components up to 7 kHz. At the time of writing, the G.722.1 standard recommends operating the codec at output bitrates of 16 kbps, 24 kbps, or 32 kbps, generating output frame lengths of 320, 480, or 640 bits per 20 ms, respectively, for which the codec was optimized. The total delay encountered by an audio frame, when passing through the codec (consisting of encoder and decoder) can be estimated to be on the order of about 60 ms. This is a result of the time-domain frame overlapping technique and the computational delay inherent in the codec.

Since the PTC employs Huffman coding for encoding the frequency domain coefficients, the decoding is very sensitive to bit errors. Hence, a single-bit error can render the whole audio frame undecodable. The PTC's standard reaction to such a frame error is simply to repeat the previous frame of coefficients, as long as they occur relatively rarely. For bursts of frame errors, the output signal is gradually muted after decoding the first erroneous frame.

9.5.9 Detailed Description of the Audio Codec

In the following we briefly describe the signal processing stages incorporated in the PTC, which is supported by the block diagram of the encoder, depicted in Figure 9.21. In the first processing step, the PCM input signal is mapped from the time domain into the frequency domain, using the modulated lapped transform (MLT), a derivative of the DCT [275]. It is well-known that the MLT can be effectively employed in applications, where blocking effects can cause severe signal distortion. The latest 320 time-domain samples form a block, which is fed together with the previous block of 320 coefficients into the MLT. As an output, the MLT then produces a block of 320 frequency-domain samples, which yields a frequency resolution of $8000\,\mathrm{Hz}/320 = 25\,\mathrm{Hz}$. As mentioned previously, only signal components with frequencies up to 7 kHz are encoded, which correspond to frequency coefficients with an index lower than 280. The other coefficients are discarded.

The remaining MLT coefficients are further grouped into 14 equal-width regions, each representing a frequency range of 500 Hz and hosting $280/14 = 20$ coefficients. For each frequency region, the square root of the mean power (RMS) is calculated in Figure 9.21, which gives an estimate of the spectral envelope. By the help of these RMS values, which are

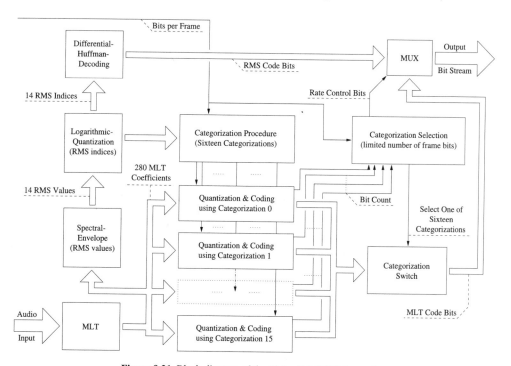

Figure 9.21 Block diagram of the PictureTel G.722.1 encoder.

transformed to the logarithmic domain in Figure 9.21, the MLT coefficients are then quantized using different step-sizes according to a perceptual model. This task is performed by calculating an initial categorization, in which a certain set of quantization and coding parameters referred to as the category is assigned to each region. As portrayed in Figure 9.21, a total of 16 tentative categorizations and bit allocations are calculated, of which finally only that one is used which makes use of the available bits in the most efficient way. After the best bit allocation has been determined, the MLT coefficients are quantized and Huffman coded along with the parameters of the associated categories. During the last computational step, the output data of the described signal processing stages is multiplexed into a data frame. The macroscopic bit allocation, which we encounter in a typical data frame at the output of the PTC encoder, is illustrated in Figure 9.22 for the case of 320 frame bits, that is, 16 kbps. As shown in Figure 9.21, the multiplexer (MUX) arranges the rms code bits, the rate-control bits, and finally the MLT code bits into a bit stream. The exact frame structure is given in Figure 9.22, together with the typical number of bits needed for encoding the spectral envelope and the transform coefficients. In every frame, the first 5 bits are occupied by the value rms_index(0), followed by the Huffman codes of the differentially coded rms indices 1, ..., 13 in spectral frequency order. The next 4 bits of every frame are occupied by the rate-control bits. Then the MLT code vector indices are transmitted, beginning with frequency region 0. Directly after a vector index's variable-length code, the associated MLT coefficient sign bits are transmitted, in spectral frequency order.

The signal processing stages that constitute the G.722.1 decoder [272] are essentially the inverse operations of the encoder shown in Figure 9.21. The decoding of a frame starts with reconstruction of the spectral envelope. Next, the 4 rate-control bits are decoded in order to determine which of the 16 possible categorizations has been used for encoding the MLT coefficients. In the same way as 16 categorizations are generated in the encoder, they are now

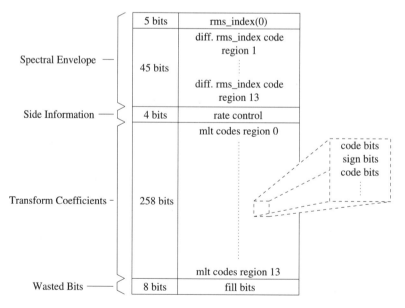

Figure 9.22 Structure of a typical coded frame at a bitrate of 16 kbps (320 bits per 20 ms frame); the number of bits needed for encoding the differential rms indices and the transform coefficients varies in each frame together with the number of wasted bits.

generated also in the decoder. Finally, the particular categorization used at the encoder is also employed by the decoder. The frequency regions, where category 7 has been applied, are treated differently. Since no MLT coefficients have been transmitted for these frequency regions, a specific technique, referred to as *noise-filling*, is used to prevent the associated MLT coefficients to be set to zero. This technique is also applied to categories 5 and 6, since most of their coefficients are quantized to zero. The coefficients, which were quantized to nonzero values, are reconstructed using a predetermined decoding table. After denormalization by multiplying all coefficients of a frequency region by their decoded RMS values, the MLT coefficients are rearranged to blocks of 320 coefficients, where the upper 40 coefficients are set to zero, since they belong to frequencies above 7 kHz. Then, the IMLT is applied to the coefficients, generating 320 time-domain samples at the output. Both the MLT and the IMLT can be decomposed into a computationally efficient discrete cosine transform (DCT) *DCT type-IV* and Inverse DCT (IDCT) *IDCT type-IV* implementation [276], followed by a window, overlap, and add operation [275]. Due to the Huffman coding, which is applied to the values of the spectral envelope as well as to the MLT coefficients, the information carried by those codewords is extremely sensitive to bit errors. If the channel decoder is unable to correct all transmissions errors, the PTC decoder's recommended behavior [272] is to repeat the MLT coefficients of the previous frame in case of a single erroneous frame, or to set the MLT coefficients to zero, which corresponds to muting the output signal, provided also that the previous frame had been contaminated by channel errors. For further details concerning the G.722.1 Transform Codec, the interested reader is referred to [272].

9.5.10 Wideband Adaptive System Performance

Our discussions related to the associated system design tradeoffs and the impact of an automatic bitrate selection scheme on the audio quality of the system are based mainly on measurements performed around channel SNR values of 5 dB, since for very low SNRs of around 0 dB the frame dropping rate is excessive, yielding an unacceptable audio quality. By contrast, for high-channel SNRs around 10 dB the FER is too low to enable us to illustrate the tradeoffs between the audio quality and FER effectively. A tentative estimate of the average quality of the reconstructed audio signal is provided by audio segmental SNR calculations, which provide an approximate measure of the subjectively perceived audio quality, especially in the presence frame dropping in conjunction with perceptual masking assisted transform-based audio coding.

9.5.11 Audio Frame Error Results

The basic tradeoff between the system throughput and audio frame dropping rate is illustrated for the 5 dB channel SNR scenario with the aid of our four fixed-bitrate modes in Table 9.13. The first column reflects for each bitrate the associated frame dropping rate that we will encounter.

As expected, by increasing the required throughput bitrate, the frame error rate (FER) will also increase, since a high proportion of reduced-quality subcarriers has to be used to convey the increased number of audio bits, although the performance of the turbo channel codec improves. Experiments have shown that a frame dropping rate of around 5% in conjunction with the 16 kbps fixed-bitrate mode is still sufficiently low in order to provide a perceptually acceptable audio quality. In the second to fourth columns of Table 9.13, the relative frequency of encountering error-free audio frames for the different audio bitrates is portrayed. Also observe in the table that the same performance figures were summarized for two different transmission schemes denoted by *Switch I* and *Switch II*. These schemes invoked a system philosophy allowing the bitrate to become time-variant and controlling the audio source codec and channel codec on a time-variant basis, in order to take this time-variant behavior into account.

Specifically, both of our experimental switching regimes, namely, *Switch I* and *Switch II*, employed the same switching algorithm, as described in Section 9.5.7, with the only difference that, in addition to the three standard bitrates of 16, 24, and 32 kbps, proposed by the PictureTel company, Switch I incorporated a 10 kbps mode, with the intention of lowering the frame dropping rate further due to the more modest "loading" of the OFDM symbols. For these switching schemes, the 5 dB SNR-related results in Table 9.13 underline that, for example, in comparison to the 16 kbps fixed-rate mode the system throughput was very much improved, conveying $(11.59 + 48.20)\%$ of the audio frames in the 24 and 32 kbps mode rather than in the 16 kbps mode, while maintaining the same frame dropping rate of 5.58% as the 16 kbps mode. Although exhibiting a slightly lower frame dropping rate, the Switch I scheme was shown to produce an audio quality inferior to that of the Switch II scheme. This was due to employment of the 10 kbps bitrate mode in the Switch I scheme, which produced a relatively low subjective audio quality. In this context, it is interesting to see that although the Switch I scheme assigns about 22% of all frames to the 10 kbps transmission mode, the frame dropping rate was increased only by about 1.1%, when disabling this subjectively low-quality but error-resilient mode in the Switch II scheme. This is an indication of the conservative decision regime of our bitrate selector. The relative frequency of invoking the different bitrates in conjunction with the Switch II scheme has also been evaluated for channel SNRs of 0 dB and 1 dB, which again characterizes the operation of the bitrate selector. The associated results are presented in Table 9.14, which become plausible in the light of our previous discussions.

9.5.12 Audio Segmental SNR Performance and Discussions

In addition to our previous results, Figure 9.23 displays the cumulative density function (CDF) of the segmental SNR (SEGSNR) of consecutive 20 ms duration audio segments obtained from the reconstructed signal of an audio test signal at the output of the PTC decoder for the schemes described above. These CDFs were recorded at a channel SNR of 5 dB. The step-function-like CDF discontinuity at a SEGSNR of 0 dB corresponds to the frame dropping rate of the associated transmission scheme, which is summarized in Table 9.14 for the various systems. As expected, for any given SEGSNR value it is desirable to maintain as low a proportion of the audio frames' SEGSNRs below a given abscissa value as possible. Hence, we concluded that the best SEGSNR CDF was attributable to the Switch II scheme, and the worst to the fixed 10 kbps arrangement, as suggested before. In the range of high audio SEGSNRs, the preference order of the various fixed schemes followed our expectations;

TABLE 9.14   FER and Relative Frequency of Different Bitrates in the Switch II Scheme (successfully transmitted frames) for Channel SNRs of 0, 5, and 10 dB

Scheme	FER[%]	Rel.Fr.: [%] 10 kbps	Rel.Fr.: [%] 16 kbps	Rel.Fr.: [%] 24 kbps	Rel.Fr.: [%] 32 kbps
0	37.69	0.0	37.79	14.42	10.10
5	5.58	0.0	34.63	11.59	48.20
10	0.34	0.0	7.81	5.61	86.24

that is, the fixed 32 kbps scheme performed best in SEGSNR terms, when neglecting frame drops. The tradeoff was that, although due to its highest audio bitrate of 32 kbps the scheme exhibited the inherently highest SEGSNR, because of its high throughput requirement, this scheme was forced to invoke a high proportion of partially impaired, low-quality OFDM subcarriers, which often resulted in corrupted and dropped audio frames. Since the fixed 10 kbps scheme exhibited the lowest audio SEGSNR performance, this scheme was excluded from the Switch II arrangement. However, FERs in excess of 10% result in distinctively audible artifacts, which—despite their high error-free SEGSNRs—virtually rendered the fixed-rate 24 kbps and 32 kbps modes unacceptable. *Hence, our proposed switching scheme—Switch II—which is based on the 16, 24, and 32 kbps bitrates, achieved at medium SNRs the best compromise between average error-free audio quality and frame dropping rate, which has been verified by our informal listening tests.*

As outlined in Section 9.5.7, the mode switching algorithm operates on the basis of statistically evaluated experimental results for the prediction of the FER. A robust, channel–independent switching regime on the basis of the turbo coder's quality perceptions can overcome this dependence. Furthermore, a target–FER driven switching scheme instead of the

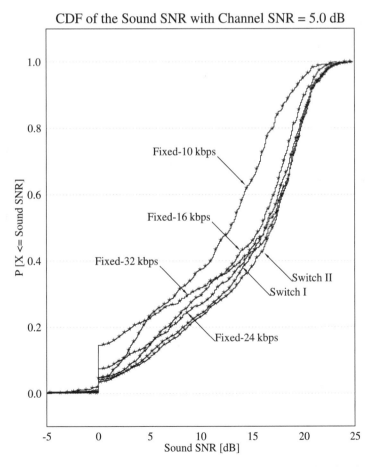

Figure 9.23 Typical CDF of the segmental SNR of a reconstructed audio signal transmitted over the fading time dispersive channel.

minimal–FER algorithm employed for this series of experiments will be investigated in future.

9.5.13 G.722.1 Audio Transceiver Summary and Conclusions

The design tradeoffs of turbo-coded burst-by-burst adaptive orthogonal frequency division multiplex (AOFDM) wideband speech transceivers were analyzed. A constant throughput adaptive OFDM transceiver was designed and benchmarked against a time-variant rate scheme. The proposed joint adaptation of source-codec, channel-codec, and modulation regime resulted in attractive, robust, high-quality audio candidate systems, capable of conveying near-unimpaired wideband audio signals over fading dispersive channels for signal-to-noise ratios (SNR) in excess of about 5 dB.

9.6 CHAPTER SUMMARY

This chapter described a range of wideband speech codecs commencing with the mature G722 SB-ADPCM standard codecs, which employed QMFs in order to guarantee sufficient flexibility in allocating channel capacity to the low and high bands. The codec operated at 48, 56, and 64 kbps, which was achieved by dropping one or two of the six low-band ADPCM bits. This allowed the codec to incorporate an 8 or 16 kbps data channel. The basic codec features are summarized in Table 9.11. As an interesting alternative technique, an FFT-based 32 kbps transform-codec was reviewed in Section 9.2.

A similar sub-band-split philosophy was suggested by Black et al. [137] in Section 9.3.2, but the authors advocated a higher-complexity backward- and forward-adaptive CELP-type codec for the encoding of the lower- and higher-sub-band signals respectively,. They achieved a similar performance to the 48 kbps mode of the G722 codec at 16 kbps. It was necessary to opt for a split-band scheme, since the full-band CELP optimization places most of the emphasis on the representation of the higher-energy low-frequency band. The Sherbrooke team invoked their ubiquitous ACELP codecs in a variety of attractive wideband schemes in both forward- and backward-adaptive arrangements in the bitrate range of 32–9.6 kbps. Their forward-adaptive 14 kbps codec [139] has a similar performance to the G722 codec operated at 56 kbps.

As an important recent development in the field, the G.722.1 speech codec's basic features were also highlighted. This codec then was employed in our system design example, characterizing the achievable performance of an intelligent adaptive OFDM transceiver.

Following our discussions on wideband coding, the next part of the book is concerned with various techniques applicable to speech compression at rates below 4.8 kbps.

PART IV
VERY LOW-RATE
CODING
AND
TRANSMISSION

10

Overview of Low-Rate Speech Coding

10.1 LOW-BITRATE SPEECH CODING

This chapter endeavors to give a rudimentary overview of low-rate speech coding, with a special emphasis on coding rates below 4.8 kbit/s and the techniques adopted in the associated codec design. This allows our readers to delve directly into the intricacies of sophisticated low-rate techniques instead of having to work their way through the previous chapters. Further related information can be found in the excellent books edited by Kleijn [32] and by Atal, Cuperman, and Gersho [28], as well as in the monographs authored, for example by Kondoz [31] and Jayant and Noll [10]. We begin with a historical perspective on the development of low-rate speech codecs, which is followed by a more in-depth review of 2.4 kbps speech coders. A brief glimpse at speech coders operating beneath 2 kbps is also offered. In addition, the methods of assessing a speech coder's performance are examined, with most emphasis placed on subjective speech quality measures. Finally, the speech database used throughout this low-rate coding-oriented part of the book is introduced.

Historically, the first speech coders were based on the now well-established waveform coding techniques [4]–[221], such as delta modulation (DM) [3] and sub-band coding (SBC) [221], which operate by directly quantizing the speech waveform. However, their operating bitrate range is restricted, since they fail to produce communications quality speech at rates below 16 kbps. Instead, this niche is filled by the class of hybrid vocoders that employ LPC [279].

These hybrid vocoders operate by parameterizing the speech signal and transmitting these parameters to the decoder. In the ubiquitous LPC schemes, this is performed through simulation of the human vocal system. Thus, an understanding of the human speech production mechanism is desirable. The stylized human voice production system is shown in Figure 10.1. The human lung forces air through the glottis to the vocal tract, where quasi-periodic vocal fold vibration or constriction of the vocal tract creates voiced and unvoiced speech, respectively. Vowel sounds, like the front vowel /ɛ/ as in "bed" are voiced sounds, whereas the fricative /s/ as in "see" is an example of an unvoiced utterance. Examples of the time- and frequency-domain representation for 20 ms or 160 samples of voiced and unvoiced

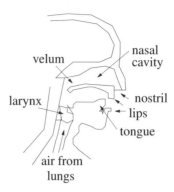

Figure 10.1 Human speech product system.

speech can be seen in Figure 10.2. The vocal tract can be alternatively labeled as the supra-glottal resonator because it is the interaction of the air with the vocal tract that determines the spectrum of the speech. This resultant speech spectrum also depends on the vocal tract shape, which itself depends on the vocal tract articulators in Figure 10.1, namely, the velum, lips, nostrils, and tongue. Further explanation of speech processing and synthesis can be found in the book by Deller, Proakis, and Hansen [19], together with the book by O'Shaughnessy [17].

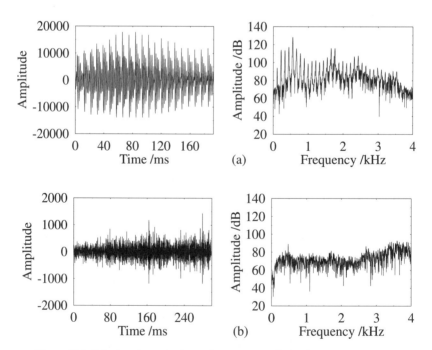

Figure 10.2 Voiced and unvoiced speech segments in both the time and frequency domain. In a) the voiced sound is the front vowel /ɛ/ as in 'b*e*d', while for b) the unvoiced sound is the fricative /s/ as in '*s*ee'.

In their traditional form [172, 279], LPC schemes synthesize speech by passing an excitation signal through a spectral shaping filter to model the frequency-domain shape produced by the vocal tract. The excitation signal mimics the glottal waveform using random noise for unvoiced speech and a sequence of periodic pulses for voiced speech. These periodic pulses are spaced according to the fundamental frequency of the speech waveform. Fundamental frequencies typically vary from 50 Hz to 300 Hz for adult male speakers and up to 500 Hz for adult female and child speakers, although the fundamental frequency can reach 1.5 kHz [20]. However, LPC schemes generally permit a fundamental frequency range of 54 Hz to 400 Hz, because at an 8 kHz sampling rate this range covers 20 to 147 samples and can be quantized with 7 bits. Increasing the permitted fundamental frequency range to cover more of the female and child fundamental frequencies would increase both the coder's bitrate and complexity. For instance, if we would allow the fundamental frequencies 400 Hz to 800 Hz, or 20 to 10 samples, the potential for the extra pitch period in each speech frame increases the associated bitrate.

In the speech coding community, the term *pitch* is often considered synonymous with fundamental frequency. Here it is noted that pitch actually refers to a perceived value and therefore is not a measure of the speech waveform. However, within this low-rate coding part of the book, the terms pitch and fundamental frequency are interchangeable, following the trend of the speech coding community.

The classical LPC vocoder schematic, where either a periodic pulse train or a Gaussian noise source is connected to the synthesis filter, is shown in Figure 10.3. The spectral shaping arrangement is an all-pole filter that fully parameterizes, and is analogous to, the shape of the vocal tract, albeit with the velum raised excluding the nasal cavity of Figure 10.1. This traditional vocoder form was proposed by Atal and Hanauer [279], and it is capable of encoding speech at 2 kbps. However, the resultant speech is of synthetic quality with a frequent "buzziness" which will be explained in Section 10.1.2.1.

10.1.1 Analysis-by-Synthesis Coding

A significant advance within the speech coding field was achieved with the introduction of analysis-by-synthesis methods, such as the multi-pulse excitation (MPE) LPC developed by Atal and Remde [9]. For MPE-LPC a selection of synthesized speech segments, of typically 5 ms length, are compared with the original segment, and the best version is selected, where the criterion is the minimum mean squared error distance. In essence, this is the

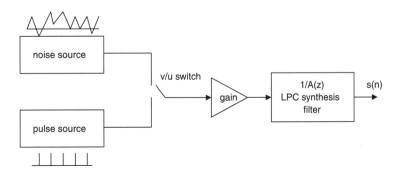

Figure 10.3 Schematic of the traditional LPC vocoder.

addition of a synthetic ear to the voice production scheme, with many different versions of the utterance created and the one that is deemed to be most like the original selected. MPE-LPC locates several pulses with varying amplitudes in the optimum positions for each speech segment. The pulse positions and quantized amplitudes are subsequently sent to the decoder.

MPE-LPC preceded regular pulse excitation (RPE) developed by Kroon et al. [11], where the distance between excitation pulses was constrained to be regular. If each pulse is separated by a regular distance D_R, then for each frame only one pulse position is required as the other pulses are $D_R, 2D_R, 3D_R \ldots$ further away. The lower number of parameters required results in a reduction in bitrate. This method of speech coding is used as the full-rate coder for the pan-European mobile radio system, known as GSM [73].

In order to achieve a further bitrate reduction, Schroeder and Atal introduced codebook excitation linear prediction (CELP) [16], where a codebook filled with vectors representing the excitation source is searched to find the best match for the speech segment, as demonstrated in Figure 10.4. Together with the spectral envelope parameters, the index for the relevant codebook entry is sent to the decoder. In order to allow the possible excitation signals to consist of random vectors, the periodicity of the speech signal must be removed, a task performed by the long-term predictor (LTP). The LTP is frequently used in an adaptive codebook format, where each entry contains a past excitation signal with a particular delay. The excitation signal corresponds to the input of the LPC synthesis filter, where at each calculation of the LTP the adaptive codebook is filled with different overlapping vectors. The LTP parameters require updating more frequently than the LPC parameters, normally every 2.5–7.5 ms as opposed to every 20–30 ms. This frequent updating means that a large proportion of the CELP bitrate is consumed by the LTP parameters.

Methods to reduce the complexity of CELP coders generally involve the use of structured codebooks allowing efficient search procedures. However, to synthesize higher quality speech than the standard LPC-10 2.4 kbps [172] scheme at least 4 kbps is required, where this assumes that about 20 bits per 20 ms speech frame are dedicated to coding the LPC filter coefficients. For coding rates below 4 kbps, the excitation vector must be 8 to 10 ms in length to achieve the desired bitrate. It is then possible for several pitch period peaks to occur within a subframe; subsequently, the random code vectors lack the required pitch periodicity, and the quality of the synthesized speech is degraded. A coder that can produce good quality speech at 3.6 kbps is the Pitch Synchronous Innovative CELP (PSI-CELP) scheme, which has been adopted for the Japanese mobile radio system's half-rate coder [186]. The PSI-CELP system produces quality speech for bitrates less than 4 kbps through exploitation of the periodicity which occurs in voiced speech. Use of periodicity in the CELP model reduces the

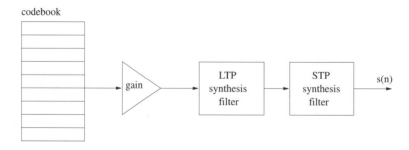

Figure 10.4 Schematic of a CELP arrangement.

overall signal-to-noise ratio (SNR) achieved by the coder; however, it improves the subjective quality of the speech signal.

Mano et al. [186] utilize pitch synchronous repetition of the random excitation to produce good quality voiced sections of speech. For the unvoiced, silent, and transient sections of speech, a fixed codebook of random excitation is used in the coder instead of an adaptive codebook. This approach is returning to the classical LPC schemes where different excitation signals were used for voiced and unvoiced speech, as shown in Figure 10.5. The PSI-CELP excitation signal is of the form $G_1 v_1(n) + G_2 v_2(n)$ for $n = 1 \cdots N$, where $v_1(n)$ is equivalent to either an adaptive codebook or a fixed random codebook entry, and $v_2(n)$ is a combination of two random codebook vectors that for voiced speech contain a repetitious random vector synchronized to the pitch period. The superposition of two codebooks in generating $v_2(n)$ reduces the memory requirements of the coder. Initially, to select the excitation signal the most appropriate $v_1(n)$ vector is determined. If the $v_1(n)$ vector came from the adaptive codebook, then the speech was deemed to be voiced. Hence, in constructing $v_2(n)$, the random vector will be repeated at pitch period intervals. If the fixed codebook is selected for $v_1(n)$, implying an unvoiced decision, then when selecting $v_2(n)$ unmodified random vectors are used.

Mano et al. [186] implement a post-filter at the decoder, which enhances the pitch harmonics for the adaptive codebook and enhances the higher frequencies for the fixed codebook. The waveform shaping post-filter decreases the SNR of the coder but improves the subjective quality of the speech. The PSI-CELP coder still requires over 3 kbps to produce good quality speech.

10.1.2 Speech Coding at 2.4 kbps

In order to produce good quality speech at less than 3 kbps, different approaches must be pursued. An interesting review of these methods can be found in the recent selection procedure for the new U.S. Department of Defense (DoD) 2.4 kbps standard. The new

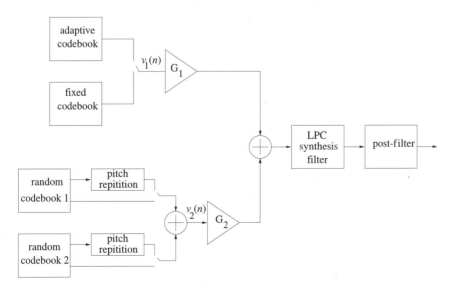

Figure 10.5 Schematic of the PSI-CELP arrangement © IEEE.

standard is designed to replace the old 2.4 kbps LPC-10 vocoder [172] and the 4.8 kbps DoD Federal Standard FS1016 CELP coder [76].

10.1.2.1 Background to 2.4 kbps Speech Coding. Historically, the first standard for 2.4 kbps speech coding using LPC was the LPC-10 recommendation, which has been in use since the late 1970s. The LPC-10 was later renamed the Federal Standard FS1015. An enhanced version of this standard, LPC-10e, was developed by the mid-1980s. However, even the enhanced version produces only synthetic quality speech, although state-of-the-art improvements in speech technology should allow significantly improved quality speech transmission. Thus, in May 1993 the United States Department of Defense Digital Voice Processing Consortium (DDVPC) began the process of selecting a successor to the LPC-10e 2.4 kbps speech coding algorithm. Kohler et al. [280] describe the workshops and general progression made in the selection process up to 1995. In May 1996, at the International Conference on Acoustics, Speech and Signal Processing (ICASSP) held in Atlanta, the winning speech compression algorithm was announced.

In this section, the seven candidate speech coding algorithms are described, and the successful candidate is revealed. The winning candidate selection method, employed by the DDVPC, is described in Section 10.3.3 where general information on speech coding performance is given. The seven candidate coders fell, disproportionately, into two categories. The first group is the category of harmonic coders, which can be further subdivided into Multiband Excitation (MBE) [79, 281] and sinusoidal coders [282, 283]. Harmonic coders consider the frequency spectrum of a speech signal and encode the amplitudes of the harmonic frequencies. MBE speech coders have voiced-unvoiced excitation waveforms, which are harnessed to represent different bands or harmonics within the frequency spectrum. Sinusoidal speech coders use an appropriate sinusoid with amplitude and phase parameters for each harmonic, where the transmitted phase defines the sinusoid as either voiced or unvoiced. Four of the DoD candidate speech coders fell into the harmonic coder category, while two other coders [284, 285] also created a frequency spectrum consisting of voiced and unvoiced excitation. The primary aim of harmonic coders is to eliminate the frequent "buzziness" of synthesized vocoder speech. This "buzziness" will be inherent to any scheme that divides time-domain speech segments into the distinct categories of voiced and unvoiced.

The buzziness occurs because speech is frequently a composite of voiced and unvoiced excitation sources, as demonstrated by voiced fricatives. In a voiced fricative, such as /v/ in "*v*alve," vocal cord vibration is accompanied by turbulence at a constriction in the vocal tract. Hence, a realistic harmonic excitation source must contain several voiced-unvoiced decisions in various frequency bands of the speech spectra [79]. Griffin and Lim [79] justify this principle with the observation that "buzzy" speech, where the speech has been synthesized from a voiced source, tends to contain a spectrum with some regions dominated by the harmonics of the fundamental frequency and other regions dominated by noise. The introduction of harmonic excitation will degrade the waveform match between the original and reconstructed speech; this is indicative of low-bitrate speech coders that tend to neglect the objective speech quality and concentrate on the subjective quality of the reconstructed speech.

The final speech coder candidate [286] belonged to the waveform interpolation category. This class can also be further subdivided, namely, into time- and frequency-domain interpolation. In waveform interpolation a characteristic, or prototype, waveform is found periodically in the original speech signal, which is then parameterized and transmitted, with interpolation between the selected prototypes producing a continuous synthesis signal.

The interpolation can be performed in either the frequency or time domain, hence the two subcategories. More explicitly, the aim of interpolation-based coders is to represent a small portion of the waveform accurately and then perform interpolation to reproduce the complete speech signal, thus, decreasing the required bitrate while maintaining the speech quality. Following this introduction, each individual candidate coder is now described in a little more detail.

10.1.2.2 Frequency Selective Harmonic Coder. The first candidate coder, the frequency selective harmonic coder (FSHC) [282], was proposed by the Communication Research Centre, in Ontario, Canada. This candidate implements a harmonic coder, which extracts and encodes only the sections of the spectral envelope that are perceptually important. Selective encoding permits the reduction of the bitrate to 2.4 kbps while maintaining good quality speech.

For harmonic coders the frequency spectrum is divided into different bands, with each band being classed as voiced or unvoiced. The speech signal is modeled by a set of sinewaves representing the fundamental frequency and its harmonics, as follows:

$$\bar{s}(n; \omega_0; \theta) = \sum_{l=1}^{L(\omega_0)} A(l\omega_0) \exp[\,j(nl\omega_0 + \theta_l)], \qquad (10.1)$$

where n is the time sample domain index, $L(\omega_0)$ is the number of harmonics in the speech bandwidth, $A(l\omega_0)$ is the vocal tract envelope, ω_0 is the fundamental frequency, and $\theta = \{\theta_1, \theta_2, \ldots \theta_{L(\omega_0)}\}$ represents the phases of the harmonics. In order to achieve a bitrate as low as 2.4 kbps the phases θ are regenerated as minimum or zero phase at the decoder.

The extraction of perceptually important harmonics is performed by dynamic frequency band extraction (DFBE). The DFBE technique extracts the harmonics that are located at the spectral envelope peaks, but it discards the harmonics situated in the spectral valleys. Elimination of the harmonics in the spectral envelope valleys by the DBFE exploits the human ear's reduced sensitivity in these regions. The overall structure of the FSHC speech coder is given in Figure 10.6, and it is described next.

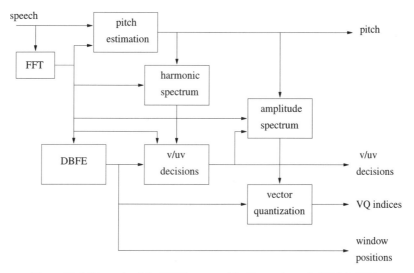

Figure 10.6 Schematic of the FSHC, proposed by Hassanein et al. [282] © IEEE.

The selected pitch period of the speech frame is that which will generate a harmonic spectrum with the minimum mean squared distance from the original spectrum. The DFBE algorithm is used to select the fraction of the spectrum whose spectral amplitudes are quantized for transmission. The window positions define the frequency bands that have been selected by the DFBE algorithm as containing perceptually important harmonics. The pitch-based amplitude spectrum, or spectral envelope, is then vector-quantized and transmitted to the decoder.

10.1.2.3 Sinusoidal Transform Coder.

The second candidate speech compression algorithm was the sinusoidal transform coder (STC) [283, 287], developed by the Lincoln Laboratory at MIT. Similarly, to the FSHC scheme, for this candidate a sinusoidal model is used to synthesize the speech signal, where the sinusoidal model is defined by amplitude, frequency, and phase parameters. These components are determined by an analysis of the short-time fourier transform (STFT) of the speech signal. Bitrate reduction is achieved through forcing the sinusoidal model to have zero-phase. Consequently, the speech signal is defined in [287] by:

$$\bar{s}(n) = \sum_{l=1}^{L(\omega_0)} A(l\omega_0) \cos[(n - n_0)l\omega_0 + \theta_l], \qquad (10.2)$$

where $A(l\omega_0)$ is the vocal tract envelope, ω_0 is the fundamental frequency, n_0 is the onset time that determines the location of the excitation pulse, and θ_l represents the voicing dependent phases, which are here set to zero.

Thus, the encoder parameters are the pitch period, voicing, and the sinewave amplitudes and they are highlighted with reference to Figure 10.7. The sinewave amplitudes were obtained from the magnitude of the harmonics of the fundamental frequency, where they determine the shape of the vocal tract spectral envelope. The amplitudes are encoded by fitting a set of cepstral coefficients to the envelope of the sinewave amplitudes, which proved to be advantageous over an all-pole speech model [283]. Following unsuccessful attempts at encoding the cepstral coefficients directly at a low bitrate, instead they were passed through a cosine transformer and quantized for coding before transmission. Use of a cosine transformer permitted a simple pulse coded modulation (PCM) scheme to be used for the encoding of the cosine transformed cepstral coefficients. The output of the cosine transformer is a set of channel gains, where these channels divide up the frequency spectrum. Before the channel gains were encoded, a perceptually based scale was used to increase the efficiency of the encoding process by placing emphasis on the perceptually relevant lower frequencies. The STC candidate was found to produce good quality speech over the 2.4–4.8 kbps range.

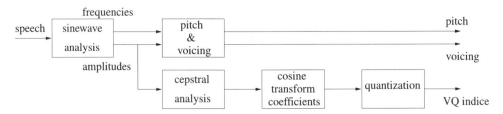

Figure 10.7 Schematic of the STC, proposed by McAulay and Quatieri [283].

10.1.2.4 Multiband Excitation Coders.   The next two candidate speech coders were the advanced multiband excitation (AMBE) Coder, developed by Digital Voice Systems Inc., and the enhanced multiband excitation (EMBE) Coder [281, 288], developed by Oklahoma State University. Both MBE models were based on the original MBE coder [79], which represents the spectral envelope $H_w(\omega)$ of a speech signal by a smoothed version of the original spectrum $\bar{S}_w(\omega)$. The excitation signal $|E_w(\omega)|$ is determined by a series of voiced-unvoiced decisions, either one decision for each harmonic or a decision for certain frequency bands spanning several harmonics. The synthesized speech signal is given by:

$$\bar{S}_w(\omega) = H_w(\omega)|E_w(\omega)|. \tag{10.3}$$

The excitation spectrum $|E_w(\omega)|$ consists of a combination of a periodic spectrum $|P_w(\omega)|$ and a random noise spectrum $|U_w(\omega)|$, where the periodic spectrum $|P_w(\omega)|$ can be viewed as the Fourier Transform of the periodic pulse train used in the LPC-10 speech coder.

The quality of speech, synthesized from the MBE model, is dependent on the correct voiced-unvoiced decisions and on the accurate fundamental frequency calculation, where both were determined using methods similar to those employed by the FSHC of Section 10.1.2.2. The original MBE model [79] was designed to operate at 8 kbps. Hence, the EMBE and AMBE models must reduce the operating bitrate, while preserving the speech quality as much as possible.

For the EMBE coder [288], a 30 ms frame structure containing two subframes was introduced to help decrease the bitrate requirements. Some parameters are transmitted every subframe, while others are transmitted only once per frame and interpolated over the other subframe. The schematic of the EMBE encoder is given in Figure 10.8 which is described below.

The speech waveform is divided into subframes of 15 ms, with the frequency spectrum calculated for each subframe. An integer estimate of the pitch period of the speech waveform from each subframe is also determined, with this initial estimate assessed for evidence of pitch doubling and halving. Since MBE coders use the pitch decision to search for evidence of voicing at the frequency harmonics, this pitch-period estimation is further refined to subsampled accuracy. A voiced-unvoiced decision is made concerning every harmonic in the speech spectrum, based on the closeness of match between the original spectrum and a fully voiced synthesized spectrum, created from the refined pitch-period estimate. Four unequal voiced-unvoiced bands are created, based on the voiced-unvoiced decision for every harmonic within each band, with the division of the four unequal bands influenced by the pitch period and perceptual importance of different frequencies. For the EMBE coder the harmonic spectrum is not represented by the harmonic amplitudes, as in the original MBE. Instead, a

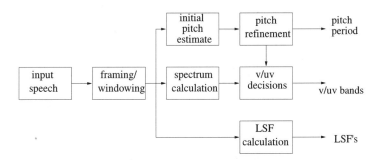

Figure 10.8 Schematic of the EMBE, proposed by Teague et al. [281] © IEEE.

detailed eighteenths order LP model is used. The LPC coefficients are transmitted once every 30 ms frame, while all other parameters are sent every 15 ms subframe.

10.1.2.5 Sub-Band Linear Prediction Coder.

The next speech coder candidate was the SBC LPC Coder [285], suggested by Thomson CSF in France. This candidate uses the frequency division aspect of MBE coders. However, it also employs LPC techniques to produce an LPC synthesis filter. The schematic of this speech coder is given in Figure 10.9.

The SBC divides the frequency spectrum into five sub-bands with each sub-band being assigned a voicing strength, based on Autocorrelation measurements. For speech that contains voicing in any of the sub-bands, the fundamental frequency of the speech signal is found, again employing autocorrelation measures. For synthesis, the fundamental frequency and sub-band voicing strengths are utilized to create an excitation signal constructed of several excitation sources. This mixed excitation is used since it has been hypothesized that such a mixed excitation source, with combined pulses and noise, will remove much of the buzziness of LPC speech [289]. Thus, the excitation signal in each sub-band can be voiced, unvoiced, a mixture of voiced and unvoiced, or transitory, as shown in the stylized Figure 10.10, where the transitory excitation is used for speech with rapidly changing characteristics, which is typical of segments at the onset of voicing. The employment of a variety of excitation sources assists in synthesizing improved quality speech at 2.4 kbps.

10.1.2.6 Mixed Excitation Linear Prediction Coder.

The penultimate short-listed candidate speech coder was the mixed excitation linear prediction (MELP) Coder [284, 290, 291], developed by Texas Instruments.

Similarly to the SBC LPC Coder of Section 10.1.2.5, this speech coder also uses different combinations of voiced and unvoiced sources, which are determined for a series of frequency bands. The scheme proposed harnesses between four and ten frequency bands, with

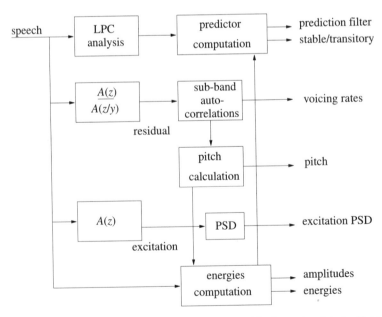

Figure 10.9 Schematic of the SBC LPC coder, proposed by Laurent and de La Noue [285]. © IEEE.

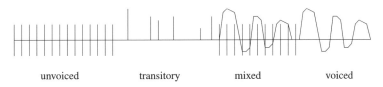

unvoiced   transitory   mixed   voiced

Figure 10.10 Excitation sources for the SBC LPC coder. © IEEE.

Figure 10.11 displaying its schematic. This candidate coder is based on the traditional LPC model, but it features many functions designed to mimic elements of the human speech generation mechanism, which have previously been employed in formant vocoders [292].

A multiband element is harnessed where voiced pulse excitation and Gaussian random excitation are passed through time-varying spectral shaping filters. These spectral shaping filters are combined to give the complete excitation as seen in Figure 10.11. The extent of voicing in a frequency band is determined by the strength of periodicity in that frequency band, while the amount of unvoiced excitation is chosen to keep the excitation power constant in each band.

In the scheme proposed by McCree and Barnwell [284], the vocoder model includes aperiodic pulses in order to simulate voicing transitions, which is similar to the excitation introduced by the SBC LPC of Section 10.1.2.5. These aperiodic pulses were created using a pulse position jitter uniformly distributed over ±25% of the pitch period, which was included only when weak correlation is apparent in the speech signal.

After the voiced and unvoiced excitation sources have been combined, adaptive spectral enhancement is performed, which helps the synthesized speech to match the spectrum of the original speech in the formant regions and is the short-term post-filter described later in Section 11.6. This enhancement is required since synthesized speech tends to reach a lower spectral valley between the time-domain formant resonances than natural speech. The excitation signal is then passed to the LPC synthesis filter and finally to a pulse dispersion filter based on a typical male glottal pulse.

The pulse dispersion filter attempts to spread the excitation energy away from the periodic pulses of the speech coder in Figure 10.11. It models the occurrences when a fraction of the original excitation is concentrated away from the instant of glottal closure. Thus, the pulse dispersion filter simulates the effect of this time-domain spread. This pulse dispersion filter has a time-domain spread based on a typical male pitch period.

10.1.2.7 Waveform Interpolation Coder. The final candidate speech coder was the Waveform Interpolation Coder (WI) [81, 286, 293], developed by AT&T, which is portrayed in Figure 10.12 and described next.

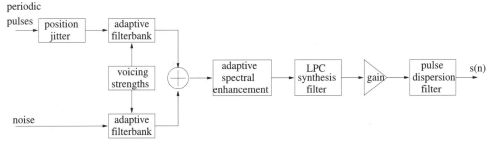

Figure 10.11 Schematic of the MELP coder arrangement, proposed by McCree and Barnwell [284] © IEEE.

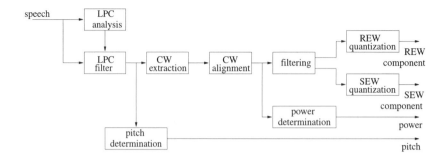

Figure 10.12 Schematic of the WI coder, proposed by Kleijn and Haagen © IEEE.

The waveform interpolation method periodically selects a prototype, or characteristic waveform (CW), which characterizes the speech signal over a given duration. The length of the CW is the pitch period of the input signal. The remainder of the encoding process is performed in the frequency domain; thus, the fast fourier transform (FFT) of the CW is calculated. The current CW is then circularly time-shifted to ensure that the CW is aligned with the previous characteristic waveforms, producing smoothly evolving CWs. The power of the CW is encoded and transmitted, enabling the CW to be normalized to unit magnitude. If the CW is determined at least once per pitch period, then the speech signal can be perfectly reconstructed, but as the CW refreshing interval is extended in an attempt to reduce the bitrate, reconstruction errors will appear. The CW is divided into two signals: the slowly evolving waveform (SEW) for voiced speech; and the rapidly evolving waveform (REW) for unvoiced speech. These two types of waveforms have different characteristics and can be most efficiently encoded separately.

To divide the speech signal into its voiced and unvoiced components, filtering is employed, where high-pass filtering reveals the components of the REW and low-pass filtering produces the SEW. In order to encode the REW, a high sampling rate is used. However, only a rough description of the waveform is encoded, for unvoiced speech is not perceptually important. The SEW signal is initially downsampled to the prototype, which is then accurately described, with the decoder reconstructing the complete signal using interpolation between consecutive CWs.

Following this review of the seven candidate speech coders for the new 2.4 kbps DoD standard we can now reveal that the winning candidate speech coder was the MELP scheme, developed by Texas Instruments. As mentioned before, the MELP is a basic vocoder with many additional features in order to more closely model the human speech production mechanism. The procedure by which the MELP coder was selected is described in Section 10.3.3.

10.1.3 Speech Coding Below 2.4 kbps

Speech coding is also progressing below 2.4 kbps, with an example given here which employs waveform interpolation in the time domain and was introduced by Hiotakakos and Xydeas [294]. It has a small portion, typically a pitch period, of the speech segment encoded in each frame, which is referred to as a prototype segment. The coder schematic is demonstrated in Figure 10.13. Smooth interpolation between the prototypes is performed at the decoder, producing a slow evolution of the excitation signal. The time-domain

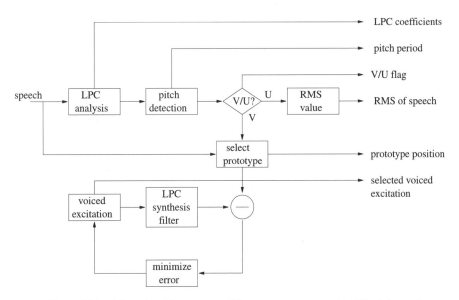

Figure 10.13 Schematic of the prototype WI arrangement, proposed by Hiotakakos and Xydeas [294].

interpolation scheme represents the unvoiced speech separately using random Gaussian noise; thus, as in any vocoder a robust pitch detector will be integral to the model.

For the model proposed by Hiotakakos and Xydeas [294], an orthogonal excitation model that utilizes zinc-basis functions [295] was employed. The excitation model's typical shape is shown by Figure 10.14, where the coefficients A and B describe the function's amplitude and λ defines its position. Sukkar et al. [295] compared the zinc functions to other excitation models, notably to the Fourier series description of the excitation. They found the zinc functions to be superior at modeling the LPC residual, which is partially due to their pulse-like shape being able to mimic the pitch-related residual pulses of voiced speech that remain after LPC analysis. The zinc functions are found to remove some of the buzziness described in Section 10.1.2.1 from the synthesized speech.

The 2 kbps interpolated zinc function prototype excitation (IZFPE) scheme proposed by Hiotakakos and Xydeas [294], detailed also in Chapter 13.2, uses a closed-loop analysis-by-synthesis model that encodes voiced and unvoiced speech separately. It processes 20 ms speech frames. As seen in the block diagram of Figure 10.13, after the LPC analysis the pitch period of the speech frame is determined, where for voiced frames a pitch prototype segment

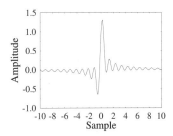

Figure 10.14 Typical shape of a zinc basis function, using the expression $z(n) = A_1 \cdot \mathrm{sinc}(n - \lambda_1) + B_1 \cdot \mathrm{cosc}(n - \lambda_1)$.

is located. The voiced speech frames have their prototype segments modeled with the zinc-basis functions characterized by Figure 10.14, and detailed in Section 13.2, while unvoiced speech frames are modeled by a Gaussian random process. At the voiced-to-unvoiced boundaries, individual pitch periods are examined for evidence of voicing, which improves the coder's performance during rapidly evolving voicing onsets.

During a voiced sequence of frames the phase of the zinc function excitation (ZFE) must be constant, thus permitting the interpolation to be performed in the time domain, which is explained in more detail in Section 13.3.4. The phase of the ZFE sequence is determined by the second voiced frame in a sequence, since according to Hiotakakos and Xydeas [294] the second voiced frame typically represents the voiced sequence better than the first voiced frame. The selection of the phase by the second voiced frame and the consideration of voicing during the last unvoiced frame implies that a delay of 60 ms can be encountered in the coder.

This introductory section has reviewed some of the milestones in low-rate speech coding, dedicated to reducing the bitrate requirements while improving the synthesized speech quality. Special attention has been afforded to the lower bitrates, with particular interest paid to the recent developments at 2.4 kbps.

Following this overview of speech coders, particularly low-bitrate speech coders, a short discussion on the LPC model is presented.

10.2 LINEAR PREDICTIVE CODING MODEL

Linear predictive coding has become a standard model for speech coders. Typically, it uses an all-pole filter to describe the transfer function of the vocal tract. The derivation of this approach can be found in Rabiner and Schafer [6]. An all-pole filter is generally an adequate model of the vocal tract, although the introduction of zeros refines the accuracy of the model, notably when the velum is lowered to introduce nasal coupling. However, the introduction of zeros in the model would prevent the separate optimization of the synthesis filter coefficients and excitation model. It would also increase the bits for encoding and transmission; hence, in practical schemes it is avoided. It is claimed that the appropriate positioning of the poles can mimic the effect of the neglected zeros, making the necessity for zeros redundant and thus reducing the complexity of the filter.

The all-pole or autoregressive model [6] represents the transfer function of the vocal tract by

$$H(z) = \frac{1}{1 - \sum_{k=1}^{p} a_k z^{-k}} = \frac{1}{A(z)} \tag{10.4}$$

where a_k are the coefficients from linear prediction and p is the order of the predictor filter. This modeling of the vocal tract shape is designated short term prediction (STP) as described next.

10.2.1 Short-Term Prediction

The schematic of STP within a basic analysis-by-synthesis model is given in Figure 10.15. Within the encoder section, the input speech is compared with the output of an LPC synthesis filter, whose coefficients a_k have previously been optimized. The excitation source that results in the minimum error between the original and synthesized speech is selected. During the decoder, the excitation source and LPC synthesis filter coefficient parameters are

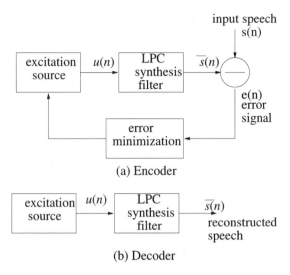

(a) Encoder

(b) Decoder

Figure 10.15 The analysis-by-synthesis approach to LPC.

received from the encoder. Passing the excitation source through the LPC synthesis filter produces the synthesized speech.

Linear prediction is useful as it predicts the next sample based on a weighted sum of previous samples, thus,

$$\bar{s}(n) = \sum_{k=1}^{p} a_k s(n - k), \tag{10.5}$$

where $\bar{s}(n)$ is the predicted sample, and $s(n - k)$ is the kth previous sample. Appropriate values for a_k produce good predictions for $\bar{s}(n)$. Any inaccuracies in the prediction result in an error signal when the synthesized and original speech are compared.

The selection of the number of past speech samples, or filter-order p, is a compromise between a low bitrate and high spectral accuracy. There should be a sufficient number of poles to represent the speech formants with an extra 2–4 poles to simulate the effect of possible zeros and for general spectral shaping. Typically, p is 8–16, and so for a sampling frequency of 8 kHz 1–2 ms of the past speech history is used. Thus, the analysis is referred to as short term prediction (STP).

From Rabiner and Schafer [6], the prediction error of Figure 10.15a is expressed as

$$e(n) = s(n) - \bar{s}(n) \tag{10.6}$$

$$e(n) = s(n) - \sum_{k=1}^{p} a_k s(n - k). \tag{10.7}$$

Taking the z transform, it can be seen that

$$E(z) = S(z)A(z), \tag{10.8}$$

where $E(z)$ is the error, $S(z)$ is the speech, and $A(z)$ is the predictor filter. Thus, the predictor filter is

$$A(z) = 1 - \sum_{k=1}^{p} a_k z^{-k}, \tag{10.9}$$

which is the inverse transform of the vocal tract. Hence, passing the original speech through this inverse transform filter, $A(z)$, produces the residual $E(z)$ which is the ideal excitation source.

The excitation signal is selected through minimization of the mean squared prediction error over a quasi-stationary 10–30 ms or 80–240 sample speech segment. The expression to be minimized is given by

$$\sum_n e^2(n) = \sum_n \left[s(n) - \sum_{k=1}^{p} a_k s(n-k) \right]^2. \tag{10.10}$$

Upon setting the partial derivatives of this expression, with respect to a_k, to zero, we arrive at a set of p equations, delivering the p filter coefficients.

The filter coefficients a_k need to be quantized before transmission to the decoder, while the stability of the LPC STP synthesis filter should be maintained. Due to the need for stability, the filter coefficients must remain within the unit circle. The quantization of any filter coefficients near the unit circle may result in a quantized value outside the circle and hence will be prone to instability problems. In order to maintain stability, the coefficients, a_k, are usually transformed into another parameter before quantization. A more appropriate parameter is the log area ratio (LAR);

$$LAR_i = \log \frac{1 - k_i}{1 + k_i} \tag{10.11}$$

where the parameters k_i are the reflection coefficients taken from vocal tract analysis [6]. A sufficient and necessary condition for the stability of $A(z)$ is $|k_i| < 1$, which can be artificially enforced when arriving at values violating this condition. When transforming k_i to LAR_i, using Equation 10.11, a reduced quantization sensitivity is achieved, facilitating their quantization using a lower number of bits. However, the most commonly used transformed spectral parameters are the line spectrum pairs [LSP] [120] or line spectrum frequencies [LSF] [94]. LSFs have well-behaved statistical properties, and if their ordering property is observed, they will ensure the stability of the filter. The ordering property of the LSFs is expressed as $f_0 < f_1 < f_2 \cdots < f_N$, where f_n are the LSFs.

10.2.2 Long-Term Prediction

The STP process will remove the short-term redundancy of the speech signal but in certain circumstances will typically result in a high prediction residual peak. For instance, when an increasing sample is predicted on the basis of the previous 8–16 samples, but the speech waveform passes its peak and starts to decrease, a high prediction residual peak will occur. This typically occurs at the start of a new pitch period, resulting in a long-term periodicity in the residual. This long-term periodicity corresponds in the spectral domain to a fine needle-structure. In order to remove the corresponding long-term residual periodicity and to model this fine spectral structure, LTP can be performed. However, for the vocoder structure of the coders described in this report, the pulse-like voiced excitation sources remove the necessity of the LTP. Thus, this report never considers the LTP.

10.2.3 Final Analysis-by-Synthesis Model

The LPC model described thus far determines the best excitation signal by minimizing the mean squared difference between the original and synthesized speech. However, the theory of auditory masking can be used to further reduce the perceived signal distortion [296]. The perceived distortion at the output of the decoder will be greatest in areas of low signal strength; therefore, warping or shaping the noise spectrum so that most energy occurs in the formant regions will reduce the subjective effect of the noise. The error weighting filter is defined by:

$$W'(z) = \frac{A(z)}{A\left(\dfrac{z}{\gamma}\right)} = \frac{1 - \sum_{k=1}^{p} a_k z^{-k}}{1 - \sum_{k=1}^{p} a_k \gamma^k z^{-k}} \tag{10.12}$$

where γ is a weighting factor between 0 and 1 that represents the degree of weighting of the error spectrum. A good choice for γ is between 0.8 and 0.9. A computationally more efficient method is to weight the original and synthesized speech, before they are subtracted, as seen in Figure 10.16. This is because the filters $A(z)$ and $\frac{1}{A(z)}$ cancel each other in the synthesis loop, where $\bar{s}_w(n)$ is synthesized a large number of times. The synthesized filter becomes:

$$W(z) = \frac{1}{A\left(\dfrac{z}{\gamma}\right)} = \frac{1}{1 - \sum_{k=1}^{p} a_k \gamma^k z^{-k}}. \tag{10.13}$$

This concludes an overview of LPC modeling, where STP, LTP and error weighting have been introduced. Next, an introduction into measuring the quality of the reconstructed speech is given.

10.3 SPEECH QUALITY MEASURES

Once the speech coder is implemented, it is imperative that speech quality be appropriately assessed. Hence, the speech quality measure must be chosen carefully. Measuring the quality of the synthesized speech can be performed using both objective and subjective measures [61]. Objective measures compare the original and reconstructed waveform and calculate a measure of the distortion between the two signals. Subjective measures involve listening tests, where judgment is passed on the intelligibility and quality of the reconstructed speech. Objective measures are simpler to repeatedly implement and evaluate, allowing the speech

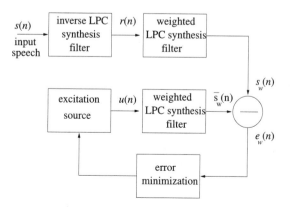

Figure 10.16 Analysis-by-Synthesis with error weighting.

quality to be continually assessed during a coder's development phase. However, subjective measures are always important to assess the human perception of the quality of a speech coder. Before continuing our disscussions we note here that a more detailed discourse on speech quality assessment methods will be provided in Chapter 17.

10.3.1 Objective Speech Quality Measures

The most frequently used objective speech quality measures are the segmental signal-to-noise ratio (SEGSNR) and the cepstral distance (CD) measures. They require the reconstructed speech to be a waveform replica of the input speech. Thus, representing unvoiced frames with a random sequence prevents the use of objective measures on unvoiced frames.

The SEGSNR is a waveform distortion measure that is defined as the distortion between the original and synthesized speech. The SEGSNR is calculated over a quasi-stationary interval of a speech frame. Thus, the distortion from a high-energy portion of speech will not overwhelm the distortion evaluation from a low-energy portion of speech. The SEGSNR measure is defined by:

$$SEGSNR = 10 \log \frac{\sum_{n=1}^{FL} s(n)^2}{\sum_{n=1}^{FL} (\bar{s}(n) - s(n))^2} \tag{10.14}$$

where FL always refers to the frame length.

The CD measure is a frequency spectrum distortion measure that determines the logarithmic spectral envelope distortion between the original and synthesized speech. The CD is given by:

$$CD = \frac{1}{\ln 10} \sqrt{2 \sum_{i=1}^{N_l} (C_{orig}(i) - C_{synth}(i))^2} \tag{10.15}$$

where C_{orig} and C_{synth} are the cepstrum coefficients and N_l is the number of the coefficients.

Because of the low bitrates considered in this report, the synthesized waveform is never a close enough match to the original waveform to utilize the described objective measures. Instead, the techniques used at low bitrates concentrate on retaining the perceptual quality of the reconstructed speech rather than reproducing a waveform match with the original signal.

10.3.2 Subjective Speech Quality Measures

Subjective measures involve listening tests using different quality ratings and test conditions. They include speech quality tests, speech intelligibility tests, pairwise comparison tests, and informal listening tests.

The two main speech quality tests are the mean opinion scores (MOS) and the diagnostic acceptability measure (DAM) [297]. The DAM test is traditionally used when selecting speech coding standards. The listeners assess how the speech coder affects various communication quality attributes. The MOS is the test traditionally implemented for commercial purposes, where the speech quality is classified by listeners into the categories excellent, good, fair, poor, and bad. The speech is then assigned a score from 5 to 1, respectively.

The diagnostic rhyme test (DRT) [298] examines a speech coder's intelligibility and rates the decoded speech on a scale from 0 to 100, where 100 represents no intelligibility errors. Rhyming words are commonly used as the test data, where the word pairs vary only in the first consonant.

Less formally, the pairwise comparison test is a test in which the listener hears two examples of a sentence, usually created by two different coders, and is asked to indicate the preference. The preference percentages are then represented as a measure of the speech coder's quality.

Lastly, informal listening tests are often used under loosely specified experimental conditions. For low-bitrate speech coders, at bitrates less than 4 kbps, it is often difficult to apply objective measures and time consuming to perform subjective tests. Thus, informal listening tests can be used as a quick measure of coder improvement. Specifically, an informal listening test might involve commenting on whether a specified distortion has been reduced by a slight change in the coder's operation.

10.3.3 2.4 kbps Selection Process

The selection of the new DoD speech coding standard for 2.4 kbps provides an informative insight into the methods used for determining the quality of a speech coder. The types of environment under which a speech coder must operate efficiently are also examined, along with the compromise between the requirements desired by the multitude of potential applications that may use the current 2.4 kbps standard.

The basic system requirements were that its performance must match the quality produced by the FS1016 CELP speech coder at 4.8 kbps. It was also important that the speech coder have a low power consumption and produce high-quality speech in a range of environments. The quality of the speech must be sufficiently high to enable speaker recognition. The model must also have the ability to operate in tandem with other systems working at different rates. The Terms of Reference were described by Tremain et al. [299] and consisted of a list of parameters, with their respective minimum acceptable values, and the objective measurements which users would wish the coder to achieve. The selected algorithm was the speech coder which provided the best overall performance to meet the Terms of Reference in a variety of noisy environments.

The test structures for the 2.4 kbps selection process [300] measured intelligibility, voice quality, talker recognizability, and communicability. The tests performed were the DRT for intelligibility and the DAM, the MOS, and the degradation mean opinion score (DMOS) for voice quality. For recognizability, a test was developed at the Naval Research Laboratory (NRL), while communicability was measured using a test designed for the U.S. Air Force Rome Laboratory (USAF-RL). The tests were performed in a number of noise environments, including an office, a car, and aircraft. The tests were also performed in a quiet environment.

The DRT for intelligibility was briefly described in Section 10.3.2. For test data the DDVPC have a large lexicon of DRT word pairs in a variety of acoustic environments.

As regards the voice quality measure the DAM, MOS and DMOS tests were examined [301] and were briefly highlighted in Section 10.3.2. It was found [301] that the MOS tests provided the most reliable set of performance measures. Thus, the traditional DAM measure was neglected.

As mentioned earlier, a test for speaker recognizability was developed by the NRL [302]. A new test was required since virtually no testing for recognizability had previously been performed. There are two levels of recognizability: the highest, in which the phone users are recognizable as themselves and the lower level of recognition; where speakers can be distinguished as different.

The testing approach employed involves a "same-different" decision being performed on the basis of whether pairs of sentences were spoken by either the same or different people. Both male and female speakers were used, with the sentence pairs constructed from processed and unprocessed speech. The term *processed* implies that the speech has been passed through the test speech coder, while the term *unprocessed* means that the speech is unmodified. The sentence pairs contained processed-processed pairs and processed-unprocessed pairs.

The employed communicability test was designed on behalf of the NASF-RL by the ARCON Corporation [303]. Communicability is a measure of the speech coder's quality under simulated operational use, including a variety of environments and different numbers of people interacting. A specified task requiring the combined effort of at least two people, linked via the speech coder, is performed. Each person involved in the test rates various aspects of the speech coder on a seven-point scale. The aspects are the level of the effort required in communicating using the coder, the quality of the received speech, the effect of the scheme on communication and task performance, and the overall acceptability of the model.

The overall selection of the new 2.4 kbps coding standard involved combining the performance scales of the algorithm in each test. The encoder's speech quality was assigned 30% of the marks, the intelligibility was assigned 35%, and the recognizability was given 15%, while the communicability was assigned 20%. The combined test total was assigned 85% of the overall marks, with the algorithm's complexity being given the remaining 15%.

The evaluation procedure initially concentrated on finding the speech coder candidates that exceeded the minimum requirements in the Terms of Reference. Subsequently, the remaining candidates were examined in more depth to find the coder that best achieved the objectives described in the Terms of Reference. As mentioned earlier, in Section 10.1.2, the winning candidate was the MELP coder developed by Texas Instruments. The other coders that met the minimum requirements were the AMBE developed by Digital Voice Systems Inc, the WI coder developed by AT&T, and the STC scheme developed by the Lincoln Laboratory at MIT.

This overview of speech quality measures has considered both subjective and objective speech measures. It has also given insights into the role of speech quality measures in selecting the winning candidate for speech coding standards. Finally, the speech database implemented throughout the developed speech coders is documented in the next section.

10.4 SPEECH DATABASE

The speech database used for our experimental purposes is detailed in Table 10.1 and subsequently is referred to by the speaker code.

TABLE 10.1 Details of the Speech Database

Speaker Code	Speaker Sex	Dialect of English	No. 20 ms Frames	Utterance
AM1	Male	American	114	Live wires should be kept covered.
AF1	Female	American	120	The kitten chased the dog down the street.
AM2	Male	American	152	The jacket hung on the back of the wide chair.
AF2	Female	American	144	To reach the end he needs much courage.
BF1	Female	British	123	Glue the sheet to the dark blue background.
BF2	Female	British	148	Rice is often served in round bowls.
BM1	Male	British	123	Four hours of steady work faced us.
BM2	Male	British	158	The box was thrown beside the parked truck.
Training	Mixed	American	2250	Conversation.

The database also contains 45 seconds of speech, which were used as training data for the quantizers designed within the speech coders. The speech is a mixture of American male and female utterances.

The speech database contains about 20 seconds of speech, uttered by four male and four female speakers with either American or British accents. The speech was recorded with no background noise and initially was stored in a 12-bit linear PCM representation. Figures 10.17 and 10.18 display the pitch-period track of each file, which was determined manually

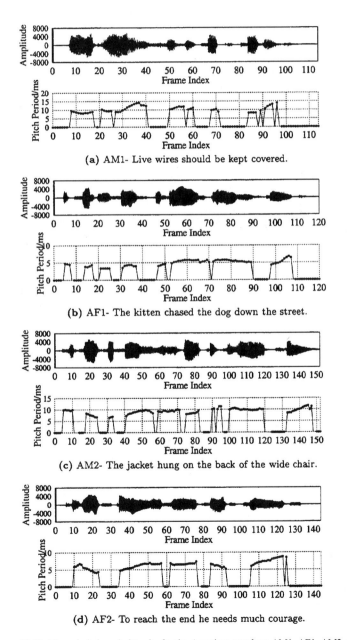

(a) AM1- Live wires should be kept covered.

(b) AF1- The kitten chased the dog down the street.

(c) AM2- The jacket hung on the back of the wide chair.

(d) AF2- To reach the end he needs much courage.

Figure 10.17 Manual pitch period tracks for the American speakers AM1, AF1, AM2 and AF2 from the speech database.

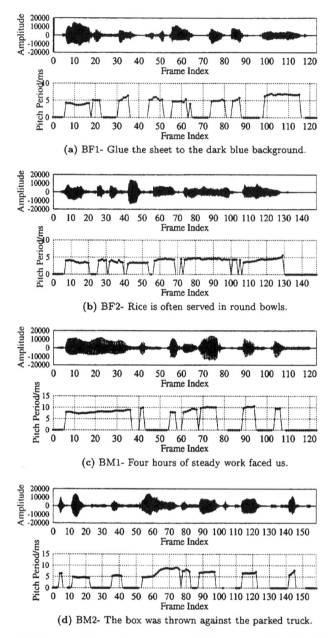

Figure 10.18 Manual pitch period tracks for the British speakers BF1, BF2, BM1 and
BM2 from the speech database.

for each speech frame. Frames that showed no visual evidence of voicing, in periodicity
terms, were set to a pitch period of zero. For some speech frames, it was difficult to determine
the pitch period or even if the frame was voiced. These frames typically occurred at the end of
voiced utterances. Later in this low-rate-coding oriented part of the book, particularly in
Chapters 11 and 12, the manually determined pitch period tracks of Figure 10.1 are used to
assess automated pitch detectors. For the speech frames where our pitch-period determination

was unreliable, the manually determined tracks were simply ignored in any assessment of the pitch-period detectors.

The pitch periods for the speakers were only permitted to be in the range of 18 ms to 2.5 ms, or 54 Hz to 400 Hz. These limits were introduced due to our bitrate constraints. Allowing pitch periods between 20 and 147 samples at a sampling rate of 8 kHz results in only 7 bits being required to transmit the 128 legitimate parameter values, while covering most expected pitch periods. It is acknowledged that some speakers will have pitch periods outside this region, particularly children. However, the pitch-period range selected permits us to use an integer pitch-period length in samples cover a wide range of expected pitch periods.

10.5 CHAPTER SUMMARY

This chapter has given a rudimentary overview of the factors influencing the development of speech coders, paying particular attention to the recent selection of a 2.4 kbps speech coder to replace the DoD FS1015 standard. Speech coding for bitrates less than 2.4 kbps was also reviewed. A brief description of LPC was given in Section 10.2, where analysis-by-synthesis was also introduced. A review of assessing the speech quality was given in Section 10.3, where particular attention was given to the speech quality measures adopted for the DoD 2.4 kbps standard. Finally, the speech database used throughout this low-rate-coding-oriented part of the book was introduced. In the next chapter we focus on the most predominant coding techniques used at coding rates below 2.4 kbps, namely, on vocoders.

11

Linear Predictive Vocoder

In this chapter we introduce a basic LPC vocoder operating on 20 ms frames, which will provide a benchmark for the low-bitrate coders developed in Chapters 13 and 14. In addition, Section 11.2 introduces the line spectrum frequency (LSF) quantizer to be used throughout the developed coders. The notion of pitch detection is introduced in Section 11.3, and the adaptive post-filter that is implemented in the developed decoders is described in Section 11.6.

11.1 OVERVIEW OF A LINEAR PREDICTIVE VOCODER

Since the topic of a linear prediction was covered in detail in Chapter 3, here only a rudimentary overview is given in order to allow readers to consult this chapter without having to read the previous chapters. The basic LPC vocoder schematic is shown in Figure 11.1. In the encoder, initially LPC STP analysis is performed in order to determine the LPC STP synthesis filter coefficients, which are then quantized into LSFs for transmission to the decoder, as described in Section 11.2. After LPC STP analysis, the short-term correlation has been removed from the speech waveform leaving the STP residual, which contains the prediction errors associated with the LPC STP analysis. This STP residual has its RMS energy determined, quantized, and sent to the decoder where it is used to scale the unvoiced excitation. The STP residual is also used in the pitch detection process, described in detail in Section 11.3, where the LPC STP residual displays more conclusive evidence of voicing due to the removal of the short-term correlation. Incorporated in the pitch detector is a voiced-unvoiced decision, which sets a flag to inform the decoder whether voiced or unvoiced excitation should be used in the synthesis process.

At the decoder, either random Gaussian noise for unvoiced excitation (detailed in Section 11.4) or a periodic pulse stream for voiced excitation (described in Section 11.5) is passed to the LPC STP synthesis filter. The subsequent output waveform is then passed to an adaptive post-filter (described in Section 11.6), which improves the perceived quality of the synthesized waveform by emphasizing the speech spectrum's formants and the spectral pitch harmonics' formants. The resultant waveform is the reconstructed speech signal.

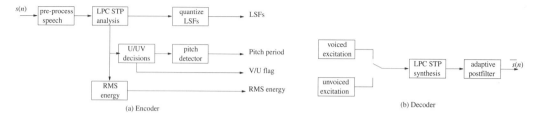

Figure 11.1 Schematic of the implemented LPC vocoder.

Following this overview of the LPC vocoder, the important implementation issues are discussed. First, the methods for quantizing the LSFs are considered.

11.2 LINE SPECTRUM FREQUENCIES QUANTIZATION

For low-bitrate speech coders, a significant portion of the available bitrate is consumed by the transmission of the LSF parameters. Thus, next we investigate an economical method of transmitting the LSFs while maintaining good perceptual speech quality. Two quantization methods are discussed here. The first is the scalar quantizer used in the DoD CELP standard FS1016 [76], which requires 34 bits/30 ms; the second is the vector quantizer from the ITU standard G.729 [123], which transmits 18 bits/10 ms. The LSF quantizer is incorporated from a speech coding standard due to the extensive training that will have been undertaken in the standardization process. To operate the quantizer at its full potential, the same pre-processing as in the standard must be employed, thereby ensuring that the quantizer is operating on speech similar to its training data.

11.2.1 Line Spectrum Frequencies Scalar Quantization

The Scalar Quantizer (SQ) from FS1016 [76] uses 34-bit nonuniform SQ for the LSFs, with the bit assignment for the LSFs given in Table 11.1. The SQ is designed to send the LSFs once every 30 ms speech frame, and they are smoothly interpolated over the 7.5 ms subframes. Since the SQ operates separately on each speech frame, the quality of the SQ will not be affected by decreasing the speech frame length to 20 ms.

The speech coding standard FS1016 [76] includes pre-processing of the speech signals in the form of a Hamming window and 15 Hz bandwidth expansion of the LPC STP filter coefficients. The Hamming window is given by:

$$w_{ham}(n) = \begin{cases} 0.54 - 0.46 \cos\left(\dfrac{2\pi n}{FL}\right), & 0 \leq n \leq FL - 1 \\ 0 & \text{elsewhere} \end{cases} \qquad (11.1)$$

where FL is the speech frame length.

TABLE 11.1 Bit Allocation for LSF Coefficients from FS1016

LSF coefficient	1	2	3	4	5	6	7	8	9	10
No. bits allocated	3	4	4	4	4	3	3	3	3	3

The 15 Hz bandwidth expansion is achieved by modifying the LPC STP filter coefficients according to the following expression:

$$\hat{a}_i = a_i \times 0.994^i \, , 1 \leq i \leq 10 \tag{11.2}$$

where a_i is the ith LPC STP filter coefficient.

Figure 11.2 demonstrates the performance of the 34-bit SQ for the speech file BF1. It shows that, while the SQ is generally good at following the unquantized LSF values, on occasion the unquantized LSF values exceed the dynamic range of the SQ.

The performance of an LSF quantizer is typically determined using the spectral distortion (SD) measure, given by:

$$S_d = \sqrt{\frac{1}{I}\sum_{i=1}^{I}[10\log(P_i) - 10\log(\hat{P}_i)]^2} \tag{11.3}$$

where S_d is the SD, P_i is the ith point in the frequency spectrum using unquantized LSF values, $\hat{P}_i$ is the ith point in the frequency spectrum using quantized LSF values, and I is the number of points in the frequency spectrum. The frequency spectra are obtained by converting the unquantized and quantized LSF values into unquantized and quantized LPC

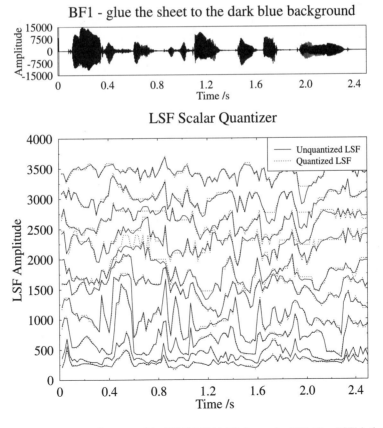

Figure 11.2 The performance of the FS1016 34-bit SQ for speaker BF1. Here LSF1 is the lowest trace, with LSF10 being the uppermost trace. Occasionally, unquantized LSF values exceed the limits of the SQ. This occurs in LSF8, LSF9, and LSF10 around the 2s mark.

STP filter coefficients, respectively. The frequency responses created by these filter coefficients are P_i and $\hat{P}_i$.

An LSF quantizer is aimed at achieving three targets in its SD measure [92]: (1) the average SD is approximately 1 dB, (2) the percentage of speech frames with an SD in the 2 dB → 4 dB range is less than 2%, and finally (3) the percentage of speech frames with an SD of greater than 4 dB is negligible.

Table 11.2 gives details about the performance of the SQ in meeting these performance criteria. It shows that the percentage of outlier frames in the 2 → 4 dB range and in the range above 4 dB are much higher than desired. However, the average SD is approximately 1 dB. Figure 11.3a displays the probability density function (PDF) of the LSF SQ SD, showing the existence of a long tail to the right. Next we introduce the LSF vector quantizer from G.729 [123].

11.2.2 Line Spectrum Frequencies Vector Quantization

The vector quantizer (VQ) from G.729 [123] is a predictive two-stage VQ which sends 18 bits/10 ms. If the LSF coefficients were transmitted every 10 ms, then with our 20 ms frame length 36 bits would be required to encode the LSFs every speech frame. This higher bitrate requirement compared to the SQ is not acceptable. A suitable alternative is to calculate the LSF coefficients for a 10 ms subframe but only send one set of quantized LSF values for the two subframes, to produce a bitrate of 18 bits/20 ms. The extra computational complexity of performing the required pre-processing on 10 ms subframes must be tolerated so that the quantizer is working on the speech data it was trained for. However, due to the predictive nature of the LSF VQ, the quantization itself is only performed once every 20 ms.

The pre-processing of the speech performed in G.729 [123] includes a high-pass input filter, windowing, and bandwidth expansion. The high-pass input filter has a cutoff frequency of 140 Hz and divides the input signal by two in order to avoid overflows in the G.729 [123] fixed point implementation. The input filter's transfer function is given by:

$$H_{h1}(z) = \frac{0.4636718 - 0.92724705z^{-1} + 0.4636718z^{-2}}{1 + 1.19059465z^{-1} + 0.9114024z^{-2}}. \tag{11.4}$$

Similarly, at the output of the decoder, a high-pass output filter with a cutoff frequency of 100 Hz is introduced. The signal must also be multiplied by two, restoring the correct amplitude level. The output filter's transfer function is given by [123]:

$$H_{h2} = \frac{0.93980581 - 1.8795834z^{-1} + 0.93980581z^{-2}}{1 - 1.9330735z^{-1} + 0.93589199z^{-2}}. \tag{11.5}$$

The windowing used in G.729 [123] is a hybrid window and spreads over several 10 ms speech frames. It includes 120 samples from previous speech frames, 80 samples of the

TABLE 11.2 SD Performance of the FS1016 SQ and the G.729 VQ

Quantizer	Mean SD/dB	SD % within 2→4dB	SD % ?? 4 dB
Scalar	1.16	10.85	1.10
Vector	0.78	1.09	0.00

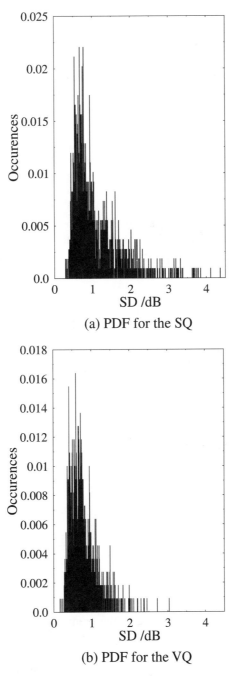

(a) PDF for the SQ

(b) PDF for the VQ

Figure 11.3 The SD PDF for the FS1016 scalar and G.729 vector quantizers. It can be seen that the SD PDF of the VQ in (b) is much more compact.

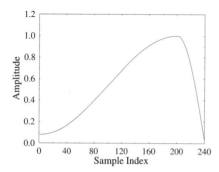

Figure 11.4 The hybrid window employed in G.729 for pre-processing the speech.

current speech frame, and 40 samples from the future speech frame. The window is displayed graphically in Figure 11.4, where the peak of the window is over the end of the current speech frame and the function created using the following expression [123]:

$$
w_p(n) = \begin{cases} 0.54 - 0.46 \cos\left(\dfrac{2\pi n}{399}\right), & 0 \leq n \leq 199 \\[2ex] \cos\left(\dfrac{2\pi(n - 200)}{159}\right), & 200 \leq n \leq 239. \end{cases} \tag{11.6}
$$

The final pre-processing performed is a 60 Hz bandwidth expansion of the LPC STP filter coefficients. This is implemented by using another windowing function on the autocorrelation coefficients, $r(k)$, from the LPC STP analysis. The autocorrelation windowing function is given by [123]:

$$
w_{lag}(k) = \exp\left[-\frac{1}{2}\left(\frac{2\pi f_o k}{8000}\right)^2\right], k = 1 \cdots 10 \tag{11.7}
$$

with a bandwidth expansion of 60 Hz, $f_o = 60$.

Figure 11.5 demonstrates the performance of the 18-bit VQ for the speech file BF1, also used in Figure 11.2. It demonstrates that the predictive nature of the VQ ensures that the unquantized values never exceed the limit of the quantizer.

The performance of the VQ was also evaluated using the SD measure. Table 11.2 shows the success of the VQ at meeting the three SD criteria. The average SD measure is less than 1 dB, and the number of outlier frames having SDs greater than 2 dB is negligible. The right-hand side of Figure 11.3b displays the PDF of the SD measure for the VQ, where it can be seen that the VQ's PDF is much more compact than the SQ, implying a better performance.

Due to the superior SD performance and the reduced bitrate, the LSF VQ was used in all speech coders developed in this low-bitrate-oriented part of the book. Next we investigate one of the most critical tasks in low-bitrate speech coders, namely, the selection of a reliable and robust pitch detector.

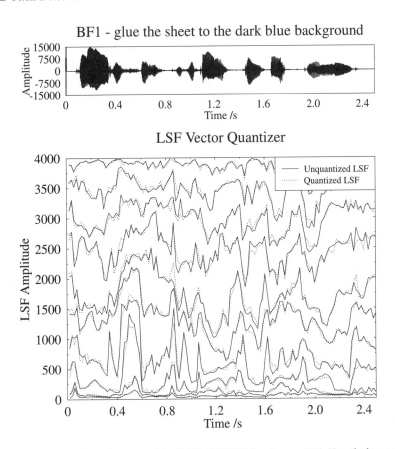

Figure 11.5 The performance of the G.729 18-bit VQ for utterance BF1. Here the lowest trace is LSF1, while LSF10 is the upper trace. The VQ performs well at quantizing the LSF values.

11.3 PITCH DETECTION

In traditional vocoders, the decision regarding the extent of voicing in a speech segment is critical. Pitch detection is an arduous task owing to a number of factors, such as the nonstationary nature of speech, the filtering effect of the vocal tract, and the presence of noise. Incorrect voicing decisions cause distortion in the reconstructed speech, and distortion is also apparent if the common phenomenon of pitch doubling occurs. Pitch doubling happens when the energy level of adjacent harmonics is higher than the energy of the fundamental frequency. Smaller pitch errors occur if the analysis window is too short, whereas with too long an analysis window a nonstationary signal may be encountered.

A detailed explanation of the considerations, together with the various approaches to pitch detection, can be found, for example, in the monograph by Hess [14]. Many different methods exist for pitch detection of speech signals, giving an indication of the difficulty involved in producing a robust pitch detector. Perhaps the most commonly used approaches are the autocorrelation-based methods, where the autocorrelation function (ACF) for a segment of speech is determined. Subsequently, the time-offset where the normalized correlation becomes maximum is deemed to be the pitch-period duration. The normalizing parameter is the autocorrelation at zero delay, namely, the signal's energy. If the maximum correlation value exceeds a certain threshold, the segment of speech is considered voiced, while beneath this threshold an unvoiced segment is indicated.

Another approach to pitch detection is to use a pattern recognizer where a selection of speech properties are assessed to make a voiced-unvoiced classification [304]. Atal and Rabiner [304] claim that voiced-unvoiced classification and pitch determination are two distinct problems that are best treated separately. The speech classification can be determined using measures such as signal energy, zero-crossing rate, and the energy of the prediction error, where each selected measure reacts differently to voiced and unvoiced speech. The output quantities of the above classifiers are assessed, and an overall decision about voicing is made. Since no decision concerning the pitch is carried out, the voiced-unvoiced decision can be performed on speech segments having a length less than a pitch period. In addition, this method lends itself to implementation with a neural network [305].

Another popular method for pitch determination is to use the cepstrum [306]. Similarly to the autocorrelation method, if the peak cepstral value exceeds a threshold, the speech is considered voiced. If the cepstral peak value did not exceed the threshold, a zero-crossing count is performed, where if the number of zero crossings exceeds a threshold the speech is deemed unvoiced; otherwise, the frame is considered voiced. For voiced segments the pitch period is again the location of the peak cepstral value. However, the calculation of the FFT of the speech segment that is required for obtaining the cepstral peak value is computationally intensive.

Recently, the wavelet transform has been applied to the task of pitch detection [307]. The wavelet approach to pitch detection is event based, which means that both the pitch period and the glottal closure instant (GCI) are located. The pitch determination methods previously mentioned are all nonevent based and assume that the pitch period is stationary within the analysis window.

Kadambe and Boudreaux-Bartels [307] used a dyadic wavelet transform denoted by $D_Y WT$. For the $D_Y WT$, the time-domain discontinuities in the speech signal, such as those at the pitch-related speech signal peaks, are represented by the corresponding local maxima in the time domain after the wavelet transform. The action of glottal closure will create a discontinuity in the speech signal. Thus, the resultant time-domain representation after the $D_Y WT$ will contain local maxima at the locations of glottal closure. However, the maxima must exceed a certain threshold for a speech segment to be labeled as voiced. The $D_Y WT$ is performed at different time-domain resolutions or scales, hence ensuring that it adequately resolves the expected fundamental frequency range (54–400 Hz). For a voiced speech segment, the local maxima at different scales will be aligned. An additional feature of the wavelet transform is that it does not require a full pitch period to operate effectively. Use of the wavelet transform is described in detail in Section 12.5.

A further review of pitch detection methods was given by Rabiner et al. [308]. In the documented LPC vocoder the ubiquitous autocorrelation approach was employed. Figure 11.6 shows the ACF for various delays for a segment of voiced speech, demonstrating that for a voiced speech segment a correlation spike occurs at the appropriate pitch period, with further peaks at the pitch harmonics.

11.3.1 Voiced-Unvoiced Decision

For the pitch predictor investigations initially a simple scheme was employed, where the speech was low-pass-filtered to 900 Hz. Noise tends to contaminate the low-energy, high-frequency speech components. Thus, by removing the high-frequency noise the prominence of the pitch-related signal components increases. The next stage involves center clipping

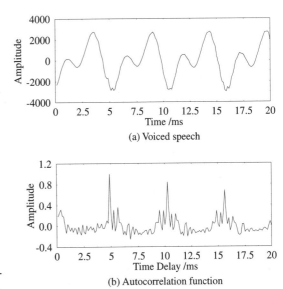

Figure 11.6 Example showing the autocorrelation for a voiced speech frame speech file AF2.

(a) Voiced speech

(b) Autocorrelation function

setting the low-magnitude signal segments to zero in order to increase the prominence of the signal's periodicity, with autocorrelation subsequently performed. This simple pitch detector failed to detect some voiced frames, particularly those near the start or end of a voiced sequence.

An alternative pitch detection technique can be constructed using the approach of the G.728 Recommendation [85]. Here the signal utilized for pitch detection is the residual signal after the LPC STP analysis, since the pitch period becomes more prominent in the residual signal due to the removal of the short-term correlation by the LPC process. The ACF selects the best candidate, A, in the current residual frame for the pitch period and the best candidate, B, around the old pitch period used in the previous frame. Preferential treatment of candidate B attempts to remove the chance of pitch doubling, through the introduction of pitch tracking as follows. If the pitch gain at delay B is more than 40% of the pitch gain for delay A, then candidate B is selected; otherwise, candidate A is selected. If the successful candidate has a pitch period gain higher than 0.7, then the frame is considered voiced, with a pitch period equal to the selected delay. Determination of unvoiced frames depends on whether the previous frame was voiced or unvoiced. If the previous frame was unvoiced, a pitch period gain of less than 0.7 indicates an unvoiced frame. Hence, the pitch period is set to zero. However, if the previous frame was voiced, then a pitch gain of greater than 0.5 would indicate a voiced frame.

The voicing strength is defined by the following normalized correlation function:

$$\frac{\sum_{n=0}^{FL} s(n) \times s(n - P)}{\sum_{n=0}^{FL} s(n - P) \times s(n - P)} \tag{11.8}$$

where FL is the frame length in samples and P is the selected pitch-period length in samples. Thus, the voicing strength is the ratio of the cross-correlation of the speech signal, $s(n)$, and pitch-period duration delayed speech signal, $s(n - P)$, to the energy of the pitch-period

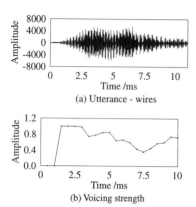

(a) Utterance - wires

(b) Voicing strength

Figure 11.7 Example showing the evolution of voicing strengths for the speech file AM1.

duration delayed speech signal. Displayed in Figure 11.7 is the evolution of the voicing strength for speech file AM1. The voicing strength can be seen to frequently fall to low levels, while the time-domain plot of the speech clearly shows that it is voiced. A pitch in the range of 54 Hz to 400 Hz is permitted because this is the typical range of human fundamental frequency. Figure 11.8 demonstrates that the current autocorrelation-based pitch detector

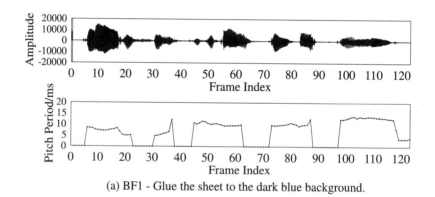

(a) BF1 - Glue the sheet to the dark blue background.

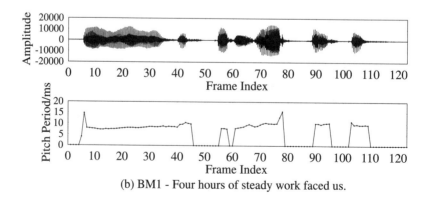

(b) BM1 - Four hours of steady work faced us.

Figure 11.8 Pitch-period decisions for (a) BF1 and (b) BM1. For BF1 a gross pitch error is visible at frame 36 in comparison to the manual track of Figure 10.18, with the rest of the speech utterance subjected to pitch halving. For BM1, gross pitch errors are visible at frames 5 and 77.

TABLE 11.3 A Comparison between the Performance of the Developed Pitch Detectors and the Manual Pitch-Period Track for the Speech Database. W_U represents the percentage of frames that are labeled voiced when they should have been identified as unvoiced. W_V indicates the number of frames that have been labeled as unvoiced when they are actually voiced. P_G represents the number of frames where a gross pitch error has occurred. The total number of incorrect frames is given by $W_U + W_V + P_G$.

Pitch Detector	$W_U\%$	$W_V\%$	$P_G\%$	Total %
ACF-based method	1.6	5.3	5.8	12.7
Oversampled ACF method	5.4	3.9	4.7	14.0
Oversampled ACF with tracking	3.1	2.3	1.8	7.2

produces both gross pitch errors and pitch halving errors, with more than half of the utterance BF1 subjected to pitch halving. The performance of this autocorrelation-based pitch detector is compared, in Table 11.3, for the entire speech database, against the manual pitch period track of Figures 10.17 and 10.18 from Section 10.4. It was found that 12.7% of frames were incorrectly labeled.

11.3.2 Oversampled Pitch Detector

A method frequently employed to improve the performance of autocorrelation-based pitch detectors is oversampling, where oversampling will increase the time-domain resolution of the search. Both the DoD 4.8 kbps standard FS1016 [173] and the Vector-Sum Excited Linear Predictive Coding (VSELP) coder utilized in the GSM half-rate coder [73] use oversampling to improve the pitch tracker's performance.

For the GSM VSELP coder, the signal is up to six times oversampled, thus allowing noninteger delays to be accepted as the pitch period. Table 11.3 shows the integer delays allowed for various pitch-period ranges.

Observing Table 11.4, we see that the GSM VSELP coder provides child and adult female speakers with the highest resolution, while adult male speakers have a lower resolution. This variable resolution produces a relative pitch error with respect to the pitch itself which is nearly constant, maintaining a similar pitch detection quality for all speakers.

In order to calculate the voicing strength of a noninteger delay l_d/D_s at a sampling frequency f_s and oversampling rate D_s, the up-sampling is performed and the equivalent integer delay l_d at sampling frequency $D_s \cdot f_s$ is found. From Figure 11.9 it can be seen that the up-sampled signal is generated by inserting $D_s - 1$ samples between every original input sample with the inserted samples being zero-valued. The resultant signal is low-pass-filtered to obey the Nyquist rate, thus producing an oversampled version of the input signal.

TABLE 11.4 Allocation of Noninteger Delays for the GSM VSELP Half-rate Coder

Delay range	Delay resolution	Number of sample points
21→22, 2/3	1/3	6
23→34, 5/6	1/6	72
35→49, 2/3	1/3	45
50→89, 1/2	1/2	80
90→142	1	52

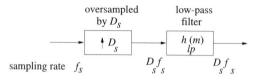

Figure 11.9 Schematic of the process of interpolation, where the inserted $D_s - 1$ values are zero samples.

For the DoD 4.8 kbps standard FS1016 [76], the up-sampling is performed with an 8-point Hamming window sinc resampling function. Thus, to oversample by a factor of:

$$w_{f_{id}}(i) = w_{ham}(12(i + f_{id})) \frac{\sin(\pi(i + f_{id}))}{\pi(i + f_{id})}, \tag{11.9}$$

where $i = -N_{ip}/2, -N_{ip}/2 + 1, \ldots, N_{ip}/2 - 1$ and $N_{ip} = 8$ is the number of interpolation points. The noninteger delays are given by $f_{id} = \frac{1}{6}, \frac{1}{3}, \frac{1}{2}, \frac{2}{3}, \frac{5}{6}$ with the integer delays given by M. The Hamming window is given by $w_{ham}(k) = 0.54 + 0.46 \cos(\pi k / 6N_{ip})$, where $k = -6N_{ip}, -6N_{ip} + 1, \ldots, 6N_{ip} = -48$ to 48.

Then for a noninteger delay $M + f_{id}$ we have:

$$r_{M+f_{id}}(i) = \sum_{k=-N_{ip}/2}^{N_{ip}/2} w_{f_{id}}(k) r_{M+f_{id}}(i - M + k) \tag{11.10}$$

where the index, i, is some point in the speech frame from which all delays are calculated and $r_{M+f_{id}}(i)$ represents a sampling point at the new sampling frequency D_s. Figure 11.10 shows a speech signal that has been oversampled by a factor of 6, where the peaks and valleys are more extreme and the signal is smoother.

The oversampled speech signal can be subjected to autocorrelation computation in order to locate the most likely pitch-period delay and to determine the voicing strengths. Figure 11.11 shows the updated pitch-period decisions for the same utterances as displayed in Figure 11.18 and 11.8, where the voiced-unvoiced threshold levels were updated from 0.7 and 0.5 to 0.8 and 0.5. Figure 11.11 shows that most of the pitch halving was removed from the pitch track; however, many more gross pitch errors were introduced. This oversampled pitch detector was compared against the manual track of Figures 10.17 and 10.18 in Table 11.4, where 14.0% of the frames were incorrectly determined.

11.3.3 Pitch Tracking

Explicit checking for pitch doubling and halving, together with pitch tracking mechanisms, are frequently employed in pitch detectors. The GSM half-rate VSELP coder [73] performs explicit checking for pitch doubling and halving. Figure 11.12 describes its operation with the flowchart followed below.

As seen in Figure 11.12, initially all the submultiples, down to 54 Hz, of the best pitch are checked. Once the submultiple has been located, the adjacent integer peaks are examined

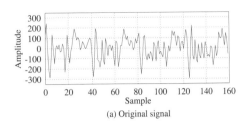

(a) Original signal

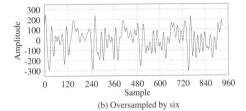

Figure 11.10 Oversampling of a speech signal by a factor of 6.

(b) Oversampled by six

to ensure that the associated prediction gain of the proposed submultiples is the highest possible value. The prediction gain is given by C_1^2/G_1, where:

$$C_1 = \sum_{n=0}^{FL} s(n) \times s(n - P) \tag{11.11}$$

$$G_1 = \sum_{n=0}^{FL} s^2(n - P) \tag{11.12}$$

with P the proposed pitch-delay, and C_1 is the correlation of $s(n)$ in the numerator of Equation 11.8, while G_1 is the energy of $s(n)$ in the denominator of Equation 11.8.

If the prediction gain at the proposed pitch submultiple is still higher than that of its neighbors, then the prediction gains of the surrounding noninteger delays are examined. The delay exhibiting the highest prediction gain is compared against a threshold as seen in Figure 11.12. If the threshold is exceeded, the associated submultiple is selected as the pitch-delay. Once all possible submultiples have been checked, all multiples of the current proposed pitch-delay are examined, up to 400 Hz. A similar procedure is followed for the pitch multiples, with the best proposed pitch-delay selected as the true pitch-delay.

The threshold for selecting a new pitch-delay is given by [73]:

$$\frac{C_1^2}{G_1} > R(0) - \frac{R(0)}{10^x}, \tag{11.13}$$

where $R(0) = \sum_{n=0}^{FL} s^2(n)$, and:

$$x = \alpha_{CG} \log_{10}\left(\frac{R(0)}{R(0) - C_{best}^2/G_{best}}\right), \tag{11.14}$$

where C_{best}^2 and G_{best} are the values for the proposed pitch-delay and the factor $\alpha_{CG} = 2.75$ was determined experimentally.

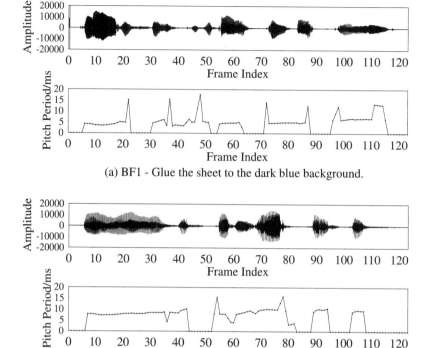

(a) BF1 - Glue the sheet to the dark blue background.

(b) BM1 - Four hours of steady work faced us.

Figure 11.11 Pitch period decisions based on the technique of Section 11.3.2 for (a) BF1 and (b) BM1. Here the ACF has been oversampled by six. For BF1 the pitch halving has been corrected, however; many gross pitch errors have been introduced. For BM1, the gross pitch errors are now located at frames 52 and 77. For comparison we refer to Figures 10.18 and 11.8.

A simple pitch tracking mechanism was also introduced into the pitch detector. The Inmarsat standard [309] performs a simple pitch tracking method whereby:

$$0.8P_{past} \leq P_{current} \leq 1.2P_{past}. \tag{11.15}$$

Thus, the pitch-delay of the current speech frame must be close to the determined pitch-delay of the previous speech frame. Our final pitch detector procedure is given in Figure 11.13.

From Figure 11.13 it can be seen that the first pitch detector task is to check whether the previous frame was voiced. If it was, then the pitch tracking mechanism described by Equation 11.15 uses the previous frame to constrain the current pitch in the vicinity of the past pitch. However, if the last frame was unvoiced, then the pitch tracking mechanism checks whether the last but one frame was voiced, and thus if it can be used in the pitch tracking. If neither frame was voiced, then no pitch tracking restrictions are imposed on the pitch detector. Subsequently, the residual signal is oversampled by six, followed by the autocorrelation function calculation. The voicing strength for the delay selected by the ACF computation is compared with a threshold, as described in Section 11.3.1 and seen in Figure 11.12. If this threshold is not exceeded, the frame is declared unvoiced. The frames that have exceeded the

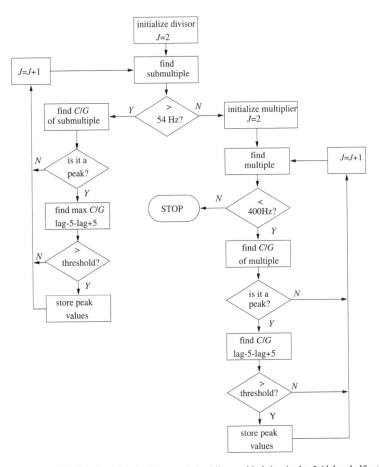

Figure 11.12 Flowchart for checking pitch doubling and halving in the 5.6 k bps half-rate
GSM VSELP coder.

threshold are voiced and have their submultiples checked, as described by Figure 11.12, and
the final pitch period is then assigned to this voiced frame.

The pitch tracks for the utterances in Figure 11.8 are given in Figure 11.14. It can be
seen that the improvements to the pitch detector have removed the gross pitch error and the
majority of the pitch halving. A performance comparison with the manual pitch track of
Figures 10.17 and 10.18 in Section 10.4 is given in Table 11.4, where 7.2% of frames were
incorrect.

11.3.3.1 Computational Complexity   The computational complexity of an operation
is measured in floating point operations per second (FLOPS), where the number of multiplications or additions performed per second is calculated, with a complexity value larger than
25 MFLOPS deemed prohibitive here. The computational complexity for the oversampled
pitch detector is given in Table 11.5, demonstrating that the use of oversampled signals in
pitch detection proportionally increases the computational complexity of the process. The first
column of Table 11.5, details the computational costs of the oversampled pitch detector in a
worst case scenario of a fundamental frequency of 400 Hz and when there is no past pitch

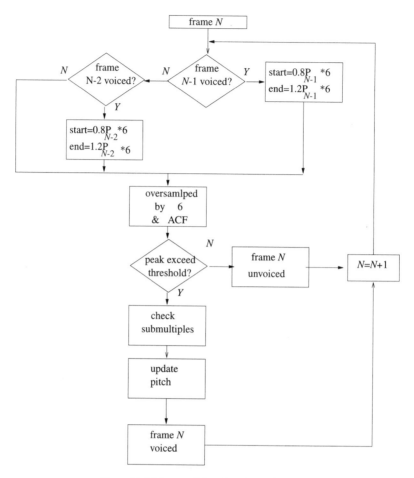

Figure 11.13 Proposed flowchart for pitch detection.

period which allows pitch tracking to be employed. As mentioned above, the complexity of 36.3 MFLOPS renders the use of a fully oversampled pitch detector prohibitively complex. It should be noted that the computational complexity values in Table 11.5 used the FFT function to reduce the complexity of the autocorrelation process. Throughout the low-bit-rate-oriented part of the book—where possible, the computational complexity values are decreased using the FFT.

11.3.4 Integer Pitch Detector

In order to reduce the computational complexity of the oversampled pitch detector, oversampling could be restricted to the final search for the nearest noninteger delay. In Figure 11.12 this is the block where the maximum value of C_1^2/G_1 for the pitch values of $lag - 5$ to $lag + 5$ is located. In the second column of Table 11.5, it can be seen that this partial oversampling procedure reduces the complexity to 27.3 MFLOPS, a value that is still prohibitive. If oversampling is removed from the autocorrelation procedure, the voiced-

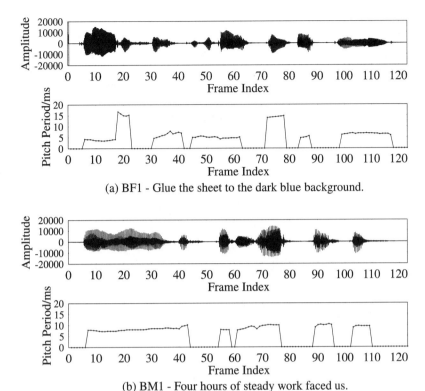

(a) BF1 - Glue the sheet to the dark blue background.

(b) BM1 - Four hours of steady work faced us.

Figure 11.14 Pitch period decisions based on the flowchart of Figure 11.13 for (a) BF1 and (b) BM1. Here the ACF has been oversampled by six and pitch tracking has been adopted. All gross pitch errors have been removed from BF1, with pitch halving occurring between frames 16 and 20, and between frames 70 and 76. For BM1 no pitch errors occur. For comparison, we refer to Figures 10.18, 11.8, and 11.11.

unvoiced thresholds should be returned to the 0.7 and 0.5 levels, used by the G.728 pitch detector [85] as described in Section 11.3.1.

In the final column of Table 11.5, the computational complexity for a pitch detector with no oversampling is given, but both pitch tracking and checking of submultiples is included. A computational complexity of 3.4 MFLOPS is more acceptable; thus, the performance of this pitch detector is considered. This pitch detector is identical to the

TABLE 11.5 Computational Complexity for Worst Case Scenario for Three Pitch Detectors with Differing Amounts of Oversampling

Procedure	Fully Oversampled by Six /MFLOPS	Partially Oversampled by Six /MFLOPS	No Oversampling/ MFLOPS
Constructing oversampled array	2.2	2.2	—
Calculating autocorrelation	5.1	1.1	1.1
Checking submultiples	29.0	24.0	2.3
Total	36.3	27.3	3.4

pitch detector described in Section 11.3.2, but all the oversampling has been removed. The new pitch detector results were surprisingly good, with results comparable to the over-sampling by the six-pitch detector, where from Table 11.4 it can be seen that 7.2% of frames had a pitch detection error. For the integer pitch detector, the parameter α_{CG} from Equation 11.14 was adjusted so $\alpha_{CG} = 3$. The quality of the integer sampling pitch detector suggests that the improvement in pitch detector quality is predominantly due to the pitch tracking and checking of pitch submultiples, rather than the oversampling process. Thus, an integer pitch detector with both pitch tracking and checking of pitch submultiples was used for implementation of the LPC vocoder.

Following this examination of pitch detection, various aspects of the LPC decoder are investigated. The first stages of the decoder are the generation of voiced and unvoiced excitation sources.

11.4 UNVOICED FRAMES

For speech frames that are classed as unvoiced, at the decoder a Gaussian random process can be used to represent unvoiced excitation. The Gaussian random process is scaled by the RMS of the LPC residual signal, as defined by:

$$RMS = \sqrt{\frac{\sum_{n=1}^{FL} r(n)^2}{FL}} \tag{11.16}$$

where FL is the frame length of the speech segment and $r(n)$ is the LPC residual signal, described in Section 10.2. The Gaussian random process was generated by applying the Box-Muller algorithm [153].

For transmission, the RMS value requires quantization. In the described LPC coder, a Max-Lloyd quantizer [10] was employed for the task, which requires knowledge of the RMS parameters' PDF. This was supplied in the form of a PDF generated from the unquantized

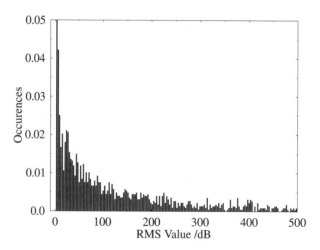

Figure 11.15 Typical PDF of the RMS of the weighted LPC residual.

TABLE 11.6 SNR Values for a Range of RMS SQs

SQ	SNR/dB
2-bit	2.79
3-bit	8.63
4-bit	19.07
5-bit	25.85
6-bit	32.28

RMS values of 45 seconds of speech from the training database, and it is portrayed in Figure 11.15. Table 11.6 displays the SNR values found for a 2-bit to 6-bit SQ. For our coder, the 5-bit quantizer was selected because this produced a similar SNR value to the SQs described later in Section 13.5.2.

11.5 VOICED FRAMES

For speech frames that are classified as voiced, the excitation source, which is passed to the LPC STP synthesis filter, is a stream of pulses. These pulses are situated a pitch period distance apart, with their energy scaled to reproduce speech of the same energy as the original speech waveform.

11.5.1 Placement of Excitation Pulses

At the beginning of a voiced sequence of frames, the first pulse is situated at the start of the frame, while subsequent pulses are placed a pitch period number of samples beyond the previous pulse. Thus, the decoder must remember the position of the last pulse in every voiced speech frame in order to calculate the position of the first pulse in the next voiced speech frame. The positioning of the pulses bears no resemblance to the position of pitch periods in the original speech; thus, it is highly probable that the synthesized and original speech; waveform will not be time aligned. This prevents any of the objective measures described in Section 10.3.1 from being utilized to determine the speech coder's performance. Instead, subjective measures will have to be relied on.

11.5.2 Pulse Energy

In order to recreate speech of the same energy as the original, the energy of the periodic pulses must be scaled. Explicitly, the energy of the reconstructed speech must equal the energy of the original speech, formulated as:

$$\sum_{n=0}^{FL} [\rho_a \delta(n) * h(n)]^2 = \sum_{n=0}^{FL} [s(n) - m(n)]^2 \tag{11.17}$$

where $s(n)$ is the original speech, $m(n)$ is the memory of the LPC STP synthesis filter, $h(n)$ is the impulse response of the LPC STP synthesis filter, ρ_a is the amplitude of each pulse, and the Kronecker delta function $\delta(n)$ represents the location of the pulses. The energy of the excitation signals must also be equal; thus:

$$\sum_{n=0}^{FL} [\rho_a \cdot \delta(n)]^2 = \sum_{n=0}^{FL} r(n)^2 \qquad (11.18)$$

where $r(n)$ is the LPC STP residual. Since the RMS energy of the LPC STP residual has already been calculated in Equation 11.16, the combined energy of the pulses, ρ_a^2, can be calculated with the RMS value of the speech. Thus, the same RMS value can be transmitted for both voiced and unvoiced frames.

For I_p pulses per frame, the energy of each pulse will be ρ_a^2/I_p, and the amplitude of each pulse becomes $\sqrt{\rho_a^2/I_p}$. Hence, the RMS of the LPC STP residual is sent to the decoder for both voiced and unvoiced frames.

The final stage of the LPC decoder is the adaptive post-filter, which is used to improve the perceived quality of the synthesized speech, and described next.

11.6 ADAPTIVE POST-FILTER

For a speech coder that operates at bitrates less than 16 kbps, the coding noise becomes a problem. A lower coding rate raises the quantization noise level, thus potentially increasing the complexity of preventing the noise from exceeding the audibility threshold. For analysis-by-synthesis speech coding models, the weighting of the LPC STP filter at the encoder can help reduce coding noise, while at the output of the decoder an adaptive post-filter can be implemented. For the basic LPC vocoder described here, the adaptive post-filter can also be used to improve the output speech.

An adaptive post-filter [86] is a series of filters whose parameters alter every frame in an attempt to conceal the coding noise. The principle of post-filtering is that, at the output of the decoder, the formant and pitch peaks of the synthetic speech are emphasized and the valleys, which are contaminated by quantization noise, are attenuated in order to render their effect less audible.

An adaptive post-filter consists of three distinct sections: a short-term post-filter, a long-term post-filter, and automatic gain control (AGC), as demonstrated in Figure 11.16.

The short-term postfilter follows the peaks and valleys of the spectral envelope, emphasizing the formants while attenuating the spectral valleys. The weighted LPC STP synthesis filter creates the shape of the spectral envelope. Thus, the short-term post-filter is

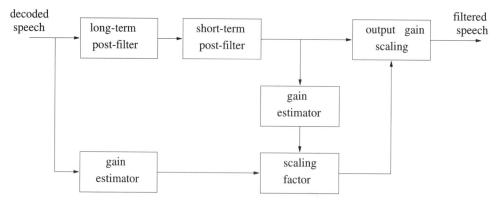

Figure 11.16 An adaptive post-filter.

based on the weighted synthesis filter. However, the weighted synthesis filter introduces a spectral tilt in the high-frequency regions, hence influencing the energy of the formants. Subsequently, an all-zero filter is introduced in the numerator of Equation 11.19 to remove the spectral tilt, together with an additional first-order filter to further reduce the tilt, namely, the bracketed term of Equation 11.19, as seen in

$$H_{spf}(z) = \frac{1 - \sum_{k=1}^{10} \beta_{pf}^k a_k z^{-k}}{1 - \sum_{k=1}^{10} \alpha_{pf}^k a_k z^{-k}} [1 - \mu_{pf} z^{-1}] \tag{11.19}$$

where $\mu_{pf} = 0.5k_1$ and k_1 is the first reflection coefficient from the LPC STP analysis, detailed in Section 10.2, with α_{pf}^k and β_{pf}^k controlling the amount of short-term post-filtering. This reduces the spectral tilt most dramatically for voiced speech, since voiced speech has previously been exposed to low-pass filtering due to the spectral tilt present in the weighted synthesis filter.

The long-term post-filter follows the peaks and valleys of the pitch harmonics, again emphasizing the peaks and attenuating the valleys. It is based on the 1-tap pitch predictor, $(1 - g_{pf} z^{-P})$, used in LTP analysis, which was described in Section 10.2. However, an all-zero filter is cascaded with it to allow more flexibility and greater control over the frequency response. The long-term post-filter is switched off during unvoiced speech since there are no pitch harmonics. Thus, the long-term post-filter must have unity power gain to ensure that voiced speech is not amplified over unvoiced speech. Hence, the long-term post-filter's transfer function is given by:

$$H_{lpf}(z) = G_{lpf} \frac{1 + \gamma_{pf} z^{-P}}{1 - g_{pf} z^{-P}} \tag{11.20}$$

where $0 < \gamma_{pf}, g_{pf} < 1$, G_{lpf} is the adaptive gain of the filter, P is the pitch period of the speech frame, and γ_{pf} and g_{pf} control the extent of the long-term post-filtering. The amount of long-term post-filtering is proportional to the voicing strength in a speech frame, thus:

$$\gamma_{pf} = \gamma_{1pf} f(x) \tag{11.21}$$

$$g_{pf} = g_{1pf} f(x) \tag{11.22}$$

where:

$$f(x) = \begin{cases} 0 & \text{if } v_s < U_{th} \\ v_s & \text{if } U_{th} < v_s \leq 1 \\ 1 & \text{if } v_s > 1 \end{cases} \tag{11.23}$$

with U_{th} being the threshold for enabling the long-term post-filter and v_s is the voicing strength indicator, generally based on the tap weight of the single-tap long-term predictor. Thus:

$$v_s = \frac{\sum_{n=0}^{FL} \bar{s}(n) \times \bar{s}(n - P)}{\sum_{n=0}^{FL} \bar{s}(n - P)^2} \tag{11.24}$$

where P is the pitch period, FL is the frame length, and $\bar{s}(n)$ is the decoded speech. This is equivalent to the pitch detector voicing strength of Equation 11.8. Chen and Gersho have shown [86] that the gain of the long-term post-filter can be controlled by:

$$G_{lpf} = \frac{1 - g_{pf}/v_s}{1 + \gamma_{pf}/v_s}. \tag{11.25}$$

When selecting the parameters of the long-term post-filter in Equation 11.20, typically g_{pf} in Equation 11.25 is set to a low value, thus decreasing the interframe memory effects in the long-term post-filter.

The final section of the adaptive post-filter is the AGC, which attempts to prevent the time-variant amplification of the speech signal. The AGC operates by estimating the magnitude of the input and output signals of the post-filter, where subsequently the output signal is adjusted on a sample-by-sample basis. The action of the AGC is described by a scaling factor of:

$$G_{pf} = \frac{\sigma_{1pf}(n)}{\sigma_{2pf}(n)}$$ (11.26)

where:

$$\sigma_{1pf}(n) = \xi_{pf} \cdot \sigma_{1pf}(n-1) + (1 - \xi_{pf}) \cdot |\bar{s}(n)|$$ (11.27)

$$\sigma_{2pf}(n) = \xi_{pf} \cdot \sigma_{2pf}(n-1) + (1 - \xi_{pf}) \cdot |\hat{s}(n)|$$ (11.28)

and $\bar{s}(n)$ is the input to the post-filter, $\hat{s}(n)$ is the output from the post-filter, while ξ_{pf} determines the rate of change for the AGC. Equations 11.27 and 11.28 constitute a weighted sum of the current signal magnitudes $|\bar{s}(n)|$ and $|\hat{s}(n)|$ together with the previous values $\sigma_{1pf}(n-1)$ and $\sigma_{2pf}(n-1)$.

Typical adaptive post-filter responses are shown in Figure 11.17. The short-term post-filter frequency response shows how the introduction of the first-order filter and the spectral tilt filter, both shown in Equation 11.19, remove the spectral tilt introduced by the all-pole filter. The post-filter frequency response demonstrates how the post-filter attenuates both the spectral envelope valleys and the pitch harmonic valleys. Following the subjective optimization of various post-filter parameters, the optimized selected parameters are given in Table 11.7.

11.7 PULSE DISPERSION FILTER

In the MELP Coder [284] described in Section 10.1.2.6, which was selected for the Department of Defense standard at 2.4 kbps, one of the novel features employed was the pulse dispersion filter. In essence it helps to spread some of the excitation pulse energy away from the main excitation impulse by following the principle of glottal pulse shaping [292, 310, 311], as highlighted in this section.

11.7.1 Pulse Dispersion Principles

A typical glottal waveform is given in Figure 11.18 and is used to spread the excitation pulse energy away from the main excitation pulse. Its shape is defined by the glottal opening time, T_P, and the closure time, T_N. Rosenberg [311] investigated the ratio of opening and closure time with respect to the pitch period P. An opening time of $T_P/P = 0.40$ and a closing time of $T_N/P = 0.16$ were found to produce the most natural sounding speech.

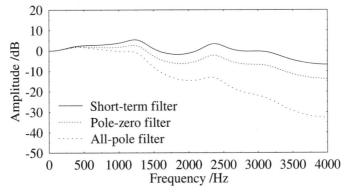

(a) Short-term post-filter frequency responses

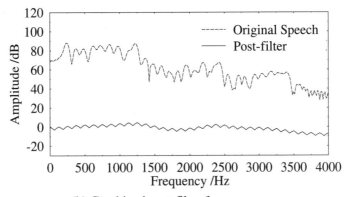

(b) Combined post-filter frequency response

Figure 11.17 Post-filter frequency responses from AM1 for the dipthong /ɑɪ/ in "live", showing (a) the short-term post-filter and (b) the long-term and combined post-filter. The speech frame had a fundamental frequency of 125 Hz. The selected postfilter parameters were $\alpha_{pf} = 0.70$, $\beta_{pf} = 0.45$, $\mu_{pf} = 0.50$, $\xi_{pf} = 0.99$, $\gamma_{pf} = 0.15$, and $g_{pf} = 0$.

TABLE 11.7 Appropriate Adaptive Post-filter Values for the LPC Vocoder, Described in Equations 11.19 to 11.27

Parameter	Value
α_{pf}	0.70
β_{pf}	0.45
μ_{pf}	0.50
γ_{pf}	0.50
g_{pf}	0.00
ξ_{pf}	0.99

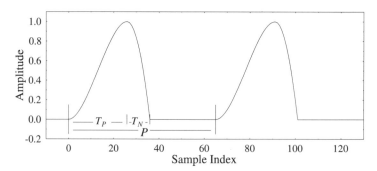

Figure 11.18 The typical shape of a glottal wave for human speech according to Equation 11.29, with $T = 65$ samples, $T_P = 26$ samples and $T_N = 10$ samples, which is used to spread the excitation pulse energy away from the main excitation pulse. The action of the filter is demonstrated in Figure 11.20.

Rosenberg also investigated which specific glottal pulse shape produced the most natural synthesized speech, with the polynomial expression:

$$f(t) = \begin{cases} \alpha\left[3\left(\dfrac{t}{T_P}\right)^2 - 2\left(\dfrac{t}{T_P}\right)^3\right], & 0 \leq t \leq T_P \\[4mm] \alpha\left[1 - \left(\dfrac{t - T_P}{T_N}\right)^2\right], & T_P \leq t \leq T_P + T_N \end{cases} \tag{11.29}$$

found to be best, where α controlled the amplitude of the glottal pulse. This polynomial expression was used to create the glottal pulse shape of Figure 11.18, where we had $P = 65$ samples, $T_P = 26$ samples, and $T_N = 10$ samples.

The principle of glottal pulse shaping was exploited by Holmes [292] to shape the excitation of a formant vocoder and by Sambur et al. [310] to form the excitation for a LPC vocoder. They both found that the introduction of the glottal pulse shaping improved the naturalness of the synthesized speech.

11.7.2 Pitch-Independent Glottal Pulse Shaping Filter

The pulse dispersion filter adopted by McCree and Barnwell [284] was a triangular-shaped pulse [311] that was spectrally flattened, as shown in Figure 11.19. The process of spectrally flattening the glottal pulse, of which the start and end states are shown in Figure 11.19, involves manipulating the frequency-domain representation of the triangular glottal pulse. The principle of spectral glottal pulse flattening is invoked [284] since the synthesized speech excitation waveform should be spectrally flat. Hence, this condition should also be imposed on the glottal pulse shape. The time-domain representation of this spectrally flattened pulse is shown in Figure 11.19b. The spectral flattening was performed using linear prediction (LP), where the triangular glottal pulse shape was passed through an LP filter to produce the pulse dispersion filter.

This triangular glottal pulse shape was also investigated by Rosenberg [311], who found it to produce inferior quality speech to the polynomial shape of Equation 11.29. However, for the same glottal opening and closure ratio T_P/T_N, the triangular shape spreads the waveform

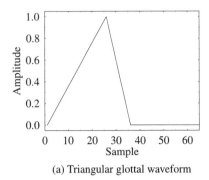

(a) Triangular glottal waveform

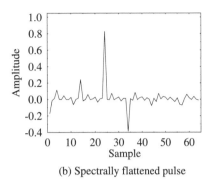

(b) Spectrally flattened pulse

Figure 11.19 A triangular glottal pulse shape together with its spectrally flattened pulse dispersion filter [284].

energy further from the impulse source than does the polynomial glottal pulse. Figure 11.20c demonstrates the time-domain energy spread achieved by the MELP coder, employing a triangular pulse shape when compared to the synthesized speech of the second trace. The delay is caused by the FIR implementation of the pulse dispersion filter.

McCree and Barnwell [284] placed their triangular pulse dispersion filter after the LPC synthesis filter, whereas previously the glottal pulse filtering was performed on the excitation [310]. Implementing the pulse dispersion filter after the LPC synthesis filter has the disadvantage that, in order to avoid imperfections, due to different filter delays, a fixed FIR filter must be employed. As seen in Figure 11.19, McCree and Barnwell used a typical male pitch period of 65 samples, at a sampling rate of 8 kHz, for the pulse dispersion filter. Their scheme was designed to benefit the male speaker most from the energy spread, since it is the longer pitch period that permits the greatest time-domain resonance decay between the pitch pulses. Thus, while the pulse dispersion filter improves the speech quality for male speakers, it does slightly reduce the speech quality of female speakers.

11.7.3 Pitch-Dependent Glottal Pulse Shaping Filter

The pitch-independent glottal pulse shape of Section 11.7.2 reduced the quality of synthesized speech for female speakers. However, a pitch-dependent glottal pulse shaping filter should avoid this effect. Following the recommendation of Rosenberg [311], the polynomial of Equation 11.29 was spectrally flattened, using linear prediction, and then it was adopted to spread the excitation waveform energy before the LPC synthesis filter. Imposing the pulse dispersion filter on the excitation followed the method of Holmes [292] and Sambur et al. [310], as opposed to McCree and Barnwell [284] who applied the pulse dispersion filter to the synthesized speech. The performance of this pitch-dependent glottal pulse shaping filter is characterized in Figure 11.20d, where it can be seen that the resultant synthesized speech contains much less energy spread than the pitch-independent triangular pulse dispersion filter of McCree and Barnwell [284] shown in Figure 11.20c.

To produce a more effective pitch-dependent glottal pulse shaping filter, the polynomial of Equation 11.29 was replaced by a spectrally flattened triangular glottal waveform pulse. The performance of this pitch-dependent triangular pulse is characterized in Figure 11.20e.

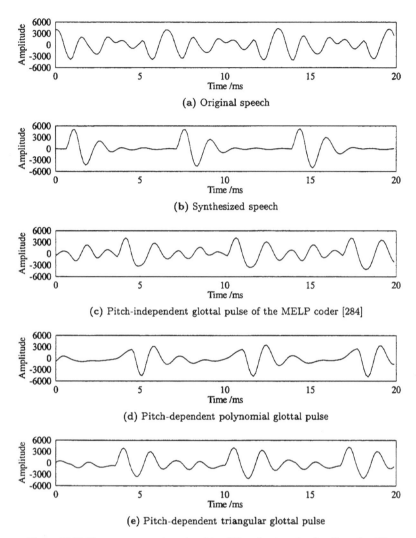

Figure 11.20 The energy spread produced by differently shaped pulse dispersion filters compared to the original synthesized speech, where the FIR implementation of the different pulse dispersion filters introduces a delay. The utterance is taken from BM2 for the back vowel /ɑ/ in "p*a*rked".

This glottal waveform shape produces good energy spread for male speakers, with its pitch-dependent nature ensuring that the speech quality of female speakers is not degraded.

However, by employing a glottal waveform shaping filter before the LPC synthesis filter, the excitation source would be constrained to be an impulse, since this is the form of excitation the glottal waveform shaping was originally designed for. Within this treatise, notably in Chapter 13, different forms of excitation are employed which would be unable to incorporate this glottal pulse shaping filter. The pulse dispersion filter, introduced by McCree and Barnwell [284], operates on the synthesized speech, thus permitting its successful operation in conjunction with the other excitations detailed later. In summary, it was found to be most perceptually beneficial to invoke the pitch-independent triangular pulse dispersion filter of McCree and Barnwell seen in Figure 11.19.

11.8 RESULTS FOR LINEAR PREDICTIVE VOCODER

The basic LPC vocoder described in this chapter was implemented, with the utterances from the speech database in Section 10.4 processed to test the LPC vocoder's performance. The time- and frequency-domain performance of the vocoder for individual 20 ms frames of speech is shown in Figures 11.21, 11.22, and 11.23, where the waveforms at different stages of the speech coder are shown. In the figures, trace (a) displays the original waveform, trace (b) shows the impulse and frequency response of the LPC STP synthesis filter, with trace (c) showing the LPC STP residual waveform. The reconstructed waveforms at different stages are also displayed, where trace (d) shows the excitation waveform, trace (e) displays the speech waveform after the LPC synthesis filter, trace (f) contains the impulse and frequency response of the adaptive post-filter, with trace (g) showing the speech waveform after the post-filter, and finally trace (h) displays the output speech following the pulse dispersion filter. The input and output speech waveforms are shown inside the pre- and post-processing filters, described in Section 11.2.2. Thus, the output speech will still be high-passed-filtered to 100 Hz. The performance of the LPC vocoder for these speech frames is described next.

Figure 11.21 displays a waveform from the testfile BM1, for the midvowel /ɜ/ in the utterance "work." From the time-domain representation of Figure 11.21a we can infer that

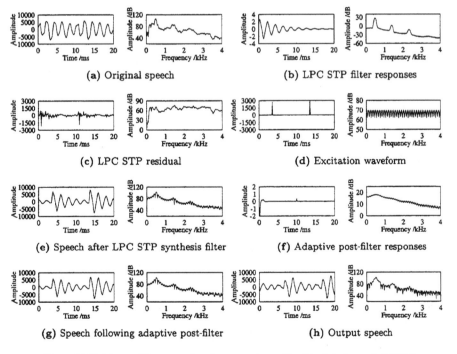

(a) Original speech

(b) LPC STP filter responses

(c) LPC STP residual

(d) Excitation waveform

(e) Speech after LPC STP synthesis filter

(f) Adaptive post-filter responses

(g) Speech following adaptive post-filter

(h) Output speech

Figure 11.21 Time and frequency domains comparison of the (a) original speech, (b) LPC STP filter impulse and frequency response, (c) LPC STP residual, (d) LPC excitation waveform, (e) speech waveform after the LPC STP filter, (f) adaptive post-filter impulse and frequency-domain response, (g) speech waveform after the adaptive postfilter, (h) output speech after the pulse dispersion filter. The 20 ms speech frame is the midvowel /ɜ/ in the utterance "work" for the testfile BM1. For comparison with the other coders developed in this study using the same speech segment, please refer to Table 16.2.

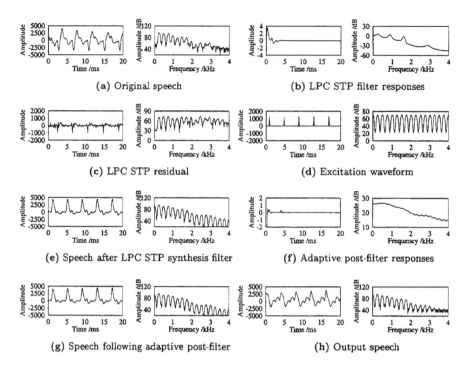

Figure 11.22 Time and frequency domains comparison of the (a) original speech, (b) LPC
STP filter impulse and frequency response, (c) LPC STP residual, (d) LPC
excitation waveform, (e) speech waveform after the LPC STP filter, (f)
adaptive post-filter impulse and frequency domain response, (g) speech
waveform after the adaptive post-filter, (h) output speech after the pulse
dispersion filter. The 20 ms speech frame is the liquid /r/ in the utterance
"*r*ice" for the testfile BF2. For comparison with the other coders developed
in this study using the same speech segment, please refer to Table 16.2.

the waveform's periodicity is approximately 80 samples, corresponding to 10 ms or to a pitch
of 100 Hz. This manifests itself in the frequency domain in terms of 100 Hz-spaced spectral
needles. From the frequency-domain representation of Figure 11.21b, we can infer that there
are spectral envelope peaks around 500 Hz, 1400 Hz, 2200 Hz, and 3500 Hz corresponding to
the formants. The LPC decoder has assigned two excitation pulses to this frame, as shown in
Figure 11.21d, which attempts to model the two pitch-related pulses in Figure 11.21c. The
reconstructed speech of Figure 11.21e contains the same type of waveform as the original;
however, the synthesized speech cannot maintain the amplitude of the original throughout the
pitch period. In the frequency domain the formants are well represented. For this particular
speech frame, the introduction of the adaptive post-filter in Figure 11.21g has little effect,
with only the first formant being emphasized and a small amount of long-term post-filtering
occurring. The pulse dispersion filter introduces a delay into the speech waveform and
reduces the amount of periodic voicing evident in the frequency spectrum, as shown in Figure
11.21h.

The utterance shown in Figure 11.22 is from the testfile BF2, for the liquid /r/ from the
utterance *r*ice." As typical, the female speaker has a much shorter pitch period than the male
speaker of Figure 11.21. From the frequency-domain representation in Figures 11.22a and
11.22b, it can be seen that the waveform has formants at 200 Hz, 800 Hz, 1500 Hz, and

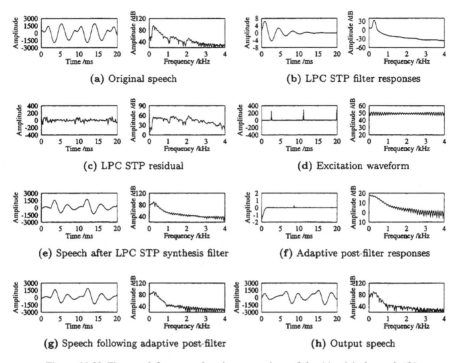

(a) Original speech
(b) LPC STP filter responses

(c) LPC STP residual
(d) Excitation waveform

(e) Speech after LPC STP synthesis filter
(f) Adaptive post-filter responses

(g) Speech following adaptive post-filter
(h) Output speech

Figure 11.23 Time- and frequency-domains comparison of the (a) original speech, (b) LPC STP filter impulse and frequency response, (c) LPC STP residual, (d) LPC excitation waveform, (e) speech waveform after the LPC STP filter, (f) adaptive post-filter impulse and frequency-domain response, (g) speech waveform after the adaptive postfilter, (h) output speech after the pulse dispersion filter. The 20 ms speech frame is the nasal /n/ in the utterance "throw$n$" for the testfile BM2. For comparison with the other coders developed in this study using the same speech segment, please refer to Table 16.2.

2800 Hz. In addition, the speech spectrum appears voiced beneath 1800 Hz and unvoiced above. Figure 11.22d shows that the LPC decoder has assigned five pitch pulses to this frame, which correspond to the five pitch periods seen in the original waveform. The pitch period is about 35 samples, corresponding to approximately 4.4 ms and a pitch frequency of about 220 Hz. In the time domain, the synthesized speech signal of Figure 11.22e contains a dominant peak in each pitch period but very little energy elsewhere. In the frequency domain the synthesized speech spectrum is visibly voiced throughout the 4 kHz. For this speech frame, the introduction of an adaptive post-filter decreases the amplitude of the pitch-period harmonics at higher frequencies. The introduction of the pulse dispersion filter in Figure 11.22h reduces the amount of periodic voicing evident in the higher frequencies of the speech spectrum, while in the time domain it spreads the energy of the waveform throughout the pitch period. It could be suggested that the pulse dispersion filter spreads too much of the energy away from the dominant peak. This highlights a disadvantage of the pitch-independent pulse dispersion filter, which was based on a typical male pitch period, and can exaggerate the dispersion of energy for a shorter pitch period.

A frame of the BM2 testfile is displayed in Figure 11.23, where characteristics of the nasal /n/ from the utterance "throw$n$" are shown. Specifically, from Figure 11.23a it can be seen that the speech waveform has a pitch period of about 70 samples, corresponding to

TABLE 11.8 Bit-allocation Table for the Investigated LPC Vocoder

Parameter	Unvoiced	Voiced
LSFs	18	18
V/U flag	1	1
RMS value	5	5
Pitch	—	7
total/20 ms	24	31
bitrate	1.2 kbps	1.55 kbps

8.75 ms or a pitch of 115 Hz. Observing the frequency spectrum, there appear to be formants at 250 Hz, 1200 Hz and 2200 Hz, although the LPC STP filter of Figure 11.23b only captures the first formant. The failure of the LPC STP filter is further evident in Figure 11.23c where the peaks at 1200 Hz and 2000 Hz are replaced by flat frequency spectrum. The frequency spectrum of Figure 11.23a shows that above 2300 Hz the spectrum is unvoiced, while in Figure 11.23e the spectrum is voiced up to 4 kHz. In addition, from Figure 11.23e it can be seen that the upper two formants are not represented. For this speech frame the addition of the post-filter decreases the amplitude of the pitch-period harmonics above 1 kHz, as shown in Figure 11.23g. As we see in Figure 11.23h, the pulse dispersion filter further reduces voicing, especially in the high-frequency region.

The perceptual quality of the synthesized speech was informally assessed, where the reproduced speech sounded slightly synthetic, with particular buzziness in the case of high-pitched female speakers.

The bit allocation for each 20 ms voiced or unvoiced speech frame is given by Table 11.8. For both voiced and unvoiced frames, the LPC STP synthesis filter coefficients were vector-quantized as LSFs for transmission, using 18 bits per frame. A voiced-unvoiced flag was also sent with both voiced and unvoiced frames. For all speech frames a 5-bit SQ, described in Section 11.4, was used to transmit the RMS of the LPC STP residual signal, indicating the energy of the synthesized excitation. For voiced frames the pitch period ranged from $20 \rightarrow 147$ samples, thus, 7 bits were required to represent its value. The total bitrate for voiced frames was 1.55 kbps, while for unvoiced frames it was 1.2 kbps.

The computational complexity for this basic LPC vocoder was dominated by the pitch detector, where in Section 11.3.2 its complexity was found to be 3.4 MFLOPS. The delay of the LPC vocoder was 60 ms, with 40 ms at the encoder required for the pitch detection and 20 ms required at the decoder.

11.9 CHAPTER SUMMARY

This Chapter has introduced a benchmark LPC vocoder, using the random Gaussian noise of Section 11.4 for unvoiced frames and the pulses of Section 11.5 for voiced frames. The chapter has detailed important aspects of low-bitrate speech coders which will be harnessed with the speech coders developed in later chapters. The investigated aspects were LSF quantization, pitch detection, adaptive post-filtering, and pulse dispersion filtering. It was found that for a 1.55 kbps speech coder the reproduced speech was intelligible but sounded distinctly synthetic. Figures 11.21, 11.22, and 11.23 illustrated that the speech can sound intelligible without faithfully reproducing the time-domain waveforms. In the next chapter we invoke various wavelet-based pitch detection techniques.

Wavelets and Pitch Detection

12.1 CONCEPTUAL INTRODUCTION TO WAVELETS

In this section we provide a simple conceptual introduction to wavelets, while a more rigorous mathematical exposure is offered in the next section.

In recent years, wavelets have stimulated substantial research interest in a variety of applications. Their theory and practice have been documented in a number of books [312]–[314] and tutorial treatises [315]–[317]. The theory of wavelets was recognized as a distinct discipline in the early 1980s. Daubechies [318] and Mallat [319] generated significant interest in the field by invoking the mathematical technique of wavelets in signal processing applications. Wavelets have many applications where previously the classical tool of Fourier theory may have been applied. Therefore, in this chapter wavelet theory is initially introduced through comparison with Fourier theory. Section 12.2 contains some of the mathematics underlying wavelet theory, while Sections 12.3, 12.4, and 12.5 describe how wavelets may be applied to the pitch detection of speech signals. Thus, for a focused discussion on the application of wavelets to pitch detection, the reader may proceed to Section 12.3.

12.1.1 Fourier Theory

Fourier theory states that any signal $f(x)$, that is 2π-periodic can be represented by an infinite series of sine and cosine functions defined by [320]:

$$f(x) = a_0 + \sum_{k=1}^{\infty} [a_k \cos(x) + b_k \sin(x)] \tag{12.1}$$

where a_k and b_k are real coefficients. Thus, we can consider the signal $f(x)$ to be constructed from a set of basis functions.

The Fourier transform is used to convert a signal between the time and frequency domain, giving a tool through which we can analyze the signal $f(x)$ in both domains, although often one domain will be more convenient than the other. The conversion between domains can occur because the coefficients of the sine and cosine functions, used to represent the time-domain signal, indicate the contribution of different frequencies to the signal $f(x)$.

Although the Fourier transform produces localized values in the frequency domain, in the time domain the sine and cosine functions have an infinite support. Thus, in order to localize the time-domain signal, windowing must be used, leading to the short-term fourier transform (STFT). Figure 12.1 displays the localization achieved by the STFT in both the time and frequency domain, yielding a uniform partitioning of the time-frequency plane since the same window is used for all frequencies. Because of localization in the frequency domain, the STFT can also be viewed as a filterbank [315], which is shown in Figure 12.2. Specifically, a filter centered at f_0 is created with a bandwidth f_b. Subsequently, this filter can be transformed to create a filter at $2 \cdot f_0$ with a bandwidth of f_b to analyze the contribution constituted by these frequencies. This process can be continued indefinitely, but the filters always have a bandwidth f_b.

Having the same localization, or resolution, in all regions of the time-frequency plane is often not ideal. The wavelet transform, which is described next, supports a variable-width localization in the time-frequency space.

12.1.2 Wavelet Theory

A wavelet is defined as a function that obeys certain conditions [318], allowing it to represent a signal $f(x)$ by a series of basis functions, in a fashion similar to the Fourier series. The wavelet decomposition of $f(x)$ can be formulated as:

$$f(x) = \sum_{ik} d_{ik}\psi_{ik}(x) \tag{12.2}$$

where d_{ik} are the coefficients of the decomposition and ψ_{ik} are the basis functions.

The wavelet transform can be used to analyze a signal $f(x)$, but unlike the short-time Fourier transform its localization resolution varies over the time-frequency space. This flexibility in resolution makes it particularly useful for analyzing signal discontinuities, where during a short time period an extensive range of frequencies is present. It can be noted that the instance of glottal closure is represented by a discontinuity in the speech waveform. As demonstrated by Figure 12.1, the wavelet transform can analyze either a large range of frequencies over a short period of time or a narrowband of frequencies over a long time period.

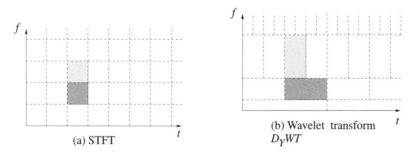

Figure 12.1 The STFT and wavelet transform time-frequency domain spaces.

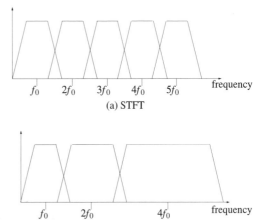

Figure 12.2 The STFT and wavelet transform filter-bank models.

Hence, it can be argued that the wavelet transform permits the analysis of a signal $f(x)$ to be viewed at different time- and frequency-domain scales. For a lengthy time window the global features of the signal $f(x)$ become prominent, while for a short time period the localized features of the signal $f(x)$ are observed. It is possible to study the frequency features in a similar manner.

The localization of the wavelet transform can also be viewed as a filterbank [315], which is shown in Figure 12.2 along with the STFT. As suggested by Figure 12.1b a filter is created at f_0 with a bandwidth of f_b, analyzing these frequencies for the signal $f(x)$ within the time duration of $[-t_{fb}, t_{fb}]$. This filter can be transformed to a filter centered at $2 \cdot f_0$ with bandwidth $2 \cdot f_b$, where the energy of these frequencies in $f(x)$ over the time duration $[-t_{fb}/2, t_{fb}/2]$ is considered. This procedure can be continued indefinitely.

Thus, our brief conceptual comparison between the Fourier and wavelet transform has been completed, highlighting the similarities and differences between the two methodologies. A study of wavelets and discontinuities is now performed.

12.1.3 Detecting Discontinuities with Wavelets

Previously, it was briefly mentioned that wavelets can be used to detect discontinuities due to the instants of glottal closure [307]. The duration between two consecutive instants of glottal closures is the pitch period of the speech signal. There have been several applications of wavelets for detecting discontinuities, with a few of them highlighted below.

In one method applying wavelet analysis to speech signals, Stegmann et al. [321] used a dyadic wavelet transform, $D_Y WT$, to distinguish between voiced, unvoiced and transient periods of speech, where the term dyadic will be elaborated on below. According to this method, the different behavior exhibited by the wavelet coefficients, at each scale, allowed the individual speech frames to be categorized into one of the above three speech classes. The different wavelet scales observe different sections of the speech spectrum. Thus, the variation in the distribution of the spectral energy in voiced and unvoiced speech segments permits discrimination between the classes. This voiced-unvoiced detector follows a philosophy similar to previous methods analyzing the statistics of voiced and unvoiced speech [304, 305].

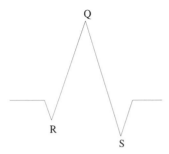

Figure 12.3 A typical QRS complex of an ECG signal.

In a second method Kadambe and Boudreaux-Bartels [307] investigated the use of the dyadic wavelet transform $(D_Y WT)$ for the pitch detection of speech signals. They used Mallat and Zhong's [322] class of spline wavelets on dyadic scales. Thresholding of located discontinuities together with their evolution across the wavelet scales were used to identify the glottal pulses, where the speech waveform's pitch period was determined by the duration between consecutive glottal pulses. Kadambe and Boudreaux-Bartels found their method to be robust to noise and more accurate than the autocorrelation-based methods of Section 11.3.

Another area where the detection of discontinuities is desirable is in biomedical signal processing [323], where their use in the analysis of electrocardiography (ECG) signals is highlighted in Reference [324]. An ECG signal has a characteristic shape termed the QRS complex, which is displayed in Figure 12.3. There it can be seen that a large positive spike is surrounded by two small negative spikes. Cardiac problems can be identified from unusual shapes of the QRS complex; thus, automatic localization of the QRS complexes would be desirable. Li et al. [324] detected the sharp discontinuities in the QRS complex with the aid of the spline wavelets suggested by Mallat and Zhong [322] and found the automatic localization of QRS complexes to be reliable.

The applications described use predominantly the class of polynomial spline wavelets introduced by Mallat and Zhong [322]. Mallat et al. [319, 322, 325] have undertaken seminal research into the task of edge detection in images, where an edge is a discontinuity in the image. Edge detection has much in common with the previously mentioned one-dimensional discontinuities, such as glottal pulses, but it occurs in two dimensions. Mallat et al. were interested in reconstructing images entirely from edge information, thereby assisting in image compression.

Following this rudimentary overview of detecting discontinuities using wavelets, the mathematics of wavelet theory is now introduced, although for a complete description the book by Koornwinder [312] is recommended. The mathematics of wavelet theory can also be found in the books by Vetterli and Kovačević [326] and Chui [313, 314].

12.2 INTRODUCTION TO WAVELET MATHEMATICS

We begin our brief introduction to the mathematics defining wavelet functions with a description of the mother wavelet, from which a class of wavelets can be derived. The mother wavelet function, ψ, is described by [312]:

$$\psi_{a,b}(t) = \frac{1}{\sqrt{a}}\psi\left(\frac{t-b}{a}\right) \tag{12.3}$$

where a is the frequency, or dilation variable and b is the position or time-domain translation parameter. Thus, wavelets exist for every combination of a and b. The Fourier transform, $\hat{\psi}(\omega)$, of the mother wavelet, $\psi(t)$, is defined by:

$$\hat{\psi}(\omega) = \int_{-\infty}^{\infty} \psi(t)e^{-jt\omega}dt \tag{12.4}$$

Any function that obeys certain constraints can be considered a mother wavelet. The reader is referred to Koornwinder [312] and Chui [313] for further clarification. The Continuous Wavelet Transform (CWT) of a function $f(t)$ is then defined by:

$$F(a, b) = \int_{-\infty}^{\infty} f(t)\psi_{a,b}(t)dt \tag{12.5}$$

which is analogous to the Fourier transform when the $e^{j\omega t}$ kernel is replaced by $\psi_{a,b}t$.

However, it is generally more useful to perform the discrete wavelet transform (DWT), or the discrete dyadic wavelet transform, $D_Y WT$, derived from the CWT by imposing the following discretization [312]:

$$F(a, b) = F(2^{-i}, 2^{-i}k), \text{ where } i, k\epsilon Z. \tag{12.6}$$

This implies that the dilation and translation indices are both elements of the discrete dyadic space Z. Within a dyadic space each time- and frequency-domain wavelet scale is down-sampled by two compared with the previous scale. Hence, from Equation 12.3 we find that the mother wavelet function takes the form of:

$$\psi(t) = 2^{i/2} \cdot \psi(2^i t - k) \tag{12.7}$$

producing a set of orthogonal basis functions exhibiting different resolutions as a function of i and k. The basic mathematics of multiresolution analysis is now introduced.

12.2.1 Multiresolution Analysis

As described in Section 12.1.2, wavelets are particularly useful when observing signals at different time or frequency scales. This technique is termed multiresolution analysis and divides the frequency space Z into a sequence of subspaces, V_m, where

$$.. \subset V_2 \subset V_1 \subset V_0 \subset V_{-1} \subset V_{-2} \subset ... \tag{12.8}$$

implying that V_m will become the space Z as $m \rightarrow -\infty$. The subspaces V_m, excluding $m = 0$, are generated through dilation of the subspace V_0. Thus, the space V_{-1} contains both the functions of V_0 and the functions that oscillate twice as fast, while only half the functions of

V_0 oscillate slowly enough to be in V_1, which is described below for a function $f(x)$ as follows:

$$f(x)\epsilon V_0 \Leftrightarrow f(2x)\epsilon V_{-1} \tag{12.9}$$

$$f(x)\epsilon V_0 \Leftrightarrow f(2^{-1}x)\epsilon V_1. \tag{12.10}$$

The subspace V_0 is generated through the father wavelet, ϕ, which is a wavelet family constructed before the mother wavelet and from which the mother wavelet is derived. The subspace V_0 contains the integer translations of the father wavelet $\phi_{0,n}$ defined by:

$$\phi_{0,n} = \phi(x - n). \tag{12.11}$$

Hence, any function $f(x)\epsilon V_0$ can be described by:

$$f(x) = \sum_{n=-\infty}^{\infty} a_n\phi(x - n) \tag{12.12}$$

where a_n are the coefficients of the decomposition and $f(x)$ is constructed from a weighted combination of integer translated father wavelets. From Equation 12.12 several statements may be inferred. First, assuming that $\phi(x)\epsilon V_0$, and since $V_0 \subset V_{-1}$ it can be said that $\phi(x)\epsilon V_{-1}$. However, Equation 12.10 states that if $\phi(x)\epsilon V_{-1}$ then $\phi(2^{-1}x)\epsilon V_0$. Hence, the two-scale difference equation can be constructed by rewriting Equation 12.12 using Equation 12.7 to arrive at:

$$\phi(x) = \sqrt{2} \sum_{n=-\infty}^{\infty} h_n\phi(2x - n) \tag{12.13}$$

where $\sqrt{2}h_n$ are a set of coefficients different from a_n, which allow h_n to be termed the filter coefficients of $\phi(x)$. Physically, this implies reconstructing the father wavelet by a weighted sum of its second harmonic components positioned at locations $-\infty \le h_n \le \infty$. The terminology "two-scale difference equation" arises from the relation of two different scales of the same function. The mother wavelet $\psi(x)$ is generated from the father wavelet, since $\psi(x)\epsilon V_0$, by the following relationship:

$$\psi(x) = \sqrt{2} \sum_{n=-\infty}^{\infty} g_n\phi(2x - n) \tag{12.14}$$

where $\sqrt{2}g_n$ are a set of coefficients constrained by $g_n = (-1)^n h_{1-n}$.

This concludes our basic introduction to wavelet mathematics, leading us to a description of the wavelets used in this chapter, namely, the polynomial spline wavelets introduced by Mallat and Zhong [322].

12.2.2 Polynomial Spline Wavelets

The family of polynomial splines is useful for practical applications since they have a compact time-domain support. Thus, they are efficient to implement in the time domain with only a few nonzero coefficients. The wavelets introduced by Mallat and Zhong [322] are constructed in detail in Appendix A, leading to the polynomial spline wavelets defined in the frequency domain by:

$$\hat{\phi}(\omega) = \left(\frac{\sin\left(\frac{\omega}{2}\right)}{\frac{\omega}{2}} \right)^3 \tag{12.15}$$

and

$$\hat{\psi}(\omega) = j\omega \left(\frac{\sin\left(\frac{\omega}{4}\right)}{\frac{\omega}{4}} \right)^4 .$$

(12.16)

These wavelets are designed so that $\hat{\phi}(\omega)$ is symmetrical with respect to 0, while $\hat{\psi}(\omega)$ is antisymmetrical with respect to 0. The filter coefficients of $\hat{\phi}(\omega)$ and $\hat{\psi}(\omega)$, namely, h_n and g_n in Equations 12.13 and 12.14, are given, respectively, by the 2π-periodic functions defined in Appendix A:

$$H(\omega) = e^{j\omega/2} \left(\cos\left(\frac{\omega}{2}\right) \right)^3$$

(12.17)

$$G(\omega) = 4je^{j\omega/2} \sin\left(\frac{\omega}{2}\right).$$

(12.18)

Appendix A also defines the filter coefficients values given in Table 12.1. Figure 12.4 displays the impulse and frequency responses for the filter coefficients, $h(n)$, of the father wavelet and the filter coefficients $g(n)$ of the mother wavelet $\hat{\psi}(\omega)$. These filter values are used in a pyramidal structure to produce a wavelet transform. This pyramidal structure is described next.

12.2.3 Pyramidal Algorithm

Mallat introduced a pyramid structure for the efficient implementation of orthogonal wavelets [319] based on techniques known from sub-band filtering. The pyramidal algorithm is illustrated in Figure 12.5.

If the input signal to the pyramidal algorithm is $A_i(\omega)$, where i represents the scale of the signal, then passing through the low-pass-filter $H(\omega)$ of Figure 12.4a and down-sampling by a factor of 2, the output is a low-pass filtered signal $A_{i+1}(\omega)$ of the input. This low-pass signal is termed the smoothed signal, because it is a smoothed version of the input signal. If the input signal $A_i(\omega)$ is passed through the high-pass filter $G(\omega)$ of Figure 12.4b and down-sampled by a factor of 2, then the output signal $D_{i+1}(\omega)$ is a high-pass-filtered version of the input signal. This high-pass output signal is termed the detail signal, for it contains the difference between the input signal $A_i(\omega)$ and the low-pass output signal $A_{i+1}(\omega)$. At the next stage of Figure 12.5, passing the smoothed signal $A_{i+1}(\omega)$ through the filters $H(\omega)$ and $G(\omega)$ and down-sampling by two results in the smoothed and detail signals $A_{i+2}(\omega)$ and $D_{i+2}(\omega)$,

TABLE 12.1 Filter Coefficients $h(n)$ and $g(n)$ Defined in Equations 12.13 and 12.14 for the Quadratic Spline Wavelet of Equations 12.17 and 12.18

n	$h(n)$	$g(n)$
-1	0.125	0
0	0.375	-2.0
1	0.375	2.0
2	0.125	0

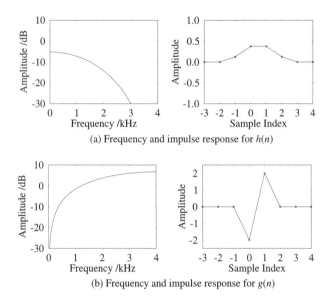

(a) Frequency and impulse response for $h(n)$

(b) Frequency and impulse response for $g(n)$

Figure 12.4 The impulse responses and frequency responses for the $h(n)$ and $g(n)$ filters described in Table 12.1 using the quadratic spline wavelets of Equations 12.17 and 12.18.

respectively. This process can be continued until the signal has been analyzed to the desired resolution.

The fundamental frequency of speech signals is assumed to vary from 54 to 400 Hz, and the sampling rate for the speech waveforms is 8 kHz. Hence, the first mother wavelet ranges from 2 to 4 kHz. This frequency band will contain higher order harmonics of the fundamental frequency, together with noise. Frequently, at the input to the speech encoder there is also present a high-pass input filter with a cutoff frequency of 100 Hz, as seen in Section 11.2.2. Thus, increasing the $D_Y WT$ scale from a mother wavelet of 2000–4000 Hz until the mother wavelet covers 125–250 Hz, namely, from scale $i + 1$ to scale $i + 5$, ensures that all frequencies passed to the speech encoder and $D_Y WT$ process are considered. Following the selection of appropriate $D_Y WT$ scales, the practical implementation of the $D_Y WT$ is discussed.

12.2.4 Boundary Effects

An important issue associated with the $D_Y WT$ is to consider the effect of time-domain boundary discontinuities due to the speech waveform's frame structure. For the computer

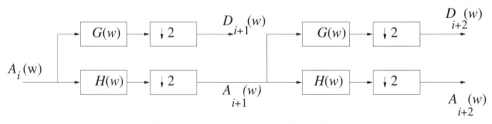

Figure 12.5 Pyramidal algorithm for multiresolution analysis.

vision problem of Mallat and Zhong [322], the discontinuities occur at the edge of the image, while in speech coding the boundaries occur at the frame edges. The method used by Mallat and Zhong to overcome the boundary discontinuity is to make the signal periodic with respect to $2T$, where T is the number of samples in the original signal and extend it symmetrically from T to $2T$.

Kadambe and Boudreaux-Bartels [307] introduced a subframe inside each speech frame to ensure that the frame boundaries did not affect the subframe under analysis, thus effectively removing any discontinuities. The approach by Stegmann et al. [321] was similar but with a lookahead to the following speech frame that introduced a delay of 8 ms.

The approach adopted was to use a lookback history into frame $N - 1$ for the frame boundary at the start of frame N, while to avoid any delay, at the end of the frame boundary, periodicity was implemented to extend the speech signal.

12.3 PRE-PROCESSING THE WAVELET TRANSFORM SIGNAL

The $D_Y WT$ of a 20 ms segment of speech is shown in Figure 12.6. For the first scale, the high-pass $D_Y WT$ of Figure 12.4b amplifies the higher frequency, predominately noisy signals, and so no periodicity is apparent. However, as the scales increase from D_{i+1} to D_{i+5}, the periodicity of the speech signal becomes more evident for both the time- and frequency-domain plots shown in Figure 12.6 for the $D_i(\omega)$ signals. We note here that, although the time-domain waveforms of Figure 12.6 are plotted on the finest scale (i.e., $i = 1$) they are waveforms subsampled by a factor of 2^i. The procedure for extracting the relevant information from Figure 12.6 is now considered.

Observing Figure 12.6, we can see that some form of pre-processing must be performed in order to determine the instants of glottal closure, and hence the fundamental frequency of the speech waveform. From Figure 12.6 the maxima and minima during each scale of the $D_Y WT$ provides most information about the speech waveform's pitch period. Thus, Figure 12.7a illustrates the initial pre-processing whereby positive impulses are placed at the maxima and negative impulses at the minima. Each of these impulses is assumed to represent possible instants of glottal closure.

The following sections describe the processes used to identify and eliminate the false glottal closure locations.

12.3.1 Spurious Pulses

Due to the presence of upper harmonics, true instants of glottal closure will manifest themselves in every $D_Y WT$ scale. Hence, all impulses that do not obey this criterion are eliminated. The remaining impulses are given in Figure 12.7b, and the elimination process is described below.

If an impulse is located in scale $i + 5$, then scale $i + 4$ is examined to look for an impulse in the vicinity of the pulse in scale $i + 5$. If a corresponding pulse exists, then scale $i + 3$ is examined. This is repeated for all scales. If any of the scales fails to contain an impulse in the correct neighborhood, the search is abandoned and the impulses are declared void. The terms *vicinity* and *neighbourhood* are used since the addition of 2^i zeros between the coefficients of the multiresolution filters $h(n)$ and $g(n)$ cause the pulses to spread out, as the scale increases.

Following the removal of superfluous impulses, the remaining impulses, which have been confirmed at all resolutions, are amalgamated to one scale. Impulses are placed

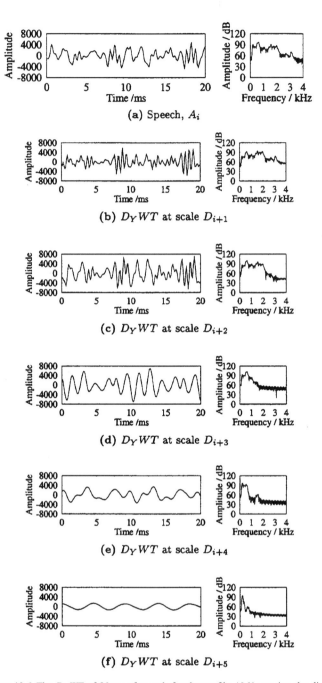

Figure 12.6 The $D_Y WT$ of 20 ms of speech for the testfile AM1 uttering the diphthong /aɪ/ as in "wires". For each scale of the $D_Y WT$, the time- and frequency-domain response are portrayed, enabling the process of the $D_Y WT$ to be clearly interpreted. The (a) speech is followed by the $D_Y WT$ scales, (b) 2000–4000 Hz, (c) 1000–2000 Hz, (d) 500–1000 Hz, (e) 250–500 Hz, and (f) 125–250 Hz, using the quadratic spline wavelet of Figure 12.4 and Table 12.1.

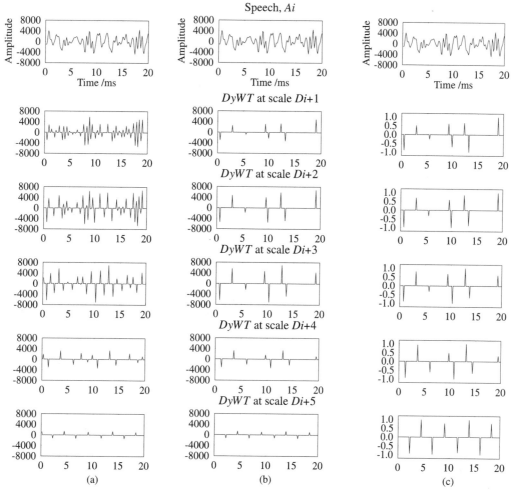

Figure 12.7 The $D_Y WT$ of 20 ms of speech for the testfile AM1 uttering the diphthong /aɪ/ as in "wires". In (a) impulses have been placed at the locations of the maxima and minima of the detail signals, $D_{i+1} \cdots D_{i+5}$. In (b) all spurious pulses have been removed. In (c) the impulses have been normalized with respect to the largest impulse at each scale.

indicating the remaining possible instants of glottal closure, which are termed candidate glottal pulses. These candidate glottal pulse locations represent the amplitudes and positions of the impulses they are combined from.

12.3.2 Normalization

Observing Figure 12.7b, we see that the amplitudes of the $D_Y WT$ decrease with scale increase, owing to the attenuation of the filter $H(\omega)$ shown in Figure 12.4. Hence, normalization of the impulses to the maximum peak at that resolution is performed, ensuring that the candidate glottal pulses are not dominated by the high-magnitude lower scale impulses. The

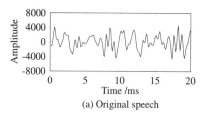

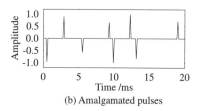

(a) Original speech   (b) Amalgamated pulses

Figure 12.8 Example of wavelet-based pre-processing for the testfile AM1 uttering the diphthong /aɪ/ as in "wires", with a combined impulse at the location of the impulse in scale $i = 1$, and then normalized to the highest pulse magnitude.

impulses after normalization are displayed in Figure 12.7c. Subsequent to this initial normalization process, the final simplification procedure is detailed below.

12.3.3 Candidate Glottal Pulses

The candidate glottal pulses sum the amplitude of the normalized pulses from each scale, but they are situated at the impulse location from scale $i = 1$, the scale with the finest time resolution. The candidate glottal pulses are renormalized to the highest pulse magnitude, as demonstrated in Figure 12.8.

The final information known about the candidate glottal pulses is that they must be at least 20 samples, or 2.5 ms apart, due to the highest expected fundamental frequency of 400 Hz. Thus, for any candidate glottal pulses within 20 samples of each other the smallest is discarded. The results from the $D_Y WT$ are now suitable for use in voiced-unvoiced classification and pitch detection.

12.4 VOICED-UNVOICED DECISION

The ability of the $D_Y WT$ to categorize speech as voiced or unvoiced has been shown previously by Stegmann et al. [321] and Kadambe and Boudreaux-Bartels [307]. The process of the $D_Y WT$ across the scales gradually removes the higher frequency components present in the speech waveform. For unvoiced speech, most energy is present in the higher frequencies, as demonstrated in Figure 10.2, while for voiced speech the energy is more evenly distributed. Thus, the voiced speech is expected to maintain its energy across the dyadic scales better than the unvoiced speech, allowing a voiced-unvoiced decision to be made. A suitable value for controlling these decisions was found to be the ratio of the RMS energy in the frequency range 2 kHz → 4 kHz, to the RMS energy in the frequency band 0 kHz → 2 kHz. This is equivalent to the ratio of the RMS energy of A_{i+1}, to the RMS energy of D_{i+1}, given by:

$$r_{th} = \sqrt{\frac{\sum_{n=0}^{FL} D_{i+1}(n)^2}{\sum_{n=0}^{FL} A_{i+1}(n)^2}} \tag{12.19}$$

where FL is the speech frame length.

A suitable threshold for the voiced-unvoiced test was found to be $r_{th} = 2$, where frames with a ratio higher than $r_{th} = 2$ are found to be unvoiced. Otherwise the frame is voiced.

Thus, for a frame of speech to be classified unvoiced it must contain twice as much energy in the $2\,\text{kHz} \rightarrow 4\,\text{kHz}$ band as in the $0\,\text{kHz} \rightarrow 2\,\text{kHz}$ frequency band.

This threshold measure performs well at distinguishing between voiced and unvoiced speech. However, it tends to classify any frames of silence as voiced speech, due to the even spread of energy across the frequency range for silent speech. A simple voiced-silence detector was implemented by considering the RMS energy of the input speech, A_i. With an RMS energy value of less than $100\,\text{dB}$, a frame is classified silent; otherwise the RMS energy level indicates a voiced speech frame. The process of pitch detection is now investigated.

12.5 WAVELET-BASED PITCH DETECTOR

Following the pre-processing procedures described in Section 12.3, the frames classified as voiced in Section 12.4 contain a group of candidate glottal pulse locations from which the pitch period of the speech frame can be deduced.

Assuming that the largest positive and negative pulses are true glottal pulse locations, we can calculate a range of possible pitch periods. The candidate pitch periods are classified on the basis of the time durations between the largest positive pulse and all other positive pulses, or the largest negative pulse and all negative pulses. Figure 12.9 displays the potential pitch periods for each speech frame in two speech files. These speech files correspond to the speech files used in Chapter 11 for the autocorrelation-based pitch detectors in Figure 11.8, 11.11, and 11.14. The resultant graphs are fairly complex, but it can be observed that the

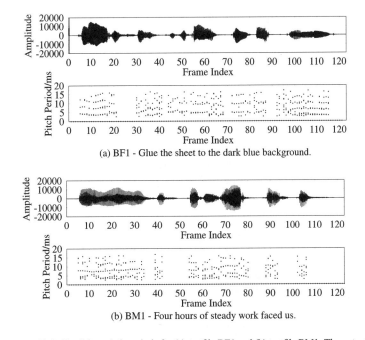

(a) BF1 - Glue the sheet to the dark blue background.

(b) BM1 - Four hours of steady work faced us.

Figure 12.9 Candidate pitch periods for (a) testfile BF1 and (b) testfile BM1. The potential pitch periods tend to consist of the true pitch and its harmonics.

candidate pitch periods are commonly placed at the true pitch period and its harmonics. Typically, the true pitch period and two or three harmonics are present.

Figure 12.9 in our scheme was interpreted with the aid of dynamic programming; first, however, some previously implemented methods are described. Specifically, Kadambe and Boudreaux-Bartels [307] used thresholding to identify the true pitch period-related pulses. However, it was found that the pulse amplitudes associated with the voiced speech waveforms varied so much that it was impossible to find a suitable threshold. Sukkar et al. [327] considered the relative amplitudes of consecutive candidate glottal pulses to determine the true instants of glottal closure, with a threshold employed for controlling which pulses to accept. Once again it was found that the excessive variation in speech waveform shapes and candidate pulse amplitude prevented the identification of a suitable threshold. A novel pitch detection method involving dynamic programming is now investigated.

12.5.1 Dynamic Programming

The implementation of a dynamic programming algorithm for determining a voiced speech frame's pitch period will introduce additional delay into a speech coder. An additional delay of 60 ms, or three speech frames, was introduced allowing the current frame N and the future two frames $N + 1$ and $N + 2$ to be examined by the dynamic program. The history of the pitch track is also examined by considering the pitch period of the previous frame. Since the minimum pitch period is 20 samples, or 2.5 ms, there can be at most seven candidate pitch periods from the positive pulses and seven candidate pitch periods, within an interval of the 20 ms frame length, from the negative pulses. Thus, the dynamic program will examine a maximum of 14 candidate pitch periods over three speech frames, namely, seven positive and seven negative pulses from each frame. Every candidate pitch period P_{N_i} in frame N is assigned a minimum cost, C_{dp_i}, defined by:

$$C_{dp_i} = |f_{N_i} - f_{N+1_j}| + |f_{N+1_j} - f_{N+2_k}| + a_{dp}|f_{N-1} - f_{N_i}| \qquad (12.20)$$

where f_{N_i}, f_{N+1_j}, and f_{N+2_k} refer to the candidate fundamental frequencies of speech frames N, $N + 1$, and $N + 2$, respectively. The fundamental frequency of frame $N - 1$ is given by f_{N-1}, and a_{dp} is a scaling function that defines the amount of pitch tracking, or the correlation between consecutive pitch values. Thus, according to Equation 12.20, the difference between the candidate fundamental frequencies, of consecutive speech frames, determines the cost of each pitch-period candidate in frame N. It is not necessarily assumed that the pitch-period candidate with the smallest cost function, C_{dp_i} is the true pitch period. Instead, the procedure given in Figure 12.10 is followed and described next.

The pitch detector design philosophy was influenced by the observations that the pitch detector performed best when strict pitch tracking was employed after a couple of consecutive pitch periods closely followed the predicted pitch-period evolution. However, the extent of pitch tracking was greatly reduced at the beginning of a voiced sequence and was removed completely after a long period, namely, 240 ms, of unvoiced speech. This is demonstrated by the sections of Figure 12.10 that consider the time elapsed from the last voiced frame, the TIME ELAPSE parameter, where if TIME ELAPSE is exceeded no pitch tracking is employed. If the TIME ELAPSE is not exceeded, then provided that the consecutive pitch values are similar, that is, $P_{N-1} \sim P_{N-2}$, extensive pitch tracking is performed, associated

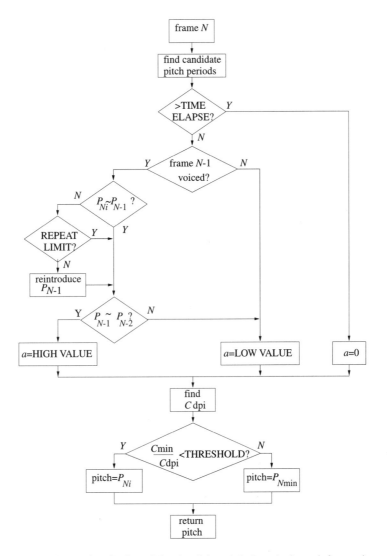

Figure 12.10 Procedure for determining the pitch period of a voiced speech frame using dynamic programming and the cost function of Equation 12.20.

with a high a_{dp} in Equation 12.20 and Figure 12.10. Otherwise weak pitch tracking is introduced.

Figure 12.10 also contains a section that reintroduces the previous pitch period, P_{N-1}, in a set number of consecutive frames. This was implemented since it was found that occasionally the true pitch period would not be among the candidate pitch periods. Thus, reintroducing the previous pitch period allowed the correct pitch track to be maintained. However, it was assumed that the true pitch period would only be missing from a certain number of consecutive voiced frames, within a voiced sequence, controlled by the REPEAT LIMIT parameter.

The final processes, at the bottom of Figure 12.10, consider whether the minimum cost function of Equation 12.20 provides the true pitch period. The higher fundamental

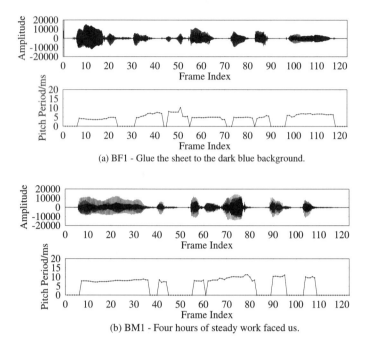

(a) BF1 - Glue the sheet to the dark blue background.

(b) BM1 - Four hours of steady work faced us.

Figure 12.11 The pitch-period decisions for (a) testfile BF1 and (b) testfile BM1. The trace from BF1 has some pitch halving between frames 45 to 50, while BM1 has a good pitch track. For comparison we refer to Figure 11.14.

frequencies have less resolution than the lower fundamental frequencies. Thus, the dynamic programming algorithm tends to favour the longer pitch periods where higher cost functions are scored. Hence, the ratio of the minimum cost C_{min} to all other costs is considered, where values higher than the THRESHOLD parameter result in the minimum cost being kept. However, a value less than the threshold results in a different pitch period being accepted.

The selected pitch periods for the two speech files shown in Figure 12.9 are portrayed in Figure 12.11. It can be seen that the wavelet-based dynamic programming pitch detector performs better than the oversampled autocorrelation-based pitch detector of Figure 11.14, since in Figure 11.14, pitch halving is observed. A comparison between the dynamic programming pitch detector and the manual track of Figure 10.18 in Section 10.4 is given in Table 12.2, where 6.8% of frames are shown to be incorrectly labeled. The computational complexity for the wavelet-based dynamic programming pitch detector is shown in Table 12.3 as 2.70 MFLOPS, a reduction from the 3.4 MFLOPS required by the autocorrelation-based pitch detector.

TABLE 12.2 A Comparison Between the Performance of the Developed Wavelet-based Pitch Detectors and a Manual Pitch Track for the Speech Database. W_U represents the percentage of frames that are labeled voiced when they should have been identified as unvoiced. W_V indicates the number of frames that have been labeled as unvoiced when they are actually voiced. P_G represents the number of frames where a gross pitch error has occurred. The total number of incorrect frames is given as $W_U + W_V + P_G$.

Pitch Detector	$W_U\%$	$V_V\%$	$P_G\%$	Total %
Wavelets and dynamic programming	1.5	1.1	4.1	6.8
Wavelets and ACF	1.3	0.3	2.3	3.9

TABLE 12.23 Computational Complexity for the Wavelet-based Dynamic Programming Pitch Detector

Operation	Complexity/MFLOPS
$D_Y WT$	1.23
Pre-processing	1.00
Dynamic programming	0.47
Total	2.70

The main disadvantage of this method of pitch-period prediction is the 60 ms delay incorporated in the coder. An additional problem is that, while the dynamic algorithm performs well for the displayed speech frames when a pitch estimation error does occur, the strong pitch tracking element propagates the error. This type of error is very audible and disconcerting. Thus, use of the ACF from Section 11.3 is investigated further in the next section.

12.5.2 Autocorrelation Simplification

An attractive alternative to using dynamic programming in the selection of the correct candidate pitch period is to employ wavelet analysis for simplifying the autocorrelation approach to pitch detection. This autocorrelation approach was investigated in Section 11.3, where the ACF was performed on the $20 \rightarrow 147$ sample range to select the correct pitch period. Harnessing the wavelet transform would permit the ACF to be computed for only 15 possible pitch periods, namely, the 14 candidate pitch periods and the reintroduced previous pitch period. This would simplify the autocorrelation procedure by more than 80%.

The process of selecting the pitch period is shown in Figure 12.12 and described next. The 14 candidate pitch periods, together with the reintroduced previous pitch period, are passed to the ACF evaluation block in Figure 12.12. The candidate pitch period which produces the highest autocorrelation value is selected for the pitch period. Finally, some simple pitch tracking is performed, where isolated voiced or unvoiced frames are removed and where pitch periods not related to either neighbor are corrected.

This simplified pitch tracking procedure introduces a delay of 40 ms into the pitch detector, an improvement on the 60 ms required by dynamic programming. The selected pitch periods for the two speech files shown in Figure 12.9 are given in Figure 12.13. Figure 12.13 can be compared with the oversampled autocorrelation-based pitch detector of Figure 11.14 and the dynamic programming algorithm of Figure 12.11. For the testfile BF1 the over-sampled autocorrelation pitch detector of Figure 11.14 produced two regions of pitch halving lasting for the complete utterances. The wavelet-based dynamic programming pitch detector of Figure 12.11 produced a small pitch error around frame 50. The most recently developed wavelet-based autocorrelation pitch detector of Figure 12.13 also contains pitch errors, which always occur at the start and end of voiced utterances, while maintaining the correct pitch track for the remainder of the utterance. The majority of these pitch errors occur at instants of low signal energy and are inaudible in the reconstructed speech signal.

For the testfile BM1, the oversampled autocorrelation pitch detector of Figure 11.14 and the wavelet-based dynamic programming pitch detector of Figure 12.11 produced no errors. In Figure 12.13, pitch doubling occurs around frame 40 and frame 80 at the low-energy termination of voiced utterances. Again, these two errors were inaudible in the reconstructed

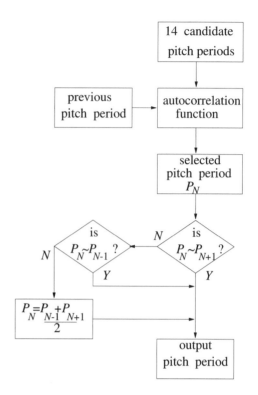

Figure 12.12 The control structure for a wavelet assisted autocorrelation-based pitch detector. Here the wavelet transform is used to reduce the possible pitch periods searched by the ACF evaluation process.

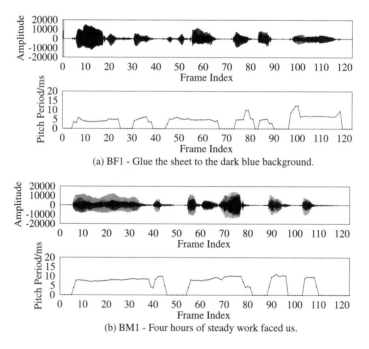

(a) BF1 - Glue the sheet to the dark blue background.

(b) BM1 - Four hours of steady work faced us.

Figure 12.13 The pitch-period decisions for (a) the testfile BF1 and (b) the testfile BM1. The BF1 trace has some pitch halving at the low-energy start and end of voiced utterances. The BM1 trace has some pitch doubling around frame 40 and 80. For a comparison we refer to Figures 11.14 and 12.11.

speech. The performance of this wavelet-based pitch detector is compared in Table 12.3 against the manual track of Section 10.4, where a total of 3.9% of the speech frames were incorrectly classified.

The computational complexity of this wavelet-assisted autocorrelation-based pitch detector is given in Table 12.4. It can be seen that at 2.67 MFLOPS the computational complexity is slightly lower than that of the wavelet-based dynamic programming approach of Table 12.2. Because of its low complexity and low delay, this wavelet-based autocorrelation pitch detector was selected for use in our speech coders.

This selected wavelet-assisted autocorrelation-based pitch detector is the fifth pitch detector investigated, all of which have been informally assessed. Their parameters are detailed in Table 12.5 along with their computational complexity, delay, and error performance details. These pitch detectors are briefly reviewed next.

The first pitch detector, described in Section 11.3.1, was used in the G.728 Recommendation [85]. It performed ACF computations with very simple pitch tracking. The results, shown in Figure 11.8, displayed excessive regions of pitch halving. The performance of this pitch detector was improved through the addition of oversampling [76] and extensive pitch tracking [73], detailed in Section 11.3.2. The pitch detector produced a substantially improved performance, as seen in Figure 11.14. However, it was excessively complex. Section 11.3.4 described a pitch detector with the oversampling removed but with the extensive pitch tracking remaining. This pitch detector had a more acceptable complexity and similar performance to Figure 11.14.

The introduction of wavelets decreased the complexity of the pitch detection procedure. With the incorporation of dynamic programming in Section 12.5.1, a good quality pitch detector was created, as shown in Figure 12.11. However, this method required a 60 ms delay. Finally, in Section 12.5.2 the ACF was incorporated into the wavelet-based scheme in order to produce a pitch detector. This pitch detector had reduced delay, low complexity, and few errors. These errors were inaudible and occurred during low-energy speech segments at utterance terminations. Hence, in our coders the wavelet-assisted autocorrelation-based pitch detector was favored.

TABLE 12.4 Computational Complexity for the Wavelet-Assisted Autocorrelation-based Pitch Detectors

Operation	Complexity/MFLOPS
$D_Y WT$	1.23
Preprocessing	1.00
Autocorrelation	0.24
Total	2.67

TABLE 12.5 Review of Considered Pitch Detectors

Pitch Detector	Complexity/MFLOPS	Delay/ms	Percent of Error Frames/%
ACF	1.1	40	12.7
ACF with pitch tracking	3.4	20	7.2
Wavelet and dynamic programming	2.7	60	6.8
Wavelet and ACF	2.7	40	3.9

12.6 SUMMARY AND CONCLUSIONS

This chapter has introduced the concept of the $D_Y WT$ in Sections 12.1 and 12.2. The resultant transformed speech signal was analyzed in Section 12.3, permitting its use in voiced-unvoiced decisions, in Section 12.4, and pitch detection, in Section 12.5.

This chapter demonstrates that a wavelet-based pitch detector incorporating autocorrelation performs better than the autocorrelation-assisted pitch detectors of Chapter 11, with errors in 3.9% of frames as opposed to 7.2% in the selected autocorrelation pitch detector from Chapter 11. The wavelet-based autocorrelation pitch detector also required a reduced computational complexity of 2.7 MFLOPS instead of the 3.4 MFLOPS of Chapter 11.

Having created a benchmark LPC vocoder, investigated the appropriate choice of LSF quantization, and selected an appropriate pitch-period detection method, we now investigate a prototype waveform interpolation coder in the next chapter.

13

Zinc Function Excitation

F.C.A. Somerville, C. Xydeas, L. Hanzo

13.1 INTRODUCTION

This Chapter introduces a prototype waveform interpolation (PWI) speech coder that uses zinc function excitation (ZFE) [295]. A PWI scheme operates by encoding one pitch-period-sized segment, a prototype segment, of speech for each frame. The slowly evolving nature of speech permits PWI to reduce the transmitted bitrates, while smooth waveform interpolation at the decoder between the prototype segments maintains good synthesized speech quality. Figure 13.1a shows two 20 ms frames of voiced speech, with a pitch period in both frames highlighted in each to demonstrate the slow waveform evolution of speech. The same pitch periods are again highlighted for the LPC STP residual waveform in Figure 13.1b, demonstrating that PWI can also be used on the residual signal. Finally, Figure 13.1c displays the frequency spectrum for both frames, showing the evolution of the speech waveform in the frequency domain. The excitation waveforms employed in this Chapter are the zinc-basis functions [295], which efficiently model the LPC STP residual while reducing the speech's "buzziness" when compared with the classical vocoders of Chapter 11 [295]. The previously introduced schematic in Figure 10.13 portrays the encoder structure for the interpolated zinc function prototype excitation (IZFPE), which has the form of a closed-loop LPC-based coding method with optimized ZFE prototype segments for the speech. A similar structure is used in the PWI-ZFE coder described in this chapter.

This Chapter follows the basic outline of the IZFPE coder introduced by Hiotakakos and Xydeas [294], but some sections of the scheme have been developed further. The Chapter begins with an overview of the PWI-ZFE scheme, detailing the operational scenarios of the arrangement. This is followed by the introduction of the zinc-basis functions, together with the optimization process at the encoder, where the wavelets of Chapter 12 are harnessed to reduce the complexity of the process. For voiced speech frames, the pitch detector employed and the prototype segment selection process are described, with a detailed discussion of the interpolation process, where the parameters required for transmission are also given. In addition, the unvoiced excitation and adaptive post-filter are briefly described. Finally, the performance of both a single ZFE and multiple ZFE arrangements is detailed.

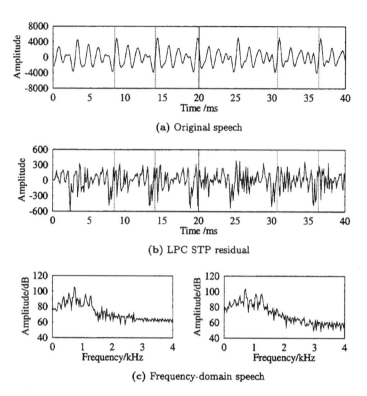

Figure 13.1 Two speech frames demonstrating the smoothly evolving nature of the speech waveform and that of the LPC STP residual in the time and frequency domain. The speech frames are from AF1, uttering the back vowel /ɔ/ in "dog".

13.2 OVERVIEW OF PROTOTYPE WAVEFORM INTERPOLATION ZINC FUNCTION EXCITATION

This section gives an in-depth description of the PWI-ZFE scheme, considering all possible operational scenarios at both the encoder and decoder. The number of coding scenarios is increased by the separate treatment of voiced and unvoiced frames, as well as by the need to accurately represent the voiced excitation.

13.2.1 Coding Scenarios

For the PWI-ZFE encoder the current, the next, and the previous two 20 ms speech frames are evaluated, as shown in Figure 13.2, which is now described in depth. Knowledge of the four 20 ms frames, namely, frames $N + 1$, N, $N - 1$ and $N - 2$, is required in order to adequately treat voiced-unvoiced boundaries. It is these transition regions which are usually the most poorly represented speech segments in classical vocoders. The parameters encoded and transmitted during voiced and unvoiced periods are summarized toward the end of the Chapter in Table 13.10, while the various coding scenarios are summarized in Tables 13.1 and 13.2.

LPC STP analysis is performed for all speech frames, and the RMS value is determined from the residual waveform. The pitch period of the speech frame is also determined.

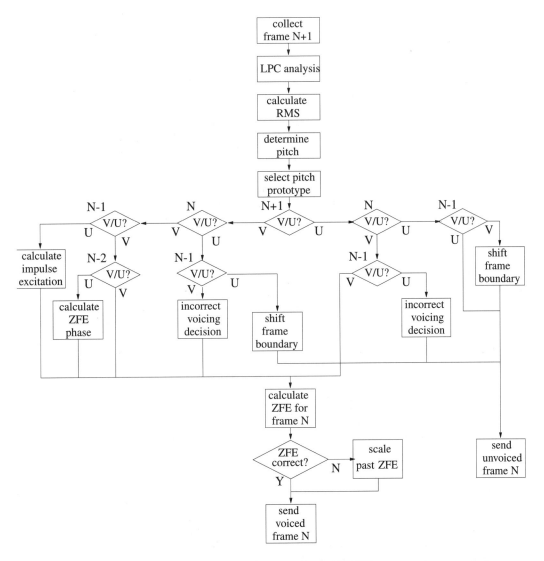

Figure 13.2 The encoder control structure for the PWI-ZFE arrangement.

However, if the speech frame lacks any periodicity, then the pitch period is assigned as zero and the speech frame is labeled as unvoiced. The various possible combinations of consecutive voiced (V) and unvoiced (U) frames are now considered.

13.2.1.1 U-U-U Encoder Scenario.   If all the speech frames $N + 1$, N and $N - 1$ are classified as unvoiced, U-U-U, then the unvoiced parameters for frame N are sent to the decoder. The unvoiced parameters are the LPC coefficients, sent as LSFs, a voicing flag that is set to *off*, and the quantized RMS value of the LPC STP residual, as described in Table 13.1.

13.2.1.2 U-U-V Encoder Scenario.   With a voicing sequence of U-U-V, where frame $N - 1$ is voiced, together with the unvoiced parameters, an extra parameter b_s, the boundary shift parameter, must be conveyed to the decoder to be used for the voicing transition regions.

TABLE 13.1 Summary of Encoder Scenarios (see text for more detail)

$N+1$	N	$N-1$	Summary
U	U	U	Frame N is located in an unvoiced sequence. Quantize and transmit the RMS value of the LPC STP residual to the decoder.
U	U	V	A voiced-to-unvoiced transition boundary has been encountered. Calculate the section of frame N that is voiced and include this boundary shift parameter, b_s, in the transmission of frame N to the decoder.
V	U	U	An unvoiced-to-voiced transition boundary has been encountered. Calculate the section of frame N that is voiced and include this boundary shift parameter, b_s, in the transmission of frame N to the decoder.
U	V	U	Assume frame N should have been classified as unvoiced; hence, treat this scenario as an U-U-U sequence.
V	V	V	Frame N is situated in a voiced sequence. Calculate the ZFE parameters, A_1 and B_1, and λ_1. Quantize the amplitude parameters A_1 and B_1, and transmit parameters to decoder.
V	U	V	Assume frame N should have been labeled as voiced; hence, treat this case as a V-V-V sequence.
U	V	V	Treat this situation as a V-V-V sequence.
V	V	U	The start of a sequence of voiced frames has been encountered. Represent the excitation in the prototype segment with an impulse.

TABLE 13.2 Summary of Decoder Scenarios (see text for more detail)

$N+1$	N	Summary
U	V	A voiced-to-unvoiced transition has been encountered. Label the portion of frame $N+1$ that is voiced, and subsequently interpolate from the pitch prototype segment in frame N to the voiced sections in frame $N+1$.
U	U	Frame N is calculated using a Gaussian noise excitation scaled by the RMS value for frame N.
V	U	An unvoiced-to-voiced transition has been encountered. Label the portion of frame N that is voiced and represent the relevant section of frame N by voiced excitation.
V	V	Interpolation is performed between the pitch prototype segments of frame N and frame $N+1$.

In order to determine the voiced to unvoiced transition point, b_s, frame N is examined, searching for evidence of voicing, in segments sized by the pitch period. The boundary shift parameter b_s represents the number of pitch periods in the frame that contain voicing. At the decoder this voiced section of the predominantly unvoiced frame N is represented by the ZFE excitation reserved for the voiced segments.

13.2.1.3 V-U-U Encoder Scenario.
The boundary shift parameter is also sent for the voicing sequence V-U-U. However, for this sequence the predominantly unvoiced frame N is examined in order to identify how many pitch-period durations can be classified as voiced. The parameter b_s in frame N then represents the number of pitch periods in the unvoiced frame N that contain voicing. At the decoder, this section of frame N is synthesized using voiced excitation.

13.2.1.4 U-V-U Encoder Scenario.
A voicing sequence U-V-U is assumed to have an incorrect voicing decision in frame N. Hence, the voicing flag in frame N is set to zero, and the procedure for an U-U-U sequence is followed.

13.2.1.5 V-V-V Encoder Scenario.   For a voiced sequence of frames V-V-V, the ZFE parameters for frame N are calculated. The ZFE is described by the position parameter λ_1, and the amplitude parameters A_1 and B_1, as shown earlier in Figure 10.14. Further ZFE waveforms are shown in Figure 13.5, which also illustrates the definition of the ZFE phase referred to below. If frame $N-2$ was also voiced, then the chosen ZFE is restricted by certain phase constraints, which will be detailed in Section 13.3. Otherwise, frame N is used to determine the phase restrictions. The selected ZFE represents a pitch-duration segment of the speech frame, which is referred to as the pitch prototype segment. If a ZFE that complies with the required phase restrictions is not found, then the ZFE parameters from frame $N-1$ are scaled, in terms of the RMS energy of the respective frames, and then they are used in frame N. This is performed since it is assumed that the previous frame parameters will be an adequate substitute for frame N, due to the speech parameters' slow time-domain evolution. The parameters sent to the decoder include the LSFs and a voicing flag set to *on*. The decoder requires the ZFE parameters A_1, B_1, and λ_1 to synthesize voiced speech, and the pitch prototype segment is defined by its starting point and the pitch-period duration of the speech segment.

13.2.1.6 V-U-V Encoder Scenario.   A voiced sequence of frames is also assumed for the voicing decisions V-U-V, with the frame N being assigned a pitch period halfway between the pitch period for frame $N-1$ and frame $N+1$.

13.2.1.7 U-V-V Encoder Scenario.   The voicing sequence U-V-V also follows the procedure of a V-V-V string, since the unvoiced decision of frame $N+1$ is not considered until the V-U-V or U-U-V scenarios.

13.2.1.8 V-V-U Encoder Scenario.   The voicing decision V-V-U indicates that frame N will be the start of a voicing sequence. Frame $N+1$, the second frame in the voicing sequence, typically constitutes a better reflection of the dynamics of the voiced sequence than the first one [294]. Hence, the phase restrictions are determined from this frame. The first voiced frame, namely N, is represented by an excitation pulse similar to that used by the LPC vocoder of Chapter 11 [81].

The speech encoder introduces a delay of 40 ms into the system, where the delay is caused by the necessity for frame $N+1$ to verify voicing decisions. In the decoder control structure, shown in Figure 13.3, only the frames $N+1$ and N are considered when synthesizing frame N. Thus, an additional 20 ms delay is introduced.

13.2.1.9 U-V Decoder Scenario.   If the sequence U-V occurs for the frame $N+1$ and N, respectively, then a voiced-to-unvoiced transition is encountered. Here the boundary shift parameter b_s, transmitted in frame $N+1$, is multiplied by the pitch period in frame N, indicating the portion of frame $N+1$ which was deemed voiced. The ZFE excitation for frame N is interpolated to the end of the voiced portion of frame $N+1$. Subsequently, the interpolation frame N is synthesized.

13.2.1.10 U-U Decoder Scenario.   When the sequence U-U occurs for frame indices $N+1$ and N, if frame $N-1$ is unvoiced, then frame N will be represented by a Gaussian noise excitation. However, if frame $N-1$ was voiced, some of frame N will already be represented by a ZFE pulse. This will be indicated by the value of the boundary shift parameter b_s. Thus, only the unvoiced section of frame N is represented by Gaussian noise.

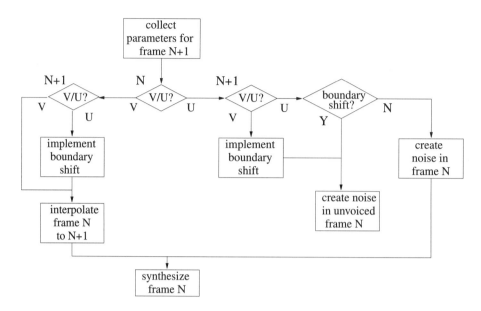

Figure 13.3 The decoder control structure for PWI-ZFE arrangement.

13.2.1.11 V-U Decoder Scenario.   The sequence V-U indicates an unvoiced-to-voiced transition. Therefore, the value of the boundary shift parameter b_s conveyed by frame N is observed. Only the unvoiced section of frame N is represented by Gaussian noise, with the voiced portion represented by a ZFE interpolated from frame $N + 1$.

13.2.1.12 V-V Decoder Scenario.   The sequence V-V directs the decoder to inter-polate the ZFE parameters between frame N and frame $N + 1$. This interpolation process is described in Section 13.6, where it occurs for the region between pitch prototype segments. Thus, each speech frame has its first half interpolated, while classified as frame $N + 1$, with its second half interpolated during the next iteration, while classified as frame N.

Following this in-depth description of the control structure of a PWI-ZFE scheme, as given by Figures 13.2 and 13.3, a deeper insight into the description of the ZFE is now given.

13.3 ZINC FUNCTION MODELING

The continuous zinc function used in the PWI-ZFE scheme to represent the LPC STP residual is defined by [295]:

$$z_k(t) = A_k \cdot \text{sinc}(t - \lambda_k) + B_k \cdot \text{cosc}(t - \lambda_k) \tag{13.1}$$

where $\text{sinc}(t) = \sin(2\pi f_c t)/2\pi f_c t$, $\text{cosc}(t) = 1 - \cos(2\pi f_c t)/2\pi f_c t$, k is the *kth* zinc function, A_k, B_k determine the amplitude of the zinc function, and λ_k determines its location. For the discrete time case with a speech bandwidth of $f_c = 4\,\text{kHz}$ and a sampling frequency of $f_s = 8\,\text{kHz}$, we have [294]:

$$z_k(n) = A_k \cdot \text{sinc}(n - \lambda_{k)} + B_k \cdot \text{cosc}(n - \lambda_{k)} = \begin{cases} A_k \,, & n - \lambda_k = 0 \\ \dfrac{2B_k}{n\pi}, & n - \lambda_k = \text{odd} \\ 0, & n - \lambda_k = \text{even} \end{cases} \qquad (13.2)$$

13.3.1 Error Minimization

From Figure 10.16, which describes the analysis-by-synthesis process, the weighted error signal $e_w(n)$ can be described by:

$$e_w(n) = s_w(n) - \bar{s}_w(n) \qquad (13.3)$$

$$= s_w(n) - m(n) - \left(\sum_{k=1}^{K} z_k(n) * h(n) \right) \qquad (13.4)$$

$$= y(n) - \left(\sum_{k=1}^{K} z_k(n) * h(n) \right) \qquad (13.5)$$

where $y(n) = s_w(n) - m(n)$, $m(n)$ is the memory of the LPC synthesis filter due to previous excitation segments, while $h(n)$ is the impulse response of the weighted synthesis filter, $W(z)$, and K is the number of ZFE pulses employed. Thus the error, $e_w(n)$, is the difference between the weighted original and weighted synthesized speech, with the synthesized speech being the ZFE passed through the synthesis filter, $W(z)$. This formulation of the error signal, where the filter's contribution is divided into filter memory $m(n)$ and impulse response $h(n)$, reduces the computational complexity required in the error minimization procedure. It is the Infinite Impulse Response (IIR) nature of the filter, which requires the memory to be considered in the error equation. For further details of the mathematics, Chapter 3 of Steele [328] is recommended. The sum of the squared weighted error signal is given by:

$$E_w^{k+1} = \sum_{n=1}^{excint} \left(e_w^{k+1}(n) \right)^2 \qquad (13.6)$$

where $e_w^{k+1}(n)$ is the kth order weighted error, achieved after k zinc-basis functions have been modeled, and $excint$ is the length over which the error signal has to be minimized—here the pitch prototype segment length.

Appendix B describes the process of minimizing the squared error signal using Figure 10.16 and Equations 13.1 to 13.6. It is shown that the mean squared error signal is minimized if the expression:

$$\zeta_{mse} = \frac{R_{es}^2}{R_{ss}} + \frac{R_{ec}^2}{R_{cc}} \qquad (13.7)$$

is maximized as a function of the ZFE position parameter λ_{k+1}, and:

$$R_{es} = \sum_{n=1}^{excint} (\text{sinc}(n - \lambda_{k+1}) * h(n)) \times e_w^k(n) \qquad (13.8)$$

$$R_{ec} = \sum_{n=1}^{excint} (\text{cosc}(n - \lambda_{k+1}) * h(n)) \times e_w^k(n) \qquad (13.9)$$

$$R_{ss} = \sum_{n=1}^{excint} (\text{sinc}(n - \lambda_{k+1})h(n))^2 \qquad (13.10)$$

$$R_{cc} = \sum_{n=1}^{excint} (\text{cosc}(n - \lambda_{k+1}) * h(n))^2 \qquad (13.11)$$

where $*$ indicates convolution.

Due to bitrate limitations it is now assumed that a single ZFE is used, that is, $k = 1$. Furthermore, it is assumed that the value *excint* becomes equivalent to the pitch-period duration, with λ_k controlling the placement of the ZFE in the range [1 to *excint*].

The ZFE amplitude coefficients are given by Equations B.14 and B.15 of Appendix B, repeated here for convenience:

$$A_k = \frac{R_{es}}{R_{ss}} \qquad (13.12)$$

$$B_k = \frac{R_{ec}}{R_{cc}}. \qquad (13.13)$$

The optimization involves computing ζ_{mse} in Equation 13.7 for all legitimate values of λ_1 in the range [1 to *excint*], subsequently finding the corresponding values for A_1 and B_1 from Equations 13.12 and 13.13. The computational complexity for this optimization procedure is now assessed.

13.3.2 Computational Complexity

The associated complexity is evaluated as follows and tabulated in Table 13.3. The calculation of the minimization criterion ζ_{mse} requires the highest computational complexity, where the convolution of both the sinc and cosc functions with the impulse response $h(n)$ is performed. From Equation 13.2 it can be seen that the sinc function is only evaluated when $n - \lambda_k = 0$, while the cosc function must be evaluated whenever $n - \lambda_k$ is odd. The convolved signals, involving the sinc and cosc signals, are then multiplied by the weighted error signal $e_w(n)$ to calculate R_{es} and R_{ec}, in Equations 13.8 and 13.9, respectively. Observing Equations 13.6 to 13.13, the computational complexity's dependence on the *excint* parameter can be seen. Thus, in Table 13.3 all values are calculated with the extreme values of *excint*,

TABLE 13.3 Computational Complexity for Error Minimization in the PWI-ZFE Encoder for the Extremities of the *Excint* Variable

Procedure	*excint* = 20/*MFLOPS*	*excint* = 147/*MFLOPS*
Convolve sinc and $h(n)$	0.02	1.06
Convolve cosc and $h(n)$	0.20	78.0
Calculate A_1	0.04	2.16
Calculate B_1	0.04	2.16
Total	0.3	83.38

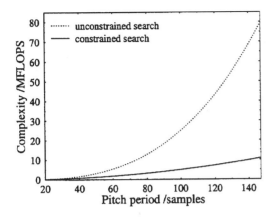

Figure 13.4 Computational complexity for the permitted pitch-period range of 20 to 147 sample duration, for both an unrestricted and constrained search.

which are 20 and 147, the possible pitch-period duration range in samples. The complexity increase is exponential, as shown in Figure 13.4 by the dashed line. The figure shows that any pitch period longer than 90 samples in duration will exceed a complexity of 20 MFLOPS.

13.3.3 Reducing the Complexity of Zinc Function Excitation Optimization

The complexity of the ZFE minimization procedure can be reduced by considering the glottal closure instants (GCI) introduced in Chapter 12. In Chapter 12 wavelet analysis was harnessed to produce a pitch detector, where the pitch period was determined as the distance between two GCIs. These GCIs indicate the snapping shut, or closure, of the vocal folds, which provides the impetus for the following pitch period. The energy peak caused by the GCI will typically be in close proximity to the position of the ZFE placed by the ZFE optimization process. This permits the possibility of reducing the complexity of the analysis-by-synthesis process. Figure 13.4 shows that as the number of possible ZFE positions increases linearly, the computational complexity increases exponentially. Hence, constraining the number of ZFE positions will ensure that the computational complexity remains at a realistic level. The constraining process is described next.

The first frame in a voiced sequence has no minimization procedure; simply put, single pulse is situated at the glottal pulse location within the prototype segment. For the other voiced frames, in order to maintain a moderate computational complexity, the number of possible ZFE positions is restricted as if the pitch period is always 20 samples. A suitable constraint is to have the ZFE located within 10 samples of the instant of glottal closure situated in the pitch prototype segment. Table 13.4 repeats the calculations of Table 13.3, for

TABLE 13.4 Computational Complexity for Error Minimization in the PWI-ZFE Encoder with a Restricted Search Procedure

Procedure	$excint = 20$/MFLOPS	$excint = 147$/MFLOPS
Convolve sinc and $h(n)$	0.02	0.15
Convolve cosc and $h(n)$	0.20	10.73
Calculate A_1	0.04	0.29
Calculate B_1	0.04	0.29
Total	0.30	11.46

complexities related to 20 and 147 sample pitch periods, for a restricted search. In Figure 13.4 the solid line represents the computational complexity of a restricted search procedure in locating the ZFE. The maximum complexity for a 147-sample pitch period is 11 MFLOPS. The degradation to the speech coder's performance, caused by restricting the number of ZFE locations, is quantified in Section 13.4.2.

13.3.4 Phases of the Zinc Functions

There are four possible phases of the ZFE produced by four combinations of positive or negative valued A_1 and B_1 parameters, which are demonstrated in Figure 13.5 for parameter values of $A_1 = \pm 1$ and $B_1 = \pm 1$. Explicitly, if $|A_1| = 1$ and $|B_1| = 1$, then the possible phases of the ZFE are the following: $A_1 = 1$ $B_1 = 1$, $A_1 = 1$ $B_1 = -1$, $A_1 = -1$ $B_1 = 1$, and $A_1 = -1$ $B_1 = -1$. The phase of the ZFE is determined during the error minimization process, where the calculated A_1, B_1 values of Equations 13.12 and 13.13 will determine the ZFE phase. For successful interpolation at the decoder, the phase of the ZFE should remain constant throughout each voiced sequence.

Following this insight into zinc function modeling, the practical formulation of an PWI-ZFE coder is discussed. Initially, the procedures requiring pitch-period knowledge are discussed, which are followed by details of voiced and unvoiced excitation considerations.

13.4 PITCH DETECTION

The PWI-ZFE coder located the voiced frame's pitch period using the autocorrelation-based wavelet pitch detector described in Section 12.5.2, which has a computational complexity of 2.67 MFLOPS. This Section investigates methods of making voiced-unvoiced decisions for pitch-sized segments and methods for identifying a pitch-period segment.

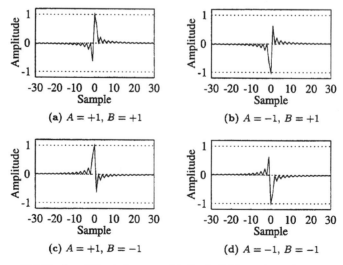

Figure 13.5 The four different phases possible for the ZFE waveform of Equation 13.1.

13.4.1 Voiced-Unvoiced Boundaries

Classifying a segment of speech as voiced or unvoiced is particularly difficult at the transition regions. Hence, a segment of voiced speech can easily become classified as unvoiced. Thus, in the transition frame, pitch-duration-sized segments are examined for evidence of voicing. In this case the autocorrelation approach cannot be used, for several pitch periods are not available for the correlation procedure. Instead, a side result of the wavelet-based pitch detector is utilized; namely, that for every speech frame candidate glottal pulse locations exist.

Therefore, if the first voiced frame in a voiced sequence is frame N, then frame $N - 1$ is examined for boundary shift. If a periodicity close to the pitch period of frame N exists over an end portion of frame $N - 1$, this end portion of frame $N - 1$ is designated as voiced. Similarly, if the final voiced frame in a voiced sequence is frame N, then frame $N + 1$ is examined for boundary shift. Any starting portion of frame $N + 1$ that has periodicity close to the pitch period of frame N is declared voiced.

In the speech decoder, the ZFE parameters must be interpolated over an integer number of pitch periods. Thus, the precise duration of voiced speech in the transition frame is not completely defined until the λ_1 interpolation process (to be described in Section 13.6) is concluded.

13.4.2 Pitch Prototype Selection

For each speech frame classed as voiced, a prototype pitch segment is located, parameterized, encoded, and transmitted. Subsequently, at the decoder interpolation between adjacent prototypes is performed. For smooth waveform interpolation the prototype must be a pitch period in duration, since this speech segment captures all elements of the pitch-period cycle, thus enabling a good reconstruction of the original speech.

The prototype selection for the first voiced frame is demonstrated in Figure 13.6. If P is the pitch period of the voiced frame, then P samples in the centre of the frame are selected as the initial prototype selection, as shown in the second trace of Figure 13.6. Following Hiotakakos and Xydeas [294], the maximum amplitude is found in the frame, as shown in the middle trace of Figure 13.6. Finally, the zero-crossing immediately to the left of this maximum is selected as the start of the pitch prototype segment, as indicated at the bottom of the figure. The end of the pitch prototype segment is a pitch-period duration away. Locating the start of the pitch prototype segment near a zero crossing helps to reduce discontinuities in the speech encoding process.

It is also beneficial in the interpolation procedure of the decoder if consecutive ZFE locations are smoothly evolving. Therefore, close similarity between consecutive prototype segments within a voiced sequence of frames is desirable. Thus, after the first frame the procedure of Hiotakakos and Xydeas [294] is no longer followed. Instead, the cross-correlation between consecutive pitch prototype segments [286] of the other speech frames is performed. These subsequent pitch prototype segments are calculated from the maximum cross-correlation between the current speech frame and previous pitch prototype segment. Figure 13.7 shows how, at the encoder, the speech waveform prototype segments can be concatenated to produce a smoothly evolving waveform.

In order to further improve the probability that consecutive ZFEs have similar locations within their prototype segments, any instants of glottal closure that are not close to the previous segment's ZFE location are discarded, with the previous ZFE location used to search for the new ZFE in the current prototype segment.

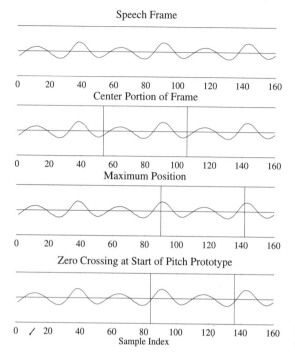

Figure 13.6 Pitch prototype selection for AM2 uttering the nasal consonant /n/ from "e*n*d".

At the encoder, the introduction of constraining the location of the ZFE pulse to within ±10 positions reduces the SEGSNR value, as shown in Table 13.5. The major drawback of the constrained search is the possibility that the optimization process is degraded through the limited range of ZFE locations searched. In addition, it is possible to observe the degradation to the mean squared error (MSE) optimization, caused by the phase restrictions imposed on the ZFEs and detailed in Section 13.3.4. Table 13.5 displays the SEGSNR values of the concatenated voiced prototype speech segments. The unvoiced segments are ignored, since these speech spurts are represented by noise. Thus a SEGSNR value would be meaningless.

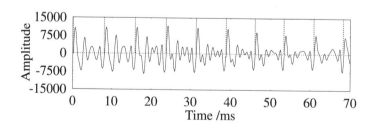

Figure 13.7 Concatenated speech signal prototype segments producing a smoothly evolving waveform. The dotted lines represent the prototype boundaries.

TABLE 13.5   SEGSNR Results for the Optimization Process with and without Phase Restrictions, or a Constrained Search

	Unconstrained search	Constrained search
No phase restrictions	3.36 dB	2.68 dB
Phase restrictions	2.49 dB	1.36 dB

Observing Table 13.5 for a totally unconstrained search, we see that the SEGSNR achieved by the ZFE optimization loop is 3.36 dB. The process of either implementing the above-mentioned ZFE phase restriction or constraining the permitted ZFE locations to the vicinity of the GCIs reduces the voiced segments' SEGSNR after ZFE optimization by 0.87 dB and 0.68 dB, respectively. Restricting both the phase and the ZFE locations reduces the SEGSNR by 2 dB. However, in perceptual terms, the ZFE interpolation procedure, described in Section 13.6, actually improves the subjective quality of the decoded speech due to the smooth speech waveform evolution facilitated, despite the SEGSNR degradation of about 0.87 dB caused by imposing phase restrictions. Similarly, the extra degradation of about 1.13 dB caused by constraining the location of the ZFEs also improves the perceived decoded speech quality due to smoother waveform interpolation.

13.5 VOICED SPEECH

For frames designated as voiced, the excitation signal is a single ZFE. For a single ZFE the equations defined in Section 13.3 and Appendix B are simplified, since the kth stage error Equation 13.5 becomes:

$$e_w^0(n) = y(n). \tag{13.14}$$

Therefore, Equation 13.6 for the weighted error of a single ZFE is given by:

$$E_w^1(n) = \left(\sum_{n=1}^{excint} e_w^1(n) \right)^2 \tag{13.15}$$

where $e_w^1(n) = y(n) - [z(n) * h(n)]$ and excinct is the interval over which the error has to be minimized, i.e., the pitch prototype segment length. Equations 13.8 and 13.9 are simplified to:

$$R_{es} = \sum_{n=1}^{excint} (\text{sinc}(n - \lambda_1) * h(n)) \times y(n) \tag{13.16}$$

$$R_{ec} = \sum_{n=1}^{excint} (\text{cosc}(n - \lambda_1) * h(n)) \times y(n). \tag{13.17}$$

Calculating the ZFE, which best represents the pitch prototype, involves locating the value of λ_1 between 0 and the pitch period that maximizes the expression for ζ_{mse} given in Equation 13.7. While calculating ζ_{mse}, $h(n)$ is the impulse response of the weighted synthesis filter $W(z)$, and the weighted error signal e_w is the LPC residual signal minus the LPC STP filter's memory, as shown by Equation 13.14. The use of prototype segments produces a ZFE determination process that is a discontinuous task. Thus, the actual filter memory is not explicitly available for the ZFE optimization process. Subsequently the filter's memory is

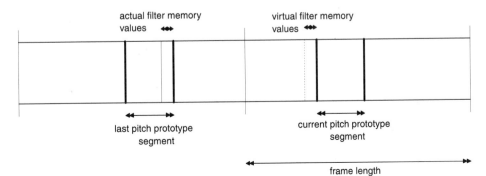

Figure 13.8 Determining the LPC filter memory.

assumed to be due to the previous ZFE. Figure 13.8 shows two consecutive speech frames, where the previous pitch prototype segment has its final p samples highlighted as LPC synthesis filter memory values, while for the current pitch prototype segment these p samples constitute virtual filter memory. Thus, for the error minimization procedure the speech between the prototype segments has been effectively removed.

Once the value of λ_1, which produces the maximum ζ_{mse} value, has been determined, the appropriate values of A_1 and B_1 are calculated using Equations 13.12 and 13.13. Figure 13.7 displayed the smooth evolution of the concatenated pitch prototype segments. If the ZFEs selected for these prototype segments are passed through the weighted LPC STP synthesis filter, the resulting waveform should be a good match for the weighted speech waveform used in the minimization process. This is shown in Figure 13.9, characterizing the analysis-by-synthesis approach used in the PWI-ZFE encoder.

The above procedure is only followed for the phase-constraining frame; for subsequent frames in a voiced sequence, the ZFE selected must have the phase dictated by the phase-constraining frame. If phase restrictions are not followed, then during the interpolation process a change in the sign of A_1 or B_1 will result in some small valued interpolated ZFEs as the values pass through zero. For each legitimate zinc pulse position, λ_1, the signs of A_1 and B_1 are initially checked, where the value of ζ_{mse} is calculated only if the phase restriction is satisfied. Therefore, the maximum value of ζ_{mse} associated with a suitably phased ZFE is selected as the excitation signal. It is feasible that a suitably phased ZFE will not be found. Indeed, with the test database 13% of the frames did not have a suitable ZFE. If this occurs, then the previous ZFE is scaled, as explained below, and used for the current speech frame.

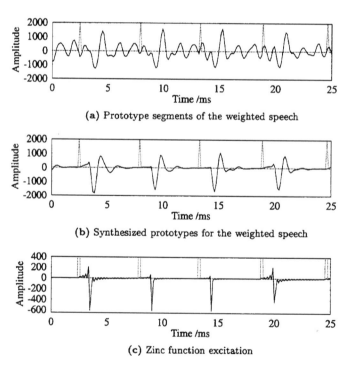

Figure 13.9 Demonstrating the process of analysis-by-synthesis encoding for prototype segments that have been concatenated to produce a smoothly evolving waveform. The dotted spikes indicate the boundaries between prototype segments.

The scaling is based on the RMS value of the LPC residual after STP analysis, which is defined by:

$$A_1(N) = \delta_s A_1(N - 1) \tag{13.18}$$

$$B_1(N) = \delta_s B_1(N - 1) \tag{13.19}$$

where

$$\delta_s = \frac{\text{RMS of LPC residual N}}{\text{RMS of LPC residual N} - 1}. \tag{13.20}$$

The value of $\lambda_1(N)$ is assigned to be the ZFE position in frame $N - 1$, becoming $\lambda_1(N - 1)$.

13.5.1 Energy Scaling

The values of A_1 and B_1 determined in the voiced speech encoding process produce an attenuation in the signal level from the original prototype signal. The cause of this attenuation is due to the nature of the minimization process described in Section 13.3, where the best waveform match between the synthesized and original speech is found. However, the minimization process does not consider the relative energies of the original weighted waveform and the synthesized weighted waveform. Thus, the values of the A_1 and B_1 parameters are scaled to ensure that the energies of the original and reconstructed prototype signals are equal, requiring that:

$$\sum_{n=1}^{excinct} (z(n) * h(n))^2 = \sum_{n=1}^{excinct} (\bar{s}_w(n) - m(n))^2 \tag{13.21}$$

where excinct is the interval over which the error has to be minimized, i.e., the pitch prototype segment length $h(n)$ is the impulse response of the weighted LPC STP synthesis filter, $\bar{s}_w(n)$ is the weighted speech signal, and $m(n)$ is the memory of the weighted LPC STP synthesis filter. Ideally, the energy of the excitation signals will also be equal, thus:

$$\sum_{n=1}^{excinct} z(n)^2 = \sum_{n=1}^{excinct} r(n)^2 \tag{13.22}$$

where $r(n)$ is the LPC STP residual.

The above equation shows that it is desirable to ensure that the energy of the synthesized excitation is equal to the energy of the LPC STP residual for the prototype segment. Upon expanding the left-hand side of Equation 13.22 to include A_1 and B_1, also introducing the scale factor S_{AB} that will ensure Equation 13.22 is obeyed, we have:

$$\sum_{n=1}^{excinct} [\sqrt{S_{AB}} A_1 \sin c(n - \lambda_1) + \sqrt{S_{AB}} B_1 \cos c(n - \lambda_1)]^2 = \sum_{n=1}^{excinct} r(n)^2 \tag{13.23}$$

where,

$$S_{AB} = \frac{\sum_{n=1}^{excinct} r(n)^2}{\sum_{n=1}^{excinct} [A_1 \sin c(n - \lambda_1) + B_1 \cos c(n - \lambda_1)]^2}. \tag{13.24}$$

Here the factor S_{AB} represents the difference in energy between the original and synthesized excitation. Thus, by multiplying both the A_1 and B_1 parameters by $\sqrt{S_{AB}}$, the energies of the synthesized and original excitation prototype segments will match.

13.5.2 Quantization

Once the A_1 and B_1 parameters have been determined, they must be quantized. The Max-Lloyd quantizer, described in Section 11.4, requires knowledge of the A_1 and B_1 parameters' PDF. These parameters are shown in Figure 13.10, where the PDF is generated from the unquantized A_1 and B_1 parameters of the training speech database, described in Section 10.4.

The Max-Lloyd quantizer was used to create 4.5- and 6-bit SQs for both the A_1 and B_1 parameters. Table 13.6 shows the SNR values for the A_1 and B_1 parameters for the various quantization schemes.

In order to gain further insight into the performance of the various quantizers, the SEGSNR and SD measures were calculated for the synthesized and original speech prototype segments. Together with the quantized A_1 and B_1 values, the SEGSNR and SD measures were calculated for the unquantized A_1 and B_1 values. Table 13.7 shows the SEGSNR values achieved. Though low, the SEGSNR values demonstrate that the 6-bit quantization produces a SEGSNR performance similar to the unquantized parameters.

Table 13.8 shows the SD values achieved, which again demonstrate that the 6-bit quantizers produce little degradation. The 6-bit A_1 and B_1 SQs were selected because of their transparency in the SEGSNR and SD tests. They have SNR values of 26.47 dB and 27.07 dB, respectively, as seen in Table 13.6.

The interpolation of the voiced excitation performed at the decoder is described next, where pitch synchronous interpolation of the ZFE and LSFs is implemented.

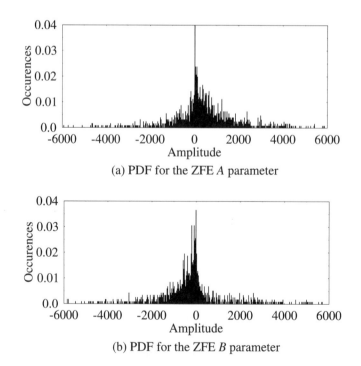

(a) PDF for the ZFE A parameter

(b) PDF for the ZFE B parameter

Figure 13.10 Graphs for the PDFs of (a) A_1 and (b) B_1 ZFE parameters, created from the combination of A_1 and B_1 parameters from 45 seconds of speech.

TABLE 13.6 SNR Values for SQ of the A_1 and B_1 Parameters

Quantizer Scheme	SNR/dB for A_1	SNR/dB for B_1
4-bit	10.45	10.67
5-bit	18.02	19.77
6-bit	26.47	27.07

TABLE 13.7 SEGSNR Values between the Original and Synthesized Prototype Segments for a Selection of SQs for the A_1 and B_1 Parameters

Quantizer Scheme	SEGSNR/dB
Unquantized	1.36
4-bit	0.21
5-bit	1.00
6-bit	1.29

TABLE 13.8 SD Values for the Synthesized Prototype Segments for a Selection of SQs for the A_1 and B_1 Parameters

Quantizer Scheme	SD/dB
Unquantized	4.53
4-bit	4.90
5-bit	4.60
6-bit	4.53

13.6 EXCITATION INTERPOLATION BETWEEN PROTOTYPE SEGMENTS

Having determined the prototype segments for the adjacent speech frames, we find that interpolation is needed to provide a continuous excitation signal between them.

13.6.1 ZFE Interpolation Regions

The associated interpolation operations will be first stated in general terms; subsequently, using the equations derived and the parameter values of Table 13.9, they will be

TABLE 13.9 Transmitted Parameters for Voiced Speech

Speech Frame	Pitch Period	Zero Crossing	A_1	B_1	λ_1
$N-1$	52	64	−431	186	16
N	52	56	−573	673	20

augmented through a numerical example. We also refer to traces three and four of Figures 13.11 and 13.12, which portray the associated interpolation operations.

Initially, we follow the method of Hiotakakos and Xydeas [294], with interpolation performed over an interpolation region d_{pit}, where d_{pit} contains an integer number of pitch periods. The provisional interpolation region, d'_{pit}, which may not contain an integer number of pitch periods, begins at the start of the prototype segment in frame $N - 1$ and finishes at the end of the prototype segment in frame N. The number of pitch synchronous intervals, N_{pit}, between the two prototype regions is given by the ratio of the provisional interpolation region to the average pitch period during this region [294]:

$$N_{pit} = nint\left\{\frac{2d'_{pit}}{P(N) + P(N - 1)}\right\} \tag{13.25}$$

where $P(N)$ and $P(N - 1)$ represent the pitch period in frames N and $N - 1$, respectively, and *nint* signifies the nearest integer. If $P(N)$ and $P(N - 1)$ are different, then the smooth interpolation of the pitch period over the interpolation region is required. This is achieved by calculating the average pitch-period alteration necessary to convert $P(N - 1)$ to $P(N)$ over N_{pit} pitch synchronous intervals, where the associated pitch interpolation factor ϵ_{pit} is defined as [294]:

$$\epsilon_{pit} = \frac{P(N) - P(N - 1)}{N_{pit} - 1}. \tag{13.26}$$

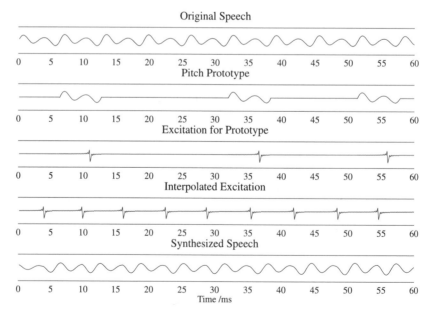

Figure 13.11 An example of the original and synthesized speech for a 60 ms speech waveform from AF2 uttering the front vowel /i/ from "he", where the frame length is 20 ms. The prototype segment selection and ZFE interpolation is also shown.

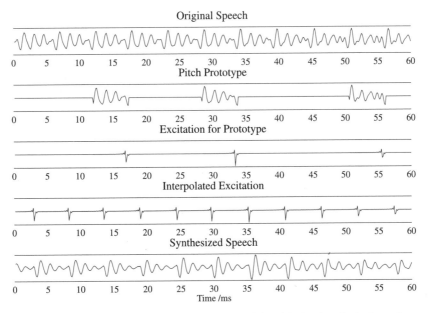

Figure 13.12 An example of three 20 ms segments of the original and synthesized speech for predominantly voiced speech from AF1 uttering the back vowel /ɔ/ "dog". The prototype segment selection and ZFE interpolation is also shown.

The final interpolation region, d_{pit}, is given by the sum of the pitch periods over the interpolation region constituted by N_{pit} number of pitch period intervals [294]:

$$d_{pit} = \sum_{np=1}^{N_{pit}} p(n_p) \tag{13.27}$$

where $p(n_p)$ are the pitch-period values between the $P(N-1)$ and $P(N)$, with $p(n_p) = P(N-1) + (n_p - 1) \cdot \epsilon_{pit}$ and $n_p = 1..N_{pit}$. In general, the start and finish of the prototype region in frame N will be altered by the interpolation process, since the provisional interpolation region d'_{pit} is generally extended or shortened, to become the interpolation region d_{pit}. To ensure correct operation between frame N and frame $N-1$, the change in the prototype position must be noted:

$$change = d'_{pit} - d_{pit} \tag{13.28}$$

and then we assign $start(N) = start(N) - change$, where $start(N)$ is the beginning of the prototype segment in frame N. Thus, the start of the prototype segment in frame N together with the position of the ZFE parameter λ_1 within the frame are altered in order to compensate for the changes to the interpolation region. Maintaining the position parameter λ_1 at the same location of the prototype segment sustains the shape within the prototype excitation but introduces a time misalignment with the original speech, where this time misalignment has no perceptual effect.

13.6.2 ZFE Amplitude Parameter Interpolation

The interpolated positions for the ZFE amplitude parameters are given by [294]:

$$A_{1,n_p} = A_1(N-1) + (n_p - 1)\frac{A_1(N) - A_1(N-1)}{N_{pit} - 1} \qquad (13.29)$$

$$B_{1,n_p} = B_1(N-1) + (n_p - 1)\frac{B_1(N) - B_1(N-1)}{N_{pit} - 1} \qquad (13.30)$$

where the formulas reflect a linear sampling of the A_1 and B_1 parameters between the adjacent prototype functions. Explicitly, given the starting value $A_1(N-1)$ and the difference $\Delta_{pit} = A_1(N) - A_1(N-1)$, the corresponding gradient is $[\Delta_{pit}/N_{pit} - 1]$, where N_{pit} is the number of pitch synchronous intervals between $A_1(N)$ and $A_1(N-1)$, allowing us to calculate the appropriate values A_{1,n_p}.

13.6.3 ZFE Position Parameter Interpolation

Interpolating the position of the ZFEs in a similar manner to their amplitudes does not produce a smoothly evolving excitation signal. Instead, the pulse position within each prototype segment is kept stationary throughout a voiced sequence. This introduces time misalignment between the original and synthesized waveforms but maintains a smooth excitation signal. In order to compensate for changes in the length of prototype segments, the normalized location of the initial ZFE position is calculated according to:

$$\lambda_r = \frac{\lambda_1(N)}{P(N)} \qquad (13.31)$$

where $P(N)$ is the pitch period of the first frame in the voiced frame sequence. For all subsequent frames in the voiced sequence, the position of the ZFE is calculated by:

$$\lambda_1(N) = nint\{\lambda_r * P(N)\} \qquad (13.32)$$

where $nint\{\cdot\}$ represents rounding to the nearest integer.

For the sake of illustration, the interpolation process is followed below for the two speech frames whose parameters are described in Table 13.9. The initial provisional interpolation region commences at the beginning of the prototype segment in frame $N-1$ and finishes at the end of the prototype segment in frame N. Since the zero crossing in frame $N-1$ is at sample index 64, the provisional interpolation region in frame $N-1$ is of duration $(160 - 64)$, while in frame N it finishes one pitch-period duration, namely 52 samples, after the zero crossing at position 56, yielding:

$$d'_{pit} = (160 - 64) + (56 + 52) = 204.$$

Using Equation 13.25, the number of pitch synchronous intervals between the two consecutive prototype segments in frames N and $N-1$ is given by d'_{pit} divided by the average pitch-period duration of $[P(N) + P(N-1)]/2$, yielding:

$$N_{pit} = nint\left\{\frac{2 \times 204}{52 + 52}\right\} = 4.$$

As $P(N)$ and $P(N-1)$ are identical, the pitch interpolation factor ϵ_{pit} of Equation 13.26 will be zero, while the interpolation region containing $N = 4$ consecutive pitch periods and defined by Equation 13.27 becomes:

$$d_{pit} = \sum_{np=1}^{4} 52 = 208.$$

The interpolated ZFE magnitudes and positions can then be calculated using the parameters in Table 13.9 and Equations 13.29 to 13.32 for frame $N - 1$, the first voiced frame in the sequence, yielding:

$$A_{1,n_p} = -431 + n_p \times \frac{-573 + 431}{3} = -478; -526; -573;$$

$$B_{1,n_p} = 186 + n_p \times \frac{673 - 186}{3} = 348; 511; 673;$$

$$\lambda_r = \frac{16}{52} = 0.308$$

$$\lambda_1(N) = 0.308 * 52 = 16.$$

Again, the associated operations are illustrated in traces three and four of Figures 13.11 and 13.12.

13.6.4 Implicit Signaling of Prototype Zero Crossing

In order to perform the interpolation procedure described above, the zero-crossing parameter of the prototype segments must be transmitted to the decoder. However, it can be observed that the zero-crossing values of the prototype segments are approximately a frame length apart; thus following the principle of interpolating between prototype segments in each frame. Hence, instead of explicitly transmitting the zero-crossing parameter, it can be assumed that the start of the prototype segments is a frame length apart. An arbitrary starting point for the prototype segments could be $FL/2$, where FL is the speech frame length.

Using this scenario, we repeat the interpolation procedure example of Section 13.6.3 with both zero crossings set to 80. The initial provisional interpolation region is calculated as:

$$d'_{pit} = (160 - 80) + (80 + 52) = 212. \tag{13.33}$$

The number of pitch synchronous intervals is given by:

$$N_{pit} = nint\left\{\frac{2 \times 212}{52 + 52}\right\} = 4. \tag{13.34}$$

Thus, the interpolation region defined by Equation 13.27 will become:

$$d_{pit} = \sum_{np=1}^{4} 52 = 208 \tag{13.35}$$

yielding the same distance as in the example of Section 13.6.3, where the zero-crossing value was explicitly transmitted. Hence, it is feasible not to transmit the zero-crossing location to the decoder. Indeed, the assumption of a zero-crossing value of 80 had no perceptual effect on speech quality at the decoder.

13.6.5 Removal of ZFE Pulse Position Signaling and Interpolation

In the λ_1 transmission procedure, although λ_1 is transmitted every frame, only the first λ_1 in every voiced sequence is used in the interpolation process. Thus, λ_1 is predictable, and it contains much redundancy. Furthermore, when constructing the excitation waveform at the decoder, every ZFE is permitted to extend over three interpolation regions—namely, its allotted region together with the previous and the next region. This allows ZFEs near the interpolation region boundaries to be fully represented in the excitation waveform, while ensuring that every ZFE will have a tapered low-energy value when it is curtailed. It is suggested that the true position of the ZFE pulse, λ_1, is arbitrary and need not be transmitted. Following this hypothesis, our experience shows that we can set $\lambda_1 = 0$ at the decoder, which has no audible degrading effect on the speech quality.

13.6.6 Pitch Synchronous Interpolation of Line Spectrum Frequencies

The LSF values can also be interpolated on a pitch synchronous basis, following the approach of Equations 13.29 and 13.30, giving:

$$LSF_{i,n} = LSF_i(N-1) + (n_p - 1)\frac{LSF_i(N) - LSF_i(N-1)}{N_{pit} - 1} \tag{13.36}$$

where $LSF_i(N-1)$ is the previous *ith* LSF and $LSF_i(N)$ is the current *ith* LSF.

13.6.7 ZFE Interpolation Example

An example of the ZFE excitation reconstructing the original speech is given in Figure 13.11, which is a speech waveform from the testfile AF2. Following the steps of the encoding and decoding process in the figure, initially a pitch prototype segment is selected at the center of the frame. Then at the encoder, a ZFE is selected to represent this prototype segment. At the decoding stage the ZFE segments are interpolated, according to Sections 13.6.1 to 13.6.5, in order to produce a smooth excitation waveform. This waveform is subsequently passed through the LPC STP synthesis filter to reconstruct the original speech. The time misalignment introduced by the interpolation process described earlier can be clearly seen, where the prototype shifting is caused by the need to have an integer number of pitch prototype segments during the interpolation region. The synthesized waveform does not constitute a strict waveform replica of the original speech, which is the reason for the coder's low SEGSNR. However, it produces perceptually good speech quality.

Figure 13.12 portrays a voiced speech section, where the same process as in Figure 13.11 is followed. The synthesized waveform portrays a similar smooth waveform evolution to the input speech, but the synthesized waveform has problems maintaining the waveform's amplitude throughout all the prototype segment's resonances. McCree and Barnwell [284] suggest that this type of waveform would benefit from the post-filter described in Section 11.6. Thus far, only voiced speech frames have been discussed; hence, next we briefly describe the unvoiced frame encoding procedure.

13.7 UNVOICED SPEECH

For frames classified as unvoiced, at the decoder a random Gaussian sequence is used as the excitation source. The same noise generator was used for the PWI-ZFE coder and the basic LPC vocoder of Chapter 11, namely, the Box-Muller algorithm. This vocoder is used to produce a Gaussian random sequence scaled by the RMS energy of the LPC STP residual, where the noise generation process was described in Section 11.4.

13.8 ADAPTIVE POST-FILTER

The adaptive post-filter from Section 11.6 was also used for the PWI-ZFE speech coder. However, the adaptive post-filter parameters were reoptimized, becoming $\alpha_{pf} = 0.75$, $\beta_{pf} = 0.45$, $\mu_{pf} = 0.60$, $\gamma_{pf} = 0.50$, $g_{pf} = 0.00$, and $\xi_{pf} = 0.99$. Finally, following the adaptive post-filter, the synthesized speech was passed through the pulse dispersion filter of Section 11.7.

Following this overview of the PWI-ZFE coder, the quality of the reconstructed speech is assessed.

13.9 RESULTS FOR SINGLE ZINC FUNCTION EXCITATION

This section assesses the performance of the PWI-ZFE speech coder described in this chapter. Figures 13.13, 13,14, and 13.15 show examples of the original and synthesized speech in the time and frequency domain for sections of voiced speech, with these graphs described in detail next. These detailed speech frames were also used to examine the LPC vocoder of Chapter 11. Hence, Figure 13.13 can be compared to Figure 11.21, Figure 13.14 to Figure 11.22, and Figure 13.15 to Figure 11.23.

The speech segment displayed in Figure 13.13 is a 20 ms frame from testfile BM1. The reproduced speech is of similar evolution to the original speech but cannot maintain the amplitude for the decaying resonances within each pitch period, which is due to the concentrated pulse-like nature of the ZFE. From Figure 13.13a and 13.13c, a time misalignment between the original and synthesized waveform is present, where the cause of the misalignment was described in Section 13.6. Specifically, the interpolation region must contain an integer number of pitch prototype segments, often requiring the interpolation region to be extended or shortened. Consequently, the later pitch prototype segments are shifted slightly, introducing the time misalignment seen in Figure 13.13c. In the frequency domain, the overall spectral envelope match between the original and synthesized speech is good, but as expected, the associated SEGSNR is low because of the waveform misalignment experienced.

The speech segment presented in Figure 13.14 shows the performance of the PWI-ZFE coder for the testfile BF2. Comparing Figure 13.14c with Figure 11.22h, we can see that the synthesized waveforms in both the time and frequency-domain are similar. Observing the frequency-domain graphs, we note that the inclusion of unvoiced speech above 1800 Hz is not modeled well by the distinct voiced-unvoiced nature of the PWI-ZFE scheme. The introduction of mixed-multiband excitation in Chapter 14 is expected to improve the representation of this signal.

The speech segment displayed in Figure 13.15 is for the testfile BM2. The synthesized speech waveform displayed in Figure 13.15c is noticeably better than the output speech in Figure 11.23h. For Figure 13.15c the first formant is modeled well. However, the upper two

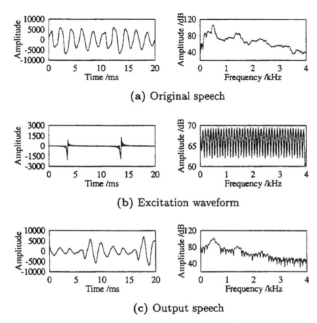

Figure 13.13 Time- and frequency-domain comparison of the (a) original speech, (b) ZFE waveform and (c) output speech after the pulse dispersion filter. The 20 ms speech frame is the midvowel /ɝ/ in the utterance "work" for the testfile BM1. For comparison with the other coders developed in this study using the same speech segment, please refer to Table 16.2.

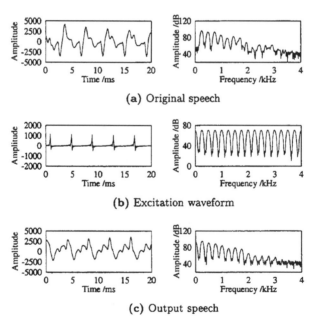

Figure 13.14 Time- and frequency-domain comparison of the (a) original speech, (b) ZFE waveform, and (c) output speech after the pulse dispersion filter. The 20 ms speech frame is the liquid /r/ in the utterance "rice" for the testfile BM2. For comparison with the other coders developed in this study using the same speech segment, please refer to Table 16.2.

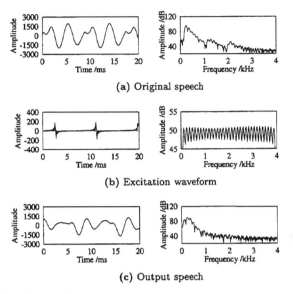

Figure 13.15 Time- and frequency-domain comparison of the (a) original speech, (b) ZFE waveform, and (c) output speech after the pulse dispersion filter. The 20 ms speech frameis the nasal /n/ in the utterance "thrown" for the testfile BM2. These signals can be compared to the basic vocoder's corresponding signals in Figure 11.23.

formants are missing from the frequency spectrum, which is a failure in the LPC STP process and will persist in all of our developed speech coders.

Informal listening tests showed that the reproduced speech for the PWI-ZFE speech coder contained less buzziness than the LPC vocoder of Chapter 11.

The bit allocation of the ZFE coder is summarized in Table 13.10. For unvoiced speech the RMS parameter requires the 5 bits described in Section 11.4, with the boundary shift parameter b_s offset requiring a maximum of:

$$\frac{\text{frame length}}{\text{minimum pitch}} = \frac{160}{20} = 8$$

values or 3 bits to encode.

TABLE 13.10   Bit Allocation table for the Investigated 1.9 kbps PWI-ZFE Coder

Parameter	Unvoiced	Voiced
LSFs	18	18
v/u flag	1	1
RMS value	5	—
b_s offset	3	—
pitch	—	7
A_1	—	6
B_1	—	6
total/20 ms	27	38
bit rate	1.35 kbps	1.90 kbps

TABLE 13.11 Total Maximum and Minimum Computational Complexity for a PWI-ZFE Coder

Operation/MFLOP	pitch period = 20	pitch period = 147
Pitch detector	2.67	2.68
ZFE minimization	0.30	11.46
Total	2.97	14.13

For voiced speech the pitch period can vary from $20 \rightarrow 147$ samples, thus requiring 7 bits for transmission. Section 13.5.2 justified the use of 6 bits to SQ the A_1 and B_1 ZFE amplitude parameters.

The computational complexity of the speech coder is dominated by the ZFE minimization loop, even when using a constrained search. Table 13.11 displays the computational complexity of the coder for a pitch period of 20 samples or 147 samples.

13.10 ERROR SENSITIVITY OF THE 1.9 kbps PWI-ZFE CODER

In this Chapter we have investigated the design of a 1.9 kbps speech coder employing PWI-ZFE techniques. However, we have not examined the speech coder's performance within a communications system, specifically its robustness to transmission errors. In this section we study how the degradation caused by a typical mobile environment affects the PWI-ZFE output speech quality.

The degradation in the PWI-ZFE speech coder's performance is caused by the hostile nature of a mobile communications environment. A mobile environment typically contains both fast and slow fading, which affects the signal level at the receiver. In addition, many different versions of the signal arrive at the receiver, each having taken different paths with different fading characteristics and different delays, thus introducing intersymbol interference. These mobile environment characteristics introduce errors into the parameters received by the speech decoder.

We begin this section by examining how possible errors at the decoder would affect the output speech quality and introduce some error correction techniques. These errors are then examined in terms of objective speech measures and informal listening tests. We then consider dividing the transmission bits into protection classes, which is a common technique that is adopted to afford the most error sensitive bits the greatest protection. Finally, we demonstrate the speech coder's performance for different transmission environments.

13.10.1 Parameter Sensitivity of the 1.9 kbps PWI-ZFE Coder

In this section we consider the importance of the different PWI-ZFE parameters of Table 13.10 in maintaining synthesized speech quality. We also highlight checks that can be made at the decoder, which may indicate errors and suggest error correction techniques. Considering the voiced and unvoiced speech frames separately, the speech coder has 10 different parameters that can be corrupted, where the vector-quantized LSFs can be considered to be four different groups of parameters. These parameters have between 7 bits, for the pitch period, and a single bit, for the voiced-unvoiced flag, which can be corrupted. In total there are 46 different bits: the 38 voiced bits of Table 13.10 and the RMS and b_s unvoiced parameters.

Finally, we note that because of the interpolative nature of the PWI-ZFE speech coder, any errors that occur in the decoded bits will affect more than just the frame where the error occurred.

13.10.1.1 Line Spectrum Frequencies. The LSF vector quantizer, described in Section 11.2.2 and taken from G.729 [123], represents the LSF values using four different parameters. The LSF VQ consists of a fourth-order moving average (MA) predictor, which can be switched on or off with the flag L0. The vector quantization is then performed in two stages. A 7-bit VQ index, L1, is used for the first stage. The second stage VQ is a split vector quantizer, using the indices L2 and L3, with each codebook containing 5 bits.

13.10.1.2 Voiced-Unvoiced Flag. The voiced-unvoiced flag is probably the most critical bit for the successful operation of the PWI-ZFE speech coder. Because of the very different excitation models employed for voiced and unvoiced speech, use of the wrong type of excitation will have a serious degrading effect.

At the decoder it is possible to detect isolated errors in the voiced-unvoiced flag, namely, V-U-V and U-V-U sequences in the $N + 1$, N, $N - 1$ frames. These sequences will indicate an error, since at the encoder they were prohibited frame combinations, as described in Section 13.2.1. However, the PWI-ZFE decoder does not operate on a frame-by-frame basis. Instead, it performs interpolation between the prototype segments of frame N and $N + 1$, as described in Section 13.2.1. Thus, without introducing an extra 20 ms delay, by performing the interpolation between frames $N - 1$ and N, it is impossible to completely correct an isolated error in the voiced-unvoiced flag.

13.10.1.3 Pitch Period. The pitch-period parameter of Table 13.10 is only sent for voiced frames, where having the correct pitch period is imperative for producing good quality synthesized speech. In Section 12.5.2 some simple pitch-period correction was already performed, where checks were made to ensure that a smooth pitch track was followed. By repeating this pitch-period correction at the decoder, the effect of an isolated pitch-period error can be reduced. However, similarly to the voiced-unvoiced flag, use of frames N and $N + 1$ in the interpolation process permits an isolated pitch period to have a degrading effect.

13.10.1.4 Excitation Amplitude Parameters. The ZFE amplitude parameters, A and B, control the shape of the voiced excitation. The A and B parameters of Table 13.10 can have both positive and negative values; however, as described in Section 13.3.4, the phase of the amplitude parameters must be maintained throughout the voiced sequence. At the decoder it is possible to maintain phase continuity for the amplitude parameter in the presence of an isolated error, with the correction that if the phase of the A or B parameter has been found to change during a voiced sequence, then the previous A or B parameter can be repeated.

13.10.1.5 Root Mean Square Energy Parameter. For unvoiced speech frames, the excitation is formed from random Gaussian noise scaled by the received RMS energy value, (see in Table 13.10 and Section 13.7). Thus, if corruption of the RMS energy parameter occurs, then the energy level of the unvoiced speech will be incorrect. However, since the speech sound is a low-pass-filtered, slowly varying process, abrupt RMS changes due to channel errors can be detected and mitigated.

13.10.1.6 Boundary Shift Parameter. The boundary shift parameter, b_s, of Table 13.10 is only sent for unvoiced frames and defines the location where unvoiced speech becomes voiced speech, or vice versa. The corruption of the boundary shift parameter will move this transition point, an event that is not amenable to straightforward error concealment.

13.10.2 Degradation from Bit Corruption

Following this discussion on the importance of the various PWI-ZFE parameters and the possible error corrections which could be performed at the speech decoder, we now investigate the extent of the degradation errors cause to the reproduced speech quality. The error sensitivity is examined by separately corrupting each of the 46 different voiced and unvoiced bits of Table 13.10, where 18 LSF plus the v/uv bits are sent for all frames; an additional 19 bits are only sent for voiced frames, and 8 bits are only sent for unvoiced frames. For each selected bit, the corruption was inflicted 10% of the time. Corrupting a bit for 10% of the time is a compromise between consistently or constantly corrupting the bit in all frames and corrupting the bit in only a single isolated frame. If the bit is constantly corrupted, then any error propagation effect is masked, while corrupting the bit in only a single frame requires that for completeness every possible frame is taken to be that single frame, resulting in an arduous process.

Figure 13.16 displays the averaged results for the speech files AM1, AM2, AF1, AF2, BM1, BM2, BF1, and BF2, described in Section 10.4. The SEGSNR and CD objective

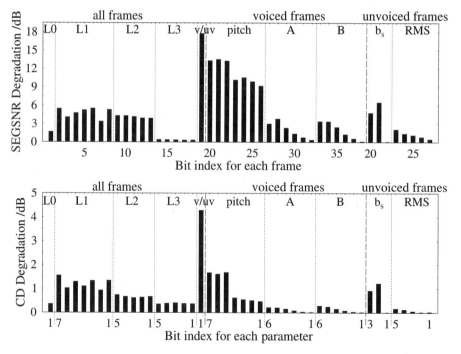

Figure 13.16 The error sensitivity of the different transmission bits for the 1.9 kbps PWI-ZFE speech coder. The graph is divided into bits sent for all speech frames, bits sent only for voiced frames and bits sent only for unvoiced frames. For the CD degradation graph containing the bit index for each parameter, bit 1 is the least significant bit.

speech measures, described in Section 10.3.1, were used to evaluate the degradation effect. In addition, the synthesized corrupted speech due to the different bit errors was compared through informal listening tests.

Observing Figure 13.16, we can see that both the SEGSNR and CD objective measures rate the error sensitivity of the different bits similarly. Both indicate that the voiced-unvoiced flag being corrected is the most critical for successful synthesis of the output speech. This was confirmed by listening to the synthesized speech, which was frequently unintelligible when there was 10% error in the voiced-unvoiced flag bit. In addition, Figure 13.6 shows that both the pitch-period and boundary shift parameters produce a significant degradation due to bit errors. However, informal listening tests do not indicate such significant quality degradation, although an incorrect pitch period does produce audible distortion. It is suggested that the time misalignment introduced by the pitch-period and boundary shift parameter errors is artificially increasing the SEGSNR and CD degradation values.

Thus, although the SEGSNR and CD objective measures indicate the relative sensitivities of the bits within each parameter, accurate interpretation of the sensitivity of each parameter has to rely more on informal listening tests.

13.10.2.1 Error Sensitivity Classes. The SEGSNR and CD objective measures, together with the informal listening tests, allow the bits to be grouped into three classes for transmission to the decoder. These classes are detailed in Table 13.12, where Class One requires the greatest protection and Class Three the least protection.

In Table 13.12 the error sensitivity classes are based on the bits sent every speech frame and the bits sent only for voiced frames, giving 38 bits. For unvoiced frames the boundary parameter shift, b_s, is given the same protection as the most significant three pitch-period bits, while the RMS value is given the same protection as the least significant four pitch-period bits and A[6].

Class One contains only the voiced-unvoiced flag, which has been identified as being very error sensitive. Class Two contains 15 bits, while Class Three contains 22 bits.

The relative bit error sensitivities have been used to improve channel coding within a GSM-like speech transceiver [329] and a FRAMES-like speech CDMA transceiver [330].

Following this analysis of the performance of a PWI-ZFE speech coder, using a single ZFE to represent the excitation, we examine the potential for speech quality improvement with extra ZFE pulses is examined.

TABLE 13.12 The Transmission Classes for the Bits of the 1.9 kbps PWI-ZFE Speech Coder, with Class 1 Containing the Most Error Sensitive Bits and Class 3 Bits Requiring Little Error Protection

Classes	Coding Bits						
1	v/uv						
2	L1[7]	L1[5]	L1[3]	L1[1]			
	pitch[7]	pitch[6]	pitch[5]	pitch[4]	pitch[3]	pitch[2]	pitch[1]
	A[6]	A[5]	B[6]	B[5]			
3	L0	L1[6]	L1[4]	L1[2]			
	L2[5]	L2[4]	L2[3]	L2[2]	L2[1]		
	L3[5]	L3[4]	L3[3]	L3[2]	L3[1]		
	A[4]	A[3]	A[2]	A[1]			
	B[4]	B[3]	B[2]	B[1]			

13.11 MULTIPLE ZINC FUNCTION EXCITATION

So far in this chapter, a single ZFE pulse has been employed to represent the voiced excitation. However, a better speech quality may be achieved by introducing more ZFE pulses [295]. The introduction of extra ZFEs will be at the expense of a higher bitrate. Thus, a dual-mode PWI-ZFE speech coder could be introduced to exploit an improved speech quality when traffic density of the system permits.

Revisiting the ZFE error minimization process of Section 13.3.1, we see that due to the orthogonality of the zinc-basis functions the weighted error signal upon using k ZFE pulses is given by:

$$E_w^{k+1} = \sum_{n=1}^{P} (e_w^{k+1}(n))^2 \qquad (13.37)$$

where P is the length of the prototype segment, over which minimization is carried out, with the synthesized weighted speech represented by:

$$\bar{s_w}(n) = \sum_{k=1}^{K} z_k(n) * h(n) \qquad (13.38)$$

where $z_k(n)$ is the kth ZFE pulse, K is the number of pulses being employed, and $h(n)$ is the impulse response of the weighted LPC synthesis filter.

13.11.1 Encoding Algorithm

The encoding process for a single ZFE was previously described in Table 13.1 and Figure 13.2. For a multiple ZFE arrangement the same process is followed, but the number of ZFE pulses is extended to K, as shown in Figure 13.17 and described next. Thus, for the phase-constrained frame, which we also refer to as the phase restriction frame, a phase is determined independently for each of the K excitation pulses. Similarly, for other voiced frames the phase of the kth pulse is based on the phase restriction for the kth pulses. Furthermore, if a suitable ZFE is not found for the kth ZFE pulse in frame N, then the kth ZFE in frame $N-1$ is scaled and reused.

For scenarios with a different number ZFE pulse per prototype segment, Table 13.13 displays the percentage of voiced frames, where some scaling from the previous frame's ZFE pulses must be performed. It can be seen that with 3 ZFE pulses employed, one-third of the voiced frames contain scaled ZFE pulses from the previous frame. In addition, some frames have several scaled ZFE pulses from the previous frame.

The implementation of the single ZFE, in Section 13.3.3, showed that for smooth interpolation it is beneficial to constrain the locations of the ZFE pulses. Constraining the K ZFE locations follows the same principles as those used in determining the single ZFE location, but it was extended to find K constrained positions. For the first voiced frame, the largest K impulses, determined by wavelet analysis according to Chapter 12 and located within the prototype segment, are selected for the positions the ZFE pulses must be in proximity to. For further voiced frames, the impulses from the wavelet analysis are examined, with the largest impulses near the K ZFE pulses in frame $N-1$ selected as excitation. If no impulse is found near the kth ZFE location in frame $N-1$, this position is repeated as the kth ZFE in frame N. It is feasible that there will be less than K wavelet analysis impulses within the prototype segment. Thus, in this situation the extra ZFEs are set to zero. They are

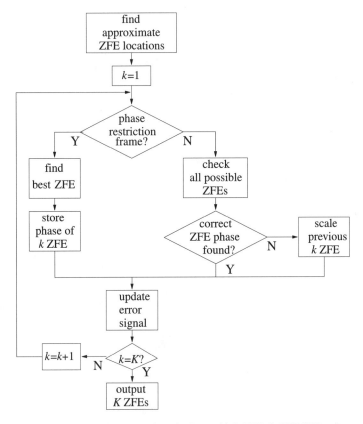

Figure 13.17 The control structure for selecting multiple ZFEs in PWI-ZFE coders.

subsequently introduced, when impulses occur within the prototype segment that are unrelated to any ZFE pulses in frame $N - 1$.

The SEGSNR values achieved for the minimization process at the encoder with different numbers of ZFE pulses per prototype segment indicate the excitation representation improvement. Figure 13.18 displays the results, showing that the improvement achieved by adding extra pulses saturates as the number of ZFE pulses increases. Thus, when eight ZFE pulses are employed, no further SEGSNR gain is achieved. The limit in SEGSNR improve-

TABLE 13.13 The Percentage of Speech Frames Requiring Previous ZFEs to be Scaled and Repeated, for K ZFE Pulses in PWI-ZFE Coders

Total K	Rescaled ZFE Needed	>1 ZFE Rescaled	>2 ZFE Rescaled	>3 ZFE Rescaled	>4 ZFE Rescaled
1	12.9%	—	—	—	—
2	20.3%	5.1%	—	—	—
3	33.0%	7.1%	1.0%	—	—
4	42.3%	16.5%	5.1%	0.6%	—
5	57.9%	28.4%	10.7%	2.5%	0.1%

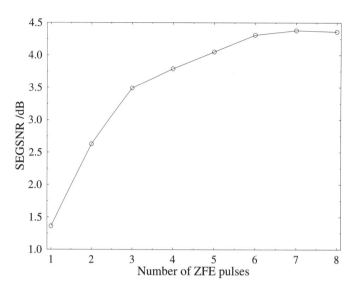

Figure 13.18 The SEGSNR achieved at the encoder minimization process for different
numbers of ZFE pulses used in the representation. The inclusion of each
new ZFE pulse requires 19 extra bits/20 ms, or 0.95 kbps extra bit rate, for
the encoding of the A_k and B_k parameters, and the additional ZFE pulse
positions λ_k, as seen in Table 13.14.

ment is due to the constraint that ZFE pulses are expected to be near the instants of glottal
closure found by the wavelet analysis. There will be a limited number of impulses within the
prototype segment. Thus, a limited number of ZFE pulses can be employed for each prototype
segment. The performance of a 3-pulse ZFE scheme at the encoder is given in Figure 13.19,
which can be compared with the performance achieved by a single ZFE, shown in Figure
13.19. It can be seen that the addition of two extra ZFE pulses improves the excitation
representation, particularly away from the main resonance.

At the decoder, the same interpolation process implemented for the single ZFE is
employed, as described in Section 13.6, and again extended to K ZFE pulses. For all ZFE
pulses, the amplitude parameters are linearly interpolated, with the ZFE pulse position
parameter and prototype segment location assumed at the decoder, as in the single-pulse
coder of earlier sections. The kth ZFE pulse position parameter is kept at the same location
within each prototype segment. For the 3-pulse PWI-ZFE scheme, the adaptive post-filter
parameters were reoptimized becoming $\alpha_{pf} = 0.75$, $\beta_{pf} = 0.45$, $\mu_{pf} = 0.40$, $\gamma_{pf} = 0.50$,
$g_{pf} = 0.00$, and $\xi_{pf} = 0.99$.

13.11.2 Performance of Multiple Zinc Function Excitation

A 3-pulse ZFE scheme was implemented to investigate the potential for improved
speech quality using extra ZFE pulses. Three excitation pulses were adopted to study the
feasibility of a speech coder at 3.8 kbps, where the bit-allocation scheme was given in Table
13.14.

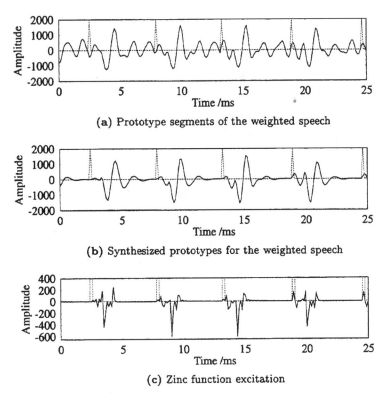

(a) Prototype segments of the weighted speech

(b) Synthesized prototypes for the weighted speech

(c) Zinc function excitation

Figure 13.19 Demonstrating the process of anlaysis-by-synthesis encoding for prototype segments that have been concatenated to produce a smoothly evolving waveform, with the excitation represented by three ZFE pulses. The dotted lines in the figure indicate the boundaries between prototype segments.

TABLE 13.14 Bit-allocation Table for Voiced Speech Frames in the 3.8 kbps Investigated PWI-ZFE Coder Employing Three ZFEs

Parameter	Voiced
LSFs	18
v/u flag	1
pitch	7
1st pulse	
A_1	6
B_1	6
2nd pulse	
λ_2	7
A_2	6
B_2	6
3rd pulse	
λ_3	7
A_3	6
B_3	6
total/20 ms	76
bitrate	3.80 kbps

Figure 13.20 displays the performance of a 3-pulse ZFE scheme for the midvowel /ɜ·/ in the utterance "wоrk" for the testfile BM1. The identical portion of speech synthesized using a single ZFE was given in Figure 13.13. From Figure 13.20b it can be seen that the second largest ZFE pulse is approximately halfway between the largest ZFE pulses. In the frequency spectrum the pitch appears to be 200 Hz, which is double the pitch from Figure 13.13b. The pitch doubling is clearly visible in the time and frequency domain of Figure 13.20c. For this speech frame the addition of extra ZFE pulses fails to improve the speech quality, where this is due to the secondary excitation pulse producing a pitch-doubling effect in the output speech.

Figure 13.21 displays the results from applying a 3-pulse ZFE scheme to a 20 ms frame of speech from the testfile BF2. The same speech frame was investigated in Figures 13.14 and 11.22. Observing Figure 13.21b, we can see that similarly to Figure 13.20b a ZFE pulse is placed midway between the other ZFE pulses. However, since this pulse has much less energy, it does not have a pitch-doubling effect. When compared with the single ZFE of Figure 13.15c, the multiple ZFEs combine to produce a speech waveform, shown in Figure 13.21c. They are much closer in both the time and frequency domain to the original, although at the cost of a higher bitrate and complexity.

Figure 13.22 portrays a 3-pulse ZFE scheme applied to a speech frame from the testfile BM2, which can be compared with Figure 13.15. From Figure 13.22b it can be seen that no pitch doubling occurs. For this speech frame the limiting factor in reproducing the original speech are the missing formants. However, observing Figure 13.22c demonstrates that 3 ZFE pulses result in an improved performance compared with a single ZFE.

Informal listening tests were conducted using the PWI-ZFE speech coder with 3 ZFE pulses, where it was found that sudden and disconcerting changes could occur in the quality of the reproduced speech. It is suggested that this effect was created by the varying success of the excitation to represent the speech. Additionally, for many speech files there was a background roughness to the synthesized speech. The problems with implementing a multiple ZFE pulse scheme are caused by the interpolative nature of the speech coder. The benefits, which are gained in improved representation of the excitation signal, are counteracted by

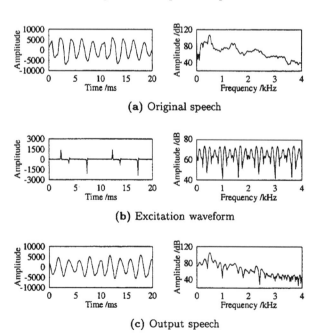

(a) Original speech

(b) Excitation waveform

(c) Output speech

Figure 13.20 Time- and frequency-domain comparison of the (a) original speech, (b) three-pulse ZFE waveform and (c) output speech after the pulse dispersion filter. The 20 ms speech frame is the midvowel /ɜ·/ in the utterance "wоrk" for the testfile BM1. For comparison with the other coders developed in this study using the same speech segment, please refer to Table 16.2.

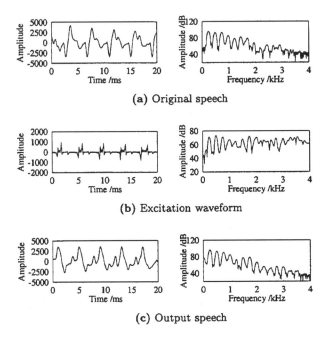

(a) Original speech

(b) Excitation waveform

Figure 13.21 Time- and frequency-domain comparison of the (a) original speech, (b) three-pulse ZFE waveform and (c) output speech after the pulse dispersion filter. The 20 ms speech frame is the liquid /r/ in the utterance "*r*ice" for the testfile BF2. For comparison with the other coders developed in this study using the same speech segment, please refer to Table 16.2.

(c) Output speech

increased problems in both obeying phase restrictions and creating a smoothly interpolated synthesized speech waveform.

For the 3.8 kbps multiple ZFE speech coder, the extra bits are consumed by the two extra ZFE pulses, with the bit allocation detailed in Table 13.14. The location of the two extra ZFE pulses, λ_2 and λ_3, with respect to the first ZFE pulse, must be transmitted to the decoder, while, similarly to the single ZFE coder, the first pulse location can be assumed at the decoder. With a permissible pitch-period range of $20 \rightarrow 147$ samples, 7 bits are required to encode each position parameter, λ. This parameter only requires transmission for the first frame of a voiced sequence, since for further frames the pulses are kept in the same location

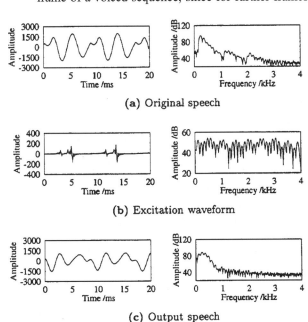

(a) Original speech

(b) Excitation waveform

Figure 13.22 Time- and frequency-domain comparison of the (a) original speech, (b) three-pulse ZFE waveform, and (c) output speech after the pulse dispersion filter. The 20 ms speech frame is the nasal /n/ in the utterance "throw*n*" for the testfile BM2. For comparison with the other coders developed in this study using the same speech segment, please refer to Table 16.2.

(c) Output speech

within the prototype region, as it was argued in Section 13.6.3. The A and B amplitude parameter for the extra ZFE pulses are scalar quantized to 6 bits.

In order to produce a dual-rate speech coder, it must be possible to change the coder's transmission rate during operation. In this multiple ZFE scheme, if a ZFE pulse were omitted from the frame, reducing the bitrate, at the decoder the ZFE pulse would be interpolated across the interpolation region to zero. Similarly, if an extra ZFE pulse was harnessed, then at the decoder the ZFE would be interpolated from zero. This interpolation from zero degrades the assumption that the previous prototype segment at the encoder is similar to the previous interpolation region at the decoder. Thus, it is prudent to only permit coding rate changes between voiced frame sequences.

13.12 A SIXTH-RATE, 3.8 kbps GSM-LIKE TRANSCEIVER[1]
F.C.A SOMERVILLE, B.L.YEAP, J.P WOODARD, L. HANZO

13.12.1 Motivation

Although the standardization of the third-generation wireless systems has been completed, it is worthwhile considering potential evolutionary paths for the mature GSM system. This tendency was hallmarked by the various GSM Phase 2 proposals, endeavoring to improve the services supported, or by the development of the half-rate and enhanced full-rate speech codecs. In this section we consider two potential improvements and their interactions in a source-sensitivity matched transceiver, namely, employing an approximately sixth-rate, 1.9 kbps speech codec and turbo coding [192, 193] in conjunction with the GSM system's Gaussian minimum shift keying (GMSK) partial response modem.

The bit allocation of the 1.9 kbps PWI-ZFE speech codec was summarized in Table 13.10, while its error sensitivity was quantified in Section 13.10. The SEGSNR and CD objective measures together with the informal listening tests allow the bit to be ordered in terms of their error sensitivities. The most sensitive bit is the voiced-unvoiced flag. For voiced frames the three most significant bits (MSB) in the LTP delay are the next most sensitive bits, followed by the four least significant LTP delay bits. For unvoiced frames, the boundary parameter shift, j, is given the same protection as the most significant three pitch-period bits, while the RMS value is given the same protection as the group of four least significant pitch-period bits and bit $A[6]$, the LSB of the ZFE amplitude A.

13.12.2 The Turbo-Coded Sixth-Rate 3.8 kbps GSM-Like System

The amalgamated GSM-like system [134] is illustrated in Figure 13.23. In this system, the 1.9 kbps speech coded bits are channel encoded with a $\frac{1}{2}$ rate convolutional or turbo encoder [192, 193] and an interleaving frame length of 81 bits, including termination bits. Therefore, assuming negligible processing delay, 162 bits will be released every 40 ms, or two 20 ms speech frames, since the 9×9 turbo-interleaver matrix employed requires two 20 ms, 38-bit speech frames before channel encoding commences. Hence, we set the data burst length at 162 bits. The channel encoded speech bits are then passed to a channel interleaver. Subsequently, the interleaved bits are modulated using Gaussian minimum shift keying (GMSK) [134] with a normalized bandwidth, $B_n = 0.3$, and transmitted at 271 Kbit/s across

---

[1]This section is based on F.C.A. Brooks, B.L. Yeap, J.P. Woodard, and L. Hanzo, *A Sixth-rate, 3.8 kbpsa GSM-like Speech Transceiver*, ACTS'98, Rhodes, Greece.

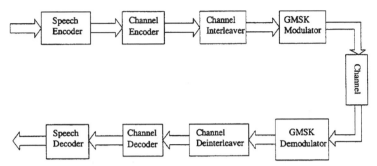

Figure 13.23 GSM-like system block diagram.

the COST 207 [273] Typical Urban channel model. Figure 13.24 is the Typical Urban channel model used, and each path is fading independently with Rayleigh statistics, for a vehicular speed of 50 km/h or 13.89 ms^{-1} and transmission frequency of 900 MHz.

The GMSK demodulator equalizes the received signal, which has been degraded by the wideband fading channel, using perfect channel estimation [134]. Subsequently, soft outputs from the demodulator are deinterleaved and passed to the channel decoder. Finally, the decoded bits are directed toward the speech decoder in order to extract the original speech information. In the following subsections, the channel coder and interleaver/deinterleaver, and GMSK transceiver are described.

13.12.3 Turbo Channel Coding

We compare two channel coding schemes, constraint-length $K = 5$ convolutional coding as used in the GSM [134] system, and a turbo channel codec [192, 193]. The turbo codec uses two $K = 3$ Recursive Systematic Convolutional (RSC) component codes, employing octally represented generator polynomials of 7 and 5, as well as eight iterations of the Log-MAP [331] decoding algorithm. This makes it approximately 10 times more complex than the convolutional codec.

It is well known that turbo codes perform best for long interleavers. However, because of the low bitrate of the speech codec, we are constrained to use a low frame length in the channel codecs. A frame length of 81 bits is used, with a 9×9 block interleaver within the turbo codec. This allows two sets of 38 coded bits from the speech codec and two termination

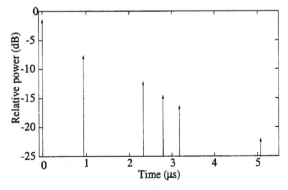

Figure 13.24 The impulse response of the COST207 Typical Urban channel used [273].

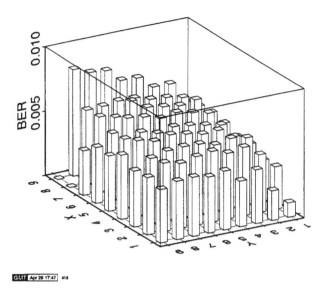

Figure 13.25 The error sensitivity of the different information bits within the 9 × 9 block interleaver used in the turbo codec.

bits to be used. The BERs of the 79 transmitted bits with the 9 × 9 block interleaver used for the turbo codec, for a simple AWGN channel at an SNR of 2 dB, is shown in Figure 13.25. It can be seen that bits near the bottom right-hand corner of the interleaver are better protected than bits in other positions in the interleaver. By placing the more sensitive speech bits here, we are able to give significantly more protection to the V/U flag and to some of the other sensitive speech bits than to the low-sensitivity bits of Figure 13.16. Currently, we are investigating the provision of more significant unequal error protection using turbo codes with irregular parity bit puncturing. Lastly, an interburst channel interleaver is used in order to disperse the bursty channel errors and to assist the channel decoders, as proposed for GSM [134].

13.12.4 The Turbo-coded GMSK Transceiver

As mentioned in Section 13.12.2, a GMSK modulator, with $B_n = 0.3$, which is employed in the current GSM [134] mobile radio standard, is used in our system. GMSK belongs to a class of continuous phase modulation (CPM) [134] and possesses high spectral efficiency and constant signal envelope, hence allowing the use of nonlinear power-efficient class-C amplifiers. However, the spectral compactness is achieved at the expense of controlled intersymbol interference (CISI). Therefore an equalizer, typically a Viterbi Equalizer, is needed. The conventional Viterbi Equalizer (VE) [134] performs Maximum Likelihood Sequence Estimation by observing the development of the accumulated metrics, which are evaluated recursively over several bit intervals. The length of the observation interval depends on the complexity afforded. Hard decisions are then released at the end of the equalization process. However, since the turbo decoders require Log Likelihood Ratios (LLRs) [332], we could use a variety of soft output algorithms instead of the VE, such as the Maximum A Posteriori (MAP) [197] algorithm, the Log-MAP [331], the Max-Log-MAP [333, 334], and the soft output Viterbi algorithm (SOVA) [27, 335, 336]. We chose to use the Log-MAP

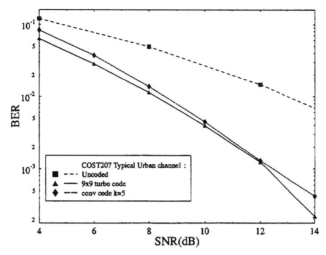

Figure 13.26 The BER performance for the turbo and convolutional coded systems over the COST 207 Typical Urban channel [273].

algorithm because it gave the optimal performance, like the MAP algorithm, but at a much lower complexity. Other schemes such as the Max-Log-MAP and SOVA are computationally less intensive but provide suboptimal performance. Therefore, for our work, we have opted for the Log-MAP algorithm in order to obtain the optimal performance, hence giving the upper-bound performance of the system.

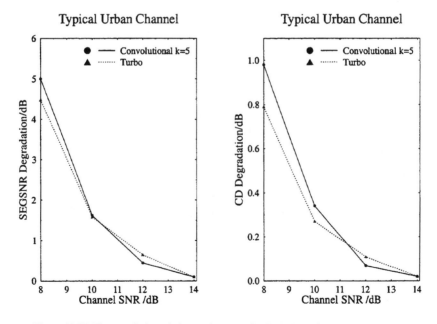

Figure 13.27 The speech degradation performance for the turbo and convolutional coded systems over the COST 207 Typical Urban channel [273].

13.12.5 System Performance Results

The performance of our sixth-rate GSM-like system was compared with an equivalent conventional GSM system using convolutional codes instead of turbo codes. The $\frac{1}{2}$ rate convolutional code [134] has the same code specifications as in the standard GSM system [134]. Figure 13.26 illustrates the BER performance over a Rayleigh-fading. COST207 Typical Urban channel [273], and Figure 13.27 shows the speech degradation, in terms of both the cepstral distance (CD) and the Segmental SNR, for the same channel. Because of the short interleaver frame length of the turbo code, the turbo- and convolutionally coded performances are fairly similar in terms of both BER and speech degradation. Hence, the investment of the higher complexity turbo codec is not justifiable, demonstrating an important limitation of short-latency interactive turbo-coded systems. However, we expect to see higher gains for higher bitrate speech codecs, such as, for example, the 260-bit/20 ms full-rate and the enhanced full-rate GSM speech codecs, which would allow us to use larger frame lengths for the turbo code, an issue currently being investigated.

13.13 CHAPTER SUMMARY

This Chapter has described a PWI-ZFE coder previously suggested by Hiotakakos and Xydeas [294]. Their work was further developed in this chapter in order to reduce the bitrate and complexity, while improving speech quality. Sections 13.2 to 13.4 gave an overview of the speech coder, with Figure 13.4 demonstrating the prohibitive complexity of the original ZFE optimization process proposed by Hiotakakos and Xydeas [294]. This prohibitive complexity was significantly reduced by introducing wavelets into the optimization process. Section 13.5 described the voiced speech encoding procedure, involving ZFE optimization and ZFE amplitude coefficient quantization. Energy scaling was also proposed to ensure that the original speech amplitude was maintained in the synthesized speech. The interpolation performed at the decoder was detailed in Section 13.6, where the justifications for not sending either the starting location of the prototype segment or the ZFE position parameter were given. The PWI-ZFE description was completed in Sections 13.7 and 13.8, which briefly described the unvoiced speech and adaptive post-filter requirements, respectively.

The PWI-ZFE speech coder at 1.9 kbps was found to produce speech with a more natural quality than the basic LPC vocoder of Chapter 11. This chapter has also shown that numerous benefits were attainable in reducing the computational complexity through use of the wavelet transform of Chapter 12 with no discernible reduction in speech quality. Particularly useful was the ability of the wavelet transform to suggest instants of glottal closure. The Chapter also outlined an interpolation method at the decoder which permitted the ZFE amplitude parameters to be transmitted without the position parameter, reducing the bitrate. Finally, in Section 13.11 multiple ZFE was considered; however, the quality of the synthesized speech was often found to be variable. In the next chapter multiband excitation will be invoked in an effort to improve the associated speech quality.

14

Mixed-Multiband Excitation

14.1 INTRODUCTION

This chapter investigates the speech coding technique of Mixed-Multiband Excitation (MMBE) [79] which is frequently adopted in very low-bitrate voice compression. The principle behind MMBE is that low-bitrate speech coders, which follow the classical vocoder principle of Atal and Hanauer [279] invoking distinct separation into voiced-unvoiced segments, usually result in speech of a synthetic quality due to a distortion generally termed "buzziness." This "buzzy" quality is particularly apparent in portions of speech which contain only voiced excitation in some frequency regions but dominant noise in other frequency bands of the speech spectrum. A classic example is the fricative class of phonemes, which contain both periodic and noise excitation sources. In low-bitrate speech coders, this type of speech waveform can be modeled successfully by combining voiced and unvoiced speech sources. Figure 14.1 shows the case of the voiced fricative /z/ as in "zoo," which consists of voiced speech up to 1 kHz and predominantly noisy speech above this frequency. Improved voiced excitation sources, such as the ZFE described in Chapter 13, can remove some of the synthetic quality of the reconstructed speech. However, the ZFE does nothing to combat the inherent problem of buzziness, which is associated with a mixed voiced-unvoiced spectrum that often occurs in human speech production.

MMBE addresses the problem of buzziness directly by splitting the speech into several frequency bands, similarly to sub-band coding [221] on a frame-by-frame adapted basis. These frequency bands have their voicing assessed individually with an excitation source of pulses, noise, or a mixture of both being selected for each frequency band. Figure 14.2 shows the PDF of the voicing strength for the training speech database of Table 10.1, where the voicing strength is defined later in Equation 14.11. It demonstrates that, although the voicing strengths have significant peaks near the values of 0.3 and 1, representing unvoiced and voiced frames, respectively, a number of frames have intermediate voicing strength. It is these frames, with about 35% having voicing strengths between 0.4 and 0.85, which will benefit from being represented by a mixture of voiced and unvoiced excitation sources.

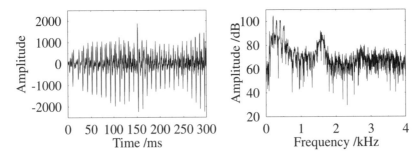

Figure 14.1 Example of a sustained voiced fricative /z/ present in "zoo". Observing the frequency domain, the phoneme is clearly beneath 1 kHz and much more noisy above 1 KHz.

This chapter begins with Section 14.2, giving an overview of an MMBE coder, and Section 14.3 details the filters that construct the multiband structure, together with the additional complexity they introduce. An augmented exposure of a MMBE encoder is given in Section 14.4, with a closer view of an MMBE decoder detailed in Section 14.5. Finally, Section 14.6 presents and examines the addition of the MMBE to the LPC vocoder of Chapter 13.

14.2 Overview of Mixed-Multiband Excitation

The control structure of an MMBE model is shown in Figures 14.3 and 14.4. The corresponding steps can also be followed with reference to the encoder and decoder schematics shown in Figure 14.5. After LPC analysis has been performed on the 20 ms speech frame, pitch detection occurs in order to locate any evidence of voicing. A frame deemed unvoiced has the RMS of its LPC residual quantized and sent to the decoder.

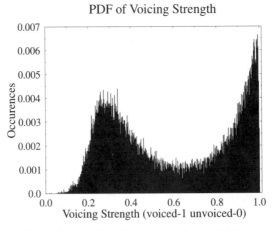

Figure 14.2 The distribution of voicing strengths for the training speech database of Table 10.1.

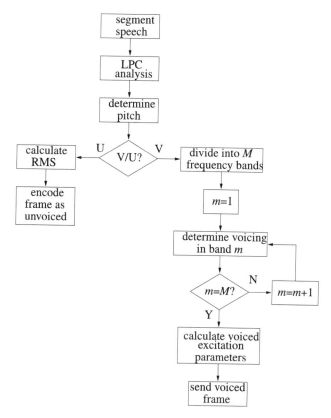

Figure 14.3 Control structure for a MMBE encoder.

Speech frames labeled as voiced are split into M frequency bands, with M constrained to be a constant value. These frequency bands generally have a bandwidth that contains an integer number of pitch-related spectral needles, where in the ideal situation each frequency band would have a width of one pitch-related spectral needle. However, in practical terms, due to coding efficiency constraints, each frequency band contains several pitch-related needles. The lower the fundamental frequency, the higher the number of pitch-related needles per frequency band. A consequence of the time-variant pitch period is the need for the time-variant adaptive filterbank, which generates the frequency bands to be reconstructed every frame in both the encoder and decoder, as shown in Figure 14.5, thus increasing the computational costs. Every frequency band is examined for voicing before being assigned a voicing strength that is quantized and sent to the decoder. Reproduction of the speech at the decoder requires knowledge of the pitch period in order to reconstruct the filterbanks of Figure 14.5b, together with the voicing strength in each band. The voiced excitation must also be determined, and its parameters have to be sent to the decoder.

At the decoder, following Figure 14.5b, both unvoiced and voiced speech frames have a pair of filterbanks created. However, for unvoiced frames, the filterbank is declared fully unvoiced with no pulses employed. For the voiced speech frames, both voiced and unvoiced excitation sources are created.

Following Figure 14.4, both the voiced and unvoiced filterbanks are created using knowledge of the pitch period and the number of frequency bands, M. For the voiced filterbanks, the filter coefficients are scaled by the quantized voicing strengths determined at

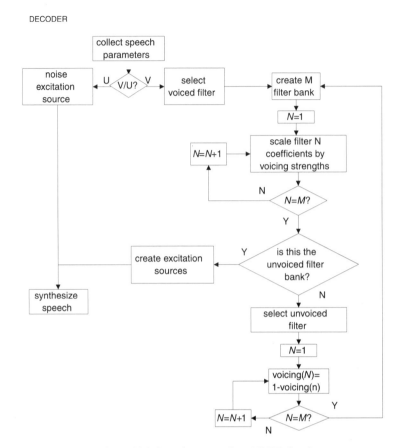

Figure 14.4 Control structure for a MMBE decoder.

the encoder. A value of 1 represents full voicing, while a value 0 signifies a frequency band of noise, with values between these extremes representing a mixed excitation source. For the unvoiced filterbank the voicing strengths are adjusted, ensuring that the voicing strengths of each voiced and unvoiced frequency band combine to unity. This constraint maintains a combined resultant from the filterbanks that is spectrally flat over the entire frequency range. The mixed excitation speech is then synthesized, as shown in Figure 14.5b, where the LPC filter determines the spectral envelope of the speech signal. Construction of the filterbanks is described in detail in Section 14.3.

14.3 FINITE IMPULSE RESPONSE FILTER

The success of MMBE is dependent on creating a suitable bank of filters. The filterbank should be capable of producing either fully voiced or unvoiced speech together with mixed speech. Two well-established techniques for producing filterbanks are finite impulse response (FIR) filters and quadrature mirror filters (QMFs), a type of FIR filter.

QMFs [223] are designed to divide a frequency spectrum in half. Thus, a cascade of QMFs can be implemented until the spectrum is divided into appropriate frequency bands. If a signal has a sampling frequency f_s, then a pair of QMFs will divide the signal into a band from 0 to $f_s/4$ and a band from $f_s/4$ to $f_s/2$. Both filters will have their 3 dB point at $f_s/4$. The filterbank of our MMBE coder was not constructed from QMFs, since the uniform division of the frequency spectrum imposes restrictions on the shape of the filterbank.

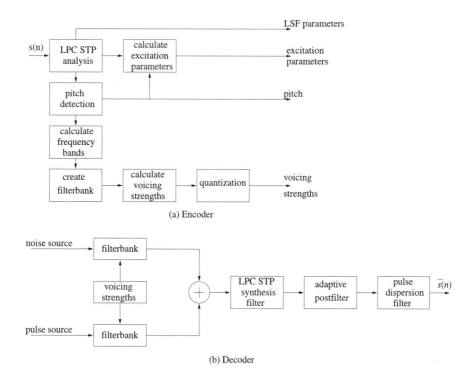

Figure 14.5 Schematic of the (a) encoder and (b) decoder for a MMBE scheme.

FIR filters contain only a finite number of nonzero impulse response taps. Thus, for a FIR filter of length K, the impulse response is given by:

$$h_T(n) = \begin{cases} b_n & 0 \le n \le K - 1 \\ 0 & \text{elsewhere} \end{cases} \tag{14.1}$$

where $h_T(n)$ is the impulse response of the filter and b_n are the filter coefficients. Using discrete convolution, we give the filter's output signal as:

$$y_T(n) = \sum_{m=0}^{K-1} h_T(m) \cdot x_T(n - m) \tag{14.2}$$

where y_T is the filter output and x_T is the filter input. Computing the Z-transform of Equation 14.2, we arrive at the following filter transfer function:

$$H(z) = \sum_{m=0}^{K-1} h_T(m) z^{-m}. \tag{14.3}$$

The impulse response of an ideal low-pass filter transfer function $H(z)$ is the well-known infinite duration sinc function given as:

$$h_T(n) = \frac{1}{\pi n r_c} \sin(2\pi n r_c) \tag{14.4}$$

where r_c is the cutoff frequency that has been normalized to $f_s/2$. In order to create a windowed ideal FIR low-pass filter, we invoke a windowing function $w(n)$, which is harnessed as follows:

$$h_T(n) = \frac{1}{\pi n r_c} w_{ham}(n) \sin(2\pi n r_c) \tag{14.5}$$

where $w_{ham}(n)$ was chosen in our implementation to be the Hamming window given by:

$$w_{ham}(n) = 0.54 - 0.46 \cos\left(\frac{2\pi n}{K}\right) \tag{14.6}$$

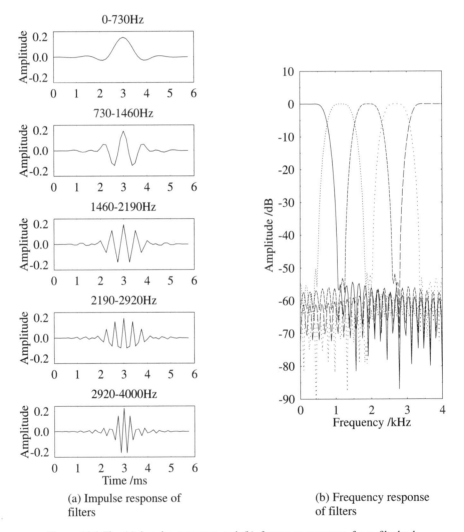

(a) Impulse response of filters

(b) Frequency response of filters

Figure 14.6 The (a) impulse responses and (b) frequency responses for a filterbank constructed from a low pass and four bandpass filters. They have frequency ranges $0 \rightarrow 730$ Hz, $730 \rightarrow 1460$ Hz, $1460 \rightarrow 2190$ Hz, $2190 \rightarrow 2920$ Hz and $2920 \rightarrow 4000$ Hz. A filter order or 47 was used.

with K being the filter length. In order to transform the low-pass filter to a bandpass filter, h_T^{BP}, the ideal windowed low pass filter, h_T^{LP} is scaled by the expression [337]:

$$h_T^{BP}(n) = h_T^{LP}(n) \cos\left(2\pi n\left(\frac{r_l + r_u}{2}\right)\right) \tag{14.7}$$

where r_l is the lower normalized bandpass frequency and r_u is the upper normalized bandpass frequency.

A filterbank consists of the low-pass filter together with the bandpass filters such that the entire frequency range is covered. Thus, as demonstrated in Figure 14.6, the filterbank contains both a low-pass filter and bandpass filters in its constitution.

14.4 MIXED-MULTIBAND EXCITATION ENCODER

At the encoder, the task of the filterbank is to split the frequency band and facilitate the determination of the voicing strengths in each frequency band. In order to accommodate an integer number of the spectral domain pitch-related needles, each frequency band's bandwidth is a multiple of the fundamental frequency. The total speech bandwidth, $f_s/2$, is occupied by a $N_n \cdot F0 \cdot M$ number of pitch-related needles, where f_s is the sampling frequency, $F0$ is the fundamental frequency, and M is the number of bands in the filterbank, while N_n is the number of needles for each sub-band, which can be expressed as [338]:

$$N_n = \frac{f_s/2}{M \cdot F0}. \tag{14.8}$$

The resultant N_n value is rounded down to the nearest integer. Any remaining frequency band between $f_s/2$ and the final filter cutoff frequency is assumed to be unvoiced.

For example, with a sampling frequency of 8 kHz and a filterbank design with five bands, the number of harmonics in each band can be determined. For a fundamental frequency of 100 Hz, it follows that:

$$N_n = \frac{4000}{100 \times 5} = 8 \tag{14.9}$$

implying that there will be eight pitch needles for each sub-band. Similarly, for a fundamental frequency of 150 Hz, we have:

$$N_n = \frac{4000}{150 \times 5} = 5.33. \tag{14.10}$$

Thus, each band will contain five pitch needles, with the frequencies 3750 to 4000 Hz being incorporated in the upper frequency band.

The method of dividing the frequency spectrum as suggested by Equation 14.8 is not a unique solution. It would be equally possible to increase the bandwidth of the higher filters due to the human ear's placing less perceptual emphasis on these regions. However, the above pitch-dependent, but even, spread of the frequency bands allows a simple division of the frequency spectrum. Since the decoder reconstructs the filter from $F0$, no extra side information requires transmission.

14.4.1 Voicing Strengths

For every voiced speech frame, the input speech is passed through each filter in the filterbank in order to locate any evidence of voicing in each band. Figure 14.7 shows the transfer function of the filterbank created and the filtered speech in both the time and frequency domain. Observing the top of Figure 14.7a, below 3 kHz the original spectrum appears predominantly voiced, whereas above 3 kHz it appears more unvoiced, as shown by the periodic and aperiodic spectral fine structure present. The corresponding time-domain signal waveforms of Figure 14.7b seem to contain substantially attenuated harmonics of the fundamental frequency $F0$, although the highest two frequency bands appear more noise-like.

The voicing strength is found in our coder using several methods [284], since if the voicing is inaccurately calculated, the reconstructed speech will contain an excessive "buzz" or "hiss," that is, too much periodicity or excessive noise, respectively. Initially, the voicing strength, v_s, is found using the normalized pitch-spaced filtered waveform correlation [284]:

$$v_s = \frac{\sum_{n=0}^{FL-1} f(n) * f(n-P)}{\sqrt{\sum_{n=0}^{FL-1} f(n)^2 \sum_{n=0}^{FL-1} f(n-P)^2}} \quad (14.11)$$

where $f(n)$ is the filtered speech of a certain bandwidth, FL is the frame length, and P is the pitch period for the speech frame. However, at the higher frequencies the correlation can be very low even for voiced speech. The time-domain envelope of the filtered speech will be a better indication of voicing [284], as demonstrated by Figure 14.8.

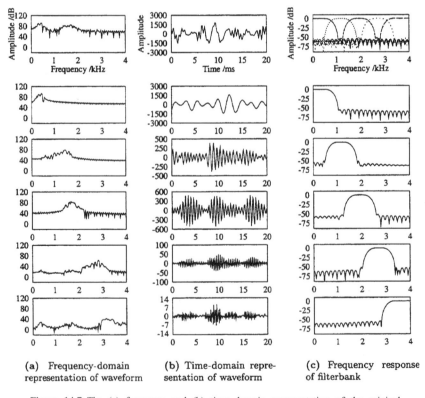

(a) Frequency-domain representation of waveform

(b) Time-domain representation of waveform

(c) Frequency response of filterbank

Figure 14.7 The (a) frequency and (b) time domain representation of the original waveform AM1 when uttering diphthong /aɪ/ in 'wi*r*es' together with the filtered waveform. (c) The frequency responses of the filterbank are also shown. A filter order of 47 was used.

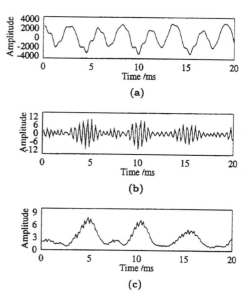

Figure 14.8 Time domain waveforms of (a) the original speech, (b) the bandpass filtered speech and (c) the envelope of the bandpass filtered speech.

The envelope of the bandpass-filtered speech is found through low-pass filtering the full-wave rectified filtered speech signal. The one-pole low-pass-filtered rectified bandpass signal is given by:

$$f(n) = \frac{1}{1 + 2\pi f_c/f_s} \cdot \left[2\pi \frac{f_c}{f_s} s(n) + f(n-1) \right] \qquad (14.12)$$

where f_c is the cutoff frequency, $s(n)$ is the input signal of the filter, $f(n)$ is the output signal of the filter, and f_s is the sampling frequency. The cutoff frequency was taken to be 500 Hz, since

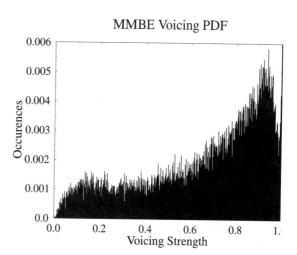

Figure 14.9 The PDF of the voicing strengths for a 20-band filterbank using the database of Table 10.1.

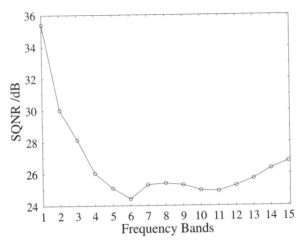

Figure 14.10 SNR values, related to the quantized and unquantized voicing strengths, achieved after the voicing levels in the respective frequency bands are 3-bit quantized for the MMBE coder.

this is just above the highest expected fundamental frequency. The voicing strength, v_s, is then calculated using Equation 14.11 for the low-pass-filtered, rectified bandpass signal.

Subsequently, each frequency band is assigned the largest calculated voicing strength achieved from the original bandpass signal or the low-pass-filtered rectified bandpass signal. The PDF of the selected voicing strengths for a 20-band filterbank is given in Figure 14.9 for the training database. The graph represents all the voicing strengths recorded in every frequency band, providing sufficient fine resolution training data for the Max-Lloyd quantizer to be used.

The PDF for the voicing strength values was passed to the Max-Lloyd quantizer described in Section 11.4. The Max-Lloyd quantizer allocates eight levels for the voicing strengths using a total of 3 bits, with level 0 constrained to be 0.2 and level 8 to be 1. If level 0 were assigned to be 0, the quantizer would be too biased toward the lower valued voicing strengths. The same quantizer is used to encode every frequency band, producing the SNR values for a 1-band to 15-band MMBE scheme given in Figure 14.10, where the speech files AM1, AM2, AF1, AF2, BM1, BM2, BF1, and BF2 were used to test the quality of the MMBE quantizer.

14.5 MIXED-MULTIBAND EXCITATION DECODER

In the MMBE scheme at the decoder of Figures 14.4 and 14.5b, for voiced speech two versions of the filterbank are constructed. Subsequently, both voiced and unvoiced excitation are passed through these filterbanks and onto the LPC synthesis filter in order to reproduce the speech waveform.

The power of the filterbank generating the voiced excitation is scaled by the quantized voicing strength, while the filterbank producing the unvoiced excitation is scaled by the difference between unity and the voicing strength. This is performed for each frequency band of the filterbank. Once combined, the resultant filterbanks produce an all-pass filter over the 0 to 4000 Hz frequency range, as demonstrated in Figure 14.11. The filterbanks are designed to allow complete voicing, pure noise, or any mixture of the voiced and unvoiced excitation. As specified in Section 14.4, any frequency in the immediate vicinity of 4 kHz which was not designated a voicing strength is included in the uppermost frequency band. From knowledge of the fundamental frequency $F0$ and the number of bands M, the decoder computes N_n, the number of pitch-related needles in each frequency band. Thus, with the normalized cutoff frequencies known the corresponding impulse response can be inferred from Equations 14.5 and 14.7.

For both voiced and unvoiced speech frames, the $(1 - v_s)$ scaled noise excitation is passed to the unvoiced filterbank. The voiced excitation is implemented with either pulses from the LPC vocoder, as detailed in Section 11.5, or the PWI-ZFE function detailed in Section 13.5. Then after scaling by v_s, the excitation is passed to the voiced filterbank. The filtered signals are combined and passed to the LPC STP filter for synthesis.

In Figure 14.12 the process of selecting the portion of the frequency spectrum that is voiced and unvoiced is shown. Figure 14.12a shows the original speech spectrum with its LPC STP residual signal portrayed in Figure 14.12b. Figure 14.12c and Figure 14.12d represent the voiced and unvoiced excitation spectra, respectively. From Figure 4.12f it can be seen that beneath 2 kHz the classification is voiced, while above 2 kHz it has been classified unvoiced. Lastly, Figure 14.12e demonstrates the synthesized frequency spectrum.

14.5.1 Adaptive Post-Filter

The adaptive post-filter from Section 11.6 was used for the MMBE speech coders, with Table 14.1 detailing the optimized parameters for each MMBE speech coder detailed in the next section. Following adaptive post-filtering, the speech is passed through the pulse dispersion filter of Figure 11.19. In the next section we consider the issues of algorithmic complexity.

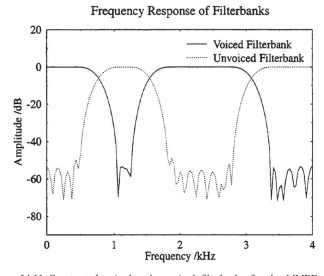

Figure 14.11 Constructed voiced and unvoiced filterbanks for the MMBE decoder. Displayed for a 5-band model, with three voiced and two unvoiced bands. Using a filter of order 47.

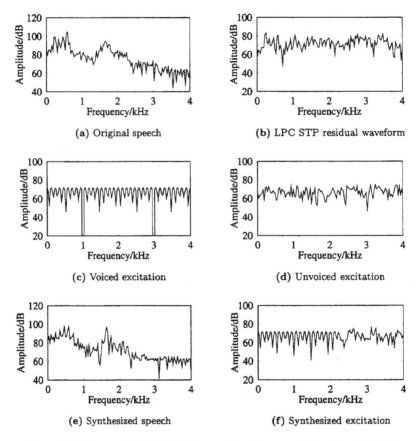

Figure 14.12 An example of the MMBE process for a 20 ms speech frame from the testfile AM1 when uttering the back vowel /u/ in "sho*u*ld". The (a) original and (e) synthesized frequency spectrum is demonstrated, along with the (b) original and (f) synthesized excitation spectra; also shown are the (c) voiced, and (d) unvoiced excitation spectra.

TABLE 14.1 Appropriate Adaptive Post-filter Values for the MMBE Speech Coders Examined in Section 14.6

Parameter	Values			
	2-band MMBE	5-band MMBE	3-band MMBE PWI-ZFE	13-band MMBE PWI-ZFE
α_{pf}	0.75	0.80	0.85	0.85
β_{pf}	0.45	0.55	0.55	0.50
μ_{pf}	0.60	0.50	0.60	0.60
γ_{pf}	0.50	0.50	0.50	0.50
g_{pf}	0.00	0.00	0.00	0.00
ξ_{pf}	0.99	0.99	0.99	0.99

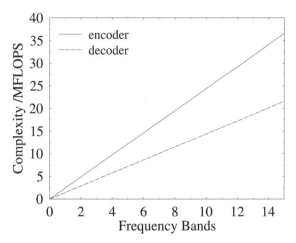

Figure 14.13 The computational complexity in the MMBE encoder and decoder for different numbers of frequency bands.

TABLE 14.2   Additional Computational Complexity Introduced at the Encoder and Decoder by the MMBE Scheme, for a 2-band and 5-band Arrangement

Procedure	2-band/MFLOPS	5-band/MFLOPS
Encoder		
Create filterbank	0.02	0.05
Filter speech into bands	1.54	3.07
Find voicing strengths	0.35	0.88
Decoder		
Create filterbank	0.02	0.05
Filter excitation sources	3.11	7.77

14.5.2 Computational Complexity

The additional computational complexity introduced by a MMBE scheme in both the encoder and decoder is given in Table 14.2 and Figure 14.13. From Table 14.2 it can be seen that at the encoder the complexity is dominated by the process of filtering the speech into different bands, while at the decoder the MMBE filtering process is dominant. In Figure 14.13, frequency band schemes between 1-band and 15-band are considered.

Following this description of the MMBE process, the reconstructed speech is examined when MMBE is added to both the benchmark LPC vocoder of Chapter 11 and the PWI-ZFE coder of Chapter 13.

14.6 PERFORMANCE OF THE MIXED-MULTIBAND EXCITATION CODER

This section discusses the performance of the benchmark LPC vocoder, in Chapter 11, and the PWI-ZFE coder in Chapter 13, with the addition of MMBE. Both a 2-band and a 5-band

MMBE were added to the LPC vocoder, as detailed in Section 14.6.1, creating speech coders operating at 1.85 kbps and 2.3 kbps, respectively. For the PWI-ZFE coder, only a 3-band MMBE was added, as detailed in Section 14.6.2, producing a 2.35 kbps speech coder.

14.6.1 Performance of a Mixed-Multiband Excitation Linear Predictive Coder

The MMBE scheme was added to the basic LPC vocoder, described in Chapter 11, with the speech database described in Table 10.1 used to assess the coder's performance. The time- and frequency-domain plots for individual 20 ms frames of speech are given in Figures 14.14, 14.15, and 14.16 for a 2-band MMBE model, while Figures 14.17, 14.18, and 14.19 display the corresponding results for a 5-band MMBE model. Both Figures 14.14 and 14.17 represent the same speech segment as Figures 11.21 and Figures 13.13, while Figures 14.15 and 14.18 represent the same speech segment as Figure 11.22 and Figure 13.14, and Figures 14.16 and 14.19 represent the same speech segment as Figure 11.23 and Figure 13.15. Initially, the performance of a 2-band MMBE scheme is studied.

Figure 14.14 displays the performance of a 20 ms speech frame from the testfile BM1. For this speech frame Figure 14.14b shows that the entire frequency spectrum is considered voiced. Thus the reproduced speech waveform is identical to Figure 11.21.

Figure 14.15 is an utterance from the testfile BF2, where observing Figure 14.15b above 2 kHz, a mixture of voiced and unvoiced excitation is harnessed. From Figure 14.15c it can be seen that the presence of noise above 2 kHz produces a better representation of the frequency spectrum than Figure 11.22c.

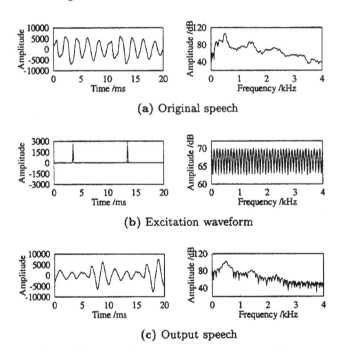

(a) Original speech

(b) Excitation waveform

(c) Output speech

Figure 14.14 Time and frequency domain comparison of the (a) original speech, (b) 2-band MMBE waveform, and (c) output speech after the pulse dispersion filter. The 20 ms speech frame is the mid vowel /ɜ/ in the utterance "work" for the testfile BM1. For comparison with the other coders developed in this study using the same speech segment, please refer to Table 16.2.

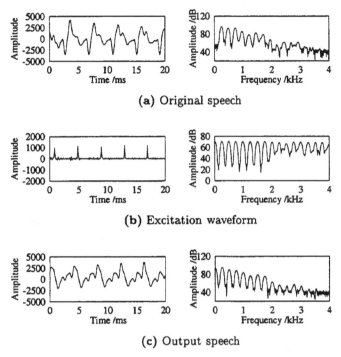

Figure 14.15 Time- and frequency-domain comparison of the (a) original speech, (b) 2-band MMBE waveform, and (c) output speech after the pulse dispersion filter. The 20 ms speech frame is the liquid /r/ in the utterance "*r*ice" for the testfile BF2. For comparison with the other coders developed in this study using the same speech segment, please refer to Table 16.2.

Figure 14.16 is a 20 ms speech frame from the testfile BM2 for the nasal /n/ in the utterance "throw*n*." Similarly to Figure 14.15, the frequency spectrum above 2 kHz is modeled by purely unvoiced excitation. Figures 14.15 and 14.16 demonstrate that many speech waveforms contain both voiced and unvoiced components. Thus, they emphasize the need for a speech coder that can incorporate mixed excitation.

Through informal listening, a comparison of the synthesized speech from an LPC vocoder with and without MMBE can be made. Introduction of the MMBE removes a significant amount of the "buzz" inherent in LPC vocoder models, producing more natural sounding speech. Occasionally, a background "hiss" is introduced into the synthesized speech, which is due to the coarse resolution of the frequency bands in a 2-band MMBE scheme. In addition, pairwise-comparison tests, detailed in Section 16.2, were conducted to compare the speech quality from the 1.9 kbps PWI-ZFE speech coder of Chapter 13 with the 2-band MMBE LPC scheme. These pairwise-comparison tests showed that 30.77% of listeners preferred the PWI-ZFE speech coder, with 23.07% of listeners preferring the 2-band MMBE LPC scheme and 46.16% having no preference.

A 5-band MMBE scheme was also implemented in the context of the LPC vocoder, which with an increased number of voicing decisions should produce better quality synthesized speech than the 2-band MMBE model.

For Figure 14.18 the addition of the extra three frequency bands is shown in Figure 14.17 for a speech frame in the testfile BM1. Figure 14.17b shows that the extra frequency

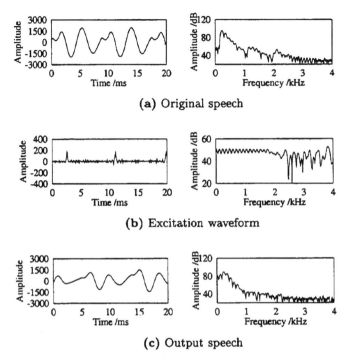

(a) Original speech

(b) Excitation waveform

(c) Output speech

Figure 14.16 Time- and frequency-domain comparison of the (a) original speech, (b) 2-band MMBE waveform, and (c) output speech after the pulse dispersion filter. The 20 ms speech frame is the nasal /n/ in the utterance "throw*n*" for the testfile BM2. For comparison with the other coders developed in this study using the same speech segment, please refer to Table 16.2.

bands produce a mixture of voiced and unvoiced speech above 3 kHz, where for the 2-band MMBE model the entire frequency spectrum was fully voiced.

· Figure 14.18 portrays the speech frame shown in Figure 14.15 from the BF2 testfile but with an extra three frequency bands. For this speech frame the additional three frequency bands have no visible effect.

Figure 14.19 displays a speech frame from the testfile BM2 with 5-band MMBE and can be compared with Figure 14.16. For this speech frame, the addition of three frequency bands produces fully unvoiced speech above 800 Hz, as shown in Figure 14.19b, with the effect on the synthesized speech visible in the frequency domain of Figure 14.19c.

Informal listening tests revealed that the addition of an extra three decision bands to the MMBE scheme has little perceptual effect. It is possible that inherent distortions caused by the LPC vocoder model are masking the improvements. The bit allocation for an LPC vocoder with either a 2-band or 5-band MMBE scheme is given in Table 14.3. The voicing strength of each decision band is quantized with a 3-bit quantizer as described in Section 14.4, thus adding 0.15 kbps to the overall bitrate of the coder. The computational complexity of the LPC speech vocoder with 2- and 5-band MMBE is given in Table 14.4, where the complexity is dominated by the MMBE function.

In the next section a 3-band MMBE scheme is incorporated into the PWI-ZFE coder of Chapter 13.

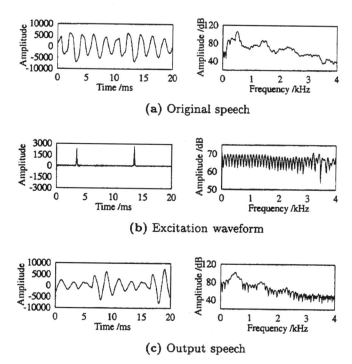

(a) Original speech

(b) Excitation waveform

(c) Output speech

Figure 14.17 Time- and frequency-domain comparison of the (a) original speech, (b) 5-band MMBE waveform, and (c) output speech after the pulse dispersion filter. The 20 ms speech frame is the midvowel /ɜ/ in the utterance "wo*r*k" for the testfile BM1. For comparison with the other coders developed in this study using the same speech segment, please refer to Table 16.2.

14.6.2 Performance of a Mixed-Multiband Excitation and Zinc Function Prototype Excitation Coder

The MMBE scheme detailed in this chapter was also added to the PWI-ZFE coder described in Chapter 13. Again, the speech database described in Table 10.1 was used to assess the coder's performance. The time and frequency-domain plots for individual 20 ms frames of speech are given in Figures 14.20, 14.21, and 14.22 for a 3-band MMBE excitation model. These speech frames are consistently used to consider the performance of the coders, and so can be compared with Figures 13.13, 13.14, and 13.15, respectively, together with those detailed during our later discourse in Table 16.2.

Figure 14.20 displays the performance of a 3-band MMBE scheme incorporated in the PWI-ZFE speech coder for a speech frame from the testfile BM1. Observing the frequency domain of Figure 14.20b, we see that a small amount of unvoiced speech is present above 2.5 kHz. The changes this noise makes to the synthesized speech are visible in the frequency domain of Figure 14.20c.

Similarly to Figure 14.20, for the speaker BF2 Figure 14.21 displays evidence of noise above 2.5 kHz. This noise is again visible in the frequency domain of Figure 14.21c.

The introduction of a 3-band MMBE scheme to the PWI-ZFE speech coder has a more pronounced effect in the context of the testfile BM2, as shown in Figure 14.22. From Figure 14.22b it can be seen that above 1.3 kHz the frequency spectrum is entirely noise. In the time

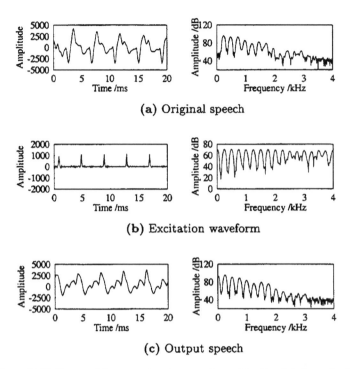

(a) Original speech

(b) Excitation waveform

(c) Output speech

Figure 14.18 Time- and frequency-domain comparison of the (a) original speech, (b) 5-band MMBE waveform, and (c) output speech after the pulse dispersion filter. The 20 ms speech frame is "*rice*" for the testfile BF2. For comparison with the other coders developed in this study using the same speech segment, please refer to Table 16.2.

domain much more noise is evident in the excitation waveform than for either Figure 14.20b or 14.21b.

Through informal listening to the PWI-ZFE coder, any audible improvements achieved by the addition of 3-band MMBE can be assessed. The MMBE removes much of the buzziness from the synthesized speech, which particularly improves the speech quality of the female speakers. Occasionally, the MMBE introduces "hoarseness," which is indicative of too much noise, especially to the synthesized speech of male speakers, but overall the MMBE improves speech quality at a slightly increased bitrate and complexity. Pairwise-comparison tests, detailed in Section 16.2, were conducted between the 2.35 kbps 3-band MMBE PWI-ZFE speech coder and the 2.3 kbps 5-band MMBE LPC scheme. These pairwise-comparison tests showed that 64.10% of listeners preferred the 3-band MMBE PWI-ZFE speech coder, with 5.13% of listeners preferring the 5-band MMBE LPC scheme and 30.77% having no preference.

As stated previously, each decision band introduces an additional 0.15 kbps to the overall bitrate of a speech coder. Hence, Table 14.5 shows that the addition of the MMBE scheme to the PWI-ZFE coder produced an overall bitrate of 2.35 kbps.

The computational complexity of the PWI-ZFE speech vocoder with 3-band MMBE is given in Table 14.6, which is dominated by the filtering procedures involved in the MMBE process and the ZFE optimization process.

In this section two schemes have been described which operate at similar bitrates: the LPC vocoder with 5-band MMBE operating at 2.3 kbps, and the PWI-ZFE coder incorporat-

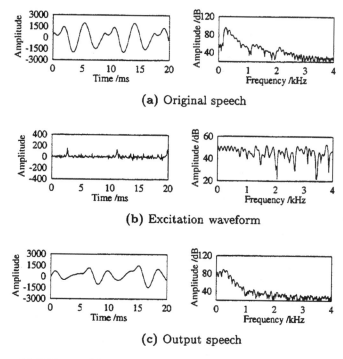

(a) Original speech

(b) Excitation waveform

(c) Output speech

Figure 14.19 Time- and frequency-domain comparison of the (a) original speech, (b) 5-band MMBE waveform, and (c) output speech after the pulse dispersion filter. The 20 ms speech frame is the nasal /n/ in the utterance "throw*n*" for the testfile BM2. For comparison with the other coders developed in this study using the same speech segment, please refer to Table 16.2.

TABLE 14.3 Bit-Allocation Table for the LPC Vocoder Voiced Frames with 2-band and 5-band MMBE

Parameter	2-band	5-band
LSFs	18	18
V/U flag	1	1
RMS value	5	5
Pitch	7	7
Voicing strengths	2×3	5×3
Total/20 ms	37	46
Bitrate	1.85 kbps	2.30 kbps

TABLE 14.4 Total Computational Complexity for a Basic LPC Vocoder Encoder with Either a 2-band or 5-band MMBE Model

Operation	2-band complexity MFLOPS	5-band complexity MFLOPS
Pitch detector	2.67	2.67
MMBE filtering	1.91	4.00
Total	4.58	6.67

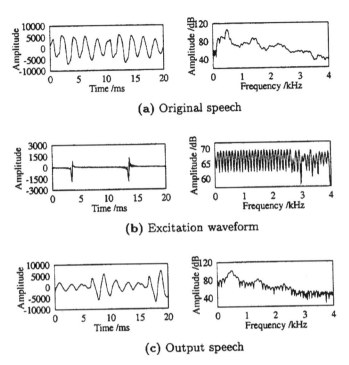

Figure 14.20 Time- and frequency-domain comparison of the (a) original speech, (b) 3-band MMBE ZFE waveform and (c) output speech after the pulse dispersion filter. The 20 ms speech frame is the midvowel /ɜ/ in the utterance "wo*r*k" for the testfile BM1. For comparison with the other coders developed in this study using the same speech segment, please refer to Table 16.2.

ing 3-band MMBE transmitting at 2.35 kbps. With informal listening tests, it was found that the PWI-ZFE coder with 3-band MMBE produced synthesized speech with slightly preferred perceptual qualities, although the quality of the reproduced speech was not dissimilar.

14.7 A HIGHER RATE 3.85 kbps MIXED-MULTIBAND EXCITATION SCHEME

In Sections 14.6.1 and 14.6.2, MMBE schemes operating at different bitrates have been investigated. The varying bitrates were achieved by either altering the excitation or by varying the number of frequency bands employed in the model. The nature of the pitch-dependent filterbank, with the filterbank being reconstructed every frame, permits simple conversion between the number of frequency bands. Following the multiple ZFE investigation of Section 13.11, an MMBE scheme operating at 3.85 kbps, incorporating a single ZFE, was implemented. The bitrate of 3.85 kbps is close to the bitrate of the PWI-ZFE speech coder with three ZFEs of Chapter 13, allowing comparisons between the two techniques at a higher bitrate. The bitrate of 3.85 kbps was achieved with the speech spectrum split into 13 bands, each scalar being quantized with 3 bits as described in Section 14.4.

The performance for an MMBE-ZFE scheme at 3.85 kbps is shown in Figures 14.23, 14.24, and 14.25, which can be compared with Figures 13.20, 13.21, and 13.22 showing the

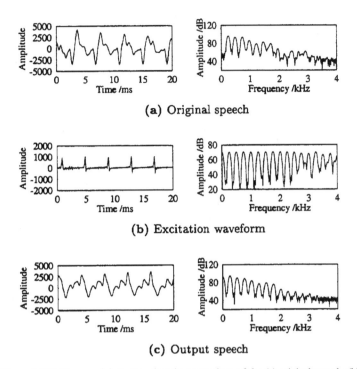

(a) Original speech

(b) Excitation waveform

(c) Output speech

Figure 14.21 Time- and frequency-domain comparison of the (a) original speech, (b) 3-band MMBE ZFE waveform, and (c) output speech after the pulse dispersion filter. The 20 ms speech frame is the mid vowel /r/ in the utterance "*r*ice" for the testfile BF2. For comparison with the other coders developed in this study using the same speech segment, please refer to Table 16.2.

TABLE 14.5   Bit Allocation Table for Voiced Frames in a 3-band and 13-bands MMBE PWI-ZFE Speech Coder

Parameter	3-band	13-band
LSFs	18	18
v/u flag	1	1
Pitch	7	7
A_1	6	6
B_1	6	6
Voicing strengths	3×3	13×3
Total/20 ms	47	77
Bitrate	2.35 kbps	3.85 kbps

TABLE 14.6   Total Computational Complexity for a PWI-ZFE Coder with a 3-band MMBE Arrangement

Operation	3-band complexity/MFLOPS
Pitch detector	2.67
MMBE filtering	2.05
ZFE minimization	11.46
Total	16.18

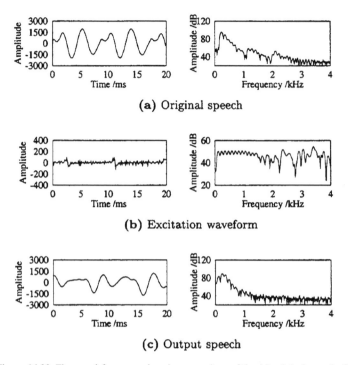

(a) Original speech

(b) Excitation waveform

(c) Output speech

Figure 14.22 Time- and frequency-domain comparison of the (a) original speech, (b) 3-band MMBE ZFE waveform, and (c) output speech after the pulse dispersion filter. The 20 ms speech frame is the nasal /n/ in the utterance "throw*n*" for the testfile BM2. For comparison with the other coders developed in this study using the same speech segment, please refer to Table 16.2.

3-pulse ZFE speech coder. Additional pertinent comparisons can be made with the figures detailed in Table 16.2.

For a speech frame from the testfile BM1 displayed in Figure 14.23, the frequency spectrum is still predominantly voiced, with noise being added only above 2.7 kHz. For this speech frame, the MMBE extension to the PWI-ZFE model performs better than adding extra ZFE pulses, since as shown in Figure 13.20 these extra ZFE pulses introduced pitch doubling.

Figure 14.24 shows a frame of speech from the testfile BF2. For this speech frame Figure 14.24b shows that up to 1 kHz the speech is voiced, between 1 and 2 kHz a mixture of voiced and unvoiced speech is present in the spectrum, between 2 and 3 kHz the speech is predominantly voiced, while above 3 kHz only noise is present in the frequency spectrum. However, when compared with Figure 13.21, it appears that the extra two ZFE pulses improve the reproduced speech more.

For a 20 ms frame from the testfile BM2, the performance is highlighted in Figure 14.25. Observing Figure 14.25b, we note that the frequency spectrum changes from voiced to unvoiced at 900 Hz. Furthermore, in the time domain it is difficult to determine the locations of the ZFE pulse.

The relative performances of the PWI-ZFE with 3-band MMBE and 13-band MMBE have been assessed through informal listening tests. Audibly, the introduction of the extra frequency bands improves the natural quality of the speech signal. However, it is debatable whether the improvement justifies the extra 1.5 kbps bitrate contribution consumed by the extra bands. Through pairwise-comparison listening tests, detailed in Section 16.2, the 13-

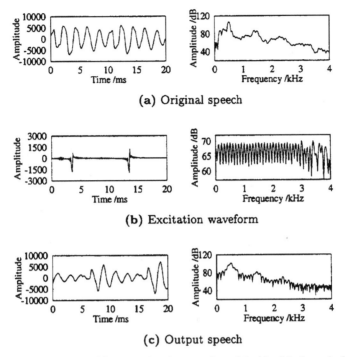

(a) Original speech

(b) Excitation waveform

(c) Output speech

Figure 14.23 Time- and frequency-domain comparison of the (a) original speech, (b) 13-band MMBE ZFE waveform and (c) output speech after the pulse dispersion filter. The 20 ms speech frame is the mid vowel /ɜ/ in the utterance 'wo*r*k' for the testfile BM1. For comparison with the other coders developed in this study using the same speech segment, please refer to Table 16.2.

band MMBE extension to the PWI-ZFE speech coder performed better than the addition of two extra ZFE pulses. Given the problems with interpolation detailed in Section 13.11, this was to be expected. The conducted pairwise-comparison tests showed that 30.77% of listeners preferred the 13-band MMBE PWI-ZFE speech coder, with 5.13% of listeners preferring the 3-pulse PWI-ZFE scheme and 64.10% having no preference. Before offering our conclusions concerning this chapter, let us in the next section consider an interesting system design example, which is based on our previously designed 2.35 kbit/s speech codec.

14.8 A 2.35 kbit/s JOINT-DETECTION-BASED CDMA SPEECH TRANSCEIVER[1]
F.C.A. SOMERVILLE, E.L. KUAN, L. HANZO

14.8.1 Background

The standardization of the third-generation wireless systems has reached a mature state in Europe, the United States, and Japan, and the corresponding system developments are well under way across the globe. All three standard proposals are based on wideband code division multiple access (W-CDMA), which optionally also supports joint multi-user detection in the up-link. In the field of speech and video source compression, similarly

[1]This section is based on F.C.A. Brooks, E.L. Kuan and L. Hanzo: *A 2.35 kbit/s Joint-detection based CDMA Speech Transceiver*, VTC'99, Houston, Texas.

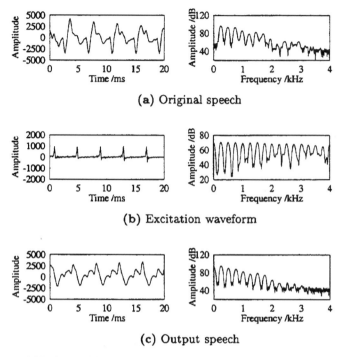

Figure 14.24 Time- and frequency-domain comparison of the (a) original speech, (b) 13-band MMBE ZFE waveform, and (c) output speech after the pulse dispersion filter. The 20 ms speech frame is the liquid /r/ in the utterance *"rice"* for the testfile BF2. For comparison with the other coders developed in this study using the same speech segment, please refer to Table 16.2.

impressive advances have been achieved. Hence, this section proposes a complete speech transceiver and quantifies its performance.

14.8.2 The Speech Codec's Bit Allocation

The codec's bit allocation was summarized in Table 14.5, where again, 18 bits were reserved for LSF vector quantization covering the groups of LSF parameters L0, L1, L2, and L3, where we used the nomenclature of the G.729 codec [123] for the groups of LSF parameters, since the G.729 codec's LSF quantizer was used. A 1-bit flag was used for the V/U classifier, while for unvoiced speech the RMS parameter was scalar-quantized with 5 bits. For voiced speech the pitch-delay was restricted to 20 → 147 samples, thus requiring 7 bits for transmission. The ZFE amplitude parameters A and B were scalar-quantized using 6 bits, since on the basis of our subjective and objective investigations we concluded that the 6-bit quantization constituted the best compromise in terms of bitrate and speech quality. The voicing strength for each frequency band was scalar-quantized, and because there were three frequency bands, a total of 9 bits per 20 ms were allocated to voicing-strength quantization. Thus, the total number of bits for a 20 ms frame became 26 or 47, yielding a transmission rate of 2.35 kbps for the voice speech segments.

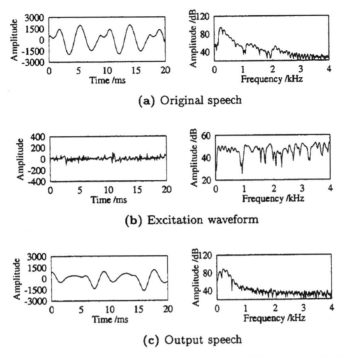

(a) Original speech

(b) Excitation waveform

(c) Output speech

Figure 14.25 Time- and frequency-domain comparison of the (a) original speech, (b) 13-band MMBE ZFE waveform, and (c) output speech after the pulse dispersion filter. The 20 ms speech frame is the nasal /n/ in the utterance "throw*n*" for the testfile BM2. For comparison with the other coders developed in this study using the same speech segment, please refer to Table 16.2.

14.8.3 The Speech Codec's Error Sensitivity

Following the above description of the 2.35 kbps speech codec, we now investigate the extent of the reconstructed speech degradation inflicted by transmission errors. The error sensitivity is examined by individually corrupting each of the 47 bits detailed in Table 14.5 with a corruption probability of 10%. Employing a less than unity corruption probability is common practice in order to allow the speech degradation caused by the previous corruption of a bit to decay, before the same bit is corrupted again, which emulates a practical transmission scenario realistically.

At the decoder, for some of the transmitted parameters it is possible to invoke simple error checks and corrections. At the encoder, isolated voiced or unvoiced frames are assumed to indicate a failure in the voiced-unvoiced decision, and corrected, an identical process can be implemented at the decoder. For the pitch-period parameter, a smoothly evolving pitch track is created at the encoder by correcting any spurious pitch-period values, and again, an identical process can be implemented at the decoder. In addition, for voiced frame sequences, phase continuity of the ZFE A and B amplitude parameters is maintained at the encoder. Thus, if a phase change is perceived at the decoder, an error occurrence is assumed, and the previous frame's parameters can be repeated.

Figure 14.26 displays the segmental signal-to-noise ratio (SEGSNR) and cepstral distance (CD) objective speech measures for a mixture of male and female speakers having British and American accents. Figure 14.26 shows that both the SEGSNR and CD

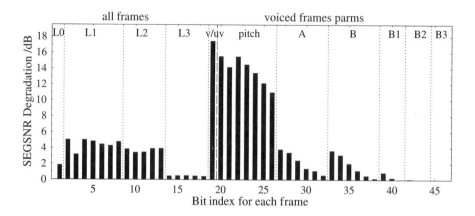

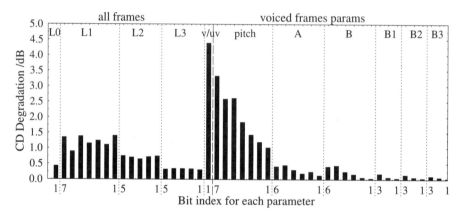

Figure 14.26 The error sensitivity of the different transmission bits for the 2.35 kbps speech codec. For the CD degradation graph, containing the bit index for each parameter, bit 1 is the least significant bit.

objective measures rate the error sensitivity of the different bits similarly. The most sensitive parameter is the voiced-unvoiced flag, followed closely by the pitch bits; the least sensitive parameters are the three voicing strengths bits of the bands $B1 - B3$, as seen in Figure 14.26.

14.8.4 Channel Coding

In order to improve the performance of the system, channel coding was employed. Two types of error correction codes were used: turbo codes and convolutional codes. Turbo coding is a powerful method of channel coding, and it has been reported to produce excellent results [192, 193]. Convolutional codes were used as the component codes for the turbo coding, and the coding rate was set to $r = 1/2$. We used a 7×7 block interleaver as the turbo interleaver. The FMA1 spread speech/data burst 1 [339] was altered slightly to fit the turbo interleaver. Specifically, the two data blocks were modified to transmit 25 data symbols in the first block and 24 symbols in the second one. In order to obtain the soft-decision inputs required by the turbo decoder, the Euclidean distance between the CDMA receiver's data estimates and each legitimate constellation point in the data modulation scheme was calculated. The set of distance values were then fed into the turbo decoder as soft inputs. The decoding algorithm used was the soft output Viterbi algorithm (SOVA) [335, 336] with eight iterations for turbo decoding. As a comparison, a half-rate, constraint-length three convolutional codec was used to produce a set of benchmark results. Note, however, that while the turbo codec used recursive systematic convolutional codecs, the convolutional codec was a nonrecursive one, which has better distance properties.

14.8.5 The JD-CDMA Speech System

The JD-CDMA speech system used in our investigations is illustrated in Figure 14.27 for a two-user scenario. The encoded speech bits generated by the 2.35 kbps prototype waveform interpolated (PWI) speech codec were channel encoded using a $\frac{1}{2}$-rate turbo encoder having a frame length of 98 bits, including the convolutional codec's termination bits, where a 7×7 turbo interleaver was used. The encoded bits were then passed to a channel interleaver and modulated using four-level quadrature amplitude modulation (4QAM). Subsequently, the modulated symbols were spread by the spreading sequence assigned to the user, where a random spreading sequence was used. The up-link conditions were investigated, where each user transmitted over a 7-path COST 207 Bad Urban channel [273], which is portrayed in Figure 14.28. Each path was faded independently using Rayleigh fading with a Doppler frequency of $f_D = 80\,\text{Hz}$ and a Baud rate of $R_b = 2.167\,\text{MBaud}$. Variations due to pathloss and shadowing were assumed to be eliminated by power control. The additive noise was assumed to be Gaussian with zero mean and a covariance matrix of $\sigma^2 \mathbf{I}$, where σ^2 is the variance of the noise. The burst structure used in our experiments mirrored the spread/speech burst structures of the FMA1 mode of the FRAMES proposal [339]. The minimum mean squared error block decision feedback equalizer (MMSE-BDFE) was used as the multiuser receiver [251], where perfect channel estimation and perfect decision feedback were assumed. The soft outputs for each user were obtained from the MMSE-BDFE and passed to the respective channel decoders. Finally, the decoded bits were directed toward the speech decoder, where the original speech information was reconstructed.

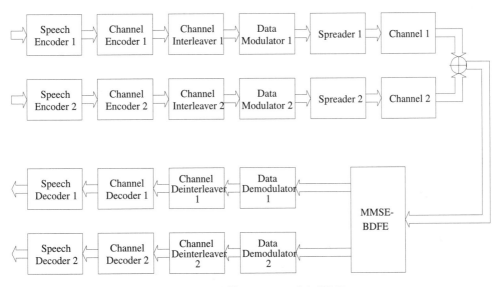

Figure 14.27 FRAMES-like two-user uplink CDMA system.

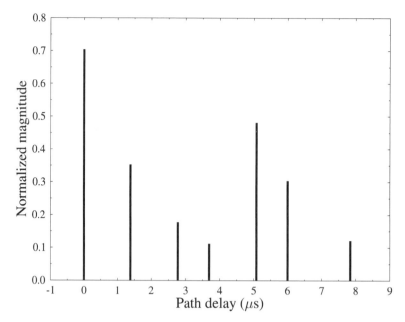

Figure 14.28 Normalized channel impulse response for a seven path Bad Urban channel [273].

14.8.6 System Performance

The BER performance of the proposed system is presented in Figures 14.29 and 14.30. Specifically, Figure 14.29 portrays the BER performance of a two-user JD-CDMA speech transceiver. Three different sets of results were obtained for the uncoded, turbo-coded, and nonsystematic convolutional-coded systems, respectively. As can be seen from the figure, channel coding substantially improved the BER performance of the system. However, in comparing the BER performances of the turbo-coded system and the convolutional-coded system, convolutional coding appears to offer a slight performance improvement over turbo coding. This can be attributed to the fact that a short turbo interleaver was used in order to maintain a low speech delay, while the nonsystematic convolutional codec exhibited better distance properties. It is well understood that turbo codecs achieve an improved performance in conjunction with long turbo interleavers. However, due to the low bitrate of the speech codec, 47 bits per 20 ms were generated. Hence, we were constrained to use a low interleaving depth for the channel codecs, resulting in a slightly superior convolutional coding performance.

In Figure 14.30, the results were obtained by varying the number of users in the system between $K = 2$ and 6. The BER performance of the system degrades only slightly, when the number of users is increased. This is due to employment of the joint detection receiver, which mitigates the effects of multiple access interference. It should also be noted that the performance of the system for $K = 1$ is also shown and that the BER performances for $K = 2$ to 6 degrade only slightly from this single-user bound.

The SEGSNR and CD objective speech measures for the decoded speech bits are depicted in Figure 14.31, where the turbo-coded and convolutional-coded systems were compared for $K = 2$ users. As expected on the basis of our BER curves, the convolutional codecs result in lower speech quality degradation compared to the turbo codes, which were

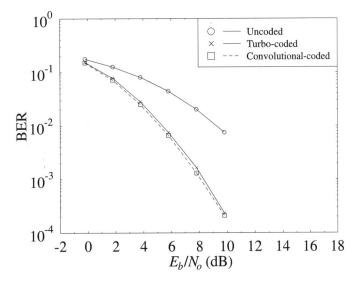

Figure 14.29 Comparison of the BER performance of an uncoded, convolutional-coded and turbo-coded two-user CDMA system, employing half-rate, constraint-length three constituent codes.

constrained to employ a low interleaver depth. Similar findings were also observed in these figures for $K = 4$ and 6 users. Again, the speech performance of the system for different number of users is similar, demonstrating the efficiency of the JD-CDMA receiver.

14.8.7 Conclusions on the JD-CDMA Speech Transceiver

The encoded speech bits generated by the 2.35 kbps prototype waveform interpolated (PWI) speech codec were half-rate channel-coded and transmitted using a DS-CDMA

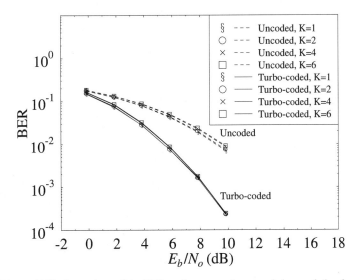

Figure 14.30 Comparison of the BER performance of an uncoded, convolutional-coded and turbo-coded CDMA system for $K = 2$, 4, and 6 users.

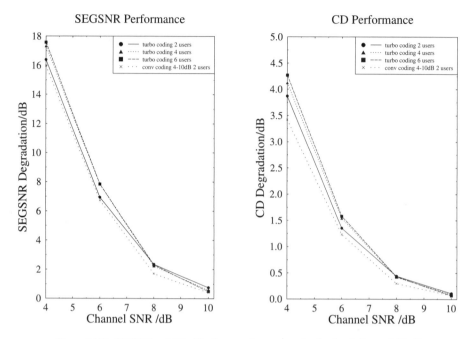

Figure 14.31 SEGSNR and CD objective speech measures for the decoded speech bits for
$K = 2$, 4, and 6 users.

scheme. At the receiver, the MMSE-BDFE multiuser joint detector was used in order to detect
the information bits, which were then channel-decoded and passed on to the speech decoder.
In our work, we compared the performance of turbo codes and convolutional codes. We found
that the convolutional codes outperformed the more complex turbo codes in terms of their
BER performance, as well as in speech SEGSNR and CD degradation terms. This was due to
the short interleaver constraint imposed by the low speech delay requirement, since turbo
codes require a high interleaver length in order to perform effectively. It was also shown that
system performance was only slightly degraded, as the number of users increased from $K = 2$
to 6, demonstrating the efficiency of the JD-CDMA scheme.

14.9 CHAPTER SUMMARY

This chapter has investigated the performance of MMBE when added to the LPC vocoder of
Chapter 11 and the PWI-ZFE coder of Chapter 13. Initially, an overview of MBE was given,
followed by detailed descriptions of the MMBE in both the encoder and decoder, given in
Section 14.4 and 14.5, respectively.

Section 14.6.1 gave a detailed analysis of 2-band and 5-band MMBE added to the LPC
vocoder, with Section 14.6.2 containing the analysis of 3-band MMBE added to the PWI-ZFE
coder. The 5-band MMBE LPC vocoder and the 3-band MMBE PWI-ZFE coder operated at
similar bitrates. Hence, they were compared through informal listening. It was found that the
3-band MMBE PWI-ZFE coder offered the best natural speech quality. The corresponding
time- and frequency-domain waveforms of our coders investigated so far were summarized
consistently using the same 20 ms speech frames. The associated figure numbers are detailed
in Table 16.2.

15

Sinusoidal Transform Coding Below 4 kbps

15.1 INTRODUCTION

In Chapters 13 and 14 the low-bitrate coding techniques of PWI and MMBE were described in detail. In this chapter we investigate a third speech coding technique, sinusoidal transform coding (STC), which similarly to PWI and MMBE is frequently employed at bitrates less than 4 kbps.

For STC it is assumed that both voiced and unvoiced speech can be represented by component frequencies that have appropriate amplitudes and phases, where these frequencies, amplitudes, and phases are determined by taking the short-term fourier transform (STFT) of a speech frame. Originally, employing the STFT of the speech waveform was proposed for the phase vocoder of Flanagan and Golden [340]. This phase vocoder synthesized speech by summing nominal frequencies, each with an associated amplitude and phase. The frequencies are taken at set intervals determined by the fixed number of channels employed in the phase vocoder. In its current most popular format, STC was first proposed by McAulay and Quatieri [341], who suggested that "peak-picking" be performed on the STFT magnitude spectra in order to determine the speech waveform's component frequencies. These frequencies have associated amplitudes and phases that can be combined to reproduce the speech waveform, with the number of frequencies determined by the number of peaks in the STFT.

Initially, STC was seen as a method for producing mid-rate speech coders [342]. However, STC speech coders typically separate the speech into voiced and unvoiced components, which, due to the complexities of determining the pitch and to carrying voiced-unvoiced decisions, can degrade the quality of the synthesized speech. With the success of the CELP coders at medium bitrates [73, 76, 123], which employ their identical synthesis scheme for both voiced and unvoiced speech, STC has never excelled in terms of quality at medium bitrates. As emphasis in speech coding is shifted to bitrates less than 4 kbps, where CELP coders do not perform well, STC coders have been adapted to operate at these lower bitrates. Low bitrate speech coders typically divide the speech into voiced and unvoiced components, which is similar to the method that STC employs. A further review of low-bitrate STC is given in Section 15.4.

Together with the ability to perform speech compression, sinusoidal analysis and synthesis has also been employed successfully for speech modification, such as frequency modification [343, 344]. Instead of peak-picking, George and Smith used analysis-by-synthesis to determine the component frequencies, amplitudes, and phases, a method they found to be more accurate than peak-picking. In addition, a sinusoidal model has been implemented successfully for pitch determination [345].

This Chapter begins by detailing, in Section 15.2 and 15.3, the methods harnessed for sinusoidal analysis and synthesis of speech waveforms, respectively. In Section 15.4 techniques required to perform STC at low bitrates are investigated. Sections 15.6, 15.7, and 15.8 described the methods required to encode the component sinewave frequencies, amplitudes and phases. In Section 15.9 the pitch synchronous interpolation performed at the decoder is detailed. Finally, in Section 15.10 the performance of the PWI-STC coder is assessed.

15.2 SINUSOIDAL ANALYSIS OF SPEECH SIGNALS

Sinusoidal coders represent the speech using sinusoidal basis functions given by [341]:

$$s(n) = \sum_{k=1}^{K} A_k \cos(\omega_k n + \phi_k) \tag{15.1}$$

where A_k represents the amplitude of the kth sinewave, ϕ_k represents the phase of the kth sinewave, with ω_k representing the frequency of the kth sinewave, and finally K is the number of component sinewaves.

15.2.1 Sinusoidal Analysis with Peak-Picking

The STFT of the speech wave is found using [346]:

$$S(\omega) = \sum_{n=-N/2}^{N/2} w(n)s(n)e^{-jn\omega} \tag{15.2}$$

where $w(n)$ is a Hamming window. In the frequency domain, the magnitude spectrum contains peaks at ω_m, with $A_m = |S(\omega_m)|$ and $\phi_m = argS(\omega_m)$.

STC operates on frames of speech, where during these frames the speech is assumed stationary. Thus the frame length must be sufficiently short to obey this assumption. The first sinusoidal speech coders [341] used speech frames of 10 ms, or 80 samples at a sampling rate of 8 kHz. The analysis window for the 512 sample-length STFT was set to 20 ms in order to incorporate the effect of a Hamming window, employed to reduce the Gibb's phenomenon. McAulay and Quatieri found that ideally the analysis window should be 2.5 × *pitch period* [346]. However, for simplicity a fixed window of 20 ms was adopted.

Figure 15.1 demonstrates the STFT for a voiced and unvoiced segment of speech. The peaks in the amplitude spectrum are highlighted by the crosses, with the corresponding phase also identified, where the phases are modulo 2π from $-\pi$ to π. The sinewaves, which constitute the speech signal, can be determined by locating the peaks in the frequency-domain magnitude spectrum [341]. In Figure 15.1 these frequencies were located using the peak-picking method.

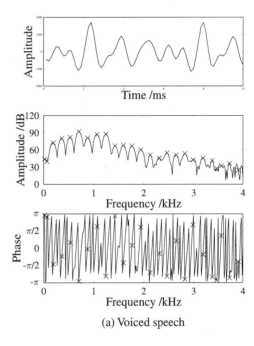

(a) Voiced speech

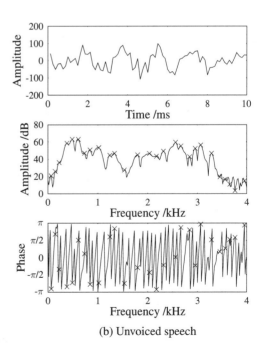

Figure 15.1 Example of sinusoidal analysis performed on (a) voiced and (b) unvoiced speech. The voiced speech segment is back vowel /ɔ/ in 'dog' for the AF1, while the unvoiced speech segment is the stop /k/ in 'kitten' also for the AF1. Together with the time-domain waveform, the amplitude and phase of the frequency spectrum are shown.

(b) Unvoiced speech

15.2.2 Sinusoidal Analysis Using Analysis-by-Synthesis

George and Smith [343, 344], proposed an alternative method to peak-picking in which the frequencies, amplitudes, and phases are determined using analysis-by-synthesis. As mentioned earlier, the peak-picking method assumes stationarity over the analysis period. If this stationarity constraint is not obeyed, then the STFT peaks will not be the optimum values to represent the speech waveform. Moreover, spectral interference in the STFT, caused by the windowing of the input speech data, can also affect the STFT peak values. Thus, the analysis-by-synthesis method improves the accuracy of the frequencies, amplitudes, and phases used to represent the speech waveform.

If the speech waveform is represented by Equation 15.1, the error signal from the modeling process is given by:

$$e(n) = s(n) - \sum_{k=1}^{K} A_k \cos(\omega_k n + \phi_k). \tag{15.3}$$

If the error signal is found iteratively, then:

$$e^{k+1} = e^k(n) - A_{k+1} \cos(\omega_{k+1} n + \phi_{k+1}) \tag{15.4}$$

where $e^0(n) = s(n)$. In order to minimize the error signal after sinusoidal modeling we take the minimum mean squared error signal given by:

$$E^{k+1} = \sum_{n=0}^{FL} [e^{k+1}(n)]^2 = \sum_{n=0}^{FL} [e^k(n) - A_{k+1} \cos(\omega_{k+1} n + \phi_{k+1})]^2. \tag{15.5}$$

A variation of this equation is minimized later in Section 15.7.2. Because of the enhanced accuracy of the analysis-by-synthesis procedure, as demonstrated later, it will be harnessed for determining the component frequencies, amplitude, and phases. In addition, the analysis-by-synthesis process should enable a weighted LPC synthesis filter to be included, when representing the LPC residual waveform by sinusoidal modeling.

Following this review of sinusoidal analysis for speech waveforms, where the component frequencies, amplitudes, and phases have been located, the process of resynthesizing the speech is now examined.

15.3 SINUSOIDAL SYNTHESIS OF SPEECH SIGNALS

Sinusoidal functions, as described in Equation 15.1, can also be employed to synthesize a speech waveform where the required constituent frequencies, amplitudes, and phases have been determined through analysis. However, if the synthesized speech frames are concatenated, with no smoothing at the frame boundaries, the resultant discontinuities will be audible in the synthesized speech.

15.3.1 Frequency, Amplitude, and Phase Interpolation

Initially, in order to overcome the discontinuity, it was proposed that every frequency, amplitude, and phase in a frame should be matched to a frequency, amplitude, and phase in the adjacent frame [341], thus performing smoothing at the frame boundaries. If an equal number of corresponding frequencies occur in all frames, the matching process is reasonably simple. However, when the number of frequencies in the sinusoidal synthesis differs between frames, then the matching process between these adjacent frames becomes more complex. In this "differing number of frequencies" scenario, the matching process involves the "birth" and "death" of sinusoids. The "birth" process occurs when an extra frequency appears in the sinusoidal representation of the speech waveform, where consequently the additional frequency must be incorporated into the matching process. The "death" process is initiated when a frequency in the current frame has no counterpart in the subsequent frame. The complexity in this interpolation process arises from the decision as to whether a frequency undergoes a birth or death process, or it is matched to a frequency in the adjacent frame. Following frequency interpolation, the corresponding amplitudes can be linearly interpolated. However, due to the modulo 2π nature of the phase values, they must be "unwrapped" before interpolation can occur.

15.3.2 Overlap-Add Interpolation

In order to circumvent the elaborate frequency matching process, sinusoidal coders typically employ an overlap-add interpolator for removing the frame boundary discontinuities [343]. For the kth frame, the speech is synthesized according to:

$$\hat{s}^m(n) = \sum_{k=1}^{K} A_k^m \cos(n\omega_k^m + \phi_k^m). \tag{15.6}$$

The synthesized speech $\hat{s}^m(n)$ is determined for the range $0 \le n \le 2 \cdot N$, where N is the frame length. The overlap-add interpolator is employed to find the reconstructed speech, given by [343]:

$$\hat{s}(n) = w_s(n)\hat{s}^{m-1}(n+N) + (1 - w_s)\hat{s}^m(n) \tag{15.7}$$

where $w_s(n)$ is typically a triangular window [346] of the form:

$$w_s(n) = 1 - \frac{n}{N}, \qquad 0 \le n \le N. \tag{15.8}$$

Thus, the synthesized speech is constructed from the windowed sinusoidal representation of the previous frame interpolated into the current frame, together with the windowed sinusoidal representation of the current frame. Figure 15.2 demonstrates the overlap-add interpolator harnessed in the sinusoidal coder to provide smoothing at the frame boundaries.

Observing Figure 15.2, the synthesized speech is a high-quality reproduction of the original speech. In addition, the previous frame's sinusoids contribute most to the synthesized speech shape at the beginning of the frame, while the current sinusoids predominantly contribute to the end of the current frame. Each set of sinusoids contributes to both the current and next speech frame. Thus, we are assuming stationarity of the speech signal over an interval of $2 \cdot N$.

For sinusoidal coders, the major assumption used is that the speech remains stationary over the analysis window, with the validity of this assumption particularly questionable for periods of voicing onset. Figure 15.3 displays a rapidly evolving voicing onset waveform, which together with the displayed synthesized speech characterizes the performance of sinusoidal analysis and synthesis at voicing onset. It can be seen that the quality of the reconstructed speech waveform is significantly degraded when compared to Figure 15.2. The reconstructed speech contains too much voicing and also produces a smoother evolution from unvoiced to voiced speech than the original waveform.

Having discussed the processes of a sinusoidal coder, we now investigate methods of implementation which will allow a low-bitrate sinusoidal coder to be constructed.

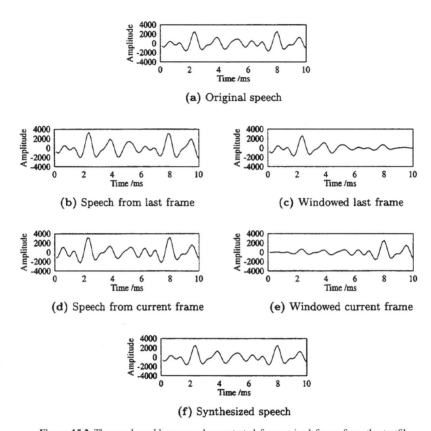

Figure 15.2 The overlap-add process demonstrated for a voiced frame from the testfile
AF1 when uttering back vowel /ɔ/ in "dog." Graph (a) represents the original
speech waveform, with graphs (b) and (d) synthesizing the speech based on
the previous and current sinusoids, respectively. Graphs (c) and (e) are the
windowed synthesized speech, with Graph (f) displaying the resultant overlap-add synthesized speech waveform.

15.4 LOW-BITRATE SINUSOIDAL CODERS

Sinusoidal coders have previously been adapted to operate at low bitrates [287, 347–349],
notably for the DoD 2.4 kbps speech coder competition [282, 283] described in Section
10.1.2, although the winning coder did not operate on the basis of sinusoidal principles. Low-
bitrate sinusoidal coders frequently employ MBE techniques [350–352]. Indeed, these two
forms of harmonic coding become conceptually rather similar at low bitrates. PWI techniques
have also been combined with STC [353].

   In low-bitrate sinusoidal coders, the bitrate is often reduced to 2.4 kbps by assuming a
zero phase sinusoidal model [287. Explicitly, at the decoder the location of the pitch pulses is
determined using the pitch period and the previous pitch pulse location, but small perturba-
tions are not encoded for transmission. The removal of phase encoding reduces the
naturalness of the synthesized speech, particularly introducing "buzziness" for unvoiced
regions. Some sinusoidal coders overcome this effect by introducing phase dispersion when
required [283, 349].

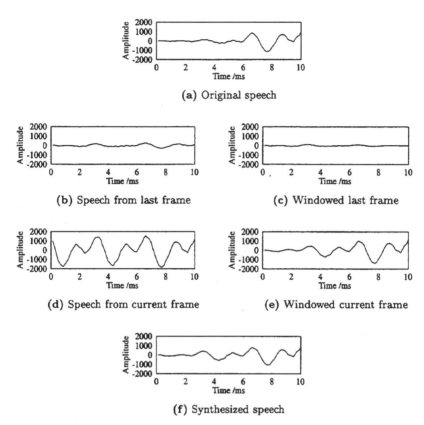

(a) Original speech

(b) Speech from last frame

(c) Windowed last frame

(d) Speech from current frame

(e) Windowed current frame

(f) Synthesized speech

Figure 15.3 Example of sinusoidal analysis and synthesis for a voicing onset from the testfile AF1 when uttering the front vowel /e/ in "chased." Graph (a) displays the original rapidly evolving voiced onset speech waveform. Graphs (b) and (c) contain the speech and windowed speech created from the previous frame's sinusoids, which are predominantly voiced. Graphs (d) and (e) display the speech and windowed speech from the current sinusoids, producing a voiced waveform. Finally, Graph (e) shows the overlap-add synthesized speech waveform.

For voiced speech, in order to reduce the required transmitted bitrate, the sinusoidal model used in the synthesis is assumed to be harmonic, which is given by [287]:

$$\hat{s}^m(n) = \sum_{k=1}^{K} A_k^m \cos(nk\omega_0^m + \phi_k^m) \tag{15.9}$$

where ω_0^m is the fundamental frequency associated with the mth frame and K represents the number of harmonics to be modeled.

In order to represent the unvoiced component of the speech, a MMBE scheme, similar to Chapter 14, can be invoked, allowing voicing information to be transmitted. It subsequently permits the decoder to mix voiced and unvoiced sounds [347].

For low-bitrate speech coders, the amplitudes associated with each frequency are typically encoded by one of two methods, namely, as LPC coefficients or with vector quantization. The number of amplitude values required depends on the pitch period. Thus, if the amplitudes are directly quantized, methods that allow using different-length amplitude vectors must be employed.

As this description makes clear, similarly to the PWI and MBE low-bitrate coders of Chapters 13 and 14, the determination of the pitch period is vital for the successful operation of the speech coder. Following this review of low bitrate sinusoidal speech coders the sinusoidal coding philosophy described in Sections 15.2 and 15.3 is adapted to become a practical low bitrate speech coder.

15.4.1 Increased Frame Length

The sinusoidal coder of Sections 15.2 and 15.3 operated on 10 ms frames in order to ensure stationarity over the analysis speech. However, for a low-bitrate scheme, this fast parameter update rate is not feasible. The low-bitrate coders explored in Chapters 13 and 14 operated on 20 ms frames. Hence the frame length of the sinusoidal coder was extended to 20 ms. Correspondingly, the analysis window was also increased, to 30 ms, for invoking the Hamming window before the STFT. As expected, audibly the increased frame length increases the background noise and reduces the naturalness of the synthesized speech.

As stated earlier the Hamming window before the STFT was extended to 30 ms; similarly, the overlap-add window must also be extended. However, a significant problem with the increased frame length is the assumption that, due to the triangular overlap-add window, the speech is stationary for twice the frame length, namely, 40 ms. Thus, an alternative overlap-add window was investigated, and the trapezoidal window as shown in Figure 15.4 was adopted for our coder, which is seen to impose less stringent stationarity requirements.

15.4.2 Incorporating Linear Prediction Analysis

For low-bitrate sinusoidal coders, typically two methods are employed for encoding the sinusoidal amplitudes. The first is to directly vector-quantize the sinusoidal amplitudes. However, the number of amplitudes depends on the pitch period of the speech waveform. Thus, initially the amplitude vector would have to be transformed to a predefined length. The second method involves performing linear prediction on the speech waveform, with the LP coefficients describing the sinusoidal amplitudes, or the associated spectral envelope. Thus, ideally the residual signal will have a constant magnitude spectrum. Since LP analysis has already been implemented for the PWI and MMBE coders of Chapters 13 and 14,

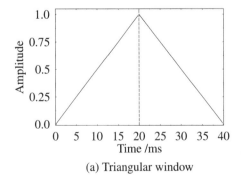

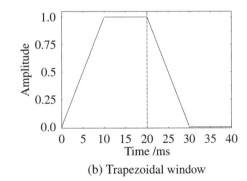

(a) Triangular window   (b) Trapezoidal window

Figure 15.4 The two overlap-add windows investigated for the STC, namely, (a) triangular window and (b) trapezoidal window.

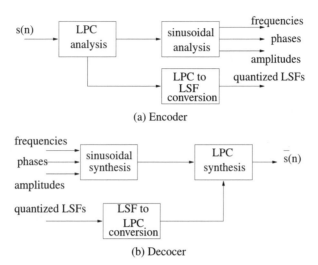

(a) Encoder

(b) Decocer

Figure 15.5 Schematic of the encoder and decoder for a STC employing LP analysis to encode the sinusoidal amplitudes.

respectively, this method was also adopted for encoding the sinusoidal amplitudes. The encoder and decoder schematics for an STC incorporating LP analysis are displayed in Figure 15.5 and they are described next.

At the encoder, the LP analysis is performed on the speech waveform, thus removing short-term redundancies and encoding the amplitudes, or the associated spectral envelope, of the sinusoids (see Figure 15.5). The LPC coefficients are transformed to LSFs (as described in Section 11.2.2), which are vector-quantized with 18 bits per frame. The remaining STP LPC residual waveform undergoes sinusoidal analysis for determining the underlying frequencies, amplitudes, and phases. At the decoder, the LSFs are converted to LP coefficients, while the frequencies, amplitudes, and phases reconstruct the LPC excitation using Equation 15.6. Finally, the excitation is passed through the LPC synthesis filter in order to synthesize the speech waveform.

Figure 15.6 demonstrates the associated waveforms together with the corresponding STFT magnitude and phase spectra. The upper trace displays the speech waveform. The second trace demonstrates the LPC STP residual, which highlights the failure in the assumption that LP analysis produces a constant-amplitude residual signal across the frequency domain.

15.5 INCORPORATING PROTOTYPE WAVEFORM INTERPOLATION

The longer frame length of 20 ms, introduced to create a low-bitrate STC coder, will reduce the accuracy of the sinusoidal amplitude and phase values, due to the increased length over which stationarity is assumed. This effect can be removed by introducing prototype waveform interpolation (PWI), so that each set of sinusoidal excitation parameters represents a pitch period. Thus, the speech is assumed to be stationary over a length of two pitch periods. The schematic for the prototype waveform interpolation sinusoidal transform coding (PWI-STC) is given in Figure 15.7. Initially, the LPC coefficients are determined for the speech frame with the LPC STP residual waveform generated. The LPC coefficients are transformed to

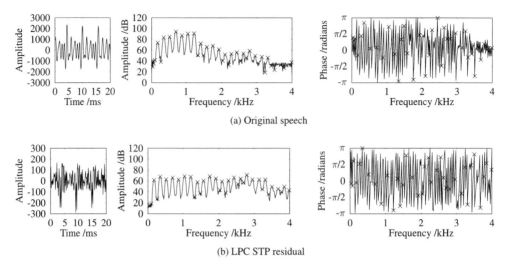

Figure 15.6 An example of STC-LPC analysis for the voiced speech utterance constituted by the back vowel /ɔ/ in "dog" for the testfile AF1, with LPC analysis incorporated to find the LPC STP residual. The amplitude peaks and corresponding phases are highlighted by crosses.

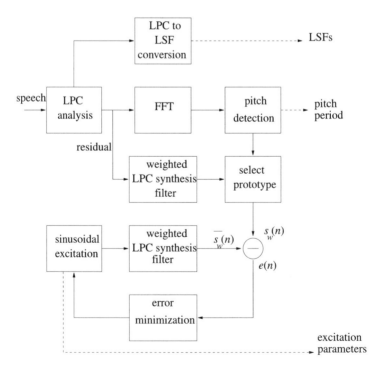

Figure 15.7 Schematic for the STC encoder.

LSFs and then vector-quantized with 18 bits/20 ms [123]. The fast fourier transform (FFT) of this residual waveform is used for pitch detection, where its preference to the wavelet assisted autocorrelation-based pitch detector of Section 12.5.2 is described in Section 15.6. The LPC STP residual is also passed through a weighted LPC synthesis filter, and a weighted speech prototype segment is determined following the principles of Section 13.4.2 and Figure 13.6. This prototype segment is then used in the analysis-by-synthesis loop, where the best sinusoidal excitation is selected by comparing the synthetic speech $\bar{s}_w(n)$ to the weighted prototype $s_w(n)$.

15.6 ENCODING THE SINUSOIDAL FREQUENCY COMPONENT

The sinusoidal frequencies are important to the successful operation of STC, since they indicate the component frequencies of the speech waveform. The most efficient way of encoding the frequencies is to constrain them to be multiples of the pitch period determined for the frame.

The pitch-period detector harnessed [287] creates the STFT magnitude spectra of the synthetic excitation for every permissible pitch period, which are the integer pitch periods from $20 \rightarrow 147$ samples. For each of these synthetic excitation magnitude spectra, the spectral distance from the original LPC STP residual waveform magnitude spectra is calculated, with the closest match selected as the pitch period. For voiced speech segments it is the true pitch period which will, typically, produce the closest spectral match to the original LPC STP residual spectrum. However, unvoiced speech segments have no pitch period. Thus, typically, a long pitch period is selected, associated with a low pitch and densely spaced pitch harmonics, since this best represents the noise-like unvoiced spectrum. This process follows the principles of LTP (described in Section 10.2) and is used in CELP coders to remove the pitch-related residual pulses. Unlike LTP, however, it is performed in the frequency domain, as highlighted below.

The pitch-period detectors investigated in Chapters 11 and 12 operated in the time domain. The most successful wavelet-based autocorrelation pitch detector described in Section 12.5.2) produced an overall pitch determination error rate of 3.9%. The frequency-domain method mentioned earlier produced an overall error rate of 4.6%, with the percentage of missed unvoiced frames $w_u = 1.2\%$, the percentage of missed voiced frames $w_v = 0.6\%$, and the percentage of gross pitch error $P_g = 2.8\%$. The frequency-domain pitch detector operates by determining which set of harmonic frequencies best represents the STFT of the LPC STP residual, thereby allowing the best harmonic sinusoidal excitation to be determined for the frame. Thus, despite its higher error rate, this frequency-domain method was adopted for pitch determination within this Chapter.

Described in more depth, the frequency-domain pitch detector selects the candidate pitch period that minimizes the error between the LPC STP residual STFT magnitude spectra and its harmonic-related pitch-based replica. However, since most noise occurs in the upper frequency regions, only the harmonics beneath 1 kHz are used in the minimization, formulated as:

$$|E(\omega)| = \sum_{\omega=0}^{\pi/4} \left[|R(\omega)| - G|P(\omega)|\right]^2 \qquad (15.10)$$

where $|E(\omega)|$ is the minimum mean squared error between the original and pitch related harmonic residual magnitude spectrum, $|R(\omega)|$ is the LPC STP residual magnitude spectrum, $|P(\omega)|$ is the magnitude spectrum of the candidate pitch related excitation whose fourier transform pair is $p(t) = \sum_{m=0}^{M} \cos(nm\omega_0 - \phi_m)$, G is the gain associated with the pitch period, ω is the normalized frequency, and $\pi/4$ represents the frequencies up to 1 kHz, since 2π corresponds to 8 kHz. In order to determine the gain, we differentiate $|E(\omega)|$ with respect to G, yielding:

$$\frac{\delta|E(\omega)|}{\delta G} = -2 \sum_{\omega=0}^{\pi/4} |P(\omega)|[|S(\omega)| - G|P(\omega)|] = 0, \tag{15.11}$$

which produces a gain value of:

$$G = \frac{\sum_{\omega=0}^{\pi/4} |P(\omega)||S(\omega)|}{\sum_{\omega=0}^{\pi/4} |P(\omega)|^2}. \tag{15.12}$$

The corresponding best pitch period is found by substituting G into Equation 15.10. Thus when:

$$G = \frac{[\sum_{\omega=0}^{\pi/4} |P(\omega)||S(\omega)|]^2}{\sum_{\omega=0}^{\pi/4} |P(\omega)|^2} \tag{15.13}$$

is maximized, $|E(\omega)|$ in Equation 15.10 is minimized.

Figure 15.8 demonstrates the pitch detector's operation for a voiced speech frame, where Figure 15.8a contains the LPC STP residual together with its STFT magnitude spectra, while Figure 15.8b displays the selected pitch period, from which it can be seen that a good match has been found.

For unvoiced frames, typically a long pitch period is selected, creating many densely spaced frequency harmonics, which produces perceptually acceptable unvoiced speech.

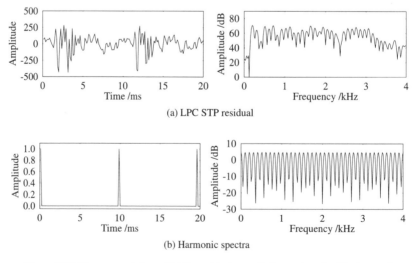

Figure 15.8 The representation of (a) the residual magnitude spectrum by (b) a harmonic pitch-period spectrum. The voiced speech segment is the glide /w/ in "*wide*" for the testfile AM2.

15.7 DETERMINING THE EXCITATION COMPONENTS

Every harmonic frequency found in Section 15.6 will have a corresponding amplitude and phase component. Based on a permissible pitch period of 20 to 147 samples, or 54 to 400 Hz, there can be between 10 and 80 corresponding amplitude and phase values in the 4 kHz range. In Section 15.2 peak-picking and analysis-by-synthesis were suggested for this task and here they are investigated in more depth.

15.7.1 Peak-Picking of the Residual Spectra

The peak-picking process described in Section 15.2.1 was implemented in order to determine the amplitudes and phases related to the selected harmonic frequencies. The performance of the peak-picking process was assessed by comparing the original prototype segment with the synthesized prototype segment, where the sinusoidal excitation components remained unquantized. For the peak-picking process, a SEGSNR of 4.8 dB was found from the above comparison.

15.7.2 Analysis-by-Synthesis of the Residual Spectrum

In Section 15.2.2 the amplitudes and phases were located by minimizing Equation 15.5, while in Section 15.4.2 LPC analysis was introduced to represent the amplitudes so that the sinusoids reconstructed the LPC STP residual waveform. Before the sinusoidal parameters are determined, the equation for representing the speech waveform is rewritten in order to include the fundamental frequency already determined. Thus, the speech waveform is synthesized by:

$$\tilde{r}(n) = \sum_{k=1}^{K} A_k \cos(nk\omega_0 + \phi_k), \tag{15.14}$$

where ω_0 is the fundamental frequency and $\tilde{r}(n)$ is the reconstructed residual waveform. Following a similar format to the ZFE minimization process in Section 13.3.1, the original speech waveform can be represented by:

$$\tilde{s}(n) = \sum_{k=1}^{K} A_k \cos(nk\omega_0 + \phi_k) * h(n), \tag{15.15}$$

where $h(n)$ is the impulse response of the LPC STP synthesis filter.

With this new representation of the speech waveform, the minimum mean squared error becomes:

$$E_{k+1} = \sum_{n=0}^{P-1} [e_{k+1}(n)]^2 = \sum_{n=0}^{P-1} [e_k(n) - A_{k+1} \cos(n(k+1)\omega_0 + \phi_{k+1}) * h(n)]^2, \tag{15.16}$$

where for $k = 0$ we have $e_0(n) = s(n) - m(n)$, with $m(n)$ being the synthesis filter memory, which similarly to Figure 13.8 in Section 13.5 is taken from the previous prototype segment and P being the analysis length, namely the length of the prototype segment which is the pitch period. This equation can be further simplified to give:

$$E_{k+1} = \sum_{n=0}^{P-1} [e_k(n) - (a_{k+1} \cos(\omega_{k+1}(n)) * h(n) + b_{k+1} \sin(\omega_{k+1}(n)) * h(n))]^2, \tag{15.17}$$

where $\omega_{k+1}(n) = n(k+1)\omega_0$, $a_{k+1} = A_{k+1}\cos\phi_{k+1}$ and $b_{k+1} = -A_{k+1}\sin\phi_{k+1}$ [343]. Differentiating Equation 15.17 with respect to a_{k+1} and setting the expression equal to zero, we find that:

$$\frac{\delta E_{k+1}}{\delta a_{k+1}} = 2 \cdot \sum_{n=0}^{P-1} [\cos(\omega_{k+1}(n)) * h(n)] \times [e_k(n)$$
$$- (a_{k+1}\cos(\omega_{k+1}(n)) * h(n) + b_{k+1}\sin(\omega_{k+1}(n)) * h(n))] = 0. \quad (15.18)$$

Similarly to George and Smith [343], we define:

$$\gamma_{11} = \sum_{n=0}^{P-1} [\cos(\omega_{k+1}(n)) * h(n)]^2 \quad (15.19)$$

$$\gamma_{22} = \sum_{n=0}^{P-1} [\sin(\omega_{k+1}(n)) * h(n)]^2 \quad (15.20)$$

$$\gamma_{12} = \sum_{n=0}^{P-1} [\cos(\omega_{k+1}(n)) * h(n)] \times [\sin(\omega_{k+1}(n)) * h(n)] \quad (15.21)$$

$$\psi_1 = \sum_{n=0}^{P-1} e_k(n) \times [\cos(\omega_{k+1}(n)) * h(n)] \quad (15.23)$$

$$\psi_2 = \sum_{n=0}^{P-1} e_k(n) \times [\sin(\omega_{k+1}(n)) * h(n)]. \quad (15.23)$$

By substituting Equations 15.19, 15.21, and 15.22 into Equation 15.18, we arrive at:

$$a_{k+1} \cdot \gamma_{11} + b_{k+1} \cdot \gamma_{12} = \psi_1 \quad (15.24)$$

Similarly, if we differentiate Equation 15.17 with respect to b_{k+1} and set it to zero, then:

$$\frac{\delta E_{k+1}}{\delta b_{k+1}} = 2 \cdot \sum_{n=0}^{P-1} [\sin(\omega_{k+1}(n)) * h(n)]$$
$$\times [e_k(n) - (a_{k+1}\cos(\omega_{k+1}(n)) * h(n) + b_{k+1}\sin(\omega_{k+1}(n)) * h(n))] = 0. \quad (15.25)$$

By substituting Equations 15.20, 15.21, and 15.23 into Equation 15.25, we achieve:

$$a_{k+1} \cdot \gamma_{12} + b_{k+1} \cdot \gamma_{22} = \psi_2. \quad (15.26)$$

Solving these, the two simultaneous equations produce values for a_{k+1} and b_{k+1}, which are given by:

$$a_{k+1} = \frac{\gamma_{22} \cdot \psi_1 - \gamma_{12} \cdot \psi_2}{\gamma_{11} \cdot \gamma_{22} - \gamma_{12}^2}. \quad (15.27)$$

$$b_{k+1} = \frac{\gamma_{11} \cdot \psi_2 - \gamma_{12} \cdot \psi_1}{\gamma_{11} \cdot \gamma_{22} - \gamma_{12}^2}. \quad (15.28)$$

From these a_{k+1} and b_{k+1} values the amplitudes and phases can be found using $A_{k+1} = \sqrt{a_{k+1}^2 + b_{k+1}^2}$ and $\phi_{k+1} = -\arctan(b_{k+1}/a_{k+1})$. Since the pitch value of ω_0 has already been found, then once Equations 15.19 and 15.23 have been constructed the a_{k+1} and b_{k+1} values can be found and A_{k+1} and ϕ_{k+1} can be determined.

Adopting the analysis-by-synthesis approach in the PWI-STC coder, when the original and synthesized prototype segments were compared, a SEGSNR of 10.9 dB was achieved.

15.7.3 Computational Complexity

The computational complexity of the PWI-STC encoder will be dominated by the analysis-by-synthesis process. In turn, the complexity of the analysis-by-synthesis process is dependent on the pitch period and the convolution processes, as suggested by Equations 15.19 to 15.23. The complexity of the analysis-by-synthesis is given in Table 15.1, where it can be seen that for a pitch period of 147 samples, corresponding to 80 harmonics, the required complexity of 140 MFLOPS is prohibitive.

15.7.4 Reducing the Computational Complexity

When George and Smith [343, 344] adopted analysis-by-synthesis for improving the STC parameter representation, they reduced the complexity of the process by incorporating the discrete fourier transform (DFT). Here the DFT can also be employed for complexity reduction However, the process is different from that of George and Smith due to the addition of the LPC synthesis filter $h(n)$, together with the lack of the slowly varying amplitude function $g(n)$ from Reference [343].

The M-point DFT for the sequence $x(n)$ is defined by:

$$X(m) \equiv \sum_{n=0}^{N-1} x(n) W_M^{mn}, \qquad 0 \leq m < M \tag{15.29}$$

where $W_M^{mn} = e^{-j(2\pi/M)mn}$, and for completeness the inverse DFT is given by:

$$x(n) = \frac{1}{M} \sum_{n=0}^{N-1} X(m) W_m^{-mn}. \tag{15.30}$$

The DFT is incorporated in order to reduce the complexity with the statements that [343]:

$$\sum_{n=0}^{N-1} x(n) \cos\left[\left(\frac{2\pi}{M}\right)mn\right] = \text{Re}[X(m)] \tag{15.31}$$

$$\sum_{n=0}^{N-1} x(n) \sin\left[\left(\frac{2\pi}{M}\right)mn\right] = -\text{Im}[X(m)]. \tag{15.32}$$

TABLE 15.1 Computational Complexity for Error Minimization in the PWI-STC Encoder

Procedure	Pitch period = 20/MFLOPS	Pitch period = 147/MFLOPS
Number of Harmonics	10	80
Colvolve sin and $h(n)$	0.01	0.55
Colvolve cos and $h(n)$	0.01	0.55
Calculate $\gamma_{11}, \gamma_{22}, \gamma_{12}, \psi_1, \psi_2$	0.01	0.10
Updating $e^{k+1}(n)$	0.02	0.59
Total for each harmonic	0.05	1.79
Overall total	0.50	140.0

Then it can be noted that the DFT of the error signal $e_k(n)$ and that of the LPC synthesis filter impulse response $h(n)$ are given by:

$$E_k(m) = \sum_{n=0}^{N-1} e_k(n) W_M^{mn} \qquad (15.33)$$

$$H(m) = \sum_{n=0}^{N-1} h(n) W_M^{mn}. \qquad (15.34)$$

With the examination of ψ_1 from Equation 15.22, a reduction in its complexity can be achieved, if initially the convolution is rewritten as a summation:

$$\psi_1 = \sum_{n=0}^{N-1} e_k(n) \times \sum_{m=0}^{M-1} h(m) \cdot \cos(\omega_{k+1}(n - m)). \qquad (15.35)$$

In order to allow the summations to become separable, the trigonometric identity $\cos(x + y) = \cos(x)\cos(y) - \sin(x)\sin(y)$ is harnessed, simplifying the equation to:

$$\psi_1 = \sum_{n=0}^{N-1} e_k(n) \cdot \cos(\omega_{k+1}(n)) \times \sum_{m=0}^{M-1} h(m) \cdot \cos(\omega_{k+1}(m))$$
$$+ \sum_{n=0}^{N-1} e_k(n) \cdot \sin(\omega_{k+1}(n)) \times \sum_{m=0}^{M-1} h(m) \cdot \sin(\omega_{k+1}(m)). \qquad (15.36)$$

Utilizing from Section 15.7.2 that $\omega_{k+1}(n) = n(k + 1)\omega_0$ together with the relationship $\omega_0 = 2\pi/P$, where P is the prototype segment length that is equal to N and M, allows $w_{k+1} = 2n\pi i/M$ to be defined. Equations 15.33 and 15.34 can then be employed to define:

$$\psi_1 = \mathrm{Re}(E_k(i)) \cdot \mathrm{Re}(H(i)) + \mathrm{Im}(E_k(i)) \cdot \mathrm{Im}(H(i)). \qquad (15.37)$$

Similarly, ψ_2 defined in Equation 15.23 can be rewritten using the trigonometric identity $\sin(x + y) = \sin(x)\cos(y) + \cos(x)\sin(y)$, giving:

$$\psi_2 = \sum_{n=0}^{N-1} e_k(n) \cdot \sin(\omega_{k+1}(n)) \times \sum_{m=0}^{M-1} h(m) \cdot \cos(\omega_{k+1}(m))$$
$$- \sum_{n=0}^{N-1} e_k(n) \cdot \cos(\omega_{k+1}(n)) \times \sum_{m=0}^{M-1} h(m) \cdot \sin(\omega_{k+1}(m)) \qquad (15.38)$$

which when described in terms of the DFTs of $E_k(m)$ and $H(m)$ becomes:

$$\psi_2 = \mathrm{Re}(E_k(i)) \cdot \mathrm{Im}(H(i)) - \mathrm{Im}(E_k(i)) \cdot \mathrm{Re}(H(i)). \qquad (15.39)$$

The term γ_{11} from Equation 15.19 can also be simplified by replacing the convolution term by a summation, becoming:

$$\gamma_{11} = \sum_{n=0}^{N-1} \left[\sum_{m=0}^{M-1} h(m) \cdot \cos(\omega_{k+1}(n - m)) \right]^2 \qquad (15.40)$$

Separating the n and m components, the expression becomes:

$$\gamma_{11} = \sum_{n=0}^{N-1} \cos^2(\omega_{k+1}(n)) \times \left[\sum_{m=0}^{M-1} h(m) \cdot \cos(\omega_{k+1}(m)) \right]^2$$

$$+ \sum_{n=0}^{N-1} \sin^2(\omega_{k+1}(n)) \times \left[\sum_{m=0}^{M-1} h(m) \cdot \sin(\omega_{k+1}(m)) \right]^2$$

$$+ \sum_{n=0}^{N-1} \cos(\omega_{k+1}(n)) \cdot \sin(\omega_{k+1}(n))$$

$$\times \left[\sum_{m=0}^{M-1} h(m) \cdot \cos(\omega_{k+1}(m)) \right] \cdot \left[\sum_{m=0}^{M-1} h(m) \cdot \sin(\omega_{k+1}(m)) \right]. \qquad (15.41)$$

This can be expressed in terms of $E_k(m)$ and $H(m)$ by:

$$\gamma_{11} = \sum_{n=0}^{N-1} \cos^2(\omega_{k+1}(n)) \cdot \mathrm{Re}(H(i))^2 + \sum_{n=0}^{N-1} \sin^2(\omega_{k+1}(n)) \cdot \mathrm{Im}(H(i))^2$$

$$- \sum_{n=0}^{N-1} \cos(\omega_{k+1}(n)) \cdot \sin(\omega_{k+1}(n)) \cdot \mathrm{Re}(H(i)) \cdot \mathrm{Im}(H(i)). \qquad (15.42)$$

Similarly, the expression for γ_{22} from Equation 15.20 can be simplified:

$$\gamma_{22} = \sum_{n=0}^{N-1} \sin^2(\omega_{k+1}(n)) \times \left[\sum_{m=0}^{M-1} h(m) \cdot \cos(\omega_{k+1}(m)) \right]^2$$

$$+ \sum_{n=0}^{N-1} \cos^2(\omega_{k+1}(n)) \times \left[\sum_{m=0}^{M-1} h(m) \cdot \sin(\omega_{k+1}(m)) \right]^2$$

$$- \sum_{n=0}^{N-1} \cos(\omega_{k+1}(n)) \cdot \sin(\omega_{k+1}(n))$$

$$- \left[\sum_{m=0}^{M-1} h(m) \cdot \cos(\omega_{k+1}(m)) \right] \cdot \left[\sum_{m=0}^{M-1} h(m) \cdot \sin(\omega_{k+1}(m)) \right], \qquad (15.43)$$

which can be rewritten as:

$$\gamma_{22} = \sum_{n=0}^{N-1} \sin^2(\omega_{k+1}(n)) \cdot \mathrm{Re}(H(i))^2 + \sum_{n=0}^{N-1} \cos^2(\omega_{k+1}(n)) \cdot \mathrm{Im}(H(i))^2$$

$$+ \sum_{n=0}^{N-1} \cos(\omega_{k+1}(n)) \cdot \sin(\omega_{k+1}(n)) \cdot \mathrm{Re}(H(i)) \cdot \mathrm{Im}(H(i)). \qquad (15.44)$$

Finally, the expression γ_{12} from Equation 15.21 can be written as:

$$\gamma_{12} = \sum_{n=0}^{N-1} \cos^2(\omega_{k+1}(n)) \cdot \mathrm{Re}(H(i)) \cdot \mathrm{Im}(H(i))$$

$$- \sum_{n=0}^{N-1} \sin^2(\omega_{k+1}(n)) \cdot \mathrm{Re}(H(i)) \cdot \mathrm{Im}(H(i))$$

$$+ \sum_{n=0}^{N-1} \cos(\omega_{k+1}(n)) \cdot \sin(\omega_{k+1}(n)) \cdot \mathrm{Re}(H(i))^2$$

$$- \sum_{n=0}^{N-1} \cos(\omega_{k+1}(n)) \cdot \sin(\omega_{k+1}(n)) \cdot \mathrm{Im}(H(i))^2. \qquad (15.45)$$

TABLE 15.2 Computational Complexity for Error Minimization in the PWI-STC Encoder

Procedure	*Pitch period = 20/MFLOPS*	*Pitch period = 147/MFLOPS*
Calculate DFT of $h(n)$	0.23	0.23
Number of harmonics	10	80
Calculate DFT of $e(n)$	0.23	0.23
Calculate $\gamma_{11}, \gamma_{22}, \psi_1, \psi_2$	—	—
Updating $e^{k+1}(n)$	0.01	0.07
Total for each harmonic	0.24	0.30
Overall total	2.63	24.23

The Equations 15.42, 15.44, and 15.45 contain summations involving only sin and cos functions. These functions can be rewritten using the identities given below:

$$\sum_{n=0}^{N-1} \cos^2(\omega_{k+1}(n)) = \frac{1}{2}\sum_{n=0}^{N-1} \cos(2\omega_{k+1}(n)) + \frac{N}{2} \tag{15.46}$$

$$\sum_{n=0}^{N-1} \sin^2(\omega_{k+1}(n)) = \frac{N}{2} - \frac{1}{2}\sum_{n=0}^{N-1} \cos(2\omega_{k+1}(n)) \tag{15.47}$$

$$\sum_{n=0}^{N-1} \cos(\omega_{k+1}(n)) \cdot \sin(\omega_{k+1}(n)) = \frac{1}{2}\sum_{n=0}^{N-1} \sin(2\omega_{k+1}(n)). \tag{15.48}$$

However, if it is taken into account that $\sum_{n=0}^{N-1} \sin(2\omega_{k+1}(n)) = 0$ for all ω_{k+1} and also that $\sum_{n=0}^{N-1} \cos(\omega_{k+1}(n)) = N$ when $\omega_{k+1} = 0$ and $\sum_{n=0}^{N-1} \cos(\omega_{k+1}(n)) = 0$ when $\omega_{k+1} \neq 0$, then Equations 15.42, 15.44, and 15.45 become:

$$\gamma_{11} = \frac{N}{2}\text{Re}(H(i))^2 + \frac{N}{2}\text{Im}(H(i))^2 \tag{15.49}$$

$$\gamma_{22} = \frac{N}{2}\text{Re}(H(i))^2 + \frac{N}{2}\text{Im}(H(i))^2 \tag{15.50}$$

$$\gamma_{12} = 0. \tag{15.51}$$

The updated computational complexity of the analysis-by-synthesis algorithm is given in Table 15.2, where the maximum complexity is the more realizable value of 24.2 MFLOPS.

15.8 QUANTIZING THE EXCITATION PARAMETERS

Following the calculation of the excitation parameters using analysis-by-synthesis, an efficient means of encoding them for transmission to the decoder is required.

15.8.1 Encoding the Sinusoidal Amplitudes

In order to encode the sinusoidal amplitudes for transmission, both vector and scalar quantization are considered. Initially, vector quantization is examined by creating a constant-length vector. Scalar quantization is also examined by dividing the amplitude parameters into frequency bands for quantization.

15.8.1.1 Vector Quantization of the Amplitudes. Vector quantization [102] is the most efficient quantization method. Thus, it is important to consider vector quantization for

TABLE 15.3 The Rational Sampling Rate Conversion Factors Required to Transform Every Amplitude Vector to a Length of 80 Samples

Amplitude Vector Length	Sampling Rate Conversion Factor M/L
10	8
15	16/3
20	4
25	16/5
30	8/3
35	16/7
40	2
45	16/9
50	8/5
55	16/11
60	4/3
65	16/13
70	8/7
75	16/15

encoding the sinusoidal amplitudes. Ideally, in order to be able to encode the remaining signal efficiently, a constant-length vector is required. Thus, for each frame the amplitude vectors have their frequency-domain spacing or sampling rate adjusted in order to produce 80 spectral lines per amplitude vector per frame. This sampling rate adjustment is performed using interpolation and decimation, where the interpolation procedure was described in Section 11.3.2.

15.8.1.2 Interpolation and Decimation. If the vector length conversion is to be performed for every possible vector size, assuming pitch-period values between 54 and 400 Hz there can be between 10 and 80 harmonics with associated amplitude values. Thus, 70 different sampling rate changes must be considered. By zero-padding each amplitude vector to the next multiple of 5, the rational sampling rate conversions can be reduced to those given in Table 15.3. These sampling rate conversions were invoked with one-stage interpolation and one-stage decimation. A sinc resampling function, as described in Section 11.3.2, was employed for interpolation, with a seventeenth order FIR low-pass filter used for decimation. Figure 15.9 shows the rational sampling rate conversion process using interpolation and decimation, with factors M and L yielding a sampling rate conversion factor of M/L.

In order to recover the original amplitude vectors, the inverse rational sampling rate conversion was performed. Figure 15.10 displays an example of a rational sampling rate conversion for the amplitude vector. For the voiced speech frame, shown in Figure 15.10, a pitch period of 64 samples was selected, corresponding to a pitch-related amplitude spacing

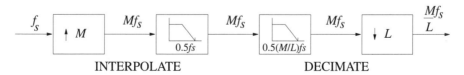

Figure 15.9 Schematic of the interpolation and decimation stages to perform a rational sampling rate conversion, which is illustrated in Figure 15.10.

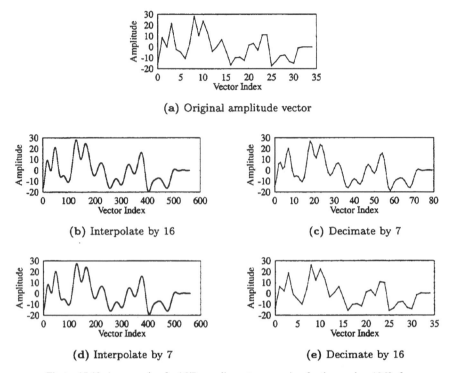

(a) Original amplitude vector

(b) Interpolate by 16   (c) Decimate by 7

(d) Interpolate by 7   (e) Decimate by 16

Figure 15.10 An example of a 16/7 sampling rate conversion for the speaker AM2, from
the front vowel utterance /æ/ in "back," where this utterance has a pitch
period of 64 samples, corresponding to a pitch-related amplitude spacing of
125 Hz. (a) shows the amplitude vectors zero padded to 35 samples. (b) and
(c) show the 16/7 sampling rate conversion to produce vectors of 80
samples. (d) and (e) show the 7/16 inverse sampling rate conversion to
reproduce the original vector.

of 125 Hz and to an amplitude vector containing 32 elements. In Figure 15.10a this amplitude
vector has been zero padded to contain 35 elements; thus, a rational sampling rate conversion
of 16/7 is performed. Figure 15.10b shows the interpolation by a factor of 16, with Figure
15.10c showing the decimation by a factor of 7. Thus, the amplitude vector now has 80
samples. The reverse sampling rate change is performed with a conversion ratio of 7/16,
shown in Figures 15.10d and 15.10e. Finally the first 32 elements are used for the amplitude
vector.

With every speech frame having an amplitude spectrum containing 80 samples, it is
possible to more easily implement vector quantization to encode each spectra.

15.8.1.3 Vector Quantization.

With the amplitude spectra at a constant vector length
of 80 samples, vector quantization (VQ) becomes relatively easy. Before the sinusoidal
amplitude spectra are vector-quantized, they are normalized by the RMS energy, which is then
scalar quantized using the Max-Lloyd algorithm described in Section 11.4.

The vector quantization was performed using a combination of the generalized Lloyd
algorithm and a pairwise nearest neighbor (PNN) design. (The reader is referred to the
monograph by Gersho and Gray [102] for further details on VQ.) Briefly, the generalized

Lloyd algorithm starts with an initial codebook, which is used to encode a set of training data. For the STC implementation, the employed quantization distortion measure is:

$$d(x, y) = \|x - y\|^2 = \sum_{i=1}^{k} (x_i - y_i)^2 \qquad (15.52)$$

where x is the training vector, y is the codebook entry, and k is the number of elements, or vector components, in the vector. The codebook entry, constituting the centroid of a cell, which has the lowest distortion when quantizing the training vector, is selected to host the training vector concerned. Once all the training vectors have been assigned to a codebook entry, every cell has its centroid recalculated on the basis of the entries in its cell, where the centroid of the cell is the codebook entry. The centroid is found using:

$$Y_j = \frac{\dfrac{1}{M}\sum_{i=1}^{M} x_i S_j(x_i)}{\dfrac{1}{M}\sum_{i=1}^{M} S_j(x_i)} \qquad \text{for } j = 1, 2, \ldots N \qquad (15.53)$$

where M is the training set size, N is the codebook size, x_i is the training vector, and $S_j(x_i)$ is a selector which indicates if $x_i \in S_j$.

The training data set is subsequently passed through the new codebook assigning all entries to the newly computed cells, after which the new codebook centroids are again recalculated. This iterative process is continued until the optimum codebook for the training data set is found.

In order to create the initial codebook for the generalized Lloyd algorithm, the PNN algorithm was used [102]. The PNN algorithm commences with an M-sized codebook, where every training vector is a codebook entry. Subsequently, codebook entries are merged, until a codebook of the required size N is arrived at. The cell merging is performed by considering the overall distortion increase upon tentatively merging every cell with every other cell, in order to find the pair inflicting the lowest overall distortion. Thus, at each iteration of the PNN algorithm, the codebook size decreases by 1.

15.8.1.4 Vector Quantization Performance. The amplitude vector describing the spectral envelope of the pitch-spaced harmonics was divided into four separate vectors for quantization, with each vector containing 20 samples. The process for the VQ of the amplitude is shown in Figure 15.11, where more efficient quantization is achieved by normalizing the amplitude vector by its RMS energy before being placed in the VQ codebook.

To demonstrate the suitability of the amplitude vector for quantization, Figure 15.12 shows the PDF of the amplitudes. In this figure the sharp peak at zero is caused by zero-padding employed in the interpolation process, as described in Section 15.8.1.2. The individual elements of the amplitude vector have their PDFs given in Appendix C, which show that the PDFs of the individual elements also support quantization.

The performance of the VQ was considered using four 8-bit VQs, together with a 5-bit scalar quantizer (SQ) for the overall RMS energy and 2-bit SQs for the codebook RMS energies, requiring a bitrate of 45 bits/20 ms or 2.25 kbps. Vector quantization produced a SEGSNR measure of 10.54 dB. However, the interpolation and decimation process required to produce a constant-length vector requires a computational complexity of 6.7 MFLOPS.

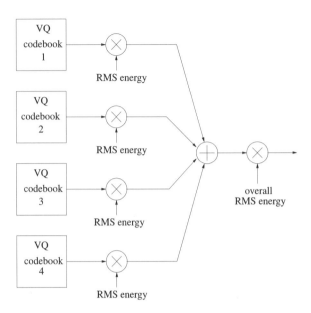

Figure 15.11 Schematic of vector quantization for the amplitude vector.

15.8.1.5 Scalar Quantization of the Amplitudes.   An alternative to VQ is to employ scalar quantization (SQ) in order to encode the sinusoidal amplitudes. For the VQ, the amplitude vector was expanded so that it would always contain 80 values, with every element represented in the VQ. However, for SQ the number of transmitted amplitude values is predetermined as M, with the pitch-spaced amplitude elements divided between the M bands. Each of the M bands is assigned to have the RMS energy of its constituent amplitude elements.

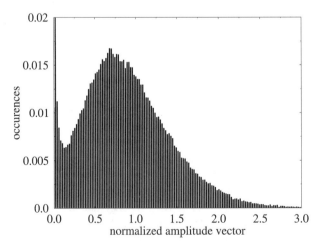

Figure 15.12 The combined PDF for the normalized amplitude vector.

Restating Equation 14.8, from the MMBE speech coder of Chapter 14, we find that the number of pitch-spaced spectral amplitude elements in each band is given by:

$$N_n = \frac{f_s/2}{M \cdot F0} \tag{15.54}$$

where f_s is the sampling frequency, $F0$ is the fundamental frequency, M is the number of spectral amplitude bands in which the transmitted amplitude values are assumed to be identical, and N_n is the number of amplitude elements in each of the M bands.

Similarly to the VQ method, the SQ process starts by quantizing the overall full-band RMS value with 5 bits. Subsequently, each of the M-transmitted amplitude parameters is assigned a value, determined as the average normalized RMS energy level of its N_n number of constituent amplitude elements. Each of the M transmitted amplitude parameters is quantized with 2-bits. The performance of the SQ was considered with $M = 20$ transmitted amplitude parameters. Thus, with the overall RMS parameter, a total of 45-bits/20 ms or 2.25 kbps is required for transmission. The scalar quantizer produced a SEGSNR value of 5.18 dB for this spectral magnitude parameter, with a negligible computational complexity.

Sections 15.8.1.1 and 15.8.1.5 show that vector quantization performs better than scalar quantization at encoding the amplitude parameters. However, they also show the significant increase in computational complexity required. Table 15.2 demonstrates that the PWI-STC speech coder is already fairly complex, where the implementation of a vector quantizer would increase the computational complexity to higher than 30 MFLOPS. Hence, due to computational complexity requirements, the scalar quantizer was selected to encode the sinusoidal amplitude parameters.

15.8.2 Encoding the Sinusoidal Phases

15.8.2.1 Vector Quantization of the Phases. It was anticipated that the phase values could be quantized in a similar manner to the amplitude values, as shown in Figure 15.11. Figure 15.13 displays the PDF of the phase values. From our detailed investigations we concluded that, in contrast to the pitch-related spectral amplitude values, the phase values are not amenable to VQ.

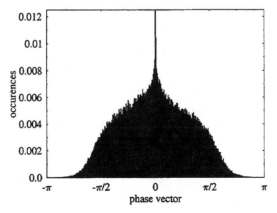

Figure 15.13 The combined PDF for the normalized phase vector.

15.8.2.2 Encoding the Phases with a Voiced-Unvoiced Switch. The traditional method for representing sinusoidal phases is to classify them as either voiced or unvoiced [346, 347], while ignoring their exact phase values. This can be justified on the basis of the knowledge that voiced phases can be represented by 0, while unvoiced phases can be represented by a uniformly distribution over the range $-\pi$ to π.

McAulay and Quatieri [346] adopted the approach of Makhoul et al. [289] in which a voicing transition frequency was determined. Beneath the voicing transition frequency the phases are assumed voiced, while above this frequency the phases are considered unvoiced. The decision as to whether a harmonic is voiced or unvoiced is typically determined by the closeness of fit between a purely voiced magnitude spectrum and the original magnitude. McAulay et al. [346, 347] based this measure on the speech spectrum itself. Since we determine the component frequencies, amplitude and phases of the LPC residual spectrum, rather than those of the speech signal, this is the most appropriate signal to employ when assessing voicing characteristics.

The extent of voicing present in each harmonic was determined using the LPC residual spectrum and the SNR applied to quantify the match between the residual and synthetic, fully voiced LPC STP residual, as suggested by McAulay et al. [346, 347], which is given by:

$$SNR = 10 \log \frac{A_{\text{orig}}(k\omega_0)^2}{[A_{\text{synth}}(k\omega_0) - A_{\text{orig}}(k\omega_0)]^2} \tag{15.55}$$

where A_{orig} refers to the original LPC residual magnitude spectrum, A_{synth} refers to a fully voiced magnitude spectrum, and $k\omega_0$ is the kth harmonic of the determined normalized fundamental frequency ω_0.

We concluded experimentally that 30 dB represented a good voiced/unvoiced threshold for each harmonic, with values above the threshold declared voiced. Following the categorization of the harmonics, the voicing transition frequency must be determined. This process examined the voiced/unvoiced state of each harmonic, where if more than two consecutive low SNR harmonics were located, which were deemed unvoiced, the rest of the harmonics were also assumed unvoiced. Otherwise all harmonics were assumed voiced. Hence, a voicing transition harmonic parameter was transmitted to the decoder, which when combined with the transmitted pitch period indicates the voicing transition frequency above which the LPC STP residual was deemed unvoiced. With a 54 Hz fundamental frequency a permissible maximum of 80 harmonics exists in the 4 kHz, thus, 7 bits will be required to transmit the voicing transition harmonic parameter.

15.8.3 Encoding the Sinusoidal Fourier Coefficients

In this section, an alternative set of parameters to the amplitude and phase values are considered for transmission. The analysis-by-synthesis approach of Section 15.7.2 operates by determining the values a_{k+1} and b_{k+1}, from Equations 15.27 and 15.28, from which the amplitude and phase values are found using $A_{k+1} = \sqrt{a_{k+1}^2 + b_{k+1}^2}$ and $\phi_{k+1} = -\arctan(b_{k+1}/a_{k+1})$. Thus, the a_{k+1} and b_{k+1} parameters are the real and imaginary Fourier coefficients, which can be encoded for transmission to the decoder as an alternative to the amplitude and phase values. The a_{k+1} and b_{k+1} values can be scalar-quantized in the same manner as the amplitude value, as described in Section 15.8.1.5. The a_{k+1} and b_{k+1} parameters have their overall RMS values scalar-quantized with 5 bits, and then the normalized values are divided into 10 frequency bands, each encoded with 2 bits producing an overall bitrate of 50 bits/20 ms or 2.5 kbps.

15.8.3.1 Equivalent Rectangular Bandwidth Scale. The SQ of the amplitude parameter divided the amplitude values into frequency bands, with Equation 15.54 employed for this purpose. However, this equation ignores the information that for human hearing lower frequencies are perceptually more important. The equivalent rectangular bandwidth (ERB) scale [354] weights the frequency spectrum in order to place more emphasis on the perceptually more important lower frequencies. Thus, it can be employed in the PWI-STC coder to produce a better SQ of the Fourier coefficients a_{k+1} and b_{k+1} of Equations 15.27 and 15.28.

The conversion between the frequency spectrum and the transformed ERB scale is given by [354]:

$$f_{ERB} = 11.17 \times \ln\left(\frac{f + 312}{f + 14675}\right) + 43.0 \qquad (15.56)$$

where f_{ERB} represents the ERB frequency scale and f is the conventional frequency spectrum.

Figure 15.14 displays the relationship between the ERB scale and frequency over 0–4 kHz. In order to divide the Fourier coefficients into M bands, the harmonic frequencies and the corresponding Fourier coefficients are converted to the ERB scale. In the ERB scale domain, the transformed frequencies are divided into M bands using Equation 14.8. Each of these M bands is then assigned a value determined as the average normalized RMS energy of its N_n constituent Fourier coefficients. In the frequency domain, the M bands will be perceptually weighted.

In our final coder, the Fourier coefficients of Equations 15.27 and 15.28 were selected for transmission to the decoder instead of the magnitude and phase parameters. This choice was due primarily to problems in determining the correct frequency point, detailed in Section 15.8.2.2, for switching from voiced to unvoiced speech in the process of coding the phases of the harmonics.

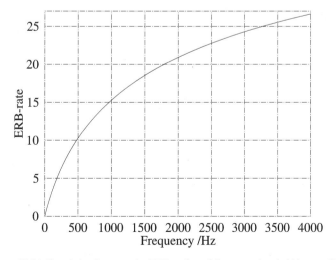

Figure 15.14 Conversion between the ERB scale and frequency bandwidth according to Equation 15.56.

15.8.4 Voiced-Unvoiced Flag

It was found that both the Fourier coefficients of Equations 15.27 and 15.28 and the amplitude and phase parameters produced too-dominant voiced excitation during the unvoiced portions of speech. For the Fourier coefficients this extra periodicity was caused by the grouping of parameters into frequency bands, removing too much of the phase signal's randomness.

In order to overcome this problem, the voiced-unvoiced decision of Section 12.4 was employed to set a voiced-unvoiced flag. At the decoder, if this flag was set to indicate a voiced frame, the Fourier coefficients of Equations 15.27 and 15.28 were used to determine the sinusoidal phases with $-\arctan(b_{k+1}/a_{k+1})$. However, if an unvoiced frame was indicated, the sinusoidal phases were uniformly distributed over $-\pi$ to π.

15.9 SINUSOIDAL TRANSFORM DECODER

The schematic of the STC decoder is shown in Figure 15.15, where the frequencies, amplitudes, and phases are determined from the transmitted parameters in order to reconstruct the residual waveform. The synthesized excitation is passed through a LPC synthesis filter together with the adaptive post-filter and pulse dispersion filter to reproduce the speech waveform.

The amplitudes are reconstructed using the transmitted quantized RMS values for the Fourier coefficients and the relationship $A_{k+1} = \sqrt{a_{k+1}^2 + b_{k+1}^2}$. For frames labeled voiced, the phases are reconstructed using $-\arctan(b_{k+1}/a_{k+1})$, while for unvoiced frames, the phases are set to random values uniformly distributed over $-\pi$ to π. The sinusoidal frequencies are reconstructed using the transmitted pitch period in order to determine the pitch-spaced excitation harmonics. These component frequencies, amplitudes, and phases are employed to reconstruct the residual waveform.

To compensate for the long analysis window of 20 ms, interpolation can be performed on a pitch synchronous basis, similarly to Chapter 13, for producing a smoothly evolving waveform. Like the PWI-ZFE speech coder and MMBE speech coders, the interpolated excitation waveform can be passed through the LPC synthesis filter, adaptive post-filter, and pulse dispersion filter to reproduce the speech waveform.

Following this overview of the STC decoder, below we concentrate on the pitch synchronous interpolation which is performed in the schematic of Figure 15.15.

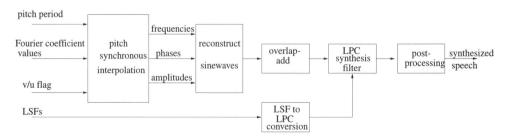

Figure 15.15 Schematic for the STC-PWI decoder.

15.9.1 Pitch Synchronous Interpolation

Performing pitch synchronous interpolation at the decoder allows the developed STC coder to be viewed as a prototype waveform interpolation sinusoidal transform coding (PWI-STC) coder, since a PWI scheme has been adopted where the excitation is implemented using STC.

With a link to PWI coders established, the interpolation process can closely follow Section 13.6. The LSFs and pitch period parameter can be interpolated as described in Sections 13.6.6 and 13.6.1, respectively. The prototype segment's zero-crossing parameter is set to zero, since the Fourier coefficient parameters contain the phase information for the prototype segment. The interpolation of the sinusoidal excitation parameters is described in detail next.

15.9.1.1 Fourier Coefficient Interpolation. The overall RMS value for the Fourier coefficient parameters can be linearly interpolated across consecutive pitch duration segments using:

$$O_{\text{RMS},n_p} = O_{\text{RMS}}(N-1) + (n_p - 1) \cdot \frac{O_{\text{RMS}}(N) - O_{\text{RMS}}(N-1)}{N_{\text{pit}} - 1} \qquad (15.57)$$

where $O_{\text{RMS}}(N)$ is the overall RMS value of the current frame and $O_{\text{RMS}}(N-1)$ is the overall RMS value of the previous frame. O_{RMS,n_p} is the overall RMS value of the n_p^{th} interpolation segment, and N_{pit} is the number of pitch synchronous intervals between $O_{\text{RMS}}(N)$ and $O_{\text{RMS}}(N-1)$.

The RMS value for each of the M number of frequency subbands must also be linearly interpolated, using:

$$\text{RMS}_{m,n_p} = \text{RMS}_m(N-1) + (n_p - 1) \cdot \frac{\text{RMS}_m(N) - \text{RMS}_m(N-1)}{N_{\text{pit}} - 1} \qquad (15.58)$$

where $\text{RMS}_m(N-1)$ is the previous {\rm RMS} value of the mth subband, while $\text{RMS}_m(N)$ is the current mth RMS value.

15.9.2 Frequency Interpolation

The frequencies of the component sinewaves are the harmonics of the fundamental frequency determined at the encoder, which is transmitted as the pitch period. Thus, interpolated frequencies are generated for each interpolation region by using the harmonics of the interpolated fundamental frequency.

15.9.3 Computational Complexity

The computational complexity of the PWI-STC decoder is dominated by the sinusoidal synthesis process described by Equation 15.6, where the transmitted Fourier coefficients have been converted into amplitude and phase parameters. Previously, in Section 15.3 it was detailed that both the present and past sinusoidal parameters were used to reconstruct the speech waveform, with a trapezoidal window used to weight the contribution of the sinusoidal parameters from each frame. The computational complexity of the sinusoidal synthesis process is dependent on the number of harmonics to be summed, where the number of harmonics can vary from 10 to 80. For a sinusoidal synthesis process that is constructed over two frame lengths or 40 ms and that contains 10 harmonics, the complexity is 1.8 MFLOPS. For a sinusoidal synthesis process that includes 80 harmonics, the computational complexity is 14.1 MFLOPS.

The pitch synchronous procedure implemented in the decoder, following the philosophy of Section 15.3.2, further increases this complexity, since the sinusoidal process must be performed for every interpolation region. For a sinusoidal process having a pitch period of 20 samples, or 400 Hz, with 10 harmonics in the 4 kHz band, up to eight 20-sample interpolation regions could be present within a 160-sample speech frame of 20 ms, producing a computational complexity of 14.4 MFLOPS. If 80 harmonics are employed in the sinusoidal synthesis process, only one interpolation region will occur within the speech frame.

The computational complexity of the PWI-STC decoder can be decreased by replacing the sinusoidal synthesis of Equation 15.6 by the inverse FFT. The inverse FFT process has a computational complexity of $N \log_2 N$, where N is the FFT length, in this case 512 samples. Thus, the associated computational complexity, using the FFT-based interpolation, for any number of harmonics will be 0.23 MFLOPS. If eight interpolation regions are present within the 20 ms speech frame, then the computational complexity will be 1.84 MFLOPS.

15.10 SPEECH CODER PERFORMANCE

Initially, the performance of a PWI-STC speech coder at 3.8 kbps was assessed, where the Fourier coefficients were divided into 10 bands for SQ and transmission to the decoder. In Chapters 13 and 14, initially a lower rate speech coder was developed before increasing the bitrate of the speech coder to 3.8 kbps. In both of these Chapters the increased bitrate did not correspond to a sufficient increase in speech quality in order to justify the added complexity and increased bitrate. By contrast, here initially a higher bitrate speech coder is developed and then its bitrate is decreased to 2.4 kbps. For the 10-band PWI-STC speech coder, the adaptive postfilter parameters, described in Section 11.6, were optimized to $\alpha_{pf} = 0.85$, $\beta_{pf} = 0.50$, $\mu_{pf} = 0.50$, $\gamma_{pf} = 0.50$, $g_{pf} = 0.00$ and $\xi_{pf} = 0.99$.

For examining the PWI-STC speech coder's performance, the speech frames used to assess the previously developed speech coders were adopted. Thus, Figures 15.16, 15.17, and 15.18 can be compared with the other 3.8 kbps speech coders demonstrated in Figures 13.20, 13.21, and 13.22 for the PWI-ZFE scheme of Section 13.11.2, together with Figures 14.23, 14.24, and 14.25 for the MMBE-ZFE scheme of Section 14.7.

Figure 15.16 displays the results for the 10-band PWI-STC speech coder for an utterance from the testfile BM1, which can be compared with the similar bitrate speech coders of Figures 13.20 and 14.23. From Figure 15.16b it can be seen that, unlike in Figure 13.20b, the reconstructed excitation has the correct pitch period. In the frequency domain, the excitation spectrum follows the shape of the formants and slightly emphasizes the higher frequencies. For the time-domain synthesized speech of Figure 15.16c it can be seen that the decay between pitch periods is more pronounced than for Figure 14.23c. However, in the frequency domain the PWI-STC speech coder better represents the formants than the MMBE-ZFE scheme of Figure 14.23c.

Figure 15.17 portrays a 10-band PWI-STC speech coder from the testfile BF2, which can be compared to Figures 13.21 and 14.24 for the PWI-ZFE and MMBE speech coders, respectively. From Figure 15.17b it can be seen that the time-domain excitation is dominated by pulses, while in the frequency domain the magnitude spectrum is flat. Figure 15.17c shows that the PWI-STC speech coder manages to reproduce both the time-domain and frequency-domain waveforms more accurately than either Figure 13.21c or 14.24c.

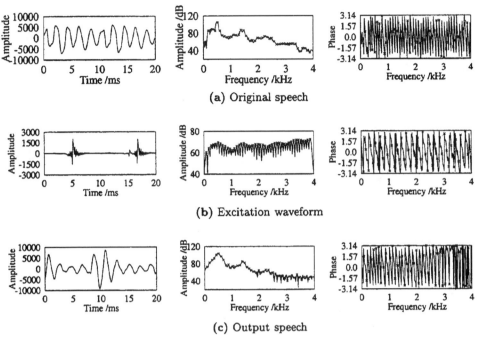

Figure 15.16 Time- and frequency-domain comparison of the (a) original speech, (b) **10-band PWI-STC** waveform, and (c) output speech after the pulse dispersion filter. The 20 ms speech frame is the mid vowel /ɜ/ in the utterance "work" for the testfile BM1. For comparison with the other coders developed in this study using the same speech segment, please refer to Table 16.2.

Figure 15.18 displays the results for a speech frame from the testfile BM2 using the 10-band PWI-STC speech coder. Here comparisons can be drawn with Figures 13.22 and 14.25. It should be noted that for this speech frame the LPC coefficients fail to represent the second and third formants. Figure 15.18b shows that the reconstructed excitation attempts to compensate for this by shaping the magnitude spectrum to follow the speech formants. Although this effect is insufficient to reproduce the missing formants in the synthesized speech spectrum of Figure 15.18c, the shape of the first formant is better represented than in Figures 13.22c and 14.25c.

The bit allocation for the 3.8 kbps PWI-STC speech coder is summarized in Table 15.4. The LPC coefficients are transformed to LSFs and quantized using 18 bits [123]. The frequencies of the component sinewaves are set to be the harmonics of the fundamental frequency, which is determined from the pitch period transmitted to the decoder employing 7 bits. The sinusoidal parameters are represented by the 10 Fourier coefficients a_k and b_k, with the overall RMS of both the a_k and b_k Fourier coefficients scalar-quantized using 5 bits. The a_k and b_k parameters are converted into the ERB scale before being split into 10 evenly spaced bands each. The average RMS value for each of these 10 bands is scalar-quantized with 2 bits. A voiced-unvoiced flag is also sent to allow random phases to be applied for unvoiced frames.

Pairwise-comparison tests, detailed in Section 16.2, were conducted between the 3.8 kbps 10-band PWI-STC speech coder and the 3.85 kbps 13-band MMBE PWI-ZFE speech coder together with the the 3.8 kbps 3-pulse PWI-ZFE scheme, where the comparison

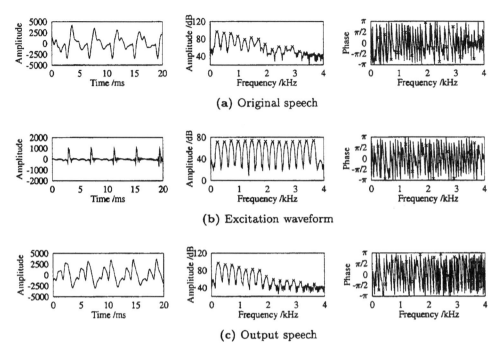

Figure 15.17 Time- and frequency-domain comparison of the (a) original speech, (b) **10-band PWI-STC** waveform, and (c) output speech after the pulse dispersion filter. The 20 ms speech frame is the liquid /r/ in the utterance "*r*ice" for the testfile BF2. For comparison with the other coders developed in this study using the same speech segment, please refer to Table 16.2.

speech coders were developed in Chapter 13 and Chapter 14, respectively. For the 3.8 kbps 10-band PWI-STC speech coder and the 3.85 kbps 13-band MMBE PWI-ZFE speech coder, these pairwise-comparison tests showed that 7.69% of listeners preferred the 10-band PWI-STC speech coder, with 30.77% of listeners preferring the 13-band MMBE PWI-ZFE scheme and 61.54% having no preference. For the 3.8 kbps 10-band PWI-STC speech coder and the 3.8 kbps 3-pulse PWI-ZFE speech coder, these pairwise-comparison tests showed that 17.95% of listeners preferred the 10-band PWI-STC speech coder, with 53.85% of listeners preferring the 3-pulse PWI-ZFE scheme and 28.20% having no preference. Thus, the 3.8 kbps 10-band PWI-STC speech coder does not perform as well as the speech coders previously developed.

The 10-band PWI-STC speech coder was also adjusted in order to produce a lower rate speech coder at 2.4 kbps, expecting that the corresponding reduction in speech quality will not be dramatic. The 2.4 kbps PWI-STC speech coder contained three frequency bands. The parameters from the adaptive postfilter of Section 11.6 were re-optimized to $\alpha_{pf} = 0.80$, $\beta_{pf} = 0.45$, $\mu_{pf} = 0.50$, $\gamma_{pf} = 0.50$, $g_{pf} = 0.00$, and $\xi_{pf} = 0.99$. The results for the same speech frames as in Figures 15.16, 15.17, and 15.18 are given in Figures 15.19, 15.20, and 15.21. These figures can also be compared with the two previously developed speech coders at 2.35 kbps; the 5-band MMBE speech coders employing simple pulse excitation and the 3-band MMBE speech coder incorporating ZFE to represent the voiced speech. The results for these speech coders were given in Figures 14.17, 14.18, and 14.19 together with Figures 14.20, 14.21, and 14.22, respectively.

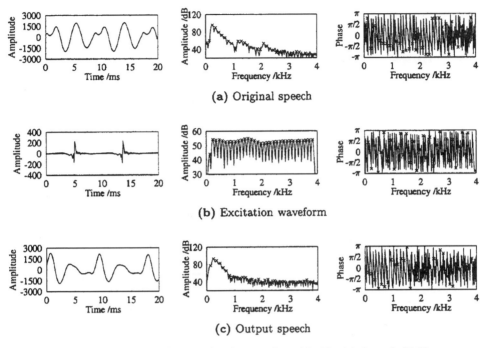

Figure 15.18 Time- and frequency-domain comparison of the (a) original speech, (b) **10-band PWI-STC** waveform, and (c) output speech after the pulse dispersion filter. The 20 ms speech frame is the nasal /n/ in the utterance "throw*n*" for the testfile BM2. For comparison with the other coders developed in this study using the same speech segment, please refer to Table 16.2.

Figure 15.19 displays the results for the 3-band PWI-STC speech coder for an utterance from the testfile BM1, which can be compared with Figure 14.17 and 14.20 generated at similar rates. Figure 15.19 can also be contrasted with the 10-band PWI-STC speech coder of Figure 15.16. The reduction to 3 frequency bands produces a flatter excitation spectrum and decreases the depth of the null between the first and second formants. When Figure 15.19c is compared with Figures 14.17c and 14.20c, it can be seen that the 3-band PWI-STC speech coder has a synthesized frequency spectrum that better represents the original spectrum.

TABLE 15.4 Bit-allocation Table for the Investigated PWI-STC Coders at 3.8 kbps and 2.4 kbps

Parameter	Required Bits	
LPC coefficients	18	18
Pitch period	7	7
v/u switch	1	1
Overall RMS for a_k	5	5
Overall RMS for b_k	5	5
a_k bands	10×2	3×2
b_k bands	10×2	3×2
Total/20 ms	76	48
Bitrate	3.8 kbps	2.4 kbps

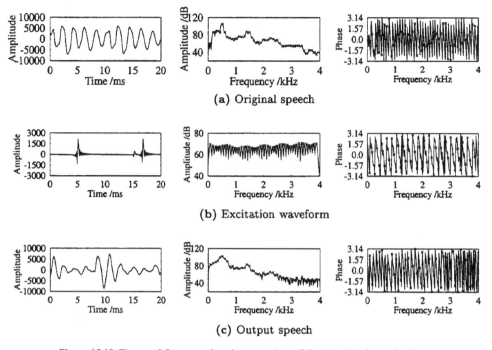

Figure 15.19 Time- and frequency-domain comparison of the (a) original speech, (b) **3-band PWI-STC** waveform, and (c) output speech after the pulse dispersion filter. The 20 ms speech frame is the midvowel /ɜ/ in the utterance "work" for the testfile BM1. For comparison with the other coders developed in this study using the same speech segment, please refer to Table 16.2.

Figure 15.20 portrays the results for a segment of speech from the testfile BF2, where the speech frame was also examined in Figures 14.18 and 14.21. Figure 15.20 related to the 3-band PWI-STC speech coder can also be compared with Figure 15.17, where the reduction in the number of frequency bands decreases the amplitude of the first resonance in each pitch period. When compared with Figures 14.18 and 14.21, the performance of the 3-band PWI-STC speech coder is deemed to be similar.

Figure 15.21 displays the performance of the 3-band PWI-STC speech coder for the testfile BM2 and can be compared with the 10-band PWI-STC speech coder of Figure 15.18. It can be seen from Figure 15.21b that the reduction in the number of frequency bands produces an excitation spectrum that places more emphasis on the lower frequencies. This results in a larger amplitude for the dominant time-domain resonance within the pitch period. The 3-band PWI-STC speech coder can also be contrasted with the 5-band MMBE speech coder and the 3-band MMBE-ZFE speech coder of Figures 14.19 and 14.22, respectively. It can be seen from Figure 15.21c that the PWI-STC speech coder produces a more dominant first formant and a larger time-domain resonance.

The 2.4 kbps PWI-STC speech coder contained only three different frequency bands. Otherwise it was identical to the 10-band PWI-STC speech coder. The bit allocation can be seen in Table 15.4.

Pairwise-comparison tests (detailed in Section 16.2) were conducted between the 2.4 kbps 3-band PWI-STC speech coder and the 2.35 kbps 3-band MMBE PWI-ZFE speech coder together with the the 2.3 kbps 5-band MMBE LPC scheme, where both speech coders were developed in Chapter 14. For the 2.4 kbps 3-band PWI-STC speech

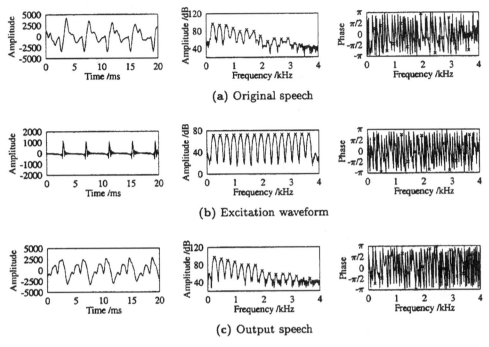

(a) Original speech

(b) Excitation waveform

(c) Output speech

Figure 15.20 Time- and frequency-domain comparison of the (a) original speech, (b) **3-band PWI-STC** waveform, and (c) output speech after the pulse dispersion filter. The 20 ms speech frame is the liquid /r/ in the utterance "*r*ice" for the testfile BF2. For comparison with the other coders developed in this study using the same speech segment, please refer to Table 16.2.

coder and the 2.35 kbps 3-band MMBE PWI-ZFE speech coder, these pairwise-comparison tests showed that 20.51% of listeners preferred the 3-band PWI-STC speech coder, with 12.82% of listeners preferring the 3-band MMBE PWI-ZFE scheme and 66.67% having no preference. For the 2.4 kbps 3-band PWI-STC speech coder and the 2.3 kbps 5-band MMBE LPC speech coder, these pairwise-comparison tests showed that 10.26% of listeners preferred the 3-band PWI-STC speech coder, with 30.77% of listeners preferring the 3-band MMBE PWI-ZFE scheme and 58.97% having no preference. These listening tests show that it was difficult to determine a preference for the speech coders developed at approximately 2.4 kbps.

15.11 CHAPTER SUMMARY

This Chapter has described sinusoidal transform coding (STC), which has recently been advocated for low-bitrate speech coding, for example, by McAulay and Quatieri [341]. The Chapter begins with a description of the STC coding algorithm developed by McAulay and Quatieri [346]. This Chapter further develops STC by incorporating the prototype waveform interpolation (PWI) philosophy in order to reduce the required bitrate. Instead of the determination of the sinusoidal parameters by peak-picking, the analysis-by-synthesis technique introduced by George and Smith was employed [343, 344]. The error minimization process was modified in order to reduce the computational complexity of the analysis-by-

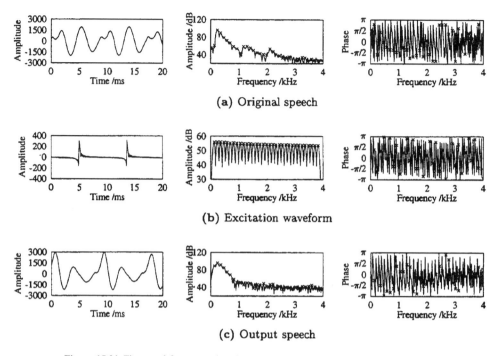

Figure 15.21 Time- and frequency-domain comparison of the (a) original speech, (b) **3-band PWI-STC** waveform, and (c) output speech after the pulse dispersion filter. The 20 ms speech frame is the nasal /n/ in the utterance "throw*n*" for the testfile BM2. For comparison with the other coders developed in this study using the same speech segment, please refer to Table 16.2.

synthesis procedure. The located sinusoidal parameters were transmitted to the decoder as Fourier coefficients. At the decoder, pitch synchronous interpolation was performed in order to improve the synthesized speech quality, with the inverse FFT harnessed for decreasing the computational complexity of the sinusoidal synthesis process.

16

Conclusions on Low-Rate Coding

15.1 OVERVIEW

This Chapter begins with an overview of the different speech coders developed throughout the low-bitrate-coding-oriented part of the book. For each speech codec the bitrate, delay, and complexity are given. Following the speech codec summary, the details of the conducted informal listening tests are given, thus assessing the quality of each speech coder. The robustness of the speech codec was only investigated for the 1.9 kbps PWI-ZFE speech coder, where the details are given in Section 13.10.

As seen in Table 16.1, for all speech codecs the requirement for a delay of less than 80 ms is met. However, for the higher bitrate PWI-ZFE speech codec, at 3.8 kbps the maximum targeted computational complexity of 25 MFLOPS is exceeded. We attempted to reduce this computational complexity, since the speech codec did not constitute a sufficiently promising design at its bitrate, in order to justify the associated research effort. The PWI-STC speech codecs narrowly exceeded the complexity limit of 25 MFLOPS.

In addition to the review of the bitrate, delay, and complexity given in Table 16.1, we also detail the location of pertinent figures and tables for each of our speech codecs, with this information summarized in Table 16.2.

16.2 LISTENING TESTS

The speech quality of the designed speech coders, detailed in Tables 16.1 and 16.2, was assessed using pairwise-comparison tests. For the pairwise-comparison test a speech utterance was passed through two speech codecs—speech codec A and speech codec B—with the listener asked to express a preference for speech coder A, speech coder B, or neither. The utterance was passed through each speech coder twice in order to give the listener more scope when selecting the best speech coder.

Thirteen listeners were used in the pairwise-comparison tests, with three different utterances passed through each speech coder A-B pair. The listening tests were conducted through headphones. Eight different utterances were employed during the listening tests,

TABLE 16.1 Summary of the Bitrate, Delay, and Computational Complexity for the Developed Speech Codecs

Speech Codec	Bitrate/kbps	Delay/ms	Complexity /MFLOPS
LPC vocoder, C1	1.55	60	3.4
PWI-ZFE, C2	1.90	70–80	14.13
PWI-ZFE, C3	3.80	70–80	37.05
2-band MMBE LPC, C4	1.85	60	4.58
5-band MMBE LPC, C5	2.30	60	6.67
3-band MMBE PWI-ZFE, C6	2.35	70–80	16.18
13-band MMBE PWI-ZFE, C7	3.85	70–80	21.30
3-band PWI-STC, C8	2.40	60	26.07
10-band PWI-STC, C9	3.80	60	26.07

where none of these utterances had been used in the design of the speech coders. The utterances were a mixture of male and female speakers with British and American accents and had differing lengths. Table 16.3 details the results of the pairwise-comparison tests.

From Table 16.3 we can see that for the top two comparison tests, where periodic pulse excitation from the LPC vocoder was compared to PWI-ZFE excitation, in both cases the PWI-ZFE excitation was preferred. The performance of the 3-band PWI-STC 2.4 kbps speech coder is variable, with it being preferred to the 2.35 kbps 3-band MMBE PWI-ZFE speech coders, but being judged inferior to the 3.8 kbps 5-band MMBE LPC speech coder. Observing Table 16.3, it can be inferred that for the speech coders operating around 2.4 kbps there was no conclusive best performer in terms of speech quality.

For the speech coders operating at approximately 3.8 kbps the 13-band MMBE PWI-ZFE speech coder performed best, followed by the 3.85 kbps 3-pulse PWI-ZFE speech coder. The 10-band PWI-STC speech coder performed least impressively.

The overall conclusion from the listening tests was that the single-pulse PWI-ZFE was the best voiced excitation source, where in order to achieve a higher bitrate the addition of MMBE was most successful.

TABLE 16.2 Summary of the Relevant Figures and Tables for Each of Our Developed Speech Codecs

Speech Codec	Characteristic Waveform Figure	Bit-Allocatioin Table	Schematic Figures	Complexity
1.55 kbps LPC vocoder, C1	11.21, 11.22, 11.23	11.8	11.1	Section 11.3.4
1.9 kbps PWI-ZFE, C2	13.13, 13.14, 13.15	13.14	10.13	Figure 13.4, Table 13.4
3.8 kbps PWI-ZFE, C3	13.20, 13.21, 13.22	13.14	10.13	Figure 13.4
1.85 kbps 2-band MMBE LPC, C4	14.14, 14.15, 14.16	14.3	14.5	Table 14.2
2.3 kbps 5-band MMBE LPC, C5	14.17, 14.18, 14.19	14.3	14.5	Table 14.2
235 kbps 3-band MMBE PWI-ZFE, C6	14.20, 14.21, 14.22	14.5	14.5	Figure 14.13
3.85 kbps 13-band MMBE PWI-ZFE, C7	14.23, 14.24, 14.25	14.5	14.5	Figure 14.13
2.4 kbps 3-band PWI-STC, C8	15.19, 15.20, 15.21	15.4	15.7, 15.15	Table 15.2, Section 15.9.3
3.8 kbps 10-band PWI-STC, C9	15.16, 15.17, 15.18	15.4	15.7, 15.15	Table 15.2, Section 15.9.3

TABLE 16.3 Details of the Listening Tests Conducted Using the Speech Codecs Detailed in Table 16.1. For the pairwise-comparison tests, the listeners were given a choice of preferring speech codec A, speech codec B or neither

Speech Codec A	Speech Codec B	Preference		
		A%	B%	Neither %
1.85 kbps 2-band MMBE LPC, C4	1.9 kbps PWI-ZFE, C2	23.07	30.77	46.16
2.3 kbps 5-band MMBE LPC, C5	2.35 kbps 3-band MMBE PWI-ZFE, C6	5.13	64.10	30.77
2.3 kbps 5-band MMBE LPC, C5	2.4 kbps 3-band PWI-STC, C8	30.77	10.26	58.97
2.35 kbps 3-band MMBE PWI-ZFE, C6	2.4 kbps 3-band PWI-STC, C8	12.82	20.51	66.67
3.8 kbps PWI-ZFE, C3	3.85 kbps 13-band MMBE PWI-ZFE, C7	5.13	30.77	64.10
3.8 kbps PWI-ZFE, C3	3.8 kbps 10-band PWI-STC, C9	53.85	17.95	28.20
3.85 kbps 13-band MMBE PWI-ZFE, C7	4.8 kbps 10-band PWI-STC, C9	30.77	7.69	61.54

16.3 SUMMARY OF VERY LOW-RATE CODING

This low-bitrate-oriented part of the book has primarily investigated three speech coding techniques frequently used at bitrate below 4 kbps; prototype waveform interpolation (PWI), multiband excitation (MBE), and sinusoidal transform coding (STC). The voiced excitation technique of zinc function excitation (ZFE) and the use of wavelets in speech coding have also been investigated.

The low-bitrate-oriented part of the book began by creating a basic LPC vocoder, which allowed decisions to be made about line spectrum frequency (LSF) quantization, pitch detection, and post-processing. It was determined that the vector quantizer (VQ) from G.729 [123] performed the LSF quantization best. Several different autocorrelation-based pitch detectors were investigated, reiterating the importance and difficulty of pitch detection, with an algorithm incorporating pitch tracking eventually selected. An adaptive post-filter and pitch-independent pulse dispersion filter were also selected.

In Chapter 12 an investigation into wavelets and pitch detection was performed. Polynomial spline wavelets introduced by Mallat and Zhong [322] were selected to produce the dyadic wavelet transform, $(D_Y WT)$, used to process the speech. Subsequent to the $D_Y WT$, a selection of possible candidate pitch periods remained, where both dynamic programming and autocorrelation were performed to determine the true pitch period. We concluded that the combination of $D_Y WT$ and subsequent autocorrelation computation produced the best pitch detector design.

Chapter 13 introduced ZFE to represent the voiced speech in a PWI scheme, creating a PWI-ZFE speech coder. The ZFE pulses were initially introduced at higher bitrates by Sukkar et al. [295] and have previously been used by Hiotakakos and Xydeas at low bitrates [294]. This Chapter introduced the principle of using glottal closure instants (GCIs) determined by the $D_Y WT$ in order to reduce the complexity of the ZFE optimization loop, where the GCIs were also used to ensure a smoothly evolving waveform within the synthesized prototype segments of the speech encoder. The Chapter also adopted the pitch prototype selection process proposed by Kleijn [81]. At the decoder the Chapter further developed the interpolation process of Hiotakakos and Xydeas [294], such that no information was required about either the pitch prototype location or the ZFE location, decreasing the number of bits requiring transmission by 6/20 ms frame.

The error sensitivity of the PWI-ZFE speech coder was also examined, where the importance of the voiced-unvoiced flag was emphasized. Finally, Chapter 13 investigated a higher bitrate speech coder with three ZFEs pulses employed to represent the voiced excitation. It was found that the extra ZFE pulses improved the representation of the voiced speech. This was counteracted by the phase restrictions imposed for the interpolation process, together with the difficulty in producing smooth interpolation at the decoder.

A mixed-MBE (MMBE) scheme was detailed in Chapter 14, where the frequency bands were based on the pitch period [338], thus requiring recalculation for every speech frame. The MMBE scheme was harnessed in both the LPC vocoder and the PWI-ZFE speech codec in order to produce a selection of different speech coders between 1.9 kbps and 3.85 kbps. It was found that at the same bitrate the MMBE scheme based on the PWI-ZFE performed best.

In Chapter 15 STC at low bitrates was investigated, where a PWI scheme was again implemented to produce PWI-STC speech coders at 2.4 kbps and 3.8 kbps. In this chapter, the analysis-by-synthesis technique used to determine the sinusoidal parameters and introduced by George and Smith [343, 344], was further developed to incorporate the weighted LPC synthesis filter. It was then modified to reduce the associated computational complexity. The sinusoidal parameters were transmitted to the decoder as scalar-quantized Fourier coefficients a_k and b_k, where these Fourier coefficients were associated with harmonics of the determined pitch.

16.4 FURTHER RESEARCH

This low-bitrate-coding-oriented part of the book demonstrated that for speech codecs operating at bitrates less than 4 kbps the principle of PWI is particularly useful. The predominant advantage of the PWI is that it allows the available bits to concentrate on encoding a single pitch period rather than three or four pitch periods as would be typical without PWI. Thus, it is suggested that further work on very low-bitrate speech coders should employ PWI as its foundation.

For the three voiced excitation methods employed in this low-bitrate-codiing-oriented part of the book—namely ZFE, MMBE, and STC-based compression—it was found that their performances were similar. When considering the most appropriate excitation for a speech coder, it is interesting to examine the winner of the U.S. Department of Defense competition for the new 2.4 kbps speech coder. This was the speech coder [284], which employed the most basic model for the excitation but incorporated additional aspects, such as a pulse dispersion filter, to model the human speech production system more closely. In the history of speech coding, understanding human speech production has produced many important developments, such as the LPC model and MBE, and should continue to provide inspiration in the future.

An area of research that is currently receiving much interest is the creation of multi-rate speech coders, which will be present in third-generation communication systems. Within this thesis for Chapters 13 and 14 we attempted to convert a speech codec to a higher bitrate, while in Chapter 15 conversion to a lower bitrate was investigated. The bitrate changes that were performed approximately doubled or halved the original bitrate, but the resultant codecs did not constitute the most attractive design tradeoff at these modified bitrates. Thus, an interesting area of further work would be to investigate how to increase or decrease a speech coder's bitrate, while maintaining an appropriate speech quality at a variable bitrate.

17

Comparison of Speech Codecs and Transceivers

17.1 BACKGROUND TO QUALITY EVALUATION

In Section 10.3 we already provided a rudimentary introduction to speech quality assessment since some of the most frequently used quality measures were invoked in our discussions. This chapter attempts to present a somewhat more detailed discussion of the topic. The major difficulty associated with the assessment of speech quality is the consequence of a philosophical dilemma. Should speech quality evaluation be based on unreliable, subjective human judgments or on reproducable objective evaluations, which may be highly uncorrelated with personal subjective quality assessments? Even high fidelity (HIFI) entertainment systems exhibit different subjective reproduction qualities, let alone low-bitrate speech codecs. It is practically impossible to select a generic set of objective measures in order to characterize speech quality because all codecs result in different speech impairments. Some objective measures appropriate for quantifying one type of distortion might be irrelevant to estimate another, just as one listener might prefer some imperfections to others. Using a statistically relevant, high number of trained listeners and various standardized tests mitigates the problems encountered but incurs cost and time penalties. During codec development, quick and cost-efficient objective preference tests are generally used, followed by informal listening tests, before a full-scale formal subjective test is embarked upon.

The literature of speech quality assessment was documented in a range of excellent treatises by Kryter [355], Jayant and Noll [10], and Kitawaki, Honda, and Itoh [61, 63].

In Reference [18] Papamichalis gives a comprehensive overview of the subject with references to Jayant's and Noll's work [10]. Further important contributions are due to Halka and Heute [356] as well as Wang, Sekey, and Gersho [357].

17.2 OBJECTIVE SPEECH QUALITY MEASURES

17.2.1 Introduction

Whether we evaluate the speech quality of a waveform codec, vocoder, or hybrid codec, objective distance measures are needed to quantify the deviation of the codec's output signal from the input speech. In this respect, any formal metric or distance measure of the mathematics, such as, for example, the Euclidean distance, could be employed to quantify the dissimilarity of the original and the processed speech signal, as long as symmetry, positive definitiveness, and the triangle inequality apply. These requirements were explicitly formulated as follows [62]:

- Symmetry: $d(x, y) = d(y, x)$,
- Positive Definitness: $d(x, x) = 0$ and $d(x, y) > 0$, if $x \neq y$,
- Triangular Inequality: $d(x, y) \leq d(x, z) + d(y, z)$.

In practice, the triangle inequality is not needed, but our distance measure should be easy to evaluate; preferably it ought to have some meaningful physical interpretation. The symmetry requires that there be no distinction between the reference signal and the speech to be evaluated in terms of distance. The positive definiteness implies that the distance is zero, if the reference and tested signals are identical.

A number of objective distance measures fulfill all criteria, some of which have waveform-related time-domain interpretations, while others have frequency-domain-related physical meaning. Often time-domain waveform codecs such as, for example, PCM are best characterized by the former, while frequency-domain codecs, like transform and sub-band codecs by the latter. Analysis-by-synthesis hybrid codecs using perceptual error weighting are the most difficult to characterize and usually only a combination of measures gives satisfactory results. Objective speech quality measures have been studied in depth by Quackenbush, Barnwell, and Clements [21]; hence, only a rudimentary overview is provided here.

The simplest and most widely used metrics or objective speech quality measures are the signal-to-noise ratios (SNR), such as the conventional SNR, the segmental SNR (SEGSNR), and the frequency-weighted SNR [358–359]. Since they are essentially quantifying the waveform similarity of the original and the decoded signal, they are most useful in terms of evaluating waveform-coder distortions. Nonetheless, they are often invoked in medium-rate codecs in order to compare different versions of the same codec, for example, during the codec development process.

Frequency-domain codecs are often best characterized in terms of the spectral distortion between the original and processed speech signal, evaluating it either on the basis of the spectral fine structure or—for example, when judging the quality of a spectral envelope quantiser—in terms of the spectral envelope distortion. Some of the often used measures are the spectral distance, log spectral distance, cepstral distance, log likelihood ratio, noise-masking ratios, and composite measures, most of which were proposed, for example, by Barnwell et al. [358–360] during the late 1970s and early 1980s. However, most of the above measures are inadequate for quantifying the subjective quality of a wide range of speech-coder distortions. They are particularly at fault predicting these quality degradations across different types of speech codecs. A particular deficiency of these measures is that when a range of different distortions are present simultaneously, these measures are incapable of evaluating the grade of the individual imperfections, although this would be desirable for codec developers.

Let us now consider some of the widely used objective measures in a little more depth.

17.2.2 Signal-to-Noise Ratios

For discrete-time, zero-mean speech signals, the error and signal energies of a block of N speech samples are given by:

$$E_e = \frac{1}{N} \sum_{u=1}^{N} (s(u) - \hat{s}(u))^2 \tag{17.1}$$

$$E_s = \frac{1}{N} \sum_{u=1}^{N} s^2(u). \tag{17.2}$$

Then the conventional signal-to-noise ratio (SNR) is computed as:

$$\text{SNR[dB]} = 10 \log_{10} (E_s/E_e). \tag{17.3}$$

When computing the arithmetic means in Equations 17.1, the gross averaging over long sequences conceals the codecs' low SNR performance in low-energy speech segments and attributes unreasonably high objective scores to the speech codec. Computation of the geometric mean of the SNR guarantees higher correlation with perceptual judgments because it gives proper weighting to the lower SNR performance in low-energy sections. This is achieved by computing the segmental SNR (SEGSNR). First, the speech signal is divided into segments of 10–20 ms, and SNR(u)[dB] is computed for $u = 1 \ldots N$, that is, for each segment in terms of dB. Then the segmental SNR(u) values are averaged in terms of dBs, as follows:

$$\text{SEGSNR[dB]} = \frac{1}{N} \sum_{n=1}^{N} \text{SNR}(u)\text{[dB]}. \tag{17.4}$$

Equation 17.4 effectively averages the logarithms of the SNR(u) values, which correspond effectively to the computation of the geometric mean. This gives proper weighting to low-energy speech segments and therefore gives values more closely related to the subjective quality of the speech codec. A further refinement is to limit the segmental SNR(u) terms to be in the range of $0 < \text{SNR}(u) < 40$[dB] because outside this interval it becomes uncorrelated with subjective quality judgments.

17.2.3 Articulation Index

A useful frequency-domain-related objective measure is the articulation index (AI) proposed by Kryter in Reference [355]. The speech signal is split into 20 sub-bands of increasing bandwidths, and the subband SNRs are computed. Their range is limited to $\text{SNR} = 30$ dB, and then the average SNR over the 20 bands is computed as follows:

$$\text{AI} = \frac{1}{20} \sum_{i=1}^{20} \text{SNR}_i. \tag{17.5}$$

The subjective importance of the sub-bands is weighted by appropriately choosing the bandwidth of all sub-bands, which then contribute 1/20th of the total SNR. An important observation is that Kryter's original band-splitting table stretches to 6100 Hz, and when using a bandwidth of 4 kHz, the two top bands falling beyond 4 kHz are therefore neglected, limiting AI inherently to 90%. When using $B = 3$ kHz, AI $\leq 80\%$. The evaluation of the AI is rather complex due to the band-splitting operation.

17.2.4 Cepstral Distance

The cepstral distance (CD) is the most highly correlated objective measure, compared to subjective measures. It maintains its high correlation over a wide range of codecs, speakers, and distortions while being reasonably simple to evaluate. It is defined in terms of the cepstral coefficients of the reference and tested speech, as follows:

$$CD = \left[\left(c_0^{in} - c_0^{out}\right)^2 + 2 \sum_{f=1}^{\infty} \left(c_j^{in} - c_j^{out}\right)^2 \right]^{1/2}. \tag{17.6}$$

The input and output cepstral coefficients are evaluated by the help of the linear predictive (LPC) filter coefficients a_j of the all-pole filter [62], which is elaborated on below.

The cepstral coefficients can be determined from the filter coefficients $a_i (i = 1 \ldots p)$ by the help of a recursive relationship, derived as follows. Let us denote the stable all-pole speech model by the polynomial $A(z)$ of order M in terms of z^{-1}, assuming that all its roots are inside the unit circle. It has been shown in Reference [361] that the following relationship holds for the Taylor series expansion of $\ln[A(z)]$:

$$\ln[A(z)] = -\sum_{k=1}^{\infty} c_k \cdot z^{-k}; \qquad c_0 = \ln(E_p/R_0), \tag{17.7}$$

where the coefficients c_k are the cepstral coefficients and c_0 is the logarithmic ratio of the prediction error and the signal energy. By substituting

$$A(z) = 1 + \sum_{k=1}^{\infty} a_k \cdot z^{-k} \tag{17.8}$$

or by exploiting that $a_0 = 1$:

$$A(z) = 1 + \sum_{k=0}^{M} a_k \cdot z^{-k}. \tag{17.9}$$

Upon differentiating the left-hand side of Equation 17.7 with respect to z^{-1}, we arrive at:

$$\frac{\delta[\ln A(z)]}{\delta z^{-1}} = \frac{1}{A(z)} \frac{\delta A(z)}{\delta z^{-1}} \tag{17.10}$$

$$\frac{\delta[\ln A(z)]}{\delta z^{-1}} = \frac{1}{\sum_{k=0}^{M} a_k \cdot z^{-k}} \sum_{k=1}^{M} k \cdot a_k \cdot z^{-(k-1)}. \tag{17.11}$$

Differentiating the right-hand side of Equation 17.7 as well as equating it to the differentiated left-hand side according to Equation 17.9 yields:

$$\left(\sum_{k=0}^{M} a_k \cdot z^{-k}\right)^{-1} \sum_{k=1}^{M} k \cdot a_k \cdot z^{-(k-1)} = -\sum_{k=1}^{\infty} k \cdot c_k \cdot z^{-(k-1)}. \tag{17.12}$$

Rearranging Equation 17.12 and multiplying both sides by z^{-1} result in Equation 17.13:

$$\sum_{k=1}^{M} k \cdot a_k \cdot z^{-k} = -\left(\sum_{k=0}^{M} a_k \cdot z^{-k}\right) \cdot \sum_{k=1}^{\infty} k \cdot c_k \cdot z^{-k}. \tag{17.13}$$

By expanding the indicated sums and performing the necessary multiplications, the following recursive equations result, which is demonstrated by an example in the next section, Section 17.2.5:

$$c_1 = -a_1 \tag{17.14}$$

$$c_j = -\frac{1}{j}\left(j \cdot a_j + \sum_{i=1}^{j-1} i \cdot c_i \cdot a_{j-i} \right); \qquad j = 2\ldots p \tag{17.15}$$

and by truncating the second sum in Equation 17.13 on the right-hand side at $2p$, since the higher order terms are of diminishing importance, we arrive at:

$$c_j = -\frac{1}{j}\sum_{k=1}^{p}(j-i) \cdot c_{j-i} \cdot a_i; \qquad j = (p+1)\ldots 2p. \tag{17.16}$$

Now, in possession of the filter coefficients the cepstral coefficients can be derived.

Having computed the cepstral coefficients $c_0 \ldots c_{2p}$, we can determine the CD as repeated below for convenience:

$$\mathrm{CD} = \left[(c_0^{\mathrm{in}} - c_0^{\mathrm{out}})^2 + 2 \cdot \sum_{j=1}^{2p}(c_j^{\mathrm{in}} - c_j^{\mathrm{out}})^2 \right]^{1/2} \tag{17.17}$$

$$c_1 = a_1$$

$$c_j = a_j - \sum_{r=1}^{j-1}\frac{r}{j} \cdot c_r \cdot a_{j-r} \qquad \text{for } j = 2 - p$$

$$c_j = -\sum_{r=1}^{p}\frac{j-r}{j}c_{j-r} \cdot a_r \qquad \text{for } j = p+1, p+2, -3p \tag{17.18}$$

where p is the order of the all-pole filter $A(z)$. The optimum predictor coefficients a_r are computed to minimize the energy of the prediction error residual:

$$e(u) = s(u) - \hat{s}(u). \tag{17.19}$$

This requires the solution of the following set of p equations:

$$\sum_{E=1}^{p} a_r \cdot R(|i-r|) = R(i) \qquad \text{for } i = 1\ldots p, \tag{17.20}$$

where the autocorrelation coefficients are computed from the segmented and Hamming-Windowed speech, as follows. First, the speech $s(u)$ is segmented into 20 ms or $N = 160$ sample long sequences. Then $s(u)$ is multiplied by the Hamming window function:

$$w_{(n)} = 0.54 - 0.45 \cos\frac{2\pi u}{N} \tag{17.21}$$

in order to smooth the frequency-domain oscillations introduced by the rectangular windowing of $s(u)$. Now the autocorrelation coefficients $R(i)i = 1\ldots p$ are computed from the windowed speech $s_w(n)$ as

$$R(i) = \sum_{n=0}^{N-1-i} s_w(u) - s_w(u+i) \qquad i = 1\ldots p. \tag{17.22}$$

Finally, Equation 17.20 is solved for the predictor coefficients $a(i)$ by the Levinson-Durbin algorithm [53]:

$$E(0) = R(0)$$

$$\epsilon_i = \left[\sum_{j=1}^{i-1} a_j^{i-1} \cdot R(i-j) \right] / E^{(i-1)} \qquad i = 1 \ldots p$$

$$a_{(i)}^{(i)} i = r_i \qquad (17.23)$$

$$a_j = a_j^{(i-1)} - k_i \cdot a_{i-j}^{(i-1)} \qquad j = 1 \ldots (i-1)$$

$$E^{(i)} = (1 - \epsilon_i^2)$$

where r_i, $i = 1 \ldots p$ are the reflection coefficients. After p iterations ($i = 1 \ldots p$), the set of LPC coefficients is given by:

$$a_j = a_j^{(p)} \qquad j = 1 \ldots p. \qquad (17.24)$$

and the prediction gain is given by $G = E(i)/E^{10}$. The computation of the CD is summarized in the flowchart of Figure 17.1.

It is plausible that the CD measure is a spectral domain parameter, since it is related to the LPC filter coefficients of the speech spectral envelope. In harmony with our expectations, it is shown in Reference [62] that the CD is identical to the logarithmic root mean square spectral distance (LRMS-SD) between the input and output spectral envelopes often used in speech quality evaluations:

$$\mathrm{LRMS - SD} = \left[\int_{-\pi}^{\pi} |\ln |b_{\mathrm{in}}/A_{\mathrm{in}}(f)||^2 - \ln |G_{\mathrm{out}}/A_{\mathrm{out}}(f)|^2 df \right]^{1/2}. \qquad (17.25)$$

In the next section we consider a simple example.

17.2.5 Example: Computation of Cepstral Coefficients

Let us make the derivation of Equations 17.14–17.17 plausible by expanding the sums in Equation 17.13 and by computing the multiplications indicated. Therefore, let us assume $p = 4$ and compute $c_1 \ldots c_{2p}$:

$$\begin{aligned}
a_1 z^{-1} &+ 2a_2 z^{-2} + 3a_3 z^{-3} + 4a_4 z^{-4} \\
&= -(c_1 z^{-1} + 2c_2 z^{-2} + 3c_3 z^{-3} + 4c_4 z^{-4} + 5c_5 z^{-5} + 6c_6 z^{-6} \\
&\quad + 7c_7 z^{-7} + 8c_8 z^{-8}) \cdot (1 + a_1 z^{-1} + a_2 z^{-2} + a_3 z^{-3} + a_4 z^{-4}) \qquad (17.26)
\end{aligned}$$

By computing the product at the right-hand side, we arrive at:

$$\begin{aligned}
a_1 z^{-1} &+ 2a_2 z^{-2} + 3a_3 z^{-3} + 4a_4 z^{-4} \\
&= c_1 z^{-1} + 2c_2 z^{-2} + 3c_3 z^{-3} + 4c_4 z^{-4} + 5c_5 z^{-5} + 6c_6 z^{-6} \\
&\quad + 7c_7 z^{-7} + 8c_8 z^{-8} + c_1 a_1 z^{-2} + 2c_2 a_1 z^{-3} + 3c_3 a_1 z^{-4} \\
&\quad + 4c_4 a_1 z^{-5} + 5c_5 a_1 z^{-6} + 6c_6 a_1 z^{-7} + 7c_7 a_1 z^{-8} + 8c_8 a_1 z^{-9} \\
&\quad + c_1 a_2 z^{-3} + 2c_2 a_2 z^{-4} + 3c_3 a_2 z^{-5} + 4c_4 a_2 z^{-6} + 5c_5 a_2 z^{-7} \\
&\quad + 6c_6 a_2 z^{-8} + 7c_7 a_2 z^{-9} + 8c_8 a_2 z^{-10} + c_1 a_3 z^{-4} + 2c_2 a_3 z^{-5} \\
&\quad + 3c_3 a_3 z^{-6} + 4c_4 a_3 z^{-7} + 5c_5 a_3 z^{-8} + 6c_6 a_3 z^{-9} \\
&\quad + 7c_7 a_3 z^{-10} + 8c_3 a_3 z^{-11}. \qquad (17.27)
\end{aligned}$$

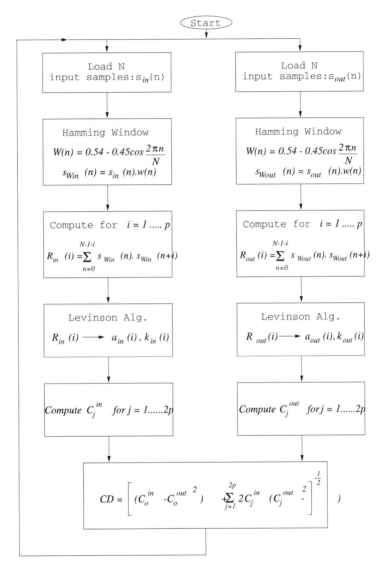

Figure 17.1 Cepstrum distance computation flowchart.

Now by matching the terms of equal order in z^{-1} on both sides:
z^{-1}:

$$c_1 = -a_1 \tag{17.28}$$

z^{-2}:

$$2a_2 = 2c_2 + a_1 c_1$$
$$2c_2 = -a_1 c_1 - 2a_2 \tag{17.29}$$

z^{-3}:

$$3c_3 = -3a_3 - 2a_1 c_2 - a_2 c_1 \tag{17.30}$$

z^{-4}:

$$4c_4 = -4a_4 - 3a_1c_3 - 2a_2c_2 \tag{17.31}$$

In general:

$$jc_j = -ja_j - \sum_{i=1}^{j-1} ic_ia_{j-i}; \qquad j = 1 \ldots p. \tag{17.32}$$

But also, there exists a number of terms with an order of higher than p that must cancel each other on the right-hand side of 17.27:

$$5c_5 + 4c_4a_1 + 3c_3a_2 + 2c_2a_3 + c_1a_4 = 0 \tag{17.33}$$

z^{-5}:

$$5c_5 = -4c_4a_1 - 3c_3a_2 - 2c_2a_3 - c_1a_4 \tag{17.34}$$

z^{-6}:

$$6c_6 = -5c_5a_1 - 4c_4a_2 - 3c_3a_3 - 2c_2a_4 \tag{17.35}$$

z^{-7}:

$$7c_7 = -6c_6a_1 - 5c_5a_2 - 4c_4a_3 - 3c_3a_4 \tag{17.36}$$

z^{-8}:

$$8c_8 = -7c_7a_1 - 6c_6a_2 - 5c_5a_3 - 4c_4a_4 \tag{17.37}$$

In general:

$$jc_j = -\sum_{i=1}^{p} (j-i)c_{j-i}a_i; \qquad j = p+1 \ldots \tag{17.38}$$

Let us now continue our review of various objective speech quality measures in the spirit of Papamichalis's discussions [18] in the next section.

17.2.6 Logarithmic Likelihood Ratio [18]

The likelihood ratio (LR) distance measure introduced by Itakura also uses the LPC coefficients of the input and output spectral envelope to quantify the spectral deviation introduced by the speech codec. The LR is defined as the ratio of the LPC residual energy before and after speech coding. Since the LPC coefficients $\underline{a}_{rin} = [a_0, a_1, \ldots a_p]$ are computed by Durbin's algorithm to minimize the LPC residual's energy, replacing $\underline{a}_{rin}$ by another LPC coefficient vector $\underline{a}_{rout}$ computed from the decoded speech certainly increases the LPC residual energy, therefore, LR ≥ 1.

The formal definition of the LR is given by

$$LR = \frac{\underline{a}_{out}^T \underline{R}^{out} \underline{a}_{out}}{\underline{a}_{in}^T \underline{R}^{in} \underline{a}_{in}} \tag{17.39}$$

where $\underline{a}_{in}, \underline{R}^{in}$ and $\underline{a}_{out}, \underline{R}^{out}$ represent the LPC filter coefficient vectors and autocorrelation matrices of the input as well as output speech, respectively. The LR measure defined in Equation 17.39 is nonsymmetric, which contradicts our initial requirements. Fortunately, this can be rectified by the symmetric transformation:

$$LRS = \frac{LR + 1/LR}{2} - 1. \tag{17.40}$$

Finally, the symmetric logarithmic LR (SLLR) is computed from:

$$\text{SLLR} = 10 \log_{10}(\text{LRS}). \tag{17.41}$$

The computational complexity incurred is significantly reduced if LR is evaluated instead of the matrix multiplications required by Equation 17.39 exploiting the following relationship:

$$\underline{a}^T \cdot \underline{Ra} = R_a(0)R(0) + 2\sum_{i=1}^{P} R_a(i) \cdot R(i), \tag{17.42}$$

where $R(i)$ and $R_a(i)$ represent the autocorrelation coefficients of the signal and that of the LPC filter coefficients $\underline{a}$, respectively.

17.2.7 Euclidean Distance

If any comprehensive set of spectral parameters closely related to the spectral deviation between input and output speech is available, the Euclidean instance between the sets of input and output speech parameters gives useful insights into the distortions inflicted. Potentially suitable sets are the LPC coefficients, the reflection coefficients computed in Equation 17.23, the autocorrelation coefficients given in Equation 17.22, the line spectrum frequencies (LSF) most often used recently, or the highly robust logarithmic area ratios (LAR). LARs are defined as

$$\text{LAR}_i = \ln \frac{1 + r_i}{1 - r_i} \qquad i = 1 \ldots p. \tag{17.43}$$

They are very robust against channel errors and have a fairly limited dynamic range, which alleviate their quantization. With this definition of LARs the Euclidean distance is formulated as:

$$D_{\text{LAR}} = \left[\sum_{i=1}^{p} \left(\text{LAR}_i^{\text{in}} - \text{LAR}_i^{\text{out}} \right)^2 \right]^{1/2}. \tag{17.44}$$

17.3 SUBJECTIVE MEASURES [18]

Once the development of a speech codec is finalized, objective and informal subjective tests are followed by formal subjective tests. Depending on the type, bitrate, and quality of the specific codec, different subjective tests are required to test quality and intelligibility. Quality is usually tested by the diagnostic acceptability measure (DAM), paired preference tests, or the most widespread mean opinion score (MOS). Intelligibility is tested by consonant-vowel-consonant (CVC) logatoms or by the dynamic rhythm test (DRT). Formal subjective speech assessment is generally a lengthy investigation carried out by a specially trained unbiased crew using semistandardized test material, equipment, and conditions.

17.3.1 Quality Tests

In diagnostic acceptability measure tests, the trained listener is asked to rate the speech codec tested using phonetically balanced sentences from the Harvard list in terms of both speech and background quality. Some terms used at Dynastat (USA) to describe speech imperfections are listed following Papamichalis in Table 17.1 [18]. As regards background qualities, the sort of terms used at Dynastat were summarized following Papamichalis in Table 17.2 [18]: The speech and background qualities are rated in the listed categories on a

TABLE 17.1   Typical Terms to Characterize Speech Impairments in DAM Tests © Papamichalis [18], 1987

Speech Impairment	Typical of
Fluttering	Amplitude modulated speech
Thin	High-pass filtered Speech
Rasping	Peak clipped speech
Muffled	Low-pass-filtered Speech
Interrupted	Packetized speech
Nasal	Low-bitrate vocoders

TABLE 17.2   Typical Terms for Background Qualities in DAM Tests © Papamichalis [18], 1987

Background	Typical of
Hissing	Noisy speech
Buzzing	Tandemed dig systems
Babbling	Low-bitrate codecs with bit errors
Rumbling	Low-frequency-noise marked speech

100-point scale by each listener, and then their average scores are evaluated for each category, giving also the standard deviations and standard errors. Before averaging the results of various categories, appropriate weighting factors can be used to emphasize features that are particularly important for a specific application of the codec.

In *pairwise preference tests*, the listeners always compare the same sentence processed by two different codecs, even if a high number of codecs has to be tested. To ensure consistency in the preferences unprocessed and identically processed sentences can also be included. The results are summarized in the preference matrix. If the comparisons show a clean preference order for differently processed speech and an approximately random preference (50%) for identical codecs in the preference matrix's main diagonal, the results are accepted. However, if no clear preference order is established, different tests have to be deployed.

17.4 COMPARISON OF SUBJECTIVE AND OBJECTIVE MEASURES

17.4.1 Background

An interesting comparison of the objective articulation index (AI) described in Section 17.2.3 and of various subjective tests was given by Kryter [362], as shown in Figure 17.3. Observe that the lower the size of the test vocabulary used, the higher are the intelligibility scores for a fixed AI value, which is due to the less subtle differences inherent in a smaller test vocabulary.

The modulated noise reference unit (MNRU) proposed by Law and Seymour [147] to relate subjective quality to objective measures is widely used by the CCITT as well. The MNRU block diagram is shown in Figure 17.2.

The MNRU is used to add noise, amplitude modulated by the speech test material, to the reference speech signal, rendering the noise speech-correlated. The SNR of the reference signal is gradually lowered by the listener using the attenuaters in Figure 17.2 to perceive

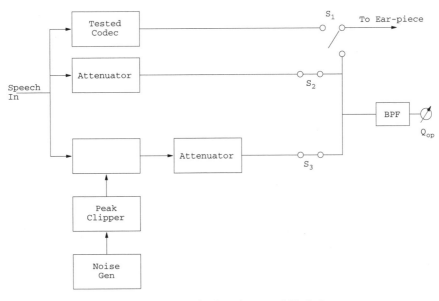

Figure 17.2 Modulated noise reference unit block diagram.

identical loudness and subjective qualities, when comparing the noisy reference signal and the tested codec's output speech. During this adjustment and comparison phase, the switches 52 and 53 are closed, and 51 is being switched between the reference and tested speech signals. Once both speech signals make identical subjective impressions, switches 52 and 53 are used to measure the reference signal's and noise signal's power and hence the opinion equivalent Q [dB] (Q_{op}) expressed in terms of the SNR computed. Although the Q_{op}, [dB] value appears to be an objective measured value, it depends on various listeners' subjective judgments and therefore is classified as a subjective measure. The Q_{op} [dB] value is easily translated into the more easily interpreted MOS measure using the reference speech's MOS vs. Q_{op} characteristic depicted in Figure 17.3.

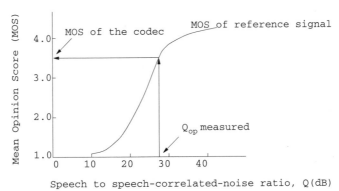

Figure 17.3 Stylized conversion function for translating Q_{op} into MOS.

1.4.2 Intelligibility tests

In intelligibility tests, the listeners are asked to recognize which one of a pair of words is uttered, where the two words only differ in one phoneme, which is a consonant [363]. Alternatively, consonant-vowel-consonant (CVC) logatoms can also be used. According to Papamichalis in the diagnostic rhyme test (DRT) developed by Dynastat [18] a set of 96 rhyming pairs of words are utilized, some of which are meat-beat, pear-tear, saw-thaw, and bond-pond. The pairs are specially selected to test the following phonetic attributes: voicing, nasality, sustention, sibilation, graveness, and compactness. If, for example, the codec under test consistently fails to distinguish between vast-fast, zoo-sue, goat-coat—that is, to deliver clear voiced sounds such as v, z, and g—it points out for the designer that the codec's long-term predictor responsible for the spectral fine structure, or voicing information in the spectrum, does not work properly. By consistently grouping and evaluating the recognition failures, vital information can be gained about the codecs' shortcomings. Typical DRT values are between 75 and 95 and for high intelligibility DRT > 90 is required.

In similar fashion, most objective and subjective measures can be statistically related to each other, but the goodness of match predicted for new codecs varies over a wide range. For low-bitrate codecs, one of the most pertinent relationships devised is [63]:

$$MOS = 0.04CD_2 - 0.80CD + 3.565. \qquad (17.45)$$

This formula is the best second-order fit to a high number of MOS-CD measurements carried out over a variety of codecs and imperfections.

In summary, speech quality evaluation is usually based on quick objective assessments during codec development, followed by extensive formal subjective tests, when the development is finalized. A range of objective and subjective measures was described, where the most popular objective measures are the simple time-domain SEG SNR [dB] and the somewhat more complex, frequency-domain CD [dB] measure. The CD objective measure is deemed to have the highest correlation with the most widely applicable subjective measure, the MOS, and their relationship is expressed in Equation 17.45. Having reviewed a variety of objective and subjective speech quality measures, let us now compare a range of previously considered speech codecs in the next section.

17.5 SUBJECTIVE SPEECH QUALITY OF VARIOUS CODECS

In previous chapters we have characterized many different speech codecs. Here we attempt a rudimentary comparison of some of the previously described codec schemes in terms of their subjective and objective speech quality as well as error sensitivity. We will conclude this chapter by incorporating some of the codecs concerned in various wireless transceivers and portray their SEGSNR versus channel SNR performance. Here we refer back to Figure 1.6, and with reference to Cox's work [1, 2] we populate this figure with actual formally evaluated mean opinion score (MOS) values, which are shown in Figure 17.4. Observe that over the years a range of speech codecs have emerged, which attained the quality of the 64 kbps G.711 PCM speech codec, although at the cost of significantly increased coding delay and implementational complexity. The 8 kbps G.729 codec is the most recent addition to this range of ITU standard schemes, which significantly outperform all previous standard ITU codecs in terms of robustness. The performance target of the 4 kbps ITU codec (ITU4) is also to maintain this impressive set of specifications. The family of codecs, which were designed for various mobile radio systems, such as the 13 kbps RPE GSM scheme, the 7.95 kbps IS-54, the IS-96, the 6.7 kbps JDC, and 3.45 kbps half-rate JDC arrangement (JDC/2), exhibits

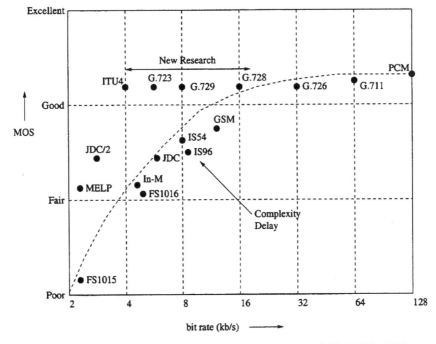

Figure 17.4 Subjective speech quality of various codecs. Cox *et al.* [1] © IEEE, 1996.

slightly lower MOS values than the ITU codecs. Let us now consider the subjective quality of these schemes in a little more depth.

The subjective speech quality of a range of speech codecs is characterized in Figure 17.4. While during our introductory discussions we portrayed the waveform coding, vocoding, and hybrid coding families in a similar, but more inaccurate, stylized illustration, this figure is based on large-scale formal comparative studies.

The 2.4 kbps Federal Standard codec FS-1015 is the only vocoder in this group, and it has a rather synthetic speech quality associated with the lowest subjective assessment in the figure. The 64 kbps G.711 PCM codec and the G.726/G.727 ADPCM schemes are waveform codecs. They exhibit a low implementational complexity associated with a modest bitrate economy. The remaining codecs belong to the hybrid coding family and achieve significant bitrate economies at the cost of increased complexity and delay.

Specifically, the 16 kbps G.728 backward-adaptive scheme maintains a similar speech quality to the 32 and 64 kbps waveform codecs, while also maintaining an impressively low 2 ms delay. This scheme was standardized during the early 1990s. The similar quality, but significantly more robust, 8 kbps G.729 codec was approved in March 1996 by the ITU. This activity overlapped with the G.723.1 developments. The 6.4 kbps mode maintains a speech quality similar to the G.711, G.726, G.727, G.728, and G.728 codecs, while the 5.3 mode exhibits a speech quality similar to the cellular speech codecs of the late 1980s. Work is under way at the time of writing to standardize a 4 kbps ITU scheme, which we refer to here as ITU4.

In parallel to the ITU's standardization activities, a range of speech coding standards have been proposed for regional cellular mobile systems. The standardization of the 13 kbps RPE-LTP full-rate GSM (GSM-FR) codec dates back to the second half of the eighties, representing the first standard hybrid codec. Its complexity is significantly lower than that of the more recent CELP-based codecs. Observe in the figure that there is also an identical-rate enhanced full-rate GSM codec (GSM-EFR), which matches the speech quality of the G.729 and G.728 schemes. The original GSM-FR codec's development was followed a little later by the release of the 8 kbps VSELP IS-54 American cellular standard. Through advances in the field, the 7.95 kbps IS-54 codec achieved a similar subjective speech quality to the 13 kbps GSM-FR scheme. The definition of the 6.7 kbps Japanese JDC VSELP codec was almost coincident with that of the IS-54 arrangement. This codec development was also followed by a half-rate standardization process, leading to the 3.2 kbps Pitch Synchronous Innovation CELP (PSI-CELP) scheme. The IS-96 American CDMA system also has its own standardized CELP-based speech codec, which is a variable-rate scheme, allowing bitrates between 1.2 and 14.4 kbps, depending on the prevalent voice activity. The perceived speech quality of these cellular speech codecs contrived mainly during the late 1980s was found to be subjectively similar to each other under the perfect channel conditions of Figure 17.4. Lastly, the 5.6 kbps half-rate GSM codec (GSM-HR) also met its specification in terms of achieving a similar speech quality to the 13 kbps original GSM-FR arrangements, although at the cost of quadruple complexity and higher latency.

17.6 ERROR SENSITIVITY COMPARISON OF VARIOUS CODECS

As a rudimentary objective speech quality measure-based bit-sensitivity comparison, in Figure 17.5 we portrayed the SEGSNR degradations of a number of speech codecs for a range of bit error rates (BER), when applying random errors. The SEGSNR degradation is in general not a reliable measure of speech quality. Nonetheless, it indicates adequately how rapidly this objective speech quality measure decays for the various codecs, when exposed to a given fixed BER. As expected, the backward-adaptive G.728 and the forward-adaptive G.723.1 schemes, which have been designed mainly for benign wireline connections, have the fastest SGSNR degradation upon increasing the BER. By far the best performance is exhibited by the G.729 scheme, followed by the 13 kbps GSM codec. In the next section

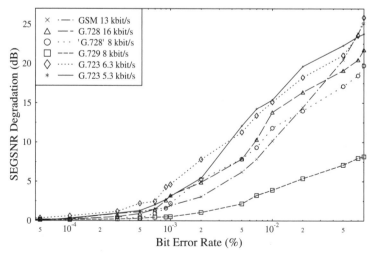

Figure 17.5 SEGSNR degradation versus BER for the investigated speech codecs.

we highlight how these codecs perform over Gaussian and Rayleigh-fading channels using three different transceivers.

17.7 OBJECTIVE SPEECH PERFORMANCE OF VARIOUS TRANSCEIVERS

In this section we embarked on a comparison of the previously analyzed speech codecs under identical experimental circumstances, when used in identical transceivers over both Gaussian and Rayleigh channels. These results are portrayed in Figures 17.6–17.11, which are detailed

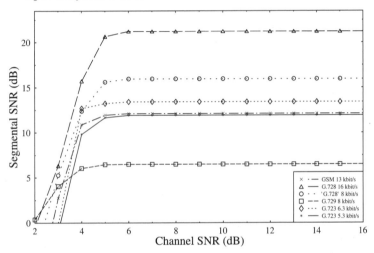

Figure 17.6   SEGSNR versus channel SNR performance of various speech codecs using the BCH(254,130,18) code and BPSK over Gaussian channels.

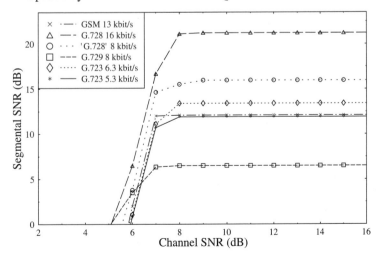

Figure 17.7   SEGSNR versus channel SNR performance of various speech codecs using the BCH(254,130,18) code and 4QAM over Gaussian channels.

Speech System Performance with 16QAM over Gaussian Channel

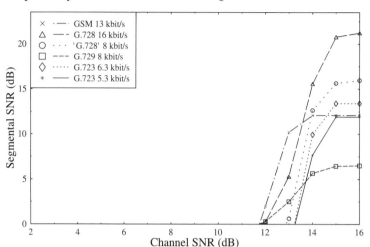

Figure 17.8   SEGSNR versus channel SNR performance of various speech codecs using
the BCH(254,130,18) code and 16QAM over Gaussian channels.

during our further discourse. Three different modems—1, 2, and 4 bits/symbol binary phase
shift keying (BPSK), 4-level quadrature amplitude modulation (4QAM), and 16QAM—were
employed in conjunction with the six different modes of operations of the four speech codecs
that were protected by the BCH(254,130,18) channel codec. Note that here no specific
source-sensitivity matched multiclass channel coding was invoked in order to ensure identical
experimental conditions for all speech codecs. Although in general the SEGSNR is not a

Speech System Performance with BPSK over Rayleigh Channel

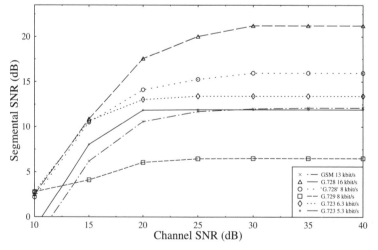

Figure 17.9   SEGSNR versus channel SNR performance of various speech codecs using
the BCH(254,130,18) code and BPSK over Rayleigh channels.

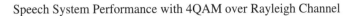

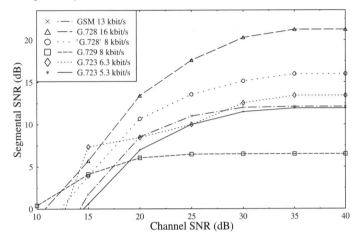

Figure 17.10   SEGSNR versus channel SNR performance of various speech codecs using the BCH(254,130,18) code and 4QAM over Rayleigh channels.

good absolute measure, when comparing speech codecs operating on the basis of different coding algorithms, it can be used as a robustness indicator, exhibiting a decaying characteristic for degrading channel conditions and hence allowing us to identify the minimum required channel SNRs for the various speech codecs and transceiver modes. Hence here we opted to use the SEGSNR in these comparisons, giving us an opportunity to point out its weaknesses on the basis of our a priori knowledge as regards the codecs' formally established subjective quality.

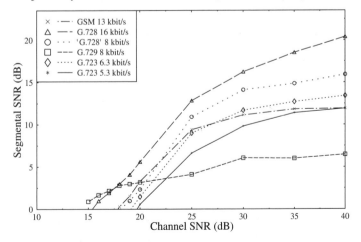

Figure 17.11   SEGSNR versus channel SNR performance of various speech codecs using the BCH(254,130,18) code and 16QAM over Rayleigh channels.

TABLE 17.3 Minimum Required Channel SNR for Maintaining Less Than 1 dB SEGSNR Degradation for the Investigated Speech Transceivers Using the BCH(254,130,18) Code and BPSK, 4QAM as well as 16QAM over both Gaussian and Rayleigh Channels

Codec	Rate (kbps)	BPSK		4QAM		16QAM	
		AWGN	Ray.	AWGN	Ray.	AWGN	Ray.
GSM	13	4	20	7	27	13	34
G.728	16	5	26	8	30	15	40
'G.728'	8	5	25	7	31	15	35
G.729	8	4	19	7	20	14	28
G.723.1	6.4	4	18	8	31	15	35
G.723.1	5.3	4	19	7	29	15	35

Under error-free transmission and no background-noise conditions, the subjective speech quality of the 16 kbps G.728 scheme, the 8 kbps G.729 codec, and the 6.4 kbps G.723.1 arrangement is characterized by a mean opinion score (MOS) of approximately 4. In other words, their perceived speech quality is quite similar, despite their different bitrates. Their similar speech quality at such different bitrates is a ramification of the fact that they represent different milestones in the evolution of speech codecs, since they were contrived in the above chronological order. They also exhibit different implementational complexities. The 13 kbps GSM codec and the 5.3 kbps G.723 arrangements are slightly inferior in terms of their subjective quality, both of which are characterized by an MOS of about 3.5. We note here, however that there exists a recently standardized enhanced full-rate, 13 kbps GSM speech codec, which also has an MOS of about four under perfect channel conditions.

These subjective speech qualities are not reflected by the corresponding SEGSNR curves portrayed in Figures 17.6–17.11. For example, the 8 kbps G.729 codec has the lowest SEGSNR, although it has an MOS similar to G.728 and the 6.4 kbps G.723.1 schemes in

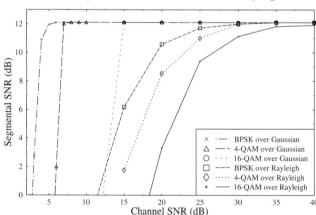

Figure 17.12   SEGSNR degradation versus channel SNR performance of the 13 kbps RPE-LTP GSM speech codec using the BCH(254,130,18) code and BPSK, 4QAM as well as 16QAM over both Gaussian and Rayleigh channels.

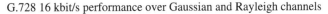

G.728 16 kbit/s performance over Gaussian and Rayleigh channels

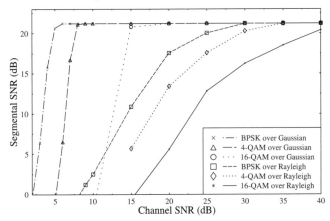

Figure 17.13 SEGSNR degradation versus channel SNR performance of the 16 kbps backward-adaptive G.728 speech codec using the BCH(254,130,18) code and BPSK, 4QAM as well as 16QAM over both Gaussian and Rayleigh channels.

terms of subjective speech quality. As expected, this is due to the high-pass filtering operation at its input, as well as a consequence of the more pronounced perceptually motivated speech quality optimisation, as opposed to advocating high-quality waveform reproduction. A further interesting comparison is offered by the 8 kbps G.728-like nonstandard codec, which exhibits a higher SEGSNR than the identical bitrate G.729 scheme but sounds significantly inferior to the G.729 arrangement. These differences become even more conspicuous when they are exposed to channel errors in the low-SNR region of the curves. In terms of error resilience, the G.729 scheme is by far the best in the group of codecs tested. The minimum required

G.729 8 kbit/s performance over Gaussian and Rayleigh channels

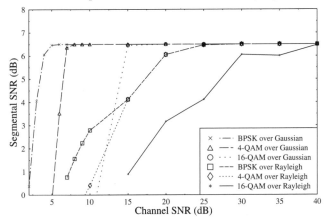

Figure 17.14 SEGSNR degradation versus channel SNR performance of the 8 kbps forward-adaptive G.729 speech codec using the BCH(254,130,18) code and BPSK, 4QAM as well as 16QAM over both Gaussian and Rayleigh channels.

G.728 16 kbit/s performance over Gaussian and Rayleigh channels

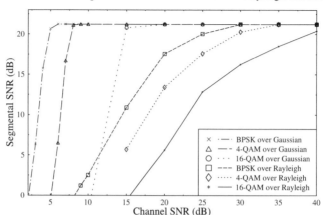

Figure 17.15   SEGSNR degradation versus channel SNR performance of the 5.3 kbps
              G.723.1 speech codec using the BCH(254,130,18) code and BPSK,
              4QAM as well as 16QAM over both Gaussian and Rayleigh channels.

channel SNR values for the various transceivers over the Gaussian and Rayleigh channels are
summarized in Table 17.3. Observe in the Rayleigh-channel curves of Figures 17.9–17.11
that the backward-adaptive codecs have a rapidly decaying performance curve, whereas, for
example, the G.729 forward-adaptive ACELP scheme exhibits a more robust behavior. Lastly,
in Figures 17.12–17.15 we organized our previous results in a different way, plotting all the
different SEGSNR versus channel SNR curves related to a specific speech codec in the same
figure, allowing a direct comparison of the expected speech performance of the various
transceivers over various channel conditions.

Constructing the Quadratic
Spline Wavelets

Previously, in Equation 12.4 the Fourier transform of a wavelet was determined. Hence, the Fourier transform of a father wavelet $\phi(x)$ is given by:

$$\hat{\phi}(\omega) = \int_{-\infty}^{\infty} \phi(x)e^{-jx\omega}dx. \tag{A.1}$$

Using the two-scale difference equation of 12.13, we can rewrite the above equation to achieve:

$$\hat{\phi}(\omega) = \sqrt{2} \sum_{n=-\infty}^{\infty} h_n \int_{-\infty}^{\infty} \phi(2x-n)e^{-jx\omega}dx. \tag{A.2}$$

Here we introduce the beneficial substitution $y = 2x - n$, allowing Equation A.2 to be rewritten as:

$$\hat{\phi}(\omega) = \frac{1}{\sqrt{2}} \sum_{n=-\infty}^{\infty} h_n e^{-jn\omega/2} \int_{-\infty}^{\infty} \phi(y)e^{-jy\omega/2}dy. \tag{A.3}$$

Employing the identity:

$$H(\omega) = \frac{1}{\sqrt{2}} \sum_{n=-\infty}^{\infty} h_n e^{-jn\omega} \tag{A.4}$$

Equation A.3 becomes:

$$\hat{\phi}(\omega) = H\left(\frac{\omega}{2}\right)\phi\left(\frac{\omega}{2}\right) \tag{A.5}$$

Iterating this result, we achieve:

$$\hat{\phi}(\omega) = \left[\prod_{i=1}^{I} H(2^{-i}\omega)\right]\phi(2^{-I}\omega) \qquad \text{for } I = 1, 2, \ldots \tag{A.6}$$

Observing that as $I \Rightarrow \infty \hat{\phi}(0) = \int \phi(x)dx = 1$, leaving:

$$\hat{\phi}(\omega) = \prod_{i=1}^{\infty} H(2^{-i}\omega). \tag{A.7}$$

Similarly, for the mother wavelet if we use the identity:

$$G(\omega) = \frac{1}{\sqrt{2}} \sum_{n=-\infty}^{\infty} g_n e^{-jn\omega} \tag{A.8}$$

we find that:

$$\hat{\psi}(\omega) = \prod_{i=1}^{\infty} G(2^{-i}\omega). \tag{A.9}$$

Mallat and Zhong [322] defined the coefficients of the father wavelet $\hat{\phi}(\omega)$ as a spline function given by:

$$H(\omega) = \left[\cos\left(\frac{\omega}{2} \right) \right]^{2n+1} \tag{A.10}$$

where $2n + 1$ is the order of the spline function. Thus, the father wavelet of Equation A.7 is given by:

$$\hat{\phi}(\omega) = \prod_{i=1}^{\infty} \left[\cos\left(2^{-i}\frac{\omega}{2} \right) \right]^{2n+1}. \tag{A.11}$$

Expanding this equation becomes:

$$\hat{\phi}(\omega) = \left[\frac{e^{j\omega/2} + 1}{2e^{j\omega/4}} \times \frac{e^{j\omega/4} + 1}{2e^{j\omega/8}} \times \cdots \right]^{2n+1}. \tag{A.12}$$

Considering the denominator as a series, we find that:

$$e^{j\omega/4} \times e^{j\omega/8} \times \cdots \Rightarrow e^{j\omega/2} \text{ as } i \Rightarrow \infty. \tag{A.13}$$

Considering the numerator as a series, we find that:

$$\frac{e^{j\omega/2} + 1}{2} \times \frac{e^{j\omega/4} + 1}{2} \times \cdots \Rightarrow \frac{1 - e^{j\omega}}{2^I(1 - e^{-j\omega/2^I})} \text{ as } i \Rightarrow I. \tag{A.14}$$

Using L'Hôpital's rule, we see that:

$$\frac{2^I(1 - e^{j\omega})}{1 - e^{-j\omega/2^I}} \Rightarrow \frac{e^{j\omega} - 1}{j\omega} \text{ as } I \Rightarrow \infty. \tag{A.15}$$

Hence, the father wavelet is given by:

$$\hat{\phi}(\omega) = \left[\frac{\sin\left(\frac{\omega}{2} \right)}{\frac{\omega}{2}} \right]^{2n+1}. \tag{A.16}$$

Mallat and Zhong [322] defined the coefficients of the mother wavelet $\hat{\psi}(\omega)$ by:

$$G(\omega) = 4j \sin\left(\frac{\omega}{4} \right). \tag{A.17}$$

The mother wavelet $\hat{\psi}(\omega)$ can now be calculated using the Fourier domain version of Equation 12.14, namely:

$$\hat{\psi}(\omega) = G\left(\frac{\omega}{2}\right)\hat{\phi}\left(\frac{\omega}{2}\right). \tag{A.18}$$

Substituting in $G(\omega)$ and $\hat{\phi}(\omega)$, we achieve:

$$\hat{\psi}(\omega) = 4j\sin\left(\frac{\omega}{4}\right) \cdot \left(\frac{\sin\left(\frac{\omega}{4}\right)}{\frac{\omega}{4}}\right)^{2n+1}, \tag{A.19}$$

producing this mother wavelet:

$$\hat{\psi}(\omega) = j\omega\left(\frac{\sin\left(\frac{\omega}{4}\right)}{\frac{\omega}{4}}\right)^{2n+2}, \tag{A.20}$$

Mallat and Zhong [322] implemented a polynomial where $2n + 1 = 3$, and subsequently they introduced a shifting constant w_{sc} that ensures $\psi(x)$ is antisymmetrical with regards to 0 and $\phi(x)$ is symmetrical about 0. This shifting constant w_{sc} is set to $\frac{1}{2}$ and added to the filter coefficients producing the following filter coefficients:

$$H(\omega) = e^{j\omega/2}\left[\cos\left(\frac{\omega}{2}\right)\right]^3 \tag{A.21}$$

$$G(\omega) = 4je^{j\omega/2}\sin\left(\frac{\omega}{2}\right) \tag{A.22}$$

and the following father and mother wavelets, respectively:

$$\hat{\phi}(\omega) = \left[\frac{\sin\left(\frac{\omega}{2}\right)}{\frac{\omega}{2}}\right]^3 \tag{A.23}$$

$$\hat{\psi}(\omega) = j\omega\left[\frac{\sin\left(\frac{\omega}{4}\right)}{\frac{\omega}{4}}\right]^4. \tag{A.24}$$

Using the identities of Equations A.4 and A.8 we can determine the time-domain filter coefficients for $h(n)$ and $g(n)$. Thus:

$$H(\omega) = e^{j\omega/2}\left[\frac{1 + e^{j\omega}}{2e^{j\omega/2}}\right]^3, \tag{A.25}$$

simplifying to become:

$$H(\omega) = \tfrac{1}{8}(e^{-j\omega} + 3 + 3e^{j\omega} + e^{2j\omega}), \tag{A.26}$$

producing the coefficients

$$h_{-1} = \frac{\sqrt{2}}{8}, \quad h_0 = \frac{3\sqrt{2}}{8}, \quad h_1 = \frac{3\sqrt{2}}{8} \quad \text{and} \quad h_2 = \frac{\sqrt{2}}{8}.$$

Similarly, for $g(n)$:

$$G(\omega) = 4je^{j\omega/2}\left[\frac{e^{j\omega/2} - e^{-j\omega/2}}{2j}\right], \tag{A.27}$$

simplifying, to become:

$$G(\omega) = 2e^{j\omega} - 2, \tag{A.28}$$

producing the coefficients $g_0 = -2\sqrt{2}$ and $g_1 = 2\sqrt{2}$.

B

Zinc Function Excitation

This appendix details the approach required to minimize the weighted error signal using ZFE, from Section 13.3.1. Following the approach of Hiotakakos and Xydeas [294] and Sukkar et al. [295], the noise weighted error signal is given by:

$$E_w^{k+1}(n) = \sum_{n=1}^{\text{excint}} [e_w^{k+1}(n)]^2 \tag{B.1}$$

and using Equations 13.1 to 13.6

$$E_w^{k+1}(n) = \sum_{n=1}^{\text{excint}} \left[e_w^k(n) - [A_{k+1} \sin c(n - \lambda_{k+1}) + B_{k+1} \cos c(n - \lambda_{k+1})] * h(n) \right]^2 \tag{B.2}$$

where A_{k+1} and B_{k+1} are the amplitude parameters for the $(k + 1)$ ZFE, and λ_{k+1} is the position parameter for the $(k + 1)$ ZFE.

In order to minimize the above expression as a function of A_{k+1}, we differentiate it with respect to A_{k+1}, giving:

$$\frac{\delta E_w^{k+1}(n)}{\delta A_{k+1}} = -2 \sum_{n=1}^{\text{excint}} [\sin c(n - \lambda_{k+1}) * h(n)]$$

$$\times [e_w^k(n) - [A_{k+1} \sin c(n - \lambda_{k+1}) + B_{k+1} \cos c(n - \lambda_{k+1})] * h(n)].$$

$$= 0 \tag{B.3}$$

Expanding the above expression yields:

$$A_{k+1} \cdot \sum_{n=1}^{\text{excint}} [\sin c(n - \lambda_{k+1}) * h(n)]^2 = \sum_{n=1}^{\text{excint}} e_w^k(n)[\sin c(n - \lambda_{k+1}) * h(n)]$$

$$- B_{k+1} \cdot \sum_{n=1}^{\text{excint}} \cos c(n - \lambda_{k+1}) * h(n)$$

$$\times \sum_{n=1}^{\text{excint}} \sin c(n - \lambda_{k+1}) * h(n) \qquad \text{(B.4)}$$

and upon introducing the shorthand:

$$R_{ss} = \sum_{n=1}^{\text{excint}} [\sin c(n - \lambda_{k+1}) * h(n)]^2 \qquad \text{(B.5)}$$

$$R_{es} = \sum_{n=1}^{\text{excint}} [\sin c(n - \lambda_{k+1}) * h(n)] \times e_w^k(n) \qquad \text{(B.6)}$$

$$R_{cs} = \sum_{n=1}^{\text{excint}} [\sin c(n - \lambda_{k+1}) * h(n)] \times [\cos c(n - \lambda_{k+1}) * h(n)] \qquad \text{(B.7)}$$

$$R_{cc} = \sum_{n=1}^{\text{excint}} [\cos c(n - \lambda_{k+1}) * h(n)]^2 \qquad \text{(B.8)}$$

$$R_{ec} = \sum_{n=1}^{\text{excint}} [\cos c(n - \lambda_{k+1}) * h(n)] \times e_w^k(n) \qquad \text{(B.9)}$$

we have:

$$A_{k+1} = \frac{R_{es} - B_{k+1} \times R_{cs}}{R_{ss}}. \qquad \text{(B.10)}$$

Similarly, if we differentiate Equation B.3 with respect to B_{k+1} we arrive at:

$$\frac{\delta E_w^{k+1}(n)}{\delta B_{k+1}} = -2 \sum_{n=1}^{\text{excint}} [\cos c(n - \lambda_{k+1}) * h(n)]$$

$$\times [e_w^k(n) - [A_{k+1} \sin c(n - \lambda_{k+1}) + B_{k+1} \cos c(n - \lambda_{k+1})] * h(n)].$$

$$= 0 \qquad \text{(B.11)}$$

Expanding this yields:

$$B_{k+1} \cdot \sum_{n=1}^{\text{excint}} [\cos c(n - \lambda_{k+1} * h(n)]^2 = \sum_{n=1}^{\text{excint}} e_w^k(n) \cdot [\cos c(n - \lambda_{k+1}) * h(n)]$$

$$- A_{k+1} \cdot \sum_{n=1}^{\text{excint}} \cos c(n - \lambda_{k+1}) * h(n)$$

$$\times \sum_{n=1}^{\text{excint}} \sin c(n - \lambda_{k+1}) * h(n) \qquad \text{(B.12)}$$

and upon introducing the shorthand of Equations B.7 to B.8, we arrive at:

$$B_{k+1} = \frac{R_{ec} - A_{k+1} \times R_{cs}}{R_{cc}}. \qquad \text{(B.13)}$$

Since the terms $\cos c(n - \lambda_{k+1})$ and $\sin c(n - \lambda_{k+1})$ are orthogonal, their cross-correlation term R_{cs} will be zero. Hence, from Equations B.10 and B.13 we have:

$$A_{k+1} = \frac{R_{es}}{R_{ss}} \tag{B.14}$$

$$B_{k+1} = \frac{R_{ec}}{R_{cc}}. \tag{B.15}$$

If we now substitute Equations B.14 and B.15, and Equations B.5 to B.9 back into the original error expression of Equation B.3, then we arrive at:

$$E_w^{k+1} = \sum_{n=1}^{\text{excint}} [e_w^k(n) - [A_{k+1} \sin c(n - \lambda_{k+1}) + B_{k+1} \cos c(n - \lambda_{k+1})] * h(n)]^2. \tag{B.16}$$

Expanding with the squared term yields:

$$E_w^{k+1} = \sum_{n=1}^{\text{excint}} e_w^k(n)$$

$$- \sum_{n=1}^{\text{excint}} 2e_w^k(n)[A_{k+1} \sin c(n - \lambda_{k+1}) + B_{k+1} \cos c(n - \lambda_{k+1})] * h(n)$$

$$+ \sum_{n=1}^{\text{excint}} [[A_{k+1} \sin c(n - \lambda_{k+1}) + B_{k+1} \cos c(n - \lambda_{k+1})] * h(n)]^2 \tag{B.17}$$

which is constituted by three distinct expressions. The first expression is given by:

$$X = \sum_{n=1}^{\text{excint}} e_w^k(n)^2 \tag{B.18}$$

with the second by:

$$Y = \sum_{N=1}^{\text{excint}} 2e_w^k(n)[(A_{k+1} \sin c(n - \lambda_{k+1}) + B_{k+1} \cos c(n - \lambda_{k+1})) * h(n)] \tag{B.19}$$

which can be further simplified using Equations B.5 to B.9, yielding:

$$Y = \sum_{n=1}^{\text{excint}} 2A_{k+1} \cdot e_w^k(n) \cdot [\sin c(n - \lambda_{k+1}) * h(n)]$$

$$+ \sum_{n=1}^{\text{excint}} 2B_{k+1} \cdot e_w^k(n) \cdot [\cos c(n - \lambda_{k+1}) * h(n)]$$

$$= 2A_{k+1} \cdot R_{es} + 2B_{k+1} \cdot R_{ec}. \tag{B.20}$$

The third term from Equation B.17 is given by:

$$Z = \sum_{n=1}^{\text{excint}} [[A_{k+1} \sin c(n - \lambda_{k+1}) + B_{k+1} \cos c(n - \lambda_{k+1})] * h(n)]^2 \tag{B.21}$$

which can be expanded to:

$$Z = \sum_{n=1}^{\text{excint}} A_{k+1}^2 \cdot [\sin c(n - \lambda_{k+1}) * h(n)]^2$$

$$+ \sum_{n=1}^{\text{excint}} 2A_{k+1} \cdot B_{k+1} \cdot [\sin c(n - \lambda_{k+1}) * h(n)] \cdot [\cos c(n - \lambda_{k+1}) * h(n)]$$

$$+ \sum_{n=1}^{\text{excint}} B_{k+1}^2 \cdot [\cos c(n - \lambda_{k+1}) * h(n)]^2. \tag{B.22}$$

If this expression is simplified using Equations B.5 to B.9 and remembering that R_{cs} was equal to zero, we get:

$$Z = A_{k+1}^2 \cdot R_{ss} + B_{k+1}^2 \cdot R_{cc}. \tag{B.23}$$

Upon using Equations B.14 and B.15, we arrive at:

$$Z = A_{k+1} \cdot R_{es} + B_{k+1} \cdot R_{ec}. \tag{B.24}$$

If we reconstruct Equation B.17, then:

$$E_w^{k+1} = X - Y + Z \tag{B.25}$$

which upon using Equations B.18, B.20, and B.24 leads to:

$$E_w^{k+1} = \sum_{n=1}^{\text{excint}} e_w^k(n)^2 - A_{k+1}R_{es} - B_{k+1}R_{ec}. \tag{B.26}$$

The expression $E_w^k(n)^2 = \sum_{n=1}^{\text{excint}} e_w^k(n)^2$ will always be positive, and it is independent of A_{k+1} and B_{k+1}. Thus the error E_w^{k+1} will be minimized when $[A_{k+1}R_{es} + B_{k+1}R_{ec}]$ is maximized. So,

$$\zeta_{mse} = \frac{R_{es}^2}{R_{ss}} + \frac{R_{ec}^2}{R_{cc}} \tag{B.27}$$

where ζ_{mse} must be maximized over the range $n = 1$ to $n = \text{excint}$.

Probability Density Function for Amplitudes

This appendix presents the probability density functions (PDFs) for the normalized amplitude residual vector from the STC speech coder of Chapter 15. Section 15.8.1.4 describes the vector quantization process for the normalized amplitude residual vector, and this appendix indicates the suitability of the amplitude residual vector for quantization.

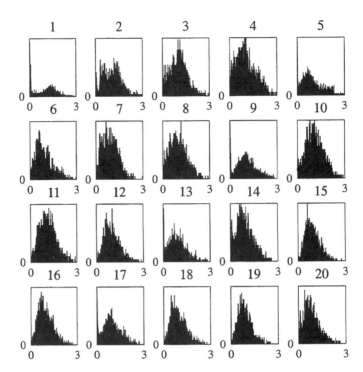

Figure C.1 The PDF for the normalized amplitude residual vector, elements 1 to 20. The abscissa represents the value of the amplitude residual element, with the ordinate representing the value's occurrence

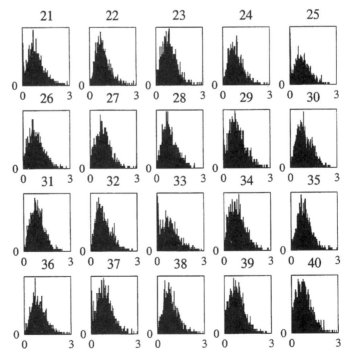

Figure C.2 The PDF for the normalized amplitude residual vector, elements 21 to 40. The abscissa represents the value of the amplitude residual element, with the ordinate representing the value's occurrence.

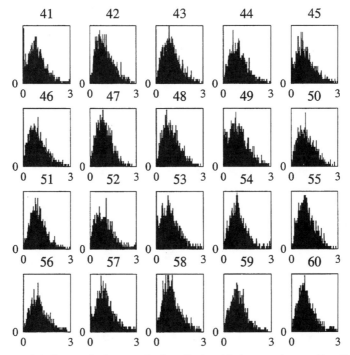

Figure C.3 The PDF for the normalized amplitude residual vector, elements 41 to 60. The abscissa represents the value of the amplitude residual element, with the ordinate representing the value's occurrence.

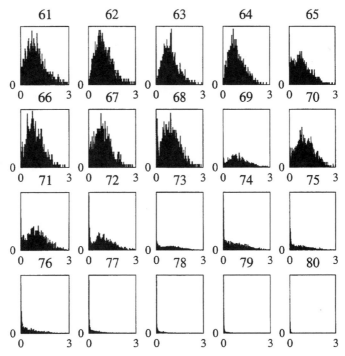

Figure C.4 The PDF for the normalized amplitude residual vector, elements 61 to 80. The abscissa represents the value of the amplitude residual element, with the ordinate representing the value's occurrence.

Bibliography

[1] R. Cox and P. Kroon. "Low bit-rate speech coders for multimedia communications." *IEEE Comms. Mag.*, pp. 34–41, December 1996.

[2] R. Cox. "Speech coding and synthesis." In *Speech coding standards* (W. Kleijn and K. Paliwal, eds.), ch. 2, pp. 49–78. Elsevier, 1995.

[3] R. Steele. *Delta Modulation Systems*. London: Pentech Press, 1975.

[4] K. Cattermole. *Principles of Pulse Code Modulation*. London: Hiffe Books, 1969.

[5] J. Markel and A. Gray, Jr. *Linear Prediction of Speech*. New York: Springer-Verlag, 1976.

[6] L. Rabiner and R. Schafer. *Digital Processing of Speech Signals*. Englewood Cliffs, NJ: Prentice-Hall, 1978.

[7] B. Lindblom and S. Ohman. *Frontiers of Speech Communication Research*. New York: Academic Press, 1979.

[8] J. Tobias, ed. *Foundations of Modern Auditory Theory*. New York: Academic Press, 1970.

[9] B. Atal and J. Remde. "A new model of LPC excitation for producing natural-sounding speech at low bit rates." In *Proceedings of International Conference on Acoustics, Speech, and Signal Processing, ICASSP'82*, pp. 614–617, IEEE, May 1982.

[10] N. Jayant and P. Noll. *Digital Coding of Waveforms, Principles and Applications to Speech and Video*. Englewood Cliffs, NJ: Prentice-Hall, 1984.

[11] P. Kroon, E. Deprettere, and R. Sluyter. "Regular pulse excitation—a novel approach to effective efficient multipulse coding of speech." *IEEE Transactions on Acoustics, Speech and Signal Processing*, vol. 34, pp. 1054–1063, October 1986.

[12] P. Vary and R. Sluyter. "MATS-D speech codec: Regular-pulse excitation LPC." In *Proc. of the Nordic Seminar on Digital Land Mobile Radio Communications (DMRII)*, (Stockholm, Sweden), pp. 257–261, October 1986.

[13] P. Vary and R. Hoffmann. "Sprachcodec für das europäische Funkfernsprechnetz." *Frequenz 42 (1988) 2/3*, pp. 85–93, 1988.

[14] W. Hess. *Pitch Determination of Speech Signals: Algorithms and Devices*. Berlin: Springer Verlag, 1983.

[15] G. Gordos and G. Takacs. *Digital Speech Processing (Digitalis Beszed Feldolgozas)*. Budapest, Hungary: Technical Publishers (Muszaki Kiado), 1983. (In Hungarian.)

[16] M. Schroeder and B. Atal. "Code excited linear prediction (CELP): High-quality speech at very low bit rates." In *Proceedings of International Conference on Acoustics, Speech, and Signal Processing, ICASSP'85* (Tampa, Florida), pp. 937–940, IEEE, March 26–29, 1985,

[17] D. O'Shaughnessy. *Speech Communication: Human and Machine*. Reading, MA: Addison-Wesley, 1987.

[18] P. Papamichalis. *Practical Approaches to Speech Coding*. Englewood Cliffs, NJ: Prentice-Hall, 1987.

[19] J. Deller, J. Proakis, and J. Hansen. *Discrete-time Processing of Speech Signals*. Englewood Cliffs, NJ: Prentice-Hall, 1987.

[20] P. Lieberman and S. Blumstein. *Speech Physiology, Speech Perception, and Acoustic Phonetics*. Cambridge: Cambridge University Press, 1988.

[21] S. Quackenbush, T. Barnwell III, and M. Clements. *Objective Measures of Speech Quality*. Englewood Cliffs, NJ: Prentice-Hall, 1988.

[22] S. Furui. *Digital Speech Processing, Synthesis and Recognition*. New York: Marcel Dekker, 1989.

[23] R. Steele, C.-E. Sundberg, and W. Wong. "Transmission of log-PCM via QAM over Gaussian and Rayleigh fading channels." *IEE Proc.*, vol. 134, Pt. F, pp. 539–556, October 1987.

[24] R. Steele, C.-E. Sundberg, and W. Wong. "Transmission errors in companded PCM over Gaussian and Rayleigh fading channels." *AT&T Bell Laboratories Tech. Journal*, pp. 991–995, July–August 1984.

[25] C.-E. Sundberg, W. Wong, and R. Steele. "Weighting strategies for companded PCM transmitted over Rayleigh fading and Gaussian channels." *AT&T Bell Laboratories Tech. Journal*, vol. 63, pp. 587–626, April 1984.

[26] W. Wong, R. Steele, and C.-E. Sundberg. "Soft decision demodulation to reduce the effect of transmission errors in logarithmic PCM transmitted over Rayleigh fading channels." *AT&T Bell Laboratories Tech. Journal*, vol. 63, pp. 2193–2213, December 1984.

[27] J. Hagenauer. "Source-controlled channel decoding." *IEEE Transactions on Communications*, vol. 43, pp. 2449–2457, September 1995.

[28] B. Atal, V. Cuperman, and A. Gersho, eds. *Advances in Speech Coding*. Boston: Kluwer Academic Publishers, 1991.

[29] A. Ince, ed. *Digital Speech Processing: Speech Coding, Synthesis and Recognition*. Boston: Kluwer Academic Publishers, 1992.

[30] J. Anderson and S. Mohan. *Source and Channel Coding—An Algorithmic Approach*. Boston: Kluwer Academic Publishers, 1993.

[31] A. Kondoz. *Digital Speech: Coding for Low Bit Rate Communications Systems*. New York: John Wiley, 1994.

[32] W. Kleijn and K. Paliwal, eds. *Speech Coding and Synthesis*. Elsevier Science, 1995.

[33] C. Shannon. *Mathematical Theory of Communication*. University of Illinois Press, 1963.

[34] J. Hagenauer. "Quellengesteuerte kanalcodierung fuer sprach- und tonuebertragung im mobilfunk." *Aachener Kolloquium Signaltheorie*, pp. 67–76, March 23–25, 1994.

[35] A. Viterbi. "Wireless digital communications: A view based on three lessons learned." *IEEE Communications Magazine*, pp. 33–36, September 1991.

[36] S. Lloyd. "Least square quantisation in PCM." *Institute of Mathematical Statistics Meeting*, Atlantic City, NJ, September 1957.

[37] S. Lloyd. "Least squares quantisation in PCM." *IEEE Trans. on Information Theory*, vol. 28, no. 2, pp. 129–136, 1982.

[38] J. Max. "Quantising for minimum distortion." *IRE Trans. on Information Theory*, vol. 6, pp. 7–12, 1960.

[39] W. Bennett. "Spectra of quantised signals." *Bell System Technical Journal*, pp. 446–472, July 1946.

[40] H. Holtzwarth. "Pulse code modulation und ihre verzerrung bei logarithmischer quanteilung." *Archiv der Elektrischen Uebertragung*, pp. 227–285, January 1949.

[41] P. Panter and W. Dite. "Quantisation distortion in pulse code modulation with non-uniform spacing of levels." *Proc. of the IRE*, pp. 44–48, January 1951.

[42] B. Smith. "Instantaneous companding of quantised signals," *Bell System Technical Journal*, pp. 653–709, 1957.

[43] P. Noll and R. Zelinski. "A contribution to the quantisation of memoryless model sources." *Technical Report*, Heinrich Heinz Institute, Berlin, 1974. (In German.)

[44] M. Paez and T. Glisson. "Minimum mean squared error quantisation in speech PCM and DPCM systems." *IEEE Trans. on Communications*, pp. 225–230, April 1972.

[45] A. Jain. *Fundamentals of Digital Image Processing*. Englewood Cliffs, NJ: Prentice-Hall, 1989.

[46] R. Salami. *Robust Low Bit Rate Analysis-by-Synthesis Predictive Speech Coding*. Ph.D. thesis, University of Southampton, 1990.

[47] R. Salami, L. Hanzo, R. Steele, K. Wong, and I. Wassell. "Speech coding." In Steele [328], ch. 3, pp. 186–346.

[48] S. Haykin. *Adaptive Filter Theory*. Englewood Cliffs, NJ: Prentice Hall, 1996.

[49] W. Webb. "Sizing up the microcell for mobile radio communications." *IEE Electronics and communications Journal*, vol. 5, pp. 133–140, June 1993.

[50] K. Wong and L. Hanzo. "Channel coding." In Steele [328], ch. 4, pp. 347–488.

[51] A. Jennings. *Matrix Computation for Engineers and Scientists*. New York: J. Wiley and Sons, 1977.

[52] J. Makhoul. "Stable and efficient lattice methods for linear prediction." *IEEE Trans. on ASSP*, vol. 25, pp. 423–428, October 1977.

[53] J. Makhoul. "Linear prediction: A tutorial review." *Proceedings of the IEEE*, vol. 63, pp. 561–580, April 1975.

[54] N. Jayant. "Adaptive quantization with a one-word memory." *Bell System Technical Journal*, vol. 52, pp. 1119–1144, September 1973.

[55] R. Steedman. "The common air interface MPT 1375." In *Cordless Telecommunications in Europe: The Evolution of Personal Communications* (W. Tuttlebee, ed.). London: Springer-Verlag, 1990.

[56] L. Hanzo. "The British cordless telephone system: CT2." In Gibson [364], ch. 29, pp. 462–477.

[57] H. Ochsner. "The digital European cordless telecommunications specification, DECT." In *Cordless Telecommunications in Europe: The Evolution of Personal Communications* (W. Tuttlebee, ed.), pp. 273–285, London: Springer-Verlag, 1990.

[58] S. Asghar. "Digital European Cordless Telephone." In Gibson [364], ch. 30, pp. 478–499.

[59] "Personal handy phone (PHP) system." RCR Standard, STD-28, Japan.

[60] "CCITT recommendation G.721."

[61] N. Kitawaki, M. Honda, and K. Itoh. "Speech-quality assessment methods for speech coding systems." *IEEE Communications Magazine*, vol. 22, pp. 26–33, October 1984.

[62] A. Gray and J. Markel. "Distance measures for speech processing." *IEEE Transactions on ASSP*, vol. 24, no. 5, pp. 380–391, 1976.

[63] N. Kitawaki, H. Nagabucki, and K. Itoh. "Objective quality evaluation for low-bit-rate speech coding systems." *IEEE Journal on Selected Areas in Communications*, vol. 6, pp. 242–249, February 1988.

[64] P. Noll and R. Zelinski. "Bounds on quantizer performance in the low bit-rate region." *IEEE Transactions on Communications*, pp. 300–304, February 1978.

[65] T. Thorpe. "The mean squared error criterion: Its effect on the performance of speech coders." In *Proceedings of International Conference on Acoustics, Speech, and Signal Processing, ICASSP'89* (Glasgow, Scotland), pp. 77–80, IEEE, May 23–26, 1989.

[66] J. O'Neal. "Bounds on subjective performance measures for source encoding systems." *IEEE Transactions on Information Theory*, pp. 224–231, May 1971.

[67] J. Makhoul, S. Roucos, and H. Gish. "Vector quantization in speech coding." *Proceedings of the IEEE*, pp. 1551–1588, November 1985.

[68] B. Atal and M. Schroeder. "Predictive coding of speech signals and subjective error criteria." *IEEE Transactions on Acoustics, Speech and Signal Processing*, pp. 247–254, June 1979.

[69] R. Steele. "Deploying personal communications networks." *IEEE Comms. Magazine*, pp. 12–15, September 1990.

[70] J.-H. Chen, R. Cox, Y. Lin, N. Jayant, and M. Melchner. "A low-delay CELP codec for the CCITT 16 kb/s speech coding standard." *IEEE Journal on Selected Areas in Communications*, vol. 10, pp. 830–849, June 1992.

[71] D. Sen and W. Holmes. "PERCELP-perceptually enhanced random codebook excited linear prediction." In *Proc. IEEE Workshop on Speech Coding for Telecommunications*, pp. 101–102, 1993.

[72] S. Singhal and B. Atal. "Improving performance of multi-pulse LPC coders at low bit rates." In *Proceedings of International Conference on Acoustics, Speech and Signal Processing*, ICASSP'84 (San Diego, California), pp. 1.3.1–1.3.4, IEEE, March 19–21, 1984.

[73] "Group speciale mobile (GSM) recommendation." April 1988.

[74] L. Hanzo and J, Stefanov. "The Pan-European Digital Cellular Mobile Radio System—known as GSM." In Steele [328], ch. 8, pp. 677–765.

[75] S. Singhal and B. Atal. "Amplitude optimization and pitch prediction in multi-pulse coders." *IEEE Trans. on Acoustics, Speech and Signal Processing*, pp. 317–327, March 1989.

[76] Federal standard 1016—telecommunications: Analog to digital conversion of radio voice by 4,800 bits/second code excited linear prediction (CELP)." February 14, 1991.

[77] S. Wang and A. Gersho. "Phonetic segmentation for low rate speech coding." In Atal et al. [28], pp. 257–266.

[78] P. Lupini, H. Hassanein, and V. Cuperman. "A 2.4 kbit/s CELP speech codec with class-dependent structure." In *Proceedings of the IEEE International Conference on Acoustics, Speech and Signal Processing (ICASSP'93)*, vol. 2 (Minneapolis, MN), pp. 143–246, IEEE, April 27–30, 1993.

[79] D. Griffin and J. Lim. "Multiband excitation vocoder." *IEEE Trans. on Acoustics, Speech and Signal Processing*, pp. 1223–1235, August 1988.

[80] M. Nishiguchi, J. Matsumoto, R. Wakatsuki, and S. Ono. "Vector quantized MBE with simplified v/uv division at 3.0 Kbps." In *Proceedings of the IEEE International Conference on Acoustics, Speech and Signal Processing (ICASSP'93)*, vol. 2 (Minneapolis, MN), pp. 151–154, IEEE, April 27–30, 1993.

[81] W. Kleijn. "Encoding speech nsing prototype waveforms." *IEEE Transactions on Speech and Audio Processing*, vol. 1, pp. 386–399, October 1993.

[82] V. Ramamoorthy and N. Jayant. "Enhancement of ADPCM speech by adaptive postfiltering." *Bell Syst Tech Journal*, vol. 63, pp. 1465–4475, October 1984.

[83] N. Jayant and V. Ramamoorthy. "Adaptive postfiltering of 16 kb/s-ADPCM speech." In *Proceedings of International Conference on Acoustics, Speech, and Signal Processing, ICASSP'86* (Tokyo, Japan), pp. 829–832, IEEE, April 7–11, 1986.

[84] J.-H. Chen and A. Gersho. "Real–time vector APC speech coding at 4800 bps with adaptive postfiltering." In *Proceedings of International Conference an Acoustics, Speech, and Signal Processing, ICASSP'87*, (Dallas, TX), pp. 2185–2188, IEEE, April 6–9, 1987.

[85] ITU-T. *CCITT Recommendation G.728: Coding of Speech at 16 kbit/s Using Low-Delay Code Excited Linear Prediction.* 1992.

[86] J.-H. Chen and A. Gersho. "Adaptive postfiltering for quality enhancement of coded speech." *IEEE Transactions on Speech and Audio Processing*, vol. 3, pp. 59–71, January 1995.

[87] F. Itakura and S. Saito. "Analysis–synthesis telephony based upon the maximum likelihood method." In *Proc. of the 6th International Congress on Acoustic* (Tokyo), pp. C17–20, 1968.

[88] F. Itakura and S. Saito. "A statistical method for estimation of speech spectral density and formant frequencies." *Electr. and Comms. in Japan*, vol. 53-A, pp. 36–43, 1970.

[89] N. Kitawaki, K. Itoh, and F. Itakura. PARCOR speech analysis synthesis system." *Review of the Electronics and Communications Laboratory, Nippon TTPC*, vol. 26, pp. 1439–1455, November–December 1978.

[90] R. Viswanathan and J. Makhoul. "Quantization properties of transmission parameters in linear predictive systems." *IEEE Trans. on ASSP*, pp. 309–321, 1975.

[91] N. Sugamura and N. Farvardin. "Quantizer design in LSP analysis–synthesis." *IEEE Journal on Selected Areas in Communications*, vol. 6, pp. 432–440, February 1988.

[92] K. Paliwal and B. Atal. "Efficient vector quantization of LPC parameters at 24 bits/frame." *IEEE Transactions on Speech and Audio Processing*, vol. 1, pp. 3–14, January 1993.

[93] F. Soong and B.-H. Juang. "Line spectrum pair (LSP) and speech data compression." In *Proceedings of International Conference on Acoustics, Speech, and Siqnal Proces-

sing, ICASSP'84 (San Diego, California), pp. 1.10.1–1.10.4, IEEE, March 19–21, 1984.

[94] G. Kang and L. Fransen. "Low–bit rate speech encoders based on line-spectrum frequencies (LSFs)." Tech. Rep. 8857, NRL, November 1984.

[95] P. Kabal and R. R,amachandran. "The computation of line spectral frequencies using chebyshev polynomials." *IEEE Trans. ASSP*, vol. 34, pp. 1419–1426, December 1986.

[96] M. Omologo. "The computation and some spectral considerations on line spectrum pairs (LSP)." In *Proc. EUROSPEECH*, pp. 352–355, 1989.

[97] B. Cheetham. "Adaptive LSP filter." *Electronics Letters*, vol. 23, pp. 89–90, January 1987.

[98] K. Geher. *Linear Circuits*. Budapest, Hungary: Technical Publishers, 1972. (In Hungarian.)

[99] N. Sugamiira and F. Itakura. "Speech analysis and synthesis methods developed at ECL in NTT—from LPC to LSP." *Speech Communications*, vol. 5, pp. 199–215, June 1986.

[100] A. Lepschy, G. Milan, and U. Viaro. "A note on line spectral frequencies." *IEEE Trans. ASSP*, vol. 36, pp. 1355–1357, August 1988.

[101] B. Cheetham and P. Huges. "Formant estimation from LSP coefficients." In *Proc. IERE 5th Int. Conf. on Digital Processing of Signals in Communications*, pp. 183–189, September 20–23, 1988.

[102] A. Gersho and R. Gray. *Vector Quantization and Signal Compression*. Boston: Kluwer Academic Publishers, 1992.

[103] Y. Shoham. "Vector predictive quantization of the spectral parameters for low rate speech coding." In *Proceedings of International Conference on Acoustics, Speech, and Signal Processing, ICASSP'87* (Dallas, TX), pp. 2181–2184, IEEE, April 6–9, 1987.

[104] R. Ramachandran, M. Sondhi, N. Seshadri, and B. Atal. "A two codebook format for robust quantisation of line spectral frequencies." *IEEE Trans. on Speech and Audio Processing*, vol. 3, pp. 157–168, May 1995.

[105] C. Xydeas and K. S. "Improving the performance of the long history scalar and vector quantisers." In *Proceedings of the IEEE International Conference on Acoustics, Speech and Signal Processing (ICASSP'93)*, vol. 2 (Minneapolis, MN), pp. 1–4, IEEE, April 27–30, 1993.

[106] K. Lee, A. Kondoz, and B. Evans. "Speaker adaptive vector quantisation of LPC parameters of speech." *Electronic Letters*, vol. 24, pp. 1392–1393, October 1988.

[107] B. Atal. "Stochastic gaussian model for low-bit rate coding of LPC area parameters." In *Proceedings of International Conference on Acoustics, Speech, and Signal Processing, ICASSP'87* (Dallas, TX), pp. 2404–2407, IEEE, April 6–9, 1987.

[108] R. Salami, L. Hanzo, and D. Appleby. "A fully vector quantised self-excited vocoder." In *Proceedings of International Conference on Acoustics, Speech, and Signal Processing, ICASSP'89* (Glasgow, Scotland), pp. 124–128, IEEE, May 23–26, 1989.

[109] M. Yong, G. Davidson, and A. Gersho. "Encoding of LPC spectral parameters using switched-adaptive interframe vector prediction." In *Proceedings of International Conference on Acoustics, Speech, and Signal Processing, ICASSP'88* (New York), pp. 402–405, IEEE, April 11–14, 1988.

[110] J. Huang and P. Schultheis. "Block quantization of correlated Gaussian random variables." *IEEE. Trans. Commun. Sys.*, vol. 11, pp. 289–296, September 1963.

[111] R. Salami, L. Hanzo, and D. Appleby. "A computationally efficient CELP codec with stochastic vector quantization of LPC parameters." In *URSI Int. Symposium on Signals, Systems and Electronics* (Erlangen, West Germany), pp. 140–143, September 18–20, 1989.

[112] B. Atal, R. Cox, and P. Kroon. "Spectral quantization and interpolation for CELP coders." In *Proceedings of International Conference on Acoustics, Speech, and Signal Processing, IC4SSP'89* (Glasgow, Scotland), pp. 69–72, IEEE, May 23–26, 1989.

[113] R. Laroia, N. Phamdo, and N. Farvardin. "Robust and efficient quantisation of speech LSP parameters using structured vector quantisers." In *Proceedings of International Conference on Acoustics, Speech, and Signal Processing, ICASSP'91* (Toronto, Onjtario, Canada), pp. 641–644, IEEE, May 14–17, 1991.

[114] H. Harbor , J. Knudson, A. Fudseth, and F. Johansen. "A real time wideband CELP for a videophone application." In *Proceedings of the IEEE International Conference on Acoustics, Speech and Signal Processing (ICASSP'94)* (Adelaide, Australia), pp. II121–II124, IEEE, April 19–22, 1994.

[115] R. Lefebvre, R. Salami, C. Lafearnme, and J. Adoul. "High quality coding of wideband audio signals using transform coded excitation (TCX)." In *Proceedings of the IEEE International Conference on Acoustics, Speech and Signal Processzng (ICASSP'94)*, (Adelaide, Australia), pp. 1193–1196, IEEE, April 19–22, 1994.

[116] J. Paulus and J. Schnitzler. "l6 kbit/s wideband speech coding based on unequal subbands." In *Proceedings of the IEEE International Conference on Acoustics, Speech and Signal Processing (ICASSP'96)* (Atlanta, GA), pp. 255–258, IEEE, May 7–10, 1996.

[117] J. Chen and D. Wang. "Transform predictive coding of wideband speech signals." In *Proceedings of the IEEE International Conference on Acoustics, Speech and Signal Processing (ICASSP'96)*, (Atlanta, GA), pp. 275–278, IEEE, May 7–10, 1996.

[118] A. Ubale and A. Gersho. "A multi-band CELP wideband speech coder." In *Proceedings of the IEEE International G'onference on Acoustics, Speech and Signal Processing (ICASSP'97)*, (Munich, Germany), pp. 1367–1370, IEEE, April 21–24, 1997.

[119] P. Combescure, J. Schnitzler, K. Fischer, R. Kirchherr, C. Lamblin, A. L. Guyader, D. Massaloux, C. Quinquis, J. Stegmann, and P. Vary. "A 16, 24, 32 Kbit/s wideband speech codec based on ATCELP." In *Proceedings of International Conference on Acoustics, Speech, and Signal Processing, ICASSP'99*, IEEE, 1999.

[120] F. Itakura. "Line spectrum representation of linear predictive coefficients of speech signals." *Journal of the Acoustic Society of America*, vol. 57, p. S35, 1975.

[121] L. Rabiner, M. Sondhi, and S. Levinson. "Note on the properties of a vector quantizer or LPC coefficients." *The Bell System Technical Journal*, vol. 62, pp. 2603–2616, October 1983.

[122] "7 khz audio coding within 64 kbit/s." CCITT Recommendation G.722, 1988.

[123] "Recommendation G.729: Coding of speech at 8 kbit/s using conjugate-structure algebraic-code-excited linear-prediction (CS-ACELP)." CCITT Study Group XVIII, June 30, 1995. Version 6.31.

[124] T. Eriksson, J. Linden, and J. Skoglung. "A safety-net approach for improved exploitation of speech correlation." In *Proceedings of the IEEE International Conference on Acoustics, Speech and Signal Processing (ICASSP'95)*, (Detroit, MI), pp. 96–101, IEEE, May 9–12, 1995.

[125] T. Eriksson, J Linden, and J. Skoglung. "Exploiting interframe correlation in spectral quantization—a study of different memory VQ schemes." In *Proceedings of the iEEE International Conference on Acoustics, Speech and Signal Processing (ICASSP'96)*, (Atlanta), pp. 765–768, IEEE, May 7–10, 1996.

[126] H. Zarrinkoub and P. Mermelstein. "Switched prediction and quantization of LSP frequencies." In *Proceedings of the IEEE International Conference on Acoustics, Speech and Signal Processing (ICASSP'96)*, (Atlanta), pp. 757–764, IEEE, May 7–10, 1996.

[127] J. Natvig. "Evaluation of six medium bit–rate coders for the pan-European digital mobile radio system." *IEEE Journal on Selected Areas in Communications*, pp. 324–331, February 1988.

[128] J. Schur. "Über potenzreihen, die im innern des einheitskreises beschränkt sind." *Journal für die reine und angewandte Mathematik, Bd 14*, pp. 205–232, 1917.

[129] W. Webb, L. Hanzo, R. Salami, and R. Steele. "Does 16–QAM provide an alternative to a half–rate GSM speech codec?." In *Proceedings of IEEE Vehicular Technology Conference (VTC'91)*, (St. Louis, MO), pp. 511–516, IEEE, May 19–22, 1991.

[130] L. Hanzo, W. Webb, R. Salami, and R. Steele. "On QAM speech transmission schemes for microcellular mobile PCNs." *European Transactions on Communications*, pp. 495–510, September/October 1993.

[131] J. Williams, L. Hanzo, R. Steele, and J. Cheung. "A comparative study of microcellular speech transmission schemes." *IEEE Th. on Veh. Technology*, vol. 43, pp. 909–925, November 1994.

[132] "Cellular system dual-mode mobile station-base station compatibility standard IS-54B." Telecommunications Industry Association, Washington, DC, 1992. EIA/TIA Interim Standard.

[133] Research and Development Centre for Radio Systems, Japan. *Public Digital Cellular (PDQ) Standard, RCR STD-27*.

[134] R. Steele and L. Hanzo, eds. *Mobile Radio Communications*, 2nd ed. New York: IEEE Press-John Wiley, 1999.

[135] L. Hanzo, W. Webb, and T. Keller. *Single- and Multi-carrier Quadrature Amplitude Modulation*. Piscataway, NJ: IEEE Press, Chichester, UK: John Wiley, April 2000.

[136] R. Salami, C. Laflamme, J.-P. Adoul, and D. Massaloux. "A toll quality 8 kb/s speech codec for the personal communications system (PCS)." *IEEE Transactions on Vehicular Technology*, pp. 808–816, August 1994.

[137] A. Black, A. Kondoz, and B. Evans. "High quality low delay wideband speech coding at 16 kbit/sec." In *Proc of 2nd Int Workshop on Mobile Multimedia Communications*, April 11–14, 1995. Bristol University, UK.

[138] C. Laflamme, J.-P. Adoul, R. Salami, S. Morissette, and P. Mabilleau. "16 Kbps wideband speech coding technique based on algebraic CELP." In *Proceedings of International Conference on Acoustics, Speech, and Signal Processing, ICASSP'91*, (Toronto, Ontario, Canada), pp. 13–16, IEEE, May 14–17, 1991.

[139] R. Salami, C. Laflamme, and J.-P. Adoul. "Real-time implementation of a 9.6 kbit/s ACELP wideband speech coder." In *Proc GLOBECOM '92*, 1992.

[140] I. Gerson and M. Jasiuk. "Vector sum excited linear prediction (VSELP)." In Atal et al. [28], pp. 69–80.

[141] M. Ireton and C. Xydeas. "On improving vector excitation coders through the use of spherical lattice codebooks (SLC's)." In *Proceedings of International Conference on Acoustics, Speech, and Signal Processing, ICASSP'89*, (Glasgow, Scotland, UK), pp. 57–60, IEEE, May 23–26, 1989.

[142] C. Lamblin, J. Adoul, D. Massaloux, and S. Morissette. "Fast CELP coding based on the barnes-wall lattice in 16 dimensions." In *Proceedings of International Conference on Acoustics, Speech, and Signal Processing, ICASSP'89*, (Glasgow, Scotland), pp. 61–64, IEEE, May 23–26, 1989.

[143] C. Xydeas, M. Ireton, and D. Baghbadrani. "Theory and real time implementation of a CELP coder at 4.8 and 6.0 kbit/s using ternary code excitation." In *Proc. of IERE 5th Int. Conf. on Digital Processing of Signals in Comms*, pp. 167–174, September 1988.

[144] J. Adoul, P. Mabilleau, M. Delprat, and S. Morissette. "Fast CELP coding based on algebraic codes." In *Proceedings of International Conference on Acoustics, Speech, and Signal Processing, IGASSP'87* (Dallas, TX), pp. 1957–1960, IEEE, April 6–9, 1987.

[145] L. Hanzo and J. Woodard. "An intelligent multimode voice communications system for indoor communications." *IEEE Transactions on Vehicular Technology*, vol. 44, pp. 735–748, November 1995.

[146] A. Kataol, J.-P. Adoul, P. Combescure, and P. Kroon. "ITU-T 8-kbits/s standard speech codec for personal communication services." In *Proceedings of International Conference on Universal Personal Communications 1985* (Tokyo, Japan), pp. 818–822, November 1995.

[147] H. Law and R. Seymour. "A reference distortion system using modulated noise." *IEE Paper*, pp. 484–485, November 1962.

[148] P. Kabal, J. Moncet, and C. Chu. "Synthesis filter optimization and coding: Applications to CELP." In *Proceedings of International Conference on Acoustics, Speech, and Signal Processing, ICASSP'88*, vol. 1, (New York), pp. 147–150, IEEE, April 11–14, 1988.

[149] Y. Tohkura, F. Itakuxa, and S. Hashimoto. "Spectral smoothing technique in PARCOR speech analysis-synthesis." *IEEE Trans. on Acoustics, Speech and Signal Processing*, pp. 587–596, 1978.

[150] J.-H. Chen and R. Cox. "Convergence and numerical stability of backward-adaptive LPC predictor." In *Proceedings of IEEE Workshop on Speech Coding for Telecommunications*, pp. 83–84, 1993.

[151] S. Singhal and B. Atal. "Optimizing LPC filter parameters for multi-pulse excitation." In *Proceedings of International Conference an Acoustics, Speech, and Signal Processing, ICASSP'83* (Boston, Mass.), pp. 781–784, IEEE, April 14–16, 1983.

[152] M. Fratti, G. Miani, and G. Riccardi. "On the effectiveness of parameter reoptimization in multipulse based coders." In *Proceedings of International Conference on Acoustics, Speech, and Signal Processing, ICASSP'92*, vol. 1, pp. 73–76, IEEE, March 1992.

[153] W. Press, S. Teukolsky, W. Vetterling, and B. Flannery. *Numerical Recipes in C.* Cambridge: Cambridge University Press, 1992.

[154] G. Golub and C. V. Loan. "An analysis of the total least squares problem." *SIAM Journal of Numerical Analysis*, vol. 17, no. 6, pp. 883–890, 1980.

[155] M. A. Rahham and K.-B. Yu. "Total least squares approach for frequency estimation using linear prediction." *IEEE Transactions on Acoustics, Speech and Signal Processing*, pp. 1440–1454, 1987.

[156] R. Degroat and E. Dowling. "The data least squares problem and channel equalization." *IEEE Transactions on Signal Processing*, pp. 407–411, 1993.

[157] F. Tzeng. "Near-optimum linear predictive speech coding." In *IEEE Global Telecommunications Conference*, pp. 508.1.1–508.1.5, 1990.

[158] M. Niranjan. "CELP coding with adaptive output-error model identification." In *Proceedings of International Conference on Acoustics, Speech, and Signal Processing, ICASSP'90* (Albuquerque, New Mexico, pp. 225–228, IEEE, April 3–6, 1990.

[159] J. Woodard and L. Hanzo. "Improvements to the analysis-by-synthesis loop in CELP codecs." In *Proceedings of IEE Conference on Radio Receivers and Associated Systems (RRAS'95)*, (Bath, UK), pp. 114–118, lEE, September 26–28, 1995.

[160] L. Hanzo, R. Salami, R. Steele, and P. Fortune. "Transmission of digitally encoded speech at 1.2 kbaud for PCN." *IEE Proceedings, Part I*, vol. 139, pp. 437–447, August 1992.

[161] R. Cox, W. Kleijn, and P, Kroon. "Robust CELP coders for noisy backgrounds and noisy channels." In *Proceedings of International Conference on Acoustics, Speech, and Signal Processing, IGASSP'89*, (Glasgow, Scotland), pp. 739–742, IEEE, May 23–26, 1989.

[162] J. Campbell, V. Welch, and T. Tremain. "An expandable error-protected 4800 bps CELP coder (U.S. federal standard 4800 bps voice coder)." In *Proceedings of International Conference on Acoustics, Speech, and Signal Processing, ICASSP'89*, (Glasgow, Scotland), pp. 735–738, IEEE, May, 23–26, 1989.

[163] S. Atungsiri, A. Kondoz, and B. Evans."Error control for low-bit-rate speech communication systems," *IEE Proceedings-I*, vol. 140, pp. 97–104, April 1993.

[164] L. Ong, A. Kondoz, and B. Evans. "Enhanced channel coding using source criteria in speech coders." *IEE Proceedings-I*, vol. 141, pp. 191–196, June 1994.

[165] W. Kleijn. "Source-dependent channel coding and its application to CELP." In Atal et al. [28], pp. 257–266.

[166] J. Woodard and L. Hanzo, "A dual-rate algebraic CELP-based speech transceive." In *Proceedings of IEEE VTC '94*, vol. 3 (Stockholm, Sweden), pp. 1690–1694, IEEE, June 8–10, 1994.

[167] C. Laflamme, J.-P. Adoul, H. Su, and S. Morissette. "On reducing the complexity of codebook search in CELP through the use of algebraic codes." In *Proceedings of International Conference on Acoustics, Speech, and Signal Processing, ICASSP'90*, (Albuquerque, New Mexico), pp. 177–180, IEEE, April 3–6, 1990.

[168] S. Nanda, D. Goodman, and U. Timor. "Performance of PRMA: A packet voice protocol for cellular systems." *IEEE Tr. on VT*, vol. 40, pp. 584–598, August 1991.

[169] M. Frullone, G. Riva, P. Grazioso, and C. Carciofy. "Investigation on dynamic channel allocation strategies suitable for PRMA schemes," *1993 IEEE Int. Symp. on Circuits and Systems, Chicago*, pp. 2216–2219, May 1993.

[170] J. Williams, L. Hanzo, and R. Steele. "Channel-adaptive voice communications." In *Proceedings of IEE Conference on Radio Receivers and Associated Systems (RRAS'95)*, (Bath, UK), pp. 144–447, IEE, September 26–28, 1995.

[171] W. Lee. "Estimate of channel capacity in Rayleigh fading environment." *IEEE Trans. on Vehicular Technology*, vol. 39, pp. 187–189, August 1990.

[172] T. Tremain. "The government standard linear predictive coding algorithm: LPC-10." *Speech Technology*, vol. 1, pp. 40–49, April 1982.

[173] J. Campbell, T. Tremain, and V. Welch. "The DoD 4.8 kbps standard (proprosed federal standard 1016)." In Atal et al. [28], pp. 121–133.

[174] J. Marques, I. Trancoso, J. Tribolet, and L. Almeida. "Improved pitch prediction with fractional delays in CELP coding." In *Proceedings of International Conference on Acoustics, Speech, and Signal Processing, ICASSP'90* (Albuquerque, New Mexico), pp. 665–668, IEEE, April 3–6, 1990.

[175] W. Kleijn, D. Kraisinsky, and R. Ketchum. "An efficient stochastically excited linear predictive coding algorithm for high quality low bit rate transmission of speech." *Speech Communication*, pp. 145–156, October 1988.

[176] Y. Shoham. "Constrained-stochastic excitation coding of speech at 4.8 kb/s." In Atal et al. [28], pp. 339–348.

[177] A. Suen, J. Wand, and T. Yao. "Dynamic partial search scheme for stochastic codebook of FS1016 CELP coder." *IEE Proceedings*, vol. 142, no. 1, pp. 52–58, 1995.

[178] I. Gerson and M. Jasiuk. "Vector sum excited linear prediction (VSELP) speech coding at 8 kbps." In *Proceedings of International Conference on Acoustics, Speech, and Signal Processing, ICASSP'90* (Albuquerque, New Mexico), pp. 461–464, IEEE, April 3–6, 1990.

[179] I. Gerson and M, Jasiuk. "Techniques for improving the performance of CELPtype speech codecs." *IEEE JSAC*, vol. 10, pp. 858–865, June 1992.

[180] I. Gerson "Method and means of determining coefficients for linear predictive coding." US Patent No 544,919, October 1985.

[181] A. Cumain. "On a covariance-lattice algorithm for linear prediction." In *Proceedings of International Conference on Acoustics, Speech, and Signal Processing, ICASSP'82*, pp. 651–654, IEEE, May 1982.

[182] W. Gardner, P. Jacobs, and C. Lee. "QCELP: a variable rate speech coder for CDMA digital cellular." In *Speech and Audio Coding for Wireless and Network Applications* (B. Atal, V. Cuperman, and A. Gersho, eds.), pp. 85–92. Boston: Kluuwer Academic Publishers, 1993.

[183] Telcomm. Industry Association (TIA), Washington, DC. *Mobile station—Base station compatibility standard for dual-mode wideband spread spectrum cellular system, EIA/TI4 Interim Standard IS,-95*, 1993.

[184] T. Ohya, H. Suda, and T. Miki. "5.6 kbits/s PSI-CELP of the half-rate PDC speech coding standard." In *Proc. of Vehicular Technology Conference*, vol. 3, (Stockholm), pp. 1680–1684, IEEE, May 1994.

[185] T. Ohya, T. Miki, and B. Suda. "Jdc half-rate speech coding standard and a real time operating prototype." *NTT Review*, vol. 6, pp. 61–67, November 1994.

[186] K. Mano, FF. Moriya, S. Mild, H. Ohmuro, K. Ikeda, and J. Ikedo. "Design of a pitch synchronous innovation CELP coder for mobile communications," *IEEE Journal on Selected Areas in Communications*, vol. 13, no. 1, pp. 31–41, 1995.

[187] I. Gerson, M. Jasiuk, J.-M. Muller, J. Nowack, and E. Winter, "Speech and channel cqiding for the half-rate GSM channel," *Proceedings ITG-Fachbericht*, vol. 130, p. 225–233, November 1994.

[188] A. Kataoka, T. Moriya, and S. Hayashi. "Implementation and performance of an 8-kbits/s conjugate structured CELP speech codec." In *Proceedings of the IEEE International Conference on Acoustics, Speech and Signal Processing (TCASSP'94)*, vol. 2 (Adelaide, Australia), pp. 93–96, IEEE, April 19–22, 1994.

[189] R. Salami, C. Laflammne, and J.-P. Adoul. "8 kbits/s ACELP coding of speech with 10 ms speech frame: A candidate for CCITT standardization." In *Proceedings of the IEEE International Conference on Acoustics, Speech and Signal Processing (ICASSP'94)*, vol. 2 (Adelaide, Australia), pp. 97–100, IEEE, April 19–22, 1994.

[190] J. Woodard, T. Keller, and L. Hanzo. "Turbo-coded orthogonal frequency division multiplex transmission of 8 kbps encoded speech." In *Proceeding of ACTS Mobile Communication Summit '97*, (Aalborg, Denmark), pp. 894–899, ACTS, October 7–10, 1997.

[191] T. Ojanpare et al. "FRAMES multiple access technology." In *Proceedings of IEEE ISSSTA'96*, vol. 1 (Mainz, Germany), pp. 334–338, IEEE, September 1996.

[192] C. Berrou, A. Glavieux, and P. Thitimajshima. "Near shannon limit error-correcting coding and decoding: Turbo codes." In *Proceedings of the International Conference on Communications*, (Geneva, Switzerland), pp. 1064–1070, May 1993.

[193] C. Berrou and A. Glavieux. "Near optimum error correcting coding and decoding: turbo codes," *IEEE Transactions on Communications*, vol. 44, pp. 1261–1271, Octcber 1996.

[194] J. Hagenauer, E. Offer, and L. Papke. "Iterative decoding of binary block and convolutional codes." *IEEE Transactions on Information Theory*, vol. 42, pp. 429–445, March 1996.

[195] P. Jung and M. Nasshan. "Performance evaluation of turbo codes for short frame transmission systems." *IEE Electronic Letters*, pp. 111–112, January 1994.

[196] A. Barbulescu and S. Pietrobon. "Interleaver design for turbo codes." *IEE Electronic Letters*, pp. 2107–2108, December 1994.

[197] L. Bahl, J. Cocke, F. Jellnek, and J. Raviv. "Optimal decoding of linear codes for minimising symbol error rate." *IEEE Transactions on Information Theory*, vol. 20, pp. 284–287, March 1974.

[198] "COST 207: Digital land mobile radio communications, final report." Luxembourg, Office for Official Publications of the European Communities, 1989.

[199] R. Salami, C. Laflamme, B. Bessette, and J.-P. Adoul. "Description of ITU-T recommendation G.729 annex A: Reduced complexity 8 kbits/s CS-ACELP codec." In *Proceedings of the IEEE International Conference on Acoustics, Speech and Signal Processing (ICASSP'97)*, (Munich, Germany), pp. 775–778, IEEE, April 21–24, 1997.

[200] R. Salami, C. Laflamme. B. Bessette, and J.-P. Adoul. "ITU-T recommendation G.729 annex A: Reduced complexity 8 kbits/s CS-ACELP codec for digital simultaneous voice and data (DVSD)." *IEEE C'ommunications Magazine*, vol. 35, pp. 56–63, September 1997.

[201] R. Salami, C. Laflamme, B. Besette, J...P. Adoul, K. Jarvinen J. Vainio, P. Kapanen, T. Hankanen, and P. Haavisto. "Description of the GSM enhanced full rate speech codec." In *Proc. of ICC'97*, 1997.

[202] "PCS1900 enhanced full rate codec US1." SP-3612.

[203] "IS-136.1A TDMA cellular/PCS—radio interface—mobile station—base station compatibility digital control channel." Revision A, August 1996.

[204] T. Honkanen, J. Vainio, K. Jarvinen, P. Haavisto, R. Salami, C. Laflamme, and J Adoul. "Enhanced full rate speech codec for 1S-136 digital cellular system." In *Proceedings of the IEEE International Conference on Acoustics, Speech and Signal Processing (ICASSP'97)*, vol. 2, (Munich, Germany), pp. 731–734, IEEE, April 21–24, 1997.

[205] "TIA/EIA/1S641, interim standard, TDMA cellular/PCS radio interface—enhanced full-rate speech codec." May 1996.

[206] "Dual rate speech coder for multimedia communications transmitting at 5.3 and 6.3 kbit/s." CCITT Recommendation G.723.1, March 1996.

[207] C. Hong, *Low Delay Switched Hybrid Vector Ezcited Linear Predictive Coding of Speech*. Ph.D. thesis, National University of Singapore, 1994.

[208] J. Zhang and H.-S. Wang. "A low delay speech coding system at 4.8 kb/s," In *Proceedings of the IEEE International Conference on Communications Systems*, vol. 3, pp. 880–883, November 1994.

[209] J.-H. Chen, N. Jayant, and R. Cox. "Improving the performance of the 16 kb/s LD-CELP speech coder." In *Proceedings of International Conference on Acoustics, Speech, and Signal Processing, ICASSP'92*, IEEE, March 1992.

[210] J.-H. Chen and A. Gersho. "Gain-adaptive vector quantization with application to speech coding," *IEEE Transactions on Communications*, vol. 35, pp. 918–930, September 1987.

[211] J.-H. Chen and A. Gersho. "Gain-adaptive vector quantization for medium rate speech coding." In *Proceedings cf IEEE International Conference on Communications 1985*, (Chicago, IL), pp. 1456–1460, IEEE, June 23–26, 1985.

[212] J.-H. Chen, Y.-C. Lin, and R. Cox. "A fixed-point 16 kb/s LD-CELP algorithm." In *Proceedings of International Conference on Acoustics, Speech, and Signal Processing, ICASSP'91*, vol. 1 (Toronto, Ontario, Canada), pp. 21–24, IEEE, May 14–17, 1991.

[213] J.-H. Chen. "High-quality 16 kb/s speech coding with a one-way delay less than 2 ms. " In *Proceedings of International Conference on Acoustics, Speech, and Signal Processing, ICASSP'90*, vol. 1 (Albuquerque, New Mexico), pp. 453–456, IEEE, April 3–6, 1990.

[214] J. D. Marca and N. Jayant. "An algorithm for assigning binary indices to the codevectors of a multi-dimensional quantizer." In *Proceedings of IEEE International Conference on Communications 1987*, (Seattle, WA), pp. 1128–1132, IEEE, June 7–10, 1987.

[215] K. Zeger and A. Gersho. "Zero-redundancy channel coding in vector quantization." *Electr. Letters*, vol. 23, pp. 654–656, June 1987.

[216] J. Woodard and L. Hanzo. "A low delay multimode speech terminal." In *Proceedings of IEEE VTC '96*, vol. 1 (Atlanta, GA), pp. 213–217, IEEE, 1996.

[217] Y. Linde, A. Buzo, and R. Gray. "An algorithm for vector quantiser design." *IEEE Transactions on Communications*, vol. Com-28, January 1980.

[218] W. Kleijn, D. Krasinski, and R. Ketchum. "Fast methods for the CELP speech coding algorithm." *IEEE Trans. on Acoustics, Speech and Signal Processing*, pp. 1330–1342, August 1990.

[219] S. D.'Agnoi, J. D. Marca, and A. Alcaim. "On the use of simulated annealing for error protection of CELP coders employing LSF vector quantizers." In *Proceedings of IEEE VTC '94*, vol. 3 (Stockholm, Sweden), pp. 1699–1703, IEEE, June 8–10, 1994.

[220] X. Maitre. "7 khz audio coding within 64 kbit/s." *IEEE-JSAC*, vol. 6, pp. 283–298, February 1988.

[221] R. Crochiere, S. Webber, and J. Flanagan. "Digital coding of speech in sub-bands." *Bell System Tech. Journal*, pp. 1069–1085, October 1976.

[222] R. Crochiere. "An analysis of 16 kbit/s sub-band coder performance: dynamic range, tandem connections and channel errors." BSTJ, vol. 57, pp. 2927–2952, October 1978.

[223] D. Esteban and C. Galand. "Application of quadrature mirror filters to split band voice coding scheme." In *Proceedings cf International Conference on Acoustics, Speech, and Signal Processing, ICASSP'77* (Hartford, CT), pp. 191–195, IEEE, May 9–11, 1977.

[224] J. Johnston. "A filter family designed for use in quadrature mirror filter banks." In *Proceedings of International Conference on Acoustics, Speech, and Signal Processing, ICATSP'80* (Denver, CO), pp. 291–294, IEEE, April 9–11, 1980.

[225] H. Nussbaumer. "Complex quadrature mirror filters." In *Proceedings of International Conference on Acoustics, Speech, an Signal Processing, ICASSP'83* (Boston, MA), pp. 221–223, IEEE, April 14–16, 1983.

[226] C. Galand and H, Nussbaumer. "New quadrature mirror filter structures." *IEEE Trans. on ASSP*, vol. ASSP-32, pp. 522–531, June 1984.

[227] S. Quackenbush. "A 7 khz bandwidth, 32 kbps speech coder for ISDN." In *Proceedings of International Conference on Acoustics, Speech, and Signal Processing, ICASSP'91* (Toronto, Ontario, Canada), pp. 1–4, IEEE, May 14–17, 1991.

[228] J. Johnston. "T nsform coding of audio signals using perceptual noise criteria." *IEEE-JSAC*, vol, 6, no. 2, pp. 314–323, 1988.

[229] E. Ordentlich and Y. Shoham. "Low-delay code-excited linear-predictive coding of wideband speech at 32 kbps." In *Proceedings of International Conference on Acoustics, Speech, and Signal Processing, ICASSP'91*, (Toronto., Ontario, Canada), pp. 9–12, IEEE, May 14–17, 1991.

[230] R. Soheili, A. Kondoz, and B. Evans. "New innovations in multi-pulse speech coding for bit rates below 8 kb/s." In *Proc. of Eurospeech*, pp. 298–301, 1989.

[231] V. Sanchez-Calle, C. Laflamme, R. Salami, and J.-P. Adoul. "Low-delay algebraic CELP coding of wideband speech." in *Signal Processing VI: Theories and Applications* (J. Vandewalle, R. Boite, M. Moonen, and A. Oosterlink, eds.), pp. 495–498. Elsevier Science Publishers, 1992.

[232] G. Roy and P. Kabal. "Wideband CELP speech coding at 16 kbit/sec." In *Proceedings of International Conference on Acoustics, Speech, and Signal Processing, ICASSP'91* (Toronto, Ontario, Canada), pp. 17–20, IEEE, May 14–17, 1991.

[233] R. Steele and W. Webb. "Variable rate QAM for data transmission over Rayleigh fading channels." In *Proceedings of Wireless '91* (Calgary, Alberta), pp. 1–14, IEEE, 1991.

[234] W. Webb and R. Steele. "Variable rate QAM for mobile radio," *IEEE Transactions on Communications*, vol. 43, no. 7, pp. 2223–2230, 1995.

[235] Y. Kamio, S. Sampei. H. Sasaoka, and N. Morinaga. "Performance of modulation-level-control adaptive-modulation under limited transmission delay time for land mobile communications." In *Proceedings of IEEE Vehicular Technology Conference (VTC'95)*, (Chicago), pp. 221–225, IEEE, July 15–28, 1995.

[236] K. Arimochi, S. Sampei, and N. Morinaga. "Adaptive modulation system with discrete power control and predistortion-type non-linear compensation for high spectral efficient and high power efficient wireless communication systems." In *Proceeding of IEEE International Symposium on Personal, Indoor and Mobile Radio Commu-*

nications, PIMRC'97, (Marina Congress Centre, Helsinki, Finland), pp. 472–477, IEEE, September 1–4, 1997.

[237] M. Naijoh, S. Sampei, N. Morinaga, and Y. Kamio. "ARQ schemes with adaptive modulation/TDMA/TDD systems for wireless multimedia communication systems." In *Proceedings of IEEE International Symposium on Personal, Indoor and Mobile Radio Communications, PIMRC'97*, (Marina Congress Centre, Helsinki, Finland), pp. 709–713, IEEE, September 1–4, 1997.

[238] A. Goldsmith. "The capacity of downlink fading channels with variable rate and power." *IEEE Tr. on Veh. Techn.*, vol. 46, pp. 569–580, August 1997.

[239] A. Goldsmith and S. Chua. "Variable-rate variable-power MQAM for fading channels." *IEEE Trans. on Comrnunications*, vol.. 45, pp. 1218–1230, October 1997.

[240] M.-S. Alouini and A. Goldsmith. "Area spectral efficiency of cellular mobile radio systems." *To appear IEEE Tr. on Veh. Techn.*, 1999. http://www.systems.caltech.edu.

[241] A. Goldsmith and P. Varaiya. "Capacity of fading channels with channel side information." *IEEE Tr on Inf Theory*, vol. 43, pp. 1986–1992, November 1997.

[242] C. Wong, T. Liew, and L. Hanzo. "Turbo coded burst by burst adaptive wide-band modulation with blind modem mode detection." In *Proceeding of ACTS Mobile Communication Summit '99* (Sorrento, Italy), pp. 303–308, ACTS, June 8–11, 1999.

[243] C. Wong and L. Hanzo. "Upper-bound of a wideband burst-by-burst adaptive modem." In *Proceeding of VTC'99 (Spring)*, (Houston, TX), pp. 1851–1855, IEEE, May, 16–20, 1999.

[244] T. Liew, C. Wong, and L. Hanzo. "Block turbo coded burst-by-burst adaptive modems." In *Proceeding of Microcoll'99, Budapest, Hungary*, pp. 59–62, March 21–24, 1999.

[245] C. Wong, T. Liew., and L. Hanzo. "Blind-detection assisted, block turbo coded, decision-feedback equalised burst-by-burst adaptive modulation." Submitted to IEEE JSAC, 1999.

[246] M. Yee, T. Liew, and L. Hanzo. "Radial basis function decision feedback equalisation assisted block turbo burst-by-burst adaptive modems." In *Proceeding of VTC'99 (Fall)*, (Amsterdam, Netherlands), pp. 1600–1604, IEEE, September 19–22, 1999.

[247] H. Matsuoka, S. Sampei, N. Morinaga, and Y. Kamio. "Adaptive modulation system with variable coding rate concatenated code for high quality multi-media communications systems." In *Proceedings of IEEE VTC '96* (Atlanta, GA, USA), pp. 487–491, IEEE, 1996.

[248] V. Lau and M. Macleod. "Variable rate adaptive trellis coded QAM for high bandwidth efficiency applications in rayleigh fading channels." In *Proceedings of IEEE Vehicular Technology, Conference (VTC'98)*, (Ottawa, Canada), pp. 348–352, IEEE, May 18–21, 1998.

[249] A. Goldsmith and S. Chua. "Adaptive coded modulation for fading channels." *IREE Tr. on Communications*, vol. 46, pp. 595–602, May 1998.

[250] T. Keller and L. Hanzo. "Adaptive orthogonal frequency division multiplexing schemes." In *Proceeding of ACTS Mobile Communication Summit '98* (Rhodes, Greece), pp. 794–799, ACTS, June 8–11, 1998.

[251] E. Kuan, C. Wong, and L. Hanzo. "Burst-by-burst adaptive joint detection CDMA." In *Proceeding of VTC'99 (Spring)*, (Houston, TX), IEEE, May 16–20, 1999.

[252] R. Chang. "Synthesis of band-limited orthogonal signals for multichannel data transmission." *BSTJ*, vol. 46, pp. 1775–1796, December 1966.

[253] L. Cimini,. "Analysis and simulation of a digital mobile channel using orthogonal frequency division multiplexing." *IEEE Transactions on Communications*, vol. 33, pp. 665–675, July 1985.

[254] K. Fazel and G. Fettweis, eds. *Multi-Carrier Spread-Spectrum*. Boston: Kluwer, 1997. p. 260.

[255] T. May and H. Rohling. "Rednktion von Nachbarkanastörungen in OFDM-Funküber-tragungssystemen" In *2. OFDM-Fachgespräch in Braunschweig*, 1997.

[256] S. Müller and J. Huber. "Vergleich von OFDM-Verfahren mit reduzierter Spitzenleis-tung." In *2. OFDM-Fachgespräch in Braunschweig*, 1997.

[257] F. Classen and H. Meyr. "Synchronisation algorithms for an ofdm system for mobile communications." In *Codierung für Quelle, Kanal und Übertragung*, no. 130 in ITG Fachbericht (Berlin), pp. 105–113, VDE-Verlag, 1994.

[258] F. Classen and H. Meyr. "Frequency synchronisation algorithms for ofdm systems suitable for communication over frequency selective fading channel." In *Proceedings of IEEE VTC '94* (Stockholm, Sweden), pp. 1655–4659, IEEE, June 8–10, 1994.

[259] S. Shepherd, P. yan Eetvelt, C. Wyatt-Millington, and S. Barton. "Simpie coding scheme to reduce peak, factor in QPSK multicarrier modulation." *Electronics Letters*, vol. 31, pp. 1131–1132, July 1995.

[260] A. Jones, T. Wilkinson, and S. Barton. "Block coding scheme for reduction of peak to mean envelope power ratio of multicarrier transmission schemes." *Electronics Letters*, vol. 30, pp. 2098–2099, 1994.

[261] M. D. Benedetto and P. Mandarini. "An application of MMSE predistortion to OFDM systems." *IEEE Trans. on Comm.*, vol. 44, pp. 1417–1420, November 1996.

[262] P. Chow, J. Cioffi, and J. Bingharn "A practical discrete multitone transceiver loading algorithm for data transmission over spectrally shaped channels." *IEEE Trans. on Communications*, vol. 48, pp. 772–775, 1995.

[263] K. Fazel, S. Kaiser, P. Robertson, and M. Ruf. "A concept of digital terrestrial television broadcasting." *Wireless Personal Communications*, vol. 2, pp. 9–27, 1995.

[264] H. Sari, G. Karam, and I. Jeanclaude. "Transmission techniques for digital terrestrial tv broadcasting." *IEEE Communications Magazine*, pp. 100–109, February 1995.

[265] J. Borowski, S. Zeisberg, J. Hübner, K. Koora, E. Bogenfeld, and B. Kull. "Perfor-mance of OFDM and comparable single carrier system in MEDIAN demonstrator 60 GHz channe." In *Proceeding of ACTS Mobile Communication Summit '97*, (Aalborg, Denmark), pp. 653–658, ACTS, October 7–10, 1997.

[266] I. Kalet, "The multitone channel." *IEEE Tran. on Comms*, vol. 37, pp. 119–124, February 1989.

[267] Y. Li and N. Sollenberger. "Interference suppression in OFDM systems using adaptive antenna arrays." In *Proceeding of Globecom '98* (Sydney, Australia), pp. 213–218, IEEE, November 8–12, 1998.

[268] F. Vook and K. Baum. "Adaptive antennas for OFDM. In *Proceedings of IEEE Vehicular Technology Conference (VTC'98)*, vol. 2, (Ottawa, Canada), pp. 608–610, IEEE, May 18–21, 1998.

[269] T. Keller, J. Woodard, and L. Hanzo. "Turbo-coded parallel modem techniques for personal communications." In *Proceedings of IEEE VTC '97* (Phoenix, AZ), pp. 2158–2162, IEEE, May 4–7, 1997.

[270] T. Keller and L. Hanzo. "Blind-detection assisted sub-band adaptive turbo-coded OFDM schemes." In *Proceeding of VTC'99 (Spring)*, (Houston, TX), pp. 489–493, IEEE, May 16–20, 1999.

[271] "Universal mobile telecommunications system (UMTS); UMTS terrestrial radio access (UTRA); concept evaluation," tech. rep., ETSI, 1997. TR 101 146.

[272] tech. rep. http://standards.pictel.com/ptelcont.htm#Audio or ftp://standard.pictel.com/sg16_q20/1999_09_Geneva/.

[273] M. Failli. "Digital land mobile radio communications COST 207," tech. rep., European Commission, 1989.

[274] J. Proakis. *Digital Communications*. 3rd ed. New York: McGraw Hill, 1995.

[275] H. Malvar. *Signal Processing with Lapped Transforms*. Boston: Artech House, 1992.

[276] K. Rao and P. Yip, *Discrete Cosine Transform: Algorithms, Advantages and Applications*. Academic Press Ltd., 1990.

[277] B. Atal and M. Schroeder. "Predictive coding of speech signals." *Bell System Technical Journal*, pp. 1973–1986, October 1970.

[278] I. Wassel, D. Goodinan, and R. Steel. "Embedded delta modulation." *IEEE Transactions on Acoustics, Speech and Signal Processing*, vol. 36, pp. 1236–1243, August 1988.

[279] B. Atal and S. Hanauer. "Speech analysis and synthesis by linear prediction of the speech wave." *The Journal of the Acoustical Society of America*, vol. 50, no. 2, pp. 637–655, 1971.

[280] M. Kohler, L. Supplee, and T. Tremain. "Progress towards a new government standard 2400bps voice coder." In *Proceedings of the IEEE International Conference on Acoustics, Speech and Signal Processing (ICASSP'95)*, (Detroit, MI), pp. 488–491, IEEE, May 9–12, 1995.

[281] K. Teague, B. Leach, and W. Andrews. "Development of a high-quality MBE based vocoder for implementation at 2400bps." In *Proceedings of the IEEE Wichita Conference on Communications, Networking and Signal Processing*, pp. 129–133, April 1994.

[282] H. Hassanein, A. Brind'Amour, S. Déry, and K. Bryden. "Frequency selective harmonic coding at 2400bps." In *Proceedings of the 37th Midwest Symposium on Circuits and Systems*, vol. 2, pp. 1436–1439, 1995.

[283] R. McAulay and T. Quatieri. "The application of subband coding to improve quality and robustness of the sinusoidal transform coder." In *Proceedings of the IEEE International Conference on Acoustics, Speech and Signal Processing (ICASSP'93)*, vol. 2 (Minneapolis, MN), pp. 439–442, IEEE, April 27–30, 1993.

[284] A. McCree and T. Barnwell III. "A mixed excitation LPC vocoder model for low bit rate speech coding." *IEEE Transactions on Speech and Audio Processing*, vol. 3, no. 4, pp. 242–250, 1995.

[285] P. Laurent and P. L. Noue. "A robust 2400bps subband LPC vocoder," in *Proceedings of the IEEE International Conference on Acoustics, Speech and Signal Processing (ICASSP'95)*, (Detroit, MI), pp. 500–503, IEEE, May 9–12, 1995.

[286] W. Kleijn and J. Haagen. "A speech coder based on decomposition of characteristic waveforms." In *Proceedings of the IEEE International Conference on Acoustics, Speech and Signal Processing (ICASSP'95)*, (Detroit, MI), pp. 508–511, IEEE, May, 9–12, 1995.

[287] R. McAulay and T. Champion. "Improved interoperable 2.4 kb/s LPC using sinusoidal transform coder techniques." In *Proceedings of International Conference on Acoustics, Speech, and Signal Processing, ICASSP'90*, (Albuquerque, New Mexico), pp. 641–643, IEEE, April 3–6, 1990.

[288] K. Teague, W. Andrews, and B. Walls. "Harmonic speech coding at 2400 bps," In *Proc. 10th Annual Mid-America Symposium on Emerging Computer Technology*, (Norman, Oklahoma), 1996.

[289] J. Makhou , R. Viswanathan, R. Schwartz, and A. Huggins. "A mixed-source model for speech compression and syntbesis." *The Journal of the Acoustical Society of America*, vol. 64, no. 4, pp. 1577–1581, 1978.

[290] A. McCree, K. Truong, E. George, T. Barnwell, and V. Viswanathan. "A 2.4kbit/s coder candidate for the new U.S. federal standard." In *Proceedings of the IEEE International Conference on Acoustics, Speech and Signal Processing (ICASSP'96)*, (Atlanta, GA), pp. 200–203, IEEE, May 7–10, 1996.

[291] A. McCree and I. Barnwell III. "Improving the performance of a mixed excitation LPC vocoder in acoustic noises." In *Proceedings of International Conference on Acoustics, Speech, and Signal Processing, ICASSP'92*, vol. 2, pp. 137–140, IEEE, March 1992.

[292] J. Holmes. "The influence of glottal waveform on the naturalness of speech from a parallel formant synthesizer." *IEEE Transaction on Audio and Electroacoustics*, vol. 21, pp. 298–305, June 1973.

[293] W. Kleijn, Y. Shoham, D. Sen, and R. Hagen. "A low-complexity waveform interpolation coder." In *Proceedings of the IEEE International Conference on Acoustics, Speech and Signal Processing (ICASSP'96)*, vol. 1 (Atlanta, GA), pp. 212–215, IEEE, May 7–10, 1996.

[294] D. Hiotakakos and C. Xydeas. "Low bit rate coding using an interpolated zinc excitation model." In *Proceedings of the ICCS 94*, pp. 865–869, 1994.

[295] R. Sukkar, J. LoCicero, and J. Picone. "Decomposition of the LPC excitation using the zinc basis function." *IEEE Transactions on Acoustics, Speech and Signal Processing*, vol. 37, no. 9, pp. 1329–1341, 1989.

[296] M. Schroeder, B. Atal, and J. Hall. "Optimizing digital speech coders by exploiting masking properties of the human ear." *Journal of the Acoustical Society of America*, vol. 66, pp. 1647–1652, December 1979.

[297] W. Voiers. "Diagnostic acceptability measure for speech communication systems." In *Proceedings of International Conference on Acoustics, Speech, and Signal Processing, ICASSP'77*, (Hartford, CT), pp. 204–207, IEEE, May 9–11, 1977.

[298] W. Voiers. "Evaluating processed speech using the diagnostic rhyme test." *Speech Technology*, January/February 1983.

[299] T. Tremain, M. Kohler, and T. Champion. "Philosophy and goals of the D.O.D 2400bps vocode selection process." In *Proceedings of the IEEE International Conference on Acoustics, Speech and Signal Processing (ICASSP'96)*, (Atlanta, GA), pp. 1137–1140, IEEE, May 7–10, 1996.

[300] M. Bielefeld and L. Supplee. "Developing a test program for the DoD 2400bps vocoder selection process." In *Proceedings of the IEEE International Conference on Acoustics, Speech arid Signal Processing (ICASSP'96)*, (Atlanta, GA), pp. 1141–1144, May 7–10, 1996.

[301] J. Tardeili and E. W. Kreamer. "Vocoder intelligibility and quality test methods." In *Proceedings of the IEEE International Conference on Acoustics, Speech and Signal Processing (ICASSP'96)*, (Atlanta, GA), pp. 1145–1148, IEEE, May 7–10, 1996.

[302] A. Schmidt-Nielsen and D. Brock. "Speaker recognizability testing for voice coders." In *Proceedings of the IEEE International Conference on Acoustics, Speech and Signal Processing (ICASSP'96)*, (Atlanta, GA), pp. 1149–1152, IEEE, May 7–10, 1996.

[303] E. Kreamer and J. Tardelli. "Communicability testing for voice coders." in *Proceedings of the IEEE International Conference on Acoustics, Speech and Signal Processing (ICASSP'96)*, (Atlanta, GA), pp. 1153–1156, IEEE, May 7–10, 1996.

[304] B. Atal and L. Rabiner. "A pattern recognition approach to voiced-unvoiced silence classification with applications to speech recognition." *IEEE Transactions an Acoustics, Speech and Signal Processing*, vol. 24, pp. 201–212, June 1976.

[305] T. Ghiselli-Crippa and A. El-Jaroudi. "A fast neural net training algorithm and its application to speech classification." *Engineering Applications of Artificial Intelligence*, vol. 6, no. 6, pp. 549–557, 1993.

[306] A. Noll. "Cepstrum pitch determination," *Journal of the Acoustical Society of America*, vol. 41, pp. 293–309, February 1967.

[307] S. Kadambe and G. Boudreatux-Bartels. "Application of the wavelet transform for pitch detection of speech signals." *IEEE Transactions on Information Theory*, vol. 38, pp 917–924, March 1992.

[308] L. Rabiner, M. Cheng, A. Rosenberg, and C. McGonegal. "A comparative performance study of several pitch detection algorithms." *IEEE Transactions on Acoustics, Speech, and Signal Processing*, vol. 24, no. 5, pp. 399–418, 1976.

[309] DVSI. *Inmarsat -M Voice Codec*. Issue 3.0 ed., August 1991.

[310] M. Sambur, A. Rosenberg, L. Rabiner, and C. McGonegal. "On reducing the buzz in LPC synthesis." *Journal of the Acoustical Society of America*, vol. 63, pp. 918–924, March 1978.

[311] A. Rosenberg. "Effect of glottal pulse shape on the quality of natural vowels." *Journal of the Acoustical Society of America*, vol. 49, no. 2, pt. 2, pp. 583–590, 1971.

[312] T. Koornwinder. *Wavelets: An Elementary Treatment of Theory and Applications*. World Scientific, 1993.

[313] C. Chui. *Wavelet Analysis and Its Applications*. Vol. I: An Introduction to Wavelets. New York: Academic Press, 1992.

[314] C. Chui. *Wavelet Analysis and its Applications*. Vol. II: Wavelets: A Tutorial in Theory and Applications. New York: Academic Press, 1992.

[315] O. Rioul and M. Vetterli. "Wavelets and signal processing." *IEEE Signal Processing Magazine*, pp. 14–38, October 1991.

[316] A. Graps. "An introduction to wavelets." *IEEE Computational Science & Engineering*, pp. 50–61, Summer 1995.

[317] A. Cohen and J. K. Kovačević. "Wavelets: The mathematical background." *Proceedings of the IEEE*, vol. 84, pp. 514–522, April 1996.

[318] I. Daubechies. "The wavelet transform, time-frequency localization and signal analysis." *IEEE Transactions on Information Theory*, vol. 36, pp. 961–1005, September 1990.

[319] S. Mallat. "A theory for multiresolution signal decomposition: the wavelet representa- tion." *IEEE Transactions on Pattern Analysis and Machine Intelligence*, vol. 11, pp. 674–693, July 1989.

[320] H. Baher. *Analog & Digital Signal Processing*. New York: John Wiley & Sons, 1990.

[321] J. Stegmann, G. Schröder, and K. Fischer. "Robust classification of speech based on the dyadic wavelet transform with application to CELP coding." In *Proceedings of the IEEE International Conference on Acoustics, Speech and Signal Processing (ICASSP'96)*, (Atlanta, GA), pp. 546–549, IEEE, May 7–10, 1996.

[322] S. Mallat nd S. Zhong. "Characterization of signals from multiscale edges." *IEEE Transactions on Pattern Analysis and Machine Intelligence*, vol. 14, pp. 710–732, July 1992.

[323] M. Unser and A. Aldroubi. "A review of wavelets in biomedical applications." *Proceedings of the IEEE*, vol. 84, pp. 626–638, April 1996.

[324] C. Li, C. Zheng, and C. Tai. "Detection of ECG characteristic points using wavelet transforms." *IEEE Transactions in Biomedical Engineering*, vol. 42, pp. 21–28, January 1995.

[325] S. Mallat and W. Hwang. "Singularity detection and processing with wavelets," *IEEE Transactions on Information Theory*, vol. 38, pp. 617–643, March 1992.

[326] M. Vetterli and J. Kovacević. *Wavelets and Subband Coding*. Englewood Cliffs, NJ: Prentice Hall, 1995.

[327] R. Sukkar, J. LoCicero, and J. Picone. "Design and implementation of a robust pitch detector based on a parallel processing technique." *IEEE Journal on Selected Areas in Communications*, vol. 6, pp. 441–451, February 1988.

[328] R. Steele, and L. Hanzor, eds. *Mobile Radio Communications*. Piscataway, NJ: IEEE Press: John Wiley, 1999.

[329] F. Brooks, B. Yeap, J. Woodard, and L. Hanzo. "A sixth-rate, 3.8kbps gsm-like speech transceiver." In *Prooceeding of ACTS Mobile Communication Summit '98*, (Rhodes, Greece), pp. 644–652, ACTS, June 8–11, 1998.

[330] F. Brooks, E. Kuan, and L. Hanzo. "A 2.35kbps joint-detection CDMA speech transceiver." In *Proceeding of VTC'99 (Spring)*, (Houston, TX), pp. 2403–2407, IEEE, May 16–20, 1999.

[331] P. Robertson, E. Villebrun, and P. Hoeher. "A comparison of optimal and sub-optimal MAP decoding algorithms operating in the log domain." In *Proceedings of the International Conference on Communications*, pp. 1009–4013, June 1995.

[332] P. Robertson. "Illuminating the structure of code and decoder of parallel concatenated recursive systematic (turbo) codes." *IEEE Globecom*, pp. 1298–1303, 1994.

[333] W. Koch and A. Baier. "Optimum and sub-optimum detection of coded data disturbed by time varying inter-symbol interference." *IEEE Globecom*, pp. 1679–1684, Decem- ber 1990.

[334] J. Erfanian, S. Pasupathy, and G. Guiak. "Reduced complexity symbol dectectors with parallel structures for ISI channels." *IEEE Transactions on Communications*, vol. 42, pp. 1661–1671, 1994.

[335] J. Hagenauer and P. Hoeher. "A viterbi algorithm with soft-decision outputs and its applications." In *IEEE Globecom*, pp. 1680–1686, 1989.

[336] C. Berrou, P. Adde, E. Angui, and S. Faudeil. "A low complexity soft-output viterbi decoder architecture." In *Proceedings of the International Conference on Communications*, pp. 737–740, May 1993.

[337] L. Rabiner, C. McGonegal, and D. Paul. *FIR Windowed Filter Design Program—WINDOW*, ch. 5.2, Piscataway, NJ: IEEE Press, 1979.

[338] S. Yeldner, A. Kondoz, and B. Evans, "Multiband linear predictive speech coding at very low bit rates." *IEE Proceedings in Vision, Image and Signal Processing*, vol. 141, pp. 284–296, October 1994.

[339] A . Klein, R. Pirhonen, J. Skoeld, and R. Suoranta "FRAMES multiple access mode 1—wideband TDMA with and without spreading." In *Proceedings of IEEE International Symposium on Personal, Indoor and Mobile Radio Communications, PIMRC'97*, vol. 1 (Marina Congress Centre, Helsinki, Finland), pp. 37–41, IEEE, September 1–4, 1997.

[340] J. Flanagan and R. Golden, "Phase vocoder." *The Bell System Technical Journal*, pp. 1493–1509, November 1966.

[341] R. McAulay and T. Quatieri. "Speech analysis/synthesis based on sinusoidal representation." *IEEE Tansactions on Acoustics, Speech and Signal Processing*, vol. 34, pp, 744–754, August 1986.

[342] L. Almeid and J. Tribolet. "Nonstationary spectral modelling of voiced speech." *IEEE Transactions on Acoustics, Speech and Signal Processing*, vol. 31, pp. 664–677, June 1983.

[343] E. George and M. Smith. "Analysis-by-synthesis/overlap-add sinusoidal modelling applied to the analysis and synthesis of musical tones." *Journal of the Audio Engineering Society*, vol. 40, pp. 497–515, June 1992.

[344] E. George and M. Smith. "Speech analysis/synthesis and modification using an analysis-by-synthesis/overlap-add sinusoidal model." *IEEE Transaction on Speech and Audio Processing*, vol. 5, pp. 389–406, September 1997.

[345] R. McAulay and T. Quatieri. "Pitch estimation and voicing detection based on a sinusoidal speech model." In *Proceedings of ICASSP 90*, pp. 249–252, 1990.

[346] R. McAulay and T. Quatieri. "Sinusoidal coding." In *Speech Coding and Synthesis* (W. B. Keijn and K. K. Paliwal, eds.), ch. 4. Elsevier Science, 1995.

[347] R. McAulay, T. Parks, T. Quatieri, and M. Sabin. "Sine-wave amplitude coding at low date rates." In *Advances in Speech Coding* (V. B. S. Atal and A.Gersho, eds.), pp. 203–214, Boston: Kluwer Academic Publishers, 1991.

[348] M. Nishiguchi and.J. Matsumoto. "Harmonic and noise coding of LPC residuals with classified vector quantization. In *Proceedings of the IEEE International Conference on Acoustics, Speech and Signal Processing (ICASSP'95)*, (Detroit, MI), pp. 484–487, IEEE, May 9–12, 1995.

[349] V. Cupernian, P. Lupini, and B. Bhattacharya. "Spectral excitation coding of speech at 2.4kb/s." In *Proceedings of the IEEE International Conference on Acoustics, Speech and Signal Processing (ICASSP'95)*, (Detroit, MI), pp. 496–499, IEEE, May 9–12 1995.

[350] S. Yeldner, A. Kondoz, arid B. Evans. "High quality multiband LPC coding of speech at 2.4kbit/s." *Electronics Letters*, vol. 27, no. 14, pp. 1287–1289, 1991.

[351] H. Yang, S.-N. Koh, and P. Sivaprakasapillai. "Pitch synchronous multi-band (PSMB) speech coding," In *Proceedings of the IEEE International Conference on Acoustics,*

Speech and Signal Processing (ICASSP'95), (Detroit, MI), pp. 516–518, IEEE, May 9–12, 1995.

[352] E. Erzin, A. Kumar, and A. Gersho. "Natural quality variable-rate spectral speech coding below 3.0kbps." In *Proceedings of the IEEE International Conference on Acoustics, Speech and Signal Processing (ICASSP'97)*, (Munich, Germany), pp. 1579–1582, IEEE, April 21–24, 1997.

[353] V. Papanastasiou and C. Xydeas. "Efficient mixed excitation models in LPC based prototype interpolation speech coders." In *Proceedings of the IEEE International Conference on Acoustics, Speech and Signal Processing (ICASSP'97)* (Munich, Germany), pp. 1555–1558, IEEE, April 21–24, 1997.

[354] O. Ghitza. "Auditory models and human performance in tasks related to speech coding and speech recognition." *IEEE Transactions on Speech and Audio Processing*, vol. 2, pp. 115–132, January 1994.

[355] K. Kryter. "Methods for the calculation of the articulation index." Tech. Rep., American National Standards Institute, 1965.

[356] U. Halka, and U. Heute, "A new approach to objective quality-measures based on attribute matching." *Speech Communications*, vol. 11, pp. 15–30, 1992.

[357] S. Wang. A. Sekey, and A. Gersho. "An objective measure for predicting subjective quality of speech coders." *Journal on Selected Areas in Communications*, vol. 10, pp. 819–829, June 1992.

[358] T. Barnwell III and A. Bush. "Statistical correlation between objective and subjective measures for speech quality." In *Proceedings of International Conference on Acoustics, Speech, and Signal Processing, ICASSP'78*, (Tulsa, OK), pp. 595–598, IEEE, April 10–12, 1978.

[359] T. Barnwell III. "Correlation analysis of subjective and objective measures for speech quality." In *Proceedings of International Conference on Acoustics, Speech, and Signal Processing, IGASSP'80* (Denver, CO), pp. 706–709, IEEE, April 9–11, 1980.

[360] P. Breitkopf and T. Barnwell III. "Segmental preclassification for improved objective speech quality measures." In *IEEE Proc. of Internal. Conf. Acoust. Speech Signal Process.*, pp. 1110–1104, 1981.

[361] L. Hanzo and L. Hinsenkamp. "On the subjective and objective evaluation of speech codecs." *Budavox Telecommunications Review*, no. 2, pp. 6–9, 1987.

[362] K. Kryter. "Masking and speech communications in noise." In *The Effects of Noise on Man*, ch. 2. New York: Academic Press, 1970.

[363] A. House, C. Williams, M. Hecker, and K. Kryter, "Articulation testing methods: Consonated differentiation with a closed-response set." *J Acoust. Soc. Am.*, pp. 158–166, January 1965.

[364] J. Gibson, ed. *The Mobile Communications Handbook*. Piscataway, NJ: CRC Press and IEEE Press, 1996.

Index

Authors Index

About the Authors

Lajos Hanzo graduated in Electronics in 1976, and in 1983 he was conferred a doctorate in the field of Telecommunications. During his 24-year career in communications, he has held various research and academic posts in Hungary, Germany, and the UK. Since 1986, he has been with the Department of Electronics and Computer Science, University of Southampton, UK, and has been a consultant to Multiple Access Communications Ltd., UK. He currently holds a Chair in Telecommunications. He has co-authored five books on mobile radio communications, published over 300 research papers, and been awarded a number of distinctions. His current teaching and research interests cover the range of Mobile Multimedia Communications, including voice, audio, and graphical source compression, channel coding, modulation, networking, as well as the joint optimization of these system components.

Dr. Hanzo is managing a research group in the wide field of wireless multimedia communications funded by the Engineering and Physical Science Research Council (EPSRC) by the Virtual Centre of Excellence in Mobile Communications and by the Commission of European Communities (CEC). He is a member of the IEE and a senior member of the IEEE.

Dr. Clare Somerville received the M.Eng. and Ph.D. degrees in Electronic Engineering in 1995 and 1999, respectively, from the University of Southampton, Southampton, UK. From 1995 to 1998, she performed research into low-bit-rate speech coders for wireless communications. She is currently with the Global Wireless System Research Department, Bell Laboratories, Swindon, UK. Her current research involves real-time services over wireless networks with a packet air interface. The focus is on the techniques for transmission of voice over GPRS and the resultant speech quality attained.

Dr. Jason Woodard received an MA in Physics from the University of Oxford in 1991, and an MS.c. in Electronics and a Ph.D. in Speech Coding from the University of Southampton in 1992 and 1995, respectively. From 1995 to 1998, he worked as a Research Fellow at the University of Southampton, researching error correction coding, especially turbo codes.

In 1998, he joined the Wireless Technology Practice at the PA Consulting Group in Cambridge. At the start of 1999, he was transferred from PA to UbiNetics Ltd., where he is responsible for the development and implementation of bit-rate processing algorithms, including turbo codes, for third-generation mobile communications products.